ANATOMY &
PHYSIOLOGY
Lab Manual

Elizabeth Mack Co
Boston University

Hilary Engebretson
Whatcom Community College

Cengage

Australia • Brazil • Canada • Mexico • Singapore • United Kingdom • United States

Anatomy and Physiology Lab Manual,
First Edition
Elizabeth Mack Co, Hilary Engebretson

SVP, Product: Cheryl Costantini

VP, Product: Thais Alencar

Portfolio Product Director: Maureen McLaughlin

Portfolio Product Manager: Diana Baniak

Product Assistant: Jeanette Ames

Senior Learning Designer: Paula Dohnal

Senior Content Manager: Brendan Killion

Digital Project Manager: Jessica Witczak

Developmental Editor: Michael Zierler

Senior Director, Product Marketing: Danae April

Product Marketing Manager: Adam Kiszka

Content Acquisition Analyst: Ann Hoffman

Production Service: MPS Limited

Designer: Chris Doughman

Cover Image Source: PrModels/Turbosquid

For product information and technology assistance, contact us at
**Cengage Customer & Sales Support, 1-800-354-9706
or support.cengage.com**.

For permission to use material from this text or product, submit all requests online at **www.copyright.com**.

Library of Congress Control Number: 2023934984

ISBN: 978-0-357-90916-4, 979-8-214-19018-1

Cengage
200 Pier 4 Boulevard
Boston, MA 02210
USA

Cengage is a leading provider of customized learning solutions. Our employees reside in nearly 40 different countries and serve digital learners in 165 countries around the world. Find your local representative at **www.cengage.com**.

To learn more about Cengage platforms and services, register or access your online learning solution, or purchase materials for your course, visit **www.cengage.com**.

Printed in the United States of America
Print Number: 01   Print Year: 2023

Dedication

EC: To my first biology teacher, Mrs. Marion Voorheis, and to teachers everywhere.
On some lives you leave a print more permanent than the pages of this book.
Your work is the most important. Thank you.

HE: To all passionate learners who are willing to get uncomfortable
in the pursuit of answering the question "Why does it work that way?"

Brief Contents

Table of Contents

About the Authors

Courtesy of Elizabeth Co

Elizabeth Mack Co is a Senior Lecturer and Director of Inclusive Pedagogy in the Department of Biology at Boston University. She teaches Human Anatomy, Human Physiology, and Physiology of Reproduction. As a professor, Dr. Co is renowned for her passion—both about the human body and about learning itself. In 2018 she was nominated by students and members of the faculty at BU and received the Metcalf Award for Excellence in Teaching, Boston University's highest teaching award. Dr. Co's current research focuses on learning, particularly critical-thinking skills development. Within her large introductory courses, she integrates an active learning and study skills curriculum. One of the focal points she investigates is how "student awareness about their learning impacts their assessment performance." She serves on the Physiology learning objective panel of the Human Anatomy and Physiology Society and works with the Howard Hughes Medical Institute developing educational resources for use in Anatomy and Physiology and Biology courses. Dr. Co holds a PhD in Biomedical Sciences from the University of California, San Francisco. She is the author of *Anatomy & Physiology 1e*, recently published by Cengage.

Courtesy of Hilary Engebretson

Hilary Engebretson is an Associate Professor in the Science and Engineering Department at Whatcom Community College in Washington State. She teaches both the lecture and lab components of Human Anatomy and Physiology 1, Human Anatomy and Physiology 2, and Human Biology. She approaches teaching these classes as a collaborative effort between instructor and students. As an academic, Hilary's first love and primary focus is teaching, so it is no surprise that her research is about teaching and learning in higher education anatomy and physiology courses. She explores how students seek help in their A&P classes and how instructors can facilitate and improve students' effective help-seeking actions at college and university.

Preface

Introduction

Like the companion text, *Co's Anatomy & Physiology 1ˢᵗ edition*, this lab manual was written with a focus on the learner. We aim to support learner metacognition and particularly the development of critical thinking skills.

The section, "How to Learn in Anatomy and Physiology Lab," aims to help each student focus their time in A&P lab and structure their study time outside the lab for preparation and extension of learning. We provide reminders throughout the lab manual of ways to continue to study the material. These learning connections aim to break the student habit of just trying to "cram" the material into their brains (which is anatomically not possible, and just doesn't work), and instead provide steady and consistent metacognitive reminders in every lab.

In our hyper-connected digital world, information is everywhere. We know that our learners need to grow their skills of critical thinking to become the most effective consumers of information that they can be. But critical thinking is only grown through regular practice. In A&P, critical thinking practice can be difficult but rewarding to incorporate. This lab manual incorporates higher-order thinking questions in the lab quizzes to provide low-stakes opportunities for students to practice these skills with every lab. Most labs include a clinical case study, which offers an extended opportunity for critical thinking skill building.

Case studies offer opportunities for students to work together and discuss their answers. These moments offer more than just critical thinking skill building, but also the chance to practice other skills such as listening, explaining, and teamwork. Case studies can also be assigned as homework or as a substitute or extension of the material when an in-person class cannot meet. There is no reason these cases need to stay in the lab room, either; instructors should feel free to bring them into the classroom. Students, if your instructor doesn't assign the case studies, try using them as optional study practice!

This lab manual is intended to be welcoming for all learners. Written in an approachable voice with an eye toward clarity, we aim to engage students in a lab experience that is cooperative among the authors, instructors, and learners. We believe that every A&P student should be able to participate in hands-on

discovery as a mechanism for deep learning. Thus, we provide a variety of laboratory experiences to choose from in each section of the manual. You will find lab activities that require no supplies, those that require readily available materials such as colored pencils or sticky notes, those that require anatomy models, and those that utilize more specialized tools. We expect instructors to choose from the available activities to build a lab experience that fits their student population, the available laboratory time, and the models and materials at their disposal.

Features and Organization

Each lab is built around a similar format. Features of each lab include:

Pre-Lab Quiz The purpose of the pre-lab quiz is to help students strengthen their understanding of background material in preparation for the lab activities. To incentivize background preparation, the first five questions of the quiz are based on content from the relevant textbook chapter. To encourage prereading the lab manual, the second five questions of the quiz are based on content from the introductions to the lab activities within the lab manual itself.

Individual Lab Activities Each of the labs is made up of several individual lab activities. These activities are intended to provide a variety of ways for students to engage with the content. Instructors may choose to use some or all of the available individual lab activities depending on the models, lab materials, and time that they have for the subject area.

Case Studies Most of the labs contain one or more case studies. Case studies present possible scenarios so that students can engage in real-world application of what they are learning in the lab. Critical thinking is expected as students are asked to apply what they learned in the lab to a novel situation.

Digging Deeper Digging Deeper features are present in most labs. These brief descriptions cover interesting facets of the material covered in the lab. Some digging deeper features describe tools and techniques that are used to help us investigate the human body. Others describe diseases, cultural connections, or diverse perspectives on the topics.

Learning Connections Every student has a different brain and therefore learns differently! Throughout each lab we provide study suggestions to help students expand the diversity of tools that they have in their learning toolbox.

Post-Lab Quiz The post-lab quiz that is provided at the end of each lab allows students another opportunity to think about the material. Students can solidify their understanding of the presented topics by having a final opportunity to challenge themselves and apply critical thinking to the material.

Course Solutions

Today's leading online learning platform, MindTap for *Co/Engebretson's Anatomy & Physiology Lab Manual 1e*, gives you complete control of your course to craft a personalized, engaging learning experience that challenges students, builds confidence, and elevates performance.

MindTap introduces students to core concepts from the beginning of your course using a simplified learning path that progresses from understanding to application and delivers access to the eText, 3D human anatomy models, auto-graded assessments, and performance analytics.

Students in A&P lab can use MindTap to submit their work from each lab activity online. Multiple-choice questions are auto-graded to save time for the instructor; they also provide explanatory feedback to the student. Open-ended questions can be graded by an instructor or TA through the MindTap Gradebook.

Use MindTap for *Co/Engebretson's Anatomy and Physiology Lab Manual 1e* as is, or personalize it to meet your specific course needs. You can also easily integrate MindTap into your Learning Management System (LMS).

MindTap: https://www.cengage.com/mindtap

Instructor Resources

Additional instructor resources for this product are available online. Instructor assets include an Instructor's Manual with answers, PowerPoint slides with images, an Educator Guide, and a test bank powered by Cognero®. Sign up or sign in at www.cengage.com to search for and access this product and its online resources.

Acknowledgments

At Cengage, we would like to thank our forever cheerleader, resource manager, and visionary, Diana Baniak. Brendon Killion is the person with a spreadsheet for a brain that made sure that all the pieces of this lab manual came together and on time. Brendan, we promise not to miss any more deadlines. Michael Zierler is our outstanding editor. Michael's knowledge, patience, and attention to details were instrumental in the writing of this book. Michael, your suggestions are always on point, we are so lucky to have gotten to work with you. This lab manual would not have been possible without the work of a team of artists, photographers, researchers, and copy editors who shared their expertise along the way.

Liz Co would like to thank her coauthor, Hilary. Writing the text was a solo expedition that was lonely at times, I so appreciated your partnership, insight, and laughter in this lab manual voyage! Hilary Engebretson would like to thank her friend-colleagues (also known as "frolleagues") Liz, Lauren, and Tealia. Your willingness to simultaneously support my wild ideas and bring me back to reality represents friendship at its best.

We would like to thank MPS for their work on the production of this lab manual. We would also like to thank our copyeditor, Calum Ross. A special thanks to our photographer Michael Gallitelli and the Cengage Creative Studio team.

How to Learn in an A&P Lab

As an A&P student you will find yourself learning in two different environments. In the classroom or lecture hall, your time will be spent listening, taking notes or perhaps engaged in a structured worksheet or case study. But when you arrive in the A&P laboratory, your time is your own. The difference between these two spaces can be striking. In lecture you might feel as if you are a rider in the backseat of a car. At the beginning of a lecture, you buckle in and experience each turn as your instructor guides you through the material. But in lab, you're the driver. You need to plan your route, know the turns, and have enough fuel for the journey. A&P instructors often observe they can tell which students will be most successful in the course by observing their behavior during lab. In this section of the lab manual, we describe the science of learning and distill the observations of many A&P lab instructors to set you off on a successful journey through the A&P lab.

The Science of Learning

Foreign, Familiar, and Mastery-Level Understanding In order to understand the process of learning, let's take an example of a person learning about the sport of baseball. Let's say our learner's name is Eliot. Eliot is taken to a baseball game for the first time by a friend. He watches intently throughout the game but doesn't understand much about what's going on. By the end of the game, he is familiar with the basic rules of the game. Eliot continues watching baseball on TV and continues to gain familiarity with both the rules and the strategy. Months later, Eliot is invited to play baseball. Though he has watched many games, he has never held a bat, thrown a ball, or worn a glove.

Considering all the time Eliot has spent *watching* baseball, will he be successful as a baseball *player*? Maybe, but probably not. In this example it's easy to see that playing baseball requires both the knowledge gained about the sport through watching as well as the skills involved in hitting, catching, running, and throwing that are gained through practice. Examine **Figure 1**. When he first attended the baseball game, Eliot was in Stage 1, the concepts of baseball were foreign to him. Over the time he spent watching games he moved to Stage 2, he became familiar with baseball. However, Eliot may need to practice the skills involved in playing baseball so that he can move to Stage 3: Baseball Mastery.

Figure 1

Learning in A&P or any other subject is no different. We must first understand the material, but then we need to build the skills involved in *using* that material. In A&P these skills are visual analysis, application of knowledge in a new environment, and making connections among seemingly disparate topics (**Figure 2**).

Figure 2

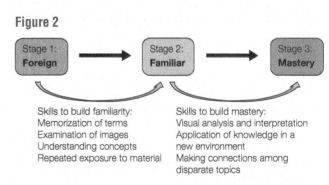

Skills to build familiarity:
Memorization of terms
Examination of images
Understanding concepts
Repeated exposure to material

Skills to build mastery:
Visual analysis and interpretation
Application of knowledge in a new environment
Making connections among disparate topics

Approaching Lab in the Right Mindset

Because of the exploratory nature of the lab environment, students are often afraid of making mistakes. Taking that analogy of moving from the backseat of the car to the driver's seat, it can be overwhelming to be in charge of your own learning. In 2008, Stanford Professor Carol Dweck proposed a way of thinking about how we learn. She describes that some learners view learning as a set of abilities that can be grown through effort and experience. She contrasted these learners, who she described as having "growth mindset," with a set of learners who believed that they either were smart and capable or were not, but that no amount of effort would change these things about them. These learners, who Dweck described as having a "fixed mindset," would become frustrated in environments where they weren't immediately successful and give up easily. For example, imagine a student who scores less than average on their first quiz in a math class. If that student had a growth mindset, they may think "I didn't learn this material well, this is my opportunity to strengthen my knowledge of this math concept." Whereas a fixed mindset learner may think, "Here is evidence that I'm just not a math person." In the time since this insightful paradigm was published, many educational researchers have discovered that students with growth mindsets were much more likely to succeed in difficult courses than students with fixed mindsets. Further, students were more likely to succeed in classes taught by instructors with growth mindsets rather than instructors with fixed mindsets. When you first try something new, it's understandable that you will make mistakes. But what growth mindset teaches us is that mistakes are actually important. Mistakes guide learning by telling you what areas you need to focus on. As you embark on becoming "the driver" of your learning in A&P lab, encourage yourself to try new things, make mistakes and learn from them. The science of learning tells us that this is important work in the process of learning and will lead you to mastery of the material.

Before Lab, During Lab, After Lab

Unless you have skeletons, models, and dissection specimens at home, it's fair to say that your time in lab is critical, because it is your only opportunity to explore the lab materials. Unlike your text, you cannot bring a dissected heart back to your home and continue learning from it, so preparation for your time in lab is essential to make the most of this limited resource. In this section we will discuss what to do before you go to lab to make the most of your time there, and what to do in lab to both maximize your time and prepare materials that you can use afterwards to continue your learning.

Lab is one of your best opportunities to work on transition from Stage 2 of learning to Stage 3 of learning (**Figure 3**). But we know that to move *beyond* Stage 2 you must first be *in* Stage 2. In other words, before arriving in lab, you want to make sure that you are familiar with the content that will be covered that day. Otherwise, you will spend most of your lab time learning what you could have learned before and miss out on the critical time with the lab materials to work on the skills of content mastery. Think of the time you spend outside of lab as an investment to streamline lab time and make it as effective as possible for you to learn.

Figure 3

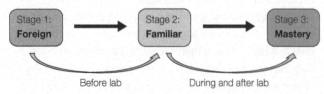

Stage 1: Foreign → Stage 2: Familiar → Stage 3: Mastery

Before lab   During and after lab

Before Lab Read the lab thoughtfully. As you read, you may want to annotate, or write notes to yourself along the way. Ask yourself, what is the purpose of each lab activity? Write down your learning goals.

What are the structures, terms, and ideas you should know before arriving on lab day? Make a list of bolded terms that appear throughout the labs and see which ones you are unfamiliar with. Examine the figures throughout the lab. Which figures in the lab represent background knowledge and which are there to guide you during the lab? For example, **Figure 4** shows figures of the kidney that you might see in a lab on the urinary system. Figure 4A is an illustration designed to help you learn structures clearly. Figure 4B is a photo designed to help you identify structures during your dissection. It would be a waste of lab time to find yourself working on learning Figure 4A. Learn it ahead of time so that while you're in lab you can work on the skill of translating this image to the kidney dissection in front of you. Remember, you want your lab time focused on skill building, not content familiarity.

Use the prelab quiz as a guide to what you should know when you come to lab. Which questions did you get wrong? Which questions did you struggle with? This is feedback highlighting the areas you need to continue working on before you come to lab.

Read with reflection. Ask yourself: what areas do you feel confident about, and which areas are confusing? Come to lab with the intent of clarifying your understanding on these topics.

During Lab Stay focused and organized. You may not be able to spend as much time as you like with the materials, but be sure to make enough time for clarifying the confusing topics and learning goals that you identified when you prepared for lab.

Once you think you've understood or even mastered a concept, ask your instructor to confirm your understanding. Do this especially around concepts you were confused about when you came to lab. Take feedback from your lab instructor, remember that mistakes are growth opportunities, feedback is encouragement to keep working in certain areas. Your mistakes are a roadmap to where you need to focus your study time.

Remember that one of the skills we build in Anatomy & Physiology is visual analysis. You build familiarity with structures by learning them on one image, for example Figure 4A, and then you strengthen your skill of visual analysis by trying to understand the same structure(s) in a different environment or representation, for example Figure 4B. When you are first faced with a new model/image/dissection, find one thing that is familiar to you. Perhaps in Figure 4 it might be the kidney pelvis.

Figure 4

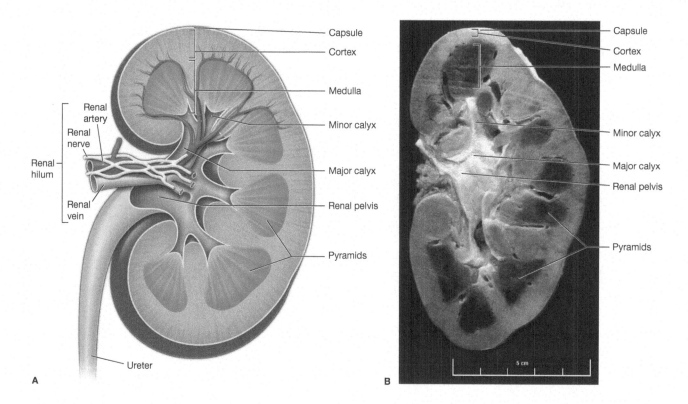

A

Capsule
Cortex
Medulla
Renal artery
Renal nerve
Renal hilum
Renal vein
Minor calyx
Major calyx
Renal pelvis
Pyramids
Ureter

B

Capsule
Cortex
Medulla
Minor calyx
Major calyx
Renal pelvis
Pyramids
5 cm

From that structure, build your understanding outward. If you learned the kidney pelvis is on the medial side of the kidney, what structures are lateral to it? Can you tell the anterior and posterior?

As you move through the lab activities, write notes to your future self. What are things you found confusing or difficult in the lab?

You instructor may have advice for you about whether you are allowed to take photographs of dissection specimens and models in your lab. If your instructor allows photographs, you may be tempted to take pictures to rush through your work or to substitute for time in the lab examining these items. Resist this temptation. Photos cannot convey the nuances of position, texture, and other attributes that you can see when looking at the real thing. There is no substitute for the time you spend examining the three-dimensional material in the lab. Use your photos as back-up references to remind you of what you learned while in the lab.

After Lab As you leave lab, or just after, write a list of the parts of the content that you're still missing or still confused on. What do you need to especially focus on learning on your own? If there were parts of an experiment, or models,

or dissections that you didn't get to examine during lab, ask your instructor if there are opportunities to go back to the lab materials, for example during office hours or an open lab session.

Make notes on what you accomplished in lab, and what the big takeaways or main themes of the lab were.

Make connections among the activities you completed in lab. For example, if your lab had both an anatomy and a physiology component, how do they connect? What anatomical structures were at work in the physiology experiments?

Complete any post-lab assignments, but don't memorize the answers to these questions. You probably will never see those exact questions again, but you may see similar questions on the same topics. Instead, use the questions and your answers as a map for where to review. Which concepts did you struggle with? These are the topics you want to go back and review.

Preparing for Assessments

Test yourself. Write a list of terms or structures from each lab. Ask yourself where you would find them, what would the model or the activity look like? What does that structure do?

An Example: **Structure:** cranial nerve 1 (CN1)

The what and the where. On what model or dissection would you find that?

Brain models and/or sheep brain dissection

Describe it. What does it look like, where is it located?

You would find CN1 by examining the inferior side of the brain. CN1 is the olfactory nerves, which are parallel structures in the anterior of the brain, on the inferior side of the frontal lobe.

What does it do? Would you be able to test for its function? What would that test look like?

CN1, the olfactory nerves, is the nerves that carry smell information to the brain. We would be able to test its function by having a subject perform smell tests of different substances.

Lab Safety

General Safety Rules

The following general safety rules should always be followed while you are in the A&P laboratory. In addition, your instructor may advise you of additional safety rules or modifications to these safety rules.

1 Wash your hands thoroughly with soap and water upon entering the lab and before leaving the lab.

2 Disinfect your lab space at the start and end of each lab session.

3 Wear closed shoes that completely cover your feet. Sandals, open-toe, or open-heel shoes are not acceptable.

4 Restrain all loose clothing, long hair, and dangling jewelry during lab activities. Brimmed hats should be turned up or backward.

5 Do not eat, drink, chew, smoke, apply cosmetics, or handle contact lenses in the lab.

6 If you are pregnant, are immunocompromised, are caring for someone who is immunocompromised, know you are allergic to a lab supply, or are uncomfortable with a lab procedure, then notify your instructor prior to starting the lab.

7 If you are allergic to latex, consult with your instructor about the disposable gloves used in your lab. If you experience any allergic reaction during the lab, notify your instructor immediately.

8 Familiarize yourself with the location of room exits, first aid kit, eyewash station, safety shower, fire extinguisher, fire blanket, and Material Safety Data Sheets (MSDSs).

9 Personal protective equipment (PPE) includes a lab coat or gown, gloves, face mask, and eye protection. Wear PPE as directed by your instructor.

10 If you are wearing PPE and must leave the lab, remove all PPE and wash your hands with soap and water before leaving.

11 Use additional precaution when using dissection equipment. Use cutting blades only in a position facing away from you and your lab partners.

12 Report accidents or injuries to your instructor immediately, no matter how small.

13 Report damaged equipment and broken slides or other broken glassware to your instructor immediately, no matter how small.

14 Return all equipment, models, and other materials used during a lab to their original **location.**

Universal Precautions for Handling Bodily Fluids

Some laboratory procedures require additional safety approaches. These Universal Precautions are standard guidelines that help prevent undue exposure to pathogens or their transmission when working with bodily fluids and specimens. In an educational laboratory setting your instructor will advise you when to employ the following Universal Precautions, but you can assume that you should practice these guidelines when working with all bodily fluids and dissection specimens. If in doubt, practice these precautions.

1 Wash your hands with soap and water before beginning laboratory procedures.

2 Wear disposable gloves and cover any exposed wounds.

3 Wear a face mask and eye protection or a face shield.

4 Wear a clean lab coat, gown, or apron.

5 Handle bodily fluid samples and equipment with care to prevent transfer to other surfaces.

6 Discard waste materials in designated biohazard containers.

7 Your instructor will demonstrate the proper way to remove disposable gloves; follow your instructor's example of this procedure and dispose of the gloves as directed.

8 Wash your hands thoroughly with soap and water immediately after removing disposable gloves.

9 Clean your work area thoroughly with disinfectant and wash your hands with soap and water when finished with lab procedures.

Disposal Guidelines

1 Dispose of sharp items (lancets, needles, etc.) in the designated sharps disposal container.

2 Dispose of broken glassware in the designated broken glass container. Always use the provided broom and dustpan.

3 Dispose of used, unbroken slides in the designated used slide container.

4 Dispose of biohazardous waste in the designated bio-hazard container.

5 If in doubt about how to dispose of any type of waste, ask your instructor.

Additional Safety Guidelines Provided by Your Instructor

Your instructor may have additional safety rules or modifications to the rules provided here. Use this space to take notes as necessary.

Lab Equipment

General Guidelines

The equipment in your lab will be used by hundreds of students. It requires all students to use care and thoughtful attention to ensure that this equipment remains in working order. The following guidelines apply generally to the equipment in the lab. Your instructor may provide additional directions on the care of certain specialized lab equipment, such as microscopes.

1 Use fixed or large pieces of equipment in the location in which they are found. Some larger pieces of equipment have a designated location and should not be carried to another location.

2 Report any broken or damaged equipment to your instructor, no matter how small the damage or problem.

3 Return any item you use in the lab to its designated location so that other students can find it.

4 Do not remove any models from the lab room.

5 Your instructor may have a picture-taking policy; consult with your instructor before taking photographs of any equipment or dissection specimens.

Anatomy and Physiology Models

The many models that are available in an anatomy and physiology lab require student care, as well.

1 Before lifting a model to carry it, identify whether there are loose parts that need to be restrained so that they do not fall.

2 Carefully carry all models with two hands.

3 Never use a pen or pencil to point to a feature of a model. It is much too easy to leave accidental marks on models even if you think you are being careful. Your instructor will provide thin sticks to use to point on models.

4 Never make additional marks or notes on a model with any sort of writing instrument.

5 When you are finished using a model, re-assemble any loose parts and return the model to its proper location.

6 Carefully close drawers and cupboard doors when returning models, ensuring that the model fits the location so that you do not bang or scratch the model.

Books, Atlases, Model Keys, and Other Written Resources

It is likely that your anatomy & physiology lab has books, atlases, model keys, and other written resources for you to use.

1 Do not remove any written resources from the lab.

2 Be careful not to tear pages in these items.

3 Do not write in these items.

4 When you are finished using a book, atlas, or model key, return it to where it belongs.

Disposable Materials in the Lab

Some disposable materials may be regularly available to you. In general, you should feel free to use these items as necessary.

1 Disposable materials that will likely be available include disposable gloves, face masks, alcohol wipes, paper towels, glass slides, and cover slips.

2 Keep containers of disposable materials neat and organized.

3 While you should use these items as necessary, do not waste disposable materials if they aren't needed.

4 If you are unsure of where to dispose of these materials once they are used, consult the Lab Safety section of this lab manual or ask your instructor.

Additional Guidelines Provided by Your Instructor

Your instructor may have additional guidelines and directions. Use this space to take notes as necessary.

Units of Measurement

Measurement	Unit and Abbreviation	Metric Equivalent	Conversion Factor to Imperial units
Length	1 meter (m)		1 m = 1.1 yards = 3.3 feet = 39.4 inches
	1 centimeter (cm)	0.01 m	1 cm = 0.4 inches (1 inch = 2.54 cm)
	1 millimeter (mm)	0.001 m	1 mm = 0.04 inches
	1 micrometer (μm)	0.001 mm or 0.000001 m	
Mass	1 kilogram (kg)	1,000 g	1 kg = 2.2 lbs
	1 gram (g)	1,000 mg	1 g = 0.04 ounces
	1 milligram (mg)	0.001 g	
Volume of Liquid or Gas	1 liter (L)	1,000 mL	1 L = 1.06 quarts
	1 milliliter (mL)	0.001 L	1 mL = 0.03 fluid ounces = 1/5 teaspoon
Time	1 second	1/60 min	
	1 millisecond	0.001 sec	
Temperature	Degrees Celsius (°C)		°F = (9/5) °C + 32 °C = (5/9) °F − 32

Lab 1   Introduction to the Human Body

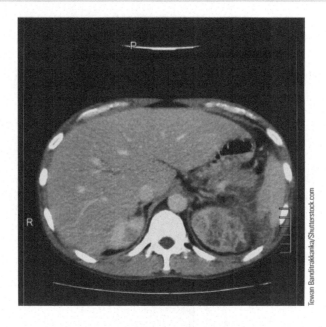

Tewan Bandirakkanka/Shutterstock.com

Learning Objectives: After completing this lab, you should be able to:

1.1   Describe the human body in anatomical position.

1.2   Describe how to use the terms right and left in anatomical reference.

1.3   Describe the location of body structures, using appropriate directional terminology.

1.4   Identify and define the anatomical planes in which a body might be viewed.

1.5   Identify and describe the locations of the body cavities and the major organs found in each cavity.

1.6*   Describe the anatomy of the serous membranes and the cavities they form.

1.7   Identify and describe the location of the four abdominopelvic quadrants and the nine abdominopelvic regions, and the major structures found in each.

* Objective is not a HAPS Learning Goal.

Introduction

Anatomical terminology is both specialized and specific. It is *specialized* in the sense that many anatomical terms are used only within anatomical and medical fields; these are not the everyday terms you may know for body structures. It is *specific* in the sense that the anatomical terms have very precise and detailed meanings. Therefore, when we use correct anatomical terminology we are able to communicate clearly with colleagues without generating confusion or a lack of understanding.

The Human Anatomy and Physiology Society includes more than 1,700 educators who work together to promote excellence in the teaching of this subject area. The HAPS A&P Learning Outcomes measure student mastery of the content typically covered in a two-semester Human A&P curriculum at the undergraduate level. The full Learning Outcomes are available at https://www.hapsweb.org.

Lab 1 Pre-Lab Quiz

This quiz will strengthen your background knowledge in preparation for this lab. For help answering the questions, use your resources to deepen your understanding. The best resource for help on the first five questions is your text, and the best resource for help answering the last five questions is to read the introduction section of each lab activity.

1. Anatomical planes. Label the three anatomical planes represented in this image.

   sagittal plane, frontal (coronal) plane, transverse plane

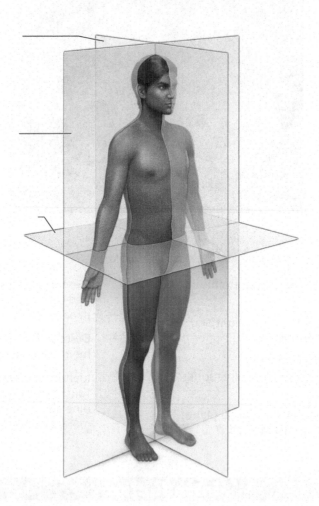

2. Write the root next to its definition:

Root	Meaning
	self or same
	against or opposite
	beside
	all around
	middle or between

List of Roots

contra-
ipsi-
mid-
para-
peri-

3. Which two cavities are located in the dorsal body cavity? Choose two correct answers.

   a. Cranial cavity
   b. Abdominal cavity
   c. Pelvic cavity
   d. Vertebral cavity

4. The serous membrane associated with the lungs is called the:

   a. Pericardium
   b. Pleura
   c. Peritoneum

5. Where is the spleen located?

   a. Left hypochondriac region
   b. Right hypochondriac region
   c. Epigastric region
   d. Hypogastric region

6. When in anatomical position, a person is _____.

   a. standing
   b. sitting
   c. lying face up
   d. lying face down

7. Directional terms are applied to the body in:

   a. The position in which it is being viewed
   b. Anatomical position

8. What is a plane of section that divides the body into right and left halves?

   a. Transverse plane
   b. Frontal plane
   c. Sagittal plane

9. The serous membranes of the body are _____ layer/s thick.

   a. one
   b. two
   c. three

10. Why is the abdominopelvic cavity divided into regions or quadrants?

   a. To reference where each organ and structure is found in the cavity
   b. To separate the structures in the cavity using serous membranes
   c. To provide a guide for incision lines during surgery

(page intentionally left blank)

Activity 1.1 Anatomical Position

This activity targets LOs 1.1 and 1.2.

One of the first steps in communicating clearly with others about the human body is to establish a standard perspective. This is done using the standard of **anatomical position**. In anatomical position a person is standing, feet slightly apart, and in line with the person's hips. Arms are at the person's side, with the palms facing forward. The head and eyes are also facing forward. Take special care with the terms right and left as they apply in anatomy. These terms always refer to the patient's or cadaver's right and left, never to the observer's right and left.

Body position during an examination or surgical procedure is usually in a lying position rather than standing. In these contexts, the body position can be described as prone or supine. **Prone** describes a face-down orientation and **supine** describes a face-up position.

Imagine that a patient is lying supine on the operating table with the surgeon to the patient's left, the anesthesiologist at the patient's head, the surgery technician to the patient's right, and a nurse at the patient's feet. The surgeon says, "we are going to remove the left foot." If the term left refers to the observer's left, such a statement could lead to confusion and increase the chances of error. In contrast, since the term left always refers to the patient's left, there is less room for confusion or mistakes.

Materials
• A lab partner

Instructions
1. Have your lab partner sit or stand in a position other than anatomical position.

2. Determine what components of your lab partner's stance are not in anatomical position and adjust their position to be in anatomical position.

3. Circle all of the components in the figure that you adjusted on your lab partner. For each component that you repositioned, write the important features of what make it correct anatomical position. For example, if your lab partner's left arm was raised over their head, you would circle the left arm in the figure and write something like, "arm at side of body with palm facing forward."

4. Reverse roles so that your lab partner can place you into correct anatomical position.

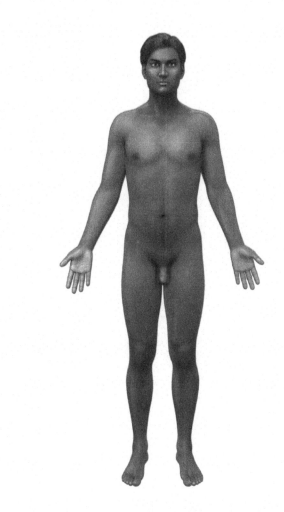

(page intentionally left blank)

Activity 1.2 Directional Terms

This activity targets LO 1.3.

Directional terms are essential for describing the relative locations of different body structures. As you consider the definitions of each directional term listed in Table 1.1 think about how these terms apply to the structures of the human body when it is in anatomical position.

Materials
- List of directional terms in Table 1.1
- Stack of blank sticky notes
- Three-dimensional model of the human body or your lab partner

Learning Connection

Chunking

Think of the directional terms as pairs of opposite terms, like right and left. Learning these terms as pairs allows you to recall both terms together.

Anterior–posterior

Ventral–dorsal

Superior–inferior

Lateral–medial

Proximal–distal

Table 1.1 Directional Terms and Definitions		
Directional Term	**Definition**	**Check When Used in Steps 1 & 2**
Superior	Describes a position above or higher than another part of the body on the head, neck, or trunk	
Inferior	Describes a position below or lower than another part of the body on the head, neck, or trunk	
Anterior	Describes a position toward the front of the body	
Posterior	Describes a position toward the back of the body	
Ventral	Describes a position toward the front of the body	
Dorsal	Describes a position toward the back of the body	
Proximal	Describes a position on a limb that is nearer to the point of attachment or the trunk of the body	
Distal	Describes a position on a limb that is farther from the point of attachment or the trunk of the body	✓
Medial	Describes a direction toward the middle of the body	
Lateral	Describes a direction toward the side of the body	
Ipsilateral	Occurring on the same side of the body	
Contralateral	Occurring on opposite sides of the body	
Superficial	Describes a position closer to the surface of the body	
Deep	Describes a position farther from the surface of the body	

Instructions

1. Write each of the following statements on a sticky note, filling in the blank. Then stick each sticky note to the body structure being described on the model or on your lab partner. The first statement has been completed for you.

   a. The left wrist is *distal* to the left elbow. *(note stuck to the left wrist)*

   b. The right knee is _____ to the right ankle.

   c. The left ear is _____ to the mouth.

d. The nose is _____ to the right eye.

e. The skin of the knee is _____ to the patella (knee cap). And the patella is _____ to the skin of the knee. (use one sticky note for this pair of terms)

2. Complete a sticky note for each directional term in Table 1.1 that you did not already use in step 1. You should have a total of 14 sticky notes stuck to your model or lab partner once you have completed this activity.

Consider why we use anatomical position and these directional terms by engaging in the following scenarios.

Scenario 1

You are working with a new patient, and a previous caregiver has written down the following instructions relaying health issues for this patient:

"The patient's arm closest to me is painful to the touch above their elbow. Just below that painful spot, on the other side of the same arm, they have a circular rash."

1. Mark all of the areas of these instructions that are unclear or for which you would be concerned that you may mistake the wrong area or structure.

2. Re-write these instructions in a way that would be clear to the caregiver on the shift following yours.

Scenario 2

Examine the figure, which shows an injured person at the scene of an accident. How would you accurately describe the location of the arm injury over the phone so that the arriving medical personnel will know specifically where it is?

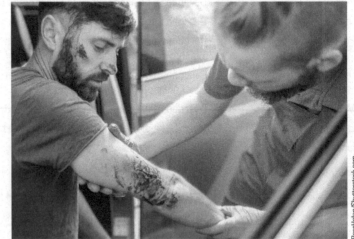

RossHelen/Shutterstock.com

Name: _____ Section: _____

Activity 1.3 Planes of Section

This activity targets LOs 1.3 and 1.4.

Anatomists often use slices through the body or through individual organs to visualize the placement of various structures. A **section** is a slice of a three-dimensional structure such as an organ; a **plane** is an imaginary slice through the body used in imaging.

In the human body a **frontal plane** of section is a slice that divides the body into anterior and posterior sections. A **transverse plane** of section is a slice that divides the body into superior and inferior sections. A **sagittal plane** of section is a slice through the body that divides the body into right and left halves. If the sagittal plane slices through the midline of the body, then it is considered a **midsagittal plane**. Otherwise, if the sagittal plane divides the body unequally, it is considered a **parasagittal plane**. Note that like a frontal or transverse plane, there are an infinite number of parasagittal planes that can be sliced through a body.

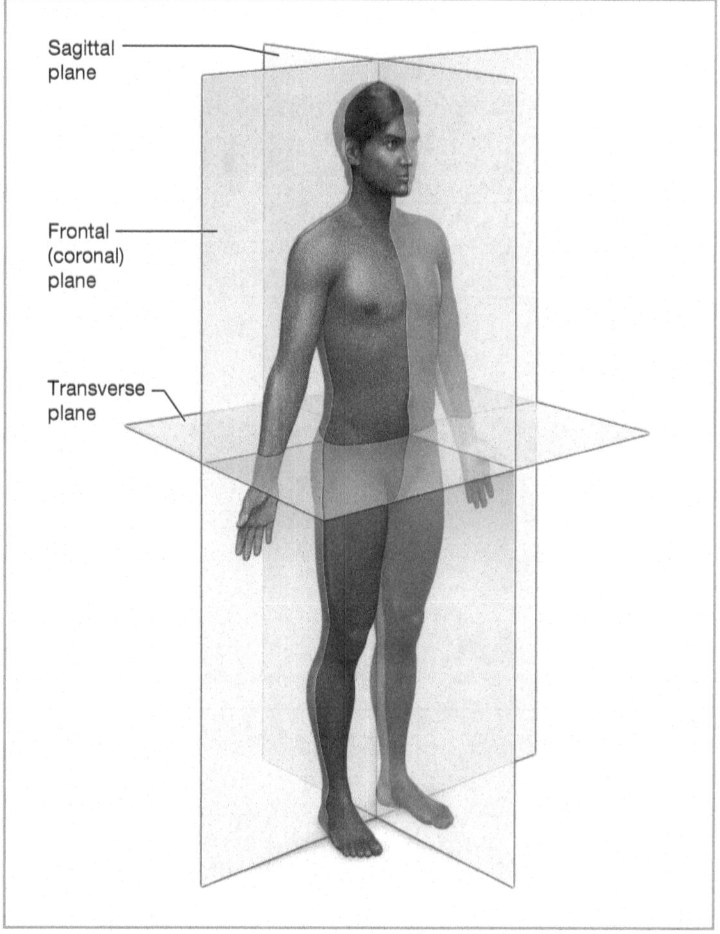

Materials
- Dissection tray or surface to cut on
- Forceps
- Gummy bears (not for human consumption), 4 per student
- Scalpel
- Whole and pre-sectioned cow, pig, or sheep brains for viewing

Figure 1.1 The Three Primary Anatomical Planes of Section Are Shown

Instructions

1. Examine Figure 1.1, which shows each of the anatomical planes of section.

2. Set the gummy bears in the dissecting tray. Use the scalpel to section each gummy bear with a different plane of section, holding the gummy bear with the forceps and not your fingers.

3. Compare your sectioned gummy bears to your lab partner's. Do you agree on the various planes of section?

4. In each of the empty spaces in the Sketches column of Table 1.2, lay out your two sections of gummy bear and trace the pieces, identifying key features (such as ear, nose, arm, etc.) if you are concerned that you might not be able to interpret your tracings later.

5. In the final column, write the directional terms that are associated with the resulting sections of gummy bear. For example, a midsagittal plane will result in equal *left* and *right* sections of gummy bear. You may have more than one pair of directional terms that are associated with a plane of section.

6. When finished, dispose of the sectioned gummy bears in the trash.

7. Carefully clean the scalpels, forceps, and dissecting trays as directed by your instructor.

Table 1.2 Anatomical Planes of Section of the Body		
Anatomical Plane of Section	**Sketches of Resulting Gummy Bear Sections**	**Associated Directional Terms**
Midsagittal plane		*Left and right*
Parasagittal plane		
Frontal plane		
Transverse plane		

Apply what you know about anatomical sections of the whole body to interpret the sections of just one organ of the body, in this case a brain.

8. For each computed tomography scan of the brain in Table 1.3, identify which plane of section has been used to produce it.

9. Examine the whole brain and anatomical sections of preserved brains labeled A through D. Using what you know about the four planes of section and the provided whole brain, match a sample section to each CT scan in Table 1.3.

10. If your instructor provides gloves so that you may handle these preserved brain specimens, dispose of your gloves as directed and wash your hands before proceeding with the rest of the lab.

Table 1.3 Anatomical Planes of Section of the Brain		
Computed Tomography Scan of Brain	Plane of Section	Which Preserved Brain Sample in Lab Shows This Plane of Section? Enter the Letters A, B, C, or D.
Zephyr/Science Source		
Zephyr/Science Source		
Zephyr/Science Source		

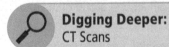

Digging Deeper:
CT Scans

Computed tomography scans are commonly referred to as CT scans in medical settings. A CT scan is actually a series of x-rays that have been taken one after another as the x-ray machine slowly shifts along the body. In this way, each x-ray is taken on the same plane of section, but incrementally shifted. Once all the x-rays have been taken, they can be stacked together by a computer to generate a more complex three-dimensional visualization of the area.

Because a CT scan records structures by using x-rays that bounce off them, it is much better at showing dense structures within the body such as bone rather than the less dense organs. Soft tissue structures such as organs will allow more of the x-rays to pass through and can be harder to see on the resulting scan. To solve this problem, clinicians may inject contrast agents that will reflect the x-rays and show these soft structures more clearly on the scan. Contrast agents are especially useful when imaging blood vessels and the heart. Contrast agents can also be ingested to allow for CT scans of various sections of the digestive system.

Some of the most common reasons for ordering a CT scan are for investigating complex bone fractures, screening for tumors or lesions in the abdominopelvic cavity, and for imaging the head or lungs to look for injuries, tumors, or problems with blood flow.

Activity 1.4 Body Cavities and Serous Membranes

This activity targets LOs 1.5 and 1.6.*

The human body is made up of several cavities, each containing organs surrounded by fluid. These cavities protect the organs within. Note that some cavities contain smaller cavities within them.

Explore these body cavities by labeling the figure with the provided terms. Some cavities are shown on both views of the body. For each body cavity in the figure, list at least one organ that is found there.

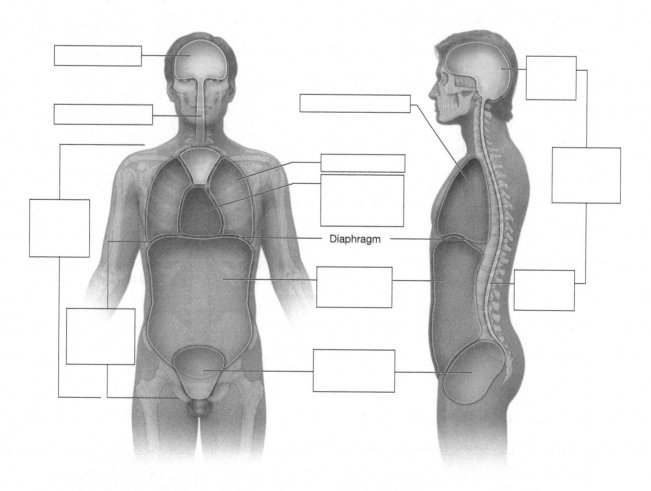

Diaphragm

A. Abdominal cavity
B. Abdominopelvic cavity
C. Anterior body cavity
D. Cranial cavity
E. Pelvic cavity

F. Pericardial cavity
G. Pleural cavity
H. Posterior body cavity
I. Thoracic cavity
J. Vertebral cavity

Many organs of the body are lined and separated from each other by a double layer of thin membranes that wrap around the organ or organs. These **serous membranes**, which can be imagined as thin sheets of tissue similar to the plastic wrap you might use in the kitchen, are two layers with their own specialized serous fluid found in a thin serous cavity between the two layers. The **parietal** layer of the serous membrane faces outward toward the cavity, while the **visceral** layer faces inward touching the surface of the organ or organs it surrounds; between the parietal and visceral layer is the serous cavity containing the serous fluid. There are three serous membranes in the body. Examine Figure 1.2, showing the **pleura**, which is the serous membrane that surrounds the lungs within the thoracic cavity. Similarly, the **pericardium** surrounds the heart within the mediastinum, and the **peritoneum** is the serous membrane that surrounds many of the organs in the abdominopelvic cavity.

Materials

- Bin or dissection tray to work over
- Human heart model
- Large sealable plastic bag such as a 1-gallon zippered bag
- Permanent marker
- Water, 50 ml

Instructions

1. Perform this activity over a large bin or dissection tray so that spilled water is easier to clean up.

2. You will be using the plastic bag to represent serous membranes and the serous cavity they create.

3. Use the permanent marker to write "parietal layer of pericardium" on one side of the bag and "visceral layer of pericardium" on the other side.

4. Fill the plastic bag with 50 ml of water.

5. Seal the bag, pushing out all the air as you close it.

6. Place the apex of the heart model (the more pointed, inferior end; see Figure 1.3) in the center of the closed bag and lift the edges so that it wraps the heart. Be sure that the visceral and parietal layers are oriented correctly.

7. Using Figure 1.3, draw a representation of the way you have wrapped the serous membranes around the heart model. Label your additions as the visceral layer of the pericardium, the parietal layer of pericardium, and the pericardial cavity on the figure.

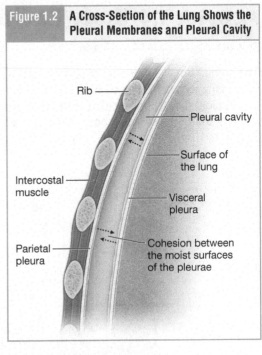

Figure 1.2   A Cross-Section of the Lung Shows the Pleural Membranes and Pleural Cavity

Rib

Pleural cavity

Surface of the lung

Intercostal muscle

Visceral pleura

Parietal pleura

Cohesion between the moist surfaces of the pleurae

8. Answer the following questions.

1. What would you call the space inside the bag where you put the water?

   a. Pleural cavity   c. Pericardial cavity
   b. Mediastinum

2. What does the water inside the bag represent?

   a. Serous fluid   c. Spinal fluid
   b. Blood

3. Is the heart found in the pericardial cavity?

   a. Yes
   b. No

4. Considering the answers to Questions 2 and 3, what is found in the pleural cavity? And in the peritoneal cavity? Explain your answer.

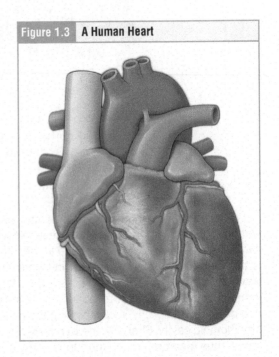

Figure 1.3   A Human Heart

Activity 1.5 Abdominopelvic Regions and Quadrants

This activity targets LO 1.7.

Health care providers divide up the abdominopelvic cavity into either nine regions or four quadrants, thus providing a reference grid for the location of the internal organs of this large space. Dividing up the abdominopelvic cavity in this manner allows a practitioner to localize a patient's pain, suspicious mass, or other concern by narrowing down the possible organs that might be affected.

The nine abdominopelvic regions are created with four lines to create a grid, like a tic-tac-toe game grid. Two vertical lines create three equal columns. The superior horizontal line is along the inferior border of the ribs and the inferior horizontal line is at the superior border of the hip bones. The superior center square is the **epigastric region**, it is bordered laterally by the **right and left hypochondriac regions**. The center square is the **umbilical region** (containing the belly button) and is bordered laterally by the right and left lumbar regions. The inferior center square is the **hypogastric region** and is bordered laterally by the **right and left iliac regions** (also called the right and left inguinal regions).

A simpler division scheme creates four abdominopelvic quadrants: the right upper quadrant, the left upper quadrant, the right lower quadrant, and the left lower quadrant. The vertical plane is along the midsagittal line of the body and the horizontal plane is just superior to the umbilicus (belly button).

Note that these regions and quadrants only pertain to the abdominopelvic cavity and do not apply to the structures of the cranial, vertebral, or thoracic cavities.

Materials
- Lab coat or apron (optional, to protect clothing from tape)
- Masking tape
- Your own body

Instructions

1. Using the description in the introduction, draw lines on the figure on page 16 to represent the nine abdominopelvic regions. Write in the names of each region on the figure.

2. Use your text to locate and name the major organs in each region; write these on the figure.

3. Use masking tape to mark these same lines on your own torso.

4. Using your notes on the figure, and your marked torso, find the location of the following structures on yourself. Note that some structures may span more than one region:

   a. Appendix
   b. Urinary bladder
   c. Spleen
   d. Gall bladder
   e. Small intestine
   f. Stomach
   g. Cecum

   h. Ascending colon
   i. Transverse colon
   j. Descending colon
   k. Sigmoid colon
   l. Rectum
   m. Liver

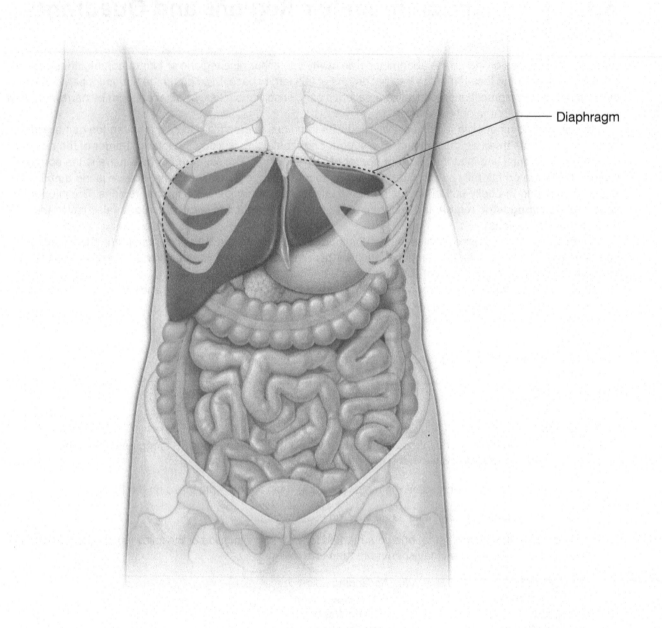

Diaphragm

5. Using the description in the introduction, draw lines on the figure below to represent the four abdominopelvic quadrants. Write in the names of each quadrant on the figure.

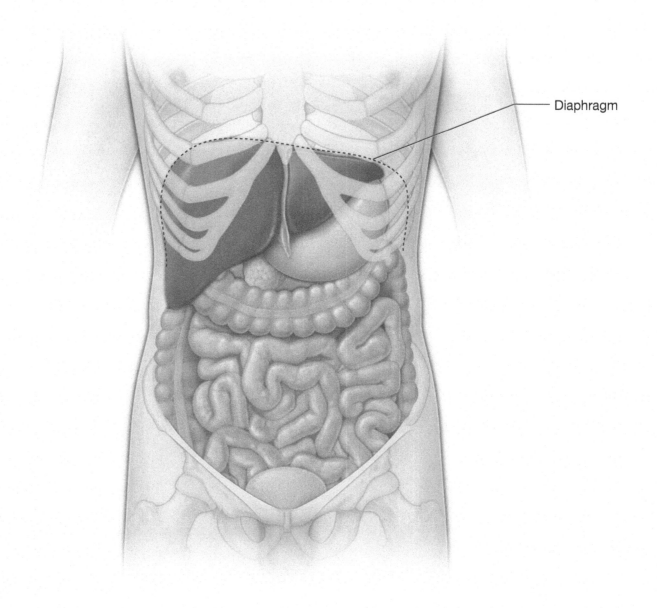

Diaphragm

6. Use masking tape to mark these same lines on your own torso.

7. Use your text to locate and name the major organs in each quadrant; write these on the figure.

8. Using your notes on the figure, and your marked torso, find on yourself the location of the structures listed in step 4. Note that some structures may span more than one quadrant.

(page intentionally left blank)

Activity 1.6 Case Study: Appendicitis

This activity targets LOs 1.1, 1.2, 1.3, 1.6, and 1.7.

Case Study

Yan is at home having dinner with his family when he experiences a sudden sharp pain in his abdomen. Yan also feels nauseous and vomits several times. Yan's spouse drives him to the emergency room. When asked at the emergency room where the pain is located, Yan points to where his abdomen hurts, as shown on Figure 1.4.

1. Yan is experiencing pain in which abdominopelvic quadrant? _____ (fill in blank)

2. Yan is experiencing pain is which abdominopelvic region? _____ (fill in blank)

3. An emergency room nurse wants to give Yan a physical examination and asks him to lie in a supine position. When Yan asks what this means, the nurse says, "please lie _____ on the examination table."
   a. face up
   b. face down

The nurse palpates Yan's abdomen and finds rebound tenderness, pain when pressure is removed from pressing on the abdomen, at a location on Yan's abdomen called McBurney's point. The nurse explains that McBurney's point is located superficially to the point in the body where the appendix attaches to the cecum, which is the first section of the large intestine.

4. Which of the following best explains where McBurney's point is located?
   a. On the nurse's right
   b. On the nurse's left
   c. On Yan's right
   d. On Yan's left

The nurse suspects an inflamed appendix and orders several lab tests. In addition, a CT scan is ordered to visualize the appendix. The results of both the lab tests and the CT scan confirm the nurse's suspicions. Yan's appendicitis is considered acute and the CT scan indicates the appendix may have already ruptured, so emergency surgery is performed.

Figure 1.4	Person Exhibiting Abdominal Discomfort

New Africa/Shutterstock.com

5. Appendicitis is considered emergent because a ruptured appendix can lead to an infection of the serous membranes that wrap the appendix and other structures within the abdominal cavity. Which term would best describe an infection of the serous membranes of the abdominal cavity?

a. Peritonitis

b. Pericarditis

c. Pleurisy

The surgeon decides to perform an open appendectomy because of concern that the appendix may have already ruptured. Once Yan is under general anesthesia, the surgeon makes a small incision in Yan's right lower quadrant. After splitting the muscles in the area, the surgeon will be able to visualize the cecum and appendix. Since both the cecum and appendix are wrapped in the serous membranes of the peritoneum, the surgeon will have to also make an incision through these membranes before being able to operate on the appendix.

6. In what order will the surgeon encounter the peritoneal membranes?

a. The visceral peritoneum will be superficial to the parietal peritoneum.

b. The parietal peritoneum will be superficial to the visceral peritoneum.

c. The parietal peritoneum will be lateral to the visceral peritoneum.

d. The visceral peritoneum will be lateral to the parietal peritoneum.

7. Yan's appendix had ruptured, so after removing the appendix and cleaning the area, the surgeon examines nearby structures in the abdominal cavity for abscesses due to the exposure to pathogens in the contents of the appendix. Which other structures are also found in the right lower quadrant of the abdomen and are therefore the most likely to have been affected? Choose all that apply.

a. Spleen

b. Lungs

c. Ascending colon

d. Small intestine

Lab 1 Post-Lab Quiz

1. The standard anatomical position for the human body will place the thumbs _____ to the other four fingers.

   a. medial
   b. lateral
   c. anterior
   d. posterior

2. The plane of section through the human body shown in the CT scan is a _____ plane.

   a. transverse
   b. frontal
   c. sagittal

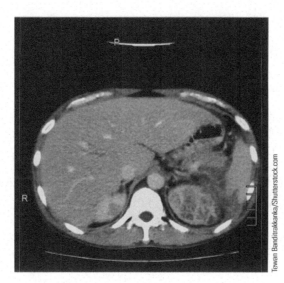

Tewan Banditrukkanka/Shutterstock.com

3. The abdominopelvic cavity is _____ to the thoracic cavity.

   a. lateral
   b. ipsilateral
   c. inferior
   d. distal

4. The fingers are _____ to the wrist.

   a. inferior
   b. superior
   c. distal
   d. proximal

5. The mediastinum contains which of the following structures? Choose all that apply.

   a. Heart
   b. Pericardium
   c. Lungs
   d. Diaphragm

6. The _____ separates the thoracic cavity from the abdominal cavity.

   a. stomach
   b. ribs
   c. diaphragm

7. What layer of serous membrane adheres to the inner wall of the thoracic cavity?

   a. Parietal pericardium
   b. Visceral pericardium
   c. Parietal pleura
   d. Visceral pleura

8. Pericardial effusion occurs when the pericardial membranes produce excess pericardial fluid. What would you expect to be the result of excess pericardial fluid?

   a. The pericardial cavity will expand, putting pressure on the heart.
   b. The heart will contain too much pericardial fluid.
   c. The thoracic cavity will fill with too much pericardial fluid.

9. If a patient presents with pain in the quadrant highlighted in this image, a likely concern might be:

   a. Appendicitis
   b. Stomach ulcer
   c. Inflammation of the gall bladder

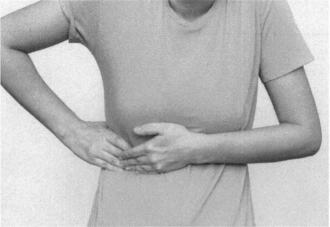

Marina Demeshko/Shutterstock.com

10. What structure marks the superior border of the epigastric region of the abdomen?

    a. The heart
    b. The diaphragm
    c. The stomach

11. A clinician who needs to examine a patient's spleen would do so by palpating (touching or pressing down on) which region?

    a. Right iliac region
    b. Left hypochondriac region
    c. Umbilical region

Lab 2 Surface Anatomy

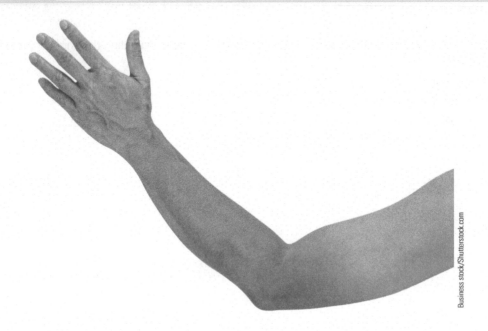

Business stock/Shutterstock.com

Learning Objectives: After completing this lab, you should be able to:

2.1 List and describe the locations of the major anatomical regions of the body.

2.2* Locate and identify soft tissue features on the surface of the human body.

2.3* Locate and identify surface bony landmarks of the human body.

* Objective is not a HAPS Learning Goal.

Introduction

The study of surface anatomy allows us to identify body structures and specific areas that can be seen on the external surface of the human body. Moreover, it also allows us to identify internal structures that are visible as a shape under the skin, can be located by proximity to another surface anatomy structure, or because they are known to be present within a specific regional area. Some surface anatomy structures may not be directly visible, but can be identified through **palpation**. Often these are bony structures that can be palpated due to their hardness relative to other softer internal structures.

Just as there are specific regional terms to allow professionals to clearly discuss various areas of the body, there are specific terms for many surface features that are visible on the human body or ones that can be palpated.

The terms **superficial** and **deep** are often used in describing surface anatomy structures as they allow for a way to clarify how various body structures are layered over or under other structures.

The Human Anatomy and Physiology Society includes more than 1,700 educators who work together to promote excellence in the teaching of this subject area. The HAPS A&P Learning Outcomes measure student mastery of the content typically covered in a two-semester Human A&P curriculum at the undergraduate level. The full Learning Outcomes are available at https://www.hapsweb.org.

Some of the body regions that you locate in Activity 2.1 share their names with surface anatomy structures that you will locate in Activities 2.2 through 2.5. Recognizing the slight differences in terminology between the regions/areas and their associated structures is important to learn. The regions are named as adjectives followed by the word "region" or "area" (which mean the same thing in these cases), while the surface anatomy structures are nouns. For example, the olecranal region refers to the posterior surface of the elbow, while the olecranon is the name of the piece of bone found deep to the skin in the olecranal region.

Why is the study of surface anatomy so important? Most medical professionals will first assess their patient by examining visible characteristics that include surface anatomy features. Being familiar with surface anatomy and what it can reveal below the surface provides considerable information about the patient. A physical exam begins with assessment of features on the surface of the body, including taking a pulse, listening to heart sounds, and palpating organs of the body at very specific surface anatomy locations.

Initially, the list of surface anatomy terms listed in this lab can seem long. But you will be using your pre-lab, time on the lab exercises, and post-lab to learn and practice each of these terms so that you can recall and use them. Beyond exams in this course, you will be expected to continue to use this terminology throughout the rest of the course, in future courses, and in your future career. Moreover, establishing good study techniques for learning and recalling terminology will serve you well for the rest of this course and future courses. Jump to the end of this lab content to read the learning connection titled "Quiz Yourself" for helpful suggestions about how to approach studying terminology.

Lab 2 Pre-Lab Quiz

This quiz will strengthen your background knowledge in preparation for this lab. For help answering the questions, use your resources to deepen your understanding. The best resource for help on the first five questions is your text, and the best resource for help answering the last five questions is to read the introduction section of each lab activity.

1. Anatomy of body regions. Label the image: with the terms listed below

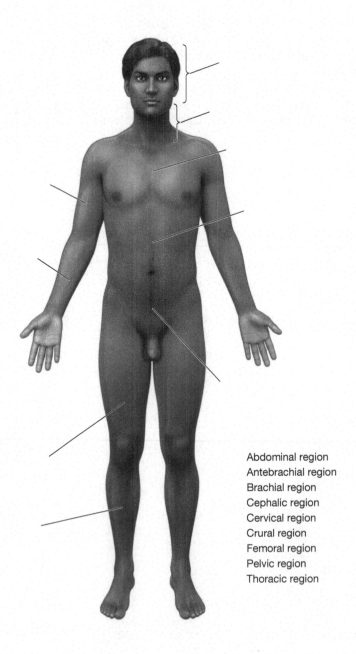

Abdominal region
Antebrachial region
Brachial region
Cephalic region
Cervical region
Crural region
Femoral region
Pelvic region
Thoracic region

2. Write the root next to its definition:

Root	Meaning
	above or on top of
	around
	before
	wing or wing-like structure
	neck or neck-like structure

List of Roots

ala
ante
cervix
peri
super

3. The anatomical region called the brachial region refers to the:

   a. Entire arm
   b. Upper arm only
   c. Lower arm only

4. The lower limb includes which of the following regions? Choose all that apply.

   a. Femoral region
   b. Crural region
   c. Pedal region
   d. Acromial region

5. The antecubital region refers to the area at the _____ of the elbow.

   a. back
   b. front

6. Feeling the surface of the body for a structure that is under the skin is called:

   a. Palpation
   b. Auscultation
   c. Percussion

7. Regional terms are always:

   a. Adjectives
   b. Nouns
   c. Verbs

8. The skin of the human body is _____ to the structures underneath.

   a. superficial
   b. deep

9. The larynx is found in the _____ region.

   a. cranial
   b. cervical
   c. facial

10. The foot makes up the _____ region.

   a. pedal
   b. patellar
   c. umbilical

(page intentionally left blank)

Activity 2.1 **Regional Terms**

This activity targets LO 2.1.

Many areas of the body are named to help anatomists and medical professionals be as specific as possible when they are referencing regions of the human body. Each region has specific boundaries or edges so that they can be carefully defined on a body of any shape or size. Note that all the regional terms listed here are adjectives and not nouns. For example, if you were describing inflammation around a person's nose, you would not say "inflammation to the nasal" but "inflammation to the nasal region" or inflammation to the nasal area."

While the list of regional terms below may appear quite long at first, you are probably familiar with many of them. Use this familiarity as a tool to help you carefully learn each term and the boundaries that define them. There are many more regional terms than what are listed in this figure; in addition, you may come across more than one regional term that applies to the same area, such as "palmar" and "volar."

Materials
• Sticky notes

Instructions
1. Use the figure on the next page to match the regional terms listed with the correct highlighted area. You may need to use your text or other resources to find all the regional terms.
2. Choose ten of the regional terms you identified in the figure and write them on sticky notes. Give the sticky notes to your partner so that they can correctly label themselves. Then switch roles with your partner choosing ten regional terms that have not been used yet.
3. Answer the following questions.

Questions
1. How can you distinguish among the cephalic region, the cranial region, and the cervical region?

2. How are the lumbar region and the abdominal region different from each other?

Regions of the Human Body

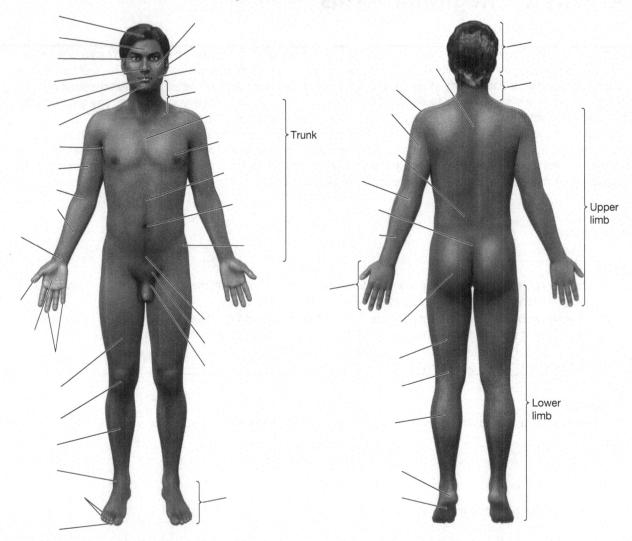

Trunk

Upper limb

Lower limb

Anterior view

Posterior view

A. Axillary region	**R.** Inguinal region	**A.** Acromial region
B. Abdominal region	**S.** Mammary region	**B.** Antebrachial region
C. Antecubital region	**T.** Mental region	**C.** Brachial region
D. Antebrachial region	**U.** Nasal region	**D.** Calcaneal region
E. Brachial region	**V.** Ocular region	**E.** Cephalic region
F. Buccal region	**W.** Oral region	**F.** Cervical region
G. Cranial region	**X.** Otic region	**G.** Dorsal region
H. Carpal region	**Y.** Pubic region	**H.** Femoral region
I. Coxal region	**Z.** Pedal region	**I.** Gluteal region
J. Crural region	**A1.** Pelvic region	**J.** Lumbar region
K. Cervical region	**B1.** Pollex region	**K.** Manual region
L. Digital (or phalangeal) region	**C1.** Palmar region	**L.** Olecranal region
M. Digital (or phalangeal) region	**D1.** Patellar region	**M.** Popliteal region
N. Frontal region	**E1.** Tarsal region	**N.** Plantar region
O. Femoral region	**F1.** Thoracic region	**O.** Sural region
P. Facial region	**G1.** Umbilical (or navel) region	**P.** Sacral region
Q. Hallux region		

Activity 2.2 Surface Anatomy of the Head and Neck

This activity targets LOs 2.2 and 2.3.

The head is made up of both the cranial area and the facial area, while the neck is found in the cervical area. You might have heard the noun cervix before, which means "neck or neck-like structure." Because more than one structure in the body is neck-like, there is more than one cervix in the body.

Because the face has many detailed structures on it, there are many specific surface anatomical structures on the head and neck. Take note of the small differences in these terms so that you can distinguish among them. The list presented here is not a comprehensive list; for example, there are at least five more specific anatomical terms for various surface anatomical features of the ear in addition to the three presented in this lab manual.

Materials
• Model of the human head (to be shared if there are not enough for each lab group)
• Skull

Regions of the Head and Neck

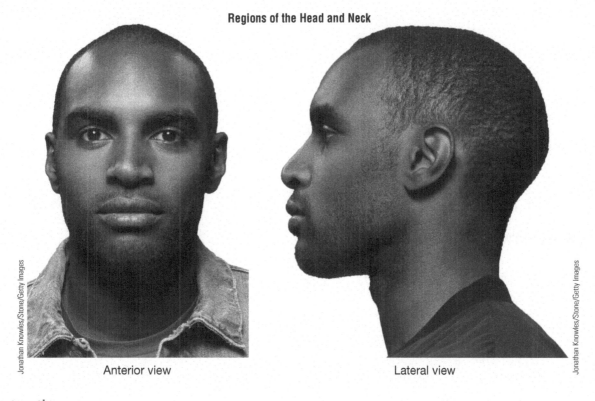

Anterior view   Lateral view

Jonathan Knowles/Stone/Getty Images

Instructions

1. Identify all of the surface anatomy structures listed in Table 2.1 on the figure.

2. Write a definition for each structure in Table 2.1 that will assist you in remembering each term.

3. Determine which regional terms from Activity 2.1 are also represented on the figure. Mark and identify each region on the figure.

4. Use your marked figure to find the requested locations on your own body, your lab partner's body, and on provided models.

5. Locate surface bony landmarks on provided skull, human head model, or images of the skull and head. Then locate and identify these bony landmarks on a living human body covered in skin, such as yourself or your lab partner.

Surface Anatomy Structure	Definition
Table 2.1 Soft Tissue Features and Bony Landmarks of the Head and Neck	
Orbit	
Superciliary arch	
Zygomatic arch	
Maxillary arch	
Mandibular arch	
Dorsum of nose	
Apex of nose	
Ala of nose	
Nares (nostrils)	
Philtrum	
Mental protuberance	
Angle of mandible	
Auricle (pinna)	
Tragus	
External acoustic meatus	
Mastoid process	
External occipital protuberance	
Laryngeal prominence	
Trachea	

Find the following locations on yourself using the model of the human head and the skull as guides.

• Finding the temporomandibular joint

   On your mandible, palpate the skin just anterior to the external acoustic meatus. Open and close your mouth several times and feel for a moving bony structure. This is the condyle of the mandible, which articulates with the temporal bone of the skull at this location and makes up the temporomandibular joint. You may have heard the acronym TMJ used to refer to this joint.

• Finding the carotid pulse point

   Place your fingers (digits 2 and 3) of either your left or right hand just anterior to the ipsilateral external acoustic meatus. Slide your fingers inferiorly until you are able to palpate the angle of the mandible. Continue to slide your fingers approximately 2 to 3 cm more inferiorly until they rest in the depression just inferior to the angle of the mandible. Pause, press gently, and feel for your pulse on the external carotid artery that is found deep to the skin at this location.

Activity 2.3 **Surface Anatomy of the Trunk**

This activity targets LOs 2.2 and 2.3.

The term trunk refers to the thorax, abdomen, pelvis, and perineum. You will find that there are many more surface anatomy features associated with the ventral surface of the trunk than with the dorsal surface.

Materials
• Articulated skeleton
• Human torso model

Regions of the Trunk

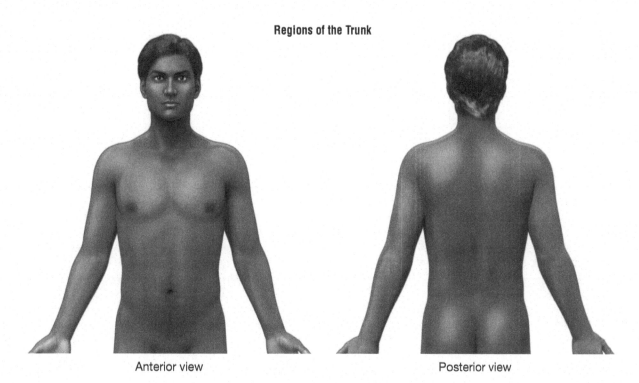

Anterior view   Posterior view

Instructions

1. Identify all surface anatomy structures listed in Table 2.2 on the figure.

2. Write in a definition for each structure in Table 2.2 that will assist you in remembering each term.

3. Determine which regional terms from Activity 2.1 are also represented on this figure. Mark and identify each region on the figure.

4. Use your marked figure to find the requested locations on your own body and/or on provided models.

5. Locate surface bony landmarks on provided skeleton, human torso model, or images of a human skeleton and torso. Then locate and identify these bony landmarks on a living human body covered in skin, such as yourself or your lab partner.

Table 2.2 Soft Tissue Features and Bony Landmarks of the Trunk	
Surface Anatomy Structure	**Definition**
Acromion	
Clavicle	
Jugular notch (suprasternal notch)	
Sternum	
Xiphoid process	
Axilla	
Axillary fold	
Areola	
Nipple	
Umbilicus	
Costal margin	
Iliac crest	
Anterior superior iliac spine	
Inguinal ligament	
Pubic symphysis	
Sacrum	
Perineum*	
Anus*	

(*)These two terms will not be directly visible in the figure. The other items in this activity will clarify these two terms.

Find the following locations on yourself, using the articulated skeleton as a guide.

• <u>Finding the spine of the scapula</u>

Use either hand to find the acromion of the contralateral side. Slide your fingers posteriorly from the acromion; you should feel a ridge of bone that extends medially at an inferior angle. This is the spine of the scapula.

• <u>Finding the spinous processes of the L5 vertebra</u>

Use both hands to locate the iliac crests. Palpate to follow these crests posteriorly. Draw your hands medially to reach each other on your back. Where your hands meet is the location of your most inferior vertebra. You may be able to just palpate the spinous process of this L5 vertebra. Directly inferior to this vertebra your sacrum begins.

Activity 2.4 Surface Anatomy of the Upper Limb

This activity targets LOs 2.2 and 2.3.

The upper limbs of the human body are also called the **pectoral appendages**. These limbs extend from the pectoral girdle of the shoulders. The brachial region, antebrachial region, and manual region are all found within the upper limb.

Materials

• Articulated skeleton

• Human arm model or your own or your lab partner's arm

Regions of the Upper Limb

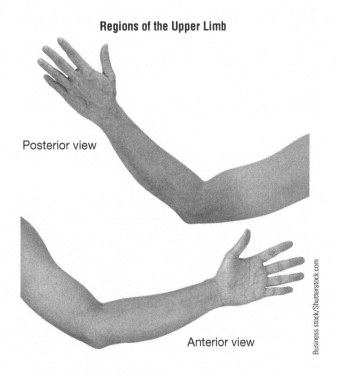

Posterior view

Anterior view

Business stock/Shutterstock.com

Instructions

1. Identify all surface anatomy structures listed in Table 2.3 on the figure.

2. Write in a definition for each structure in Table 2.3 that will assist you in remembering each term.

3. Determine which regional terms from Activity 2.1 are also represented on this figure. Mark and identify each region on the figure.

4. Use your marked figure to find the requested locations on your own body, your lab partner's body, and on provided models.

5. Locate surface bony landmarks on provided skeleton, human arm model, or images of a skeleton and human arm. Then locate and identify these bony landmarks on a living human body covered in skin, such as yourself or your lab partner.

Table 2.3 Soft Tissue Features and Bony Landmarks of the Upper Limb	
Surface Anatomy Structure	**Definition**
Brachium	
Antebrachium	
Cubital fossa (antecubitus)	
Olecranon	
Ulnar styloid process	
Radial styloid process	
Dorsum of hand	
Palm of hand	
Pisiform bone	
Pollex (digit 1 of the hand)	
Thenar eminence	

Find the following locations on yourself, using the articulated skeleton as a guide.

• Finding the radial pulse point

With the palmar surface of your hand facing superiorly, use digits 2 and 3 of your contralateral hand to palpate your carpal region. Locate the radial styloid process. Slide your fingers approximately 1 cm medially along the skin until your fingers are located proximal to the thenar eminence. Pause, press gently, and feel for your pulse on the radial artery that is found deep to the skin at this location.

• Identifying the digits and phalanges of the hand

The pollex, commonly known as the thumb, can also be called "digit 1 of the hand." The remaining four digits of your hand are numbered in sequence moving medially from the pollex. Name, with this numbering system, all five digits on your own hand now. Now examine the sections of each digit. Each section contains a single bone, called a phalanx (plural = phalanges).

How many phalanges does digit 1 of the hand contain?

How many phalanges do digits 2 through 5 each contain?

Digging Deeper:
The Anatomical Snuff Box

The anatomical snuff box is a surface anatomy feature of the dorsal aspect of your hand. This triangular space, shown in the figure, gets its name from the historical use of this depression as a place to pour a small amount of ground tobacco, called snuff, before sniffing it into the nose.

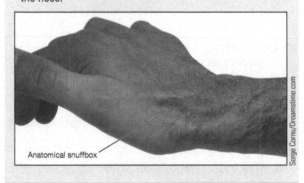

Anatomical snuffbox

Serge Cornu/Dreamstime.com

Activity 2.5 Surface Anatomy of the Lower Limb

This activity targets LOs 2.2 and 2.3.

The lower limbs of the human body are also called the **pelvic appendages**. These limbs extend from the pelvic girdle of the hips. The femoral region, crural region, and pedal region are all found within the lower limb.

Materials
• Articulated skeleton
• Human leg model or your own or your lab partner's leg

Regions of the Lower Limb

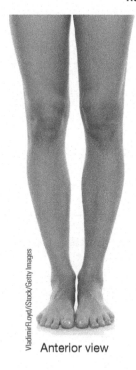

Anterior view

Posterior view

Instructions

1. Identify all surface anatomy structures listed in Table 2.4 on the figure.

2. Write in a definition for each structure in Table 2.4 that will assist you in remembering each term.

3. Determine which regional terms from Activity 2.1 are also represented on this figure. Mark and identify each region on the figure.

4. Use your marked figure to find the requested locations on your own body, your lab partner's body, and on provided models.

5. Locate surface bony landmarks on provided skeleton, human leg model, or images of a skeleton and leg. Then locate and identify these bony landmarks on a living human body covered in skin, such as yourself or your lab partner.

Table 2.4 Soft Tissue Features and Bony Landmarks of the Lower Limb	
Surface Anatomy Structure	Definition
Femoris	
Crus	
Greater trochanter	
Gluteal cleft	
Gluteal fold	
Ischial tuberosity	
Patella	
Tibial tuberosity	
Popliteal fossa	
Lateral malleolus	
Medial malleolus	
Calcaneal tendon	
Calcaneus	
Hallux (digit 1 of the foot)	
Dorsum of foot	
Plantar surface of foot	

Find the following location on the articulated skeleton.

• Finding the ventral gluteal injection site

Place the palmar surface of your non-dominant hand on the greater trochanter of the skeleton. Choose the side of the skeleton that allows your pollex to point anteriorly as you do this. If you are using your right hand, you will be using the left side of the skeleton. With your palm still in place, palpate the anterior superior iliac spine with digit 2 of the same hand. With your palm and digit 2 in place, spread digit 3 posteriorly as if you were making a "peace" sign. The space between digits 2 and 3 marks the injection site. An intramuscular (IM) injection is often given here because of the presence of significant muscle mass that is free from major nerves.

Activity 2.6 Case Study: Car Accident

This activity targets LOs 2.1, 2.2, and 2.3.

Maria was driving to work when she was involved in a car accident. When paramedics arrive on the scene, Maria tells them that she feels some numbness in the area indicated on Figure 2.1.

Case Study

| Figure 2.1 | A Car Accident Injury |

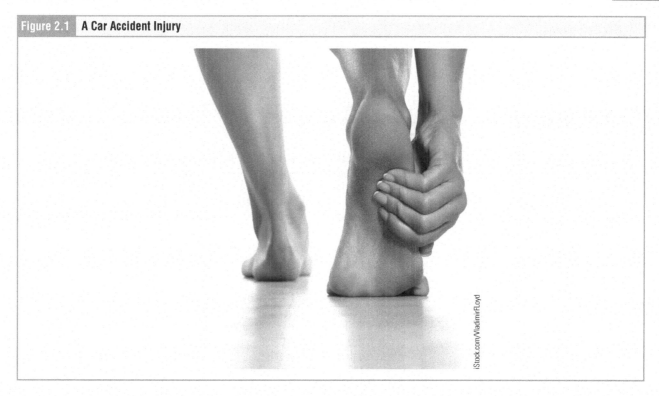

iStock.com/VladimirFLoyd

Questions

1. In their report, the paramedic should note that Maria has numbness in the:
   a. Left patellar region
   b. Right patellar region
   c. Left pedal region
   d. Right pedal region

To assess the situation, the paramedics will determine whether blood flow to Maria's left leg has been compromised by the accident. This is done by palpating the femoral triangle superior to the affected area. The femoral triangle is a triangular-shaped area of skin and the structures deep to that region of skin.

2. The femoral triangle is named for the body region in which it is found. Which body region contains the femoral triangle?
   a. Femoral region
   b. Inguinal region
   c. Frontal region

3. The superior border of the femoral triangle is made up of the inguinal ligament. The inguinal ligament reaches from a medial attachment point just lateral to the pubic symphysis to the:
   a. Tibial tuberosity
   b. Anterior superior iliac spine
   c. Lateral malleolus

The paramedics palpate the skin in the femoral triangle because the femoral artery travels deep to the skin in this area. If the femoral artery has been damaged it may not be delivering enough blood to supply the leg with the necessary oxygen and nutrients.

4. In addition to Maria's foot, what other regions or structures of the body might be affected if the right femoral artery has been damaged? Choose all correct answers.

   a. Crural region
   b. Calcaneus
   c. Clavicle

Learning Connections

Try Writing It

Learning the correct spelling of anatomical terms in this course can be difficult. As an aid, when you are studying, write down the name of each term as you recall it. Then check your notes, text, or lab manual to confirm that you have correctly spelled the term before moving on. Pay special attention to the vowels.

Use mnemonics

In addition, when you find terms that share similar spellings (such as axilla and areola) address the possible confusion head-on by writing both terms next to each other with their definitions or small sketches of each structure. Then find a clue to help you keep the terms separated. For example, you could imagine a large X in your armpit to remember the term axilla.

Learning Connections

Quiz Yourself

This lab introduces you to a significant amount of new terminology. While the pre-lab, lab activities, and post-lab quiz will help you learn and practice these terms, it will likely take more practice to be able to easily and accurately use all of the listed terms. What can you do to review these terms to prepare for tests, use them in this and other classes, and apply them in your future career? Try one of the following approaches:

- Stick blank sticky notes to each surface anatomy structure on your own body. Then, without the aid of your lab notes or a textbook, write in the correct term on each sticky note.

- Find unlabeled images of human bodies and print out two copies of each image. On the first copy, use your lab notes and your textbook to identify all the surface anatomy structures. Then on the second copy, draw lines to the same structures but do not label them yet. Close your notes and textbook and fill in all of the structures that you can on this second copy. Are there structures you could not identify? Check them against your completed first copy to determine which terms need more attention.

- Working with a study partner, have them point to a surface anatomy structure on themselves; name the structure. When you get to a structure you cannot name, write it down. Once you have a list of ten terms that you could not answer, switch roles and repeat the task. For each of you, the list of ten terms is your top priority for study. After a few minutes of study, have your lab partner quiz you again on those top ten structures, then switch roles.

Remember, your goal is to learn, practice, and then check your understanding of these terms. The best way to check your understanding is to quiz yourself in a new way, so after you learn and practice the terms with one of the above-listed approaches, choose a different approach to quiz yourself; the extra challenge of a new approach will help you find the weak spots in your understanding of the material.

Lab 2 Post-Lab Quiz

1. If you wanted to be able to refer to a person's thoracic region and abdominal region at the same time, you could call this their _____.

   a. trunk
   b. upper limb
   c. umbilical region

2. If you received a blow to the kneecap, you could indicate that you have an injury to the:

   a. Popliteal region
   b. Crural region
   c. Patellar region

3. The _____ regions make up the inferior border of the pelvic region.

   a. sacral
   b. inguinal
   c. femoral

4. Identify the bony structure found deep to the skin in the location marked on the image:

   a. Mandibular arch
   b. Maxillary arch
   c. Zygomatic arch

Inside Creative House/Shutterstock.com

5. Identify the structure marked on the image:

   a. axilla
   b. areola
   c. acromion

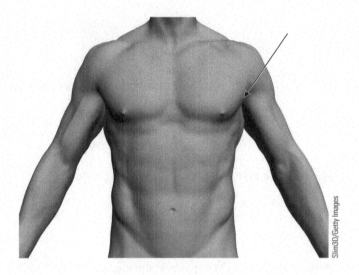

Slim3D/Getty Images

6. Identify the bony structure found deep to the skin in the location marked on the image.

   a. Thenar eminence
   b. Pisiform bone
   c. Ulnar styloid process

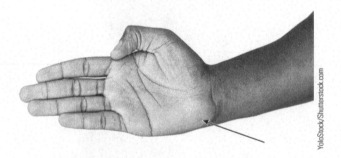

YoloStock/Shutterstock.com

7. Identify the bony structure found deep to the skin in the location marked on the image.

   a. Popliteal fossa

   b. Patella

   c. Hallux

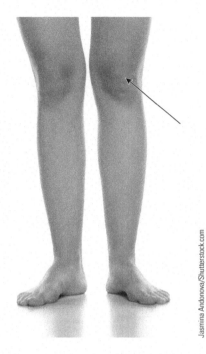

Jasmina Andonova/Shutterstock.com

8. Identify the bony structure found deep to the skin in the location marked on the image.

   a. Calcaneus

   b. Lateral malleolus

   c. Medial malleolus

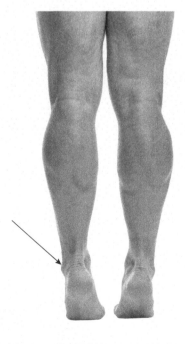

Deman/Shutterstock.com

9. Digit 1 of the human foot is _____ to digits 2–5 of the foot.

   a. medial
   b. lateral

10. How many total phalanges are found in a human's left foot?

    a. 3
    b. 5
    c. 14
    d. 15

Lab 3 Cellular Anatomy

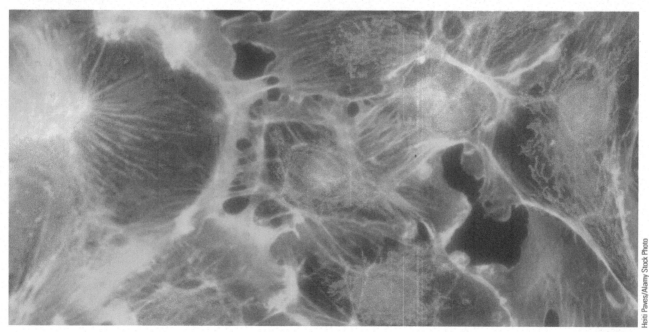

Heiti Paves/Alamy Stock Photo

Learning Objectives: After completing this lab, you should be able to:

3.1 Describe the three main parts of a cell (plasma [cell] membrane, cytoplasm, and nucleus), and explain the general functions of each part.

3.2 Compare and contrast cytoplasm and cytosol.

3.3 Describe the structure and roles of the cytoskeleton.

3.4 Describe the structure of the plasma (cell) membrane, including its composition and arrangement of lipids, proteins, and carbohydrates.

3.5 Describe the functions of different plasma membrane proteins (e.g., structural proteins, receptor proteins, channels).

3.6 Define the term organelle.

3.7 Describe the structure and function of the various cellular organelles.

3.8 Describe the general phases (e.g., G phases, S phase, cellular division) of the cell cycle.

3.9 Compare and contrast chromatin, chromosomes, and chromatids.

3.10 Describe the events that take place during mitosis and cytokinesis.

3.11* Predict the effect of various chemicals on the cell cycle.

* Objective is not a HAPS Learning Goal.

The Human Anatomy and Physiology Society includes more than 1,700 educators who work together to promote excellence in the teaching of this subject area. The HAPS A&P Learning Outcomes measure student mastery of the content typically covered in a two-semester Human A&P curriculum at the undergraduate level. The full Learning Outcomes are available at https://www.hapsweb.org.

Introduction

From tiny red blood cells to extremely long neurons, the 40 trillion cells of your body are as diverse as they are plentiful. Each cell in the body is adapted to its function and its environment. Though they are very small, cells are intricate and complex, possessing a number of different organelles. Each **organelle** is a membrane-enclosed structure with a unique function. The representation of organelles differs among cell types and is based on what the cells require in order to fulfill their functions. All cells have a **cell membrane**, an elegantly flexible phospholipid and protein structure that separates each cell from its outer environment. In this lab we will explore the cell membrane, the cytoplasm including the different organelles and protein structures within it, and the cell nucleus that exists in animal and human cells. We will also consider the diverse appearances of our cells and the mechanism by which cells replicate.

Lab 3 Pre-Lab Quiz

This quiz will strengthen your background knowledge in preparation for this lab. For help answering the questions, use your resources to deepen your understanding. The best resource for help on the first five questions is your text, and the best resource for help answering the last five questions is to read the introduction section of each lab activity.

1. Anatomy of the nucleus. Label the image with the terms listed below:

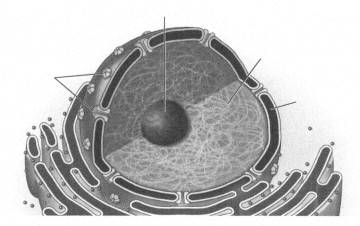

Chromatin
Nuclear envelope
Nuclear pores
Nucleolus

2. Anatomy of a chromosome: Place the following labels on their structures in the diagram:

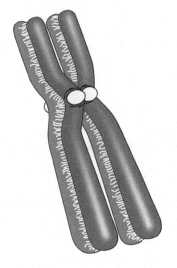

centromere
homologous chromosomes
sister chromatids

3. Write the root next to its definition:

Root	Meaning
	cell
	thread
	body
	color

List of Roots

chromo
cyto
filament
soma

4. In which organelle is the majority of ATP generated during aerobic cellular respiration?

   a. Mitochondria
   b. Lysosomes
   c. Nuclei
   d. Ribosomes

5. How do vesicles move around the cell?

   a. Propelled by flagella
   b. On the cytoskeleton
   c. By diffusion

6. Among the following cell types, which undergoes mitosis most frequently in an otherwise healthy adult?

   a. Skin cells
   b. Skeletal muscle cells
   c. Bone cells (osteoblasts)

7. What is the function of ribosomes?

   a. To break down waste products
   b. To make proteins
   c. To convert nutrients to usable cellular energy

8. What is the difference between chromatin and chromosomes?

   a. Humans have chromosomes and bacteria have chromatin
   b. Chromosomes are condensed packages of chromatin, and chromatin is a complex of DNA and histone proteins
   c. Chromatin is RNA and chromosomes are DNA

9. Chemotherapy drugs are effective against cancer cells because they inhibit _____.

   a. cellular respiration
   b. cellular reproduction
   c. cellular movement

10. A friend hands you a slide of liver cells taken from a healthy patient. Before you even examine the slide under the microscope, what phase of the cell cycle would you anticipate that the majority of cells would be in?

    a. Interphase
    b. Prophase
    c. Metaphase
    d. Anaphase
    e. Telophase

Activity 3.1 Parts of the Cell

This activity targets LOs 3.1, 3.2, 3.3, 3.4, 3.5, 3.6, and 3.7.

The interior of a cell is delimited by its cell membrane. In animals, the cell membrane is composed of phospholipids, cholesterol, and proteins. Phospholipids are the largest contributor to the membrane, forming a lipid bilayer that encapsulates the cell (Figure 3.1). The aqueous inside of the cell along with all its compartments and organelles (minus the nucleus) is called the **cytoplasm**.

The largest of the cell's organelles is its nucleus, which contains the cell's DNA inside of a double membrane (Figure 3.2). The **endoplasmic reticulum (ER)** is a system of flattened sacs that is continuous with the outer membrane of the nucleus. The ER is made of the same lipid bilayer as the nuclear membrane and the cell membrane. The ER provides membranous passages throughout much of

| Figure 3.1 | **Phospholipid Bilayer of Cell Membranes** |

The phospholipid bilayer consists of two sheets of phospholipids. The hydrophobic tails cluster together and the hydrophilic heads orient themselves toward the watery environments on either side.

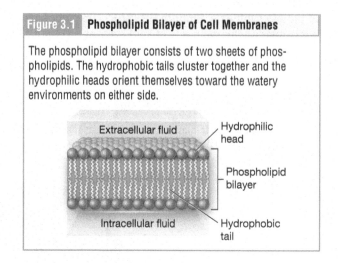

| Figure 3.2 | **A Model Human Cell** |

While this image is not representative of any one particular human cell (i.e., skin cells and muscle cells, for example, do not look like this!), it is a model of a cell containing the primary organelles and internal structures.

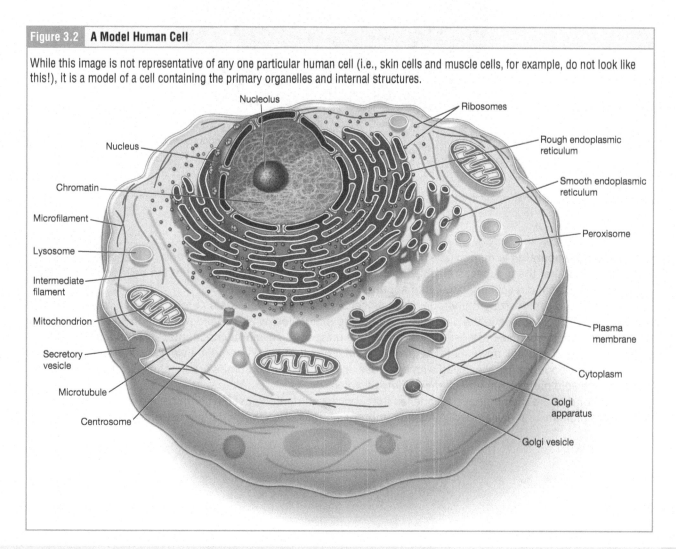

the cell that function in transporting, synthesizing, and storing materials. Endoplasmic reticulum occurs in two forms: rough ER and smooth ER. Each form performs different functions and can be found in different amounts depending on the functions of the cell. **Rough ER (RER)** is so called because it associates with numerous ribosomes attached to the cytosolic side of the ER. **Ribosomes** are large protein–RNA complexes that are the sites of protein synthesis. **Smooth ER (SER)** lacks ribosomes and therefore does not contribute to protein synthesis but is instead the main site of lipid synthesis (Figure 3.3).

Figure 3.3 | Endoplasmic Reticulum (ER)

The ER is a network of thin membrane-bound sacs that surround the nucleus of the cell. The ER can be studded with ribosomes, in which case it is found closer to the nuclear membrane and named *rough ER*, or it can be ribosome-less, in which case it is called *smooth ER*.

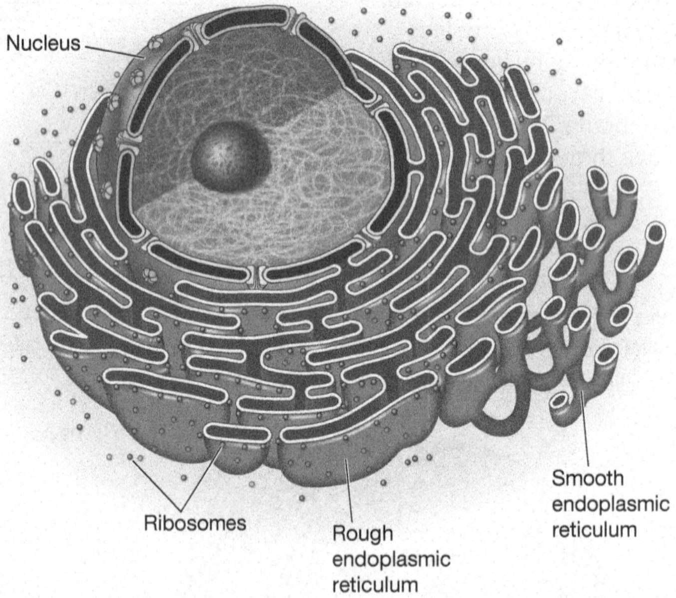

The **Golgi apparatus** is responsible for storing, modifying, and exporting the products that come from the rough ER. Note that the processes involved in creating and exporting products out of the cell require that the nucleus, endoplasmic reticulum, and Golgi apparatus work together (Figure 3.4).

In addition to synthesizing numerous products cells require, a cell must also break down or detoxify components. The SER contributes to the detoxification of some molecules, while breakdown of molecules often occurs in the lysosome. **Lysosomes** are organelles that digest unneeded cellular components, such as a damaged organelle. They also break down material taken in from the external environment. Like lysosomes, **peroxisomes** are membrane-bound organelles filled with enzymes. In function, peroxisomes are most similar to SER, performing both lipid metabolism and chemical detoxification.

Mitochondria are membranous organelles that convert nutrients to usable chemical energy within the cell. A series of proteins and enzymes within the membranes of mitochondria perform most of the biochemical

Figure 3.4 | Golgi Apparatus

The Golgi apparatus modifies the products made by the rough ER. The modified products are packaged in vesicles and sent to the cell membrane for export (exocytosis) or transported to other areas of the cell for use in cellular processes. The Golgi apparatus also produces new organelles called lysosomes.

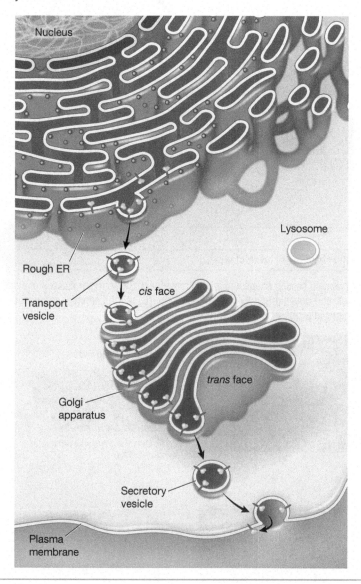

reactions of cellular respiration. These reactions convert energy stored in nutrient molecules (such as glucose) into **adenosine triphosphate (ATP)**, which provides usable cellular energy to the cell (Figure 3.5).

The cell's **cytoskeleton** is made up of fibrous proteins that provide structural support for cells. These fibers are also critical for cell motility, cell reproduction, and transportation of substances and organelles within the cell. The cytoskeleton forms a complex, threadlike network throughout the cell consisting of three different kinds of protein-based filaments: **microfilaments**, **intermediate filaments**, and **microtubules** (Figure 3.6). **Centrioles** and the centrosome serve as an organization point for microtubules, one of the types of cytoskeleton fibers.

There are three types of appendages that can be found on the surface of human cells: microvilli, cilia, and flagella. **Microvilli** are tiny and numerous projections of the cell membrane that serve to expand the cell's surface area where

Figure 3.5 | **Mitochondrion**

The main function of mitochondria is to produce ATP, the cell's major energy currency, through the process of cellular respiration.

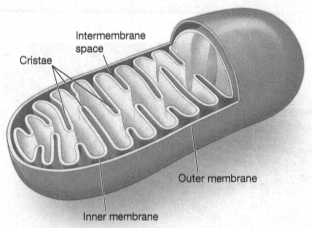

Cristae

Intermembrane space

Outer membrane

Inner membrane

Figure 3.6 | **The Three Components of the Cytoskeleton**

The cytoskeleton functions to maintain cellular shape and enable movement. Three major classes of cytoskeletal fibers are common in human cells: (A) microtubules, composed primarily of tubulin protein, (B) microfilaments, composed primarily of actin protein, and (C) intermediate filaments, composed primarily of keratin protein.

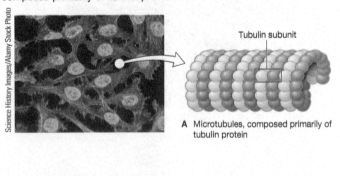

Tubulin subunit

A Microtubules, composed primarily of tubulin protein

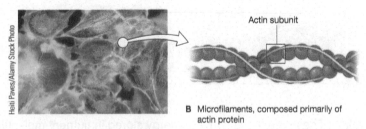

Actin subunit

B Microfilaments, composed primarily of actin protein

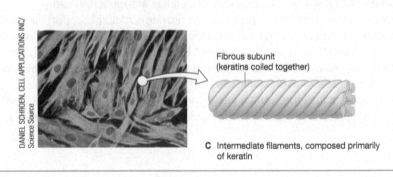

Fibrous subunit (keratins coiled together)

C Intermediate filaments, composed primarily of keratin

it interacts with the extracellular environment (Figure 3.7A). **Cilia** are found on many cells of the body, including the epithelial cells that line the airways of the respiratory system. Like microvilli, there are typically many cilia on the surface of a cell, resembling a hairbrush (Figure 3.7B). But the key difference between microvilli and cilia is that cilia contain extensions of the cytoskeleton, specifically microtubules, and are therefore capable of movement, while microvilli lack microtubules (but do contain microfilaments). Cilia move rhythmically to sweep materials off the surface of the cell. A **flagellum** is an appendage, larger than a cilium or microvillus, that is specialized for propelling the cell through fluid (Figure 3.7C). Cytoskeletal elements (made from microtubules, like cilia) within the flagellum enable its movement.

| Figure 3.7 | Cellular Appendages |

There are three types of human cell surface appendages: (A) Microvilli are associated with microfilaments, but incapable of movement. Their purpose is to expand the surface area of the cell. (B) Cilia are composed of microtubules and are capable of movement. Their function is to sweep material off the surface of the cell. (C) Flagella are very long; the only human cell with a flagellum is a sperm cell. Its function is to propel the sperm toward the egg.

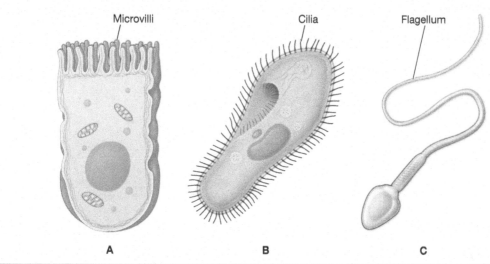

Microvilli   Cilia   Flagellum

A   B   C

Materials
- Models or diagrams of the cell and the plasma membrane
- Sticky notes

Instructions
1. Identify each of the structures listed in Table 3.1 on your model or diagram.

2. If you're working with a model in the lab, write the name of each organelle on a sticky note and label the parts on the model.

3. Complete Table 3.1 with descriptions and functions of each organelle.

4. Complete Table 3.2 with information about cellular appendages.

Learning Connection

Broken Process

For each of the cell parts you've identified, ask yourself what would happen to a cell, an organ, or an individual if that organelle or cell component couldn't perform its function. For example, what processes might be inhibited if the cytoskeleton didn't function?

Table 3.1 Cellular Structures

	Structure	Description of Appearance	Function
Membrane	Phospholipid bilayer		
	Integral membrane proteins		
	Peripheral proteins		
Cytoskeleton	Actin filaments		
	Intermediate filaments		
	Microtubules		
Cytoplasmic organelles	Ribosomes		
	Peroxisomes		
	Mitochondria		
	Rough endoplasmic reticulum		
	Smooth endoplasmic reticulum		
	Golgi apparatus		
	Lysosomes		
	Centrioles		
Nucleus	Nuclear membrane		
	Nuclear pores		
	Chromatin		
	Nucleolus		

Table 3.2 Cellular Appendages

Cellular Appendage	Description of Appearance	Function	Example in the Body
Microvilli			
Cilia			
Flagella			

Digging Deeper:
Antibiotics

Antibiotics are a category of drugs that affect bacterial cells. Bacterial cells are similar to human cells in many ways, but have certain features that are unique to bacteria and not the same as human cells. These differences are targets for drugs to treat bacterial infections because they will disproportionately impact the bacterial cells while leaving the human cells alone. Bacterial cells and human cells both have ribosomes, but the ribosome is different in bacterial cells. Many types of antibiotics, including azithromycin and chloramphenicol, attack the bacterial ribosome and prevent it from functioning. What would be the result of inhibiting the bacterial ribosome? What would these cells be able to do and not be able to do? Other drugs target other cell parts. Daptomycin, for example, disrupts the bacterial cell membrane, and pokes holes in it. What would be the result of introducing holes into a cellular membrane?

Activity 3.2 Mitosis

This activity targets LOs 3.8, 3.9, and 3.10.

All of us originated from a single cell, a fertilized egg. From this one cell, trillions and trillions of rounds of cell replication are needed in order to grow the body you are sitting in right now. Inside your body you are constantly making new white blood cells to fight infections, replacing the epithelial cell lining of your stomach, producing new skin cells, and healing from damage. Cell replication is a constant and lifelong pursuit.

Gametes (egg and sperm), red blood cells, most neurons, and some muscle cells are incapable of replicating themselves, while most somatic cells reproduce often. The **cell cycle** is the sequence of events in the life of a cell from the moment it is created to the moment it divides itself, generating two new cells. One "turn" of the cell cycle consists of three general phases: interphase, mitosis, and cytokinesis (Figure 3.8). **Interphase** is the period of the cell cycle during which the cell is not replicating. Most cellular life is spent in interphase. During interphase a cell preparing to replicate will copy its DNA so that it enters mitosis with one copy for each of the eventual cells. **Mitosis** is the division of the nucleus and its genetic material, during which the nucleus breaks down and two new, fully functional nuclei are formed. **Cytokinesis** divides the cytoplasm (including the daughter nuclei) into two separate cells. Mitosis is separate from meiosis, the process in which germ cells, which give rise to the egg and sperm reproductive cells, are made.

Figure 3.8	Life Cycle of a Cell

Cell replication can be divided into mitosis and cytokinesis. Nonreplicating life, or interphase can be subdivided into G1 (when the cell grows and performs all of its normal functions), S (in which the cell replicates its DNA), and G2 phase (in which the cell prepares its intracellular contents for replication).

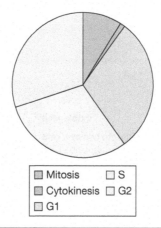

☐ Mitosis　☐ S
☐ Cytokinesis　☐ G2
☐ G1

Interphase

A cell grows and carries out all its metabolic functions and processes in a period called G_1. **G_1 phase** is the phase in which cells are growing, making proteins, and carrying out their functions. Cells that have permanently stopped reproducing are said to be in G_0 phase. For cells that will divide again, G_1 phase is followed by **S phase**, the phase in which a cell replicates all of the DNA in its genome. After S phase the cell proceeds through **G_2 phase** during which the cell prepares for mitosis.

Mitosis

In mitosis, chromosomes are organized so that they can be distributed equally between the two new cells. Remember that DNA was replicated in interphase; within mitosis the two copies of the genomes are distributed in an organized fashion. Mitosis is divided into four major stages: prophase, metaphase, anaphase, and telophase (Figure 3.9). Mitosis is followed by cytokinesis.

During **prophase** the cell's centrioles move to opposite poles and begin to weave a net of microtubules called the **mitotic spindle**. During prophase the cell's duplicated DNA condenses and the **chromosomes** (units of condensed DNA) become easily recognizable. The nuclear membrane begins to disintegrate during prophase, which will allow the DNA to move throughout As **metaphase** begins, the chromosomes have attached to the microtubules of the

Figure 3.9 | **Cell Division: Mitosis Followed by Cytokinesis**

The stages of cell division oversee the separation of identical genetic material into two new nuclei, followed by the division of the cytoplasm.

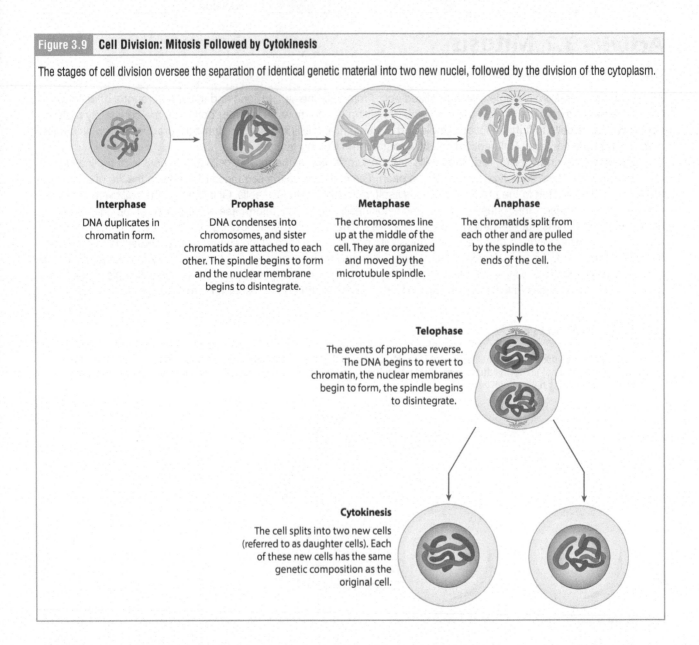

Interphase

DNA duplicates in chromatin form.

Prophase

DNA condenses into chromosomes, and sister chromatids are attached to each other. The spindle begins to form and the nuclear membrane begins to disintegrate.

Metaphase

The chromosomes line up at the middle of the cell. They are organized and moved by the microtubule spindle.

Anaphase

The chromatids split from each other and are pulled by the spindle to the ends of the cell.

Telophase

The events of prophase reverse. The DNA begins to revert to chromatin, the nuclear membranes begin to form, the spindle begins to disintegrate.

Cytokinesis

The cell splits into two new cells (referred to as daughter cells). Each of these new cells has the same genetic composition as the original cell.

spindle. The spindle microtubules begin to pull the chromosomes into a line at the center of the cell. In **anaphase**, the chromosomes, each of which contains two copies of its DNA, are pulled by the microtubules so that the chromatids, the DNA copies, are pulled apart from each other and toward the poles of the cell.

In many ways, telophase is the reverse of the events of prophase. In **telophase** the nuclear membranes begin to reform around the chromosomes. Those chromosomes are decondensing, looking more and more like tangles of DNA instead of discrete DNA bodies. The mitotic spindle begins to break apart. At the end of telophase, each new cell has a full complement of DNA, organelles, and centrioles. In **cytokinesis** the membrane pinches off in the center of the cell (pulled together by cytoskeletal components) and eventually the two cells separate.

Materials

- Classroom models of mitosis
- Colored pencils or pens
- Interlocking beads
- Pipe cleaners in three colors
- Strands of string

Notes to Instructor: It may be helpful to make pre-made kits and put into plastic bags; pipe cleaner pairs could vary by size; there are also commercial chromosome kits

Instructions

We will model the events of mitosis in a theoretical cell with six chromosomes. During S phase, these six chromosomes duplicate, becoming six chromosomes each with two chromatids. Each pipe cleaner represents a chromatid.

1. String one interlocking bead onto each of the pipe cleaners. And snap together the matching pairs of pipe cleaners. This action represents DNA replication in interphase and chromosome condensation in prophase. Examples of pipe cleaners with interlocking beads are shown below.

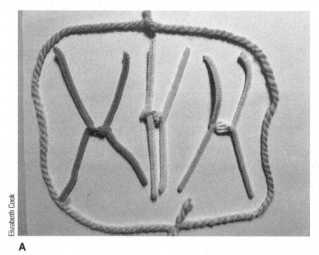 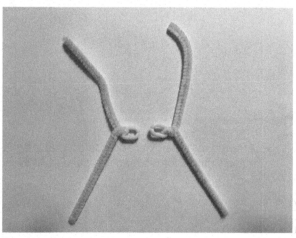

A   B

2. Place two strings together to form a circle around the cell. These strings represent the cell membrane.

3. Arrange the pipe cleaner chromosomes as you would expect to find them in prophase. Use the classroom mitosis model for reference as needed. Draw a cell in prophase in the space provided below.

4. Arrange the pipe cleaner chromosomes as you would expect to find them in metaphase. Use the classroom mitosis model for reference as needed. Draw a cell in metaphase in the space provided below.

5. Arrange the pipe cleaner chromosomes as you would expect to find them in anaphase. Use the classroom mitosis model for reference as needed. Draw a cell in anaphase in the space provided below.

6. Arrange the pipe cleaner chromosomes as you would expect to find them in telophase. Use the classroom mitosis model for reference as needed. Draw a cell in telophase in the space provided below.

7. Now separate the two strings of the membrane, and arrange them around the newly formed cells. Draw the two cells as you would see them at the end of cytokinesis.

Activity 3.3 Case Study: Chemotherapy

This activity targets LO 3.11.

Cancer is an extremely complex condition that has a number of contributing factors. In the US, one in two women and one in three men will develop some form of cancer in their lifetimes. At its core, cancer occurs when cell replication, mitosis, occurs at times and in cells when it shouldn't. As a result, many of the drugs that treat cancer, collectively called chemotherapy, inhibit some portion of the cell cycle. Below a few different chemotherapy drugs are described.

1. One class of chemotherapy drugs is alkylating agents. These drugs damage DNA by pulling nucleotides together (specifically guanine molecules are pulled together and covalently bonded). Which phase(s) of the cell cycle would not be able to go forward in a cell treated with an alkylating agent?

   a. Interphase
   b. Prophase
   c. Metaphase
   d. Anaphase
   e. Telophase
   f. All of these

2. What cells in the body would be targeted if alkylating agents were delivered into the bloodstream?

   a. All cells
   b. All actively replicating cells
   c. Only replicating cancer cells

3. Antimetabolites substitute nucleotides in DNA during DNA replication. Replication proceeds but the cell becomes non-functional. During which phase do these drugs act?

   a. Interphase
   b. Prophase
   c. Metaphase
   d. Anaphase
   e. Telophase
   f. All of these

4. What cells in the body would be targeted if antimetabolites were delivered into the bloodstream?

   a. All cells
   b. All actively replicating cells
   c. Only replicating cancer cells

5. Taxol is an extremely common chemotherapy drug that prevents microtubules from moving within the cell (specifically it prevents microtubule depolymerization). During which phase would a cell become stuck in the cell cycle?

   a. Prophase
   b. Anaphase
   c. Telophase

6. What cells would be targeted by Taxol if it was delivered into the bloodstream?

   a. All cells
   b. All actively replicating cells
   c. Only cancer cells

7. Doxil inhibits an enzyme (topoisomerase) involved in DNA replication. During which phase does this drug act?

   a. Interphase
   b. Prophase
   c. Metaphase
   d. Anaphase
   e. Telophase
   f. All of the stages

8. What cells in the body would be targeted if Doxil was delivered into the bloodstream?

   a. All cells
   b. All actively replicating cells
   c. Only cancer cells

Lab 3 Post-Lab Quiz

1. Neurons make large quantities of proteins called neurotransmitters. We expect that, due to their emphasis on protein manufacturing, they would have lots of _____.
   a. nuclei
   b. ribosomes
   c. mitochondria
   d. lysosomes

2. A stem cell undergoes mitosis frequently. Between one round of mitosis (M phase) and the next, we expect DNA to be in the form of _____ in stem cells.
   a. chromatin
   b. chromosomes
   c. chromatids

3. In mitosis, chromosomes move around the cell. How?
   a. They are highly negatively charged, so they are drawn to the positively charged regions of the cell.
   b. They are pulled through magnetic forces.
   c. They are pulled by the microtubule spindle.

4. Cyanide is a poison that interferes with events that take place inside the mitochondrion. Which of the following is the most likely issue during cyanide poisoning?
   a. Waste products build up in the cell because they cannot be broken down.
   b. Cell functions cease because there is no energy to fuel them.
   c. Cell replication cannot occur because DNA cannot replicate.

5. Liver cells detoxify the blood by breaking down toxins and drugs. In order to perform this breakdown function, we'd expect them to have lots of _____.
   a. nuclei
   b. ribosomes
   c. mitochondria
   d. peroxisomes

6. In what phase of the cell cycle is DNA replicated?
   a. Interphase
   b. Prophase
   c. Metaphase
   d. Anaphase
   e. Telophase
   f. All of these

7. When comparing a cell in prophase versus a cell in G1, which has more DNA?

   a. The cell in prophase has more DNA.
   b. The cell in G1 has more DNA.
   c. The two cells have the same amount of DNA.

8. Are chromosomes attached to the spindle in metaphase?

   a. Yes, chromosomes are attached to the spindle in the beginning of metaphase.
   b. Yes, chromosomes attach to the spindle at the end of metaphase.
   c. No, chromosomes do not attach to the spindle until after metaphase.
   d. No, chromosomes have already detached from the spindle by the beginning of metaphase.

9. Which statement describes an event in anaphase?

   a. The nuclear membrane disappears.
   b. The cytoskeleton causes a pinching of the membrane, leading to the separation of the newly formed cells.
   c. DNA replicates.
   d. The sister chromatids separate because the spindle fibers pull them apart.

10. You are working in a lab that studies cells. You add cells to the center of a Petri dish and place the dish under a microscope to view the cells. As you begin to focus the microscope and the cluster of cells you added comes into view, you feel your phone vibrate in your pocket. You remove your gloves, reach into your pocket to turn the phone to silent, put your gloves back on and then look back in the microscope. All the cells are gone! You scan around and find that they have rapidly spread throughout the dish due to the action of their cellular appendages. Which appendage is responsible for this movement?

    a. Microvilli
    b. Cilia
    c. Flagella

Lab 4   Introduction to the Light Microscope

Dani Kristiani/Shutterstock.com

Learning Objectives: After completing this lab, you should be able to:

4.1* Identify the parts of a compound light microscope and describe their functions.

4.2* Demonstrate the ability to locate an object and bring it into focus.

4.3* Describe the use of stains in histology.

4.4* Produce your own slide for viewing under the microscope.

4.5* Calculate the total magnification and field diameter for one objective lens and use it to estimate the size of the object being viewed.

4.6 Apply an understanding of the magnification power of a compound microscope to its uses in research and medicine.

* Objective is not a HAPS Learning Goal.

Introduction

The microscope was invented by a father–son team who made eyeglasses in 1590. Microscopes were initially used to examine insects, but the first A&P application of microscope use was when Antonie van Leeuwenhoek used a microscope to answer a question he'd wondered for a long time: *what is all this gooey white stuff on my teeth?* Because he examined his own dental plaque Leeuwenhoek became the first person to ever see bacteria! Since Leeuwenhoek's observations were published in the 1680s microscopes have been an integral part of nearly all medical discoveries from identifying the causative agents of diseases like influenza and tuberculosis to studying cancer, reproduction, and Alzheimer's disease. In this lab we will learn (or refresh our knowledge of) how to use the microscope to more closely examine anatomical structures in A&P.

The Human Anatomy and Physiology Society includes more than 1,700 educators who work together to promote excellence in the teaching of this subject area. The HAPS A&P Learning Outcomes measure student mastery of the content typically covered in a two-semester Human A&P curriculum at the undergraduate level. The full Learning Outcomes are available at https://www.hapsweb.org.

Lab 4 Pre-Lab Quiz

This quiz will strengthen your background knowledge in preparation for this lab. For help answering the questions, use your resources to deepen your understanding. The best resource for help on the first five questions is your text, and the best resource for help answering the last five questions is to read the introduction section of each lab activity.

1. Place the following in order from smallest to largest:

    a. Cell

    b. Tissue

    c. Organelle

    d. Organ

2. What type of molecule makes up the majority of the cellular membrane?

    a. DNA

    b. Phospholipids

    c. Cholesterol

    d. Hydrogen ions

3. Methylene blue is a dye that is commonly used to stain cells for viewing under the microscope (in fact, you may use it in the exercises in this lab). Methylene blue is a positively charged molecule. Which of the following statements are TRUE about methylene blue (choose all that are correct):

    a. It will be soluble in an aqueous solution.

    b. It will bind to the fatty acid tails of phospholipids.

    c. It will bind to negatively charged DNA.

    d. It will be specifically attracted to the acidic interiors of lysosomes.

4. What type of cell will you be able to isolate by gently scraping the inside of your cheek?

    a. Muscle cell

    b. Epithelial cell

    c. Nervous cell

5. Of the following types of cells, which do you think you might not be able to see in their entirety in one field of view using this microscope?

    a. Epithelial cells

    b. Fibroblasts

    c. Smooth muscle cells

    d. Neurons

6. Anatomy of a microscope. Complete each sentence in the figure using the words in the following list: arm, coarse, condenser, eyepieces, fine, illuminator, objective lenses, stage, stage control.

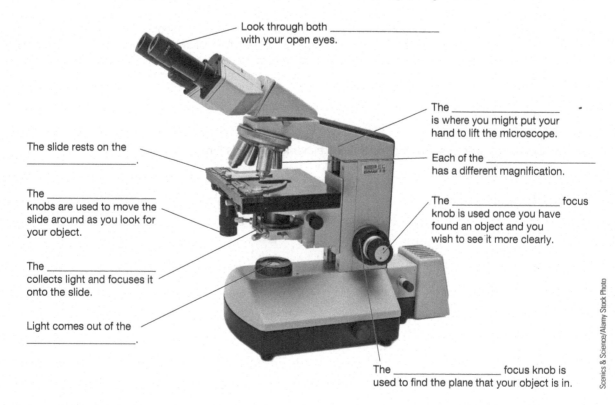

Look through both _____ with your open eyes.

The _____ is where you might put your hand to lift the microscope.

Each of the _____ has a different magnification.

The slide rests on the _____.

The _____ knobs are used to move the slide around as you look for your object.

The _____ focus knob is used once you have found an object and you wish to see it more clearly.

The _____ collects light and focuses it onto the slide.

Light comes out of the _____.

The _____ focus knob is used to find the plane that your object is in.

7. Write the root next to its definition:

Root	Meaning	List of Roots
	small	lumen
	inner part	macro
	large	micro
	light	nuc/nucleus

8. What material should you use to clean the lenses of a microscope?

a. Tissues

b. Fabric or clothing

c. Lens paper

d. Paper towels

9. True or False: If you remove an objective from the microscope and look through it with your eye, it will have the same magnification power as when you look through the microscope with that objective in place.

10. Another student in the lab has found something fascinating under the microscope. They invite you over to their station to view the object. As you approach, you notice that their microscope has the 40× objective in place. As you look through the eyepieces, you are not able to see clearly, you need to adjust the focus for your own eyes. You reach for the focus knobs, which one should you use at **this** moment?

    a. Fine focus.
    b. Coarse focus.
    c. Either knob is fine.
    d. You should not need to adjust the focus using either knob at this time.

Activity 4.1 Parts of the Microscope

This activity targets LO 4.1.

A **compound microscope** is a microscope that uses multiple lenses to magnify an object or specimen. The microscopes in your lab are delicate, expensive, and shared among many students. Therefore, you should carry it carefully and clean it before and after use. Specifically:

- Carry the microscope with one hand under the base supporting the microscope's weight and one hand on the arm lifting the microscope (Figure 4.1). Always carry the microscope with two hands even if you are strong enough to carry it with one.

- Always clean your microscope before and after use, remembering that it was used by another student prior to you and will be used by another student after. We use a special paper called **lens paper** to clean the eyepieces and objectives (Figure 4.2). Lens paper usually comes in booklets, not in a box or a roll. If you're unsure of what to use to clean your microscope, be sure to ask.

- Never remove any parts from the microscope or try to fix it yourself if it appears broken. Always alert your instructor to the issue.

- At the beginning and end of microscope use, always turn the rotating nosepiece so that the lowest power (and shortest!) objective is in place over the stage.

There are at least two lenses within your microscope, each of these lenses magnifies the image of objects on the slide. Your microscope will have one or more than one **objective lens** and each of these should have a number on it with a symbol × next to the number, for example 10×. This number signifies the magnification power of the lens. A 10× objective will magnify

Figure 4.1 | Proper Way to Carry a Microscope

The safest way to carry a microscope is with one hand under the base and another gripping the arm.

Figure 4.2 | Lens Paper Cleans Lenses Safely

The lenses of your microscope are delicate and prone to scratches. Specialized lens paper is used to clean them.

Nithin T A/Shutterstock.com

Table 4.1 The Lenses of the Microscope

Magnification	Name	Used for
4×	Scanning power	Locating the area of the slide to view more closely
10×	Low power	Viewing tissues
40×	High power	Examining cells more closely
100×	Oil immersion	Requires the use of oil between the slide and the lens, infrequently used in A&P but helpful for viewing bacteria

the image of the object(s) on the slide to make them appear 10 times larger than they are. The common objective lenses and their names are listed in Table 4.1.

The other lens present in all microscopes is the **ocular lens**. In each of the eyepieces sits an ocular lens. Almost all ocular lenses are 10× magnifiers, you can assume your ocular lens is as well unless your instructor tells you otherwise. The image that reaches the ocular lens has already been magnified by the objective lens. If you were viewing your slide using a 10× objective lens and you account for the 10× ocular lens, then you are seeing an image of the slide that is 100× larger than the object actually is (Figures 4.3 and 4.4). As you look through your microscope lenses the area of the slide that you can see is called the field of view. As you increase magnification, the field of view progressively narrows.

Materials
• Compound microscope
• Magnifying lens

Instructions
1. Record the magnification of each of the objectives on your microscope and calculate the total magnification of an object viewed with each objective lens and a 10× ocular lens:

   Objective magnification: _____×   Total magnification: _____×

   Objective magnification: _____×   Total magnification: _____×

   Objective magnification: _____×   Total magnification: _____×

   Objective magnification: _____×   Total magnification: _____×

2. Another important part of the microscope is the focus knobs. To understand how focus knobs contribute to microscopy, first take a magnifying lens and hold it between your eyes and this lab manual (or another object).

   a. Choose an object (for example one word or image) that you'd like to focus on.

   b. Find the distance that you need to hold the magnifying lens so that the object appears larger and in crystal-clear focus and you can see the letters or parts of the image clearly.

   c. While keeping your eyes on the focus object, move the magnifying glass closer to your eyes. Does the image become clearer or blurrier?

   d. Hold the lens farther away. The image blurs if you are too far or too close, there is a perfect distance for the lens to view this object.

3. Find something else to focus on that is a different size and shape, such as an anatomical model or poster in the room. Repeat this process of finding the sweet spot in which the object you are viewing is perfectly in focus. You'll notice that the optimal distance for the magnifying lens changes with each object based on their three-dimensional depth and shape.

Figure 4.3 | Total Magnification Is the Product of the Lens Powers

The compound microscope is named compound because it provides two opportunities for magnification. Here we see the letter e magnified only with the objective lens (center), or magnified with both the objective and ocular lenses (right).

The letter e, when written in 10-point font is so small it can barely be seen with the naked eye

The same letter e, magnified 10 times, as an object would be if viewed with just the 10X objective lens

The same letter e, magnified 100 times, as an object would be if viewed with the 10X objective lens and the 10X ocular lens

4. Now turn to your microscope. Make sure there is no slide on the stage. Find the focus knobs. Use the larger knob, the **coarse focus knob**, and turn it while watching the stage.

5. Use the smaller knob, the **fine focus knob**, and turn it while watching the stage.

While you held a magnifying glass and moved that lens in order to adjust the focus, the lenses of your microscope are fixed in place, therefore the stage can move in order to bring the object closer to or farther away from the lens and adjust its focus. The coarse focus knob noticeably moves the stage and can achieve big changes in focus. The fine focus knob moves the stage so slightly that it is hard to see with your eyes.

The last component of microscopy to discuss is the light. Light enables us to view the object on the stage and is especially important as we increase the magnification. The light comes out of the illuminator, which is more-or-less a fancy flashlight incorporated in your microscope. As you turn on any light source (a lamp, a flashlight, or your microscope's illuminator) the light travels outward in all directions (Figure 4.5). The job of the **condenser** is to collect and direct the light that passes through it. It prevents the light from being cast widely but instead directs it to the slide. It does not contribute to magnification. Within the condenser, a **diaphragm**, similar to a camera's iris, can widen or narrow to regulate the amount of light passing through the condenser (Figure 4.6).

Figure 4.4 | Magnification by a Compound Microscope

A compound microscope magnifies an object using two different lenses, the ocular lens and the objective lens.

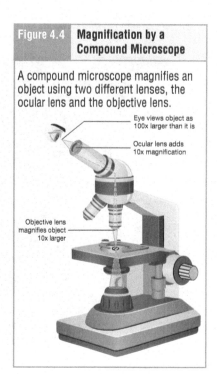

Eye views object as 100x larger than it is

Ocular lens adds 10x magnification

Objective lens magnifies object 10x larger

Figure 4.5 Role of the Condenser

The condenser captures light, which naturally dissipates and spreads out, and focuses it.

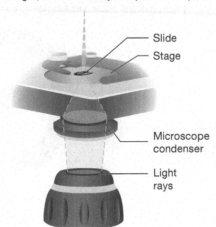

- Slide
- Stage
- Microscope condenser
- Light rays

Figure 4.6 Role of the Diaphragm

Within the condenser is a diaphragm which can widen (left) or narrow (right) to adjust the amount of light passing through.

Dani Kristiani/Shutterstock.com

Name: _____   Section: _____

Activity 4.2 Using the Microscope

This activity targets LO 4.2.

Materials
• Compound microscope
• Ruler in millimeters
• Slide prepared with a letter e or thread on it

Instructions

1. Start with the microscope light source on, and the lowest power objective in place over the stage.

2. Place the slide on the stage, and secure it with the stage clips or slide holder (if your microscope has these). Use the stage control knobs to move the slide so that the letter e or thread is directly below the objective. The light beam should be passing directly through the letter e.

3. While looking at the microscope from the side, use the coarse focus knob to bring the slide and stage as close to the objective as possible.

4. With both eyes open, look through the ocular lenses. If the light is too bright or not bright enough, adjust the diaphragm to allow more or less light through.

5. Once you've adjusted the light, look through the oculars again and use the coarse focus knob to slowly move the slide and stage away from the objective until it comes into focus.

6. Use the fine focus knob to make fine adjustments to the height of the stage to achieve the optimal focus.

7. Measure the working distance using the mm ruler. The **working distance** is the vertical distance between the objective and the slide (Figure 4.7).

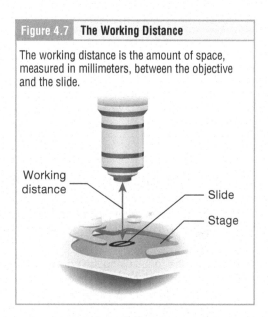

Figure 4.7 **The Working Distance**

The working distance is the amount of space, measured in millimeters, between the objective and the slide.

Working distance

Slide

Stage

8. In the circle below, draw what you see in the field and in the spaces provided, record the total magnification and the working distance at this magnification.

Appearance of the field:

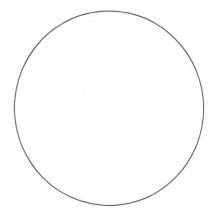

Total magnification used with the lowest power objective: _____.

Working distance with the lowest power objective: _____.

9. Rotate the objectives so that the 10× objective is in place.

10. With both eyes open, look through the ocular lenses. If the light is too bright or not bright enough, you can adjust the diaphragm to allow more or less light through.

11. Adjust the focus using the fine focus knob. DO NOT USE THE COARSE FOCUS KNOB. After you have increased your magnification, the objectives are too large to use the coarse focus knob. You run the risk of hitting and breaking the slide with the objective. Using the fine focus knob only, adjust the focus so that you can see the letter e perfectly.

12. Measure this working distance using the mm ruler.

13. In the space below, draw what you see in the field and record the working distance at this magnification.

Appearance of the field:

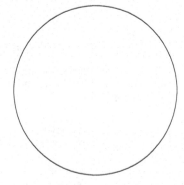

Total magnification used with 10× power objective: _____.

Working distance with the 10× power objective: _____.

14. Rotate the objectives so that the 40× objective is in place.

15. With both eyes open, look through the ocular lenses. If the light is too bright or not bright enough, you can adjust the diaphragm to allow more or less light through.

16. Adjust the focus using the fine focus knob. DO NOT USE THE COARSE FOCUS KNOB. After you have increased your magnification, the objectives are too large to use the coarse focus knob. You run the risk of hitting and breaking the slide with the objective. Using the fine focus knob only, adjust the focus so that you can see the letter e perfectly.

17. Measure this working distance using the mm ruler.

18. In the space below, draw what you see in the field and record the working distance at this magnification.

Appearance of the field:

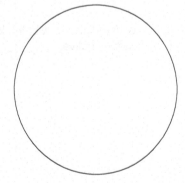

Total magnification used with 40× power objective: _____.

Working distance with the 40× power objective: _____.

Activity 4.3 Slide Preparation

This activity targets LOs 4.3 and 4.4.

Materials
- Clean slide
- Coverslip
- Dropper bottle of methylene blue stain
- Dropper bottle of physiological saline solution
- Filter paper or paper towels
- Toothpicks

Instructions

1. Place a small drop of saline on the slide.

2. Gently scrape the inside of your cheek using a toothpick.

3. Swish the end of the toothpick gently in the saline to release the cells from the toothpick. Discard the toothpick in a biohazard waste receptacle.

4. Your cheek epithelial cells on the slide are nearly ready for viewing, but human cells are fairly transparent. To increase their visibility, we add a stain, in this case **methylene blue**. Add a tiny drop of the methylene blue stain to the saline drop on the slide and use a second toothpick to stir gently and carefully. Discard the toothpick in a biohazard waste receptacle.

5. Use Figure 4.8 as a guide for this step.
   a. Hold a coverslip lightly between your index finger and thumb and lower it slowly to the liquid on the slide. **One tip for keeping your hand steady is to place your elbow or wrist firmly on the countertop.**
   b. Once the coverslip makes contact with the liquid, release it slowly to try to eliminate or reduce air bubbles.
   c. If there is excess fluid around the edges, absorb it using the filter paper or paper towel. The easiest way to do this is to fold the paper towel and use its folded edge to dab at the liquid gently.
   d. Throw this paper towel in the biohazard waste as it may contain cheek cells.

6. Use the directions outlined in Activity 4.2, beginning with step 1 to view your slide.

| Figure 4.8 | Preparing a Wet Mount |

Placing a coverslip on a specimen from an angle reduces the number of air bubbles trapped beneath the coverslip. Blot away any excess liquid with a paper towel.

7. While you always need to start with the lowest magnification, choose the magnification that you felt was optimal for viewing your cheek cells to complete the following:

Appearance of the field:

Total magnification used: _____.

Working distance at this magnification: _____.

8. Follow the directions provided by your instructor for disposal of your slide. You probably need to place the slide in a beaker of bleach solution and dispose of the coverslip in sharps waste.

Digging Deeper:
Staining Cells

Histology is the study of tissues using a microscope. In order to visualize the architecture of the tissue, different stains and preparations can be used to highlight various features of the tissue. Two of the most common stains are hematoxylin and eosin. Hematoxylin is a slightly basic (alkaline) and positively charged stain. It is attracted to DNA and some negatively charged proteins. When it binds, it stains these items a deep blue color. Eosin is acidic and negatively charged. It is attracted to some proteins and organelles and stains them pink. Because these two stains have different net charges, they are attracted to and stain unique sets of molecules. Therefore, they are

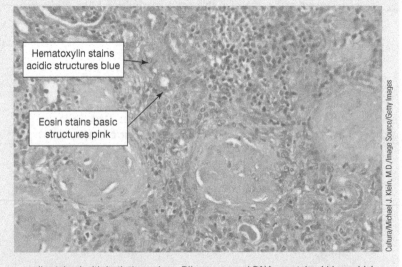

Hematoxylin stains acidic structures blue

Eosin stains basic structures pink

Cultura/Michael J. Klein, M.D./Image Source/Getty Images

frequently used together. For example, in the figure are cells stained with both these dyes. Ribosomes and DNA are stained blue, which allows the nucleus to be clearly differentiated. The cytoplasm stains pink with eosin.

Other commonly used histological stains include Giemsa stain, which is commonly used for blood and bone marrow, and Masson's trichrome which can help highlight collagen fibers.

Learning Connection

Chunking

What structures that you've learned about so far in Anatomy and Physiology would you be able to view under the microscope? Make a list of them. What have you learned about that would be too small to be seen using this type of compound microscope? What are some structures that are too large?

Activity 4.4 Estimate Diameter of Field

This activity targets LO 4.5.

Materials
• Compound microscope
• Grid slide or a slide prepared with graph paper that is ruled in millimeters

Instructions

1. Start with the microscope light source on, and the lowest power objective in place over the stage.

2. Place the slide on the stage and secure it with the stage clips or slide holder (if your microscope has these). Use the stage control knobs to move the slide so that the grid lines are directly below the objective.

3. Use the coarse focus knob to bring the slide and stage as close to the objective as possible.

4. With both eyes open, look through the ocular lenses. If the light is too bright or not bright enough, you can adjust the diaphragm to allow more or less light through.

5. Once you've adjusted the light, look through the oculars again and use the coarse focus knob to slowly move the slide and stage away from the objective until the grid lines come into clearer focus.

6. Now use the fine focus knob to move back and forth until you've found the optimal focus.

7. Measure the working distance using the mm ruler. The **working distance** is the vertical distance between the objective and the slide (see Figure 4.7).

8. Move the slide using the stage control knobs so that one grid line is at the left edge of the field. Now, using Figure 4.9 as a guide, count the number of squares you can see across the diameter of the field. If you can only see part of a square, count it as 0.5.

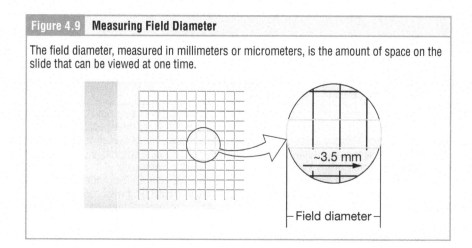

Figure 4.9 | **Measuring Field Diameter**

The field diameter, measured in millimeters or micrometers, is the amount of space on the slide that can be viewed at one time.

~3.5 mm

Field diameter

Table 4.2 Magnification Power of Microscope Lenses

	Scanning Power (4×)		Low Power (10×)		High Power (40×)		Oil Immersion (If Using, 100×)	
Magnification of objective lens	×		×		×		×	
Magnification of ocular lens	10×		10×		10×		10×	
Total magnification	×		×		×		×	
Working distance	mm		mm		mm		mm	
Field diameter	mm	μm	mm	μm	mm	μm	mm	μm

9. Record the field diameter in Table 4.2. List the field diameter in both millimeters (mm) and micrometers (μm). You can use Table 4.3 for help converting to micrometers.

10. Repeat steps 1–9 for each of the objectives.

Table 4.3 Units of Measurement

Metric Unit	Abbreviation/Symbol	Equivalent
Meter	m	
Centimeter	cm	10^{-6}m
milimeter	mm	10^{-7}m
Micrometer	μm	10^{-8}m

| Clinical Correlation | Finding the Causative Agent of Disease |

Pathology is the study of disease and is among the most ancient of sciences. However, it was not until the late 1800s, 150 years after the microscope was invented, that clinicians began using the microscope to understand the changes at the cellular level that occur during disease. Today the microscope is a central instrument to the pursuit of understanding and diagnosing disease. Have you or someone you know ever had a biopsy? The tissue removed during a biopsy procedure probably was examined in several ways, but one of those is to stain the cells and examine the tissue using a compound microscope, much as you have done in this lab. Microscopy has leant some limitations to the study of pathology as well. In 1918, as the worst influenza (flu) pandemic in history was whipping around the globe infecting one out of every three people worldwide, scientists were unable to discover its causative agent. The microscopes at the time, much like the one you're using today, were unable to show viruses, only bacteria could be seen at the resolution of these microscopes. And so the pandemic came and went without anyone understanding what had caused it. It wasn't until a more powerful microscope, the electron microscope, was invented in 1931 that scientists were able to take preserved samples of tissue from patients of the 1918 flu and understand what the causative agent was. Today microscopes are some of our most powerful and widely used tools in understanding disease.

Name: _____ Section: _____

1. As you increase magnification, you probably need to _____ the light intensity.
   a. increase
   b. decrease
   c. not change

2—5. Fill in the blanks to identify the parts of the microscope in the image. Choose from this set of words: arm, coarse focus knob, condenser, eyepieces, fine focus knob, illuminator, objectives, stage, stage control.

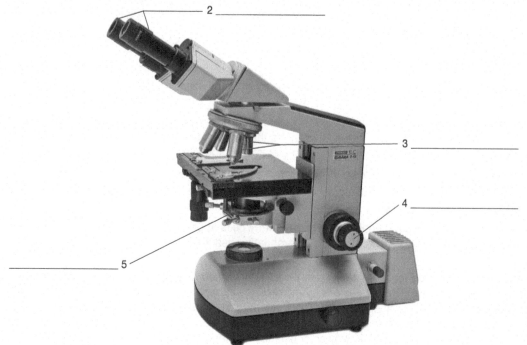

2 _____

3 _____

4 _____

5 _____

6. You're participating in a study abroad program in a different country and as you walk into lab on the first day, you notice that their microscopes have different objectives from what you are used to, though the eyepieces are the same. The instructions in your lab manual tell you to choose the 60× objective. What will the total magnification be of the specimen you're examining?

   a. 6×

   b. 60×

   c. 100×

   d. 600×

7. You've been asked to serve as a peer instructor in a biology class this week. The students are learning how to use a microscope. When called over to a group, you notice that they have a slide on the stage and the 4× objective in place. When you look through the eyepieces, everything is completely out of focus, just a blur. Which focus knob is it appropriate to reach for at this time.

   a. Coarse focus.

   b. Fine focus.

   c. The problem is probably not one that can be solved with adjusting the focus.

8. Think about the features and cell types of the four major tissue types (connective, epithelial, muscle, nervous). Would you be able to differentiate among these types using a compound microscope or would you need a higher-powered microscope such as an electron microscope?

   a. Their individual features are too small to be differentiated using the compound microscope.

   b. The resolution of the compound microscope is too small, tissues are larger, and you would not be able to see more than one cell at a time.

   c. The compound microscope enables me to see many cells at once including some of their features and would be an ideal tool for visualizing and differentiating tissues.

9. As you read in the Introduction, microscopes were first invented over 400 years ago and bacteria were some of the first objects viewed. Tuberculosis is a lung disease caused by bacteria that has been a dominant infectious disease for at least 9,000 years. The bacterium that causes tuberculosis, however, was not discovered until about 140 years ago. If we had the power to see bacteria using microscopes for 400 years, which of the following could explain why scientists took so long to discover this type of bacteria?

   a. These bacteria have highly permeable cell membranes, allowing many substances to pass in and out.

   b. These bacteria have large fats in their cell walls, preventing water-soluble dyes from entering.

   c. These bacteria are too large to be seen clearly using microscopes.

10. The sizes of each of the following objects are listed in parentheses. Using your calculations from Table 4.1, which of these objects would you be able to see in its entirety in the compound microscope using the 40× objective?

    a. Coronavirus particle (0.08 μm)

    b. *E. coli* bacterium (1 μm)

    c. Cardiac muscle cell (100 μm)

    d. Cross-section of the esophagus (30 mm)

Lab 5 Movement Across Biological Membranes

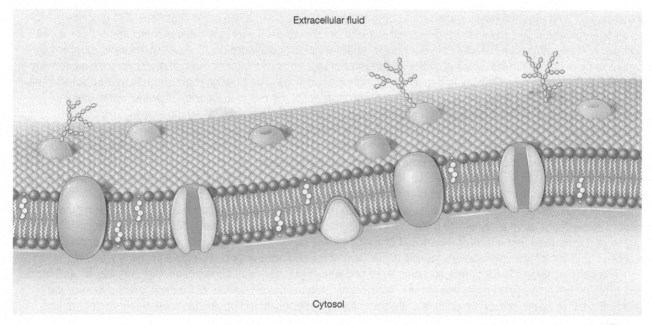

Extracellular fluid

Cytosol

Learning Objectives: After completing this lab, you should be able to:

5.1* Define the terms solution, solute, and solvent.

5.2* Using the properties of a given molecule, predict its relative rate of diffusion.

5.3 Define osmosis and explain how it differs from simple diffusion across membranes.

5.4* Compare solutions and predict the direction of osmosis using the terms hyperosmotic, hypoosmotic, and isosmotic.

5.5 Compare and contrast osmolarity and tonicity of solutions.

5.6 Describe the effects of hypertonic, isotonic, and hypotonic solutions on cells.

5.7* Predict the effects of osmosis and diffusion on different body fluids in physiological circumstances.

5.8 Describe the structure of the plasma (cell) membrane, including its composition and arrangement of lipids, proteins, and carbohydrates.

5.9* Define the terms intracellular fluid and extracellular fluid.

* Objective is not a HAPS Learning Goal.

The Human Anatomy and Physiology Society includes more than 1,700 educators who work together to promote excellence in the teaching of this subject area. The HAPS A&P Learning Outcomes measure student mastery of the content typically covered in a two-semester Human A&P curriculum at the undergraduate level. The full Learning Outcomes are available at https://www.hapsweb.org.

Introduction

Each 40 trillion of the cells that make up the human body is surrounded by a membrane. Just as the outer layer of your skin separates your body from its environment, the cell membrane (also known as the plasma membrane) separates the inner contents of a cell from its external environment. Within the body, cells are living in tissues where they are anchored in place by an extracellular matrix and bathed in watery extracellular fluid. The inside of cells is a water-based fluid too, but ideally a cell would benefit from being able to keep some materials (like glucose) inside and some materials (like the waste products of its neighbors) outside. Therefore, the cell membrane gives a cell some autonomy much like the walls of your house, apartment, or dorm room do. The cell membrane can prevent passage of some molecules, while allowing passage of others; thus, we refer to the cell membrane as **selectively permeable**. Selective permeability is based on molecular properties. The plasma membrane is **hydrophobic**, meaning water-repelling, and since the fluid both inside and outside of the cell is aqueous, meaning water-based, the membrane resists movement through the plasma membrane of water and dissolved molecules. The cell membrane is hydrophobic because it is composed mostly of phospholipids. A single phospholipid molecule has a phosphate group on one end, often called the head of the molecule, and two fatty acid tails (Figure 5.1). The phosphate group is negatively charged, making the head **polar** and **hydrophilic** (water-loving). Polar refers to any molecule that has different charges on one side compared to another. Water, for example, is slightly positive on one end and slightly negative on another, therefore water is polar. The phosphate heads are drawn toward the aqueous environments on the inside and outside of the cell. The fatty acid tails are hydrophobic, and they are drawn toward each other, and tend to exclude, or push away, water molecules. Within the aqueous environment of the body, the phospholipids organize themselves into a bilayer (Figure 5.2) so that their hydrophilic heads face the watery insides and outsides of the cells, and their fatty acid tails are tucked into a core of the membrane and protected from the water.

The cell membrane is not composed entirely of phospholipids, however. Other lipids, like cholesterol, and numerous proteins are also part of a cell membrane. Membrane proteins have a variety of functions, some membrane proteins provide paths across the plasma membrane that allow specific molecules to move in and out of the cell.

The **intracellular fluid (ICF)** on the inside of cells and the **extracellular fluid (ECF)** on the outside of cells is composed of water with a wide variety of dissolved molecules in it. These dissolved molecules and the water they

---

| Figure 5.1 | **Phospholipid Structure** |

A phospholipid molecule consists of a polar phosphate "head," which is hydrophilic, and a nonpolar lipid "tail," which is hydrophobic. Unsaturated fatty acids result in kinks in the hydrophobic tails.

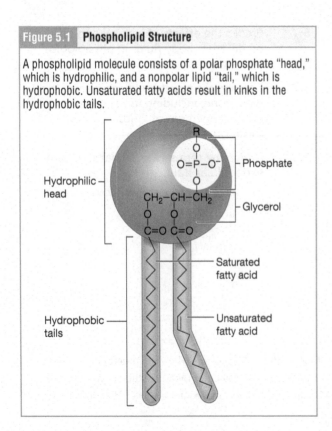

| Figure 5.2 | **Phospholipid Bilayer of Cell Membranes** |

The phospholipid bilayer consists of two sheets of phospholipids. The hydrophobic tails cluster together and the hydrophilic heads orient themselves toward the watery environments on either side.

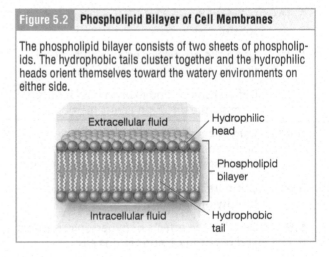

Figure 5.3   **Kinetic Energy of Molecules**

The random kinetic motion of molecules causes collisions and drives the molecules apart from each other.

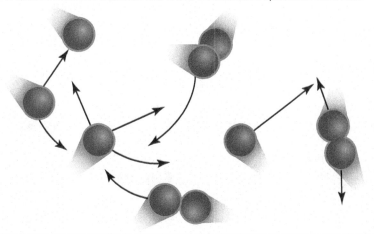

are in move within the ICF and ECF, and some exchange between the two fluids. In this lab we explore these different types of molecular movement.

First, let us consider why molecules move at all. All molecules possess **kinetic energy** and are thus moving constantly. Their motion is random, and molecular collisions occur all the time (Figure 5.3). The closer the molecules are to each other, the more collisions will occur, just as the more crowded a room is, the more likely it is to accidentally bump into another person. If molecules are more closely packed in one place, and more spread out in another, over time and due to their constant random motion, the molecules will spread out until they are evenly distributed throughout the space. This movement of molecules from an area where they are at a higher concentration to an area where they are at a lower concentration is called **diffusion** (Figure 5.4). The difference in concentration between two locations is termed a **concentration gradient**. The driving force for diffusion is kinetic energy, so any environmental factor that impacts kinetic energy will impact the rate at which diffusion occurs. The two most impactful factors in diffusion rate are size and temperature. Kinetic energy is faster at higher temperatures so the rate of diffusion increases as temperature rises. Smaller molecules can move faster than can larger ones, so the size of the molecule will impact the rate of diffusion as well.

Diffusion continues until no gradient remains because molecules are always in motion. Their movement never stops but when there are more of them in a closed space, they bounce off each other more, producing more net movement. Once the molecules are relatively equally distributed, this movement is less noticeable until it finally

Figure 5.4   **Diffusion in Liquids**

(A) Sugar molecules diffuse throughout a hot cup of coffee. They begin in a group at the bottom of the cup, but slowly diffuse, down their concentration gradient, until they are evenly spread out (equilibrium) throughout the coffee liquid. (B) Molecules of dye spread throughout a container of water, down their concentration gradient.

Before diffusion   After diffusion

A   B

The properties of the cell membrane allow small, uncharged molecules such as oxygen and carbon dioxide, and hydrophobic molecules such as lipids, to pass through—down their concentration gradient—by simple diffusion.

Extracellular fluid

Small uncharged molecules

Lipid bilayer (cell membrane)

Time

Time

Cytosol

---

reaches **equilibrium**, the state at which there is no net movement in any direction. Diffusion can occur in a watery environment or in air, it can also occur across biological membranes for molecules that are able to permeate the membrane's hydrophobic core (**Figure 5.5**). CO_2 and O_2 are examples of small, non-polar molecules that can pass through the plasma membrane. When molecules can diffuse across the membrane by dissolving within the hydrophobic membrane core, their movement is called **simple diffusion**. Many molecules, like ions, water, and glucose, cannot easily diffuse across the membrane because they are too large or their properties, such as their polarity, prevent them from diffusing among the membrane's fatty acid tails. These molecules can still move down their concentration gradient across the membrane but they require a channel that allows them to move (**Figure 5.6**). This type of diffusion, which requires a channel, is called **facilitated diffusion**.

Figure 5.6 | **Facilitated Diffusion**

When a molecule's properties prevent it from diffusing through the lipid bilayer, a membrane protein may provide a channel through which the molecule can diffuse.

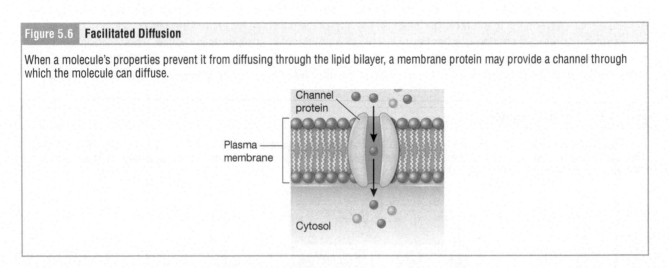

Channel protein

Plasma membrane

Cytosol

Water molecules will also move from areas of greater water molecule concentration to areas of lesser water molecule concentration. **Figure 5.7** illustrates how water molecule concentration is related to solute concentration. The more solute molecules, represented by red dots in Figure 5.7, that are in a solution, the fewer water molecules are in that solution. One way to describe this is that the water molecule concentration and the solute concentration are inversely proportional. If there is a difference in water molecule concentration on either side of a membrane that is permeable to water, the water will move down its concentration gradient to the side of the membrane that has fewer water molecules and more solute molecules (**Figure 5.8**). This movement of water molecules down their concentration gradient is called

Figure 5.7 **Water Molecule Concentration Is Related to Solute Concentration**

Think of a solution as being made up of a combination of solute and water molecules; the more solute molecules in the solution, the lower the concentration of water molecules.

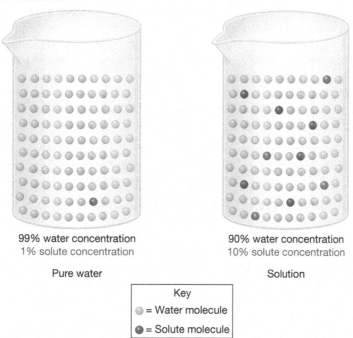

99% water concentration
1% solute concentration

Pure water

90% water concentration
10% solute concentration

Solution

Key
○ = Water molecule
● = Solute molecule

Figure 5.8 **Osmosis**

Osmosis is the diffusion of water across a membrane down its concentration gradient. The membrane illustrated here is permeable to water, but not to the solute. (A) Depicts the two solutions at the moment they are placed in the beaker. Over time, water molecules will move toward the area where they are in a lower concentration (in other words, where the solute is in a higher concentration) on the right side of the beaker. (B) Illustrates the volumes on each side of the membrane after the water molecules have reached a dynamic equilibrium. Both solutions are now at equal water and solute concentrations, but have different volumes.

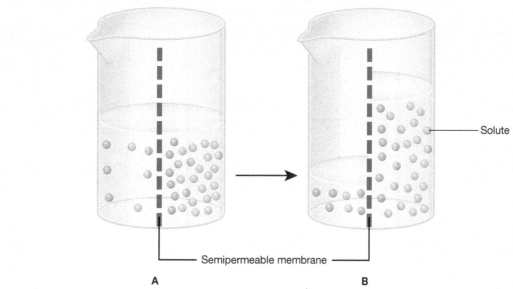

Solute

Semipermeable membrane

A   B

osmosis. Water molecules can move across cell membranes very slowly, but the membrane is largely impermeable to water. Instead, efficient osmosis occurs through water molecule channels called **aquaporins** that are embedded in the plasma membrane.

The rate at which molecules are able to cross the membrane depends on four different factors. First, the degree of the concentration gradient will determine the rate of diffusion. That is to say, if there is a significant difference between one location and another, the rate will be faster than if there is only a slight difference. Second, the amount of surface area across which diffusion can take place will affect the rate of diffusion, with a larger surface area permitting more molecules to diffuse at any one time, therefore increasing the rate of diffusion. Third, the permeability of the membrane to the diffusing molecule influences the rate at which diffusion may occur. This is true for both simple and facilitated diffusion. And fourth, the rate of diffusion is dependent on membrane thickness, which can impede molecular movement. This applies only to simple diffusion. The contributions of these four factors on the rate of diffusion of a molecule through a membrane are expressed in Fick's law, which is:

$$\text{Rate of diffusion} = \frac{\text{membrane permeability} \times \text{surface area} \times \text{concentration gradient}}{\text{membrane thickness}}$$

As far as a cell is concerned, diffusion across its plasma membrane and osmosis are forms of **passive transport**; that is to say, they do not require the cell to invest any energy to move molecules. The kinetic energy of the molecules is all that's required.

Lab 5 Pre-Lab Quiz

This quiz will strengthen your background knowledge in preparation for this lab. For help answering the questions, use your resources to deepen your understanding. The best resource for help on the first five questions is your text, and the best resource for help answering the last five questions is to read the introduction section of each lab activity.

1. Anatomy of a cell membrane. Write the following labels on the image:

channel protein
transmembrane protein
peripheral protein
phospholipid heads
phospholipid tails

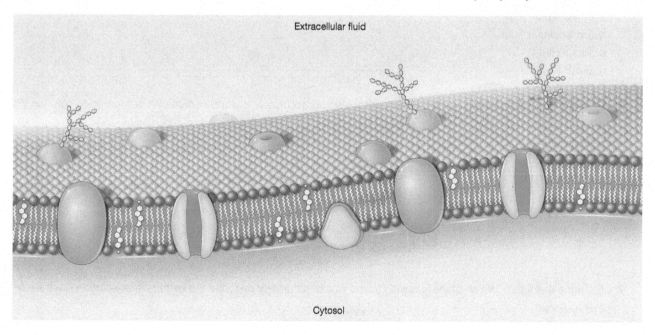

2. Write the root next to its definition:

Root	Meaning
	beloved
	water
	fear
	push
	equal
	over or excessive
	under or less

List of Roots

hydro-
hyper-
hypo-
iso-
osmo-
-philic
-phobic

3. How do water molecules most efficiently cross cell membranes?

   a. Simple diffusion
   b. Osmosis
   c. Active transport using pumps

4. Water is a _____ molecule, and therefore _____ molecules dissolve in water.

   a. polar; polar
   b. non-polar; non-polar
   c. polar; non-polar
   d. non-polar; polar

5. What is the term for the fluid found outside of cells?

   a. Intracellular fluid
   b. Extracellular fluid
   c. Tissue fluid
   d. Serum

6. Examine this figure. Which of the following terms would accurately describe solution X compared to solution Y?

   a. Hypertonic
   b. Isosmotic
   c. Hypoosmotic
   d. Hyperosmotic

Solution X

Solution Y

7. Examine this figure. Which of the following terms would accurately describe the extracellular solution illustrated?

   a. Hypotonic
   b. Isosmotic
   c. Hypoosmotic
   d. Hypertonic

8. What do you expect might happen to the cell illustrated in question 7 after a few minutes of exposure to the solute concentration outside the cell?

   a. The cell should be perfectly fine, no change in shape or size.
   b. The cell will likely crenate (shrivel).
   c. The cell will swell, it may even burst.

9. What is the correct term to describe the green dots in the figure in question 7?

    a. Solution

    b. Solvent

    c. Solute

    d. None of these terms are correct

10. Which of the following is TRUE of kinetic energy?

    a. Only a few molecules move by kinetic energy.

    b. All molecules move by kinetic energy.

    c. All molecules except water move by kinetic energy.

    d. Water is the only type of molecule that moves by kinetic energy.

(page intentionally left blank)

Activity 5.1 Diffusion through Two Media

This activity targets LOs 5.1 and 5.2.

In the introduction to this lab, we discussed that molecular size influences the rate or ability of the molecule to diffuse. You can visualize this diffusion as a person trying to squeeze through a crowd. Smaller people would have an easier time weaving through the crowd and would get through faster. But it's not just the size of the molecule that matters, the medium that the molecule is trying to diffuse through impacts the rate of diffusion as well, just as the density of the crowd would impact a person's ability to get from one place to another. When examining diffusion, we use the words solute and solvent. The **solute** is the material that is dissolving in the liquid, its molecules are spreading out from the area of greater concentration to areas of lesser concentration. The **solvent** is the substance in which the solute is diffusing. In the illustrative example of a person moving through a crowd, the person is the solute, the crowd is the solvent.

In this lab we will examine the rate of diffusion of two different molecules, potassium permanganate and methylene blue. The properties of these two molecules are listed in Table 5.1.

Table 5.1 Properties of Molecules Used in Activity 5.1			
Solute or Solvent?	**Substance**	**Description**	**Molecular Weight**
Solute	Potassium permanganate ($KMnO_4$)	Purple dye	158 amu
Solute	Methylene blue ($C_{16}H_{18}ClN_3S$)	Blue dye	320 amu
Solvent	Agar	1% w/v semi-solid polysaccharide mixture	336 amu
Solvent	Water (H_2O)	Universal solvent	18 amu

We will examine the diffusion of these two molecules through two different substances. The first is water, which is considered the universal solvent. The ICF and ECF are aqueous fluids, so this rate of diffusion will resemble diffusion in cellular fluids. The other medium we will examine diffusion in is agar, a semi-solid polysaccharide that resembles the texture of Jello. The properties of these two substances are listed in Table 5.1.

Materials
- Distilled water (dH_2O)
- Forceps
- Medicine dropper or pipette
- Methylene blue with dropper or in dropper bottle
- Millimeter-ruled graph paper
- Petri dish, empty
- Petri dish with 1% agar
- Potassium permanganate crystals
- Potassium permanganate with dropper or in dropper bottle
- Sharpie marker or wax pencil

Instructions

1. Formulate a prediction about the rates of diffusion of the solutes. Which solute do you think will diffuse the fastest? Will the rates of solute diffusion differ among the solvents? Predict the relative diffusion rates in the following match-ups: potassium permanganate in water versus potassium permanganate in agar, potassium permanganate in agar versus methylene blue in agar. Record your predictions here:

2. Place the empty Petri dish over your graph paper and add dH_2O until the Petri dish is half-full.

3. Wait until the surface of the water is completely still and, using forceps, add one crystal of potassium permanganate near the center of the dish, placing it just over a location where two lines of the graph paper intersect.

4. Measure the diameter of the purple potassium permanganate dye in millimeters using the graph paper. Use a timer or the clock in your lab room to measure elapsed minutes, and record the diameter each minute for 10 minutes. Record your observations in Table 5.2.

Table 5.2 Observations	
Time Interval	**Diameter of Purple Dye**
0 min (start)	
1 min	
2 min	
3 min	
4 min	
5 min	
6 min	
7 min	
8 min	
9 min	
10 min	

5. Using your marker or wax pencil, draw a line across the bottom of your agar-filled Petri dish, dividing it in half. Place the Petri dish on top of the graph paper as shown in Step 1 of Figure 5.9.

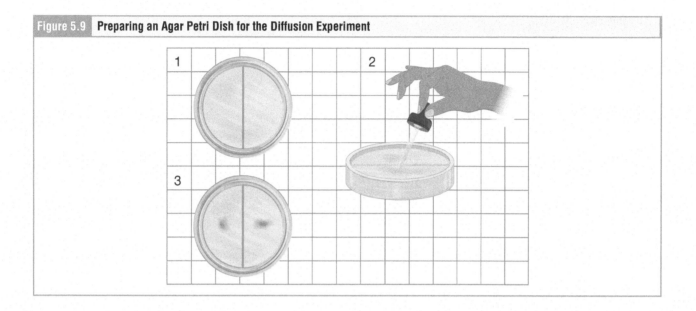

Figure 5.9 | **Preparing an Agar Petri Dish for the Diffusion Experiment**

6. Examine Step 2 of Figure 5.9. You will use the medicine dropper to carve two depressions in the agar surface. These depressions will serve as wells to hold the solutes. Squeeze the bulb of the medicine dropper, place the tip of the medicine dropper against the surface of the agar and release the bulb. The sucking action should suck up a small amount of agar leaving a depression behind. Expel the agar from the dropper into the appropriate waste container indicated by your lab instructor.

7. Repeat this step on the agar on the other side of the line you've drawn. When finished your agar plate should resemble Step 3 of Figure 5.9.

8. Fill one well with methylene blue and the other with potassium permanganate.

9. Record the time in the clock time column next to 0 min (start) in Table 5.3. Now calculate the time it will be on the clock at the 15-, 30-, 45-, and 60-minute intervals. You may want to set a timer or alarm on your watch or phone, or one member of your group can be the timekeeper.

Table 5.3 Data Collection			
Time Interval	Clock Time	Diameter of Methylene Blue	Diameter of Potassium Permanganate
0 min (start)			
15 min			
30 min			
45 min			
60 min			

10. At each interval, measure the distance the dye has diffused from the edge of the well.

11. After you have collected all your data, reread your hypotheses from Step 1 and describe how your hypotheses matched up with your observations. Was anything surprising to you? Reflect on your hypotheses and observations here:

Activity 5.2 Diffusion and Osmosis Across Membranes

This activity targets LOs 5.3 and 5.4.

As described in the introduction to this lab, diffusion can occur in a wide-open space, as we explored in Activity 5.1, or we can measure diffusion across a membrane. In both cases molecular properties will determine the rate of diffusion. When we measure diffusion across a membrane, the permeability of the solute in the membrane determines the rate of diffusion.

In this lab we will observe the movement of water and solutes between two solutions. When comparing solutions, we can use a set of terms to describe their relative solute concentration. If solution A has a higher concentration of solutes than solution B, we would describe solution A as being **hyperosmotic** to solution B. Solution B is **hypoosmotic** to solution A. If the two solutions have the same concentration of solutes, we refer to them as being **isosmotic** to each other.

Dialysis bags are selectively permeable plastic sacs that have been developed to mimic some permeability properties of the cell membrane. You can't see them, but the plastic sacs are covered with tiny holes that will allow molecules below a particular molecule size to move freely from one side of the plastic to the other.

In this lab we will observe both osmosis and diffusion; you may want to begin by reminding yourself of the difference between these two processes.

Materials
- Beakers, 4 @ 250 mL
- Benedict's solution
- Dialysis tubing clamps or string
- Dialysis tubing, 4 pieces
- Distilled H_2O
- Glucose solution, 40%
- Graduated cylinder, 25 mL
- Laboratory scale (also called a balance)
- NaCl solution, 10%
- Paper towels
- Pipettes, 5 ml and pipettors
- Silver nitrate
- Small funnel
- Sucrose solution, 40%
- Test tube holder
- Test tube rack
- Test tubes, 4
- Water bath or hot plates with beakers of water @ 100°C
- Wax pencil or lab tape and marker

Learning Connection

Micro to Macro

In this lab we imagine the microscopic movement of molecules and demonstrate that movement by observing weight changes in the dialysis bags. What if the dialysis bags were blood vessels? What would happen if there was net osmosis into a blood vessel? What if there was net osmosis out?

Instructions

1. Set the strips of dialysis tubing to soak in a beaker of water for 2–5 minutes while you set up the other materials.

2. Using the wax pencil or lab tape and a marker, number the beakers 1 through 4.

3. Fill beakers 1 through 3 halfway (125 mL) with distilled water.

4. Fill beaker 4 with 125 mL of 40% glucose solution.

Table 5.4 Weight Changes

Beaker #	Liquid in Beaker	Liquid in Sac	The Solution in the Sac Is ____ (Hypoosmotic, Hyperosmotic, Isosmotic) to the Solution in the Beaker	Initial Weight	Final Weight	Weight Change (+ if Gain and – if Lost)
1	dH₂O	40% glucose				
2	dH₂O	10% NaCl				
3	dH₂O	40% sucrose				
4	40% glucose	40% glucose				

5. Fill each strip of dialysis tubing by folding and clamping (or tying) one end. Take the other end and rub it between your thumb and index finger to open it. Place the funnel into the tube and add 20 mL of the liquid specified in Table 5.4 and Figure 5.10. Once filled, carefully press any air out, fold the unsealed end of the tube and clamp or tie it. Dry off the outside of the tubing using paper towels and weigh it on the scale. Record its initial weight in Table 5.4.

Figure 5.10	Experimental Set-Up for Activity 5.2

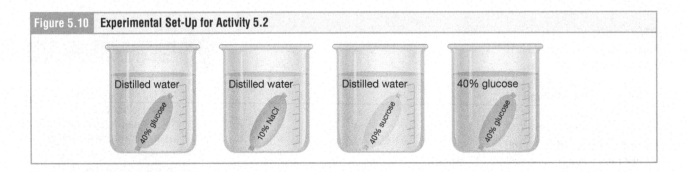

Distilled water — 40% glucose

Distilled water — 10% NaCl

Distilled water — 40% sucrose

40% glucose — 40% glucose

6. After recording the weight of each dialysis bag, lower them into the beakers specified in Table 5.4. Be sure that the bags are submerged in liquid, if a portion of the bag is exposed, add more liquid until it is covered.

7. Allow the bags to sit in the beakers for 60 min. At the end of that time, take them out, blot them dry and weigh them. Record their final weights in Table 5.4.

8. Think about your observations. If osmosis occurred and water moved down its concentration gradient, would you expect the sac to gain or lose weight? If solute diffusion occurred would the sac gain or lose weight or would its weight stay the same?

9. When diffusion occurs, the concentration of the solute will change in the beakers. Beakers 1 through 3 started with distilled water. We can now test to see if solute diffused into the beaker. Benedict's solution tests for the presence of sugars such as sucrose and glucose. Adding silver nitrate to a solution that contains salt (NaCl) will cause a reaction, the silver (Ag) will combine with the chloride of salt to form silver chloride (AgCl). In Table 5.5, decide and record which beakers you'd like to test for the presence of sugars by using Benedict's solution, and which beakers you'd like to test for NaCl by using silver nitrate.

Table 5.5 Experiment Planning

Beaker #	I Want to Test for the Presence of (Sugar or Salt?)	So, I Will Add (Benedicts or Silver Nitrate)	Test Results
1			
2			
3			
4			

10. Prepare four test tubes by labeling them with the wax pencil and adding either five drops of Benedict's solution or two drops of silver nitrate (use your work in Table 5.5 as your guide). To the tube marked #1, add 5 mL of beaker solution from beaker 1 using a pipet. For the tubes that you are testing with Benedict's solution, grab them with a test tube holder and place them in a 100°C water bath or beaker of boiling water for 5 minutes.

11. Sugar is present if a green, yellow, or red precipitate forms in a Benedict's test. Silver nitrate shows a positive result, the presence of salt, if a white precipitate forms.

12. Which beakers showed that diffusion had occurred out of the sacs and into the beaker liquid?

13. Did diffusion occur in beaker 4? Why or why not?

14. Based on your observations, put the molecules, sucrose, glucose, and NaCl, in order of increasing molecular size.

Digging Deeper:
Sweat

Sweat is a critical mechanism for maintaining body temperature homeostasis. Sweat cools us down through evaporative cooling. As a liquid evaporates and moves off a surface and into the air, it carries heat with it. If you put any water onto your skin (think about how you feel after swimming or even walking in the rain) it will carry heat away with it as it evaporates and you dry off. You may even have had the experience of putting a liquid on your skin that evaporates faster than water. Ethanol (rubbing alcohol) and acetone (nail polish remover) both evaporate very quickly, and in doing so, they feel quite cool on your skin even though the liquid itself is not cooler than water. As the surface of the skin cools, the blood flowing beneath that skin is losing heat, returning to the circulation at a lower temperature. When our bodies become overly warm, we dilate (widen) blood vessels headed toward the skin, to maximize the loss of heat. You may experience this as flushing or swelling of your skin when you're hot.

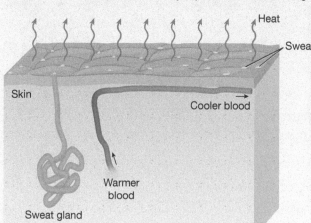

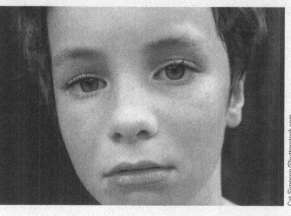

Cat Simpson/Shutterstock.com

(page intentionally left blank)

Activity 5.3 Osmosis and Cells

This activity targets LO 5.5, 5.6, and 5.7.

In Activity 5.2 we introduced three terms—hypoosmotic, hyperosmotic, and isosmotic—to compare the solute concentration in two solutions. If the two solutions are on either side of a membrane, then these terms predict the likelihood of solute or water movement across the membrane.

When we examine the likelihood that water will move across a cell membrane based on the solute concentration inside and outside that cell, we use a related set of terms. A solution that has a higher solute concentration than what is found inside the cell is called a **hypertonic** solution. A solution that has a lower solute concentration than what is found inside the cell is called a **hypotonic** solution. A solution that has the same solute concentration as what is found inside the cell is called an **isotonic** solution (Figure 5.11). These terms, collectively called **tonicity**, predict the degree to which water will move in or out of the cell, since most solutes cannot freely diffuse across the cell membrane. Just like you saw in Activity 5.2, if water moves by osmosis into or out of a cell, the cell will gain or lose volume. If a cell gains water volume, it will become swollen and more round, and if it gains too much water, the volume of water can cause the membrane to burst, killing the cell. If a cell loses water volume, it shrinks. The membrane may sag against the rigid cytoskeleton, causing the cell to take on a ragged, crinkled appearance (Figure 5.11c). The process of the cell shrinking is called **crenation**.

Figure 5.11 | **Cells in Isotonic, Hypotonic, and Hypertonic Solutions**

The term tonicity is used to compare the concentration of solute in the solution surrounding a cell to the concentration of solute in the intracellular fluid. (A) A red blood cell in an isotonic solution will have no net movement of water across its membrane and no change in size. (B) A red blood cell in a hypotonic solution will have a higher solute concentration, and therefore lower water molecule concentration, on the inside of the cell. Water will travel through aquaporins into the cell, causing the cell to swell in size. (C) A red blood cell in a hypertonic solution will have a lower solute concentration, and therefore a higher water molecule concentration, on the inside of the cell. Water will travel through aquaporins out of the cell, causing the cell to shrink and shrivel.

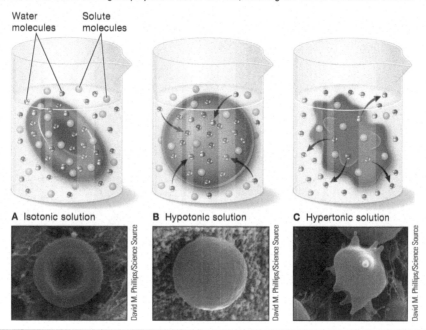

Water molecules   Solute molecules

A Isotonic solution   **B** Hypotonic solution   **C** Hypertonic solution

David M. Phillips/Science Source

Materials

- An unknown solution
- Clean slides and coverslips
- Disposable gloves
- Distilled water (dH$_2$O)
- Dropper
- Mammalian blood
- Microscope
- NaCl solution (saline), 0.9%
- NaCl solution, 5.0%
- Paper towels

Instructions

1. Wear disposable gloves throughout this procedure. Once you have gloves on, take care to use clean techniques, do not touch your phone, lab manual, personal items, or classroom items such as doorknobs or shared reagents. You may want to work together with your lab partner to make sure one person, who has gloves on, works with the blood in this lab, and the other person navigates the lab space.

2. Place a drop of 0.9% NaCl solution on the first slide, add a drop of blood and place a coverslip on top using the technique illustrated in Figure 5.12. Blot any excess liquid from the sides of the coverslip using a paper towel. Then place the slide on the microscope stage with the scanning (4×) objective in place. Focus until the cells come clearly into view. Now rotate the objectives so that the low-power (10×) objective is in place and focus again, using only the fine focus knob. Now rotate the objectives so that the high-power (40×) objective is in place and focus again, using only the fine focus knob. Notice the cells' shape and appearance. This solution is the isotonic solution, so we can consider these cells to illustrate the normal shape. Draw what you see here:

Figure 5.12 **Coverslips**

To add a coverslip onto your wet mount slide, hold the coverslip by the edges (take care not to touch the surface of the glass or you will add fingerprints) and hold the coverslip against the wet area at a 45° angle. Gently and slowly lower the coverslip until it covers the wet mount. Blot excess fluid with a paper towel.

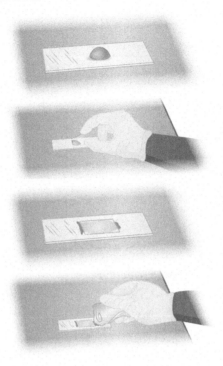

3. Repeat the procedure you followed in Step 2 but with a drop of 5% NaCl solution and a drop of blood. Notice the shape and appearance of the cells. Draw what you see here:

4. Repeat the procedure you followed in Step 2 but with a drop of distilled water and a drop of blood. Notice the cells' shape and appearance. Draw what you see here:

5. Which solution was hypotonic and which was hypertonic?

6. Now place a drop of the unknown solution prepared by your instructor on the last slide, add a drop of blood and place a coverslip on top using the technique illustrated in Figure 5.12. Blot any excess liquid from the sides of the coverslip using a paper towel. Then place the slide on the microscope stage with the scanning (4×) objective in place. Focus until the cells come clearly into view. Now rotate the objectives so that the low-power (10×) objective is in place and focus again, using only the fine focus knob. Now rotate the objectives so that the high-power (40×) objective is in place and focus again, using only the fine focus knob. Notice the cells' shape and appearance. Draw what you see here:

7. Was the unknown solution hypotonic, isotonic, or hypertonic?

Activity 5.4 Case Study: Heat Stroke

This activity targets LO 5.7.

Case Study

It's a hot summer day in a central Florida town. George is a 64-year-old, fair-skinned man who is the coach of his granddaughter's softball team. After several hours in the sun coaching he began to feel dizzy and confused. The players noted that his words were a little slurred and hard to understand, and he couldn't remember which inning they were playing. Although he had earlier been sweating profusely, he wasn't sweating now. His granddaughter approached him after the game to offer him some water and she was alarmed to see that his skin had turned a pale blue color. She started walking with him to his car but halfway across the field he fainted. A nearby parent called for an ambulance, and George was taken to a hospital and diagnosed with a condition called *heat stroke*.

Heat stroke is a condition of elevated body temperature. Usually, when the body temperature rises, sweat is sufficient to cool the temperature back down to the homeostatic range. Let's discuss sweat for a moment. Sweat is slightly more dilute than interstitial fluid. The production of sweat is illustrated in Figure 5.13.

Figure 5.13 | **Where Sweat is Produced**

Sweat production in a sweat gland involves the transport of ions across the epithelium to create an osmotic gradient, water follows by osmosis.

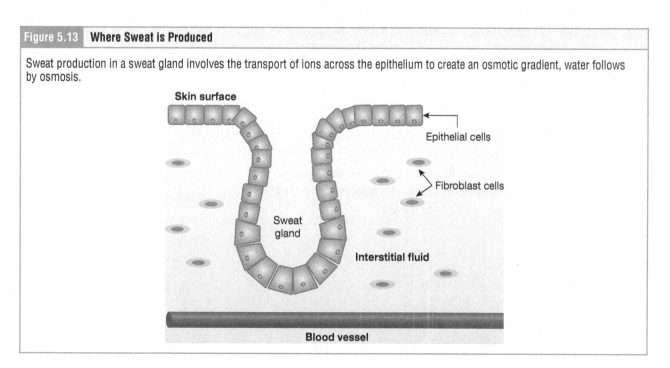

---

1. If the final sweat released from sweat glands onto the skin is **more dilute** than body fluids we would consider this sweat to be _____ to body fluids.

   a. hypoosmotic

   b. isosmotic

   c. hyperosmotic

2. How does water move across the sweat gland epithelium to reach the lumen of the sweat gland?

   a. It is pumped using cellular energy (active transport).

   b. It diffuses through the membrane (simple diffusion).

   c. It diffuses through aquaporins (facilitated diffusion).

3. While sweat is more dilute than body fluids, it does still contain dissolved solutes. Since the purpose of sweating is to cool the body via the evaporation of water, why isn't sweat just composed of pure water?

   a. There is no way to facilitate the movement of water alone into the sweat gland lumen.
   b. Since water and ions move across plasma membranes through the same channels, there is no way to move one without the other.
   c. Active transport of pure water would generate too much heat, counteracting the cooling gains of sweating.
   d. Active transport of pure water would be too energetically expensive, causing a greater physiological problem than just allowing body temperature to increase.

4. After a long period of sweating, what would happen to the volume of interstitial fluid?

   a. It would not change.
   b. It would increase.
   c. It would decrease.

5. Since sweat is more dilute than interstitial fluid, more water than solutes is lost from the body. After a long period of sweating, what would happen to the osmolarity of interstitial fluid?

   a. It would increase.
   b. It would not change.
   c. It would decrease.

6. If the temperature outside did not change, why would this man stop sweating?

   a. A decrease in available ions
   b. A decrease in body temperature (back to homeostatic levels)
   c. The brain region that controls sweating has been compromised.
   d. A decrease in blood circulation to the skin

Lab 5 Post-Lab Quiz

1. Hemolysis is a specific term that refers to the death of red blood cells. Hemolysis occurs when blood is exposed to which type of solution?

   a. Hypotonic

   b. Hypertonic

   c. Isotonic

2. Gatorade and electrolyte drinks were developed with the thinking that it might be good to provide ions in the water that a person drinks. The design was for the drink to be _____ compared to human cells.

   a. hypotonic

   b. hypertonic

   c. isotonic

3. When you stir sugar into your coffee, we can consider the coffee to be the _____ and the sugar to be the _____.

   a. solution, solvent

   b. solute, solvent

   c. solvent, solute

   d. solute, solution

4. Which part of the plasma membrane resists the movement of water molecules due to its chemical properties?

   a. Fatty acid tails of phospholipids

   b. Heads of phospholipids

   c. Protein channels

5. Which of the following accurately describes the difference between a water molecule moving across the membrane by osmosis and a molecule of oxygen O_2 moving across the membrane by diffusion?

   a. Both types of molecules need to travel through channels in the membrane.

   b. Both types of molecules can move by wiggling in between phospholipids in the membrane.

   c. Water molecules can move by wiggling in between phospholipids in the membrane but O_2 molecules require a channel.

   d. O_2 molecules can move by wiggling in between phospholipids in the membrane but water molecules require a channel.

6. Abby is an athlete and has been competing in the hot sun for hours. They collapsed on the field at the end of the game and was brought to the emergency room for treatment. A medical student says, "Abby is desperately dehydrated, let's start an IV of distilled water immediately!" and the nurse replies, "Are you trying to kill Abby?!?" Why is the medical student wrong to order an IV of distilled water?

   a. Distilled water is not sterile, it may contain dangerous microbes.

   b. Distilled water is hypotonic, it may cause Abby's red blood cells to burst.

   c. Distilled water does not contain any nutrients, and clearly Abby needs calories.

7. Urea is a waste product that is produced as cells metabolize protein. It has a molecular weight of 60 amu and is non-polar. What do you think is the most common method for urea to cross the cell membrane?

   a. Simple diffusion

   b. Facilitated diffusion

   c. Active transport

8. Examine the two solutions separated by a membrane in the figure. In which direction will net osmosis occur?

   a. From right to left
   b. From left to right
   c. No net osmosis will occur

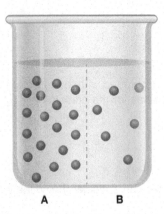

A   B

9. Examine the solutions in the figure. If the membrane is not permeable to water molecules, but is permeable to the solute, will either side change in volume after osmosis and diffusion have had time to occur?

   a. A will gain volume
   b. B will gain volume
   c. Neither side will
      gain volume

A   B

10. Cerebrospinal fluid is the fluid found within the brain and spinal cord. It circulates around neurons (nervous system cells) bathing them, supplying them with nutrients, and carrying away their wastes. How would we classify cerebrospinal fluid?

    a. Intracellular fluid
    b. Extracellular fluid
    c. Neither of these terms applies

Lab 6 Introduction to Tissues

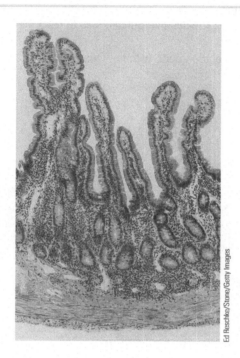

Ed Reschke/Stone/Getty Images

Learning Objectives: After completing this lab, you should be able to:

6.1   Compare and contrast the general features of the four major tissue types.

6.2   Classify different types of tissues based on structural characteristics.

6.3*   Describe the microscopic anatomy of each tissue type.

6.4*   Classify different types of tissue based on functions.

6.5*   Identify the locations in the body where each tissue type is found.

6.6   Identify examples of each type of tissue.

6.7   Compare and contrast neurons and glial cells with respect to cell structure and function.

6.8*   Compare and contrast the types of cells found in connective tissues.

6.9*   Explain how the functions of each tissue are determined by its form (structure), and how the function can alter structure.

*Objective is not a HAPS Learning Goal.

Introduction

Your body's cells are organized into functional groups called tissues. In multicellular organisms such as humans, organizing like-functioning cells into tissues allows for a division of labor within the body. Individual tissues have specific functions that serve the more general function of the organs in which they are found. Histology, the study of these various tissues, requires the use of microscopes so that we can see the detail of the various cells and extracellular matrix that make them up.

The Human Anatomy and Physiology Society includes more than 1,700 educators who work together to promote excellence in the teaching of this subject area. The HAPS A&P Learning Outcomes measure student mastery of the content typically covered in a two-semester Human A&P curriculum at the undergraduate level. The full Learning Outcomes are available at https://www.hapsweb.org.

Lab 6 Pre-Lab Quiz

This quiz will strengthen your background knowledge in preparation for this lab. For help answering the questions, use your resources to deepen your understanding. The best resource for help on the first five questions is your text, and the best resource for help answering the last five questions is to read the introduction section of each lab activity.

1. Anatomy of epithelial tissue. Label the image with the following terms:

    apical surface
    basal surface
    lumen

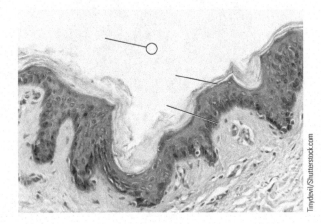

2. Match the root next to its definition:

Root	Meaning
	web or tissue
	fiber
	cell
	upon or outside
	cartilage
	bone

List of Roots

chondro
cyte
epi
fibro
histo
osteo

3. Arrange the following terms in order of least complex to most complex.
    Tissue, Molecule, Organ, Cell, Organelle

4. The large, usually dark-staining structure found in almost all cells of the body that contains the cell's DNA is the:

    a. Nucleus

    b. Mitochondria

    c. Cytoskeleton

5. When using a microscope, you should always first view the slide on:

   a. The highest power objective, then reduce to a lower power.

   b. A moderate power objective, then increase to a higher power.

   c. The lowest power objective, then increase to a higher power.

6. All cells in the human body are surrounded by a:

   a. Cell wall

   b. Cell membrane

   c. Both a cell wall and a cell membrane

7. Tissues are made of cells and extracellular matrix. What is extracellular matrix?

   a. The cytoplasm inside each cell

   b. The inside of a hollow organ

   c. The material outside of the cell membrane

8. Match the basic tissue type to its function:

Tissue	Meaning
	wraps or lines a structure
	shortens to move the tissue
	sends electrical messages within the tissue
	links or supports structures of the body

Types of Tissues

connective tissue
epithelial tissue
muscle tissue
nervous tissue

9. Which of the following tissue type(s) consist of tightly packed cells with little extracellular matrix? Mark all correct answers.

   a. Epithelial tissue

   b. Muscle tissue

   c. Nervous tissue

   d. Connective tissue

10. What general tissue type is always found deep to epithelial tissues and provides blood supply to the epithelium?

   a. Muscle tissue

   b. Nervous tissue

   c. Connective tissue

(page intentionally left blank)

Activity 6.1 Explore the General Functions of the Four Basic Tissue Types

This activity targets LO 6.1.

Anatomists have classified all tissues based on their functions into four basic types: epithelial, muscle, nervous, and connective. Remember, a core concept of anatomy and physiology is that structure and function are inherently related to each other. The structure of each tissue (its composition and appearance) determines how it functions (the job or jobs it performs in the human body). Consider how the description of the functions of each of the four basic tissue types is driven by the structure of that tissue type.

Epithelial tissues are wrappers and liners. These tissues, as tightly packed sheets of cells, must hold together as a continuous layer. Some epithelial tissues wrap around the outside of an organ; you might imagine how a house is covered in siding so that the wiring and plumbing in the walls is protected from the elements. Other epithelial tissues line the inside of a hollow structure or organ in the body like drywall lines the inside of a house.

Muscle tissue is contractile. These tissues can be stimulated to shorten in order to move the skeleton or the wall of organs in the body. Like epithelial tissue, muscle cells are packed closely together; in muscle tissues the tight connections between the cells are so that the shortening of the many cells within the tissue will cause the whole tissue to shorten.

Our body's **nervous tissue** functions to send messages, doing this through cells that extend long arm-like processes to reach each other. In contrast to epithelial and muscle tissues, when you view nervous tissue, you are likely to see quite a bit of extracellular matrix. The additional space around the cells provides room for them to reach out and connect with their neighbors.

The many types of **connective tissue** in the body have a wider range of functions than the other tissues. Broadly speaking, connective tissues serve to link (connect), support, and cushion other tissues. Most connective tissues will include significant extracellular matrix; often this matrix is full of fibers that make the tissue stronger.

Materials
• Four color images (one each of the four tissue types; Figures 6.1–6.4).

Figure 6.1 | **Tissue 1**

Histological view of the lining of the urinary bladder that provides a stretchy barrier between the urine within the bladder and the deeper tissues.

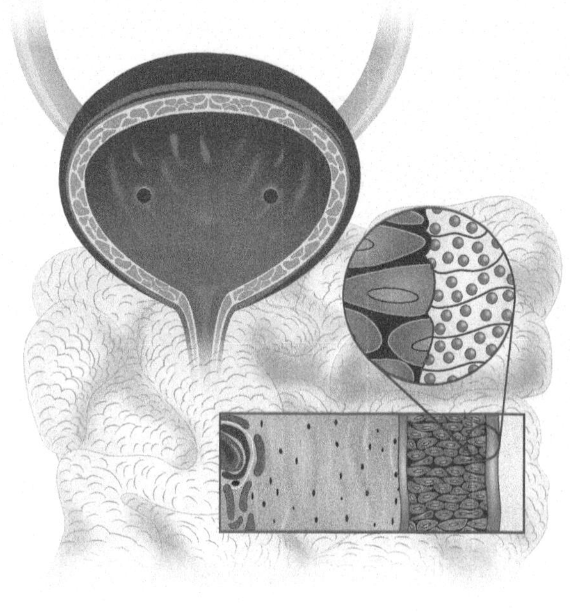

Figure 6.2   Tissue 2

Histological view of brain tissue with neurons connecting to each other to communicate.

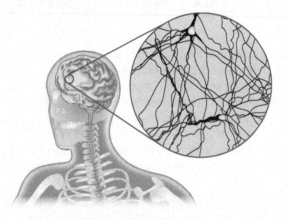

Figure 6.3   Tissue 3

Histological view of a tendon, which connects muscle to bone in the body.

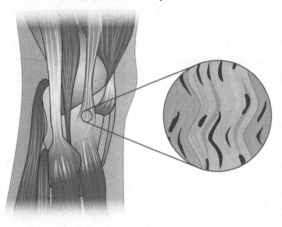

Figure 6.4   Tissue 4

Histological view of cardiac tissue, which contracts to pump blood.

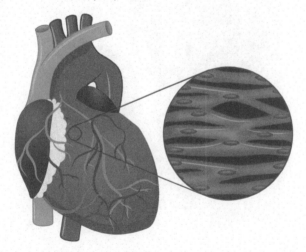

Instructions

1. View the four color images, labeled Tissue 1, Tissue 2, Tissue 3, and Tissue 4.

2. Use the information above and what you already know about the structures and functions of the parts of the human body to identify each tissue based on the structures and organization that you can see in the image. For each, give an explanation that justifies your answer. Phrase your answer as, "The structure of the tissue in the image is…which means it functions to…"

Tissue 1

Tissue type:

Justification:

Tissue 2

Tissue type:

Justification:

Tissue 3

Tissue type:

Justification:

Tissue 4

Tissue type:

Justification:

(page intentionally left blank)

Activities 6.2–6.5 Explore Each of the Four Basic Tissue Types

In the following activities you will be asked to make detailed drawings of the various tissue types of the body; however, this is not a test of your art skills. Drawing and sketching help us see the details and remember what we are seeing in unique ways; careful consideration of what you are looking at and how the structures you see are arranged will help you learn the features of each tissue. In addition, drawing helps you to see how tissues differ and aids you in recognizing those differences whenever you encounter them in the future. Combining writing and drawing is a powerful combination, so add written descriptions of what you are seeing to your sketches. The more time you take to think about the details of what you are drawing and why those details matter, the more useful and satisfying you will find these lab exercises.

An important comment about color and histology: many students find that at first all of the slides of the various tissues look very much like a sea of similar pink swirls. This is because certain stains are regularly used to dye the almost-transparent tissues so that we can see them under the light microscope. Many of these stains give dark pink, purple, and blue color to portions of the tissues. But the stains themselves do not really differentiate tissue types from each other, they are just there so we can see the distinct structures that make up each tissue. Look past the color to the shapes you see made up by the colors; it can be helpful to imagine it like you are interpreting black-and-white photographs in which it is not the color that is the clue it is the shapes themselves that help you understand the scene.

(page intentionally left blank)

Activity 6.2 Examine the Structure of Epithelial Tissue

This activity targets LOs 6.2, 6.3, 6.4, 6.5, and 6.6.

Epithelial tissues are the wrappers and liners of the body's organs. Because the cells are tightly packed together, epithelial tissues are **avascular**; they have no blood vessels weaving between the cells. Think of how this adds to the structural integrity of the sheets of cells that make up the tissue. But also think about how this means that these sheets of cells cannot be very thick, since the oxygen and nutrients needed by each cell in the epithelial tissue must diffuse into the tissue from the connective tissue that is nearby (more on that when you reach connective tissues).

Since epithelial tissues always wrap or line a structure or organ, one side of the epithelium is always facing an open space either to the outside or inside of the structure or organ. If the epithelium lines a hollow organ this space is referred to as the **lumen**. The presence of a clear empty lumen or free space (if the epithelium is a wrapper rather than a liner) in a histology slide can be a big clue that you are looking at epithelial tissue at the edge of that space. The side of the epithelium facing the space is called the **apical** surface. The opposite side of the epithelium that attaches to the connective tissue that is deep to it is called the **basal** surface. If you look carefully at the basal surface of epithelial tissue you may be able to see a thin layer of extracellular proteins making up the **basement membrane** which helps the epithelial tissue stick to the connective tissue deep to it. Examine Figure 6.5 to see these items. Can you locate other examples of a lumen, basement membrane, and the apical and basal surfaces of epithelial cells in Figure 6.5?

Figure 6.5	Epithelial Tissue with Lumen and Basement Membrane

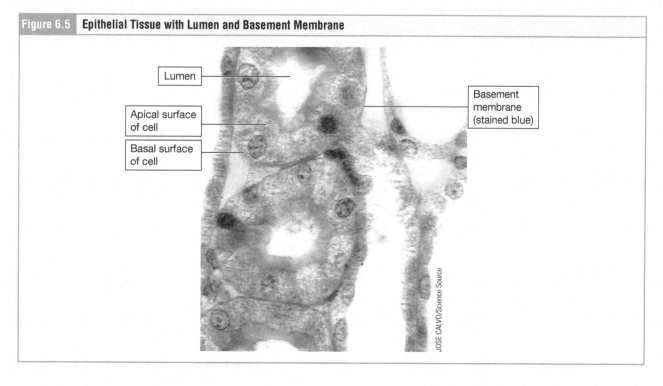

While all epithelial tissues are wrappers and liners, the various subtypes of epithelium represent all the different functions the body needs these wrappers and liners to serve. For example, some epithelia must be extra thin to allow material to pass across, while others must be specialized to withstand abrasion (scraping) of material passing across it, while still others must be able to actually move substances along within a lumen. As you examine each epithelial sub-type, you will be asked to consider how the structure of the subtype that can see and describe provides that subtype with its specific wrapping or lining functions. You will also be asked to note specific structural features of some epithelial subtypes. For example, one epithelial tissue subtype produces a thick apical layer of **keratin** protein, while other epithelial tissue subtypes contain special cells called **goblet cells** that produce mucus. As you notice and identify these structural features, consider the specialized functions they provide for those tissue subtypes.

Examine Table 6.1 to compare the related terms that are being used here to talk about this type of tissue. The term "epithelial" is an adjective, as in epithelial tissue. The term "epithelium" is a noun and can be used instead of epithelial tissue. Finally, "epithelia" is the plural of epithelium.

Table 6.1 Terminology Associated with Epithelial Tissue

Term	Part of Speech
Epithelium	Singular noun
Epithelia	Plural noun
Epithelial	Adjective

Materials
• Compound microscope
• Prepared slides of the epithelial tissue subtypes

Instructions

1. Obtain the epithelial tissue slides.

2. For each slide, examine the slide first on scanning power (4×) and use the coarse focus knob to locate the lumen or free space. Then increase the magnification by switching to the low-power (10×) objective lens and use the fine focus knob. Then switch to the high-power (40×) objective lens and again use the fine focus knob until you can clearly see the epithelial tissue.

3. Check with your instructor if you are unsure about what you are looking at.

4. Draw and describe the epithelial tissue, identifying all requested structures that are listed. In a blank area on the drawing, note the magnification used.

Slide 1: Simple Squamous Epithelium

- On your drawing, identify: lumen or free space, apical surface, basal surface, basement membrane, a cell nucleus
- How many layers of epithelial cells are present?
- What is the function of this tissue?

- How does the structure of this tissue support these functions?

- Where in the body will this tissue be found?

Slide 2: Simple Cuboidal Epithelium

- On your drawing, identify: lumen or free space, apical surface, basal surface, basement membrane, a cell nucleus
- How many layers of epithelial cells are present?
- What is the function of this tissue?

- How does the structure of this tissue support these functions?

- Where in the body will this tissue be found?

Slide 3: Simple Columnar Epithelium, Non-ciliated

- On your drawing, identify: lumen or free space, apical surface, basal surface, basement membrane, a cell nucleus, brush border (microvilli), goblet cell
- How many layers of epithelial cells are present?
- What is the function of this tissue?

- How does the structure of this tissue support these functions?

- Where in the body will this tissue be found?

Slide 4: Simple Columnar Epithelium, Ciliated

- On your drawing, identify: lumen or free space, apical surface, basal surface, basement membrane, a cell nucleus, cilia, goblet cell
- How many layers of epithelial cells are present?
- What is the function of this tissue?

- How does the structure of this tissue support these functions?

- Where in the body will this tissue be found?

Slide 5: Stratified Squamous Epithelium, Keratinized

- On your drawing, identify: lumen or free space, apical surface, basal surface, basement membrane, a cell nucleus, keratin layer
- Estimate the number of layers of epithelial cells.
- What is keratin made of?

- What is the function of this tissue?

- How does the structure of this tissue support these functions?

- Where in the body will this tissue be found?

Slide 6: Stratified Squamous Epithelium, Non-keratinized

- On your drawing, identify: lumen or free space, apical surface, basal surface, basement membrane, a cell nucleus
- Estimate the number of layers of epithelial cells.
- What is the function of this tissue?

- How does the structure of this tissue support these functions?

- Where in the body will this tissue be found?

Slide 7: Stratified Cuboidal Epithelium

- On your drawing, identify: lumen or free space, apical surface, basal surface, basement membrane, a cell nucleus
- Estimate the number of layers of epithelial cells.
- What is the function of this tissue?

- How does the structure of this tissue support these functions?

- Where in the body will this tissue be found?

Slide 8: Pseudostratified Columnar Epithelium, Ciliated

- On your drawing, identify: lumen or free space, apical surface, basal surface, basement membrane, a cell nucleus, cilia, goblet cell
- How many layers of epithelial cells are present (note the prefix "pseudo-" in the name of this tissue)?
- What is the function of this tissue?

- How does the structure of this tissue support these functions?

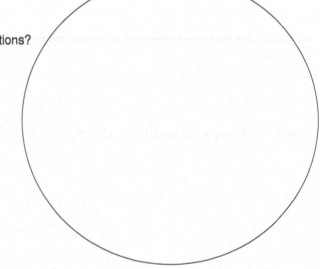

- Where in the body will this tissue be found?

Slide 9: Transitional Epithelium (Relaxed)

• On your drawing, identify: lumen or free space, apical surface, basal surface, basement membrane, a cell nucleus

• Estimate the number of layers of epithelial cells.

• What is the function of this tissue?

• How does the structure of this tissue support these functions?

• Where in the body will this tissue be found?

• You are viewing this tissue in its relaxed state, what will this same tissue look like if in its stretched (distended) state? When would this occur?

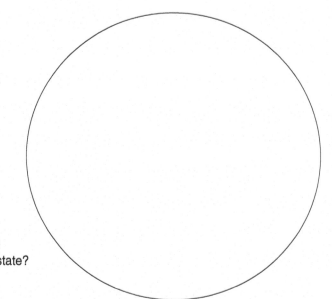

(page intentionally left blank)

Activity 6.3 Examine the Structure of Muscle Tissue

This activity targets LOs 6.2, 6.3, 6.4, 6.5, and 6.6.

Muscle tissue is **contractile** and shortens when stimulated to move the skeleton or the soft tissue of organs in the body. Muscle tissues are divided into three sub-types based on where they are found in the body and their functions. Importantly, we can see how they vary in structure because of their differing functions. As you read through the following descriptions of the features of muscle tissue, fill in Table 6.2. Skeletal muscle is probably the muscle tissue you are most familiar with. This type of muscle, which connects the bones of your body, allows you to move. This muscle subtype is **voluntary**, meaning that you have conscious control over whether it contracts or not. As you draw the skeletal muscle, note the incredible size and shape of the individual cells, the multiple nuclei per cell, and the visible striations. Depending on the stain used to prepare the skeletal muscle cell, the striations may be very faint or may be obvious; these striations are the stacked proteins within the cell that can make it contract in order to shorten the muscle. Cardiac muscle, as the name implies, is found throughout the wall of the heart, and its contractions enable the heart to function as a pump, circulating blood throughout the body. Like skeletal muscle, cardiac muscle has striations that may be more or less visible based on the staining technique that is used. In contrast to skeletal muscle, cardiac muscle cells are short and branched. Where one cardiac cell attaches to its neighbor, **intercalated disks**, which are aggregations of specialized proteins, can be seen tightly stitching the cells together so that they do not pull apart as the heart contracts. Smooth muscle cells differ from skeletal and cardiac muscle cells in that their contractile proteins are not visible as striations and therefore look smooth, but we can still see that the cells are large and full of material around a central nucleus. The cells are long and tapered at both ends like a sweet potato; this shape is called fusiform. Both cardiac muscle and smooth muscle are **involuntary**, meaning that their contraction is not under our conscious control.

Table 6.2 Differentiations Between Muscle Tissue Subtypes

Use the introduction to complete the empty cells in this table

Muscle Tissue Subtype	Function	Voluntary or Involuntary?	Cell Shape	Has Visible Striations When Stained? (Yes or No)	Has Intercalated Disks? (Yes or No)
Skeletal muscle					
Cardiac muscle					
Smooth muscle					

Materials
• Prepared slides of the muscle tissue subtypes

Instructions

1. Obtain the muscle tissue slides.

2. For each slide, examine the slide first on scanning power (4×) and use the coarse focus knob to locate the tissue. Then increase the magnification by switching to the low-power (10×) objective lens and use the fine focus knob. Then switch to the high-power (40×) objective lens and again use the fine focus knob until you can clearly see the tissue.

3. Check with your instructor if you are unsure about what you are looking at.

4. Draw and describe the muscle tissue, identifying all requested structures that are listed. In a blank area on the drawing, note the magnification used.

Slide 1: Skeletal Muscle Tissue

• On your drawing, identify: what makes up a single cell, multiple nuclei, striations

• What is the function of this tissue?

• How does the structure of this tissue support these functions?

• Where in the body will this tissue be found?

Slide 2: Cardiac Muscle Tissue

- On your drawing, identify: what makes up a single cell, nucleus, striations, branching shape of the cells, intercalated disk
- What is the function of this tissue?

- How does the structure of this tissue support these functions?

- Where in the body will this tissue be found?

Slide 3: Smooth Muscle Tissue

- On your drawing, identify: what makes up a single cell, nucleus, fusiform shape of the cells
- What is the function of this tissue?

- How does the structure of this tissue support these functions?

- Where in the body will this tissue be found?

(page intentionally left blank)

Activity 6.4 Examine the Structure of Nervous Tissue

This activity targets LOs 6.2, 6.3, 6.4, 6.5, 6.6, and 6.7.

Nervous tissue is found in our brain, spinal cord, and nerves and allows the body to communicate through long chains of connected cells. While both epithelial tissue and muscle tissue had different subtypes, nervous tissue does not. Nervous tissue contains two general types of cells: neurons and glial cells. The large **neurons** send and receive messages through their many long arm-like processes called **dendrites** and the usually longer single process called an **axon**. Surrounding the neuron are many smaller **glial cells** that provide support for these neurons through various specialized functions.

Materials
• Prepared slide of nervous tissue showing individual neurons and glial cells

Instructions

1. Obtain the nervous tissue slide.

2. Examine the slide first on scanning power (4×) and use the coarse focus knob to locate a neuron. Then increase the magnification by switching to the low-power (10×) objective lens and use the fine focus knob. Then switch to the high-power (40×) objective lens and again use the fine focus knob until you can clearly see the neuron and surrounding glial cells.

3. Check with your instructor if you are unsure about what you are looking at.

4. Draw and describe the nervous tissue, identifying all requested structures that are listed. In a blank area on the drawing, note the magnification used.

Slide: Nervous Tissue

• On your drawing, identify: a neuron, the neuron's nucleus and cell body, glial cells

• What is the function of the neurons in this tissue?

• What is the general function of the glial cells in this tissue?

• How do the structures of each of these cell types support these functions?

• Where in the body will this tissue be found?

(page intentionally left blank)

Activity 6.5 Examine the Structure of Connective Tissue

This activity targets LOs 6.2, 6.3, 6.4, 6.5, 6.6, and 6.8.

Connective tissues are present in the body to connect, support, and cushion other tissues. Connective tissues are always found deep to epithelial tissues and contain the blood vessels that supply the epithelium. Generally, connective tissues have significant space between the cells. This space is filled with **ground substance**, a somewhat clear, thick material containing various fibers. Examine Figure 6.6, which shows three possible fiber types made by **fibroblasts** that can be found in the ground substance: collagen fibers, reticular fibers, and elastic fibers. Thick collagen fibers are long, unbranched, unstretchy proteins that look like miniature stiff ropes. Somewhat thinner elastic fibers are made of proteins that branch away and rejoin so that they look wavy; these fibers are stretchy. Reticular fibers are the thinnest of the three fiber types and are named for their angular branched shape which can look boxy or net-like; just like a net, reticular fibers are strong but flexible.

Figure 6.6	Types of Connective Tissue Fibers

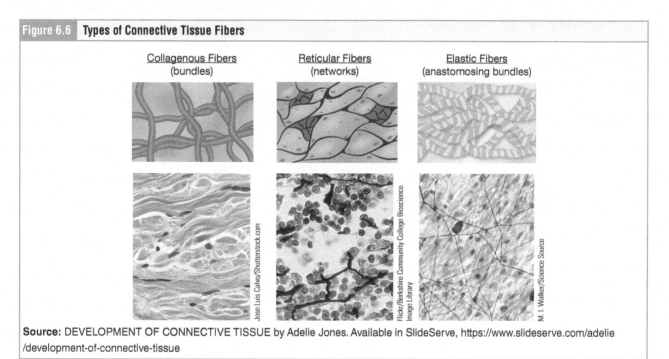

Collagenous Fibers (bundles)   Reticular Fibers (networks)   Elastic Fibers (anastomosing bundles)

Jose Luis Calvo/Shutterstock.com

Flickr/Berkshire Community College Bioscience Image Library

M. I. Walker/Science Source

Source: DEVELOPMENT OF CONNECTIVE TISSUE by Adelie Jones. Available in SlideServe, https://www.slideserve.com/adelie/development-of-connective-tissue

The many subtypes of connective tissues represent different structural solutions to address the varying ways that connective tissues function to provide connection, support, and cushioning to nearby cells and tissues. The subtypes can be grouped into three general categories: connective tissues proper that have fibers within ground substance (areolar connective tissue, adipose connective tissue, and reticular connective tissue), dense connective tissues that have more fibers and less ground substance (dense regular connective tissue, dense irregular connective tissue, elastic connective tissue), supporting connective tissues that are dense and strong (cartilage and bone), and fluid connective tissues that are, as the name implies, fluids (blood and lymph). While lymph is, technically, a connective tissue, it does not contain any cells, fragments, or fibers and is therefore not examined histologically.

The connective tissue subtypes often contain different specialized cells depending on the functions that the tissue needs to perform. **Adipocytes** are found in adipose connective tissue and store triglycerides inside themselves. **Chondrocytes** are found in the cartilage tissue subtypes and **osteocytes** are found in bone tissue. Both chondrocytes and osteocytes build the extracellular matrix of these tissues; as they do so they become trapped in small spaces in the matrix called **lacunae** (singular = lacuna). **Erythrocytes** and **leukocytes**, located in blood, carry oxygen and perform immune system functions, respectively.

Materials
• Prepared slides of the connective tissue subtypes

Instructions
1. Obtain the connective tissue slides.

2. For each slide, examine the slide first on scanning power (4×) and use the coarse focus knob to locate the tissue. Then increase the magnification by switching to the low-power (10×) objective lens and use the fine focus knob. Then switch to the high-power (40×) objective lens and again use the fine focus knob until you can clearly see the tissue.

3. Check with your instructor if you are unsure about what you are looking at.

4. Draw and describe the connective tissue, identifying all requested structures that are listed. In a blank area on the drawing, note the magnification used.

Connective Tissue Proper
Slide 1: Areolar Connective Tissue
• On your drawing, identify: fibroblasts, collagen fibers, elastic fibers, ground substance
• What are the functions of this connective tissue?

• How does the structure of this tissue support these functions?

• Where in the body will this tissue be found?

Slide 2: Adipose Connective Tissue

• On your drawing, identify: adipocyte, adipocyte nucleus, cell membrane of adipocyte

• What are the functions of this connective tissue?

• How does the structure of this tissue support these functions?

• Where in the body will this tissue be found?

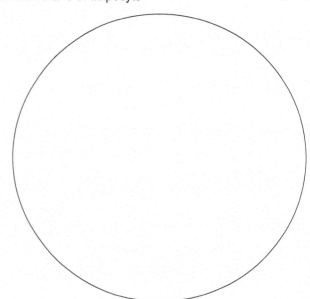

Slide 3: Reticular Connective Tissue

• On your drawing, identify: leukocytes (white blood cells), reticular fibers, ground substance

• What are the functions of this connective tissue?

• How does the structure of this tissue support these functions?

• Where in the body will this tissue be found?

Dense Connective Tissues

Slide 4: Dense Regular Connective Tissue

• On your drawing, identify: fibroblasts, collagen fibers, ground substance

• Dense regular connective tissue can sometimes be mistaken for smooth muscle tissue. Examine your drawing of smooth muscle tissue to determine how you will differentiate the two tissues from each other.

• What are the functions of this connective tissue?

• How does the structure of this tissue support these functions?

• Where in the body will this tissue be found?

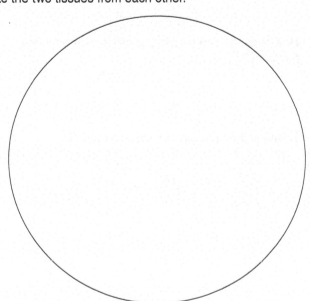

Slide 5: Dense Irregular Connective Tissue

• On your drawing, identify: fibroblasts, collagen fibers, ground substance

• What are the functions of this connective tissue?

• How does the structure of this tissue support these functions?

• Where in the body will this tissue be found?

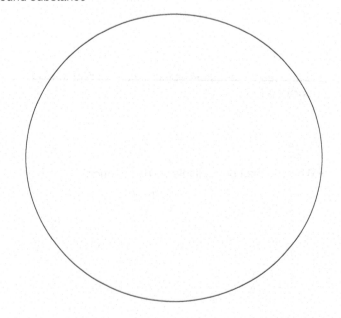

Slide 6: Elastic Connective Tissue

- On your drawing, identify: elastic fibers, ground substance
- What are the functions of this connective tissue?

- How does the structure of this tissue support these functions?

- Where in the body will this tissue be found?

Supporting Connective Tissues
Slide 7: Hyaline Cartilage

- On your drawing, identify: chondrocyte, lacuna, ground substance
- Collagen fibers are present in the ground substance of hyaline cartilage, but are they visible? If they are visible, how are the fibers arranged in the ground substance?

- What are the functions of this connective tissue?

- How does the structure of this tissue support these functions?

- Where in the body will this tissue be found?

Slide 8: Fibrocartilage

• On your drawing, identify: chondrocyte, lacuna, collagen fibers, ground substance

• Collagen fibers are present in the ground substance of fibrocartilage, are they visible? If they are visible, how are the fibers arranged in the ground substance?

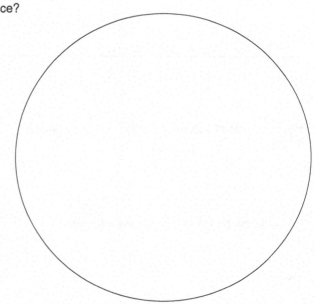

• What are the functions of this connective tissue?

• How does the structure of this tissue support these functions?

• Where in the body will this tissue be found?

Slide 9: Elastic Cartilage

• On your drawing, identify: chondrocyte, lacuna, elastic fibers, ground substance

• How are the elastic fibers arranged in the ground substance?

• What are the functions of this connective tissue?

• How does the structure of this tissue support these functions?

• Where in the body will this tissue be found?

Slide 10: Bone Tissue

• On your drawing, identify: osteocyte, lacuna, central canal, calcified ground substance

• What are the functions of this connective tissue?

• How does the structure of this tissue support these functions?

• Where in the body will this tissue be found?

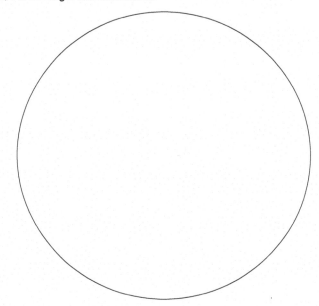

Fluid Connective Tissues

Slide 11: Blood Tissue

• On your drawing, identify: erythrocyte (red blood cell), leukocyte (white blood cell), platelet, liquid ground substance (plasma)

• What are the functions of this connective tissue?

• How does the structure of this tissue support these functions?

• Where in the body will this tissue be found?

(page intentionally left blank)

Activity 6.6 Connecting the Structure and Function of the Various Tissue Types

This activity targets LOs 6.4, 6.5, and 6.6.

You have gathered the important features of each tissue type, including each tissue type's critical structure, its function, and locations in the body where it is found. Now it is time to put it all together and practice considering how each tissue's location in the body is driven by what that tissue can do, which in turn relies on key structural features.

Instructions

1. You are going to complete Table 6.3. The first row has been filled in for you.

2. For each incomplete row, determine what belongs in the empty boxes.

3. In the bottom three rows, which are blank, create your own challenges by thinking of a tissue type and filling in only one or two of the boxes in the row. Then exchange with your lab mate to solve each other's challenges by filling in the remainder of the boxes in each row.

Table 6.3 Key Structural Elements of Tissue Types			
Tissue Type	**Key Structural Elements of This Tissue Type**	**Function of This Tissue Type**	**Location in the Body Where This Tissue Type Is Found**
Elastic cartilage	Contains stretchy elastic fibers with a lot of ground substance	To bend and return to shape	Flexible outer ear
	Collagen and elastic fibers in an open meshwork		Subcutaneous layer under the skin
		Can stretch and relax to accommodate changes to the size of the lumen	
Ciliated pseudostratified columnar epithelium			Lining larger airways of the lungs
	A single continuous layer of thin, flat cells that line a lumen		Air sacs of the lungs
	Short, branching, striated cells with no ground substance		
		Send electrical signals through the body	
	Closely packed cells containing triglycerides		
		Move material through the hollow organs of the body	
			Ligaments and tendons
Fibrocartilage			
	Apical surface of keratin with many layers of flat cells		
			Lining most glands of the body
	Scattered reticular fibers in ground substance		
			The skeleton of the body

Activity 6.7 Case Study: Barrett's Esophagus

This activity targets LOs 6.4, 6.5, and 6.9.

Case Study

We have been discussing how the structure of a tissue type allows for its function in the human body. Consider the esophagus, which is a stretchy tube connecting the mouth to the stomach. When we swallow food it passes down the esophagus to reach the stomach, which uses churning and stomach acid to help break that food up into a uniform slurry.

1. The tissue that lines the lumen of the esophagus is specialized for managing the rough abrasion of swallowed bites of food. Therefore, you can predict that the inner surface of the esophagus is lined with:

   a. Hyaline cartilage

   b. Stratified squamous epithelial tissue

   c. Smooth muscle tissue

   d. Adipose connective tissue

2. Where the esophagus connects to the stomach, a ring of muscle surrounds the passageway; this muscle is called the lower esophageal sphincter. This sphincter automatically contracts after a bite of food or a drink of liquid has passed and entered the stomach. This prevents food and liquid from refluxing back into the esophagus as the stomach churns. Which type of muscle tissue would you expect to find making up the lower esophageal sphincter?

   a. Skeletal muscle tissue

   b. Cardiac muscle tissue

   c. Smooth muscle tissue

3. The tissue that lines the lumen of the stomach is specialized for protecting the deeper tissues from the damaging effects of the low pH of the stomach acid through significant production of alkaline mucus. Therefore, you can predict that the stomach is lined with:

   a. Dense regular connective tissue

   b. Reticular connective tissue

   c. Simple columnar epithelial tissue

   d. Transitional epithelial tissue

Brad regularly experiences a recurring burning pain in his chest that usually happens after he eats a big meal, sometimes he also has trouble swallowing. Ten years ago, Brad self-diagnosed these as symptoms of gastro-esophageal reflux disease (GERD) after reading about it on the internet and began taking over-the-counter heartburn medication to treat the symptoms. Lately, the symptoms have been getting worse, and Brad complains to his doctor about the problem. Brad's doctor is concerned about the long-term effects of GERD and the worsening symptoms, so she orders an upper endoscopy, which involves a small tube inserted down the throat into the esophagus. During the endoscopy, a small biopsy (piece of tissue) is removed from the area of the esophagus where it joins the stomach, and a histological slide is made.

Figure 6.7 | Normal Tissue Where Esophagus Transitions to Stomach

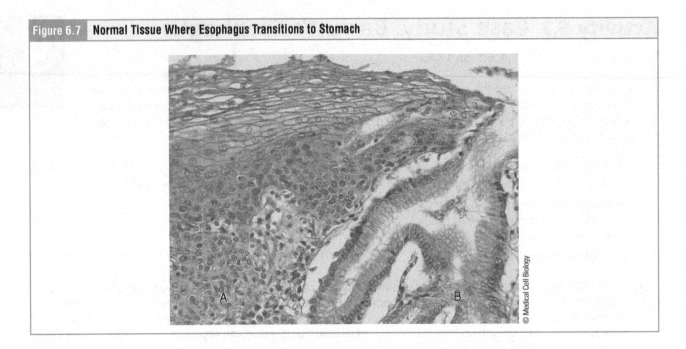

Examine Figure 6.7, which shows the histological slide of normal esophagus at the transition between esophagus and stomach. Then examine Figure 6.8, which shows a histological slide of the biopsy of Brad's esophagus.

Figure 6.8 | Brad's Esophagus

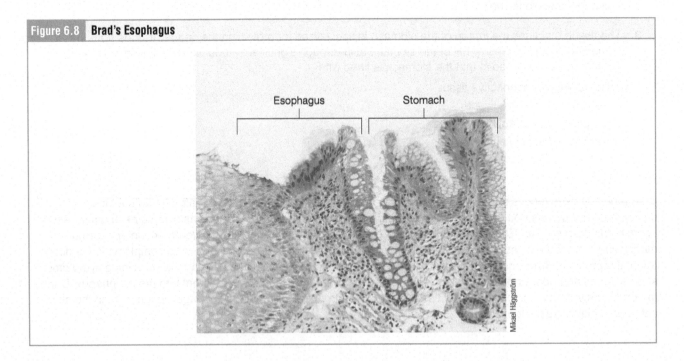

4. In Figure 6.7, the histological slide of a normal esophagus–stomach transition, the letter A is marking the _____ while the letter B is marking the _____.

   a. stomach; esophagus
   b. esophagus; stomach

5. In Figure 6.8, the histological slide of Brad's biopsy, you can see that some of the epithelial tissue of the esophagus contains goblet cells that are producing mucus. Would you expect to find goblet cells in the epithelial tissue lining a healthy esophagus?

   a. Yes
   b. No

6. Just as the function of a tissue is determined by its form, the form of a tissue can be altered by its function. What "new" function will the goblet cells in Brad's esophagus perform because of his reflux?

   a. Release of alkaline mucus to protect against stomach acid
   b. Churning of the contents of the esophagus
   c. Movement of material through the lumen

Brad's doctor tells him that Barrett's esophagus sometimes leads to esophageal cancer. To prevent the damaged cells from becoming cancerous, she suggests that Brad work to reduce the reflux that is damaging his esophagus. In addition to prescribing medication to treat the GERD and regular endoscopies to monitor the condition of his esophagus, she also suggests that he make some lifestyle changes to lose weight.

7. GERD is common in both obese and pregnant individuals. Which of the following statements could explain the link between these two conditions and GERD?

   a. The presence of a growing fetus or extra fat around the stomach pushes stomach contents up through the lower esophageal sphincter.
   b. The consumption of foods that are lower in pH contributes to the acidic nature of the stomach contents.
   c. A high-fat diet will lead the tissues of the body to increase the number of goblet cells to store triglycerides.

(page intentionally left blank)

Activity 6.8 Case Study: Marfan Syndrome

This activity targets LOs 6.4 and 6.8.

Shonda is 12 years old. Shonda has just been diagnosed with scoliosis (curvature of the spine) and during a follow-up visit the doctor notices that in addition to being tall and thin, Shonda's arms, legs, and fingers are unusually long and incredibly flexible. Shonda's doctor considers a possible diagnosis of Marfan syndrome and explains that this genetic disorder affects a gene on chromosome 15. If a person has just one abnormal version of the gene they will exhibit Marfan syndrome. This gene codes for a structure called fibrillin, which is necessary for the correct formation of elastic fibers. Without the correct directions for fibrillin, cells cannot produce elastic fibers into the correct arrangement in the extracellular matrix; this also has a secondary effect on the arrangement of collagen fibers in the matrix.

1. Which tissue type would you expect to exhibit problems in a person with Marfan syndrome?
   a. Epithelial tissues
   b. Muscle tissues
   c. Nervous tissue
   d. Connective tissues

2. Which cell type makes elastin and collagen fibers?
   a. Fibroblasts
   b. Chondrocytes
   c. Osteocytes

3. Which of the follow connective tissues would you expect to show the most drastic effects in someone with Marfan syndrome?
   a. Blood
   b. Elastic connective tissue
   c. Hyaline cartilage

4. Shonda's doctor suspects that the scoliosis diagnosis may suggest a possible Marfan syndrome diagnosis because the ligaments are not holding the vertebra together as tightly as they should, allowing the spine to curve. This implies that another connective tissue affected by Marfan syndrome is:
   a. Dense regular connective tissue
   b. Reticular connective tissue
   c. Adipose connective tissue

5. Which of the following would be a useful question for Shonda's doctor to ask to determine if she has other key signs of Marfan syndrome?
   a. Are her joints more flexible than other people's joints?
   b. Do her sweat glands over-produce?
   c. Is the surface of her skin overly dry?

Shonda's doctor would like to do further testing to determine if Shonda does have Marfan syndrome. In addition to a detailed family history and genetic testing to determine if Shonda had the abnormal gene on chromosome 15, the doctor also orders an echocardiogram to examine Shonda's heart valves and aorta, which is the large blood vessel that leaves the heart to deliver blood to the rest of the body.

6. Which connective tissue is likely to be abundant in the heart valves and aorta?
   a. Bone
   b. Elastic cartilage
   c. Elastic connective tissue

7. In a person without Marfan syndrome, what function does this type of connective tissue provide?

   a. It resists stress applied in one direction.

   b. It allows the tissue to stretch and recoil.

   c. It acts as a shock absorber.

8. The heart valve problems that would be likely in a person with Marfan syndrome are that the heart valves would:

   a. Be calcified and no longer flexible

   b. Not spring back quickly to closed, allowing blood to leak through

   c. Not contain enough adipocytes and triglyceride stores

9. Which problem with the aorta is most likely in Marfan syndrome?

   a. An aortic dissection, in which the wall of the aorta tears.

   b. Infection of the aorta, in which a bacteria or virus infects the aorta.

   c. Atherosclerosis of the aorta, in which hard fat and calcium build up in the wall of the aorta.

The results of the echocardiogram and genetic testing confirm that Shonda does have Marfan syndrome. Of all the signs and symptoms of Marfan syndrome, the ones most concerning to doctors are the effects on the heart. The doctor shares that regular monitoring of the heart and large blood vessels like the aorta is important to prevent and manage some possible problems. In addition, the doctor suggests some lifestyle changes to protect these structures.

10. A good suggestion for Shonda would be to choose:

   a. A high-fat diet that increases the amount of adipose connective tissue in her body

   b. Light activity that does not greatly increase the flow of blood through her heart and vessels

   c. High-intensity activity to increase the strength of her cardiac muscle tissue

---

Digging Deeper:
Making a Histology Slide

Preparing tissues to view under the microscope is a laborious process that takes both skill and patience. First, the tissue must be fixed by soaking it in chemicals that will prevent the cells and molecules from breaking down. Next, the tissue is processed with further chemicals to remove the water and then embedded into wax so that it can be sliced into incredibly thin pieces that are then placed on slides. Finally, since most cell and extracellular components are clear, stains are washed over the slices of tissue and allowed to adhere to various molecules in the tissue. A cover slip is placed on top of the tissue and sealed with glue to protect the finished slide.

The most commonly used stains in histology are hematoxylin and eosin. They are often used together and referred to as H&E.

Learning Connection

Let those Latin and Greek roots do some heavy lifting for you by finding connections between the names of tissue types and the cells and structures found in those tissues. For example, **adipo**cytes are found in **adipo**se connective tissue and **fibro**blasts make **fibers**. What other word connections can you make to help you learn these terms?

Lab 6 Post-Lab Quiz

Use the figure to the right to answer questions 1 through 4.

1. In the figure, the clear space marked as A is the:

   a. Extracellular matrix
   b. Lumen
   c. Basement membrane
   d. Ground substance

2. in the figure, the cells marked as C are:

   a. Goblet cells
   b. Fibroblasts
   c. Leukocytes
   d. Chondrocytes

3. In the figure, the tissue marked as D is specialized for maximum absorption of material into the body. Because of this, you can predict that a higher-magnification view of this tissue would reveal what structure at the cell surface marked as B?

   a. Brush border (microvilli)
   b. Cilia
   c. Basement membrane
   d. Goblet cells

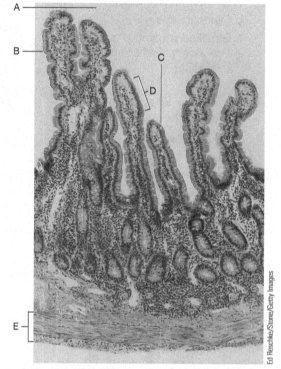

Figure for questions 1–4.

4. The tissue in this figure was taken from the small intestine; the long projections of tissue that you see are called villi. As tissue E contracts, the villi move to increase the absorption of material into the body. Therefore, tissue E is:

   a. Stratified squamous epithelium
   b. Dense irregular connective tissue
   c. Smooth muscle tissue
   d. Hyaline cartilage

5. Which cells within nervous tissues function to send electrical messages?

   a. Glial cells
   b. Neurons
   c. Nerves
   d. Fibroblasts

6. You are examining a histological slide and see that the cells you are looking at have cilia on their apical surface. Because of the presence of cilia, you can narrow down your identification of the tissue type to which specific tissues?

Use the figure to the right to answer questions 7 through 9.

7. In the figure, what tissue type is marked by the letter B?

   a. Smooth muscle tissue

   b. Fibrocartilage

   c. Stratified squamous epithelium

   d. Pseudostratified columnar epithelium

8. In the figure, what structure is marked by the letter A?

   a. Keratin layer

   b. Cilia

   c. Basement membrane

   d. Microvilli

9. In the figure, the tissue marked as C functions to prevent tears to this organ. Which tissue type is tissue C?

   a. Hyaline cartilage

   b. Nervous tissue

   c. Dense irregular connective tissue

   d. Smooth muscle tissue

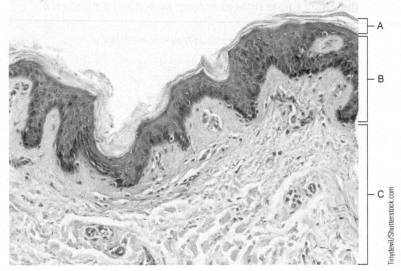

Figure for questions 7–9.

10. Reticular connective tissue contains many small cells between the reticular fibers. Sharon identifies the cells as leukocytes, while Jermaine identifies them as fibroblasts. Who is correct? Defend your answer.

11. You are provided with a slide labeled "involuntary muscle of the bladder wall." Which tissue feature will you be looking for on the slide?

    a. Fusiform-shaped cells

    b. Striations

    c. Intercalated disks

    d. Extracellular fibers

12. Cells of your body that make collagen fibers need vitamin C to do so. If you are deficient in vitamin C, which tissue type would be most affected?

    a. Epithelial tissues

    b. Nervous tissues

    c. Connective tissues

    d. Muscle tissues

Lab 7 Integument

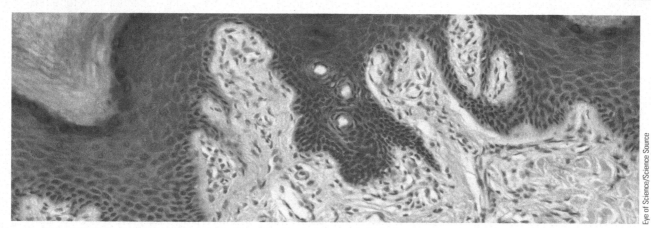

Eye of Science/Science Source

Learning Objectives: After completing this lab, you should be able to:

7.1 Describe the general functions of the integumentary system and the subcutaneous layer.

7.2 Identify and describe the tissue type making up the epidermis.

7.3 Identify and describe the layers of the epidermis, indicating which are found in thin skin and which are found in thick skin.

7.4 Describe the processes of growth and keratinization of the epidermis.

7.5 Identify and describe the dermis and its layers, including the tissue types making up each dermal layer.

7.6 Identify and describe the subcutaneous layer, including the tissue types.

7.7 Describe the functions of the epidermis.

7.8 Describe the functions of the dermis, including the specific function of each dermal layer.

7.9 Describe the functions of the subcutaneous layer.

7.10 Describe the structure and function of hair.

7.11 Describe the structure and function of nails.

7.12 Describe the structure and function of exocrine glands of the integumentary system.

7.13 Given a factor or situation (e.g., second-degree burns [partial-thickness burns]), predict the changes that could occur in the integumentary system and the consequences of those changes (i.e., given a cause, state a possible effect).

7.14 Explain how the integumentary system maintains homeostasis with respect to thermoregulation and water conservation.

7.15 Explain how the integumentary system relates to other body systems to maintain homeostasis.

7.16 Explain how each of the five layers, as well as each of the following cell types and substances, contributes to the functions of the epidermis: stem cells of stratum basale, keratinocytes, melanocytes, epidermal dendritic (Langerhans) cells, tactile (Merkel) cells and discs, keratin, and extracellular lipids.

The Human Anatomy and Physiology Society includes more than 1,700 educators who work together to promote excellence in the teaching of this subject area. The HAPS A&P Learning Outcomes measure student mastery of the content typically covered in a two-semester Human A&P curriculum at the undergraduate level. The full Learning Outcomes are available at https://www.hapsweb.org.

Introduction

The skin is the largest organ in the human body. It plays essential roles in protection from infection, thermoregulation, water regulation, and environmental sensing. Like all organs, the skin is made of tissues that work together as a single structure to perform these critical functions. The skin is made of multiple layers of cells and is held to the body with connective tissue. While the skin is very complex, we can discuss it as generally being composed of three layers: the epidermis, dermis, and hypodermis.

The **epidermis**, the outermost layer, is composed of stratified squamous epithelium. While all structures in the body that are open to the outside, such as the mouth, are lined with stratified squamous epithelium, the skin is exposed to a much drier environment. Therefore, the stratified squamous epithelium is keratinized stratified squamous. **Keratin** is an intracellular fibrous protein that gives hair, nails, and skin their hardness and water-resistant properties. Like all epithelia, the skin's epithelium is avascular (does not contain blood vessels) and is therefore supported by a vascularized connective tissue layer, called the **dermis**, just below the epidermis. Beneath the dermis is the **hypodermis**, a second connective tissue layer with a composition that differs from the dermal layer (Figure 7.1).

Figure 7.1 | Layers of Skin

The skin is composed of many layers. The epidermis, which is the major outermost layer of the skin, is composed of epithelial tissue. The dermis, which lies deep to the epidermis, is composed of connective tissue. Beneath the dermis lies the hypodermis, which has a different composition of connective tissues than the dermis. *Fascia* is the term for the connective tissue that anchors the skin to the muscle beneath.

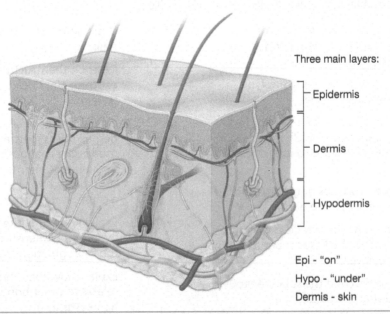

Three main layers:

— Epidermis

— Dermis

— Hypodermis

Epi - "on"

Hypo - "under"

Dermis - skin

Digging Deeper:
The Importance of Medical Illustration and Representation

There are many different infections and illnesses that can impact the skin and/or be diagnosed based on skin symptoms. Therefore, it is imperative for doctors to be able to recognize skin symptoms in skin with all varieties of pigmentation. Unfortunately, the representation of darker skin in the descriptions and images of skin conditions in textbooks is often based on lighter shades of skin. For example, The Centers for Disease Control website describes the bacterial infection impetigo as "a red, itchy sore. As it heals, a crusty, yellow or honey-colored scab forms over the sore." This can be difficult to identify on black or brown skin. Similarly, measles is a dangerous viral infection that presents with cold or flu like symptoms in addition to a rash. The rash, consisting of flat raised bumps that are red or brown on lighter skin, are very difficult to distinguish on darker skin. It is essential that science and medical education include descriptions and images of skin conditions on all shades of skin so that disease recognition, diagnosis, and treatment occurs rapidly, regardless of race.

Lab 7 Pre-Lab Quiz

This quiz will strengthen your background knowledge in preparation for this lab. For help answering the questions, use your resources to deepen your understanding. The best resource for help on the first five questions is your text, and the best resource for help answering the last five questions is to read the introduction section of each lab activity.

1. Anatomy of connective tissue and epithelia. Label panel A of the figure with the following tissue components: apical side, basal side, epithelial cells. Label panel B with these tissue features: blood vessels, collagen fibers, fibroblasts.

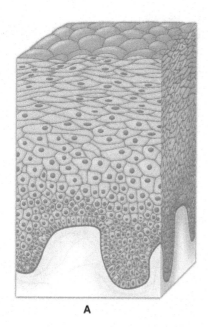

A

B

2. Write the root next to its definition

Root	Meaning
	a bump
	a foundation
	skin
	a layer

List of Roots

bas-
derm-
papill-
stratum

3. The epidermis is composed of _____, while the dermis is composed of _____.

   a. epithelium, connective tissue
   b. muscle, connective tissue
   c. muscle, epithelium
   d. connective tissue, epithelium

4. True or False: The main protein fiber of the dermis is the same as the main protein fiber of the nails and hair.
   a. True
   b. False

5. Which of the following correctly describes a difference between thick and thin skin?
   a. Thick skin is found at joints such as the knee and the elbow, thin skin is found at broad surfaces like the abdomen and back.
   b. Thick skin has five layers of epidermis and thin skin has four.
   c. Thick skin has many hair follicles, while thin skin has only a few hair follicles.

6. Which is the correct order of the layers of the epidermis of thin skin from superficial to deep?
   a. Stratum corneum, stratum granulosum, stratum spinosum, stratum basale
   b. Stratum granulosum, stratum spinosum, stratum corneum, stratum basale
   c. Stratum basale, stratum granulosum, stratum spinosum, stratum corneum
   d. Stratum corneum, stratum spinosum, stratum granulosum, stratum basale

7. In which of the following layers of skin can you find blood vessels?
   a. Stratum basale
   b. Papillary dermis
   c. Reticular dermis
   d. All of these

8. The diverse array of skin color amongst humans is due to the production and deposition of pigment molecules into the epidermis. Which type of cells produce that pigment?
   a. Merkel cells
   b. Keratinocytes
   c. Melanocytes
   d. Stem cells

9. Which of these images depicts the correct location of the bulb of a hair follicle?

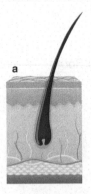 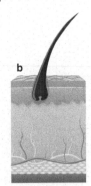 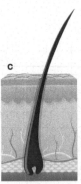

10. Which of these glands is found the deepest in the skin?
   a. Apocrine sweat glands
   b. Sebaceous glands
   c. Eccrine sweat glands

Activity 7.1 Anatomical Layers of the Skin

This activity targets LOs 7.1, 7.2, 7.3, 7.4, 7.5, 7.6, 7.7, 7.8, and 7.9.

Epidermis

The epidermis is composed of many layers (stratified) of epithelial cells. Collectively, these cells are called **keratinocytes**. The keratinocytes are loosely organized into 4–5 layers. These layers, from deep to superficial are: stratum basale, stratum spinosum, stratum granulosum, and stratum corneum (Figure 7.2). The skin on the palms of the hands and soles of the feet has a fifth layer, called the stratum lucidum, located between the stratum corneum and the stratum granulosum. Skin in these two locations, with its fifth layer of cells, is referred to as "thick" skin, and all other skin is referred to as "thin" skin (Figure 7.3). Thick and thin skin also differ in their glands, which we will discuss later.

The **stratum basale** is the deepest epidermal layer and anchors the epidermis to its basement membrane, which attaches to the dermis below. The stratum basale is a single layer of cells, most of which are stem cells that give rise to the other cells of the epidermis. All of the other keratinocytes originate from this basal layer. As new cells are produced, the existing cells of the epidermis are pushed superficially away from the basal layer, toward the surface of the skin. Two other types of cells are found dispersed among the stem cells of the stratum basale. The first is a **Merkel cell**, which is a neuroendocrine cell; it functions as a sensory receptor and secretes chemicals that can act as hormones. Merkel cells are especially abundant in the hands and feet, where we have more touch sensation. The second is a **melanocyte**, a cell that produces the pigment melanin. **Melanin**, a protein

---

Figure 7.2 | Layers of the Epidermis

The epidermis has four layers in most areas of the body: stratum basale, stratum spinosum, stratum granulosum, and stratum corneum.

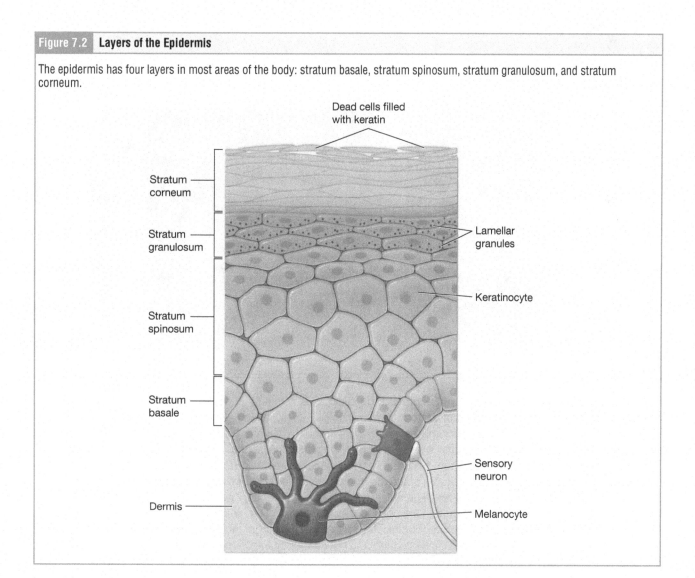

Figure 7.3 **Thin Skin Versus Thick Skin**

(A and B) Thin skin exists all over the body and has four distinguishable cell layers in the epidermis. (C and D) Thick skin, found only on the palms of the hands and soles of the feet, has an additional layer, the stratum lucidum.

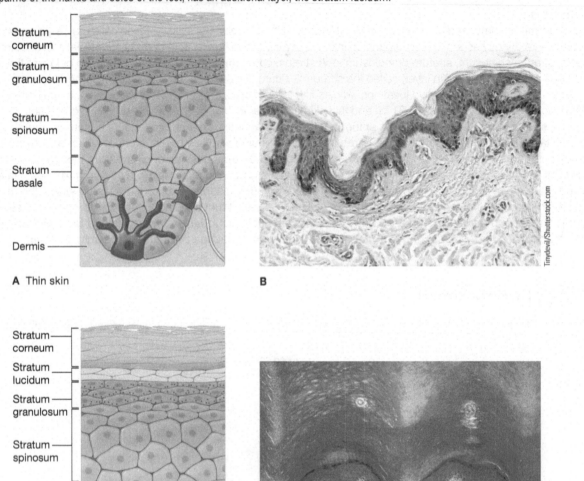

A Thin skin

B

Tinydevil/Shutterstock.com

C Thick skin

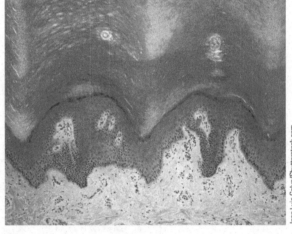

D

Jose Luis Calvo/Shutterstock.com

that can be made in two forms, functions to protect the structures within cells below the epidermis from ultraviolet light. The protein melanin lends color to both skin and hair.

The 8–10 rows of cells of the **stratum spinosum** are wide in the middle and tapered at the end, similar in shape to almonds or American footballs. The keratinocytes of this layer begin synthesis of keratin and release a water-repelling glycolipid that helps prevent water loss from the body, making the skin relatively waterproof. As new keratinocytes are produced in the stratum basale, keratinocytes of the stratum spinosum are pushed into the stratum granulosum.

The **stratum granulosum** is named for its granular, or spotted, appearance. The cells of this layer are a bit flatter than the cells of the stratum spinosum and are spotted with granules, which are intracellular protein-filled vesicles. These vesicles are filled with keratin precursor molecules and melanin pigment molecules. Melanin is made by melanocytes in the stratum basale, but is then excreted and subsequently engulfed by the keratinocytes in the stratum granulosum. The stratum granulosum is the most superficial layer that contains living cells.

The **stratum lucidum** is a feature of thick skin, and present in the skin of the palms of the hands and soles of the feet. This seemingly translucent layer of the epidermis is located just under the stratum corneum. Cells in the stratum lucidum are dead, and empty of internal organelles, which is why they appear clear when viewed with a microscope.

The **stratum corneum** is the most superficial layer of every epidermis and is the layer of skin that is exposed to the outside environment. The stratum corneum is usually composed of about 15–30 layers of dead skin cells. This dry, dead layer helps prevent the penetration of microbes and the dehydration of underlying tissues. It also protects against damage to the living cells from friction.

Dermis and Hypodermis

The **dermis** contains blood and lymph vessels, nerves, and other structures, such as the deeper portions of hair follicles and sweat glands. The dermis has two layers, both are connective tissue composed of an interconnected mesh of elastin and collagenous fibers produced by fibroblasts (Figure 7.4). Collagen provides tensile strength, and appears thicker and is stained pink by the dye eosin when viewed with the microscope. Elastin fibers provide stretchiness to skin, and appear thin and black or purple when stained and viewed with the microscope (Figure 7.5).

The **papillary layer** of the dermis is made of loose areolar connective tissue. This layer gets its name from its ridges, or papillae, that project into the epidermis to interlock and prevent separation of the epidermis from the dermis (Figure 7.6). Within the papillary layer are small blood vessels, nerve fibers, and touch receptors called tactile (Meisner's) corpuscles. The collagen fibers of the papillary layer are smaller and thinner than those of the reticular layer.

The **reticular layer** of the dermis is composed of dense, irregular connective tissue. This layer has a rich vascular and nerve supply. It makes up the majority (80%) of the thickness of the skin.

The **hypodermis** is a layer directly below the dermis and serves to connect the skin to the underlying tissues. The hypodermis consists of well-vascularized, loose areolar connective tissue and adipose tissue. Its functions include fat storage, insulation, and cushioning.

Learning Connection

Chunking

Try making a list of as many skin terms as you can think of (you might start with the bolded terms in the intro to this activity and add terms from your textbook). Then group them into epidermis, dermis and hypodermis. For example, the term keratin would be placed in the epidermis category whereas the term collagen would be placed under dermis.

| Figure 7.4 | **The Dermis** |

The two layers of the dermis—papillary and reticular—are differentiated by the thickness and weave of the collagen fibers.

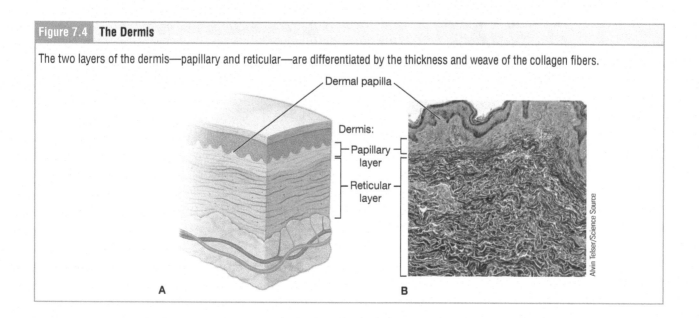

Collagen fibers

Elastin fibers

Biophoto Associates/Science Source

Figure 7.6 Dermal Papillae

(A) The bumps and ridges of the dermis and epidermis interlock to form a combined structure that rarely pulls apart when exposed to friction or shear force. (B) The interdigitation can be likened to the interlocking of fingers.

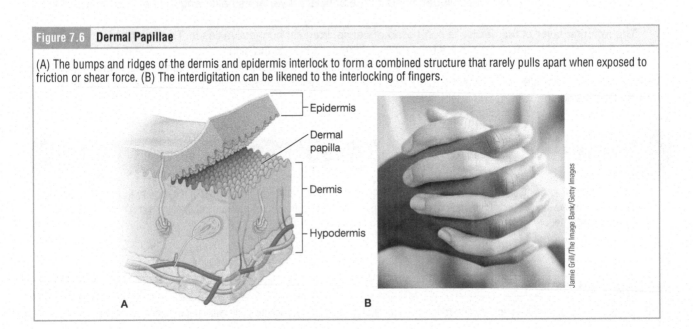

Epidermis

Dermal papilla

Dermis

Hypodermis

A

B

Jamie Grill/The Image Bank/Getty Images

Materials
- Compound microscope
- Lab tape
- Model of skin
- Slides of skin

Instructions
1. On the model of human skin, identify all of the layers and sub-layers listed in **Table 7.1**. Use lab tape to label each of them.

2. Observe a prepared slide of human skin.

   a. Using Figure 7.7 as a reference, identify the stratum corneum and stratum basale. Describe in Table 7.1 how these two layers differ.

   b. Draw the epidermis in Table 7.1.

   c. Using Figure 7.7 as a reference, compare the papillary and reticular dermis. Draw and describe these layers in Table 7.1.

Table 7.1 Layers of Human Skin			
Layer	**Sub-Layer**	**Unique Characteristics Observed**	**Drawing**
Epidermis	Stratum corneum		
	Stratum granulosum		
	Stratum spinosum		
	Stratum basale		
Dermis	Papillary dermis		
	Reticular dermis		
Hypodermis	Hypodermis		

Figure 7.7   **Layers of the Epidermis and Dermis**

The skin is composed of the epidermis and the dermis. The dermis and epidermis each has sublayers.

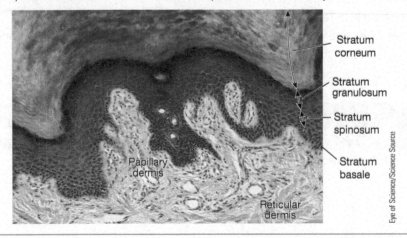

Stratum corneum

Stratum granulosum

Stratum spinosum

Stratum basale

Papillary dermis

Reticular dermis

Eye of Science/Science Source

Activity 7.2 Accessory Structures of the Skin

This activity targets LOs 7.10, 7.11, and 7.12.

Accessory structures of the skin include hair, nails, sweat glands, and sebaceous glands.

Hair

The hair that you see emerging from your skin surface is technically the **hair shaft**, a long strand of dead keratinized cells. This shaft of hair is grown within an epidermal structure called the **hair follicle**. The hair follicle is an epidermal structure but as it grows larger it pushes down into the dermis. At the base of the hair follicle dermal cells form the **dermal papilla**, a structure that contains the blood vessels and nerves associated with the hair follicle. Together the base of the hair follicle and the dermal papilla are called the **hair root** (Figure 7.8).

Figure 7.8	The Structure of Hair

Hair grows in follicles, which are mostly epidermal structures, but project into the dermis as they become larger.

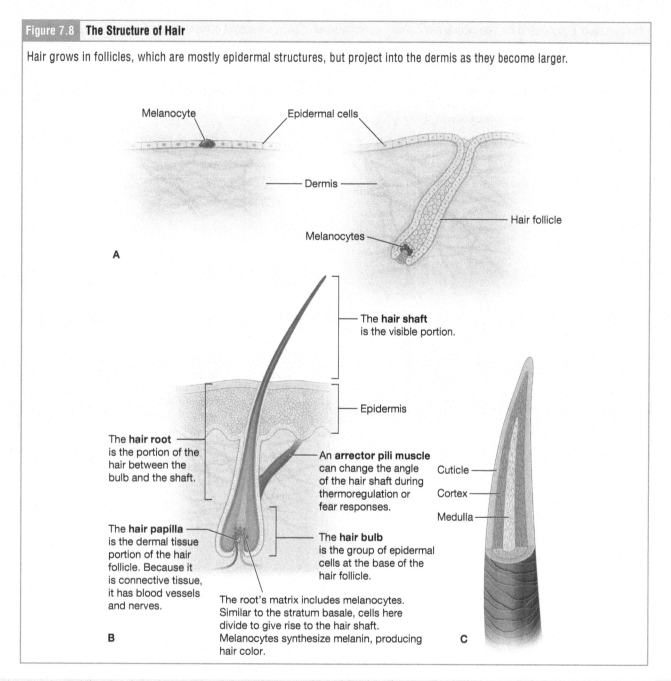

Melanocyte

Epidermal cells

Dermis

Melanocytes

Hair follicle

A

The **hair shaft** is the visible portion.

Epidermis

The **hair root** is the portion of the hair between the bulb and the shaft.

An **arrector pili muscle** can change the angle of the hair shaft during thermoregulation or fear responses.

Cuticle

Cortex

Medulla

The **hair papilla** is the dermal tissue portion of the hair follicle. Because it is connective tissue, it has blood vessels and nerves.

The **hair bulb** is the group of epidermal cells at the base of the hair follicle.

The root's matrix includes melanocytes. Similar to the stratum basale, cells here divide to give rise to the hair shaft. Melanocytes synthesize melanin, producing hair color.

B

C

Glands

Each hair follicle is associated with an oil gland known as a **sebaceous gland**. These glands produce an oily mixture called **sebum** which lubricates the hair and skin (Figure 7.9). These glands and their associated hair follicles are found all over the body except in thick skin.

The other exocrine glands of human skin are sweat glands. Eccrine sweat glands are the type of sweat gland found all over the body; they produce sweat for thermoregulation (Figure 7.9). These glands are coiled glands that lie deep in the dermis with the duct carrying the sweat to a pore on the skin surface. This type of sweat is composed mostly of water with a few ions, metabolic waste, and antibacterial compounds.

Apocrine sweat glands are usually associated with hair follicles and are found in densely hairy areas of the armpits (i.e., axilla) and groin. Apocrine sweat glands are larger than eccrine sweat glands and lie deeper in the dermis, sometimes even reaching the hypodermis (Figure 7.9). Figure 7.10 compares these two glands histologically.

Figure 7.9 The Accessory Structures of the Skin

The accessory structures of the skin include hair follicles, apocrine sweat glands, sebaceous glands, eccrine sweat glands, and nails.

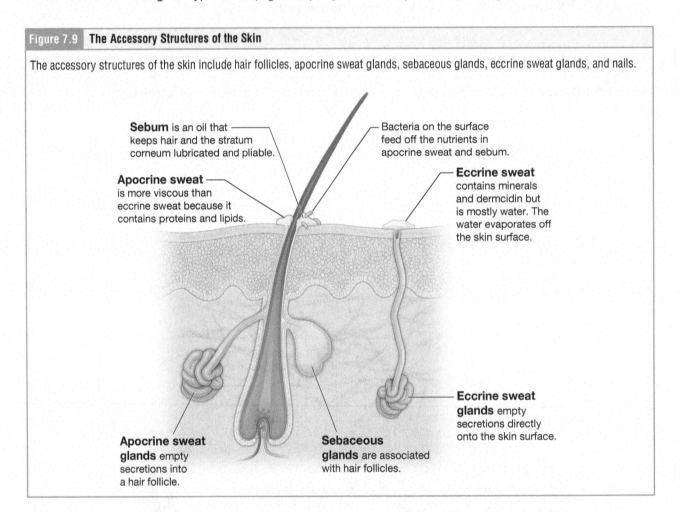

Sebum is an oil that keeps hair and the stratum corneum lubricated and pliable.

Bacteria on the surface feed off the nutrients in apocrine sweat and sebum.

Apocrine sweat is more viscous than eccrine sweat because it contains proteins and lipids.

Eccrine sweat contains minerals and dermcidin but is mostly water. The water evaporates off the skin surface.

Apocrine sweat glands empty secretions into a hair follicle.

Sebaceous glands are associated with hair follicles.

Eccrine sweat glands empty secretions directly onto the skin surface.

Figure 7.10 | Sweat Glands

There are two different kinds of sweat glands in the human body. Apocrine sweat glands (A) are found in the axilla and groin, while eccrine sweat glands (B) are found all over the body.

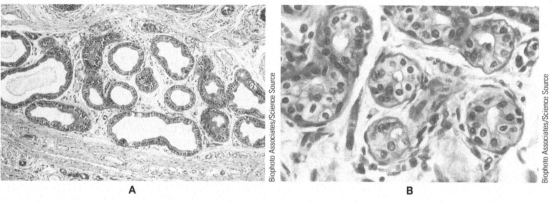

A   B

Nails

The **nail bed** is the living component of nails, a specialized structure of the epidermis, the cells of which produce the nail body. The **nail body** is the hard, bladelike structure you might paint with nail polish, though its real function is to protect the tips of our fingers and toes. The nail body is composed of densely packed dead keratinocytes. The nail body forms a **nail root** at the proximal edge of the nail which has a matrix of proliferating cells from the stratum basale that enables the nail to grow continuously. The nail cuticle is a strip of epidermis around the perimeter of the nail (Figure 7.11).

Figure 7.11 | Nails

Nails are accessory structures of the epithelium made of dead cells packed with keratin.

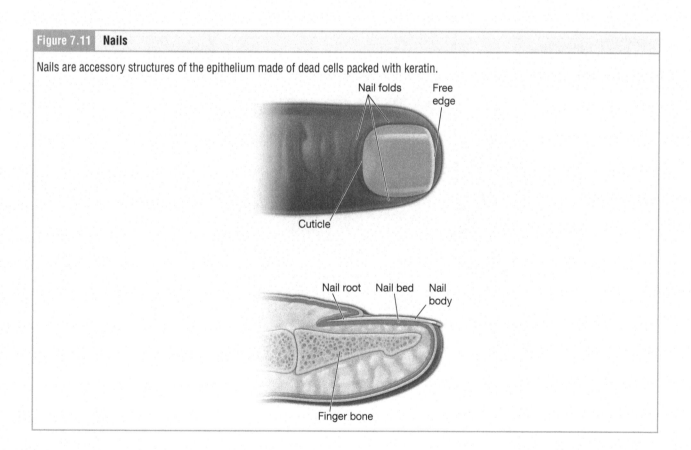

Materials

- Compound microscope
- Dissecting microscope or hand lens
- Prepared slides of human scalp, human skin, and human axilla

Instructions

1. Using Figure 7.11 as a guide, identify the parts of the nail on yourself or your lab partner using the hand lens or dissecting microscope. Take a photo or make a drawing and label the parts of the nail using Figure 7.11 as your guide.

2. Using your microscope, observe the slide of human scalp using the 40× objective. Using Figure 7.12 as your guide, identify the parts of the hair follicle. Draw an example of a hair follicle in the space below.

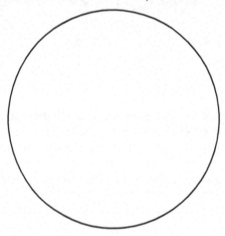

Figure 7.12   **Hair Follicles**

Hair follicles are epidermal structures that extend through the dermis consisting of a shaft, follicle, bulb, and papilla.

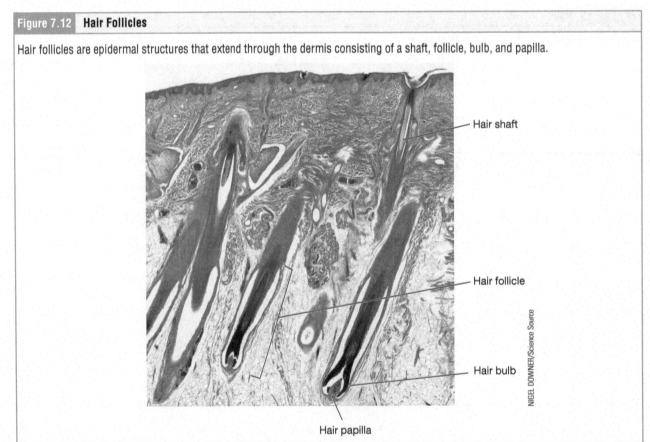

Hair shaft

Hair follicle

Hair bulb

Hair papilla

NIGEL DOWNER/Science Source

3. Observe the slide of the axilla. Find a sebaceous gland and an apocrine sweat gland. Draw each of them in the spaces below.

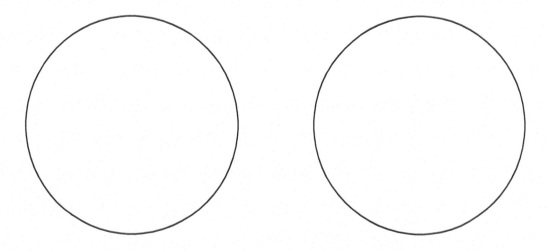

4. Observe the slide of human skin. Find an eccrine sweat gland. Draw an example in the space below.

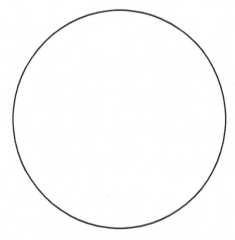

How would you describe the difference between the appearance of an eccrine sweat gland and an apocrine sweat gland to a person who couldn't see them under the microscope?

(page intentionally left blank)

Activity 7.3 Case Study: Skin Cancer

This activity targets LO 7.13.

Cancer occurs when some cells replicate uncontrollably. Often these cells can spread to other parts of the body. Cancer can occur in almost any organ, including the skin. One in five Americans will develop skin cancer in their lifetimes, in fact, more people are diagnosed with skin cancer than all other cancers combined. Skin cancer is more common in fair- or light-skinned individuals, because melanin is protective against UV radiation, but skin cancer can affect any human, regardless of skin color. In the United States, skin cancer is typically diagnosed only in later stages of disease progression in people of color and therefore has a worse prognosis. It is important, therefore, for all people to be aware of the signs and risks of skin cancer. In this activity we will explore five different varieties of skin cancer.

Kaposi's sarcoma is a type of skin cancer that occurs when endothelial cells replicate uncontrollably. Endothelial cells are the cells that line the inside of blood vessels and lymph vessels.

1. In which layer of the skin does Kaposi's sarcoma originate?

   a. Stratum corneum

   b. Stratum basale

   c. Reticular dermis

Merkel cell carcinoma is a type of skin cancer in which Merkel cells replicate in an uncontrolled manner. Melanoma is a type of skin cancer in which melanocytes replicate in an uncontrolled manner.

2. In which layer of the skin do Merkel cell carcinoma and melanoma originate?

   a. Both originate in the stratum granulosum.

   b. Both originate in the stratum basale.

   c. Merkel cell carcinoma originates in the reticular dermis, and melanoma originates in the papillary dermis.

   d. Merkel cell carcinoma originates in the stratum basale, and melanoma originates in the stratum granulosum.

3. Comparing Merkel cell carcinoma and melanoma, which tumor would tend to appear darker?

   a. Merkel cell carcinoma would appear darker.

   b. Melanoma would appear darker.

   c. Both would appear the same color.

Basal cell carcinoma and squamous cell carcinoma are two common forms of skin cancer.

4. Based on their names alone, do you think these cancers originate in the dermis or epidermis?

   a. Both originate in the dermis.

   b. Both originate in the epidermis.

   c. Basal cell carcinoma originates in the dermis and squamous cell carcinoma originates in the epidermis.

   d. Basal cell carcinoma originates in the epidermis and squamous cell carcinoma originates in the dermis.

5. Which type of cancer is most likely depicted in this image?

    a. Basal cell carcinoma
    b. Kaposi's sarcoma
    c. Squamous cell carcinoma

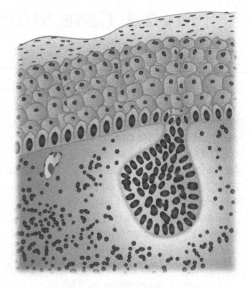

6. Cancer is often diagnosed using microscopy. A pathologist studies a section of the tumor and looks for characteristic features of specific types of cancer. Match the histological characteristics with the type of cancer for the three skin cancers below.

Type of Cancer	Histological Characteristics
Melanoma	
Kaposi's sarcoma	
Squamous cell carcinoma	

    a. Tumor cells will be in or near blood vessels
    b. Tumor cells will have a squamous (flat) shape
    c. Tumor cells will appear darker than surrounding tissue

7. Cancer is often diagnosed using microscopy. An important diagnostic observation is to note the shape of the cells. In this microscopy image, two tumors are outlined. Which type of cancer might this be?

    a. Melanoma
    b. Squamous cell carcinoma
    c. Kaposi's sarcoma

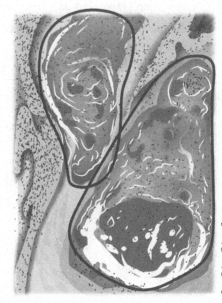

Garry DeLong/Science Source

Lab 7 Post-Lab Quiz

1. The color pigment of hair comes from _____ cells, and the nutrients that nourish the hair follicle come from blood vessels of the _____ layer.

   a. epidermis, dermis
   b. dermis, hypodermis
   c. epidermis, hypodermis
   d. epidermis, epidermis

2. Imagine a genetic condition in which thick skin covered the body. Which of the following would be absent?

   a. Eccrine sweat glands
   b. Hair follicles
   c. Sensory reception

3. The amount of hypodermal adipose tissue varies from one body region to another and one person to another. Because of this _____ also varies.

   a. thermoregulatory capacity
   b. proportion of thick versus thin skin
   c. density of hair follicle

4. A second-degree burn extends through the epidermis and into the dermis. When treating a patient with a second-degree burn, which of the following is a concern in addition to the damage to the skin?

   a. Loss of water
   b. Loss of protein
   c. Decreased capacity to store ATP

5. Examine the cell in the image. How did this cell get to be in this location?

   a. A neighboring cell went through mitosis to produce this cell.
   b. This cell originated in the stratum basale and gradually moved up to this location as more and more cells were produced below it.
   c. This cell migrated down from the corneum in response to friction.

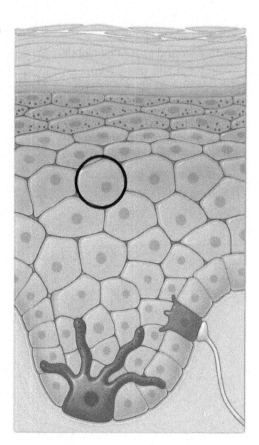

6. During exercise, which of the following glands increases its rate of exocrine secretion?

   a. Eccrine glands
   b. Sebaceous glands
   c. Endocrine glands

7. This image shows a blister, a fluid-filled skin injury resulting from friction, such as wearing shoes that don't fit well. How can we correctly describe this blister?

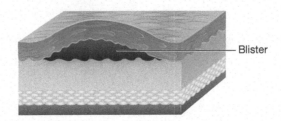

   a. As this blister formed, the reticular dermis separated from the papillary dermis.
   b. As this blister formed, the reticular dermis separated from the epidermis.
   c. As this blister formed, the papillary dermis separated from the epidermis.

8. Why are there no cancers of the stratum corneum?

   a. These cells are packed together too tightly to form tumors.
   b. These cells are already dead and cannot reproduce.
   c. Because they have no access to blood vessels, these cells cannot form tumors.

9. The _____ work together to maintain temperature homeostasis through shivering and sweating.

   a. muscles and skin
   b. liver and endocrine glands
   c. brain and spinal cord

10. Melanin pigment is produced in the stratum _____ and packaged into granules that reside in the stratum _____.

    a. granulosum, granulosum
    b. basale, basale
    c. spinosum, granulosum
    d. basale, granulosum

Lab 8 Bone Tissue

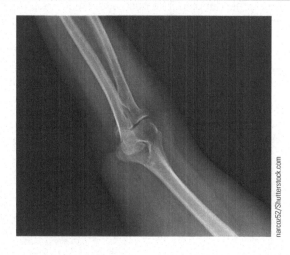

narcoz52/Shutterstock.com

Learning Objectives: After completing this lab, you should be able to:

8.1 Classify bones of the skeleton based on their shape.

8.2 Identify and describe the structural components of a long bone and explain their functions.

8.3 Describe the bone repair and remodeling process and how it changes as humans age.

8.4 Describe how the location and distribution of red and yellow bone marrow varies during a lifetime.

8.5 List and describe the cellular and extracellular components of bone tissue.

8.6 Explain the roles that specific bone cells play in the formation of bone tissue.

8.7 Identify the microscopic structure of compact bone and spongy bone.

8.8 Compare and contrast the function of osteoblasts and osteoclasts during bone growth, repair, and remodeling.

8.9 Explain the steps involved in fracture repair.

8.10* Compare and contrast the qualities of bone tissue to other tissues in the body.

8.11 Compare and contrast interstitial (lengthwise) and appositional (width or circumferential) growth.

* Objective is not a HAPS Learning Goal.

Introduction

The human skeleton is among the most dynamic structures in the body. It is under constant revision, perennially being broken down, and built again in a cycle of remodeling based on needs and use. Bone is composed of living cells, osteocytes, that are separated from each other by a rigid, mineralized matrix. Bone is a form of connective tissue. It is infused with blood vessels and nerves, wrapped in fibrous connective tissue, and many bones are capped by cartilage. In this lab we will explore the structure of bone, the bone remodeling process, and examine some of the disease states that can occur with disruptions to bone structure.

In this lab you will explore bone tissue at the gross anatomy level in Activities 8.1–8.3 and at the microanatomy level in Activities 8.4–8.6.

The Human Anatomy and Physiology Society includes more than 1,700 educators who work together to promote excellence in the teaching of this subject area. The HAPS A&P Learning Outcomes measure student mastery of the content typically covered in a two-semester Human A&P curriculum at the undergraduate level. The full Learning Outcomes are available at https://www.hapsweb.org.

Lab 8 Pre-Lab Quiz

This quiz will strengthen your background knowledge in preparation for this lab. For help answering the questions, use your resources to deepen your understanding. The best resource for help on the first five questions is your text, and the best resource for help answering the last five questions is to read the introduction section of each activity.

1. Anatomy of a long bone. Label panel A of the figure with the following parts of the bone: compact bone, endosteum, medullary cavity, periosteum, red marrow, spongy bone, yellow marrow. Label panel B of the figure with the following terms: compact bone, lamellae, osteon, periosteum, spongy bone, trabeculae.

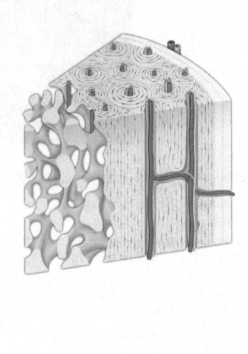

A   B

2. Write the root next to its definition:

Root	Meaning
	in
	around
	builder
	cartilage
	cell
	sheet
	bone

List of Roots

-blast
chondro-
-cyte
endo-
lamella
osteo-
peri-

3. Which of the following are functions of the skeleton (select all that apply)?

    a. Storage of minerals
    b. Production of hormones
    c. Red and white blood cell synthesis

4. The external surface of a bone is composed of compact bone/spongy bone {circle one}.

5. The fetal skeleton is composed of _____, which is slowly converted to bone.

    a. hyaline cartilage
    b. fibrocartilage
    c. fibrous connective tissue

6. The bones of your fingers are called phalanges. Each of these is a cylinder, and, depending on the size of your hands, is a little more or less than an inch in length. Which type of bone would your phalanges be classified as?

    a. Long
    b. Short
    c. Irregular

7. Which type of bone cell is responsible for breaking down bone matrix in order to add minerals to the blood?

    a. Osteoblasts
    b. Osteoclasts
    c. Osteocytes

8. Place these items in order of size from smallest to largest: lacuna, concentric lamella, osteon.

9. Which of the following components of bones are proteins?

    a. Calcium
    b. Phosphorus
    c. Collagen
    d. None of these are proteins

10. When a bone has suffered a fracture and is healing, what is the first material constructed in the area of the break?

    a. Bone tissue is built immediately.
    b. Fibrocartilage is built first, then ossified.
    c. Epithelial tissue is built first, then slowly converted to bone.

(page intentionally left blank)

Activity 8.1 Bone Classification

This activity targets LO 8.1.

At the gross anatomical level, we can learn a lot about bones. Bones are classified into four categories based on their shape (Figure 8.1 and Table 8.1). The shape of a bone tells us a lot about its functions in the body. *Flat* bones are, well, flat in shape. These bones are generally found protecting vulnerable organs like the heart and brain. *Long* bones are longer than they are wide. These bones might be small, like the bones in your fingers, or large, like the bones in your leg, but they are always longer than they are wide. *Short* bones, by contrast, are roughly the same width as length. Short bones are usually found in clusters where they form slightly movable joints, like the wrist and ankle. Any bones that don't fit neatly into these three categories are usually termed *irregular* bones. These bones have complicated shapes, usually with bumps and processes that stick out from the bone and attach to the tendons of muscles.

Materials
• An assortment of bones from a disarticulated skeleton OR an articulated skeleton

Instructions
1. Obtain an assortment of bones provided by your instructor. As you examine each bone, determine if it is a flat, irregular, long, or short bone.

2. For each bone provide one clue that justifies your answer. One of the bones has been done for you as an example.

Learning Connection

Chunking

Look at an image of the entire human skeleton. Can you group the bones into their shape categories? Where on the skeleton do you tend to find flat bones? Long bones?

Table 8.1 Types of Bone		
Shape	**Example**	**Description**
Flat	Sternum	Typically thin and flat. Found in locations where they can cover and protect delicate organs
Long	Femur	Longer than they are wide. Note that long bones can be small, like the phalanges of the fingers, but that they are always longer than they are wide
Short	Tarsals	Roughly as wide as they are long
Irregular	Os coxae	Have complex shapes. Typically have many bumps or processes where muscles attach

Figure 8.1 Bones Are Classified According to Their Shape

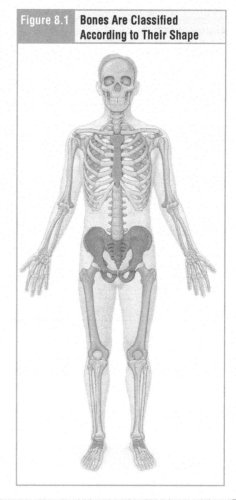

Bone 1

Name of bone: *Vertebra*

Type of bone: *Irregular*

Justification: *This bone does not have a long shaft, it is not very flat, but it does have a lot of processes that stick out*

Bone 2

Name of bone: _____

Type of bone: _____

Justification: _____

Bone 3

Name of bone: _____

Type of bone: _____

Justification: _____

Bone 4

Name of bone: _____

Type of bone: _____

Justification: _____

Bone 5

Name of bone: _____

Type of bone: _____

Justification: _____

Bone 6

Name of bone: _____

Type of bone: _____

Justification: _____

Bone 7

Name of bone: _____

Type of bone: _____

Justification: _____

Name: _____ Section: _____

Activity 8.2 Long Bone Anatomy

This activity targets LOs 8.2, 8.3, and 8.4.

Long bones are the most common bone type in the body. These bones serve as structural columns to bear the weight of the body as well as enabling us to move. Although long bones come in many sizes, their structures are all very similar. The tissue of bone is a hard, mineralized matrix, but it can be organized in two different ways. **Compact bone** is almost completely made of matrix with only small canals and microscopic caves for cells to live within. **Spongy bone** is a network of bone matrix with large spaces. It is named spongy bone for its appearance, which looks almost like a kitchen sponge, but the matrix is still hard; spongy bone is not at all squishy or soft. The matrix of spongy bone forms a network of bridges, each called a **trabecula** (plural trabeculae). Both compact and spongy bone are constantly remodeled, with matrix being leeched away or added to bone. As spongy bone is remodeled, matrix is added to areas that receive the most weight-bearing stress, therefore the architecture of the trabeculae is organized so that these matrix bridges support the weight where the most weight is carried. You can read more about this process, and the physiological factors that influence the rate of bone remodeling, in your textbook.

The ends of long bones are each called an **epiphysis**. All of the long bones of the body are found in the limbs, so we can use our limb-specific directional terms, proximal and distal, to describe each epiphysis. In children's bones, each epiphysis will contain a region of hyaline cartilage. Under the influence of growth hormone, this hyaline cartilage plate, called an epiphyseal plate, will expand, elongating the entire bone. Some of the cartilage will then ossify (be converted to bone). You can read more about the growth of bones in your textbook. In late puberty, the body will stop growing taller. Each long bone will fully convert these cartilage plates into bone and then elongation of the bone becomes impossible. As the bone matrix seals, a small line, the **epiphyseal line**, is often left behind, marking the location where the epiphyseal plate used to be.

Digging Deeper:
Why and When and How Do We Stop Growing?

LeBron James, a famous NBA player, was 6'7" when he was drafted to play for the Cleveland Cavaliers at age 19. While 6'7" is plenty tall, Lebron continued to grow to 6'9" during his career. Many basketball players continue to grow after their rookie year. Some humans can grow even into their early 20s, while others stop growing in middle school. When and how do our bodies stop growing? Growth in height is primarily determined by long bone growth. Long bones can grow as long as they have epiphyseal plates, plates of cartilage near the ends of bone. Growth of these plates is mediated by growth hormone, which is secreted by the anterior pituitary. Eventually these plates ossify or are converted to bone. Since growth hormone cannot produce elongation of bone tissue, only cartilage tissue, once there is no more cartilage within the length of the bone, the bone will stop growing and gains in height become impossible. The ossification of the growth plates is mediated primarily by the sex hormones estrogen and testosterone. During puberty the production of estrogen and testosterone ramps up. These increasing levels promote growth hormone secretion as well as the deposition of minerals in the bone matrix. The hormones also cause the growth and maturation of the gonads and reproductive organs. When estrogen or testosterone reach very high levels more consistent with adulthood, they cause so much mineralization of bone that the growth plates mineralize as well. At this point, growth in height is no longer possible. One way to think about growth and reproduction is that humans evolved through times in which food was often scarce. In fact, it's only in the last 70 years or so that much of the world has consistent food supplies throughout the year. With food and other resources limited, it was an advantage for the body to be conservative with energy use. Once the reproductive structures were mature enough for reproduction, energy is conserved for that process and growth, which is also costly in terms of energy, is stopped.

The long cylinder of bone that stretches between the proximal and distal epiphyses is called the diaphysis. While the epiphyses are composed of both compact and spongy bone, the diaphysis is typically made of compact bone only. In its interior, instead of spongy bone, is a hollow medullary cavity. This space, as well as the spaces among the trabeculae, is filled with bone marrow in a living bone. There are two kinds of bone marrow, yellow and red (Figure 8.2). **Red bone marrow** is composed of a population of stem cells that can divide and give rise to new blood cells. **Yellow bone marrow** is actually just adipose tissue! In children's bones most of the medullary spaces are filled with red bone marrow, but as we age, we slowly add more and more yellow bone marrow. Most adult long bones are filled with yellow bone marrow in the medullary cavity, and red bone marrow is found among the spaces of the spongy bone in the epiphyses. You won't see any bone marrow in the bones of this lab, because bone marrow, unlike bone matrix, degrades quickly after death.

Materials
• A femur cut longitudinally to expose the interior of the bone

Instructions
1. Examine a femur that has been cut lengthwise or a photo of a cut femur.

2. On your bone identify the structures listed below.

   Structures to identify in a long bone:
   Compact bone
   Diaphysis
   Epiphyseal line
   Epiphysis
   Medullary cavity
   Spongy bone

3. In the figure to the right, label the six structures listed in Step 2.

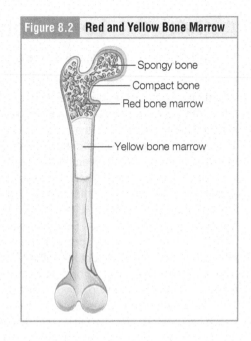

Figure 8.2 **Red and Yellow Bone Marrow**

- Spongy bone
- Compact bone
- Red bone marrow
- Yellow bone marrow

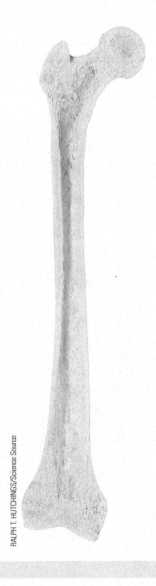

RALPH T. HUTCHINGS/Science Source

Questions

1. What do you notice about the organization and orientation of the trabeculae?

2. In life, what materials can be found among the spaces in spongy bone?

3. Bones are highly innervated and vascularized. Where are blood vessels found in bone? Can you find any of these spaces on your long bone specimen?

4. The figure shows an X-ray of one of the ends of the humerus. Compare this photo to an articulated skeleton in your lab room. Is the photo showing the proximal epiphysis or the distal epiphysis?

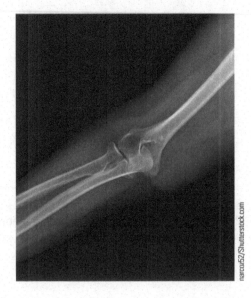

5. Is the bone that you inspected an adult bone or a child's bone? How can you determine this?

6. How would we describe the anatomical plane that the bone was cut in:
   a. Frontal plane
   b. Transverse plane
   c. Sagittal plane
   d. Oblique plane

Activity 8.3 Cow Bone Dissection

This activity targets LOs 8.2 and 8.5.

Materials
• Dissecting pan
• Forceps
• Fresh cow bone cut in longitudinal section
• PPE as per your instructor
• Probes

Instructions

1. Before starting Activity 8.3, make sure you are wearing PPE approved by your instructor.

2. First examine the exterior of the bone. Are there any muscles or tendons still attached? Examine the attachments? Draw or describe the attachments and include the periosteum in your description.

3. What are tendons and ligaments composed of? What is the periosteum composed of?

4. Using the blunt probe, clean out the medullary cavity of the bone. Is the material in the medullary cavity, yellow bone marrow or red bone marrow? What does this tell you about the age of the cow that the bone came from?

5. Continue cleaning out the medullary cavity until you locate the trabeculae at the epiphysis. Describe the locations where you find red bone marrow.

6. Observe the shiny white articular cartilage on the epiphysis. What type of cartilage is this? Is the periosteum found covering the articular cartilage or does it end at the edges of the articular cartilage?

Activity 8.4 Microanatomy of Bone

This activity targets LOs 8.2, 8.5, 8.6, and 8.7.

Bones are rigid structures and their microanatomy helps them to support a lot of weight. Bones are developed from a hyaline cartilage template. Calcium and phosphorus minerals are deposited slowly onto the framework of collagen fibers within the cartilage matrix. When we think of bones we often think of them as dead static structures, but during life, bones are incredibly dynamic. Living bones are vascularized, innervated, and comprised of many cells, meaning that we can find blood vessels, nerves, and cells within and along the inner and outer surfaces of the bones. There are three cell types within bone tissue: osteocytes, osteoclasts, and osteoblasts. Each type of cell has its own function, as listed in Table 8.2.

Learning Connection

Broken Process

What would bone look like if osteoblasts didn't function correctly? Consider the different impacts if osteoblasts were over or under functioning. What about osteoclasts?

Healthy bones are the result of a balance of the action of these three types of cells. When one type of cell is more active than the others, the bone composition and structure will change. The activity of the cells is influenced by a variety of factors that include bone use, hormones, growth factors, and response to injury. An example of bone use is if the amount of weight-bearing stress increases during a change in activity, say, taking up running for exercise.

On its internal and external surfaces bone matrix is covered by a sheet of connective tissue. The connective tissue covering on the outer surface of the bone is **periosteum**, whereas the connective tissue that lines the surfaces of the medullary cavity and trabeculae is the **endosteum**. The connective tissue hugging the bone surface is rich in osteoblasts and therefore capable of generating new bone matrix. When these osteoblasts are activated and producing a lot of matrix, the bone becomes wider. These osteoblasts are also important in the repair of broken bones. Osteoclasts also live in the periosteum and endosteum. When osteoclasts are activated, they secrete acids onto the bone surface which digest the matrix. The liberated minerals are absorbed into the bloodstream, increasing the mineral content of the blood. Other tissues throughout the body may use these minerals, for example neurons require calcium to function (you may have read about this in your textbook). When nervous tissue is running low on calcium, it takes up calcium from the blood, which can come from demineralized bone or from foods that we eat.

The osteoclasts are immobile in the bone because they are isolated within their own cave of bone matrix called a **lacuna**. The lacunae are organized in circles surrounding a **central canal** that houses blood vessels (Figure 8.3). Cutting through compact bone reveals that each central canal is surrounded by rings of organized lacunae, each ring is called a **concentric lamella**. Each central canal and its surrounding concentric lamellae is a functional weight-bearing column called an **osteon**. You can think of these osteons like the columns at the front of a large building.

Table 8.2 The Cells of Bone		
Cell Type	**Description**	**Function**
Osteocytes	Mature bone cells found in cavities within the solid bone matrix	Secrete minerals and collagen to maintain bone matrix
Osteoclasts	Found on the periphery of the bone	Secrete acid, which digests the bone matrix, freeing minerals to be used by other tissues
Osteoblasts	Found in the inner layer of the connective tissue that covers bone	Secrete bone matrix in growth and repair

Figure 8.3 | **Diagrams and Photomicrograph of Compact Bone**

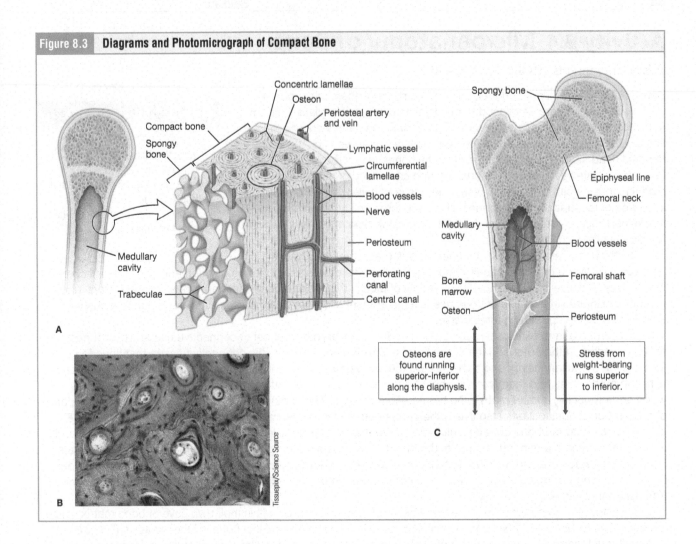

The osteons are wrapped with larger sheets of organized lacunae that encircle the entire circumference of the bone; these much wider circles are called **circumferential lamellae**. Most of the time each cell is alone in its lacuna, however the cells can share nutrients through long cellular extensions that traverse small pores in the bone matrix called **canaliculi**.

Materials
• Light microscope
• Osteon model
• Prepared microscope slide of bone

Instructions

1. On the osteon model, identify the locations in which each type of cell would be found.

2. Using just your eyes, examine the slide. Are you looking at an entire bone cross-section, or a wedge of bone that looks like a slice of pie? Can you see just compact bone? Just spongy bone? Or both? Roughly where on the slide will you find spongy bone? Compact bone?

3. Identify the zone of the slide where you expect to be able to examine the microscopic structure of the osteons.

4. Place the slide on the stage of the microscope and use the stage adjustment knobs to orient your slide so that the zone you've identified with your eyes is under the objective. Using the 4x objective on your microscope, focus on the slide until you can find an osteon. Identify the central canal and its lacunae. Can you find the small, hair-like canaliculi?

5. Does your slide have both circumferential and concentric lamellae? Where is each one located?

6. Increase the magnification by moving up to the 10x and then 40x objectives. Can you identify individual nuclei? What types of cells are these?

(page intentionally left blank)

Activity 8.5 Chemical Composition of Bone

This activity targets LOs 8.5 and 8.6.

Materials
• One bone that has been baked (for at least 2 hours at 250°F)
• One bone that has been soaked in HCl overnight and then dried

Instructions
In this lab you will have the opportunity to examine two bones that have been softened by different methods. One of the bones has been soaked in hydrochloric acid (HCl) overnight. The other bone has been heated in a hot oven.

1. Match each bone component with its definition.

Bone Component	Definition	List of Bone Components
	A protein that provides scaffolding and a structure onto which minerals are deposited.	Calcium Collagen Osteocyte
	A cell that secretes bone matrix.	
	A mineral that adds strength and rigidity to bones.	

2. Soaking the bone in HCl mimics the function of which of the following types of cells?
   a. Osteoclasts
   b. Osteoblasts
   c. Osteocytes

3. To create acid-soaked bone used in this lab, it was placed in a tub full of HCl. In addition to the acid, which of the following substances do you think would have been found in the tub after the bone had soaked overnight?
   a. Calcium ions
   b. Glucose molecules
   c. Sodium ions
   d. Red blood cells

4. In order to affect the bone matrix, what was the first layer that the acid penetrated?
   a. A layer of blood vessels
   b. A layer of connective tissue
   c. A layer of epithelium

5. Compare the acid-soaked bone to the heat-treated bone. How are they similar? How do they differ?

6. What component of bone was directly changed by exposure to heat?
   a. Minerals
   b. Collagen fibers

7. Did the heat-treated bone also lose minerals? Why or why not?

Activity 8.6 Case Study: Fracture Repair

This activity targets LOs 8.8 and 8.9.

Case
Study

Figure 8.4 illustrates different types of bone breaks. Notice that bone fractures can have a variety of orientations and also range in severity. A greenstick fracture, for example, affects only the outside of the bone, whereas a comminuted fracture affects the entire bone from one side to the other.

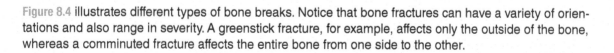

Figure 8.4   **Types of Bone Fractures**

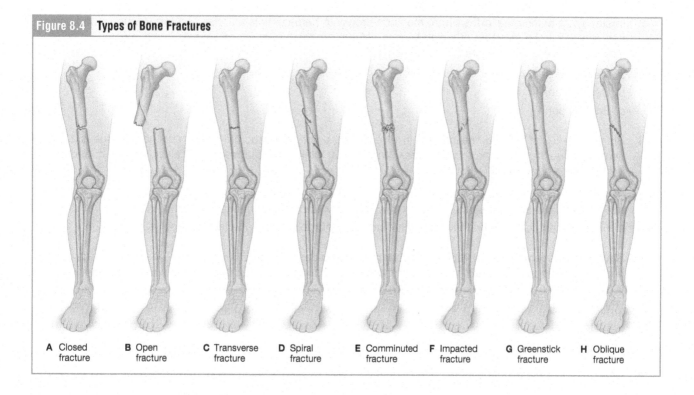

| A Closed fracture | B Open fracture | C Transverse fracture | D Spiral fracture | E Comminuted fracture | F Impacted fracture | G Greenstick fracture | H Oblique fracture |

1. In a greenstick fracture such as the one illustrated in Figure 8.4, will spongy bone or compact bone be affected?

   a. Spongy bone

   b. Compact bone

   c. Both will be affected

   d. We cannot tell without using a microscope to examine the fracture

2. In the comminuted fracture illustrated in Figure 8.4, will bone marrow be impacted by the break? If so, would it be yellow or red bone marrow?

   a. Yellow bone marrow will be affected.

   b. Red bone marrow will be affected.

   c. Neither type of bone marrow would be affected by the break.

Let's examine the greenstick fracture in more depth. Kenan was playing soccer when an opponent struck his calf from the lateral side while trying to steal the ball. Kenan sustained a greenstick fracture on his fibula. At the emergency room, the doctor set the bone and placed him in a hard cast. The doctor explained that Kenan's fracture would heal in four stages according to the following timeline:

Stage 1 (days 1–5): hematoma formation
Stage 2 (days 5–11): fibrocartilage callus formation
Stage 3 (days 11–28): bony callus formation
Stage 4 (days 18–months+): bone remodeling

3. In stage 1, a **hematoma** forms where the bone breaks. A hematoma is a bruise that forms when blood pools after a break in the structure of a blood vessel. Where are the broken blood vessels found in compact bone that are leaking the blood that forms the hematoma?

   a. Medullary cavity
   b. Central canals
   c. Spongy bone

4. Stages 2 and 3 of fracture healing both involve the formation of a structure called a callus. A **callus** is a healing tissue that forms around the break. Why do you think that a fibrocartilage callus forms first, and is followed by a bony callus?

   a. Fibrocartilage is faster to build, but bone takes more time due to the scarcity of minerals in the body.
   b. Fibrocartilage establishes the collagen framework, which is then enriched with minerals.
   c. Fibrocartilage is made first because there are more fibroblasts in bone than osteoblasts.

5. Stage 4, bone remodeling, involves both the making and destruction of bone matrix. During this stage bone is built or strengthened in locations that bear more weight, and taken away in locations where it is not needed. Which type(s) of cells are involved in this process (select all that apply).

   a. Osteoclasts
   b. Osteoblasts
   c. Osteocytes

6. What do you think might occur in an individual who was healing a fracture, but had severe mineral deficiency?

   a. The fracture site would heal, but would be misaligned, a noticeable bump would be able to be felt at the skin.
   b. The fracture site would heal to the fibrocartilage stage, but would not ossify.
   c. The fracture site would not heal past the hematoma state, because minerals are required for fibrocartilage synthesis.

Activity 8.7 Case Study: Osteopetrosis

This activity targets LOs 8.3, 8.4, 8.5, and 8.6.

Case Study

Kate is a 16-year-old biological female who has been complaining of a stuffy nose and sinus pressure for 6 months. At first, her parents assumed that she had seasonal allergies but allergy medication never alleviated the symptoms and the stuffy nose has persisted well beyond the allergy season. Her doctor orders blood tests to check for allergic responses, infection, or other abnormalities. Kate's results are presented in Table 8.3.

Table 8.3 Lab Values			
	Kate's Blood Levels	**Typical Ranges**	**High/Low/Typical**
Hematocrit	41.7%	35.5–44.9%	Typical
Eosinophils	1%	1–3%	Typical
Basophils	0.5%	0–1%	Typical
Lymphocytes	32%	20–40%	Typical
Monocytes	3%	4–8%	Low
Neutrophils	38%	40–60%	Low
Glucose (non-fasting)	103	<140mg/dL	Typical
Calcium	4.1	8.6–10.3mg/dL	Low
Phosphorus	1.8	2.8–4.5mg/dL	Low

1. If Kate was suffering from allergies, her eosinophil or basophil counts would likely be elevated. Is Kate suffering from allergies?

    a. Yes, allergies are likely.

    b. No, it is not likely that Kate is suffering from allergies.

2. If Kate had a viral infection, it is likely that her lymphocyte count would be elevated. If Kate had a bacterial infection, it is likely that her neutrophil count would be elevated. Does the blood work show that Kate is suffering from an infection?

    a. Yes, it is likely that Kate has a viral infection.

    b. Yes, it is likely that Kate has a bacterial infection.

    c. No, it does not appear that Kate has an infection.

3. Neutrophils, lymphocytes, monocytes, eosinophils, and basophils are all types of white blood cells. Where in the bone are these cells generated?

    a. Yellow bone marrow in the medullary cavity

    b. Red bone marrow in the spongy bone

    c. Periosteum

    d. Lacunae

4. The doctor takes note of the low blood calcium and blood phosphorus levels. These minerals come from the diet, but they are also added to the blood through the breakdown of bone matrix. Which type of cell is responsible for breaking down bone matrix and adding minerals to the blood?

    a. Osteoclasts

    b. Osteoblasts

    c. Osteocytes

Based on the lab values, the doctor suspects that Kate may have a disease called osteopetrosis. In this disease, bone matrix is not broken down and bones are not remodeled. One of the symptoms of this disease is frequent fractures, especially when activity levels change. The doctor notes that Kate suffered two fractures in the last year.

5. Match the cells with their function:

Bone Component	Definition	List of Bone Components
	Maintains bone density by secreting matrix.	Osteoblast
	Builds bone from the edges by secreting bone matrix.	Osteoclast
	Breaks down bone, releasing minerals into the blood.	Osteocyte

6. In osteopetrosis, bone matrix is not being broken down. Which cell type is under-functioning?
   a. Osteocytes
   b. Osteoblasts
   c. Osteoclasts

7. Low white blood cells values are also a symptom of osteopetrosis. Which of the following explanations provides the most plausible reason for the low white blood cell counts?
   a. So much of the body's energy is devoted to bone production that fewer white blood cells are produced.
   b. Without bone remodeling and bone matrix breakdown, the amount of space for bone marrow is restricted.
   c. The same acid that is used to break down bone matrix is required for white blood cell production.

A few months later, Kate returns to the doctor's office. This time she is complaining of headaches and tingling in the skin of her face. The doctor suspects that these symptoms are also related to osteopetrosis.

8. Which of the following could explain the link between osteopetrosis and pain and tingling in the head and face?
   a. Compression of cranial nerves due to narrowed bony passageways through the skull
   b. Increased blood calcium levels
   c. Inflammation due to excessive lymphocyte levels

Lab 8 Post-Lab Quiz

1. The figure shows a bone. Which kind of bone is it?

2. Put the following in order of size from smallest to largest: canaliculi, osteon, trabecula, lacuna.

3. Label the diagram with the following terms: compact bone, spongy bone, periosteum, endosteum, circumferential lamellae, concentric lamellae.

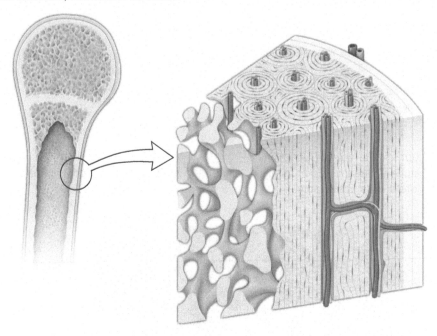

4. Osteogenesis imperfecta is a condition in which collagen does not correctly form. What role does collagen play in bone matrix?

5. In someone with osteogenesis imperfecta, without sufficient collagen what qualities would the bones exhibit?

   a. Smaller and more bendy

   b. Larger in circumference and more rough on the outside

   c. More rigid, dense and heavy

6. Which bone cells are exhibiting the issue in osteogenesis imperfecta?

   a. Fibroblasts

   b. Osteoblasts

   c. Osteoclasts

7. Which of these other tissues would be affected in this disease?

   a. Skeletal muscle

   b. Epithelium, like the surface of the skin

   c. Connective tissues like tendons

Use the following information to answer Questions 8–10.

   Eloise, an A&P student, has joined her family for dinner at a restaurant. Her mother orders bone marrow and the dish shown in the figure arrives at the table.

Brent Hofacker/Alamy Stock Photo

8. Which of the following words best describes the bone fragment on the plate?

   a. Diaphysis of a long bone

   b. Epiphysis of a long bone

   c. Transverse section through an irregular bone

   d. Frontal section through a flat bone

9. You ask the waiter what kind of animal the bone came from and they answered that it was from an adult cow. Taking this information plus your answer Question 8, is Eloise's mother eating yellow bone marrow or red bone marrow?

   a. Yellow bone marrow

   b. Red bone marrow

10. Defend your answer to Question 9 by explaining where yellow and red bone marrow are found.

11. Define osteon.

12. Which type of cell lives within a lacuna?

   a. Osteoclasts

   b. Osteoblasts

   c. Osteocytes

13. What is the function of canaliculi?

14. What minerals are involved in ossification?

15. In which of the following circumstances would you find osteoclasts activated?

a. During times of low blood calcium

b. During times of low iron levels

c. During times of low blood pressure

16. In cases of weight gain the added stress on the bones leads to appositional growth. Which type of bone cell is responsible for the additional matrix deposition in cases of appositional growth?

a. Osteoclasts

b. Osteoblasts

c. Osteocytes

Lab 9 The Axial Skeleton

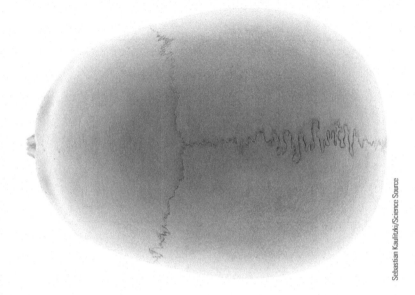

Sebastian Kaulitzki/Science Source

Learning Objectives: After completing this lab, you should be able to:

9.1 Distinguish between the axial and appendicular skeletons and list the major bones contained within each.

9.2 Identify individual bones and their locations within the body.

9.3 Identify major bone markings (e.g., spines, processes, foramina) on individual bones.

9.4 Identify the approximate locations of each of the paranasal sinuses.

Introduction

The skeletal system forms the rigid internal framework of the body and includes the bones, cartilages, and ligaments. The bones support the weight of the body, allow for body movement, and protect internal organs. In addition, they provide a location for hematopoiesis and mineral storage. The bony skeleton is subdivided into two major divisions—the axial and appendicular.

In this lab we will explore the anatomy of the bones of the axial skeleton. The **axial skeleton** forms the vertical, central axis of the body, which protects the brain, spinal cord, heart, and lungs. It also serves as the attachment site for muscles that move the head, neck, and back, the muscles of respiration, and the abdominal muscles. The axial skeleton includes the bones of the skull, vertebral column, and thoracic cage (Figure 9.1).

Bones are dynamic organs that respond to changes in muscle strength and use. Bones also take shape in a manner that creates space for soft-tissue structures such as blood vessels that travel near them. The various identifiable features of individual bones, called **bone markings**, represent points where ligaments and tendons attach, where blood vessels, nerves, and muscles travel along the surface, and openings where blood vessels and nerves pass through the bone.

The Human Anatomy and Physiology Society includes more than 1,700 educators who work together to promote excellence in the teaching of this subject area. The HAPS A&P Learning Outcomes measure student mastery of the content typically covered in a two-semester Human A&P curriculum at the undergraduate level. The full Learning Outcomes are available at https://www.hapsweb.org.

Figure 9.1 **The Axial and Appendicular Skeletons**

The axial skeleton forms the vertical axis of the body. It consists of the skull, vertebral column (including the sacrum and coccyx), and the thoracic cage, formed by the ribs and sternum. The appendicular skeleton is made up of all bones of the upper and lower limbs and the bones that connect the limbs to the axial skeleton.

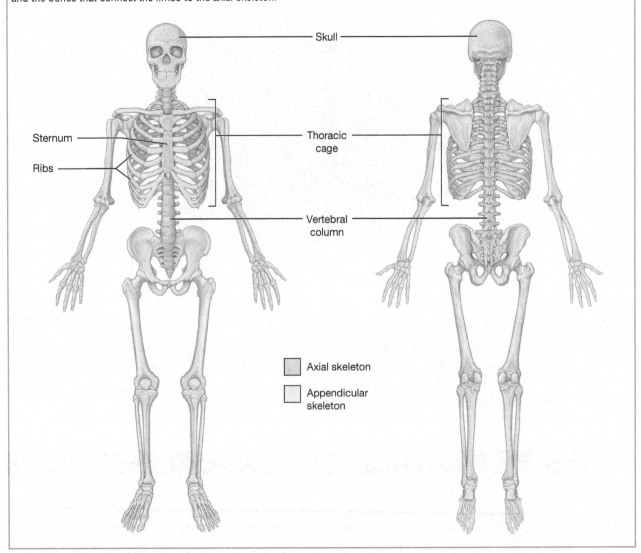

Skull

Sternum

Ribs

Thoracic cage

Vertebral column

☐ Axial skeleton

☐ Appendicular skeleton

Lab 9 Pre-Lab Quiz

This quiz will strengthen your background knowledge in preparation for this lab. For help answering the questions, use your resources to deepen your understanding. The best resource for help on the first five questions is your text, and the best resource for help answering the last five questions is to read the introduction section of each lab activity.

1. Anatomy of the axial skeleton.

   Label the skull, thoracic cage, and vertebral column of the axial skeleton.

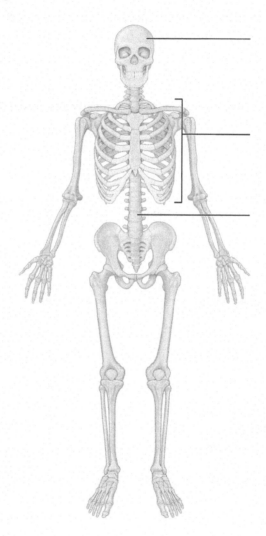

2. Match each of the roots to its definition:

Root	Meaning
	passage or path
	ditch
	knuckle
	branch
	hole or opening

List of Roots

condyle
foramen
fossa
meatus
ramus

3. Are the pectoral and pelvic bones part of the axial skeleton?

   a. Yes
   b. No

4. What bony space encloses and protects the brain?

   a. The orbit
   b. The paranasal sinuses
   c. The cranial cavity

5. Which of these bones of the skull are unpaired? Choose all correct answers.

   a. The parietal bone
   b. The occipital bone
   c. The frontal bone

6. What pair of bones forms the lateral aspect of the cranium and contains the external auditory meatus that leads to the middle and inner ear?

   a. The temporal bones
   b. The parietal bones
   c. The sphenoid bones

7. What structure forms the lateral aspect of the nasal cavity and creates a convoluted surface there?

   a. The nasal conchae
   b. The bony nasal septum
   c. The palate

8. How many individual vertebrae are found in the human spinal column?

   a. Seven
   b. Nineteen
   c. Twenty-four

9. Each of the thoracic vertebrae can be distinguished by what feature?

   a. The attachment of a pair of ribs
   b. The direct attachment to the sternum
   c. The presence of transverse foramina

10. Each rib contains a costal groove, a slight indent where blood vessels and nerves travel. On what surface is this costal groove found?

   a. The superior surface
   b. The inferior surface

Activity 9.1 **Bones of the Skull**

This activity targets LOs 9.1, 9.2, 9.3, and 9.4.

The skull consists of a total of 22 bones that lead to an intricate and complex anatomical structure. The bones of the skull can be subdivided into the **cranial bones** and the **facial bones** (Figure 9.2).

The Cranial Bones

The cranial bones form the rounded cranium that surrounds and protects the brain and houses the middle and inner ear structures. The eight cranial bones are:

• The **frontal bone** (Figure 9.3) forms the forehead. A slight depression between the eyebrows is called the **glabella**. The **supraorbital margin** marks the superior ridge of the orbit and the **supraorbital foramina** are a pair of openings for nerves and blood vessels to enter and exit the bone. Within the frontal bone directly superior to each orbit are a pair of air-filled spaces, the **frontal sinuses**. The location of the frontal sinuses is shown in Figure 9.14.

Figure 9.2 | **The Bones of the Skull Can Be Divided into Two Groups**

The bones that form the rounded brain case that houses the brain are called *cranial bones* and the bones of the upper and lower jaws, nose, orbits, and other facial structures are called *facial bones*.

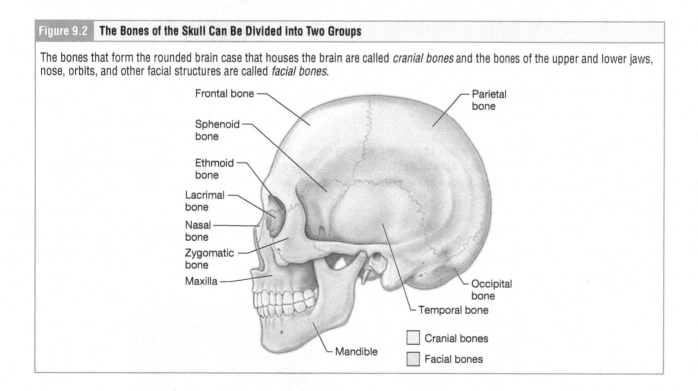

Figure 9.3 | The Frontal Bone

The single frontal bone encloses the anterior cranium and forms the upper portions of the eye orbits.

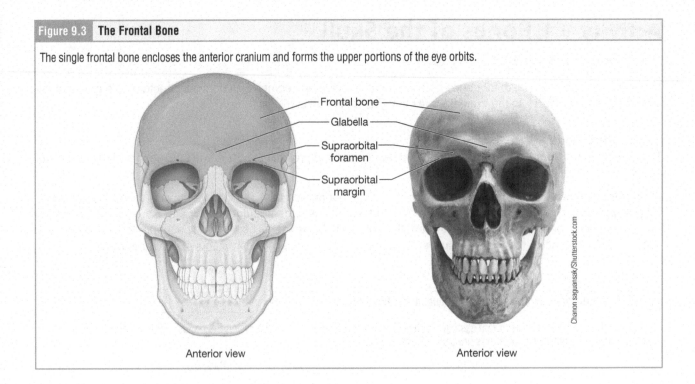

Frontal bone
Glabella
Supraorbital foramen
Supraorbital margin

Anterior view

Anterior view

Chanon saguansak/Shutterstock.com

- The **parietal bones** (Figure 9.4) are a pair of bones that form the superior portion of the skull. These two bones articulate with each other at the **sagittal suture**. They also articulate anteriorly with the frontal bone at the **coronal suture**, laterally with the temporal bones at the **squamous sutures,** and posteriorly with the occipital bone at the **lambdoidal suture**.

Figure 9.4 | The Paired Parietal Bones

The parietal bones make up the middle of the cranium.

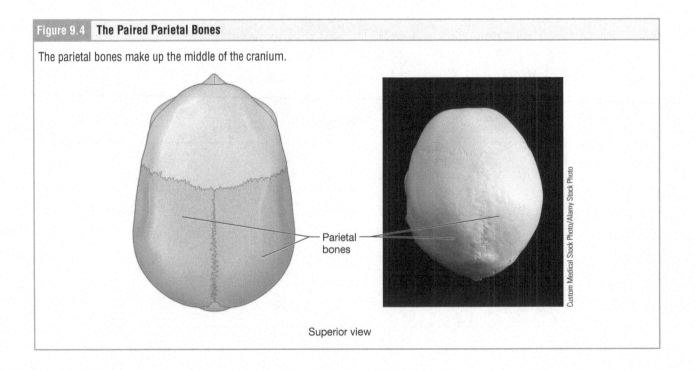

Parietal bones

Custom Medical Stock Photo/Alamy Stock Photo

Superior view

- The **occipital bone** (Figure 9.5) forms the posterior and inferior portion of the skull. A small protrusion called the **external occipital protuberance** leads to the lateral, raised **superior nuchal line** on each side. An obvious central opening, the **foramen magnum**, allows the spinal cord to exit the skull. Just lateral to the foramen magnum are two large bumps, the **occipital condyles**, that sit on the first cervical vertebra to support the skull.
- The **temporal bones** (Figure 9.6) are a pair of bones that form the lateral portion of the skull. The temporal bone articulates anteriorly with the zygomatic bone via the **zygomatic process**, which forms a portion of the *zygomatic arch*. Inferior to the base of the zygomatic process is the **mandibular fossa**, a smooth, shallow depression where the mandible articulates at the *temporomandibular joint*. Directly posterior to the mandibular fossa is the opening to the auditory canal, the **external acoustic meatus**, that allows sound waves to reach the middle and inner ear. Inferior to this opening is the thin **styloid process** of the temporal bone. The larger **mastoid process**, posterior to the external acoustic meatus and lateral to the styloid process, provides an attachment point for neck muscles. Medial to each styloid process on the inferior surface of the temporal bones are the openings to the **carotid canals** that allow the internal carotid arteries to reach the brain. On the internal view of the temporal bone from within the

---

Figure 9.5 **The Occipital Bone**

(A) The single occipital bone forms the back of the cranium. The occipital bone articulates with the first vertebra via the occipital condyles. (B) From the bottom of the skull, we can see that the occipital bone comprises much of the base of the skull, including the large foramen magnum, the opening through which the spinal cord passes.

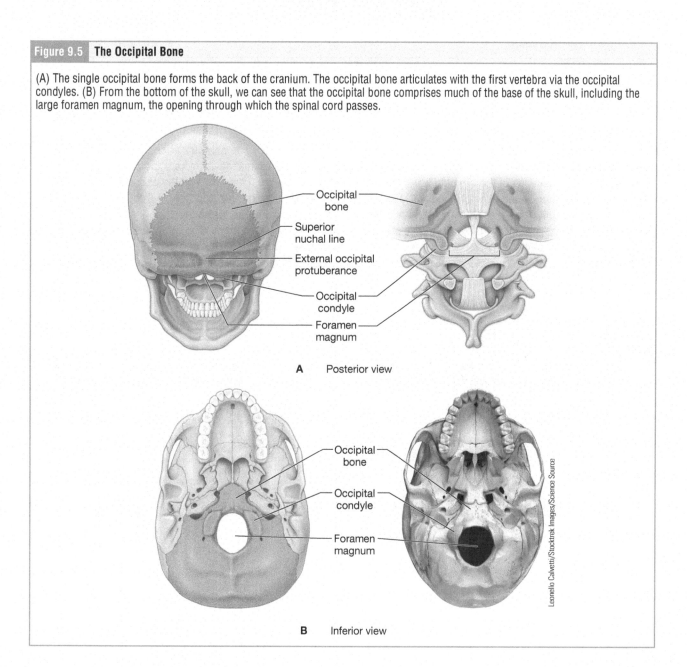

A   Posterior view

B   Inferior view

Leonello Calvetti/Stocktrek Images/Science Source

Figure 9.6 | The Temporal Bone

(A) The temporal bone contributes to the lateral walls of the cranium. (B) Zooming in on the isolated temporal bone, we can identify many bone features.

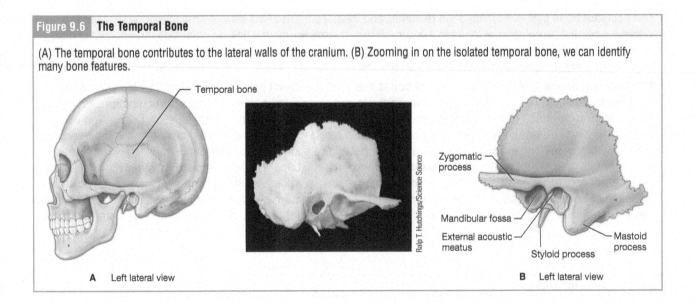

A   Left lateral view

B   Left lateral view

cranial cavity, the **internal acoustic meatus** allows nerves for hearing to reach from the brain to the middle and inner ear. Both the carotid canal and the internal acoustic meatus are shown more clearly in Figure 9.16.

- The **ethmoid bone** (Figure 9.7) is deep to the nasal bones and maxilla, making its complex shape partially viewable in the orbital and nasal cavities. Its superior surface can be seen inside the cranial cavity, where the pointed **crista galli** separates two flat sections of the ethmoid bone called the **cribriform plates**. These two plates of bone, punctuated with small holes, support the olfactory bulbs and extensions of the olfactory nerves that dangle through the holes to reach the nasal cavity. A centrally located plate of bone, the **perpendicular plate**, forms the superior portion of the bony nasal septum while lateral projections, the **superior nasal conchae** and the **middle nasal conchae**, form the irregular surfaces of each nasal cavity. Between the surfaces of the ethmoid bone that can be seen in each orbit and those nasal conchae are small air-filled spaces called the **ethmoid sinuses**. These sinuses are also shown in Figure 9.15.

Figure 9.7 | The Ethmoid Bone

The single, complex interior ethmoid bone separates the nasal cavity from the brain. (A) The ethmoid bone makes up a portion of the eye orbit. (B) The ethmoid bone lies anterior to the sphenoid bone. (C) The perpendicular plate makes up part of the nasal septum.

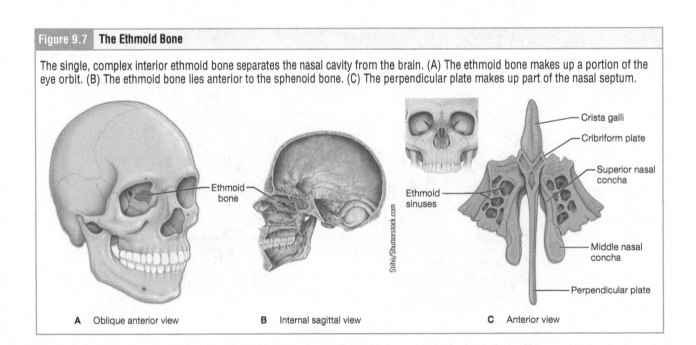

A   Oblique anterior view

B   Internal sagittal view

C   Anterior view

• The **sphenoid bone** (Figure 9.8) is partially visible from the lateral, superior, and inferior aspects of the skull. Inferiorly, a portion of the **greater wings of the sphenoid bone** is directly anterior to the temporal bones. Another portion of the greater wings of the sphenoid bone is visible in the orbits, where they form the posterior-most portions of the orbit. Viewing the sphenoid bone in the orbits you can see two long openings: the **superior orbital fissure** descends through the center of this portion of the sphenoid bone, while the **inferior orbital fissure** marks the margin between the sphenoid bone and the maxilla. The remainder of the sphenoid bone is more easily visualized inside the cranial cavity where it lies posterior to the frontal bone. The **lesser wings** of the sphenoid bone are located anteriorly and superiorly to the greater wings. Both pairs of wings come together centrally at the body of the sphenoid bone, where a saddle-like formation, the **sella turcica**, provides bony protection for the pituitary gland. Two holes, the **optic canals**, on the anterior surface of the sella turcica where it joins the lesser wings allow the optic nerves to pass between the brain and the eyes in the orbits. There are many foramina in the sphenoid bone (the **foramen rotundum, foramen ovale,** and **foramen spinosum**) which will be addressed later in this introduction and can be seen in Figure 9.16. Medial to the optic canals within the body of the sphenoid bone is another air-filled space, the **sphenoid sinus,** also shown in Figure 9.15. The inferior extensions of the sphenoid bone, the **pterygoid processes**, form the posterior edge of the nasal cavity.

Figure 9.8	The Sphenoid Bone

The complex, interior sphenoid bone articulates with almost every other bone in the skull. (A) The sphenoid bone is visible on the external aspect of the skull. (B) The sphenoid bone spans the entire width of the skull. and (C) The sphenoid bone has a unique butterfly shape and many intricate features.

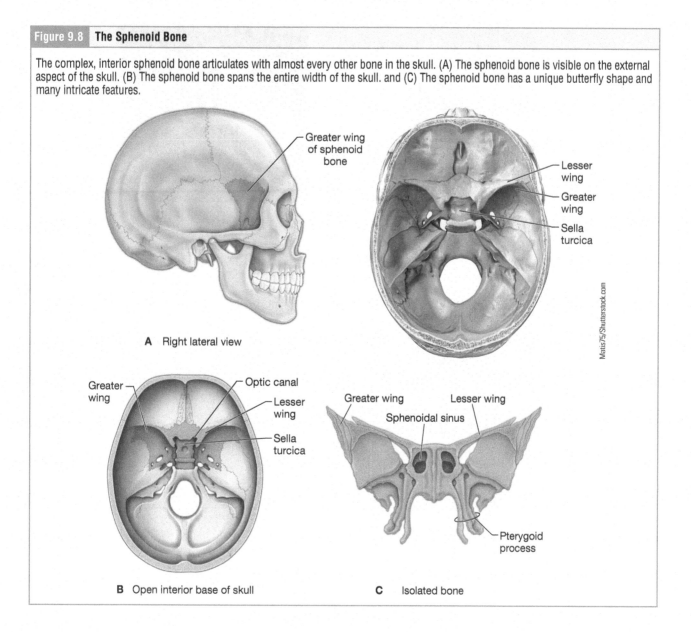

A   Right lateral view

B   Open interior base of skull

C   Isolated bone

Matis75/Shutterstock.com

The Bones of the Face

The facial bones form the structure of the face, nasal cavity, and the mouth, and enclose the eyeballs. The 14 facial bones are:

• The **mandible** (Figure 9.9) is sometimes called the jawbone. The lateral "arms" of the mandible are the **rami** (singular = ramus). The rami terminate superiorly with two bony projections: each **mandibular condyle** articulates with the temporal bone at the mandibular fossae while the **coronoid processes** provide locations for muscle attachments to move the jaw. Between these two projections is the **mandibular notch**. The rami of the mandible terminate inferiorly on the **body** of the mandible at the **mandibular angles**. The body of the mandible holds the lower teeth; the anterior-most portion of the mandible is called the **mental protuberance,** or chin. Two small but obvious holes here, the **mental foramina**, allow blood vessels and nerves to enter the bone.

Figure 9.9	The Mandible

(A) The mandible, a single movable bone, forms the lower jaw. (B) In the isolated mandible we can see many of the bone's features.

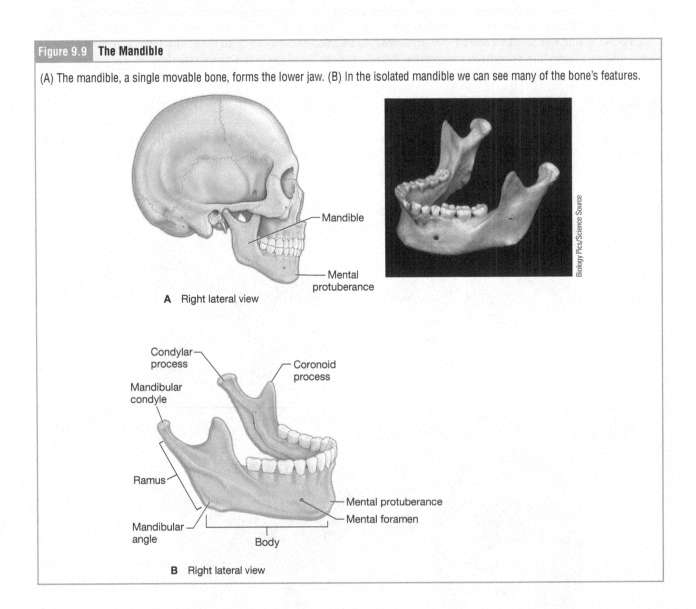

A   Right lateral view

B   Right lateral view

Figure 9.10 | **The Maxilla**

(A) The two maxillae articulate to form the upper jaw. (B) On the isolated maxilla we can observe many of its features.

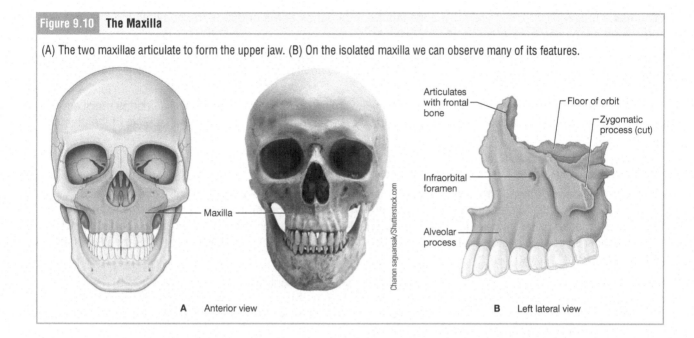

A Anterior view   **B** Left lateral view

- The fused pair of maxillary bones (Figure 9.10), also called the **maxilla**, hold the upper teeth. The maxilla forms the floor of each orbit and articulates laterally with each zygomatic bone via the zygomatic processes. Medial to the processes and inferior to each orbit are a pair of **infraorbital foramina**. A pair of air-filled spaces within the maxilla, the maxillary sinuses, are shown in Figure 9.15.
- The **zygomatic bones** (Figure 9.11), or cheek bones, form the lateral border of each orbit. These bones also form the anterior half of each zygomatic arch with their **temporal processes**, which reach to meet the zygomatic processes of each temporal bone to form each complete arch.

Figure 9.11 | **The Zygomatic Bone**

The zygomatic bone forms the main bony support for the cheek. The temporal process of the zygomatic bone merges with a process from the temporal bone to form the zygomatic arch, a supporting structure for the side of the skull.

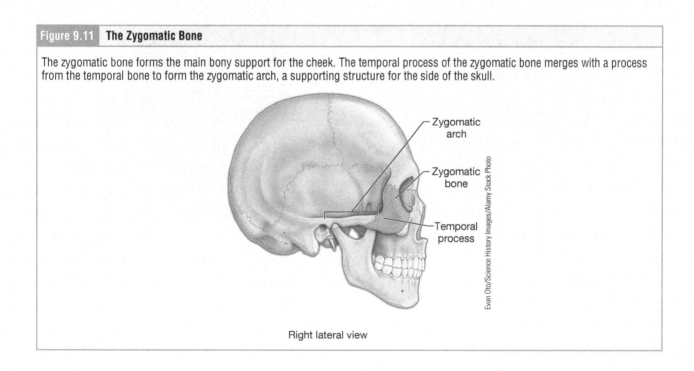

Right lateral view

- The **nasal bones** (Figure 9.12) are the medial-most pair of bones at the bridge of the nose, where they are bracketed by the vertical projections of the maxilla.
- Just lateral to the vertical projections of the maxilla are the small **lacrimal bones** (Figure 9.12). Between the maxilla and each lacrimal bone is a small groove that contains the lacrimal duct to drain tears from the eye into the nasal cavity.
- The **vomer** (Figure 9.12) is seen as a small, central bony projection from the anterior maxilla into the nasal cavity, forming the inferior portion of the bony nasal septum. The vomer meets the perpendicular plate of the ethmoid bone as it rises into the nasal cavity.
- The **inferior nasal conchae** (Figure 9.12), along with the middle and superior nasal conchae of the ethmoid bone, form the irregularly shaped lateral walls of the nasal cavity.
- The **palatine bones** (Figure 9.13) are best viewed from an inferior view of the skull. This pair of bones form the posterior portion of the hard palate (the roof of the mouth) with their **horizontal plates** and then rise laterally just posterior to the nasal conchae to follow the pterygoid processes of the sphenoid bone. A tiny portion of the superior-most tip of each palatine bone is visible in the orbit, where it is sandwiched between the meeting of the ethmoid bone and maxilla.

Figure 9.14 shows the interior of the skull that has been sectioned midsagittally. This view shows how many of the bones of the skull articulate. The **hyoid bone** shown in Figure 9.14 is occasionally categorized as a skull bone, although it does not articulate with the bones of the skull. It serves as an attachment point for muscles of the tongue and creates some of the structure of the larynx (voicebox).

Figure 9.12	The Lacrimal Bone, Nasal Bone, Vomer, and Inferior Nasal Concha

Viewing the skull from the anterior, we can identify many of the bones.

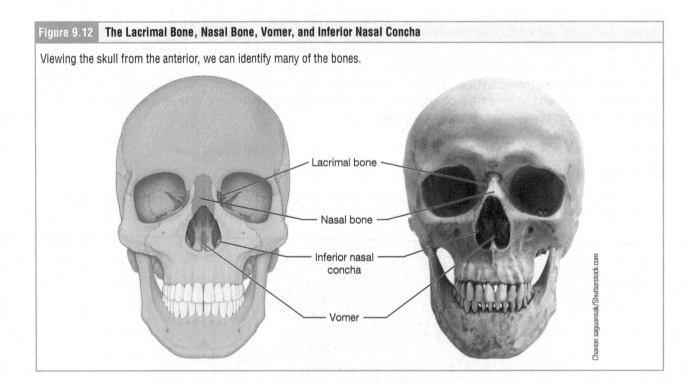

Lacrimal bone

Nasal bone

Inferior nasal concha

Vomer

Chanon saguansak/Shutterstock.com

Figure 9.13 **The Palatine Bone**

Two palatine bones form the posterior palate and nasal cavity walls. (A) The palatine bones make up a small posterior portion of the hard (bony) palate and help to form the walls of the nasal cavity. (B) The unusual J shape of the palatine bones allows them to fit among several surrounding bones.

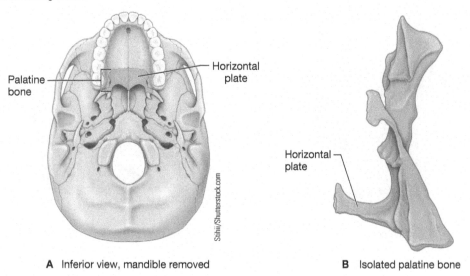

A Inferior view, mandible removed

B Isolated palatine bone

Stihii/Shutterstock.com

Figure 9.14 **Midsagittal Section of Skull**

This view shows the interior of a skull that has been sectioned sagittally.

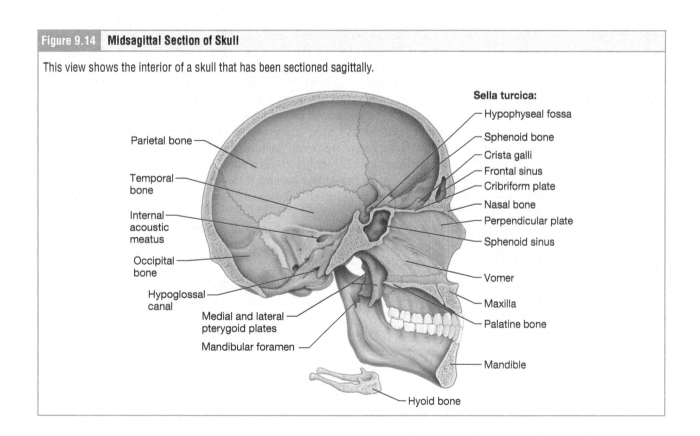

Figure 9.15 | The Paranasal Sinuses

Sinuses are hollow, air-filled spaces within bones. They are named for the skull bone that they occupy.

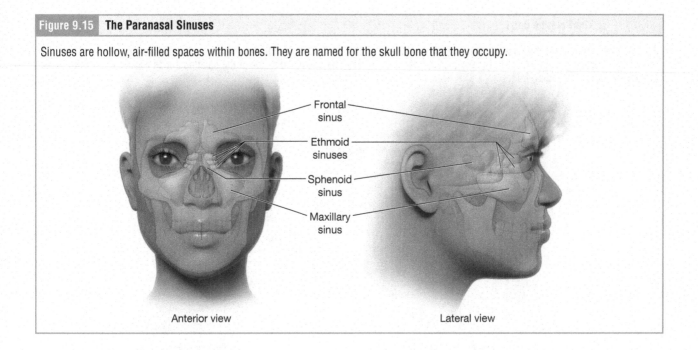

Frontal sinus

Ethmoid sinuses

Sphenoid sinus

Maxillary sinus

Anterior view   Lateral view

Sinuses and Holes in the Skull

The **paranasal sinuses** are hollow, air-filled spaces within some of the bones of the skull. They aren't visible on the exterior of a skull but their approximate locations are shown in Figure 9.15. All of the paranasal sinuses have small openings that connect them to your nasal cavity so that air can flow in and out of these spaces to equalize pressure between them and the environment around you. These openings may become blocked or filled with mucus when you have a cold or sinus congestion, preventing the easy flow of air between the sinuses and your nasal cavity. Each sinus is named for the bone in which it is found: the pair of frontal sinuses are found superior to the orbits in the frontal bones, the ethmoid sinuses are medial to the orbits, while the maxillary sinuses are inferior to the orbits and lateral to the nasal cavity. Directly posterior to the ethmoid sinuses is the single sphenoid sinus.

The many holes in the skull can be seen in Figure 9.16. Most of these openings have already been addressed as components of the bones that they occupy. However, several openings occur at the juncture between bones. The pair of **foramen lacerum** are located to each side of the sella turcica where the sphenoid bone, temporal bone, and occipital bone meet. The **jugular foramina**, which can be viewed from both the internal superior view and the external inferior view, are located just lateral to the foramen magnum where the occipital bone meets the temporal bone.

The purpose of this activity is to learn and practice the many bone names and bone markings of the skull.

Materials
- Colored pencils
- Human skulls or models of human skulls
- Small sticks or pipe cleaners for pointing to bones and bone markings

Important Note: While it is tempting to use your pen or pencil to point to the many small structures on the skull, it is critical that you do not use any writing instrument to point or touch the skull, even if you think that you are being careful. Instead, use the small sticks or pipe cleaners provided by your instructor to avoid leaving any smears or marks on the bones.

Figure 9.16   The Openings in the Skull

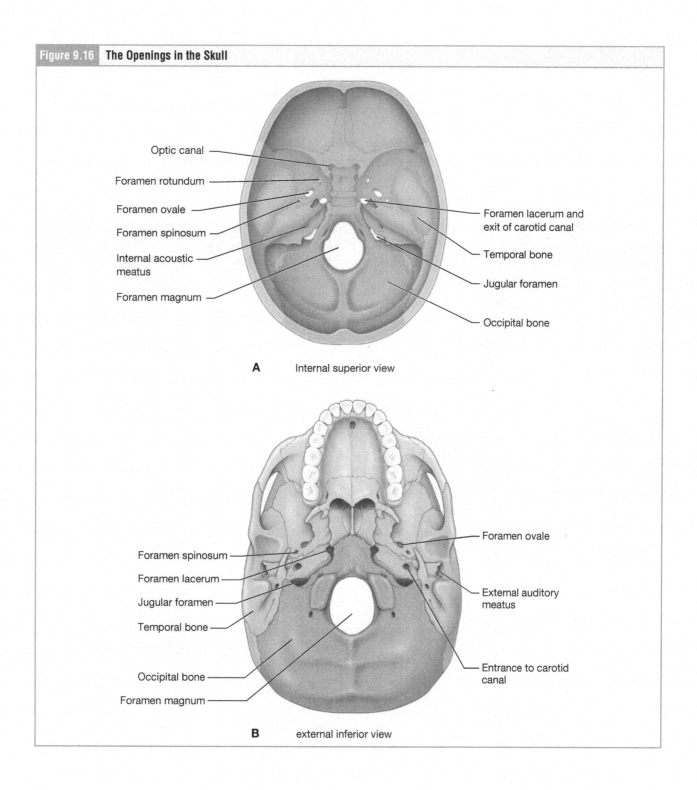

Optic canal

Foramen rotundum

Foramen ovale

Foramen spinosum

Internal acoustic meatus

Foramen magnum

Foramen lacerum and exit of carotid canal

Temporal bone

Jugular foramen

Occipital bone

A   Internal superior view

Foramen spinosum

Foramen lacerum

Jugular foramen

Temporal bone

Occipital bone

Foramen magnum

Foramen ovale

External auditory meatus

Entrance to carotid canal

B   external inferior view

Instructions

1. Work with your lab partner to learn the bones and bone markings of the skull.

   a. For each bone discussed above, find the bone on the provided skull.

   b. Then locate each bone marking identified in bold. The bones and bone markings of the skull are listed in Table 9.1.

   c. Review the many openings of the skull identified in Figure 9.16.

   d. Locate the positions of the paranasal sinuses on the skull. Remember, these sinuses won't be visible on the skull, but you should be able to approximate their locations using Figure 9.15 as a guide.

Table 9.1 Bones and Bone Markings of the Skull

Bones of the Skull	Bone Markings of the Skull	Bones of the Skull	Bone Markings of the Skull
CRANIAL BONES		sphenoid bone	greater wings
frontal bone	glabella		superior orbital fissure
parietal bones	supraorbital margin		inferior orbital fissure
	supraorbital foramina		lesser wings
	frontal sinuses		sella turcica
	sagittal suture		optic canals
	coronal suture		foramen rotundum
	squamous sutures		foramen ovale
occipital bone	lambdoidal suture		foramen spinosum
	external occipital protuberance		sphenoid sinus
	superior nuchal line		pterygoid processes
	foramen magnum	**FACIAL BONES**	
	occipital condyles	mandible	mandibular condyle
temporal bones	zygomatic process		coronoid processes
	mandibular fossa		mandibular notch
	external acoustic meatus		mandible body
	styloid process		mandibular angles
	mastoid process		mental protuberance
	carotid canals		mental foramina
	internal acoustic meatus	maxilla	infraorbital foramina
ethmoid bone	crista galli	zygomatic bones	temporal processes
	cribriform plates	nasal bones	vomer
	perpendicular plate	lacrimal bones	inferior nasal conchae
	superior nasal conchae	palantine bones	horizontal plates
	middle nasal conchae		paranasal sinuses
	ethmoid sinuses		foramen lacera
			jugular foramina

2. There are seven bones that form the orbits of the skull. Use the introduction above and the skull to locate these bones on both the anterior and oblique anterior views in the figure below.

a. Choose a different colored pencil for each bone of the orbit, then label each bone and color it in on both views:

   i. Frontal bone   v. Ethmoid bone

   ii. Sphenoid bone   vi. Palatine bone

   iii. Zygomatic bone   vii. Lacrimal bone

   iv. Maxilla

b. Label the three openings in the posterior aspect of the orbit on the image:

   i. Optic canal

   ii. Superior orbital fissure

   iii. Inferior orbital fissure

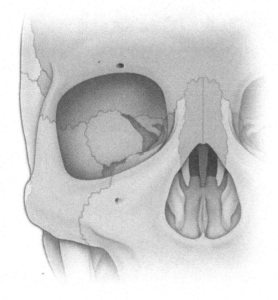

A   Anterior view

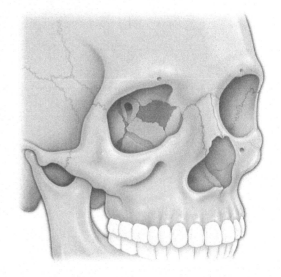

B   Oblique anterior view

Learning Connection

While there are a lot of holes in the skull, there are some tricks to help you keep them organized in your brain.

For example, there are five sets of openings in the sphenoid bone alone: the optic canal, foramen rotundum, foramen ovale, foramen spinosum, and foramen lacerum. If you place your hands in a pair of L shapes, as shown in Figure 9.17, you can trace the path of these foramina with the phrase "Open Ripe Olives to Share Lunch."

What other phrases could you develop to learn groups of openings? Some to explore might be:

- The optic canal, superior orbital fissure, and inferior orbital fissure
- The foramen magnum, the carotid canal, and the jugular foramen

---

Figure 9.17 **Learning the Openings in the Sphenoid Bone**

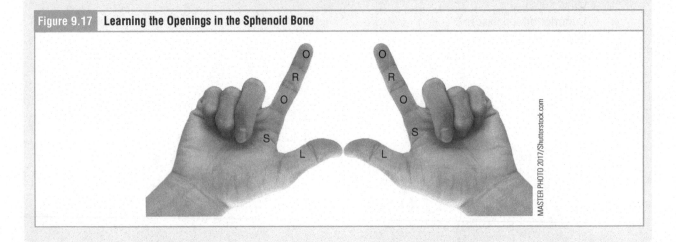

Activity 9.2 Sutures of the Skull

This activity targets LO 9.3.

The bones that make up the skull are joined together by immobile joints called sutures. The narrow gap between the bones at the suture is filled with dense fibrous connective tissue that eventually ossifies. The long sutures located between the bones of the brain case are not straight but instead follow irregular, tightly twisting paths. These twisting lines serve to tightly interlock the adjacent bones, thus adding to the strength of the skull.

These four sutures are the most prominent on the skull. The single frontal bone borders the anterior surfaces of the pair of parietal bones at the **coronal suture**. You may recall that the coronal plane through the human body would pass through the skull in the orientation of the coronal suture. Similarly, the **sagittal suture** is located along the midsagittal plane of the skull, where the two parietal bones meet each other superiorly. At the inferior border of each parietal bone where it meets with the temporal bone is the curved **squamous suture**. Finally, on the posterior aspect of the skull, the single occipital bone meets the pair of parietal bones at their posterior edges to form the **lambdoid suture**.

Materials
- Human skulls or models of human skulls
- Small sticks or pipe cleaners for pointing to bones and bone markings

Instructions
1. Close all resources, including your textbook and atlas. You should only have this lab manual and the skull in front of you.

2. With your lab partner, use the directional terms and explanations of the sutures in the introduction above to find the location of each suture on the skull or skull model.

   a. Coronal suture

   b. Sagittal suture

   c. Squamous suture

   d. Lambdoid suture

3. Label the sutures on the images below.

> **Important Note:** While it is tempting to use your pen or pencil to point to structures on the skeleton, it is critical that you do not use any writing instrument to point or touch the bones or bone models, even if you think that you are being careful. Instead, use the small sticks or pipe cleaners provided by your instructor to avoid leaving any smears or marks on the bones.

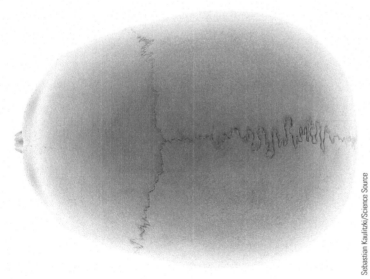

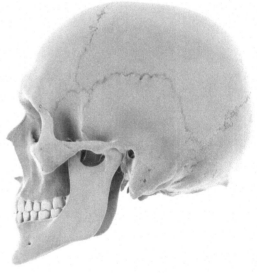

Sebastian Kaulitzki/Science Source

Sebastian Kaulitzki/Science Source

(page intentionally left blank)

Activity 9.3 Bones of the Vertebral Column

This activity targets LOs 9.1, 9.2, and 9.3.

The axial skeleton includes the vertebral column. The vertebral column, shown in Figure 9.18, is made up of 24 vertebrae, the sacrum, and the coccyx. Together, these structures form a stack of bones that protect the soft spinal cord and allow the spinal nerves to exit from this protection to reach to the body's periphery.

The **vertebrae** can be separated into seven cervical vertebrae, 12 thoracic vertebrae, and five lumbar vertebrae. We will first explore the anatomy of a typical vertebra, and then examine the unique features that distinguish the cervical, thoracic, and lumbar vertebrae from each other. A typical vertebra, shown in Figure 9.19, consists of a **body**, a pair of **pedicles** that extend posteriorly away from the body, and a pair of flat **lamina** that project from the pedicles and fuse medially to create a complete arch. The two pedicles and two lamina together form the **vertebral arch** that creates the **vertebral foramen** that encases the spinal cord. Projecting somewhat laterally from the vertebral arch are two **transverse processes** and projecting posteriorly is the **spinous process**. Rising up from the vertebral arch are two **superior articular processes** and reaching down from the base of the spinous process are a pair of **inferior articular processes**. The inferior articular processes of one vertebra meet the superior articular processes of the vertebra just superior to it to form a bracket that holds the vertebrae in position, as can be seen in Figure 9.19. Between these articulated processes, lateral passageways called **intervertebral foramina** allow the spinal nerves to leave the protection of the spinal column. In a living body, soft-tissue structures called **intervertebral discs** cushion each vertebral body from its neighbors.

Figure 9.18	Vertebral Column

The adult vertebral column consists of 24 individual vertebrae, plus the fused vertebrae of the sacrum and coccyx. The spine is divided into three regions: cervical C1–C7 vertebrae, thoracic T1–T12 vertebrae, and lumbar L1–L5 vertebrae.

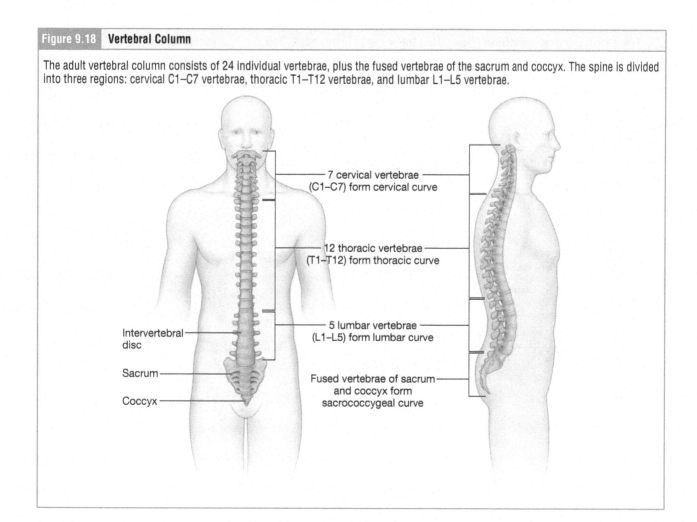

7 cervical vertebrae (C1–C7) form cervical curve

12 thoracic vertebrae (T1–T12) form thoracic curve

5 lumbar vertebrae (L1–L5) form lumbar curve

Intervertebral disc

Sacrum

Coccyx

Fused vertebrae of sacrum and coccyx form sacrococcygeal curve

Figure 9.19 | Anatomy of a Typical Vertebra

Some vertebral features are found on all types of vertebrae.

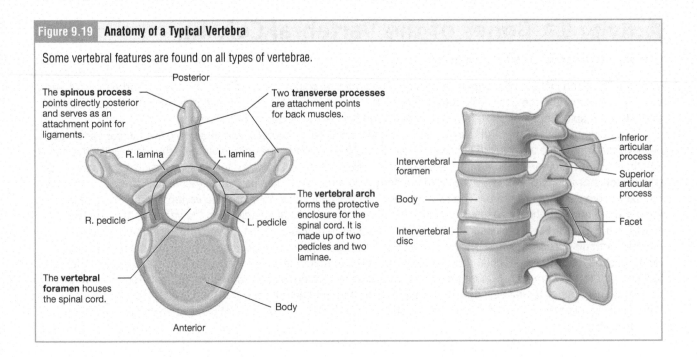

Posterior

The **spinous process** points directly posterior and serves as an attachment point for ligaments.

Two **transverse processes** are attachment points for back muscles.

R. lamina

L. lamina

The **vertebral arch** forms the protective enclosure for the spinal cord. It is made up of two pedicles and two laminae.

R. pedicle

L. pedicle

The **vertebral foramen** houses the spinal cord.

Body

Anterior

Intervertebral foramen

Body

Intervertebral disc

Inferior articular process

Superior articular process

Facet

The seven **cervical vertebrae** (Figure 9.20) are located in the neck. The cervical vertebrae have an extra pair of foramina, the **transverse foramina**, at the base of the transverse processes, through which the vertebral artery and vein pass. Most of the spinous processes of the cervical vertebrae are forked; this is called a bifid spinous process.

Figure 9.20 | Cervical Vertebrae

A typical cervical vertebra has a small body, transverse processes with a transverse foramen, and a bifid spinous process.

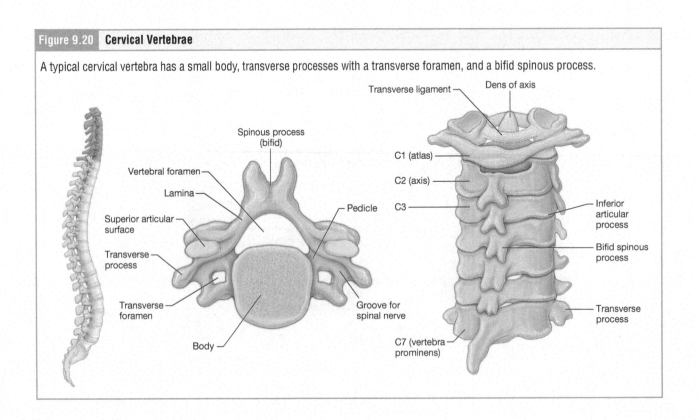

Spinous process (bifid)

Vertebral foramen

Lamina

Superior articular surface

Transverse process

Transverse foramen

Body

Pedicle

Groove for spinal nerve

Transverse ligament

Dens of axis

C1 (atlas)

C2 (axis)

C3

Inferior articular process

Bifid spinous process

Transverse process

C7 (vertebra prominens)

The first two cervical vertebrae, shown in Figure 9.21 are named separately from the other vertebrae due to their unique anatomy:

- The **atlas** (also called the C1 vertebra) does not have a body or spinous process. Instead, it is ring-shaped, consisting of an **anterior arch** and **posterior arch**.
- The **axis** (also called the C2 vertebra) has an additional structure, the **dens,** that projects upward from the vertebral body on its anterior side, allowing the axis to tuck into the inner surface of the anterior arch of the atlas.

Figure 9.21 **Atlas and Axis**

(A) The C1 vertebra (atlas) does not have a body or spinous process. It consists of an anterior and a posterior arch and elongated transverse processes. (B) The C2 vertebra (axis) contains the upward-projecting dens, which articulates with the anterior arch of atlas. (C) The anterior view of axis shows the huge upward-projecting dens. (D) The dens of C2 is held in place by the transverse ligament.

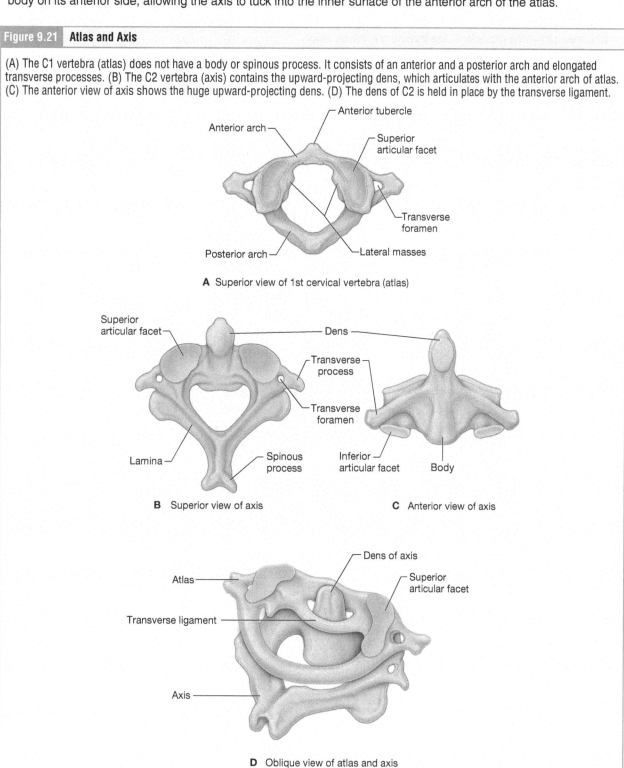

A Superior view of 1st cervical vertebra (atlas)

B Superior view of axis

C Anterior view of axis

D Oblique view of atlas and axis

The 12 **thoracic vertebrae** (Figure 9.22) have long spinous processes with a pronounced downward angle. From the side, these vertebrae look somewhat like the head of a giraffe.

They also have additional articulation sites not present in the vertebrae of other regions because each thoracic vertebra must articulate with a pair of ribs. Figure 9.23 shows these additional articulations, which appear as smooth, rounded depressions on the vertebrae. The head of each rib articulates with the body of the vertebra at a **costal facet**, which is found at the superior edge of the vertebral body. In fact, some of these ribs are positioned so superiorly on the body of their vertebra that they also create a partial demifacet on the body of the vertebra just superior to it. The rib not only attaches to the vertebra at the body, but also again at the transverse process: the articulating tubercle of each rib touches the transverse process at the **transverse costal facet**.

Figure 9.22   **Thoracic Vertebrae**

A typical thoracic vertebra is distinguished by the downward pointing spinous process.

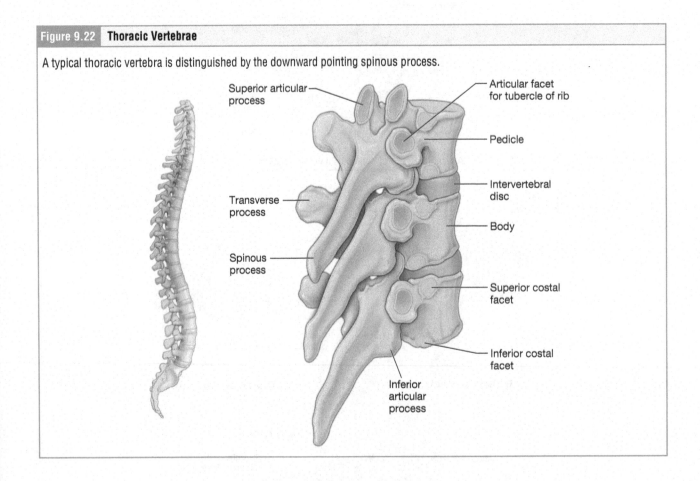

Figure 9.23 **Each Rib Articulates with a Thoracic Vertebra**

The body of a thoracic vertebra articulates with the head of a rib, and the transverse process articulates with the rib tubercle.

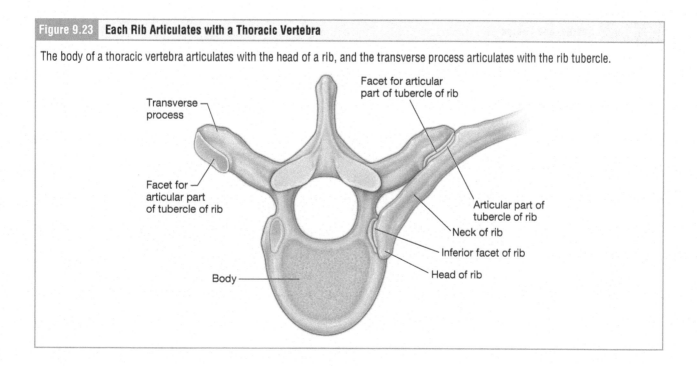

Transverse process

Facet for articular part of tubercle of rib

Facet for articular part of tubercle of rib

Articular part of tubercle of rib

Neck of rib

Inferior facet of rib

Head of rib

Body

The five **lumbar vertebrae** (Figure 9.24) have large, bulky bodies with blunt, thick spinous processes. From the side, these vertebrae look somewhat like the head of a moose. In addition to their larger size, they can be distinguished from the other groups of vertebrae because they lack the transverse foramina of the cervical vertebrae and the costal facets of the thoracic vertebrae.

Figure 9.24 **Lumbar Vertebrae**

Lumbar vertebrae are characterized by their thick bodies and short, rounded spinous processes.

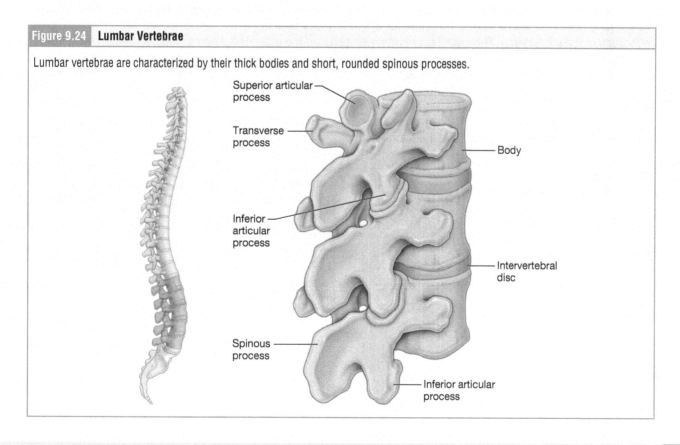

Superior articular process

Transverse process

Body

Inferior articular process

Intervertebral disc

Spinous process

Inferior articular process

Figure 9.25   **Sacrum and Coccyx**

The sacrum is formed from the fusion of five sacral vertebrae; the locations of vertebral fusion are indicated by the transverse ridges. The coccyx is formed by the fusion of four small coccygeal vertebrae.

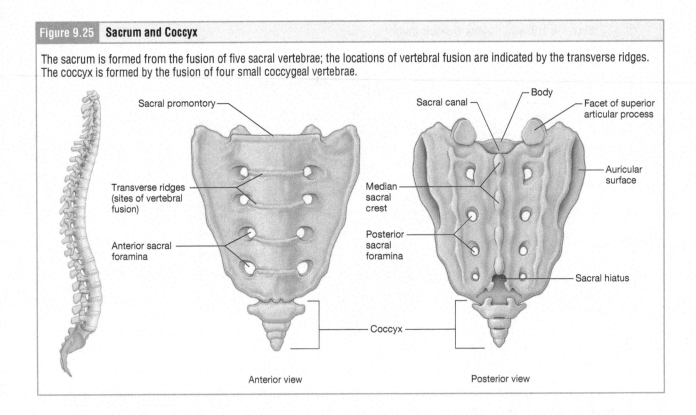

Anterior view

Posterior view

The **sacrum** (Figure 9.25) is made up of five vertebrae that have fused together. On the posterior side of the sacrum, running down the midline, is the **medial sacral crest**, a bumpy ridge that is the remnant of the fused spinous processes of these vertebrae. Deep to the median sacral crest is the **sacral canal**, which terminates at an inferior opening called the **sacral hiatus**. The **superior articular processes** of the sacrum articulate with inferior articular processes of the L5 vertebra. The **sacral promontory** projects on the anterior surface of the sacrum, marking the anterior edge of the bone. The anterior and posterior surfaces of the sacrum have a series of paired openings called **sacral foramina**, which allow nerves to exit the sacral canal. The **coccyx**, sometimes referred to as the tailbone, is made of four small fused coccygeal vertebrae. The coccyx is an attachment point for several ligaments and muscles of the pelvic floor and posterior leg.

The purpose of this activity is to allow you to distinguish between the three types of vertebrae and to learn the many structures found on vertebrae and the sacrum.

Materials
- Sacrum and coccyx (model or real bone)
- Small sticks or pipe cleaners for pointing to bones and bone markings
- Vertebrae (models or real bones)

Important Note: While it is tempting to use your pen or pencil to point to the many small structures on the vertebrae and sacrum, it is critical that you do not use any writing instrument to point or touch the bones or bone models, even if you think that you are being careful. Instead, use the small sticks or pipe cleaners provided by your instructor to avoid leaving any smears or marks on the bones.

Instructions

1. Separate the whole class into as many groups as there are full sets of vertebrae available.

2. With the classmates in your group, divide the vertebrae into the three types:

   a. Seven cervical vertebrae

   b. Twelve thoracic vertebrae

   c. Five lumbar vertebrae and the sacrum

3. Divide your group into three sub-groups. Each sub-group can take one of the piles of vertebrae to complete steps 4 through 6.

4. Identify the following general features:

   a. Body

   b. Pedicles and lamina

   c. Transverse processes

   d. Spinous process

   e. Vertebral foramen

   f. Superior articular processes

   g. Inferior articular processes

5. Stack several of the vertebrae on top of each other.

   a. While it may be difficult to choose the vertebrae that are directly next to each other (such as T8 and T9) based on visual cues alone, you should be able to find the correct order by trial and error, determining which vertebrae fit best with each other like puzzle pieces.

   b. Be careful, and do not stack more than five vertebrae on top of each other, as they tend to topple, especially when not separated from each other by the intervertebral discs present in a living body. Whether you are working with real bones or models, it is important to take care of them.

   c. Notice how the superior articular processes attach to the inferior articular processes to create the intervertebral foramina.

   d. How do the features of the vertebrae within this type change throughout the stack? Record your observations in Table 9.2, one section of the table has been completed for you.

6. Identify the additional features of the type of vertebrae that you are working with.

7. Return the vertebrae you were working on and obtain one of the other two types. Repeat steps 4 through 6.

8. Return the vertebrae you were working on and obtain the final type. Repeat the steps.

9. Practice distinguishing cervical, thoracic, and lumbar vertebrae from each other.

   a. Obtain a single vertebra and show it to your lab partner. Have them determine whether it is a cervical, thoracic, or lumbar vertebra while describing their reasoning to you using correct anatomical terminology.

   b. Repeat several times and then switch roles.

Table 9.2 Distinguishing Features of the Vertebrae Within Each Vertebral Group

Vertebral Type	Size and Shape of Spinous Process	Size and Shape of Vertebral Body	Position of Costal Facets	Orientation of Articular Processes	Other Observations
Cervical vertebrae			X (not present)		
Thoracic vertebrae					
Lumbar vertebrae	Superior-most spinous process points inferiorly a little, but the inferior ones point directly posterior Spinous processes become thicker inferiorly		X (not present)		

Activity 9.4 Bones of the Thoracic Cage

This activity targets LOs 9.1, 9.2, and 9.3.

The remainder of the axial skeleton is called the **thoracic cage**, or rib cage, and is composed of the sternum, ribs, and the costal cartilages that attach them (Figure 9.26). The thoracic cage protects the heart and lungs.

| Figure 9.26 | **The Thoracic Cage** |

The thorax, or thoracic cage, is formed by bone and cartilage to protect the heart and lungs while allowing for the movements of breathing. (A) The sternum has three separate components. (B) Ten of the 12 pairs of ribs are attached to the sternum via costal cartilages. (C) All 12 pairs of ribs are anchored posteriorly to the thoracic vertebrae.

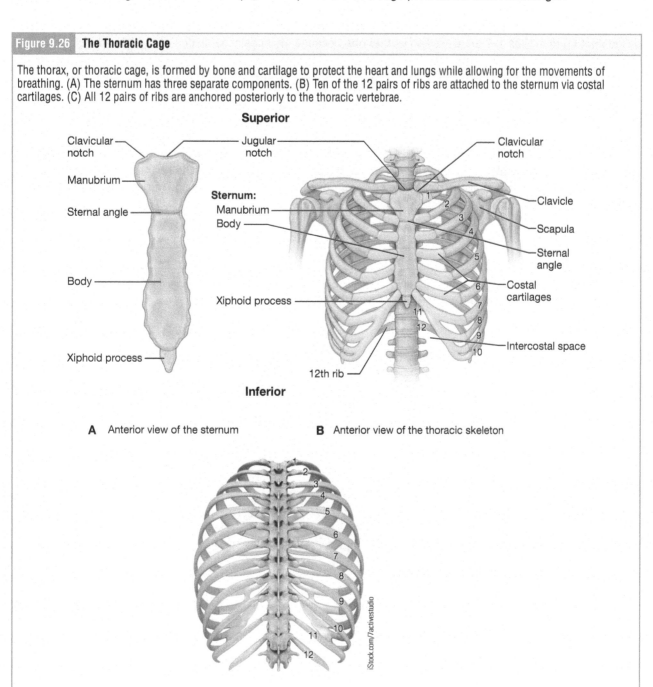

A Anterior view of the sternum

B Anterior view of the thoracic skeleton

C Posterior view of the thoracic skeleton

The **sternum**, or breastbone, consists of three fused bones: the **manubrium**, **sternal body**, and **xiphoid process**. The pair of clavicles (collar bones) articulate with the manubrium at a pair of **clavicular notches** on the manubrium, creating a shallow depression superior to the manubrium called the **jugular notch**. The manubrium articulates with the sternal body at the **sternal angle**, which creates a slight bump along the sternum that can be felt through the skin on some individuals. The inferior tip of the sternum is the xiphoid process.

Each rib is a curved, flattened bone. There are 12 pairs of ribs that articulate posteriorly with the 12 thoracic vertebrae. None of the ribs articulate directly with the sternum; rather, they attach to the sternum via hyaline cartilage strips called **costal cartilages**. The last two rib pairs, ribs 11 and 12, do not attach to the sternum at all and are called **floating ribs**. Ribs 1–7 are classified as **true ribs** because their costal cartilages each attach directly to the sternum. Ribs 8–12 are called **false ribs** because their costal cartilages don't attach directly to the sternum to create an individual connection but rather the cartilage attaches to the cartilage of the next higher rib (or not at all, in the case of ribs 11 and 12).

The parts of a typical rib are shown in Figure 9.27. The posterior end where the rib attaches to the body of the vertebra is called the **head of the rib**. Lateral from the head a short distance is the **tubercle of the rib** where it articulates with the transverse process of the vertebra; between the head and the tubercle is the **neck of the rib** (see Figure 9.23 for a reminder of how a rib articulates with a vertebra). The remainder of the rib is the **body of the rib**. Progressing laterally along the body, the rib curves dramatically at the **angle of the rib**. Along the body on the inferior edge you will find a shallow groove, the **costal groove**, where blood vessels and nerves travel out from the spinal cord and supply the thoracic wall.

Figure 9.27 **A Typical Rib**

Each rib connects with a thoracic vertebra at the head and tubercle of the rib. Along the rib's inferior edge is a groove in which blood vessels and nerves are nestled and protected. Anteriorly, most ribs connect to costal cartilages.

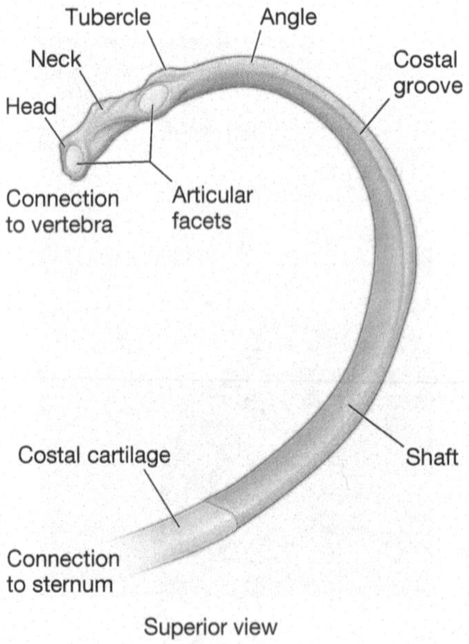

Materials

- Articulated skeleton
- Individual thoracic vertebrae and ribs (human or model)
- Small sticks or pipe cleaners for pointing to bones and bone markings

Important Note: While it is tempting to use your pen or pencil to point to structures on the skeleton, it is critical that you do not use any writing instrument to point or touch the bones or bone models, even if you think that you are being careful. Instead, use the small sticks or pipe cleaners provided by your instructor to avoid leaving any smears or marks on the bones.

Instructions

1. Examine the skeleton with your lab partner to visualize how the sternum connects to the ribs via the costal cartilages. Identify all of the bones and bone markings in Figure 9.26.

2. Obtain several ribs.

3. Work with your lab partner to determine how you can distinguish between a right and left rib; make sure that you have at least one right and one left rib.

4. Obtain two neighboring thoracic vertebrae.

5. Lay the left and right ribs and the two thoracic vertebrae out on the table in position, using Figures 9.23 and 9.27 to help you name the locations where the bones articulate with each other.

6. Sketch a drawing of these articulations in the space below.

Digging Deeper:
Chest Compressions during CPR

Cardiopulmonary resuscitation (CPR) may include chest compressions. This procedure uses the bones of the thoracic cage to push on the heart in an attempt to move blood through the heart when it is not beating on its own. The heart lies deep to the sternal body, so pressure on this area of the sternum will compress the heart and push blood forward. Directions for CPR indicate that the heel of the hand should be placed two finger widths above the xiphoid process, which can be felt through the skin (Figure 9.28). This focuses pressure directly over the likely location of the heart. Pressure directly on the small xiphoid process should be avoided during CPR because it is possible for it to break and detach from the rest of the sternum during forceful compressions, risking a puncture or laceration by the sharp bony fragments. Can you palpate the xiphoid process on yourself?

| Figure 9.28 | Feeling for the Xiphoid Process Before Starting Chest Compression During CPR |

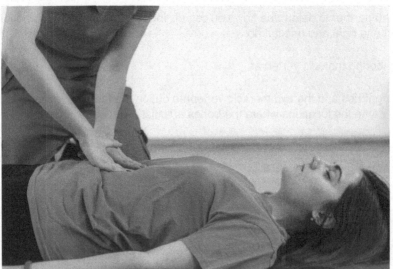

Microgen/Shutterstock.com

Lab 9 Post-Lab Quiz

1. Which of the following bones is **not** part of the axial skeleton?

   a. Sacrum
   b. Scapula
   c. Sphenoid bone

2. What bone is highlighted in this image?

   a. The temporal bone
   b. The maxillary bone
   c. The zygomatic bone

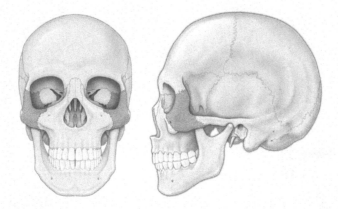

3. What two bones articulate to allow your mouth to open and close?

   a. The maxilla and the mandible
   b. The mandible and temporal bone
   c. The temporal bone and zygomatic bone

4. What portion of the orbit is formed by the ethmoid bone?

   a. The medial wall
   b. The lateral wall
   c. The floor
   d. The roof

5. As you eat a bite of food it passes along the roof of your mouth. What two bones does the food pass by, in order?

   a. The vomer and then the perpendicular plate of the ethmoid bone
   b. The mandible and then the hyoid bone
   c. The maxilla and then the horizontal plate of the palatine bone

6. Identify the space highlighted in this radiograph.

   a. Orbits
   b. Frontal sinuses
   c. Nasal cavity

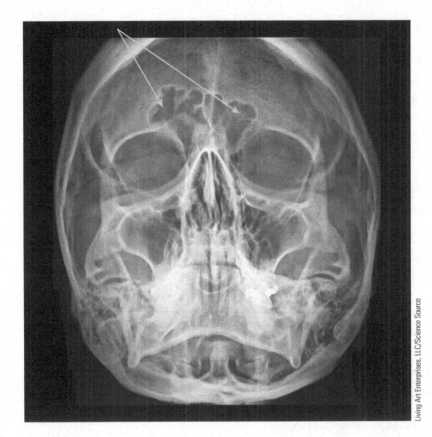

Living Art Enterprises, LLC/Science Source

7. With what structure do the occipital condyles articulate?

   a. The superior articulating processes of the atlas
   b. The mandibular condyles
   c. The dens of the axis

8. Which of the following sutures is paired?

   a. Lambdoid suture
   b. Frontal suture
   c. Squamous suture
   d. Sagittal suture

9. A blow to the nose is most likely to break which bones?

   a. Zygomatic bones
   b. Palatine bones
   c. Nasal bones

10. Which type of vertebrae are shown here?

    a. Cervical

    b. Thoracic

    c. Lumbar

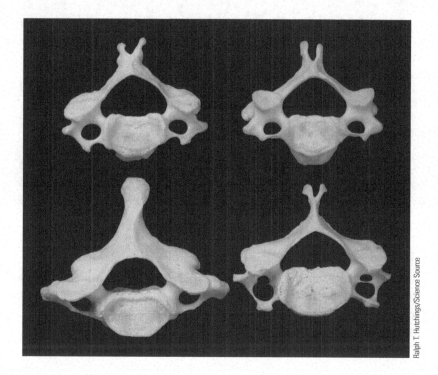

Ralph T. Hutchings/Science Source

11. How many total ribs are in the body?

    a. Eleven

    b. Twelve

    c. Twenty-three

    d. Twenty-four

Lab 10 Appendicular Skeleton

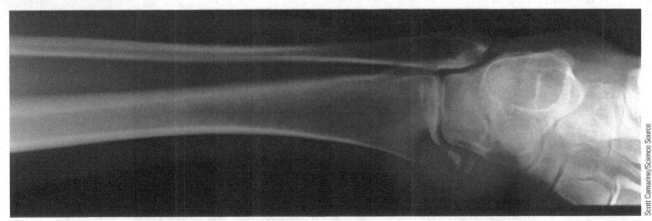

Scott Camazine/Science Source

Learning Objectives: After completing this lab, you should be able to:

10.1   Distinguish between the axial and appendicular skeletons and list the major bones contained within each.

10.2   Identify individual bones and their locations within the body.

10.3   Identify major bone markings (e.g., spines, processes, foramina) on individual bones.

Introduction

The appendicular skeleton contains a total of 126 bones that make up the appendages plus the bones that attach the appendages to the axial skeleton: the pectoral girdles and the pelvic girdle. Each upper limb attaches to the body via a **pectoral girdle** (also called a shoulder girdle), while a single **pelvic girdle** connects both lower limbs to the axial skeleton.

As you study the bones of the appendicular skeleton, consider how the two sets of limbs, your arms and legs, are similar to each other in anatomical arrangement. Consider also how the two sets of limbs differ from each other due to their differing functions in the upright human body.

Note that you may not be able to work through the lab activities in order, depending on how many sets of bones your instructor provides to the class. You may also be assigned to work on a set of bones with more than just your lab partner, perhaps working in groups of four students, to have enough bones for all of the students.

The Human Anatomy and Physiology Society includes more than 1,700 educators who work together to promote excellence in the teaching of this subject area. The HAPS A&P Learning Outcomes measure student mastery of the content typically covered in a two-semester Human A&P curriculum at the undergraduate level. The full Learning Outcomes are available at https://www.hapsweb.org.

Lab 10 Pre-Lab Quiz

This quiz will strengthen your background knowledge in preparation for this lab. For help answering the questions, use your resources to deepen your understanding. The best resource for help on the first five questions is your text, and the best resource for help answering the last five questions is to read the introduction section of each lab activity.

1. Anatomy of the appendicular. Label the following parts of the appendicular skeleton:

Femur   Pelvic girdle
Fibula   Radius
Humerus   Tibia
Pectoral girdle   Ulna

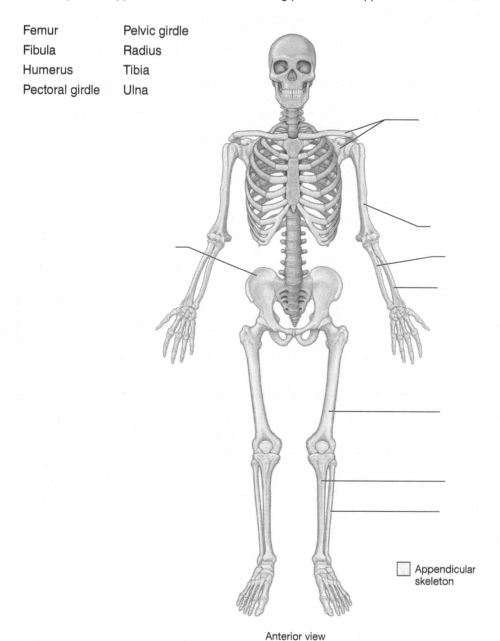

☐ Appendicular skeleton

Anterior view

2. Match the following Greek and Latin roots to their definitions:

Root	Meaning
	after or beyond
	above
	below
	upon

List of Roots

epi
infra
meta
supra

3. Which of the following bones is <u>not</u> part of the appendicular skeleton?

   a. The sacrum
   b. The femur
   c. The scapula

4. The pectoral girdle assists in the movement of what body part?

   a. The head
   b. The arms
   c. The legs

5. How many bones are found in the distal limb of the arm and the distal limb of the leg?

   a. One
   b. Two
   c. Three

6. Which of the following bones is <u>not</u> part of the pectoral girdle?

   a. The clavicle
   b. The scapula
   c. The sternum

7. The humerus of the brachium articulates with what two bones at the elbow?

   a. The radius and ulna
   b. The tibia and fibula
   c. The carpals and metacarpals

8. What paired structures together make up the pelvic girdle?

   a. A pair of pubis bones
   b. A pair of os coxae
   c. A pair of scapulae

9. Does the femur articulate directly with the fibula?

   a. Yes
   b. No

10. Digits 2–5 of the hand and 2–5 of the foot each contain how many phalanges?

   a. Two
   b. Three
   c. Four

Activity 10.1 Bones of the Pectoral Girdle

This activity targets LOs 10.1, 10.2, and 10.3.

Each pectoral girdle (Figure 10.1) is made up of two bones: the clavicle and the scapula. The complex shape of the scapula allows space and attachment for the significant number of muscles that move the arm. The clavicle supports the position of the scapula, holding the shoulder joint superiorly and laterally away from the ribcage. The clavicle also protects underlying nerves and blood vessels as they pass between the trunk of the body and the arm.

    The **clavicle** (Figure 10.2) is a small S-shaped bone that connects the sternum to the shoulder. Its medial end, called the **sternal end**, articulates with the manubrium of the sternum. Its lateral end, called the **acromial end**, articulates with the acromion of the scapula. In anatomical position the acromial end points somewhat anteriorly as it abuts the anterior-pointing acromion. The inferior surface of the clavicle is often rougher and more irregular due to the impressions made by the blood vessels and nerves that travel along the inferior aspect of the bone.

    The **scapula** (Figure 10.3) has many important landmarks. The three borders of the scapula are named for their positions within the body: the **superior border**, the **lateral border**, and the **medial border**. The two corners of the triangular scapula are the **superior angle** and **inferior angle**. The remaining corner of the scapula, between the superior and lateral borders, is the location of the **glenoid cavity** (glenoid fossa), a shallow depression that cups the head of the humerus of the arm. The scapula has two prominent projections superior to the glenoid cavity. The **coracoid process** projects anteriorly, serving as an attachment point for muscles of the chest and arm. Lateral to the coracoid process is the broad, flat **acromion** that forms the bony tip of the superior shoulder area. The acromion articulates with the acromial end of the clavicle to brace the shoulder. Rising from the posterior superior surface of the scapula and leading medially away from the acromion is a ridge of bone, the **spine of the scapula** (scapular spine). The scapula has three depressions that are named for their relative positions to the scapular spine: the **supraspinous fossa** superior to the spine, the **infraspinous fossa** inferior to the spine, and the **subscapular fossa**, deep to the scapular spine.

    The purpose of this activity is to learn the bones and bony markings of the pectoral girdle.

Materials
- Articulated skeleton, if available
- Clavicle bone
- Scapula
- Small sticks or pipe cleaners for pointing to bones and bone markings

---

Figure 10.1 | **The Shoulder Girdle**

The shoulder girdle is the anchor for the upper limb, consisting of the clavicle and the scapula. It is shown in three views: (A) Anterior view; (B) Posterior view; and (C) Superior view.

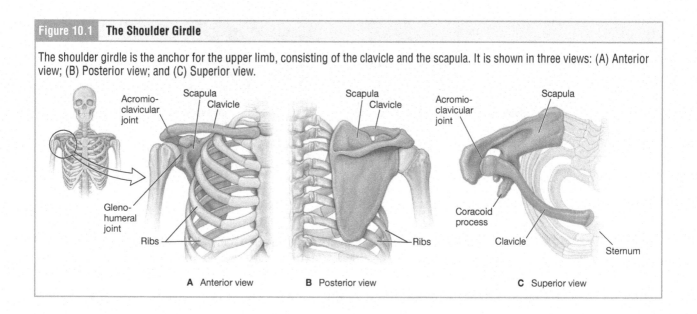

A Anterior view   B Posterior view   C Superior view

Figure 10.2 **The Clavicle**

The clavicle connects the acromion of the scapula to the sternum of the axial skeleton. Therefore, it has a lateral acromial end and a medial sternal end.

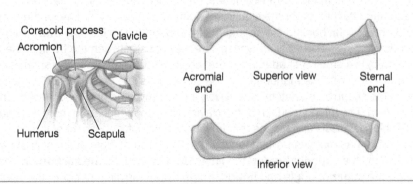

Figure 10.3 **The Scapula**

The scapula is shown here from its anterior side, which faces the ribcage, and its posterior side, which faces muscles and skin of the back.

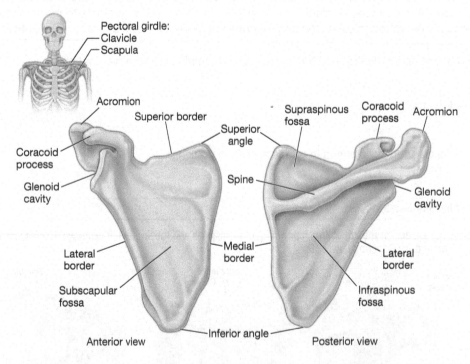

Important Note: While it is tempting to use your pen or pencil to point to the many small structures on these bones, it is critical that you do not use any writing instrument to point or touch the bones, even if you think that you are being careful. Instead, use the small sticks or pipe cleaners provided by your instructor to avoid leaving any smears or marks on the bones.

Instructions

1. Obtain either a right clavicle and scapula or a left clavicle and scapula. If you are not sure how to determine whether you have the "matching" pair of bones, work with your lab partner to identify the anatomical features that will allow you to determine whether the bones are right or left. Use an articulated skeleton, if available, to help you.

2. How were you able to distinguish between the left and right bones? Use the space below to take some notes on the anatomical features that were helpful.

3. Identify each of the bone markings listed in bold in the introduction on these bones and those listed in Table 10.1.

4. Hold or lay out the bones on the table to approximate anatomical position. Note the location where the clavicle and scapula articulate and the names of the markings that meet to form this articulation.

Table 10.1 Bones and Markings of the Pectoral Girdle		
Bone	**Bone Marking**	
Clavicle	Sternal end	
	Acromial end	
Scapula	Superior border	Coracoid process
	Lateral border	Acromion
	Medial border	Spine of the scapula
	Superior angle	Supraspinous fossa
	Inferior angle	Infraspinous fossa
	Glenoid cavity	Subscapular fossa

(page intentionally left blank)

Activity 10.2 Bones of the Upper Limb

This activity targets LOs 10.1, 10.2, and 10.3.

In humans, the upper limbs are divided into three regions (Figure 10.4): the **brachium** (upper arm), the **antebrachium** (forearm), and the **hand** or manus. The brachium contains a single bone, the **humerus**. The antebrachium contains a pair of bones, the **ulna** and **radius**. The hand contains eight **carpal bones**, five **metacarpal bones**, and 14 **phalanges** (singular = phalanx).

The proximal end of the humerus (Figure 10.5), the **head of the humerus**, faces medially to articulate with the glenoid cavity of the scapula. Laterally from the head is a large projection, the **greater tubercle**. And anterior to the greater tubercle is a somewhat smaller projection, the **lesser tubercle**. Between the two tubercles is the narrow **bicipital groove** (intertubercular groove) where a tendon of the biceps brachii muscle passes. Along the **shaft of the humerus** is an anterior projection of bone, the **deltoid tuberosity**, which is the site of attachment for the deltoid muscle.

The distal portion of the humerus articulates with the radius and ulna to form the elbow joint; several important markings are found here. The medial and lateral sides of the humerus, just proximal to the articulations at the elbow, have a pair of bony projections: the **medial epicondyle of the humerus** and **lateral epicondyle of the humerus**. The two condyles of the humerus that form the articulation points for the elbow are the **trochlea**, which articulates with the ulna, and the **capitulum**, which articulates with the radius. On the posterior surface of the humerus just superior to the trochlea is an indentation called the **olecranon fossa** where the humerus accommodates the ulna when the elbow is straight. Anteriorly in this same region the **coronoid fossa** accommodates the ulna as it pivots around the humerus to bend the elbow. Lateral to the coronoid fossa, the **radial fossa** accommodates the head of the radius when the elbow is bent.

The forearm contains two bones: the ulna and radius (Figure 10.6). The ulna is wider proximally and narrow distally. The U-shape of the proximal ulna is made up of several important markings that correspond with structures of the humerus. The scooped shape on the inside (anterior side) of the U is the **trochlear notch** that articulates with the trochlea of the humerus. The trochlear notch peaks in a point of bone called the **coronoid process of the ulna**, which tucks into the coronoid fossa of the humerus when the elbow is bent. On the posterior surface of the proximal end of the ulna, the **olecranon process** forms the large point of bone that you may think of as your elbow. Still on the proximal end of the ulna, just lateral to the trochlear notch you will find a small, smooth depression where the head of the radius will articulate with the ulna to form the *proximal radioulnar joint*, this is the **radial notch** of the ulna. The narrow distal end of the ulna terminates in the distal **head of the ulna**, which articulates with the carpals, and the **styloid process of the ulna**, also called the medial styloid process.

In contrast to the ulna, the radius is narrow proximally at the **head of the radius** and then widens distally. Just distal to the head, on the anterolateral surface, the **radial tuberosity** provides an attachment point for muscles. On the distal end of the radius, the **styloid process of the radius**, also called the lateral styloid process, forms a lateral bulge of bone. The anterior surface of the distal radius forms a shallow concave shape that can help distinguish the anterior and posterior surfaces of this simple bone from each other.

| Figure 10.4 | The Regions of the Arm |

The arm can be divided into three major regions—the brachium, the antebrachium, and the hand—with two major functional joints, the elbow and the wrist. There are also many minor joints between individual bones.

Arm (brachium) — Humerus

Forearm (antebrachium) — Ulna — Radius — Carpals — Metacarpals

Hand — Phalanges

Figure 10.5 | Humerus and Elbow Joint

(A) The humerus is the single bone of the brachium. (B) Its distal end has articulating surfaces where it forms the elbow joint with the radius and ulna.

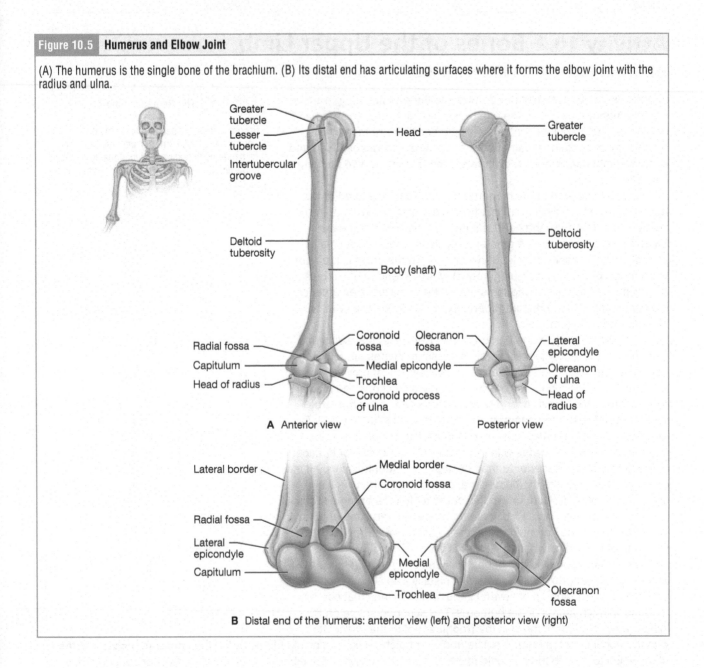

A Anterior view

Posterior view

B Distal end of the humerus: anterior view (left) and posterior view (right)

The small bones of the wrist and hand are best examined in an intact, articulated view (Figure 10.7). The wrist and base of the hand are formed by a series of eight small carpal bones. The carpal bones are arranged in two rows, forming a proximal row of four carpal bones and a distal row of four carpal bones. The bones in the proximal row, running from the lateral (thumb) side to the medial side, are the **scaphoid, lunate, triquetrum**, and **pisiform** bones. The small, rounded pisiform bone sits on the anterior surface of the triquetrum, rather than articulating with it laterally. The distal bones (lateral to medial) are the **trapezium, trapezoid, capitate**, and **hamate**. The hamate bone is easily distinguished from the other carpals by the hooked projection of bone on its anterior (palmar) surface.

Figure 10.6 **The Radius and Ulna**

The radius and ulna make up the antebrachium. (A) The medial side of the radius and lateral side of the ulna are attached to each other by an interosseous membrane. (B) This view of the ulna shows the large olecranon process that fits into the olecranon fossa of the humerus when the arm is extended at the elbow. (C) A close-up view of the proximal ulna shows the trochlear notch, which cups the distal end of the humerus in the elbow joint.

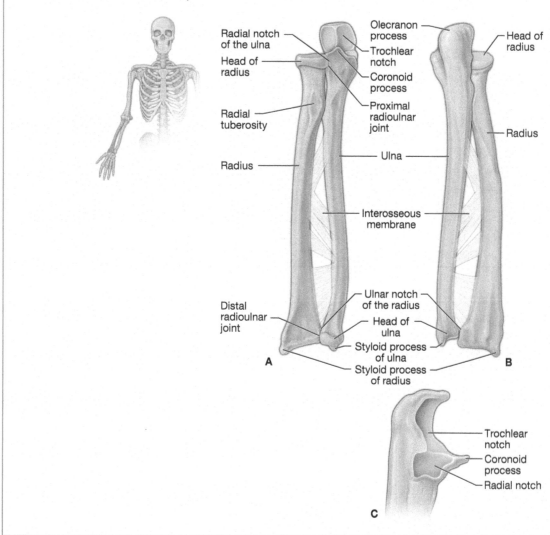

The palm of the hand contains five elongated metacarpal bones. These bones lie between the carpal bones of the wrist and the bones of the fingers and thumbs. The metacarpal bones are numbered 1 through 5, beginning laterally, in the same pattern as the numbering of the digits of the hand. The digits (fingers and thumb) contain a total of 14 bones, each of which is called a phalanx bone. Digit 1 of the hand, also called the **pollex**, has two phalanges which are called a *proximal phalanx bone* and a *distal phalanx bone*. Digits 2 (the index finger) through 5 (the pinkie or little finger) have three phalanges each; between the proximal and distal phalanges is a third bone, the *middle phalanx bone*.

Figure 10.7 | Bones of the Wrist and Hand

(A) The wrist is composed of eight carpal bones arranged in two rows. The metacarpal bones form the palm of the hand, and the phalanx bones (phalanges) compose the thumb and fingers. (B) A coronal section through the wrist illustrates the radiocarpal joint. The radius articulates with the scaphoid and lunate. The ulna is shielded from the carpals by the articular disc and does not directly articulate with them.

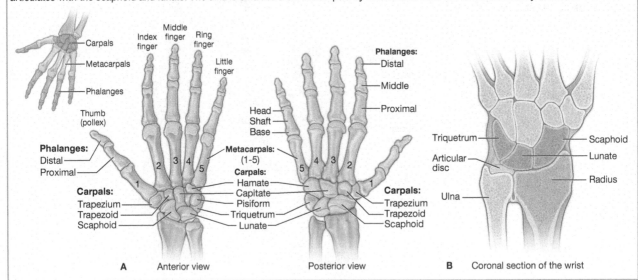

Important Note: While it is tempting to use your pen or pencil to point to the many small structures on these bones, it is critical that you do not use any writing instrument to point or touch the bones, even if you think that you are being careful. Instead, use the small sticks or pipe cleaners provided by your instructor to avoid leaving any smears or marks on the bones.

The purpose of this activity is to learn the bones and bony markings of the bones of the brachium, antebrachium, and hand.

Materials
- Articulated bones of the hand
- Articulated skeleton, if available
- Humerus
- Radius
- Small sticks or pipe cleaners for pointing to bones and bone markings
- Ulna

Instructions
1. Obtain a right or left humerus, radius, ulna, and articulated hand. If you are not sure how to determine whether you have a "matching" set of bones, work with your lab partner to identify the anatomical features that will allow you to determine whether the bones are right or left. Use an articulated skeleton, if available, to assist you.

2. How were you able to distinguish between the left and right bones? Use the space below to take some notes on the anatomical features that were helpful.

3. Identify each of the markings listed in bold in the introduction on these bones and in Table 10.2.

4. Hold or lay out the bones on the table to approximate anatomical position. Note the locations where the bones articulate and the names of the bone markings that meet to form these articulations.

Table 10.2 Bones and Markings of the Upper Limb			
Bone	**Bone Markings**	**Bone**	**Bone Markings**
BONE OF THE BRACHIUM			Radial notch
Humerus	Head of the humerus		Head of the ulna
	Greater tubercle		Styloid process of the ulna
	Lesser tubercle	Radius	Head of the radius
	Bicipital groove		Radial tuberosity
	Shaft of the humerus		Styloid process of the radius
	Deltoid tuberosity	**BONES OF THE HAND**	
	Medial epicondyle of the humerus	Scaphoid	
	Lateral epicondyle of the humerus	Lunate	
	Capitulum	Triquetrum	
	Olecranon fossa	Pisiform	
	Coronoid fossa	Trapezium	
	Radial fossa	Trapezoid	
BONES OF THE ANTEBRACHIUM		Capitate	
Ulna	Trochlear notch	Hamate	
	Coronoid process of the ulna	Metacarpals	
	Olecranon process	Phalanges	

(page intentionally left blank)

Activity 10.3 The Articulations of the Upper Appendicular Skeleton

This activity targets LOs 10.1, 10.2, and 10.3.

Together, the bones of the pectoral girdles, arms, and hands articulate to provide a wide range of motion to the upper limbs of the body. This activity will allow you to "build" a shoulder, arm, and hand to visualize how these bones come together to form the upper appendicular skeleton.

Materials
- Articulated bones of the hand
- Articulated skeleton, if available
- Clavicle
- Humerus
- Radius
- Scapula
- Ulna

Instructions

1. Obtain the bones listed here, working with your lab partner to make sure they are all either right or left. Use an articulated skeleton, if available, to assist you.

2. How were you able to distinguish between the left and right bones? Use the space below to take some notes on the anatomical features that were helpful.

3. Assemble the bones into a shoulder, arm, and hand.

4. Roughly sketch the assembled bones in the space on the facing page.

5. Add the manubrium (of the axial skeleton) to the sketch to show where the clavicle articulates with it.

6. Label the articulating features of each bone on your sketch.

(page intentionally left blank)

Activity 10.4 Bones of the Pelvic Girdle

This activity targets LOs 10.1, 10.2, and 10.3.

The adult **pelvis** (Figure 10.8) is formed by the two os coxae (singular = os coxa) or hip bones, the sacrum, and the coccyx. While the sacrum and coccyx are considered part of the axial skeleton, the os coxae, which directly articulate with and assist in the movement of the legs, are part of the appendicular skeleton. The term pelvic girdle refers to the two os coxae without their accompanying axial skeleton bones.

Some of the features of the pelvic girdle are best understood by examining the entire pelvis together, shown in Figure 10.8. Laterally on each side of the pelvis is an **acetabulum**, which provides a depression for the head of the femur of the leg to articulate. Anteriorly and medially, the two os coxae articulate with each other at the **pubic symphysis**, while posteriorly each os coxa articulates with one side of the sacrum at a **sacroiliac joint**. The large entrance to the pelvis is marked anteriorly and laterally by a ridge of bone, the **pelvic brim**. The posterior border of the pelvic brim is marked by the sacral promontory. Two smaller inferior openings, the **obturator foramina**, provide a space for nerves to pass to the anterior region of each leg.

Each adult os coxa is formed by three separate bones that fuse together during the late teenage years: the ilium, ischium, and pubic bones (Figure 10.9). Thus, you could say that the pelvic girdle is initially six bones: two ilium bones, two ischium bones, and two pubic bones.

The **ilium** is the largest of the three bones of the os coxa (Figure 10.10). The superior edge of the ilium, the **iliac crest**, terminates anteriorly at the **anterior superior iliac spine**. Just inferior to this bony point is a second, smaller point called the **anterior inferior iliac spine**. The posterior terminus of the iliac crest is the **posterior superior iliac spine**, which has a corresponding **posterior inferior iliac spine**. Between these two posterior spines on the medial surface of the broad body of the ilium is the rough **auricular surface** that marks the articulation point with the sacrum. Inferior to the posterior inferior iliac spine is an obvious indentation that marks the inferior border of the ilium: the **greater sciatic notch**, which allows a nerve to pass from the pelvis to the posterior aspect of the thigh. The pelvic brim, previously identified in Figure 10.8, is made of a sharp edge of bone that spans both the ilium and pubis bones. The portion of the pelvic brim that is on the ilium is called the **arcuate line**.

Figure 10.8	The Adult Pelvis

The adult pelvis is a solid structure made up of four bones—the right and left os coxae, the sacrum of the spine, and the coccyx.

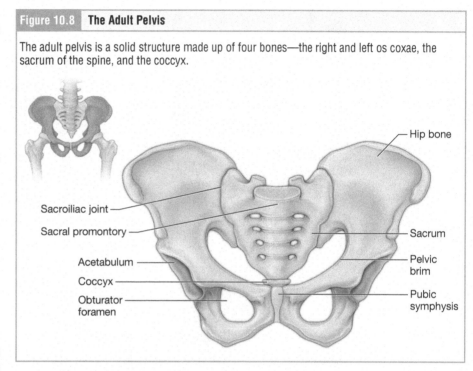

Figure 10.9	The Os Coxa Is Formed by Three Separate Bones: the Ilium, Pubis, and Ischium

In adulthood, the three pelvic bones fuse to form each of the os coxae bones. The three bones all contribute to the acetabulum, the bowl-shaped region of articulation of the hip and femur. The ischium and pubis also meet to form an arch around a large opening, the obturator foramen.

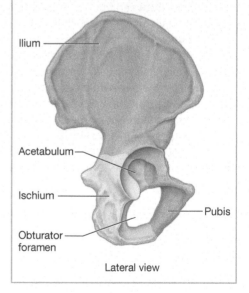

Lateral view

Figure 10.10 | The Ilium

The ilium forms the large fan-shaped superior portion of the hip bone. (A) Lateral and medial views. (B) Anterior view.

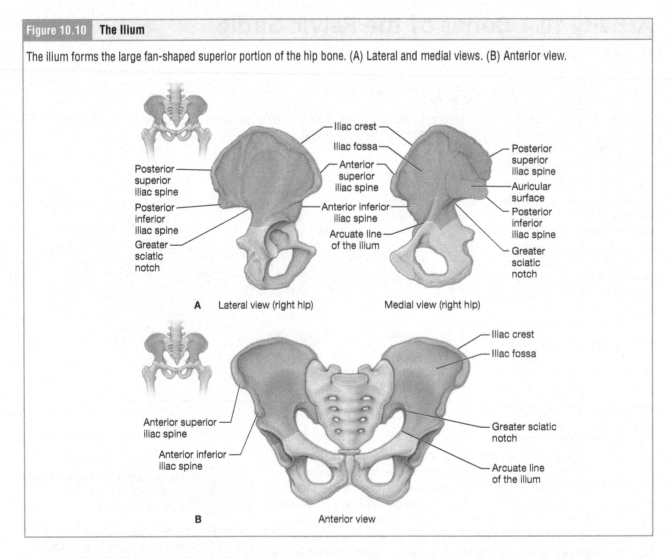

A Lateral view (right hip)   Medial view (right hip)

B   Anterior view

The **ischium** is the posterior, inferior aspect of the os coxa (Figure 10.11). The large, thick, inferior-most portion of the ischium is the **ischial tuberosity**. The branch of bone that reaches anteriorly from the ischial tuberosity to the pubis is the **ischial ramus**. Posteriorly, and superior to the ischial tuberosity, is a smaller, thin point of bone called the **ischial spine**. Between the tuberosity and the spine is a small indentation called the **lesser sciatic notch**. Note how the lesser sciatic notch relates to the greater sciatic notch of the ilium.

The **pubis**, or pubic bone, forms the anterior portion of the os coxa (Figure 10.12). The two pubic bones meet each other anteriorly at the fibrocartilage pad of the pubic symphysis. Just lateral to this articulation is a small bump on each pubis that marks the inferior attachment point of the inguinal ligaments: the **pubic tubercle**. Reaching laterally away from the pubic tubercle, the pubic portion of the sharp edge of the pelvic brim is the **pectineal line**. The pubis has two extensions: the **superior ramus of the pubis** reaches toward the ilium and superior portion of the ischium, while the **inferior ramus of the pubis** reaches toward the inferior portion of the ischium.

The purpose of this activity is to learn the bones that make up the os coxae and key bony markings of this structure that articulate with each other to form the pelvic girdle and also articulate with the sacrum of the axial skeleton to form the pelvis.

Figure 10.11 **The Ischium**

The ischium forms the posteroinferior portion of the hip bone. (A) Lateral and medial views. (B) Anterior view.

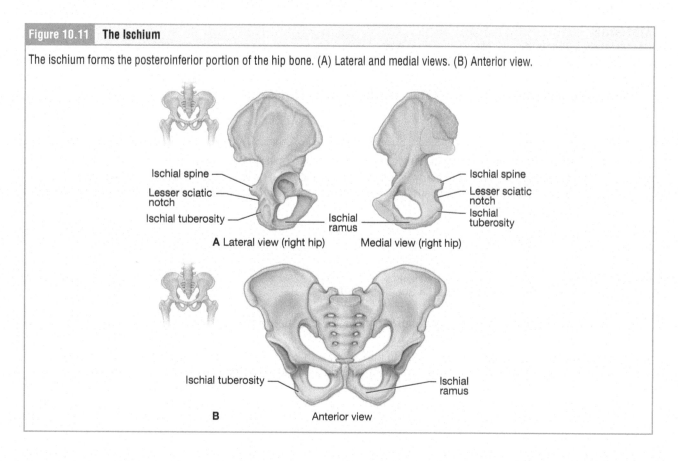

Ischial spine
Lesser sciatic notch
Ischial tuberosity
Ischial ramus

A Lateral view (right hip)

Ischial spine
Lesser sciatic notch
Ischial tuberosity

Medial view (right hip)

Ischial tuberosity
Ischial ramus

B Anterior view

Figure 10.12 **The Pubis**

The pubis forms the anteromedial portion of the hip bone. (A) Lateral and medial views. (B) Anterior view.

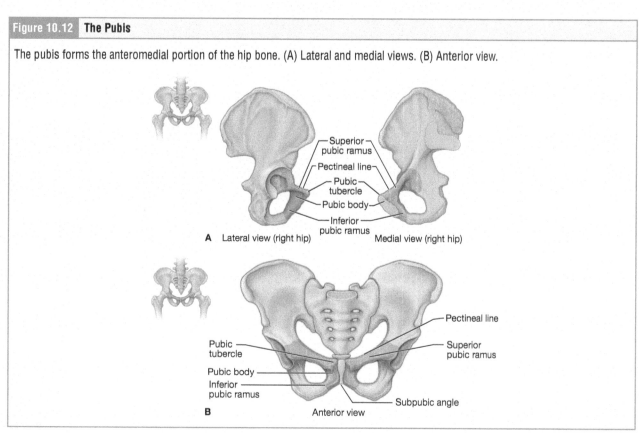

Superior pubic ramus
Pectineal line
Pubic tubercle
Pubic body
Inferior pubic ramus

A Lateral view (right hip)

Medial view (right hip)

Pubic tubercle
Pubic body
Inferior pubic ramus

Pectineal line
Superior pubic ramus
Subpubic angle

B Anterior view

Materials
- Articulated skeleton, if available
- Os coxae, both right and left
- Sacrum
- Small sticks or pipe cleaners for pointing to bones and bone markings

Important Note: While it is tempting to use your pen or pencil to point to the many small structures on these bones, it is critical that you do not use any writing instrument to point or touch the bones, even if you think that you are being careful. Instead, use the small sticks or pipe cleaners provided by your instructor to avoid leaving any smears or marks on the bones.

Instructions

1. Obtain both the right and left os coxa and the sacrum.

2. Identify each of the bone markings listed in bold in the introduction on these bones.

3. Which os coxa is the right and which is the left? Work with your lab partner to identify the anatomical features that will allow you to determine which is which. Use an articulated skeleton, if available, to help you.

4. How were you able to distinguish between the left and right bones? Use the space below to take some notes on the anatomical features that were helpful.

5. Hold or lay out the bones on the table to approximate anatomical position. Note the locations where the bones articulate and the names of the markings that meet to form these articulations.

Activity 10.5 Bones of the Lower Limb

This activity targets LOs 10.1, 10.2, and 10.3.

Similar to the upper limb, the lower limbs are also separated into three regions (Figure 10.13): the **thigh** (upper leg), the **lower leg**, and the **foot**. An additional bone, the **patella** (kneecap) is embedded in the tendon that passes across the anterior knee at the joint. The thigh contains a single bone, the **femur**. The lower leg contains a pair of bones, the **tibia** and **fibula**. The foot contains seven **tarsal bones**, five **metatarsal bones**, and 14 **phalanges** (singular = phalanx).

The femur (Figure 10.14) articulates with the acetabulum of the os coxa at the **head of the femur**. The center of the head of the femur contains a small, rough pit called the **fovea capitis**, which is the attachment point for the *ligament of the head of the femur*. The **neck of the femur** is bounded by two projections: the **greater trochanter** is a large lateral knob of bone, while the **lesser trochanter** is a posteromedial smaller bony projection. On the anterior surface between the two trochanters is the **intertrochanteric line**, and on the posterior surface between the two trochanters is a second ridge called the **intertrochanteric crest**. Inferior and lateral from the intertrochanteric crest is a rough bulge of bone, the **gluteal tuberosity**, that serves as the attachment point for the large gluteus maximus muscle. Tracing inferiorly along the long anterior shaft of the femur is a ridge of bone, the **linea aspera**, an attachment point for muscles of the thigh. At the distal end of the femur, reminiscent of similar anatomical features of the humerus, lie two bumps of bone, the **lateral epicondyle of the femur** and the **medial epicondyle of the femur**. Just distal to the epicondyles are the large, bulky condyles that articulate with the tibia at the knee joint: the **lateral condyle of the femur** and the **medial condyle of the femur**. On the posterior surface of the femur between the two condyles, the **intercondylar fossa** forms a deep pit in this area. On the distal anterior surface, the **patellar surface** is a shallow depression where the patella articulates.

The tibia (Figure 10.15) is the larger of the two bones of the crus. Its two articular surfaces, the **articular surface of the medial condyle** and the **articular surface of the lateral condyle**, (together sometimes called the tibial plateau) support the medial and lateral condyles of the femur. On the tibial plateau, lying between the two articular surfaces, are two small points of bone, the **intercondylar eminences**. Below the surface of the tibial plateau, the two large bulges of bone are the **medial condyle of the tibia** and the **lateral condyle of the tibia**. The proximal fibula sits just distal to the lateral condyle of the tibia, look for the smooth surface marking the location of the *proximal tibiofibular joint* here. On the anterior surface of the proximal tibia is a bulge of bone, the **tibial tuberosity**, which is where the patellar ligament attaches. The distal tibia is marked by an obvious inferior bulge of bone on the medial aspect, the **medial malleolus**, that marks the location where the tibia, along with the fibula, bracket the talus (a tarsal bone) to form a portion of the ankle. The lateral surface of the distal tibia curves in somewhat to accommodate the distal portion of the fibula at the *distal tibiofibular joint*.

Figure 10.13	The Regions of the Leg

The leg can be divided into three major regions—the thigh, the lower leg, and the foot—with two major functional joints, the knee and the ankle. There are also many minor joints between individual bones.

Figure 10.14 **The Femur and the Knee Joint**

The femur is the single bone of the thigh region. Its head fits into the acetabulum of the os coxae to form the hip joint. Inferiorly, the femur articulates with the tibia at the knee joint. The patella articulates with the distal end of the femur.

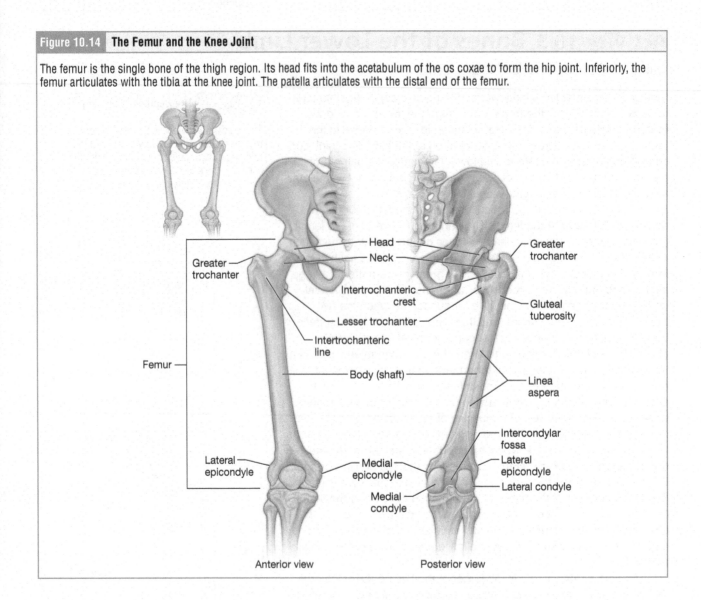

Anterior view   Posterior view

The fibula (Figure 10.15), the smaller and thinner of the two bones of the crus, is lateral to the tibia. The **head of the fibula** is proximal and the **lateral malleolus** of the ankle is distal.

The small bones of the ankle and foot are best examined in an intact, articulated view (Figure 10.16). The ankle is formed by seven tarsal bones. The largest tarsal bone that forms the heel of the foot is the **calcaneus**. The bulge on the posterior portion of the calcaneus, the calcaneal tuberosity, is the attachment point for the **calcaneal tendon** (Achilles tendon). Proximal to the calcaneus and bracketed by the distal ends of the tibia and fibula is the **talus**. Anterior to the talus is the **navicular** bone. The final four tarsal bones form a row. From medial to lateral they are the **medial cuneiform**, **intermediate cuneiform**, **lateral cuneiform**, and **cuboid**.

The foot contains five elongated **metatarsal** bones. These bones lie between the tarsal bones of the ankle and the bones of the toes. The metatarsal bones are numbered 1 through 5, beginning medially. The toes contain a total of 14 bones, each of which is called a phalanx bone. Digit 1 of the foot, also called the **hallux**, has two phalanges which are called a *proximal phalanx bone* and a *distal phalanx bone*. Digits 2 through 5 (the little toe) have three phalanges each, between the proximal and distal phalanges is a third bone, the *middle phalanx*.

The purpose of this activity is to learn the bones and bony markings of the bones of the thigh, lower leg, and foot.

Figure 10.15 **Tibia and Fibula**

The tibia is the weight-bearing bone of the lower leg. On its lateral side it articulates with the fibula, which does not bear weight but has many muscle attachments.

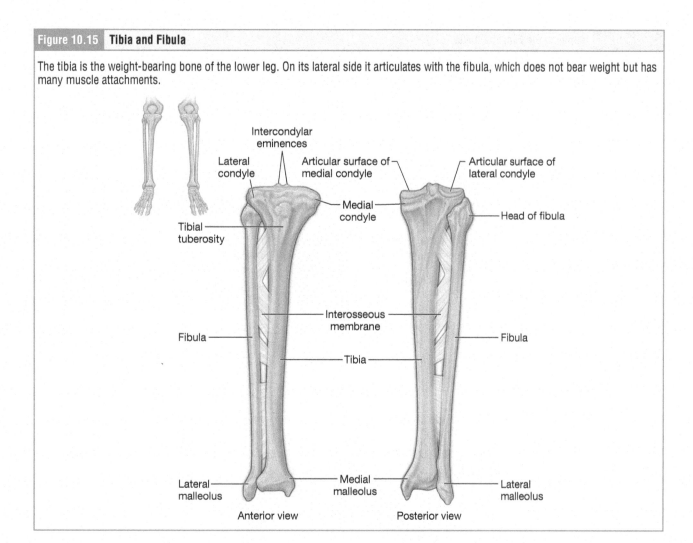

Intercondylar eminences

Lateral condyle

Articular surface of medial condyle

Articular surface of lateral condyle

Medial condyle

Head of fibula

Tibial tuberosity

Interosseous membrane

Fibula

Tibia

Fibula

Lateral malleolus

Medial malleolus

Lateral malleolus

Anterior view

Posterior view

Materials
- Articulated bones of the foot
- Articulated skeleton, if available
- Femur
- Fibula
- Patella
- Small sticks or pipe cleaners for pointing to bones and bone markings
- Tibia

Important Note: While it is tempting to use your pen or pencil to point to the many small structures on these bones, it is critical that you do not use any writing instrument to point or touch the bones, even if you think that you are being careful. Instead, use the small sticks or pipe cleaners provided by your instructor to avoid leaving any smears or marks on the bones.

Figure 10.16 **Bones of the Foot and Ankle**

The seven tarsal bones form the inferior ankle and posterior foot, shown in three views—(A) superior view, (B) medial view, and (C) lateral view. The arch of the foot is formed by the metatarsals and the toes are composed of phalanges.

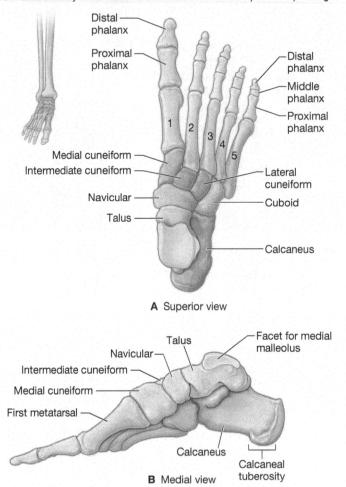

A Superior view

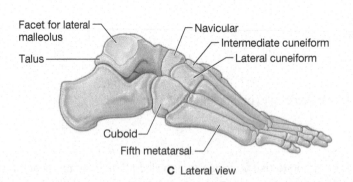

B Medial view

C Lateral view

Instructions

1. Obtain a right or left femur, patella, tibia, fibula, and articulated foot. If you are not sure how to determine whether you have the "matching" set of bones, work with your lab partner to identify the anatomical features that will allow you to determine whether the bones are right or left. Use an articulated skeleton, if available, to help you.

2. How were you able to distinguish between the left and right bones? Use the space below to take some notes on the anatomical features that were helpful.

3. Identify each of the bone markings listed in bold in the introduction on these bones and in Table 10.3.

Bone	Bone Markings	Bone	Bone Markings
Table 10.3 Bones and Markings of the Lower Limb			
	BONE OF THE THIGH		Intercondylar eminences
Femur	Head of the femur		Medial condyle of the tibia
	Fovea capitis		Lateral condyle of the tibia
	Neck of the femur		Tibial tuberosity
	Greater trochanter		Medial malleolus
	Lesser trochanter	Fibula	Head of the fibula
	Intertrochanteric line		Lateral malleolus
	Intertrochanteric crest		**BONES OF THE FOOT**
	Gluteal tuberosity	Calcaneus	
	Linea aspera	Talus	
	Lateral epicondyle of the femur	Navicular	
	Medial epicondyle of the femur	Medial cuneiform	
	Lateral condyle of the femur	Intermediate cuneiform	
	Medial condyle of the femur	Lateral cuneiform	
	Intercondylar fossa	Cuboid	
	Patellar surface	Metatarsals	
	BONES OF THE LOWER LEG	Phalanges	
Tibia	Articular surface of the medial condyle		
	Articular surface of the lateral condyle		

4. Hold or lay out the bones on the table to approximate anatomical position. Note the locations where these bones articulate and the names of the markings that meet to form these articulations.

 Learning Connection

Quick!

What are the two long bones in the lower leg? Did you pronounce them correctly? Sometimes students will inject the letter "L" into the word "tibia" when it is not there. Conversely, sometimes students will accidentally remove the letter "L" from the word "fibula." A helpful cue to remember where the L belongs and the relative position of the tibia and fibula in relation to each other is this phrase: "the fibuLa is Lateral!"

(page intentionally left blank)

Activity 10.6 The Articulations of the Lower Appendicular Skeleton

This activity targets LOs 10.1, 10.2, and 10.3.

Together, the bones of the pelvic girdle, the legs, and the feet articulate to provide a wide range of motion to the lower limbs of the body. This activity will allow you to "build" a hip, leg, and foot to visualize how these bones come together to form the lower appendicular skeleton.

Materials
- Articulated bones of the foot
- Articulated skeleton, if available
- Femur
- Fibula
- Os coxa
- Patella
- Tibia

Instructions

1. Obtain the bones listed here, working with your lab partner to make sure they are all either right or left. Use an articulated skeleton, if available, to help you.

2. How were you able to distinguish between the left and right bones? Use the space below to take some notes on the anatomical features that were helpful.

3. Assemble the bones into a hip, leg, and foot.

4. Roughly sketch the assembled bones in the space on the facing page.

5. Add in the sacrum (of the axial skeleton) to the sketch to show where the os coxa articulates with it.

6. Label the articulating features of each bone on your sketch.

(page intentionally left blank)

Activity 10.7 Case Study: A Broken Ankle

This activity targets LOs 10.1, 10.2, and 10.3.

Julia is out for a jog and steps on a small stick in the path. As the stick rolls, Julia's right foot rolls with it, twisting at the ankle as she falls. Luckily, Julia has her phone and calls her sister to come help her. While she hopes her injury isn't serious, the ankle is becoming swollen and painful; Julia's sister insists on a trip to the emergency room. While waiting to be seen in the ER, Julia asks her sister, an Anatomy and Physiology student, some questions about the bones of her ankle.

1. "What bones make up the place where my ankle pivots?" asks Julia. What is her sister's answer?

   a. The seven tarsal bones
   b. The metatarsals
   c. The tibia, fibula, and talus

2. "What is the bone of my heel called?" asks Julia. What is her sister's answer?

   a. The calcaneus
   b. The navicular
   c. The cuboid

Julia is eventually seen by an ER doctor. The doctor gently examines Julia's right ankle. The doctor explains that the ankle joint can be thought of as having three sides and a "roof" that box in the talus bone.

3. Which bone forms the "roof" and medial wall of the ankle?

   a. The tibia
   b. The fibula

4. Which bone forms the lateral and posterior wall of the ankle?

   a. The tibia
   b. The fibula

   The doctor is not able to determine whether there is only soft-tissue injury or a break to any bones of Julia's ankle. He explains that a soft-tissue injury such as a sprain can have similar symptoms to a bone break and orders a radiograph (X-ray) to determine if there is a break and where it is. Julia's radiograph, a posterior view of her right foot, is shown here.

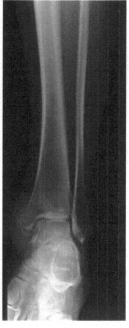

Scott Camazine/Science Source

5. Which bone marking, highlighted on the radiograph, is broken in Julia's ankle?

    a. Medial malleolus

    b. Lateral malleolus

    c. Tibial tuberosity

6. To which bones does this broken bone marking belong?

    a. The talus

    b. The tibia

    c. The fibula

The ER doctor refers Julia to an orthopedic surgeon who examines the radiographs and determines that there is enough misalignment (shift) of the broken piece of the bone that surgery is required to use surgical plates and screws to align the bones properly. This will prevent improper healing that could make it difficult for Julia to properly move her ankle.

Digging Deeper:
Chest Compressions during CPR

X-rays are a form of radiation that can pass through most objects, including the tissues of the human body. The X-rays that pass through the body can be detected on the other side to form images: the more dense the tissue is that the X-rays must pass through to reach the detector, the more are absorbed by the tissue, registering with higher contrast as a brighter "shadow" on the resulting image. Less dense structures allow more X-rays to pass and show as gray on the image; the least dense materials, such as air in the lungs, are black. Because the bones are more dense than other body tissues, they are easier to see on the resulting image, which is called a radiograph. Because of this, X-ray machines are good diagnostic tools for assessing the bones of the body.

    The radiograph in Figure 10.17 is of a left pectoral girdle and proximal humerus. Can you spot the fracture of the greater condyle of the humerus? What other bone markings are evident in this radiograph? Label these on the figure.

Figure 10.17 **Radiograph of Left Pectoral Girdle and Proximal Humerus**

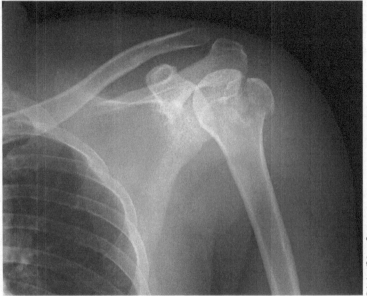

Rajaaisya/Science Source

Lab 10 Post-Lab Quiz

1. The pelvic girdle of the appendicular skeleton, is made up of which bones? Choose all correct answers.

   a. Sacrum

   b. Coccyx

   c. Os coxae

2. What structure of the scapula articulates with the clavicle?

   a. The acromion

   b. The coracoid process

   c. The glenoid cavity

3. The point of your elbow is sometimes called the "funny bone." What bone marking is this point of bone?

   a. The trochlea of the humerus

   b. The head of the radius

   c. The olecranon process of the ulna

4. Which carpal bone articulates with metatarsal 5 of the hand?

   a. Trapezium

   b. Hamate

   c. Lunate

5. Which bone, indicated with an arrow in this radiograph, is broken?

   a. Humerus

   b. Radius

   c. Ulna

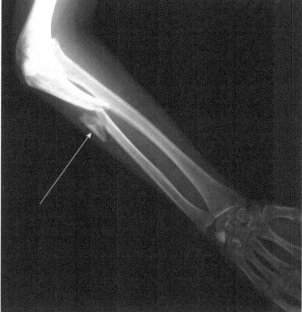

6. What bone articulates at the location shown in the image?

   a. Tibia
   b. Fibula
   c. Patella
   d. Femur

Anterior view

7. As a fetus passes into the birth canal during labor and delivery, its head must pass by the bony structures that make up the pelvic brim. What two bone markings make up the anterior and lateral borders of the pelvic brim?

   a. Lineal aspera and intertrochanteric line
   b. Arcuate line and pectineal line
   c. Superior ramus and inferior ramus of the pubis

8. What bone is the arrow pointing to in the figure

   a. Tarsal 1 of the left foot
   b. Proximal phalanx 1 of the left foot
   c. Metatarsal 1 of the left foot

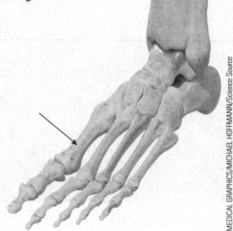

MEDICAL GRAPHICS/MICHAEL HOFFMANN/Science Source

9. A pelvic girdle that develops under the influence of estrogen will have a wider distance at location 4 in this image. The distance between what two bony points is being measured at location 4?

a. Ischial spines

b. Ischial tuberosities

c. Anterior superior iliac spines

d. Anterior inferior iliac spines

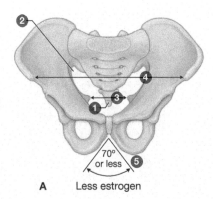

A   Less estrogen

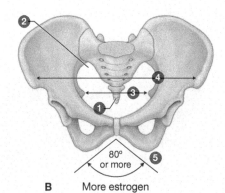

B   More estrogen

10. When we say a person has broken their hip, the break usually occurs at the location in the radiograph shown here. What bony structure is broken?

a. Ischial tuberosity

b. Acetabulum

c. Neck of the femur

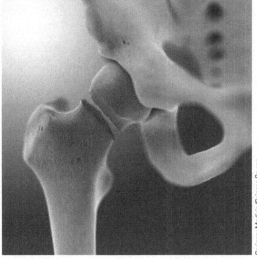

Springer Medizin/Science Source

Lab 11 | Joints

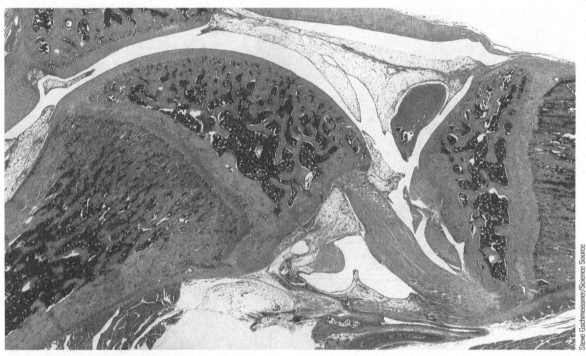

Steve Gschmeissner/Science Source

Learning Objectives: After completing this lab, you should be able to:

11.1 Describe the functional classification of joints (e.g., synarthrosis, diarthrosis) based on amount of movement permitted, and provide examples of each type.

11.2 Describe the anatomical classification of joints based on structure: fibrous (i.e., gomphosis, suture, syndesmosis), cartilaginous (i.e., symphysis, synchondrosis), and synovial (i.e., planar/gliding, hinge, pivot, condylar, saddle, ball-and-socket), and provide examples of each type.

11.3 For each of the six structural types of synovial joints, describe its anatomic features, identify locations in the body, and predict the kinds of movement each joint allows.

11.4 Explain the relationship between the anatomical classification and the functional classification of joints.

11.5 Identify and describe the major structural components of a typical synovial joint.

11.6 Define the movements that typically occur at a joint (e.g., flexion, extension, abduction, adduction, rotation, circumduction, inversion, eversion, protraction, retraction).

11.7* Perform a range of motion assessment with a goniometer at selected joints.

* Objective is not a HAPS Learning Goal.

The Human Anatomy and Physiology Society includes more than 1,700 educators who work together to promote excellence in the teaching of this subject area. The HAPS A&P Learning Outcomes measure student mastery of the content typically covered in a two-semester Human A&P curriculum at the undergraduate level. The full Learning Outcomes are available at https://www.hapsweb.org.

Introduction

A **joint** is any place in the body where adjacent bones or bone and cartilage come together to form a connection. The two tissues that come together are described as articulating with each other at that location. Between the two bones there is usually a **joint cavity**, which is filled with connective tissue. A joint cavity may be incredibly small so that the bones appear to touch each other, or it may be quite large so that several different connective tissues exist in the gap between the bones. Some joints are fixed, while others allow the bones to move in relationship to each other. Each time you move your body your muscles are pulling on your bones to produce a change in the angle of many moveable joints.

Because these joints involve bones, this lab will rely heavily on your ability to picture where the bones of the body are located, important markings on those bones, and how the bones fit together at the joints. You may want to review the bone sections of your lab manual and textbook in preparation for this lab.

Lab 11 Pre-Lab Quiz

This quiz will strengthen your background knowledge in preparation for this lab. For help answering the questions, use your resources to deepen your understanding. The best resource for help on the first five questions is your text, and the best resource for help answering the last five questions is to read the introduction section of each lab activity.

1. Anatomy of a typical synovial joint. Label the image with the following terms:

   Articular capsule
   Articular cartilage
   Ligament
   Synovial fluid
   Synovial membrane

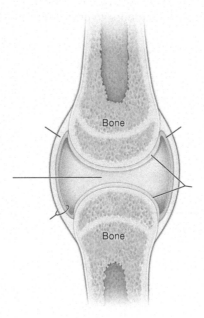

2. Match each of the roots to its definition:

Root	Meaning
	joint
	cartilage
	on both sides
	with or together
	twice

List of Roots

amphi
arthrosis
chondro
di
syn

3. Does movement occur at all joints?

   a. Yes
   b. No

4. What three bones articulate at the talocrural joint?

   a. The femur, patella, and tibia
   b. The tibia, fibula, and talus
   c. The talus, calcaneus, and navicular

5. Which of the following joints is an example of a synovial joint?

   a. The articulation between the parietal bones of the skull
   b. The articulation of the two pubic bones at the pubic symphysis
   c. The articulation between the head of the femur and the acetabulum

6. What functional classification describes all synovial joints?

   a. Diarthrotic
   b. Synarthrotic
   c. Amphiarthrotic

7. A synovial joint that produces movement in a single plane or axis is given what term?

   a. Uniaxial
   b. Biaxial
   c. Multiaxial

8. What is the most common type of joint in the body?

   a. Cartilaginous
   b. Fibrous
   c. Synovial

9. Which synovial joint is the largest in the human body?

   a. Elbow
   b. Hip
   c. Knee

10. What term describes how much movement occurs at a synovial joint?

   a. Range of motion
   b. Gomphosis
   c. Goniometer

(page intentionally left blank)

Activity 11.1 Joint Classification

This activity targets LOs 11.1, 11.2, 11.3, and 11.4.

Joints can be described and grouped either functionally or structurally. As you read through the descriptions of both classification systems, consider when it might be useful to compare joints based on how they work versus when it might be useful to compare them based on how they are anatomically built.

A functional classification of joints is primarily concerned with how the joint moves. A **synarthrosis** is a joint where the bones do not move or change position relative to each other. An **amphiarthrosis** is a joint where the bones can move slightly, while a **diarthrosis** provides a greater degree of motion at the joint. A few examples of each functional type of joint are shown in Figure 11.1.

A structural classification of joints considers what holds the bones together in the space between them. Three general types of structural joints exist: fibrous, cartilaginous, and synovial. At a **fibrous joint**, the bones are directly connected by fibrous (dense regular) connective tissue and there is no joint cavity. A **cartilaginous joint** does not have a joint cavity, either. Here, the space between the two bones is filled with cartilage. Finally, a **synovial joint** has a fluid-filled joint cavity between the two bones, each of which is covered with articular cartilage at their articulating surfaces. More detail about the complex anatomy of a synovial joint will be covered in a separate activity. These structural classifications are connected to the functional classifications: a fibrous joint may be a synarthrosis or amphiarthrosis; cartilaginous joints, also, may be synarthroses or amphiarthroses; and all synovial joints, due to their structure, are highly moveable diarthroses. Each of the three general types of structural joints has subcategories.

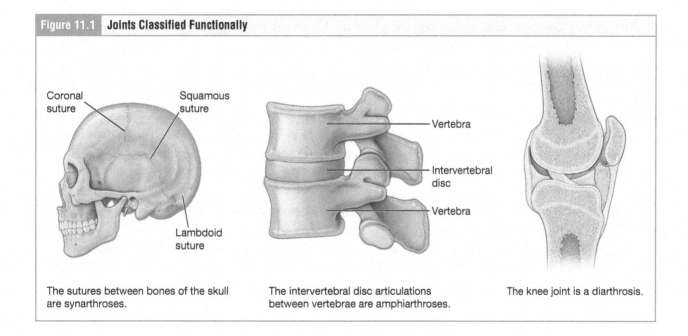

Figure 11.1 Joints Classified Functionally

Coronal suture / Squamous suture / Lambdoid suture

Vertebra / Intervertebral disc / Vertebra

The sutures between bones of the skull are synarthroses.

The intervertebral disc articulations between vertebrae are amphiarthroses.

The knee joint is a diarthrosis.

Figure 11.2 | Types of Fibrous Joints

Bones joined by a sheet of connective tissue are called fibrous joints. Examples include: (A) sutures between the skull bones, (B) an interosseous membrane, which spans the distance between the shafts of two bones, such as the radius and ulna of the forearm, and (C) a gomphosis, which is a specialized fibrous joint that ahchors a tooth in its bony socket.

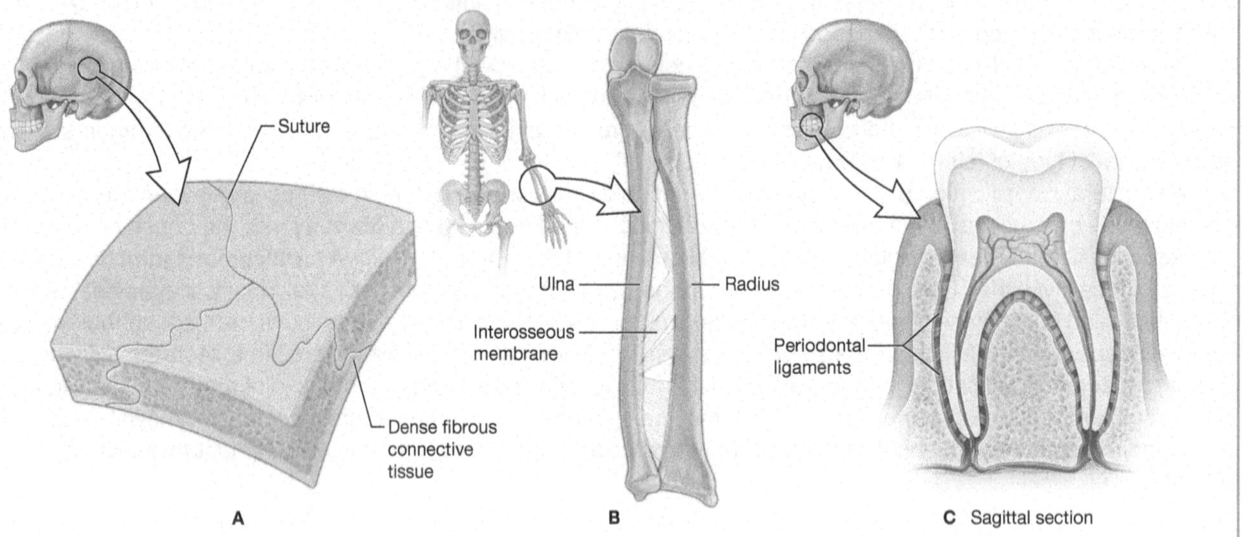

A

B

C Sagittal section

The primary function of fibrous joints is to firmly hold two bones together. There are three subcategories of fibrous joint. These subcategories are shown in Figure 11.2.

- **Sutures** are joints between the bones of the skull, as seen in both Figure 11.1 and Figure 11.2. These joints are immoveable and strong, as the sutures are wavy and there is only a narrow fiber-filled space between the adjoining bones.

- A **gomphosis** is a specialized joint that anchors the root of a tooth into its bony socket. Between the socket and the tooth are short strands of dense regular connective tissue that fix the tooth into the socket so that it does not move.

- **Syndesmoses** are formed between the long bones of the forearm and between the long bones of the lower leg. The long strands of dense regular connective tissue between the pair of bones allow for only a small amount of movement at these joints.

Cartilaginous joints are separated into two groups based on whether the tissue that joins them is hyaline cartilage or fibrocartilage.

- A **synchondrosis** is joined by hyaline cartilage and provides no opportunity for movement. These include the epiphyseal plates in children's bones, where the cartilaginous epiphyseal plate joins the epiphysis of the bone to the diaphysis. Synchondroses are also found where each rib connects to costal cartilage, as shown in Figure 11.3. These joints, called costochondral joints, are different from the synovial joints that connect the costal cartilage medially to the sternum.

- A **symphysis** contains a pad of fibrocartilage between the two bones, allowing for a small amount of movement at the joint. Both the pubic symphysis and the intervertebral joints between neighbor vertebrae are such joints.

There are six types of synovial joints. All types of synovial joints are highly moveable; their general shape and examples can be seen in Figure 11.4. Notice the arrows in the figure, which show how movement at a synovial joint may be in one plane (**uniaxial**), two planes (**biaxial)**, or multiple planes (**multiaxial**).

- At a **pivot joint** a rounded portion of a bone is enclosed within a ring formed partially by the other bone and par-tially by a ligament. The rounded bone rotates within this ring. The articulation of the atlas (C1) vertebra with the axis (C2) vertebra is a pivot joint, which allows movement in the horizontal plane when you shake your head "no." The proximal radioulnar joint is also a pivot joint.

Figure 11.3 **Cartilaginous Joints**

Cartilaginous joints include (A) synchondroses such as the epiphyseal plates and the costochondral joints, and (B) symphyses such as the intervertebral discs and pubic symphysis.

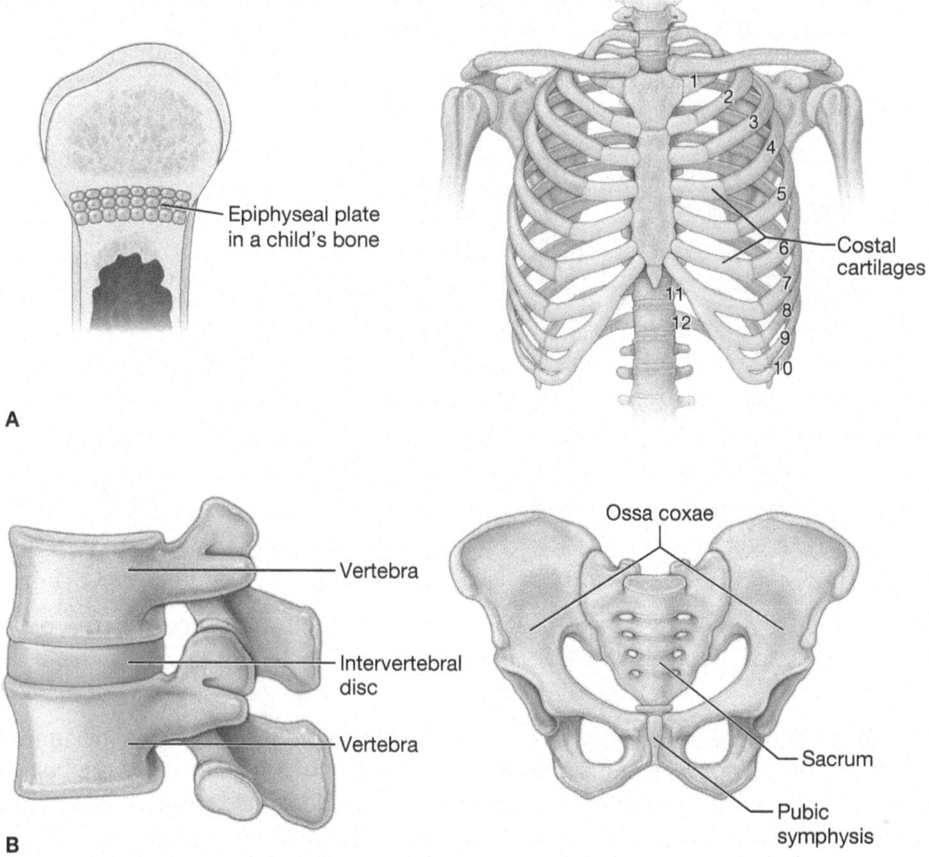

A

Epiphyseal plate in a child's bone

Costal cartilages

B

Vertebra

Intervertebral disc

Vertebra

Ossa coxae

Sacrum

Pubic symphysis

- In a **hinge joint** the convex end of one bone articulates with the concave end of the other. As for a pivot joint, this type of joint allows for movement in one plane. The trochlea of the humerus articulates with the trochlear notch of the ulna at a hinge joint. The knee and interphalangeal joints are also hinge joints.

- At a **condyloid** joint the shallow depression at the end of one bone articulates with a rounded structure on an adjacent bone or bones. The joints between each metacarpal and its proximal phalanx is a condyloid joint, which allows movement in two planes. Notice that you can both bend your digits toward your palm at these joints as well as spread your fingers apart from each other. The radiocarpal joint at your wrist is also a condyloid joint.

- At a **saddle joint** both articulating surfaces have a saddle shape which is concave in one direction and convex in the other. This allows the bones to fit together like a rider sitting on a saddle and allows movement in two planes. The articulation between metacarpal 1 and the trapezium, at the base of the thumb, is a saddle joint.

- At a **plane joint** the surfaces of the two bones are almost flat so that they can slide past each other. While it is possible to create movement in many planes at a plane joint, nearby ligaments constrain most movement to a single axis. The intercarpal joints and intertarsal joints are plane joints, as are the acromioclavicular joint, the joints between the superior and inferior articular processes of adjacent vertebrae, and the joints between the vertebral bodies and the heads of the ribs.

- A **ball-and-socket joint** has the greatest range of motion of all the synovial joint types. At these joints, the rounded head of one bone (the ball) fits into the bowl-shaped socket of another. The hip joint and the glenohumeral joint are both ball-and-socket joints.

Figure 11.4   **The Six Types of Synovial Joints**

The six types of synovial joints are specialized to accomplish different types of movement. (A) Pivot joints feature rotation around an axis, such as between the first and second cervical vertebrae, which allows for turning of the head side to side (as you might when you say "no"). (B) Hinge joints feature the rounded end of one bone that another bone moves around within one plane. (C) Condyloid joints feature one rounded bone end that is cupped within a bowl-like depression of another bone. (D) Saddle joints feature two curved bones, like saddles, that fit into each other complementarily. (E) Plane joints (such as those between the tarsal bones of the foot) feature flattened, articulating, bony surfaces on both bones, which allow for limited gliding movements. (F) Ball-and-socket joints (such as the hip and shoulder joints) feature a rounded head of a bone moving within a cup-shaped depression in another bone.

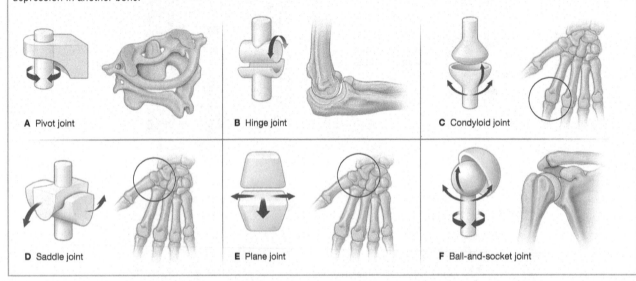

A Pivot joint   B Hinge joint   C Condyloid joint

D Saddle joint   E Plane joint   F Ball-and-socket joint

The temporomandibular joint of the jaw is a unique synovial joint that produces complex movements (Figure 11.5). As such, it can be considered a hinge joint or a plane joint depending on which movements are being examined.

The purpose of this activity is to allow you to synthesize the functional and structural classification systems, considering how each approach informs the other and practicing the terminology presented here.

Figure 11.5   **Temporomandibular Joint**

The temporomandibular joint is the hinge joint of the mouth, where the condyle of the mandible fits within the mandibular fossa of the temporal bone.

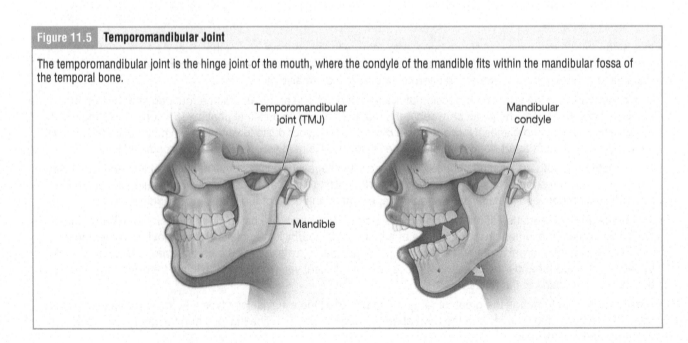

Temporomandibular joint (TMJ)

Mandibular condyle

Mandible

Instructions

1. Use the introduction, your textbook, and any other resources provided by your instructor to complete Table 11.1.

2. The first three columns will help you organize your thoughts about joints from a structural perspective.

   a. First consider each structural classification provided in the table; describe the general anatomic features that characterize that classification in the second column.

   b. In the third column, separate each type of general type of structural joint into its subcategories. Provide a description of each subcategory, as shown in the table for a suture. You will have to do some investigation in the reading of the introduction and in Figure 11.4 to complete the subcategories of synovial joints.

Learning Connection

Try Drawing It

Are you struggling to learn and connect all of the terms describing the classification of different joints? Map them out with a mind map. There are many different ways to map the classifications out. As you think about how you might represent it, you are allowing your brain to make connections and distinctions that will help you remember and understand the terms.

3. The fourth column will help you connect the functional classification system to the structural classification system. Again, you will have to do some investigation within the reading of the introduction to complete this column.

4. The fifth column is for writing in all the joints of the body that were given as examples within the reading.

5. Compare your completed table with your lab partner's, adding information as necessary to complete it.

Table 11.1 Joint Classifications

Structural Classification	Structural Features	Structural Subcategories	Functional Classification	Examples
Fibrous		*Suture—narrow, wavy fiber-filled space between bones*		
Cartilaginous				
Synovial		Uniaxial		
		Biaxial		
		Multiaxial		

Activity 11.2 Anatomy of a Synovial Joint

This activity targets LO 11.5.

Synovial joints are the most common type of joint in the body. They feature an enclosed space that contains fluid between the articulating bones. While each synovial joint of the body is somewhat unique, they share some common features, which are shown in Figure 11.6. Note that Figure 11.6 is representing these features generally and is not drawn to represent a specific joint in the human body.

The walls of the fluid-filled space are the **articular capsule** (joint capsule). The outer layer of this capsule is made of dense connective tissue. The inner layer of the capsule is lined with a **synovial membrane**. The cells of the synovial membrane secrete **synovial fluid**. This thick, slimy fluid provides lubrication to reduce friction between the bones as they move within the capsule. The synovial fluid keeps the bones in the joint from touching each other at the articulation.

Each bone in the joint is covered with **articular cartilage** at the articulating surface; usually this is hyaline cartilage. This cartilage protects the surface of the bone and makes these surfaces very smooth so that they can easily move within the joint cavity.

Ligaments tie the bones of the joint to each other. These dense regular connective tissue fibers reinforce the joint to keep it stable. At least one of the following types of ligaments will be found at a synovial joint. Ligaments that make up the outer layer of the articular capsule are called **intrinsic ligaments**. Other ligaments, external to the capsule, are called **extrinsic ligaments**. Some ligaments connect the bones within the capsule; these are **intracapsular ligaments**.

While the above features are found at every synovial joint, other anatomical features are found at some but not all synovial joints. These features are shown in Figure 11.7.

Many synovial joints are supported by more than just ligaments that tie the bones to each other. **Tendons**, which are made of dense regular connective tissue extending from the muscles near the joint, provide additional support.

Figure 11.6	Anatomical Features Present at All Synovial Joints

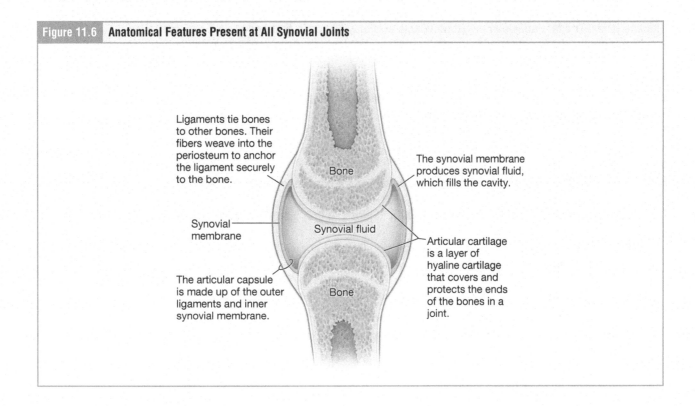

Ligaments tie bones to other bones. Their fibers weave into the periosteum to anchor the ligament securely to the bone.

The synovial membrane produces synovial fluid, which fills the cavity.

Bone

Synovial membrane

Synovial fluid

The articular capsule is made up of the outer ligaments and inner synovial membrane.

Bone

Articular cartilage is a layer of hyaline cartilage that covers and protects the ends of the bones in a joint.

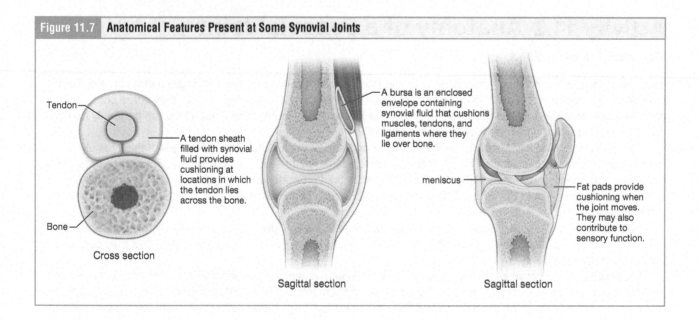

Figure 11.7 | **Anatomical Features Present at Some Synovial Joints**

Tendon

A tendon sheath filled with synovial fluid provides cushioning at locations in which the tendon lies across the bone.

Bone

Cross section

A bursa is an enclosed envelope containing synovial fluid that cushions muscles, tendons, and ligaments where they lie over bone.

meniscus

Fat pads provide cushioning when the joint moves. They may also contribute to sensory function.

Sagittal section

Sagittal section

A few synovial joints of the body have a pad of fibrocartilage between the bones within the articular capsule called a **meniscus** (or articular disc). The meniscus can smooth the movement between the two bones, provide shock absorption, and create a better fit between the shapes of the articulating surfaces of the bones.

External to the articular capsule, some joints have structures to smooth and separate the many moving parts and reduce friction. A **bursa** or bursae (plural) is a small sac of synovial fluid that creates a cushion where ligaments, muscles, or tendons might rub against bone. A **tendon sheath** is an elongated version of a bursa that wraps around a muscle tendon where the tendon crosses the joint. A **fat pad** is a small accumulation of adipose tissue that occurs at some joints. The fat pad is somewhat moveable and may fill space that arises as a joint moves.

The purpose of this activity is to locate the typical joint features described above within actual joints of the body.

Materials
• Anatomical models of joints (such as the shoulder, knee, hip, and/or elbow)
• Fresh joint specimens such as pig feet (if available)
• Protective gown, gloves, and eye protection for examining fresh joint specimens

Instructions
1. Working with your lab partner, obtain an anatomical joint model or fresh joint specimen.

2. Locate all features listed above and shown in Figure 11.6 that are typical to all synovial joints.

3. What joint features from Figure 11.7 can you also find on your joint? Use your textbook and other resources provided by your instructor to guide you in which of these features to look for.

4. Sketch the joint in the space provided on the next page.

   a. Identify which joint you are sketching in the upper left-hand corner.
   b. Use all the space provided to draw the joint as large as possible.
   c. Label all found joint structures.

5. If time permits, obtain a second, different model or specimen and repeat steps 2–4.

Joint:

Joint:

(page intentionally left blank)

Activity 11.3 The Anatomy of the Knee Joint

This activity targets LO 11.5.

We will examine the knee joint to discover how the typical joint structures look and function in a specific joint in the human body. The knee joint (Figure 11.8) is the largest joint of the body. It is an articulation between the lateral and medial condyles of the femur with the lateral and medial condyles of the tibia; the patella is also involved in the joint where it articulates with the patellar surface of the distal femur.

Figure 11.8 | **Knee Joint**

(A) The knee joint is the articulation of the femur, the patella, and the tibia. (B) The knee is supported by fibrocartilage discs called *menisci* and held in place by cruciate ligaments and (C) collateral ligaments.

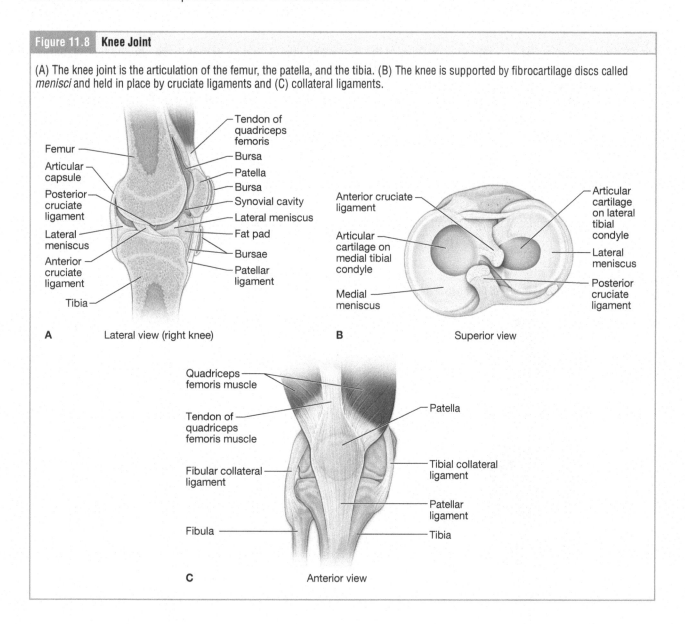

A   Lateral view (right knee)

B   Superior view

C   Anterior view

The knee is stabilized by many ligaments and tendons.

- Posteriorly within the articular capsule, the **posterior cruciate ligament** (PCL) reaches from the posterior tibial plateau to the medial femoral condyle. The PCL prevents the tibia from sliding posteriorly at the joint.

- Anteriorly within the articular capsule, the **anterior cruciate ligament** (ACL) reaches from the anterior tibial plateau to the lateral femoral condyle. The ACL prevents the knee from hyperextending. Together, the PCL and ACL form an X shape within the articular capsule.

- Laterally, the **fibular collateral ligament** attaches from the head of the fibula to the lateral surface of the lateral condyle of the femur. This ligament reinforces the lateral aspect of the joint.

- Medially, the **tibial collateral ligament** attaches from the medial surface of the medial condyle of the tibia to the medial surface of the medial condyle of the femur. This ligament provides strength to the medial aspect of the joint.

- Forming the superior, anterior portion of the articular capsule is the **quadriceps femoris tendon** attaching to the patella.

- Forming the inferior, anterior portion of the articular capsule is the **patellar ligament** reaching from the inferior base of the patella to the tibial tuberosity.

Two C-shaped menisci sit between the condyles of the femur and tibia. The **medial meniscus** and the **lateral meniscus** stabilize and cushion the joint. Several fat pads and bursae surround the patella, distributing forces and reducing the friction that occurs in this area as the joint is used.

The purpose of this activity is to locate and learn the structures of the knee joint, connecting the functions of these structures to their specific locations within the knee.

Materials
• Model of the knee joint, if available, or photo of the knee joint

Instructions
1. Examine the model of the knee joint, if available. Also examine the photograph of a model of the knee shown in Figure 11.9.

| Figure 11.9 | **Anatomical Model of the Knee Joint** |

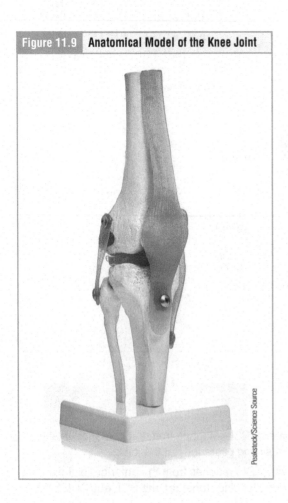

Peakstock/Science Source

2. Using Figure 11.8 and the introduction above, find the following structures on the model and label the photograph of the knee joint with the following terms.

- Femur
- Tibia
- Patella
- Fibula
- Capsular space
- Anterior cruciate ligament
- Posterior cruciate ligament
- Fibular collateral ligament

- Tibial collateral ligament
- Quadriceps femoris tendon
- Patellar ligament
- Medial meniscus
- Lateral meniscus
- Fat pads
- Bursae

3. Not all the structures will be present on the model or visible in the photograph.

   a. Some structures are not visible because they are not present. Sketch them in and label them.

   b. Some structures are not visible because the anteromedial orientation of the photograph does not show them. Sketch a posterior view of the knee joint next to Figure 11.9 and label the visible structures.

4. For each structure labeled on the photograph and your posterior view sketch, write a brief description of the function of that structure under your label.

5. Examine the light micrograph of a section through a knee joint shown in Figure 11.10. This is a lateral view. Which structures from the above list can you locate? Work with your lab partner to label as many structures as possible on the image.

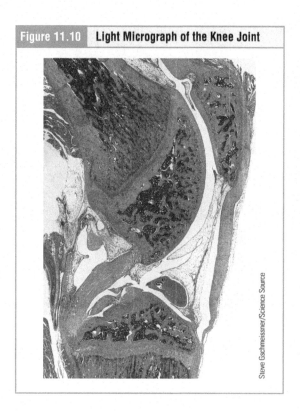

Figure 11.10   Light Micrograph of the Knee Joint

Steve Gschmeissner/Science Source

(page intentionally left blank)

Activity 11.4 Movements at Synovial Joints

This activity targets LOs 11.3 and 11.6.

Synovial joints enable the body to have a tremendous range of movements. The type of movement that can be produced at a synovial joint is determined by its structural type. In some regions of the body, several joints may work together to produce a particular movement. These movements are best described using directional terms and anatomical positions of the body; review this material in your textbook if you need a refresher. Examine Figure 11.11 and Figure 11.12 as you read through the descriptions of these movements.

Figure 11.11	Movements at the Joints of the Body

(A–B) *Flexion* decreases a joint angle and *extension* brings the joint back to its resting position (usually back to a straight line).
(C–D) Extension beyond the straight line, increasing the joint angle to greater than 180 degrees, is called *hyperextension*.
(E) *Abduction* (moving the structure away from the body, or spreading the fingers and toes) and *adduction* (bringing the structure toward the body or bringing the fingers and toes together) are motions of the limbs, hands, fingers, or toes. *Circumduction* (moving a structure in a circular pattern) combines flexion, adduction, extension, and abduction. (F) Turning of the head or twisting of the trunk is described as *rotation*. The limbs are capable of medial and lateral rotation if they turn the anterior side toward or away from the midline. (G) The radius and ulna are parallel in anatomical position, which is called *supination*. When turning the hand so that the palm faces posteriorly, the radius must rotate over the ulna, a movement called *pronation*.

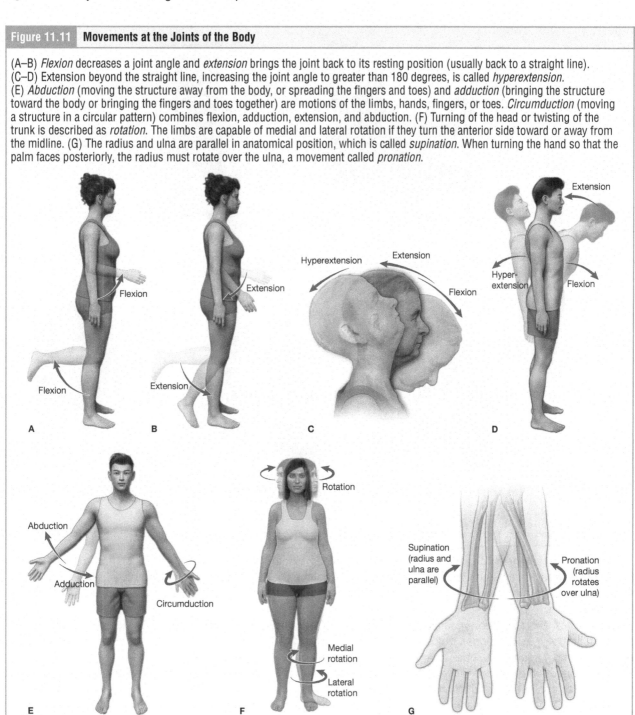

Figure 11.12   **More Movements at the Joints of the Body**

(H) The ankle is a hinge joint capable of two movements: *dorsiflexion*, in which the top of the foot moves closer to the anterior leg, and *plantar flexion*, in which the foot and leg approach a straight line, often referred to as pointing the toes. (I) Eversion and inversion of the foot are accomplished through the plane joints of the tarsal bones. *Eversion* moves the foot so that the sole of the foot is facing laterally, and *inversion* moves the foot so that the sole faces medially. (J) *Protraction* of the mandible pushes the chin forward as the mandible moves forward, and *retraction* returns the mandible to its resting position. (K) The opening and closing of the mouth is accomplished through *depression of the mandible* (opening) and *elevation* (closing). (L) *Opposition* moves the thumb so that it can contact the fingers, and *reposition* restores anatomical position.

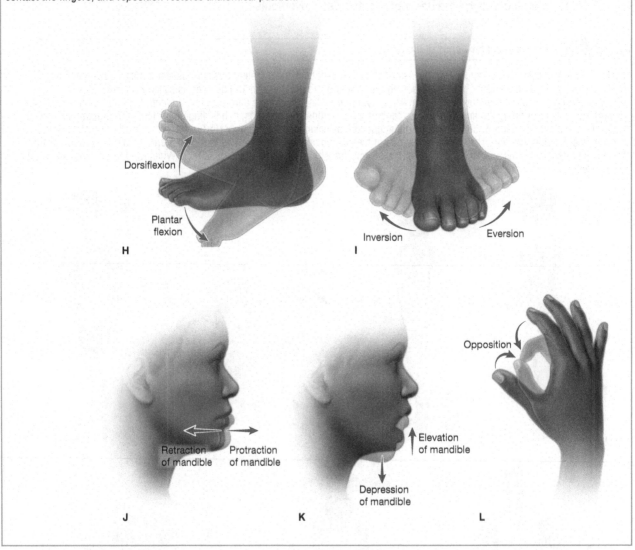

Dorsiflexion

Plantar flexion

H

Inversion   Eversion

I

Retraction of mandible   Protraction of mandible

J

Elevation of mandible

Depression of mandible

K

Opposition

L

Flexion and **extension** change the angle of a joint along the sagittal plane (anterior or posterior to the body). Flexion decreases a joint angle and extension brings the joint back to its resting position (usually back to a straight line). Extension beyond the straight line, increasing the joint angle to greater than 180 degrees, is called **hyperextension.**

Abduction and **adduction** motions occur on the limbs along the coronal plane (lateral to the body). Abduction is a motion that pulls a limb, finger, or toe away from the middle of the body. Adduction is the opposing movement that brings the limb, finger, or toe back toward the body or even past the midline of the body.

Circumduction describes movement in a circular way so that a limb is moved sequentially through flexion, adduction, extension, and abduction.

Rotation at a joint is a twisting movement. This type of movement can occur at the neck and torso due to the summating of the small movements at each vertebral joint. It can also occur at the ball-and-socket joints of the shoulder and hip as the bone rotates around the long axis to turn in the socket. At a ball-and-socket joint, movement that brings the anterior surface of the limb toward the midline is called **medial rotation** and movement that brings the anterior surface away from the midline is called **lateral rotation**.

Supination and **pronation** are specialized movements that only occur at the forearm. Supination is the motion that rotates the forearm so that the radius and ulna are parallel to each other and so that the palm faces anteriorly. Pronation is the motion that moves the radius so that it forms an X with the ulna, turning the palm to face posteriorly in the process. Notice that this pair of movements is occurring as the head and shaft of the radius rotate; the bony articulations at the wrist are not moving to any great extent.

The ankle joint is a hinge joint; two pair of movements occur only here. Lifting the front of the foot so that the top of the foot moves toward the anterior surface of the leg is **dorsiflexion**. Pointing the toes downward is **plantar flexion**.

The multiple plane joints among the tarsal bones of the foot work to produce inversion and eversion. **Inversion** is the turning of the foot so that the plantar surface (bottom of the foot) faces medially. **Eversion** is the turning of the foot so that the plantar surface faces laterally. Like the dorsiflexion and plantar flexion of the ankle joint, inversion and eversion at the intertarsal joints are specific to this location and do not occur in other areas of the body.

The thumbs of the hands have a special pair of movements, as well. **Opposition** brings the tip of the thumb in contact with the tip of the finger at the first carpometacarpal joint. Returning the thumb to its anatomical position is called **reposition**.

Two pair of movements describe the ways in which the mandible or the scapula can be moved. Protraction and retraction are anterior-posterior movements. **Protraction** moves the bone anteriorly and **retraction** moves the bone posteriorly. **Depression** and **elevation** are downward and upward movements of these two bones.

Instructions

1. Work with your lab partner to perform the actions listed below and complete Table 11.2.

2. For each action, one of you will perform the action while the other observes the action. Swap roles for every other action so that each of you gets a chance to perform each role.

3. Identify all the joints that are moving during the action.

4. Then, for each joint, identify the particular movements that are happening.

5. Use this lab manual, your textbook, and any other resources provided by your instructor to help.

6. The first action in the table has been done as an example.

7. If directed by your instructor, perform and assess other actions described by your instructor. Add them to the final rows of Table 11.2.

8. When you have completed the table, join with another lab group and compare tables. Do your tables agree? Did you miss any joints or movements? Discuss and add the missing pieces to your table.

Learning Connection

Are you using the word "flex" appropriately? In the past you may have used the phrase "flex your muscles" to refer to the shortening or contraction of a muscle. But note that this word does not refer to just any muscle in your body shortening! And those shortening muscles are causing movement at a joint! In A&P, the word flex is describing a very particular type of movement at a joint. Which of the following phrases correctly use the word and which don't? Circle the correct phrases and cross out the incorrect ones.

"I flex at the elbow joint to touch my finger to my nose."

"I flex my muscles to pick up a heavy box off the ground."

"I flex my neck to look up at the ceiling."

"I flex at the hip as I sit down."

Table 11.2 Joint Movements During Typical Body Actions		
Action	**Joints That Are Moving**	**Specific Movements at Each Joint**
With elbows at your sides, placing your palms together and raising them to your face in a praying motion	Shoulder (glenohumeral joint)	Flexion
	Elbow (humeroulnar)	Flexion
	Elbow (humeroradial)	Flexion
	Radioulnar	Pronation
	Wrist (radiocarpal)	Extension
Taking a bite and chewing (pretend you are biting and chewing a very tough piece of food, such as a piece of licorice)		
Shrugging your shoulders up and down		
Reaching toward your toes and standing back up (like a toe-touch; assume no bend at the knee or ankle)		
Walking (observe only the joints of the lower body, ignore any movements at the upper torso or arms)		

Activity 11.5 **Range of Motion**

This activity targets LOs 11.6 and 11.7.

Synovial joints vary in the range of motion (ROM) they allow. The ligaments, tendons, and the bony structures of the articulations combine to create anatomical structures with unique constraints. A goniometer is an inexpensive tool used to measure the ROM at a joint. Physical therapists use goniometry to assess a joint to determine whether ROM has been compromised. If an intervention is being applied to a joint, an initial measurement and follow-up measurements at subsequent treatment sessions can determine whether the intervention is effective.

A typical goniometer is shown in Figure 11.13. The tool has a stationary arm that is fixed to either a circle or half circle with angle degrees marked on it, this is the fulcrum. The tool also has a moving arm that can pivot at the fulcrum. As shown in the figure, the stationary arm is placed along the non-moving body part and the fulcrum is placed at the center of the joint. While the fulcrum of the goniometer is held trapped against the joint, the moveable limb is moved through its range of motion. Then the moveable arm of the goniometer is pivoted to align with the just-moved limb. The moving arm of the goniometer is now aligned with a marking on the fulcrum identifying the angle of degrees through which the limb just moved. For example, in Figure 11.13, the scale reads 110. This means that the ROM of shoulder abduction by this patient is 110 degrees. When reading the ROM, remember that 90 degrees is a right angle and 180 degrees is flat. This will help you double-check your measurement to make sure you are correctly reading the scale on the fulcrum.

Figure 11.13	A Universal Goniometer in Use to Measure Shoulder Abduction

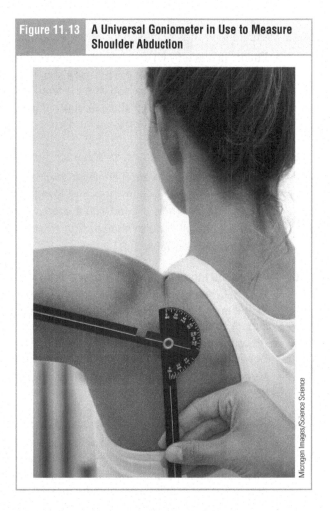

Microgen Images/Science Science

Materials
• Plastic universal goniometer for measuring joints (such as the one shown in Figure 11.13)

Instructions
1. The goniometer may or may not look exactly like the one in Figure 11.13. Examine the goniometer you have.

   a. Find the stationary arm and the moving arm on the goniometer.

   b. Examine the scale on the fulcrum. Different goniometers have slightly different scales, so take some time to become familiar with the scale on the goniometer that you are using. If you have questions about how to read the scale, ask your instructor for help.

2. Before beginning your measurements with the goniometer, consider each of the joint movements listed in the first column of Table 11.3.

   a. Review your lab manual, textbook, and/or other resources to remember how each movement occurs.

   b. Make a small sketch of the movement in column 2 for each listed movement.

   c. The first one has been done for you.

3. Working with your lab partner, one of you will be the patient and the other will be the assessor. Then you will switch roles. When you have completed measuring all of the ROMs described below and filled in Table 11.3 you will have two measurements at each joint, one for each student.

4. Measure the range of motion for *shoulder flexion*.
   a. Patient is seated with arm hanging loosely at the side. Palm faces medially.
   b. Place the goniometer on the patient's shoulder aligning the fulcrum with the center of the humeral head. The stationary arm is aligned with the torso and the moving arm is aligned with the humerus.
   c. Have the patient perform shoulder flexion as far as possible.
   d. Write the ROM in Table 11.3.

5. Measure the range of motion for *shoulder hyperextension*.
   a. Patient is seated with arm hanging loosely at the side. Palm faces medially.
   b. Place the goniometer on the patient's shoulder aligning the fulcrum with the center of the humeral head. The stationary arm is aligned with the torso and the moving arm is aligned with the humerus.
   c. Have the patient perform shoulder hyperextension as far as possible.
   d. Write the ROM in Table 11.3.

6. Measure the range of motion for *wrist flexion*.
   a. Patient is seated next to a table with their forearm resting on the table and the wrist just over the edge of the table so that the wrist and hand are hanging off the edge. Begin with the hand aligned with the arm, fingers straight but relaxed.
   b. Place the goniometer on the patient's lateral wrist, aligning the fulcrum with the triquetrum. The stationary arm is aligned with the ulna and the moving arm is aligned with metacarpal 5.
   c. Have the patient perform wrist flexion as far as possible.
   d. Write the ROM in Table 11.3.

7. Measure the range of motion for *wrist hyperextension*.
   a. Patient is seated next to a table with their forearm, wrist, and palm resting on the table. Begin with the hand aligned with the arm, fingers straight but relaxed.
   b. Place the goniometer on the patient's lateral wrist, aligning the fulcrum with the triquetrum. The stationary arm is aligned with the ulna and the moving arm is aligned with metacarpal 5.
   c. Have the patient perform wrist extension as far as possible.
   d. Write the ROM in Table 11.3.

8. If directed by your instructor, measure the ROM of other joint movements. Your instructor will provide you with directions for how to perform these assessments. Add them to the final rows of Table 11.3.

9. Answer these questions:
   a. Compare the ROM of shoulder flexion to that of shoulder hyperextension. Which has more ROM? Do some research in your textbook or other available resources. What structures in the shoulder joint reduce motion in the movement that has less ROM?
   b. How do the ROMs of you, your lab partner, and normal values compare? What are some possible explanations for differences and similarities between your measurements?

Table 11.3 ROM at Selected Joints				
Joint Movement	Sketch of Movement	Student #1 ROM	Student #2 ROM	Normal ROM for Adults at a Healthy Joint
Shoulder flexion				150 to 180 degrees
Shoulder hyperextension				50 to 60 degrees
Wrist flexion				60 to 80 degrees
Wrist hyperextension				60 to 70 degrees

(page intentionally left blank)

Activity 11.6 Case Study: Sprained Knee

This activity targets LOs 11.3 and 11.5.

Zola is playing soccer with friends when she falls and twists her knee. She gets back up, but her knee is weak and she is having trouble walking. Zola visits her doctor, hoping that the damage is not serious.

   The doctor examines Zola's swollen knee. With Zola sitting on the examining table with her legs hanging down, the doctor performs several tests to determine if there is a tear to one of the ligaments of the knee.

1. First, the doctor tugs anteriorly on Zola's tibia. Which ligament should prevent too much forward movement of the tibia at the knee joint?

   a. Anterior cruciate ligament
   b. Posterior cruciate ligament
   c. Fibular collateral ligament
   d. Tibial collateral ligament

2. Next, the doctor pushes posteriorly on Zola's tibia. Which ligament should prevent too much backward movement of the tibia at the knee joint?

   a. Anterior cruciate ligament
   b. Posterior cruciate ligament
   c. Fibular collateral ligament
   d. Tibial collateral ligament

Both tests are negative, and further assessment by the doctor confirms that all four of the important ligaments at the knee are intact. The doctor is concerned about the swelling at the knee and requests an MRI to visualize the soft tissues of the knee joint. The results are shown in Figure 11.14. The blue highlighted areas of the MRI shown here are synovial fluid.

3. What structure at the joint produces the synovial fluid that fills the articular cavity?

   a. The synovial membrane
   b. The bursae
   c. The articular cartilage

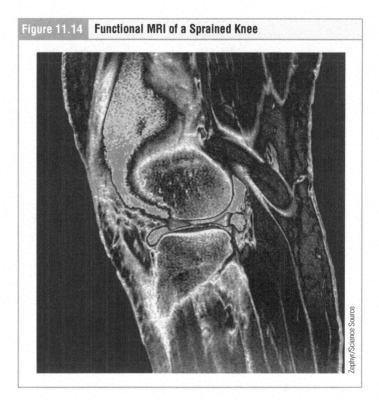

Figure 11.14 Functional MRI of a Sprained Knee

Zephyr/Science Source

4. Is the synovial fluid in the MRI contained within the articular cavity?

   a. Yes
   b. No

5. Which of the following structures must have ruptured to allow the synovial fluid to leak?

   a. A meniscus
   b. A fat pad
   c. The articular capsule

Because there was significant fluid around Zola's knee, surgical intervention was required to remove the fluid and repair the tear. The doctor told Zola that she would need to be careful and rest the knee so that the tear could heal.

 Digging Deeper:
Osteoarthritis and Physical Therapy

Osteoarthritis is inflammation at a joint as the articular cartilage at a joint deteriorates. The knee and hip joints are often affected by osteoarthritis. Osteoarthritis can make a joint stiff and painful, so physical therapists work to help patients increase mobility at the joint, choose body positions that reduce the stress at the affected joint, and choose assistive devices such as walkers or a cane to further take pressure off a joint.

Lab 11 Post-Lab Quiz

1. Your instructor hands you a skull. Which type or types of fibrous joints will you be able to locate on a skull? Choose all correct answers.

   a. Gomphosis
   b. Suture
   c. Syndesmosis

2. Which joint classification below does not have cartilage filling the entire joint cavity?

   a. Synchondrosis
   b. Symphysis
   c. Syndesmosis

3. The histological slide seen here is fibrocartilage. From which type of joint could it have been taken?

   a. An intervertebral joint
   b. An intercarpal joint
   c. A carpometacarpal joint

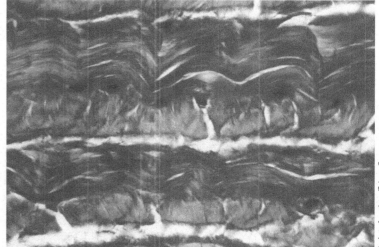

Biophoto Associates/Science Source

4. Which kind of joint will be found in a child's body but absent in an adult's body?

   a. Gomphoses
   b. Condylar joints
   c. Synchondroses

5. Which of the following correctly describes a plane joint?

   a. One bone rotates around another bone
   b. One bone slides across the surface of another bone
   c. One bone rests within a cup-shaped depression in another bone

6. What three bones articulate at the knee joint?

   a. The tibia, the fibula, and the femur
   b. The femur, the tibia, and the patella
   c. The humerus, the femur, and the calcaneus

7. Bone articulations can be held immobile by which types of tissue? Choose all correct answers.

   a. Dense regular connective tissue
   b. Cartilage
   c. Synovial fluid

8. Which of the following structures may be found outside the articular capsule of a synovial joint? Choose all correct answers.

   a. Fat pad
   b. Meniscus
   c. Bursa
   d. Ligament

9. What type of movement is shown in this exercise that moves one leg laterally away from the other leg?

   a. Flexion at the hip
   b. Extension at the hip
   c. Adduction at the hip
   d. Abduction at the hip

MedicalRF/Science Source

10. The movement called opposition occurs at what type of synovial joint?

   a. Saddle joint
   b. Plane joint
   c. Pivot joint

Lab 12 Muscle Tissue

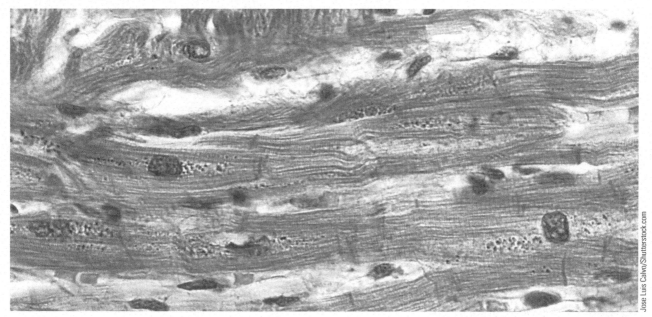

Jose Luis Calvo/Shutterstock.com

Learning Objectives: After completing this lab, you should be able to:

12.1 Describe the organization of skeletal muscle, from cell (skeletal muscle fiber) to whole muscle.

12.2 Name the connective tissue layers that surround each skeletal muscle fiber, fascicle, entire muscle, and group of muscles and indicate the specific type of connective tissue that composes each of these layers.

12.3 Describe the components within a skeletal muscle fiber (e.g., sarcolemma, transverse [T] tubules, sarcoplasmic reticulum, myofibrils, thick [myosin] myofilaments, thin [actin] myofilaments, troponin, tropomyosin).

12.4 Define sarcomere.

12.5 Describe the structure, location in the body, and function of skeletal, cardiac, and smooth muscle.

12.6 Compare and contrast the general microscopic characteristics of skeletal, cardiac, and smooth muscle.

12.7 Define the sliding filament theory of skeletal muscle contraction.

12.8 Describe the sequence of events involved in the contraction of a skeletal muscle fiber, including events at the neuromuscular junction, excitation–contraction coupling, and cross-bridge cycling.

12.9 Describe the sequence of events involved in skeletal muscle relaxation.

12.10 Given a factor or situation (e.g., muscular dystrophy), predict the changes that could occur in the muscular system and the consequences of those changes (i.e., given a cause, state a possible effect).

The Human Anatomy and Physiology Society includes more than 1,700 educators who work together to promote excellence in the teaching of this subject area. The HAPS A&P Learning Outcomes measure student mastery of the content typically covered in a two-semester Human A&P curriculum at the undergraduate level. The full Learning Outcomes are available at https://www.hapsweb.org.

Introduction

The defining quality of muscle tissue is its ability to contract, meaning that it shortens while generating force. Most often, when we think of muscle tissue, we think of skeletal muscle tissue: the muscles that are attached to our bones and allow for movements of the body. But there are three types of muscle tissue. In addition to skeletal muscle, there are smooth muscle and cardiac muscle. Smooth muscle lines the walls of hollow organs like the vagina, uterus, blood vessels, and gastrointestinal tract. Cardiac muscle is found only in the walls of the heart. In this lab, we will explore the properties of muscle by examining skeletal muscle.

The function of muscle is to contract (and then relax, so it can contract again). The ability of muscle to contract is a function of the coordinated movement of specific groups of proteins in muscle cells. All three types of muscle utilize these contractile proteins, alternating arrangements of thick and thin filaments, for contraction. The term **functional unit** means the smallest unit that is capable of the function of the whole. In this case, the smallest unit of a skeletal muscle cell that is capable of contracting is the **sarcomere**. Sarcomeres are incredibly small; you would need to stack 45 sarcomeres end-to-end to build the thickness of a sheet of printer paper. Imagine how many sarcomeres, laid end to end, it would take to reach the length of a skeletal muscle cell, which can be inches long. All three types of muscle tissue also exhibit excitability in addition to contractility. The property of excitability means that they change from a relaxed to a contracted state in response to electrical signals.

Within a muscle cell, sarcomeres are arranged in long columns called **myofibrils**. Skeletal muscle cells are packed with myofibrils (Figure 12.1). The number of myofibrils within a muscle cell is variable. As we gain strength, muscle cells add myofibrils and become larger in diameter. Losing strength occurs for the opposite reason: muscle cells decrease in both diameter and myofibril number.

Figure 12.1	The Organization of a Muscle

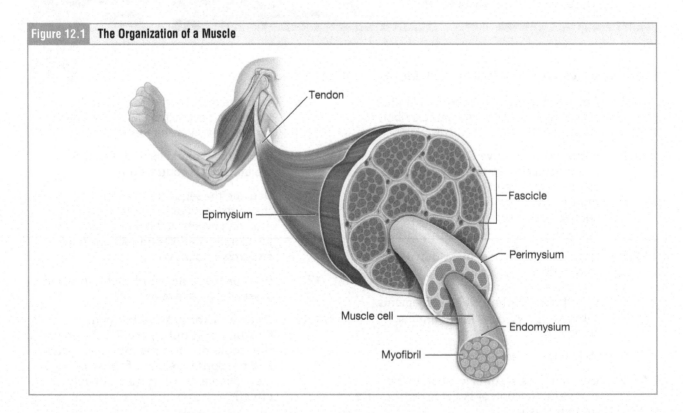

Tendon
Fascicle
Epimysium
Perimysium
Muscle cell
Endomysium
Myofibril

While all three types of muscle achieve contraction using the same contractile proteins, significant differences exist among the three muscle types in terms of how these components are organized. Skeletal muscle cells are long, multinucleated structures that comprise skeletal muscle. Skeletal muscle cells are typically as long or almost as long as the muscle organ they are arranged in; for this reason they are often referred to as *muscle fibers*, a synonym for skeletal muscle cells. These long cells result from the fusion of many individual cells, resulting in a long cell with many nuclei. The sarcomeres of neighboring muscle cells are precisely arranged, leading to patterns, like stripes, along the length of muscle tissue (Figure 12.2A).

Figure 12.2 **The Three Types of Muscle Tissue**

Three types of muscle tissue are found in the body: (A) Skeletal muscle (associated with the bones), (B) cardiac muscle (found only in the heart), and (C) smooth muscle (found in the walls of tubular organs).

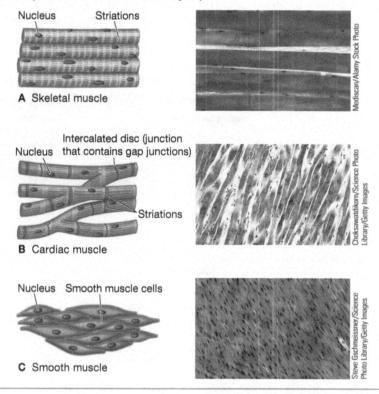

A Skeletal muscle

B Cardiac muscle

C Smooth muscle

Cardiac muscle cells have one or two nuclei per cell and are physically and electrically connected to each other via gap junctions. The gap junctions permit ions to move from one cell to another. Therefore, when one or a small group of cardiac muscle cells experiences an electrical change, the charged ions of that change will diffuse through gap junctions to the neighboring cells. Because of this, cardiac muscle cells contract as a sheet. Cardiac muscle tissue, like skeletal muscle tissue, appears striated under the microscope due to the precise arrangement of the sarcomeres (Figure 12.2B). Cardiac muscle cells are forked at one end, resulting in a shape like the letter Y.

Smooth muscle earns its name because its contractile proteins are not arranged in a pattern, and therefore this tissue does not have the same striped appearance. Smooth muscle cells are small and football-shaped. They each have a single nucleus. Most smooth muscle contracts in a coordinated, sheet-like fashion, though some does not (Figure 12.2C).

In this lab we will explore the structure and function of muscle cells and whole muscle organs.

Lab 12 Pre-Lab Quiz

This quiz will strengthen your background knowledge in preparation for this lab. For help answering the questions, use your resources to deepen your understanding. The best resource for help on the first five questions is your text, and the best resource for help answering the last five questions is to read the introduction section of each lab activity.

1. Anatomy of a sarcomere. Label the figure of a sarcomere with the following terms:

   actin

   M line

   myosin

   thick filament

   thin filament

   Z line

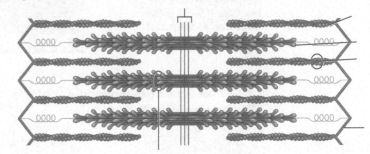

2. Match each of the roots to its definition:

Root	Meaning
	muscle
	bundle of sticks
	flesh

List of Roots

fasci-

myo-

sarco-

3. There are three types of muscle: cardiac, skeletal, and smooth. Fill in the blanks to match the type of muscle with the description:

   The wall of the esophagus (the stretchy tube that connects the throat to the stomach) _____

   The wall of the right atrium (one of the receiving chambers of the heart) _____

   Moves the humerus _____

4. Which of the following graphs resembles what you would see if you were measuring the electrical potential of a skeletal muscle cell membrane using a voltmeter?

a.

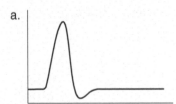

c.

b.

d.

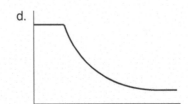

5. What is the neurotransmitter that is used by neurons to excite skeletal muscle cells?

   a. Dopamine
   b. Serotonin
   c. Acetylcholine
   d. GABA

6. Which kind of muscle is striated (select all that apply)?

   a. Smooth muscle
   b. Cardiac muscle
   c. Skeletal muscle

7. Which of the following is TRUE of a myofibril?

   a. It is composed of sarcomeres
   b. It is covered in perimysium
   c. It is larger than a muscle cell

8. Which of the following is the term for the smallest unit capable of contraction?

   a. Myofibril
   b. Myosin
   c. Sarcomere
   d. Sarcoplasm

9. The striations, or striped appearance, of some muscle cells is due to the precise arrangement of myofilaments within _____.

   a. sarcomeres
   b. sarcoplasmic reticulum
   c. T-tubules

10. Which of the following is the correct order from the surface of a skeletal muscle organ to deep within it?

    a. Epimysium, endomysium, perimysium
    b. Endomysium, perimysium, epimysium
    c. Epimysium, perimysium, endomysium

(page intentionally left blank)

Activity 12.1 Skeletal Muscle Anatomy

This activity targets LOs 12.1, 12.2, 12.3, and 12.4.

When we refer to an individual muscle of the body, such as the biceps brachii or the deltoid, we are referring to a skeletal **muscle organ**. Each muscle organ is comprised of hundreds of thousands or millions of individual muscle cells, also called muscle fibers. But it is not only contractile muscle cells within the muscle organ. Like most organs, muscle organs consist of multiple integrated tissues. These tissues include the skeletal muscle cells, blood vessels, nerve fibers, and wrappings of connective tissue. Each skeletal muscle is wrapped by an outer sheath of connective tissue called the **epimysium** (Figure 12.1).

Learning Connection

Micro to Macro

Place the following structure terms in order of size: muscle, myofibril, sarcomere, muscle cell.

The inside of each skeletal muscle organ is subdivided into small bundles of muscle cells called fascicles. Each fascicle is wrapped with a thin connective tissue wrapper called **perimysium**. The fascicle is a bundle of independent muscle cells that may contract together or separately. Each muscle cell is wrapped in an **endomysium** (pl. endomysia).

Underneath the endomysium, the plasma membrane of each muscle cell is called the sarcolemma. The sarcolemma is punctuated by deep invaginations that resemble deep pits. Each of these is called a **T-tubule**, and it functions to carry the electrical impulse that travels along the sarcolemma into the depths of the cell. As you examine Figure 12.3, you can see that the sarcolemma binds together long cylinders of contractile proteins. These intracellular bundles are called myofibrils, and each **myofibril** consists of a long line of connected sarcomeres. When these sarcomeres contract, the entire myofibril shortens, pulling on the ends of the muscle cells. The endomysia that coat the individual muscle cells weave together at the ends of the muscle, thus as the myofibrils contract, the endomysia are pulled toward the center of the muscle and the entire muscle contracts, bringing the attachment points closer together.

How does an electrical signal cause contraction? The answer is through ions of calcium. Calcium ions are stored within cellular structures called **sarcoplasmic reticulum (SR)**. These structures are modifications of endoplasmic

---

Figure 12.3   **The Sarcolemma, the T-tubules, and the Sarcoplasmic Reticulum**

Invaginations of the plasma membrane called T-tubules bring the electrical impulses that travel along the membrane into the depths of the cell. The SR, which stores calcium, abuts the T-tubules. When an electrical impulse travels along a T-tubule, the SR membrane is affected and releases calcium into the sarcoplasm.

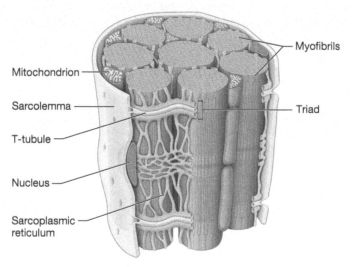

reticulum (ER), which is found in most cells. SR has slightly different function than ER, however. SR functions to store calcium between muscle contractions; when an electrical impulse is generated by nervous system input, that electrical signal travels along the sarcolemma and into the depths of the cell as it traverses the T-tubules. The nearby electrical signal causes ion channels embedded in the SR to open, releasing calcium at this moment of electrical excitement. Calcium ions are critical to contraction, enabling contractile proteins within each sarcomere to interact and initiate the contraction. Contraction ends when the calcium ions are gathered back into the SR. You can think of the presence of calcium ions permitting contraction to occur kind of like the way that a car's key is required for the engine to run.

Structures to identify in this activity:

• Endomysium
• Epimysium
• Mitochondria
• Myofibril
• Nuclei
• Perimysium
• Sarcolemma
• Sarcoplasmic reticulum
• Striations
• T-tubule

Materials
• Compound light microscope
• Muscle fiber model or images
• Prepared microscope slides of skeletal muscle tissue (longitudinal and cross section)

Instructions

1. On the model or using the figures in this lab identify the structures in the structure list above.

2. Place the slide on the microscope stage with the scanning (4×) objective in place. Focus until the cells come clearly into view. Now rotate the objectives so that the low-power (10×) objective is in place and focus again, using only the fine focus knob. Now rotate the objectives so that the high-power (40×) objective is in place and focus again, using only the fine focus knob. Notice the cells' shape and appearance. You can use Figures 12.4A and B for reference. Locate the nuclei and the striations. Notice where one cell ends and the next begins. Draw one cell, and state how many cells you can see in your field of view.

How many cells can you
see at 40×? _____

Figure 12.4 Skeletal Muscle Histology

Skeletal muscle tissue viewed through a light microscope 100×. (A) longitudinal section, (B) cross section

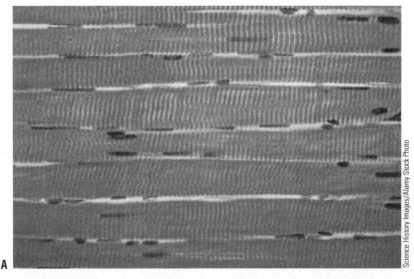

Science History Images/Alamy Stock Photo

A

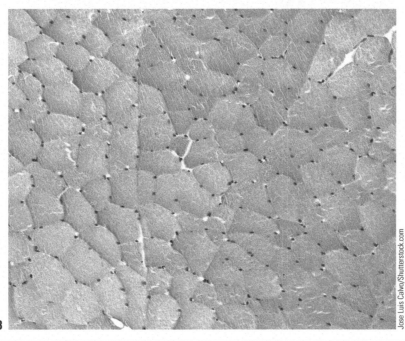

Jose Luis Calvo/Shutterstock.com

B

(page intentionally left blank)

Activity 12.2 Cardiac and Smooth Muscle Histology

This activity targets LOs 12.5 and 12.6.

Cardiac muscle cells are joined to their neighbors via junctions called intercalated discs. An intercalated disc is composed of both gap junctions and **desmosomes**. The gap junctions unite the cells functionally, allowing for ions (and therefore the excitatory electrical signal) to pass from one cell to its neighbor. This way, once one cell is stimulated electrically to contract, its neighbors will contract as well. The desmosomes unite the cells structurally, tying their membranes together so that the cells do not come apart during contraction. These two types of cellular junctions can be reviewed in Figure 12.5. With so many membrane proteins within intercalated discs, these structures are very dense, and appear as dark lines under the microscope.

Digging Deeper:
Ticker Trouble

In order to contract, cardiac muscle depends on the electrical signal traveling from one cardiac muscle cell to its neighboring cells, because they lack individual neuromuscular junctions. In a healthy heart, the heart wall is composed of continuous sheets of connected cells so that the contraction signal can spread rapidly and the heart can contract in a single, coordinated squeeze to efficiently eject blood. When cardiac muscle tissue is damaged, for example after a heart attack (myocardial infarction), some of the cardiac muscle cells die. These dead cells are unable to receive and transmit the electrical signal. The result can be arrhythmias, irregular heartbeats caused by unusual patterns of electrical signals and the resulting contractions.

Smooth muscle cells have a single, central nucleus and tapered ends. This tissue was named smooth because it lacks the striations apparent in skeletal and cardiac muscle. Smooth muscle cells contain the same contractile proteins as the other muscle types, but their physical arrangement is very different, and lacks the precise alternation of density that causes the striping of cardiac and skeletal muscle. Like cardiac muscle, some smooth muscle cells are connected by gap junctions. But smooth muscle lacks desmosomes, and therefore the sites of attachment are not as dense and noticeable as the intercalated discs of cardiac muscle.

| Figure 12.5 | Types of Cell Junctions |

The classes of cell-to-cell junctions are tight junctions, gap junctions, and desmosomes.

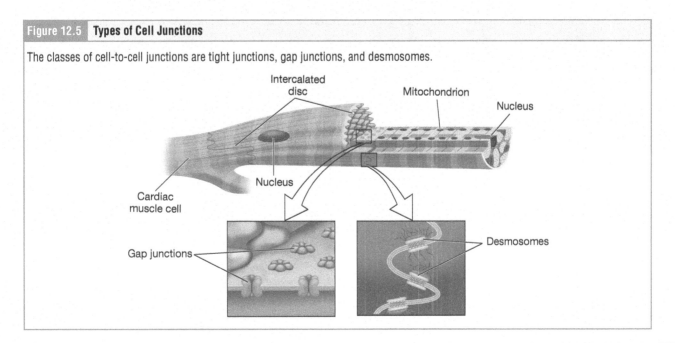

Materials

• Prepared slides of cardiac and smooth muscle (longitudinal section)

Instructions

1. Place the slide of cardiac muscle on the microscope stage with the scanning (4×) objective in place. Focus until the cells come clearly into view. Now rotate the objectives so that the low-power (10×) objective is in place and focus again, using only the fine focus knob. Now rotate the objectives so that the high-power (40×) objective is in place and focus again, using only the fine focus knob. Notice the cells' shape and appearance. You can use Figure 12.6 for reference. Locate the nuclei, intercalated discs, and striations. See if you can tell where one cell ends and the next begins. Draw one cell, and state how many cells you can see in your field of view below.

Figure 12.6	Cardiac Muscle Histology

Cardiac muscle tissue viewed under the light microscope at 100×.

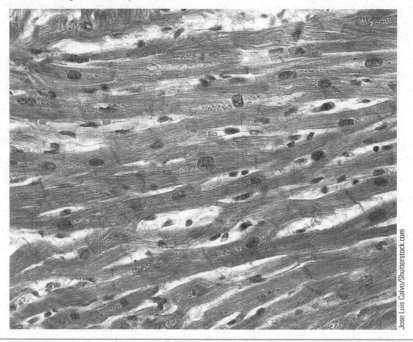

Jose Luis Calvo/Shutterstock.com

How many cells can you see at 40×? _____

2. Place the slide of smooth muscle on the microscope stage with the scanning (4×) objective in place. Focus until the cells come clearly into view. Now rotate the objectives so that the low-power (10×) objective is in place and focus again, using only the fine focus knob. Now rotate the objectives so that the high-power (40×) objective is in place and focus again, using only the fine focus knob. Notice the cells' shape and appearance. You can use Figure 12.7 for reference. Locate the nuclei. See if you can tell where one cell ends and the next begins. Draw one cell, and state how many cells you can see in your field of view below.

Figure 12.7	Smooth Muscle Histology

Smooth muscle tissue viewed under the light microscope.

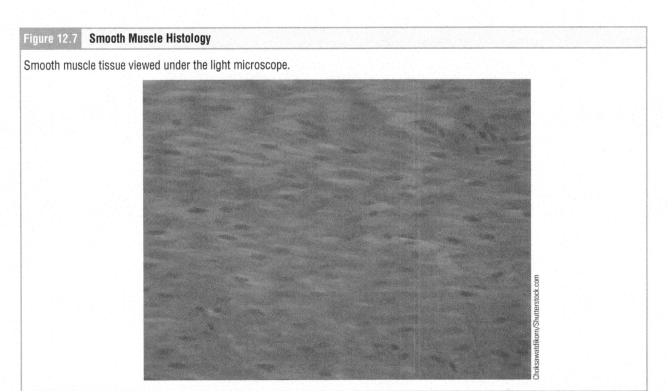

Choksawatdikorn/Shutterstock.com

How many cells can
you see at 40×? _____

(page intentionally left blank)

Activity 12.3 Steps in Skeletal Muscle Contraction

This activity targets LOs 12.7, 12.8, 12.9, and 12.10.

Materials
• Index cards or strips of paper with one of the steps of muscle contraction written on each (see list below).

Instructions
1. Mix up your cards so that they are entirely out of order.

2. Working with your lab partner or group, place the cards back in order.

3. Answer the challenge questions by finding the steps that would not occur in those scenarios.

Motor neuron steps (in alphabetical order):
- a. Acetylcholine is released
- b. Ca^{2+} enters motor neuron cytoplasm
- c. Chemically gated Na^+ channel opens on motor neuron cell body/dendrites
- d. Voltage-gated Ca^{2+} channels open on plasma (cell) membrane
- e. Voltage-gated Na^+ channel opens on motor neuron axon

End plate steps (in alphabetical order):
- a. Acetylcholine binds to nicotinic receptors on end plate
- b. Action potential proceeds down motor neuron axon
- c. Muscle cell action potential proceeds down membrane and T-tubules
- d. Muscle cell membrane returns to resting membrane potential

Steps inside the muscle cell (in alphabetical order):
- a. A new ATP molecule binds to actin
- b. Acetylcholine is broken down by acetylcholinesterase
- c. ATP is used to pump Ca^{2+} ions into sarcoplasmic reticulum
- d. Ca^{2+} concentration in the sarcoplasm decreases
- e. Ca^{2+} flows through ryanodine receptor/channel
- f. Ca^{2+} ions binds to troponin, inducing shape change
- g. Ca^{2+} ions diffuse through muscle cell sarcoplasm
- h. Myosin head releases actin
- i. Myosin head returns to original position
- j. Myosin heads bind to actin
- k. Myosin heads undergo a conformational change called "the powerstoke" (change from obtuse to acute angle)
- l. The ATP molecule bound to myosin is hydrolyzed into ADP + P_i
- m. Tropomyosin moves off the myosin-binding sites of actin filaments
- n. Tropomyosin slides back into place over actin/myosin binding sites
- o. Voltage-gated Ca^{2+} channels open on SR membrane

Challenge questions: Sarin gas is a chemical weapon used in warfare. It acts as a competitive inhibitor to acetylcholinesterase.

1. Which of the following correctly describes a competitive inhibitor?
   a. A competitive inhibitor binds to the active site of an enzyme and prevents the substrate from binding
   b. A competitive inhibitor binds to a different location on an enzyme (allosteric site) and changes the function of the enzyme
   c. A competitive inhibitor binds to the active site of an enzyme and destroys the enzyme

2. What steps will NOT occur in the presence of sarin gas poisoning?
   a. The muscle cell will never depolarize, no contraction will occur
   b. The muscle cell will never repolarize, it will be stuck in contraction
   c. The muscle cell will depolarize, but only half of the calcium ions will be released, the contraction will be weaker than usual

3. If you worked in a hospital and a person was admitted who had been exposed to sarin gas, what would their symptoms look like?
   a. The patient would experience flaccid paralysis, they would be unable to contract their muscles
   b. The patient would experience spastic paralysis, they would be unable to relax their muscles
   c. The patient would experience progressive weakening of muscle contraction

Challenge questions: Pancuronium is a nicotinic receptor antagonist commonly used as a component of anesthesia.

1. The nicotinic receptor is located on the surface of the _____ and it binds _____.
   a. motor neuron; calcium ions
   b. motor neuron; GABA
   c. muscle cell; sodium ions
   d. muscle cell; acetylcholine

2. What is the definition of a molecular antagonist?
   a. An antagonist binds to a receptor and blocks it
   b. An antagonist binds to a receptor and activates it
   c. An antagonist binds to a receptor but does not have a biological effect on it

3. What steps will NOT occur in the presence of pancuronium?
   a. The motor neuron will not release acetylcholine
   b. The muscle cell will not depolarize or contract
   c. The muscle cell will contract but not relax

4. Mario is scheduled for gall bladder surgery. Prior to intubation, Mario is given pancuronium through their IV line. What will Mario's response be to the pancuronium?
   a. The patient would experience flaccid paralysis, they would be unable to contract their muscles
   b. The patient would experience spastic paralysis, they would be unable to relax their muscles
   c. The patient would experience progressive weakening of muscle contraction

Lab 12 Post-Lab Quiz

1. Botulism is a disease that can occur when a species of bacteria, *Clostridium botulinum*, secretes a toxin, botulinum toxin (aka botox). Botulinum toxin causes local paralysis, an inability to contract muscles. If the toxin spreads throughout the body, the muscles that control breathing can be affected, causing death. Which of the following is a plausible mechanism of botulinum toxin?

   a. Botulinum toxin inhibits the action of acetylcholinesterase

   b. Botulinum toxin binds to calcium channels on the sarcoplasmic reticulum, holding them in an open position

   c. Botulinum toxin inhibits the motor neuron from releasing acetylcholinesterase

2. Imagine a fictional disease in which gap junctions do not exist in the intercalated discs. Which of the following might describe the resulting heart?

   a. Heart muscle cells would never replicate, and the heart would be too small

   b. Heart muscle cells would not be able to drain their fluid, each cell would swell and could burst

   c. Heart muscle cells would not share electrical signals, they would not contract as a unit and many would never contract at all

3. The energy for muscle contraction comes from cellular respiration. Which of the following is a byproduct of cellular respiration?

   a. Glucose

   b. Nitric oxide

   c. Heat

   d. N_2

4. Which type of ion leaves the muscle cell in large quantities through channels?

   a. Na^+

   b. K^+

   c. Ca^{2+}

5. Which of the following is TRUE about a myofibril?

   a. It is composed of sarcomeres

   b. It is covered in perimysium

   c. It is larger than a muscle cell

6. You've been invited to a cadaver lab, and you're assisting in the skin removal of the cadaver. As you peel back a layer of skin, you find muscle beneath that is covered in a thin, slightly iridescent, relatively transparent sheet of tissue. You hear the instructor telling the class that this thin sheet of tissue is rich in collagen and that it is called fascia. Just below the fascia is a thinner material called _____.

   a. endomysium

   b. perimysium

   c. epimysium

7. Your labmate has called you over to a microscope to look at muscle cells that have been separated from each other. You see the field of view in the image. You can see a single nucleus in the center of each of the cells. What kind of muscle is this?

    a. Smooth muscle

    b. Cardiac muscle

    c. Skeletal muscle

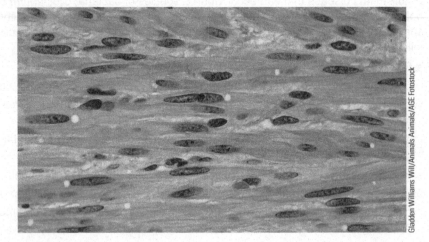

8. The tongue is composed largely of muscle. The tongue is a structure within the digestive system, but you can control its movements. What type of muscle composes the tongue?

    a. Smooth muscle

    b. Cardiac muscle

    c. Skeletal muscle

9. The pupil of your eye can dilate or constrict in response to light stimulus. This dilation regulates the amount of light coming into the inner eye. You have no control of the muscles, which are called ciliary muscles, that control the pupil. The ciliary muscles are most likely composed of which type of muscle?

    a. Smooth muscle

    b. Cardiac muscle

    c. Skeletal muscle

10. Which type of muscle has multiple nuclei?

    a. Smooth muscle

    b. Cardiac muscle

    c. Skeletal muscle

Lab 13 Axial Muscles

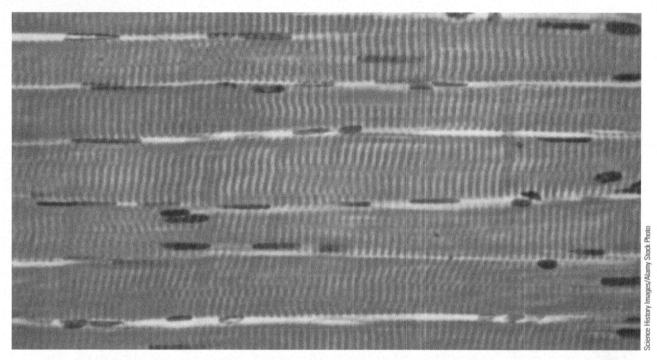

Science History Images/Alamy Stock Photo

Learning Objectives: After completing this lab, you should be able to:

13.1 Identify the location, general attachments, and actions of the major skeletal muscles.

13.2 Describe similar actions (functional groupings) of muscles in a particular compartment (e.g., anterior arm) or region (e.g., deep back).

Introduction

Skeletal muscles produce the movements of our body. They are responsible for our posture and our facial expression. They even help increase our body temperature and are used when we urinate. Tendons attach our muscles to the bones that they pull on. Each muscle is attached in at least two places. The muscle's more mobile end, where the action occurs when the muscle contracts, is called its **insertion**. Its less mobile end, where no movement occurs during contraction, is its **origin**. The word **action** describes what happens when the muscle contracts.

When we think of a movement or an exercise, like lifting your leg to take a step, multiple muscles are involved, but the principal muscle is called a **prime mover**. Other muscles may contribute by flexing a joint or stabilizing the insertion, and these are called **synergists** (Figure 13.1). Every movement has an opposite movement, for example flexing your elbow has the opposite motion of extending it. The groups of muscles that oppose each other to create these opposite actions are **antagonists** (Figure 13.2).

The Human Anatomy and Physiology Society includes more than 1,700 educators who work together to promote excellence in the teaching of this subject area. The HAPS A&P Learning Outcomes measure student mastery of the content typically covered in a two-semester Human A&P curriculum at the undergraduate level. The full Learning Outcomes are available at https://www.hapsweb.org.

Figure 13.1   Prime Movers and Synergists

The action shown in this illustration is flexion of the elbow. (A) The biceps brachii is the prime mover, the main muscle responsible for this action. (B) The brachioradialis and brachialis are both synergists that aid in this motion.

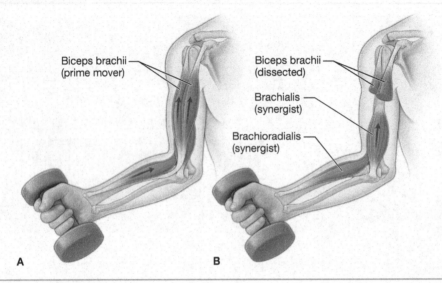

Figure 13.2   Agonists and Antagonists

Each action has an opposite. The opposite action to flexing the elbow is extending the elbow. (A) In the action of flexing the elbow, the biceps brachii is the agonist, the muscle accomplishing the action. (B) In the action of extending the elbow, the biceps brachii is the antagonist that causes the opposite action of the triceps brachii.

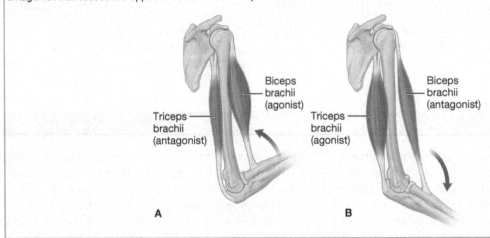

Just like the skeleton, we often consider the muscles of the axial body separately from the appendages (Figure 13.3). In this lab we will learn about the axial muscles, and in Lab 14 we will learn about the appendicular muscles. One note about this distinction, however, we classify muscles as axial or appendicular based on where their *action* is, not on where they are located. Latissimus dorsi, for example, is located on the trunk, but acts on the arm, therefore it is an appendicular muscle (Figure 13.4). Muscles that act in the region where they are located

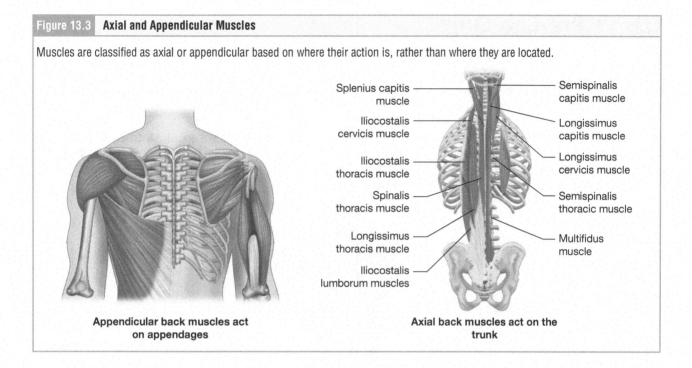

Figure 13.3 Axial and Appendicular Muscles

Muscles are classified as axial or appendicular based on where their action is, rather than where they are located.

Splenius capitis muscle

Iliocostalis cervicis muscle

Iliocostalis thoracis muscle

Spinalis thoracis muscle

Longissimus thoracis muscle

Iliocostalis lumborum muscles

Semispinalis capitis muscle

Longissimus capitis muscle

Longissimus cervicis muscle

Semispinalis thoracic muscle

Multifidus muscle

Appendicular back muscles act on appendages

Axial back muscles act on the trunk

are **intrinsic** muscles, and muscles that act outside the region of their location are **extrinsic** muscles. You can appreciate the difference by wrapping one of your hands around the forearm of your opposite arm. Now wiggle your fingers of the arm being held. Do you feel the forearm muscles rippling under your hand? These are extrinsic muscles. They're located in the forearm but connected, via long tendons, to the bones of the hand. Their action is at the hand. Do you think you also have intrinsic hand muscles?

Figure 13.4 Extrinsic Muscles

Latissimus dorsi is a great example of a muscle that is located on the back, but its action is at the arm (arm adduction).

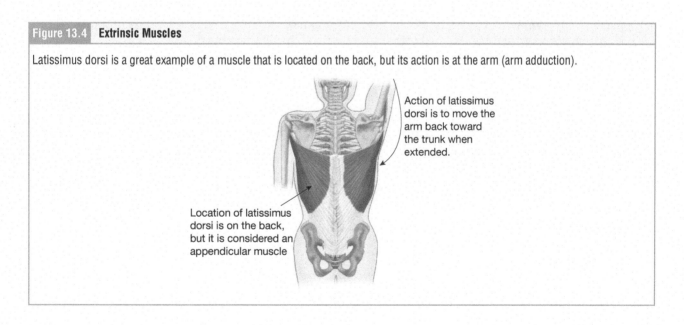

Action of latissimus dorsi is to move the arm back toward the trunk when extended.

Location of latissimus dorsi is on the back, but it is considered an appendicular muscle

Lab 13 Pre-Lab Quiz

This quiz will strengthen your background knowledge in preparation for this lab. For help answering the questions, use your resources to deepen your understanding. The best resource for help on the first five questions is your text, and the best resource for help answering the last five questions is to read the introduction section of each lab activity.

1. Anatomy of muscle layers. Muscles are often arranged in layers in a region. The topmost, or the one you would find first if you were dissecting, is the **superficial** layer. The layer closest to the bone is the **deep** layer. In some regions there is also one in between that is called the **intermediate** layer. Label the muscle layers shown in the figure below with the following terms: deep, extrinsic, intrinsic, superficial.

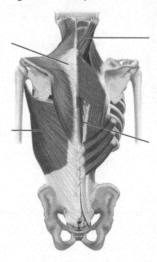

2. Match each of the roots to its definition:

Root	Meaning	List of Roots
	the neck	capit
	diagonal	cervic
	straight	oblique
	the head	orbit
	a circle	rect

3. Which of the following muscles of facial expression has a circular fiber arrangement?
   a. Buccinator
   b. Masseter
   c. Zygomaticus major
   d. Orbicularis oris

4. Zygomaticus major has which of the following fiber orientations?
   a. Fusiform
   b. Parallel
   c. Bipennate
   d. Circular

5. In the figure, which of the labels indicates the muscle belly?

   a. A
   b. B
   c. C

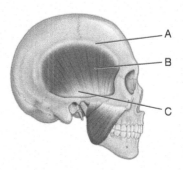

6. What is the most superficial muscle of the anterior neck?

   a. Platysma
   b. Buccinator
   c. Masseter
   d. Sternocleidomastoid

7. If you peeled back the platysma, you would be able to see the origin of which muscle more clearly?

   a. Masseter
   b. Sternocleidomastoid
   c. Zygomaticus major

8. What is the action of splenius capitis?

   a. Anterior flexion of the neck (chin to chest)
   b. Lateral flexion of the neck (ear to shoulder)
   c. Extension of the neck (looking at the ceiling)

9. Which of the axial muscles listed inserts on the scapula?

   a. Splenius cervicis
   b. Longissimus
   c. Transversospinalis
   d. None of these

10. The term *oblique* is used to describe a diagonal fiber orientation. Which layer of intercostals does <u>not</u> have oblique orientation?

   a. External intercostals
   b. Internal intercostals
   c. Innermost intercostals

(page intentionally left blank)

Activity 13.1 Muscles of Facial Expression

This activity targets LOs 13.1 and 13.2.

Most of the skeletal muscles in the body attach to and move bones. The muscles of facial expression are unique in that they originate on the bones of the skull but insert in the skin of the face. Because these muscles insert on skin rather than on bone, when they contract they produce facial expressions. These muscles are illustrated in Figure 13.5.

Figure 13.5 | **Muscles of Facial Expression**

This illustration shows (A) anterior and (B) lateral views of the muscles of facial expression, with the largest ones being labeled. These muscles insert into the skin of the face.

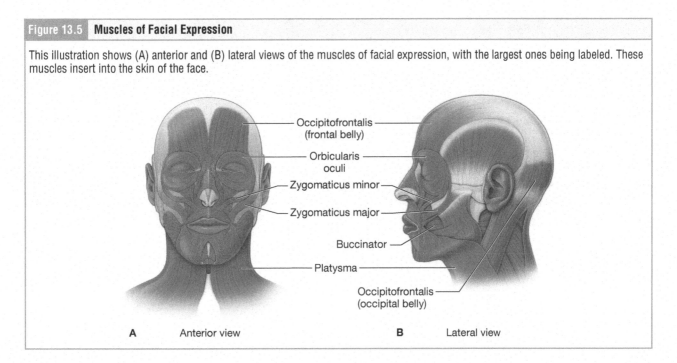

A   Anterior view   B   Lateral view

Materials
• Human cadaver or head and neck muscle model or images
• Mirror

Instructions
1. On the cadaver or model, identify the muscles found in Table 13.1.

	Table 13.1 Muscles of Facial Expression		
Muscle	**Description**	**Origin (O) and Insertion (I)**	**Action**
Orbicularis oris	Circular muscle surrounding the mouth	(O) Medial aspects of the maxilla and mandible (I) Skin and mucous membrane of the lips	Puckers the lips
Orbicularis oculi	Circular muscle surrounding the eye	(O) Nasal part of frontal bone, maxilla, and lacrimal (I) Skin of orbital region	Opens and closes the eye
Occipitofrontalis	Muscle with two bellies spanning the top and back of the head	(O) Skin of the eyebrow, muscles of the forehead, and the superior nuchal line (I) Epicranial aponeurosis	Moves the scalp and the eyebrows
Buccinator	Makes up the cheeks and contributes to facial expression	(O) Maxilla, mandible, and temporomandibular joint (I) Fibers of the orbicularis oris	Compresses the cheek

(continued)

Table 13.1 Muscles of Facial Expression (*continued*)

Muscle	Description	Origin (O) and Insertion (I)	Action
Zygomaticus major	One of the main muscles involved in facial expression, attaching at the corners of the mouth	(O) Anterior lateral surface of the zygomatic bone (I) Muscles of the upper lip	Raising the corners of the mouth
Zygomaticus minor	One of the main muscles involved in facial expression, attaching at the corners of the mouth	(O) Posterior lateral surface of the zygomatic bone (I) Muscles of the upper lip	Raising the corners of the mouth
Platysma	Very superficial muscle of the neck	(O) Fascia over the upper, lateral chest (I) Skin, fascia, and muscles of the mandible	Tenses the skin of the neck

2. Using a mirror, make the facial expressions in the first column of Table 13.2 and determine which muscle(s) is/are used. Or have your lab partner make each of the facial expressions. Write the muscle(s) in the second column.

Table 13.2 Facial Expressions for Activity 13.1	
Expression	**Muscle(s) Involved**
Smiling	
Winking at someone	
Raising eyebrows in surprise	
Blowing a kiss	

Activity 13.2 **Muscles of the Neck and Back**

This activity targets LOs 13.1 and 13.2.

The muscles of the posterior and lateral neck are involved in head movements (Figure 13.6). The muscles of the back are involved in posture (Figure 13.7).

Figure 13.6	**Lateral and Posterior Neck Muscles**

The muscles of the neck move the head, cervical vertebrae, and scapulae.

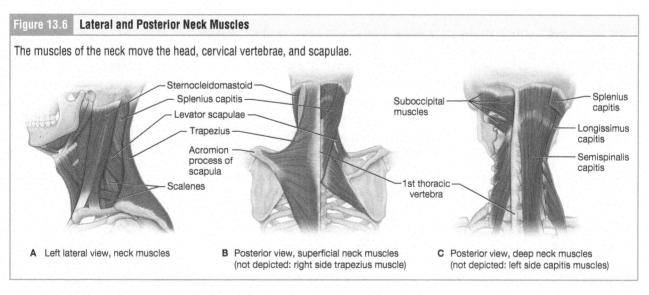

A Left lateral view, neck muscles

B Posterior view, superficial neck muscles
(not depicted: right side trapezius muscle)

C Posterior view, deep neck muscles
(not depicted: left side capitis muscles)

Figure 13.7	**Muscles of the Neck and Back**

Keeping the spinal column erect or extending it is the work of the long muscles to either side of the spine. Smaller muscles act as synergists and also contribute to lateral flexion.

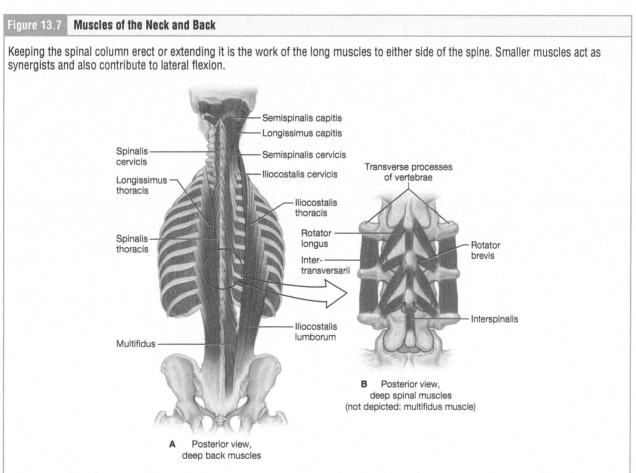

A Posterior view,
deep back muscles

B Posterior view,
deep spinal muscles
(not depicted: multifidus muscle)

Materials

• Human cadaver or neck and back muscle model or images
• Mirror

Instructions

1. On the cadaver or model, identify the muscles found in Table 13.3.

Table 13.3 Muscles of the Neck and Back			
Muscle	**Description**	**Origin (0) and Insertion (I)**	**Action**
Sternocleidomastoid	Thick muscle attaching the sternum, clavicle, and mastoid together	(0) Manubrium of the sternum and medial clavicle (I) Mastoid process of the temporal bone	Prime mover: flexion of the head and neck
Scalenes	Muscles of the neck that assist the sternocleidomastoid	(0) Transverse processes of cervical vertebrae (I) First and second ribs	Flexion of the neck
Splenius capitis	Intrinsic straplike muscle of the posterior neck working to keep the head upright	(0) Ligamentum nuchae, lower cervical and upper thoracic spinous processes (I) Occipital bone and mastoid process	Extension of the head and neck
Splenius cervicis	Intrinsic muscle of the neck superficial to splenius capitis	(0) Thoracic spinous processes (I) Atlas and axis	Bilateral extension of the head and neck, and unilateral rotation
Iliocostalis	Deep, erector spinae group	(0) Posterior ribs, thoracolumbar aponeurosis (I) Cervical transverse processes and posterior ribs	Vertebral column extension and control of flexion, unilateral flexion (side-bending), and rotation
Longissimus	Deep, erector spinae group	(0) Lower cervical and thoracic transverse processes, thoracolumbar aponeurosis (I) Mastoid process and cervical and thoracic transverse processes	Vertebral column extension and control of flexion, unilateral flexion (side-bending), and rotation
Spinalis	Deep, erector spinae group	(0) Ligamentum nuchae and spinous processes (I) Thoracic spinous processes	Vertebral column extension and control of flexion, unilateral flexion (side-bending), and rotation
Transversospinalis (group)	Small muscles that connect the vertebrae to one another	(0) Transverse processes of each vertebra (I) Spinous processes of each vertebra	Stabilizing the vertebral column contributes to rotation

2. While viewing yourself in a mirror, perform the exercises in the first column of Table 13.4 and determine which muscle(s) is/are used. Or have your lab partner perform each of the exercises. Write the muscle(s) in the second column.

Table 13.4 Positions for Activity 13.2	
Exercise	**Muscle(s) Involved**
Resting chin on chest	
Resting ear on shoulder	
Rotating to the side as in swinging a baseball bat	
Tilting your head upwards to look at the ceiling (or clouds)	
Back bend	

(page intentionally left blank)

Activity 13.3 Muscles of the Thorax and Abdomen

This activity targets LOs 13.1 and 13.2.

The muscles of the thorax are involved in breathing (Figure 13.8). The thoracic cavity and abdominal cavity are separated by the diaphragm (Figure 13.9) which is also a prime mover of breathing. The muscles of the anterior abdomen (Figure 13.10) are involved in posture primarily, but also are recruited for protection from impact (for example when someone punches you) as well as laughing, sneezing, coughing, and vomiting. The muscles of the posterior abdominal wall (Figure 13.11) contribute to both spinal stability and hip flexion.

Figure 13.8	Intercostal Muscles

The intercostal muscles span the spaces between adjacent ribs. There are three layers of intercostal muscles—external, internal, and innermost.

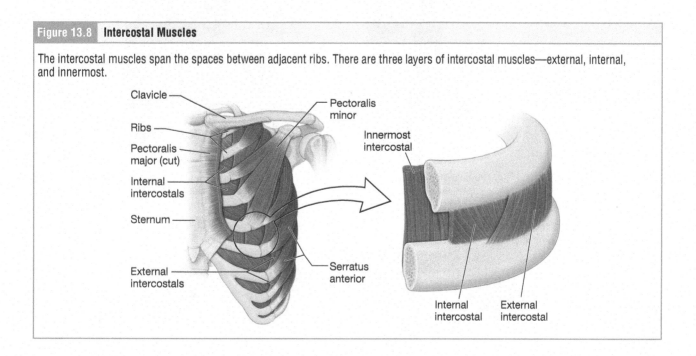

Figure 13.9	The Muscles of the Diaphragm

The diaphragm divides the thoracic and abdominal cavities and functions in respiration. (A) In the anterior view, the diaphragm is a parachute- or bell-shaped muscle. (B) The inferior view shows the large central tendon of the diaphragm.

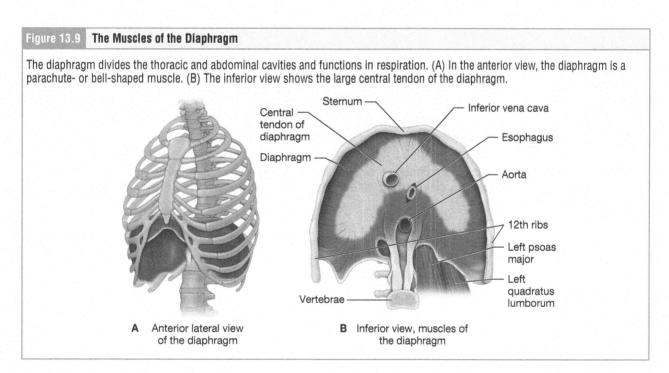

Figure 13.10 | **Muscles of the Anterior Abdomen**

(A and B) The anterior view of the abdomen dominated by the rectus abdominis, which is covered by a sheet of connective tissue called the rectus sheath. (C) On the sides of the abdomen, the external oblique muscles form the superficial layer, while the internal oblique muscles form the middle layer and the transversus abdominis forms the deepest layer.

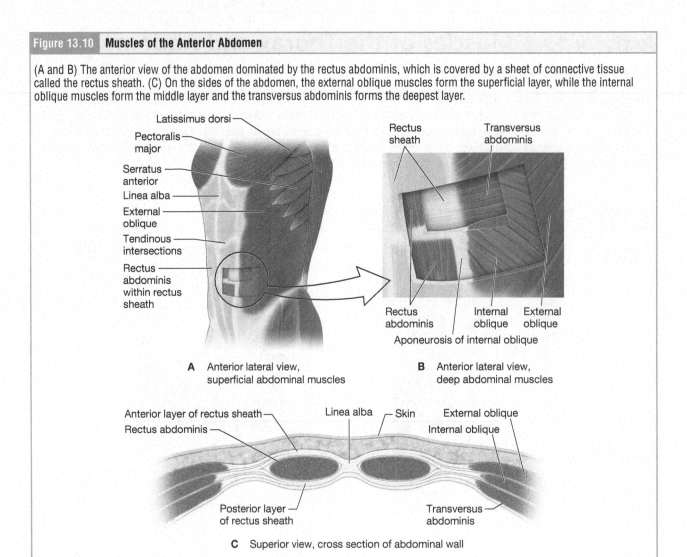

A   Anterior lateral view, superficial abdominal muscles

B   Anterior lateral view, deep abdominal muscles

C   Superior view, cross section of abdominal wall

Materials
• Human cadaver or thorax and abdomen muscle model or images
• Mirror

Instructions
1. On the cadaver or model, identify the muscles found in Table 13.5.

2. While viewing yourself in a mirror, perform the exercises in the first column of Table 13.6 and determine which muscle(s) is/are used. Or have your lab partner perform each of the exercises. Write the muscle(s) in the second column.

Figure 13.11 **Muscles of the Posterior Abdomen**

The muscles in the posterior wall of the abdomen contribute to both lumbar spine and hip movements.

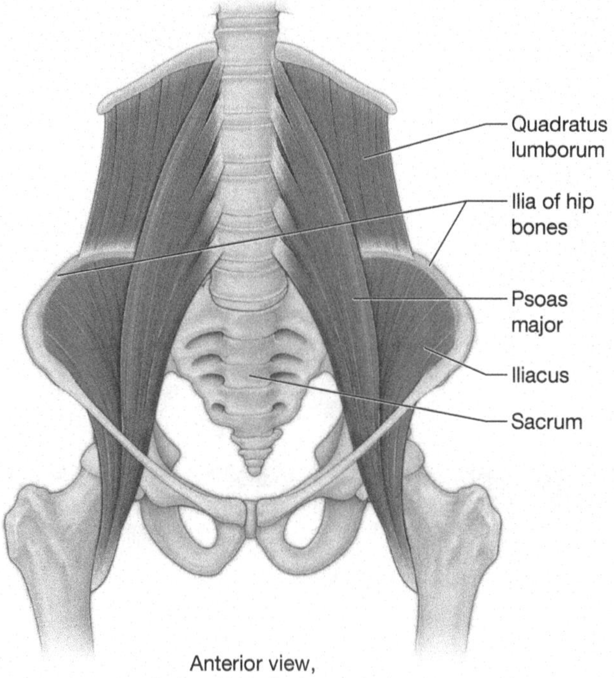

Anterior view,
posterior abdominal muscles

Table 13.5 Muscles of the Abdomen and Thorax			
Muscle	**Description**	**Origin (O) and Insertion (I)**	**Action**
External abdominal oblique	Outermost layer of abdominal muscles with fibers running inferomedially	(O) Lower ribs (I) Pubis and abdominal aponeurosis	Flexion, lateral flexion, and rotation of the trunk
Internal abdominal oblique	Intermediate layer of abdominal muscles with superomedial fibers	(O) Iliac crest and thoracolumbar aponeurosis (I) Costal cartilages and abdominal aponeurosis	Flexion, lateral flexion, and rotation of the trunk
Transverse abdominis	Deepest layer of abdominal muscles, fibers run transversely	(O) Iliac crest and thoracolumbar aponeurosis (I) Pubis and abdominal aponeurosis	Compression of abdomen
Rectus abdominis	Superficial abdominal muscle that runs along the middle of the abdomen	(O) Pubic symphysis (I) Costal cartilages	Flexion of the trunk
Psoas major	Posterior abdominal muscle, comes together with iliacus to become a major hip flexor—iliopsoas	(O) Lumbar vertebrae (I) Lesser trochanter of the femur	Flexion of the hip and trunk
Iliacus	Comes together with psoas major to become a major hip flexor—iliopsoas	(O) Anterior ilium (I) Lesser trochanter of the femur	Flexion of the hip and trunk
Quadratus lumborum	Deep layer of muscle forming the posterior wall of the abdomen	(O) Iliac crest (I) Transverse processes of lumbar vertebrae	Lateral flexion of the trunk

(continued)

	Table 13.5 Muscles of the Abdomen and Thorax (*continued*)		
Muscle	**Description**	**Origin (O) and Insertion (I)**	**Action**
Diaphragm	Most important breathing muscle, separates the thorax from the abdomen	(O) Costal cartilages and lumbar vertebrae (I) Central tendon of the diaphragm	Depresses the central tendon to create more space in the thorax during inspiration
External intercostals	Most external layer of muscles between the ribs, fibers are oriented inferomedially	(O) Inferior margins of ribs (I) Superior margins of ribs	Elevates ribs during inspiration
Internal intercostals	Middle layer of muscles between the ribs, fibers are oriented superomedially	(O) Inferior margins of ribs (I) Superior margin of ribs	Depress and draw the ribs together during expiration
Innermost intercostals	Deepest layer of muscles between the ribs, the fibers are horizontal	(O) Costal grooves of ribs (I) Superior border of rib below	Assist the internal intercostals during expiration

Table 13.6 Movements for Activity 13.3	
Exercise	**Muscle(s) Involved**
Twisting and rotating, as you would if you were reaching your right hand to the outside of your left knee	
Holding a plank position	
Performing a sit up	
Hip flexion, as you would if you were standing and raised one knee	
Taking a deep breath	

Digging Deeper:
Abdominal Muscles for Pressure

In addition to their functions in spine flexion, rotation and maintaining posture, the abdominal muscles are recruited in an involuntary fashion for increasing abdominal or thoracic pressure. Try placing your hand on your abdomen as you take a deep inhale and then let it out in an exhale. As you come to the end of the exhale force the air out with all your might squeezing out as much air as possible. What happened under your hand? Chances are, your abdominal muscles contracted to squeeze on your abdominal organs, forcing them up against the diaphragm. The pressure against the diaphragm helps force out extra air from the lungs. The same thing happens in coughing, your abdominal muscles are recruited in an involuntary fashion to help generate pressure that pushes air out of the lungs. What about vomiting, have you ever thought of that? Once again, your abdominal muscles are used in sharp, sudden contractions to pressurize your abdomen and force the contents of your stomach and upper duodenum out of your body. Defecation is one more example in which those abdominal muscles are generating pressure.

Activity 13.4 Case Study: Facial Paralysis

This activity targets LOs 13.1 and 13.2.

Facial paralysis can be caused by stroke or a viral infection. The key to telling these two conditions apart has to do with the anatomy of cranial nerve VII, one of the facial nerves. Examine Figure 13.12, which illustrates how a stroke, which is caused by an interruption in blood flow in the brain, differs from the effects of a viral infection that impacts the facial nerve.

Figure 13.12 | Facial Paralysis

Bell's palsy and strokes can both cause facial paralysis, but they have different anatomical origins.

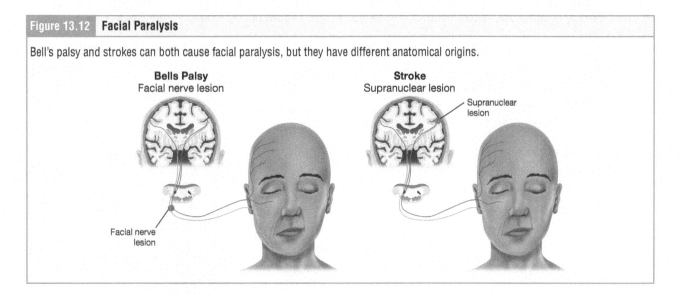

---

1. You are shadowing a neurologist when they are called to consult on a case. The patient has one-sided facial paralysis and the neurologist has been asked to evaluate if a stroke is a possible cause. Asking the patient to perform which of the following expressions (three photos below) could help differentiate between stroke or Bell's palsy?

   a. Raising eyebrows
   b. Smiling
   c. Puffing out cheeks

2. If the patient was not able to perform the task you selected in question 1, what is their diagnosis?
   a. Stroke
   b. Bell's palsy

3. Which of the following muscles would be affected in Bell's palsy (select all that apply)?

    a. Occipitofrontalis

    b. Buccinator

    c. Zygomaticus major

    d. Orbicularis occuli

    e. All of these

Bell's palsy is not usually permanent. Control over facial muscles usually returns between 2 weeks and 6 months. The affected muscles, however, can become atrophied, which means that they lose size and strength without use. Physical therapy is commonly recommended to keep the muscles active during recovery from Bell's palsy.

4. Which of the following exercises might be incorporated into physical therapy for a Bell's palsy patient?

    a. Wiggling the ears

    b. Bending the neck from side-to-side

    c. Moving the tongue inside the mouth

    d. Raising the eyebrows up and down

5. Would this expression be possible in a patient with Bell's palsy?

    a. Yes, on the unaffected side only

    b. Yes, on both sides

    c. No, on neither side

Name: _____   Section: _____

1. Examine the cross section of the abdomen carefully. What layer is rectus abdominus in?

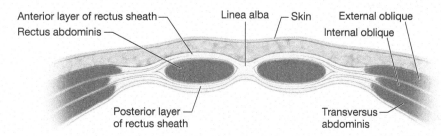

Anterior layer of rectus sheath —   Linea alba   — Skin   External oblique
Rectus abdominis —   Internal oblique

Posterior layer —   Transversus —
of rectus sheath   abdominis

Superior view, cross section of abdominal wall

   a. Deep layer
   b. Superficial layer
   c. Intermediate layer

2. Which of the following muscles is the deepest muscle of the abdominal wall?

   a. Transversus abdominis
   b. Internal oblique
   c. Rectus abdominis

3. Which of these muscles is found in the deep posterior abdominal wall?

   a. Quadratus lumborum
   b. Spinalis
   c. Interspinalis

4. Imagine that iliocostalis and longissimus are contracting on the right side but not the left. What movement do you expect to see?

   a. Hyperextension of the back
   b. Flexion of the back
   c. Lateral flexion toward the right
   d. Lateral flexion toward the left

5. Contracting the _____ muscles would cause you to squint your eyes, taking in less light

   a. orbicularis oris
   b. orbicularis oculi
   c. occipitofrontalis

6. Which of the following muscles could contribute to turning the head to the side?

   a. Zygomaticus major
   b. Sternocleidomastoid
   c. Occipitofrontalis
   d. Platysma
   e. All of these

7. Any muscle that inserts on the ribcage can assist in respiration, particularly in times when it is difficult to take deep enough breaths such as during infection or exercise. Which of the following muscles could assist in respiration when needed?

   a. Platysma

   b. Anterior scalenes

   c. Spinalis

   d. Splenius capitis

8. Which abdominal muscle is the prime mover during a sit up as shown?

   a. External abdominal oblique

   b. Internal abdominal oblique

   c. Quadratus lumborum

   d. Rectus abdominus

Patrik Giardino/Stone/Getty Images

9. Which abdominal muscle, if any, inserts on the femur?

   a. Internal abdominal oblique

   b. Rectus abdominus

   c. Psoas major

   d. Quadratus lumborum

10. What is the correct order, from lateral to medial, of the erector spinae muscles?

   a. Spinalis, iliocostalis, longissimus

   b. Longissimus, spinalis, iliocostalis

   c. Iliocostalis, longissimus, spinalis

Lab 14 Appendicular Muscles

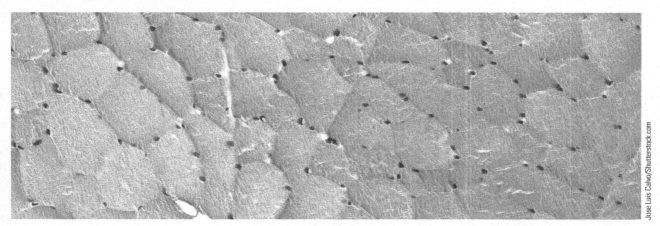

Learning Objectives: After completing this lab, you should be able to:

14.1 Identify the location, general attachments, and actions of the major skeletal muscles.

14.2 Describe similar actions (functional groupings) of muscles in a particular compartment (e.g., anterior arm) or region (e.g., deep back).

Introduction

The appendicular muscles move our appendages, that is our arms and legs. Each muscle is attached in at least two places. The more mobile end, where the action occurs, meaning the bone moves, when the muscle contracts, is called its **insertion**. The less mobile end, where no movement occurs during contraction, is its **origin**. We use the word **action** to describe what happens when the muscle contracts. We classify muscles as axial or appendicular based on where their *action* is, not where they are located, so some of the appendicular muscles we will study in this lab are located on the trunk, but they produce action at the limbs.

You may have heard the saying that every action has a reaction; and in anatomy, every muscle that produces a movement has a muscle that produces the opposite action. The **prime mover**, or **agonist**, is the muscle that produces the action. For example, the biceps brachii muscle is the prime mover in flexing the elbow (Figure 14.1). The triceps brachii is the **antagonist**, the muscle that produces the opposite action, in this case elbow extension. Other muscles may assist the prime mover in the action, and these are called **synergists**. Often synergists are located in the same compartment or location, for example, the anterior side of your upper arm contains several elbow flexors, the posterior side of your upper arm contains the elbow extensors.

Muscles that act in the region where they are located are **intrinsic** muscles, and muscles that are located outside of the place where they act, but connected to that location via long tendons are **extrinsic** muscles (Figure 14.2).

The Human Anatomy and Physiology Society includes more than 1,700 educators who work together to promote excellence in the teaching of this subject area. The HAPS A&P Learning Outcomes measure student mastery of the content typically covered in a two-semester Human A&P curriculum at the undergraduate level. The full Learning Outcomes are available at https://www.hapsweb.org.

Figure 14.1 | **Prime Movers and Synergists**

The action shown in this illustration is flexion of the elbow. (A) The biceps brachii is the prime mover, the main muscle responsible for this action. (B) The brachioradialis and brachialis are both synergists that aid in this motion.

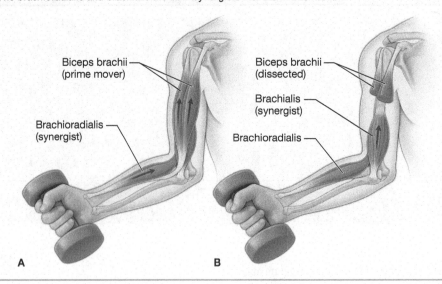

Biceps brachii (prime mover)

Brachioradialis (synergist)

Biceps brachii (dissected)

Brachialis (synergist)

Brachioradialis

A

B

In this lab we will learn a lot of muscle names and their locations in the body. To help with the sizable load of new terms, here are a few themes of how muscles are named:

- Muscle shape: muscles are often named based on their resemblance to geometric shapes, such as rhomboid or deltoid (Figure 14.3).
- Muscle size: sometimes individual muscles in a group are named for their size relative to one another. The terms maximus, medius, and minimus mean large, medium, and small, respectively. Table 14.1 lists Greek and Latin prefixes that pertain to size.
- Location: muscles can be named for their region. For example, biceps brachii is in the brachial region, whereas biceps femoris is in the femoral region.
- Orientation of fibers: muscles may be named based on whether their fibers run straight up and down (rectus), at a diagonal (oblique), or in a circle (orbicularis).
- Number of origins: anatomists often describe the origin of a muscle as heads, so a muscle with two origins would earn the name biceps, one with three origins could be named triceps. Table 14.2 summarizes Greek and Latin prefixes that indicate number.
- Action: many muscles are named based on their action, for example, many of the muscles that adduct the thigh are called adductors.
- Attachment location: the locations of attachments sometimes appear in the muscle name. Flexor carpi ulnaris, for example, is a muscle with attachments on the ulna and carpal region.

Figure 14.2 | **Intrinsic and Extrinsic Muscles**

Intrinsic muscles are located within the region where they act and extrinsic muscles are located in a different location than where they act, but connected to that location through tendons.

Extrinsic muscles

Intrinsic muscles

Figure 14.3 | **Some Muscles Are Named for Their Shape**

(A) The rhomboid muscles of the upper back are named for their shape, which is similar to that of a rhombus. (B) The deltoid muscle of the shoulder is so named because it resembles an upside-down Greek letter delta.

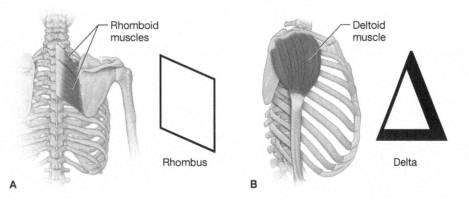

Table 14.1	Greek and Latin Words That Pertain to Muscle Size
Term	**Meaning**
Maximus	The largest of a group
Medius	Medium-sized in a group
Minimus	The smallest of a group
Brevis	Short
Longus	Long
Major	The larger of two
Minor	The smaller of two
Longissimus	The longest

Table 14.2	Greek and Latin Prefixes That Indicate Number
Prefix	**Meaning**
Uni	One
Bi / Di	Two
Tri	Three
Quad	Four
Multi	Many

(page intentionally left blank)

Lab 14 Pre-Lab Quiz

This quiz will strengthen your background knowledge in preparation for this lab. For help answering the questions, use your resources to deepen your understanding. The best resource for help on the first five questions is your text, and the best resource for help answering the last five questions is to read the introduction section of each lab activity.

1. Anatomy of muscle compartments. The muscles of a limb are often housed within compartments. The image is a cross section through a limb. Label the following structures on the image:

   extensor compartment
   flexor compartment
   lateral
   medial
   neurovascular bundle

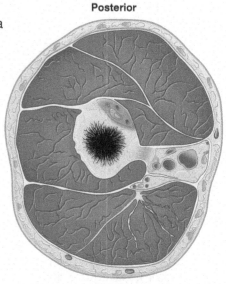

Posterior

Anterior

2. Anatomy of muscle shapes. Correctly identify each of the seven types of muscle using the following terms:

   bipennate
   circular
   convergent
   fusiform
   multipennate
   parallel
   unipennate

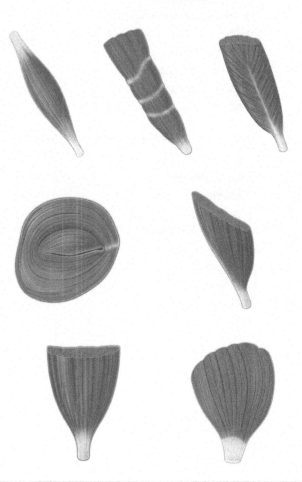

3. Match each of the roots to its definition:

Root	Meaning		List of Roots
	to bring in		abduct
	long		adduct
	to lead away		brev-
	great		longus
	short		magnus

4. Which muscle(s) is/are the prime mover of shoulder retraction (moving the shoulders back for improved posture)?

   a. Trapezius

   b. Latissimus dorsi

   c. Rhomboid major and minor

   d. Teres minor

5. Which of the following is the best description of arm abduction?

   a. If you are standing with your arms out, as a letter T and bring them back to your sides, that is abduction.

   b. If you move your arms out in front of you so that they are parallel with the floor, that is abduction.

   c. If you are standing with your arms at your sides and you move them out so that your body resembles a letter T, that is abduction.

6. Hip flexors cross the hip on the _____ side, knee flexors cross the knee on the _____ side.

   a. anterior, anterior

   b. anterior, posterior

   c. posterior, anterior

   d. posterior, posterior

7. Which of the following is a movement term that we use to describe the ankle?

   a. Abduction

   b. Plantar flexion

   c. Elevation

8. The extrinsic muscles of the hand are found where?

   a. Along the palm of the hand

   b. Along the anterior surface of the fingers

   c. At the wrist

   d. At the forearm

9. The action of the gastrocnemius is to

    a. Adduct the arm
    b. Plantarflex the toes
    c. Flex the elbow
    d. Extend the hip

10. Which of the labels in this figure correctly identifies infraspinatus?

    a. A
    b. B
    c. C
    d. D

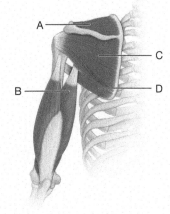

(page intentionally left blank)

Activity 14.1 Muscles of the Upper Limb

This activity targets LOs 14.1 and 14.2.

The arm attaches to the trunk at the shoulder, so many of the muscles that move the arm at the shoulder originate on the trunk. These muscles are listed in Table 14.3 and illustrated in Figure 14.4. The shoulder is the most mobile joint in the human body, its movements are illustrated in Figure 14.5.

Table 14.3 Shoulder and Brachial Muscles			
Muscle	**Description**	**Origin (O) and Insertion (I)**	**Action**
Subclavius	Muscle sitting just underneath the clavicle	(O) Costal cartilage of first rib (I) Inferior clavicle	Stabilizes the clavicle
Pectoralis minor	Smaller anterior chest muscle, deep to pectoralis major	(O) Anterior surfaces of upper ribs (I) Coracoid process of the scapula	Protraction and depression of the scapula
Serratus anterior	Thin muscle laying right on top of the anterior ribs	(O) Outer surface of upper ribs (I) Anterior surface of the scapula	Protraction and stabilization of the scapula
Trapezius	Superficial muscle of the back and shoulder	(O) Occipital bone, ligamentum nuchae, spinous processes of cervical and thoracic vertebrae (I) Lateral clavicle and spine of scapula	Elevation, retraction, and depression of the scapula
Rhomboid major	Superficial muscle just deep to traps	(O) Spinous processes of upper thoracic vertebrae (I) Vertebral border of the scapula	Retraction of the scapula
Rhomboid minor	Superficial muscle just deep to the traps and above rhomboid major	(O) Spinous processes of cervical and thoracic vertebrae (I) Root of spine of scapula	Retraction of the scapula
Pectoralis major	Anterior superficial muscle of the chest	(O) Medial clavicle, sternum, and costal cartilages (I) Proximal humerus	Prime mover: flexion, medial rotation, and adduction of the shoulder
Latissimus dorsi	Convergent, covers much of the superficial lower back	(O) Thoracolumbar aponeurosis and spinous processes of thoracic vertebrae (I) Proximal humerus	Prime mover: extension, medial rotation, and adduction of the shoulder
Deltoid	Thick muscle covering the majority of the shoulder superficially and laterally	(O) Clavicle and posterior scapula (I) Lateral and proximal humerus	Prime mover for abduction of the arm, flexion and medial rotation, extension and lateral rotation
Subscapularis	Part of the rotator cuff, the only muscle of this group that is anterior to the scapula	(O) Anterior surface of the scapula (I) Lesser tubercle of the humerus	Medial rotation and stabilization of the shoulder and arm
Supraspinatus	Part of the rotator cuff, located on the posterior scapula	(O) Surface of the scapula above the spine (I) Greater tubercle of the humerus	Abduction and stabilization of the shoulder and arm
Infraspinatus	Part of the rotator cuff, located on the posterior scapula	(O) Surface of the scapula below the spine (I) Greater tubercle of the humerus	Lateral rotation and stabilization of the shoulder and arm

(continued)

Table 14.3 Shoulder and Brachial Muscles (*continued*)

Muscle	Description	Origin (O) and Insertion (I)	Action
Teres major	Short but powerful muscle, inferior to teres minor	(O) Lateral border of the posterior and inferior scapula (I) Proximal humerus	Extension, adduction, and medial rotation of the arm
Teres minor	Part of the rotator cuff, located on the posterior scapula	(O) Lateral border of the scapula (I) Greater tubercle of the humerus	Lateral rotation, extension, and stability of the shoulder and arm
Coracobrachialis	Muscle extending from the scapula to the middle of the brachium anteriorly	(O) Coracoid process of the scapula (I) Medial surface of the humerus	Flexion and adduction of the arm

Figure 14.4 | Shoulder Muscles

Many muscles stabilize the shoulder so that it can serve as a steady base for the movements of the arm. (A) Left anterior lateral view and (B) posterior view of the shoulder muscles.

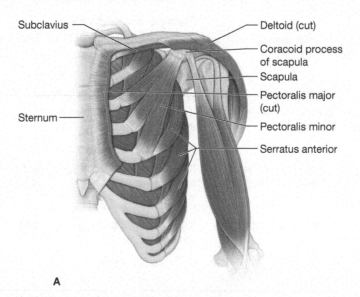

A

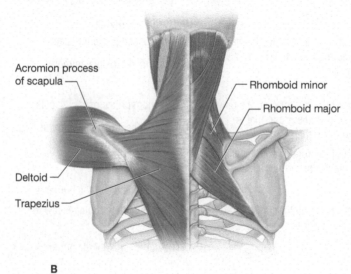

B

Figure 14.5 Shoulder Movements

The movements available at the shoulder joint include (A) retraction and protraction, (B) flexion and extension, (C) abduction and adduction, and (D) internal and external rotation.

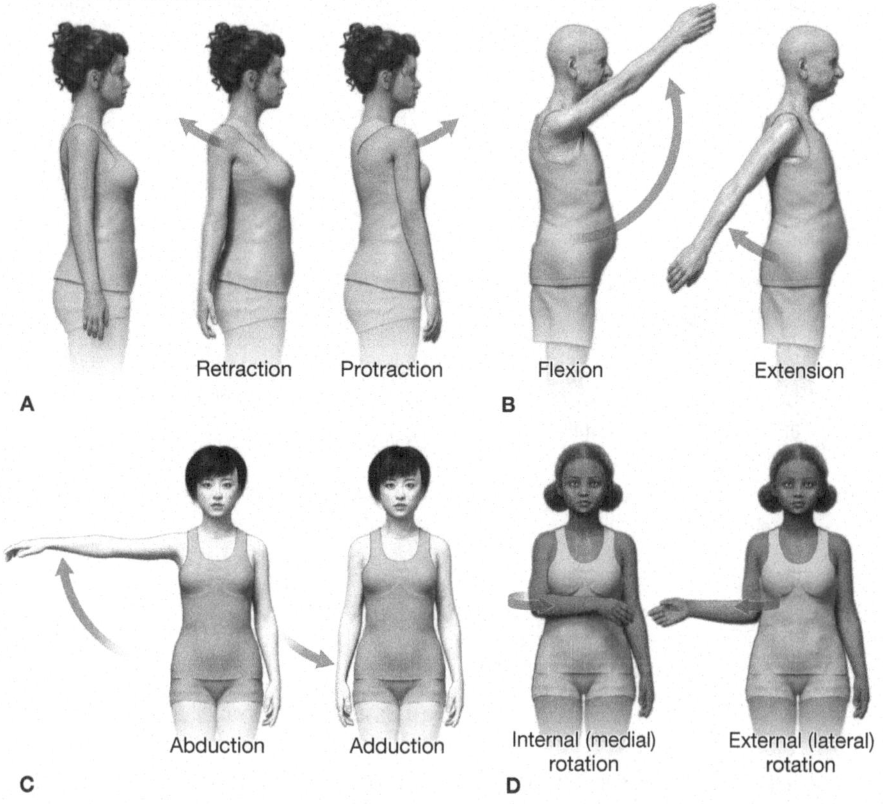

Retraction   Protraction   Flexion   Extension

A   B

Abduction   Adduction   Internal (medial)   External (lateral)
                          rotation   rotation

C   D

The muscles of the forearm are listed in Table 14.4 and illustrated in Figure 14.6. These muscles are a mix of muscles with actions on the forearm, and muscles with actions at the hand, in other words, extrinsic muscles of the hand. The movements of the forearm, wrist, and hand are illustrated in Figure 14.7. In addition to the extrinsic muscles that produce hand movements, there are also intrinsic hand muscles, illustrated in Figure 14.8.

		Table 14.4 Forearm, Wrist, and Hand Muscles	
Muscle	**Description**	**Origin (O) and Insertion (I)**	**Action**
Biceps brachii	Superficial, two-headed muscle that crosses two joints and sits on the anterior brachium	(O) Distal scapula (I) Radial tuberosity of the radius	Prime mover in forearm flexion, supination of forearm
Brachialis	Deep brachial muscle crossing the elbow joint	(O) Lower anterior humerus (I) Coronoid process of the ulna	Flexion of the forearm
Brachioradialis	More distal muscle separating the anterior and posterior compartments of the forearm	(O) Distal and lateral humerus (I) Distal radius	Flexion of the forearm with the arm in neutral position

(continued)

Table 14.4 Forearm, Wrist, and Hand Muscles (*continued*)

Muscle	Description	Origin (O) and Insertion (I)	Action
Triceps brachii	Superficial muscle spanning the posterior brachium with three heads	(O) Distal scapula and posterior, proximal humerus (I) Olecranon process of the ulna	Extension of the shoulder and elbow
Anconeus	Small muscle crossing the elbow joint	(O) Lateral epicondyle of the humerus (I) Lateral surface of the olecranon	Extension and stabilization of the forearm
Pronator teres	Superficial and proximal muscle of the anterior antebrachium	(O) Medial epicondyle of the humerus and proximal ulna (I) Middle of lateral radius	Pronation of the forearm
Pronator quadratus	Deepest layer of the anterior antebrachial muscles, this muscle is shaped like a quadrangle	(O) Distal anteromedial ulna (I) Distal anterolateral radius	Pronation of the forearm
Supinator	Deep posterior muscle of the antebrachium	(O) Lateral epicondyle of the humerus (I) Lateral surface of the proximal radius	Prime mover for supination
Flexor carpi radialis	Lateral and superficial antebrachial muscle in the anterior compartment and acting on the wrist	(O) Medial epicondyle of the humerus (I) Bases of second and third metacarpals	Flexion and radial deviation of the wrist
Palmaris longus	Very thin and long superficial muscle, not everyone has this muscle	(O) Medial epicondyle of the humerus (I) Flexor retinaculum and proximal phalanges	Weak wrist flexion
Flexor carpi ulnaris	Medial, anterior, and superficial antebrachial muscle acting on the wrist	(O) Medial epicondyle of the humerus and proximal posterior ulna (I) Carpal bones and fifth metacarpal	Flexion and ulnar deviation of the wrist
Flexor digitorum superficialis	Superficial muscle in the anterior compartment and acting on the fingers	(O) Medial epicondyle of the radius and proximal ulna (I) Middle phalanges of digits 2–5	Flexion of digits 2–5 and flexion of the wrist
Flexor pollicis longus	Deep and extrinsic anterior antebrachial muscle	(O) Anterior surface of the radius (I) Distal phalanx of the thumb	Flexion of the thumb
Flexor digitorum profundus	Deep anterior antebrachial muscle acting on the fingers	(O) Proximal anterior ulna (I) Distal phalanges of digits 2–5	Flexion of digits 2–5 and flexion of the wrist
Extensor carpi radialis longus	Superficial posterior antebrachial muscle	(O) Lateral epicondyle of the humerus (I) Posterior surface of the second metacarpal	Extension and radial deviation of the wrist

(*continued*)

Table 14.4 Forearm, Wrist, and Hand Muscles (*continued*)

Muscle	Description	Origin (O) and Insertion (I)	Action
Extensor carpi radialis brevis	Superficial posterior antebrachial muscle running just beneath ECRL	(O) Lateral epicondyle of the humerus (I) Posterior surface of the third metacarpal	Extension of the wrist
Extensor digitorum	Superficial posterior antebrachial muscle	(O) Lateral epicondyle of the humerus (I) Digits 2–5	Extension of the digits and wrist
Extensor digiti minimi	Superficial posterior antebrachial muscle	(O) Lateral epicondyle of the humerus (I) Digit 5	Extension of digit 5 and wrist
Extensor carpi ulnaris	Superficial posterior antebrachial muscle	(O) Lateral epicondyle of the humerus and proximal posterior ulna (I) Fifth metacarpal	Extension and ulnar deviation of the wrist
Abductor pollicis longus	Deep extrinsic muscle running laterally along the antebrachium to the thumb	(O) Posterior surfaces of the radius and ulna (I) First metacarpal	Abduction of the thumb and radial deviation of the wrist
Extensor pollicis brevis	Deep posterior muscle of the antebrachium	(O) Posterior radius (I) Proximal phalanx of the thumb	Extension of the thumb and radial deviation of the wrist
Extensor pollicis longus	Deep posterior muscle of the antebrachium	(O) Posterior ulna (I) Distal phalanx of the thumb	Extension and radial deviation of the wrist
Extensor indicis	Deep posterior antebrachial muscle	(O) Posterior ulna (I) Digit 2	Extension of digit 2 and wrist
Abductor pollicis brevis	Intrinsic thenar muscle, most lateral	(O) Flexor retinaculum and lateral carpal bones (I) Proximal phalanx of the thumb	Abduction of the thumb
Opponens pollicis	Intrinsic thenar muscle	(O) Flexor retinaculum and the trapezium (I) first metacarpal	Opposition of thumb
Flexor pollicis brevis	Intrinsic thenar muscle, deep to opponens pollicis and abductor pollicis brevis	(O) Flexor retinaculum and lateral carpal bones (I) Proximal phalanx of the thumb	Flexion of the thumb
Adductor pollicis brevis	Intrinsic thenar muscle, fibers run between digits 1 and 2	(O) Second and third metacarpals, lateral carpal bones (I) Proximal phalanx of the thumb	Adduction of the thumb
Abductor digiti minimi	Intrinsic hypothenar muscle	(O) Pisiform (I) Proximal phalanx of digit 5	Abduction of digit 5

(*continued*)

Table 14.4 Forearm, Wrist, and Hand Muscles (*continued*)

Muscle	Description	Origin (O) and Insertion (I)	Action
Flexor digiti minimi	Intrinsic hypothenar muscle	(O) Flexor retinaculum and the hamate (I) Proximal phalanx of digit 5	Flexion of digit 5
Opponens digiti minimi	Intrinsic hypothenar muscle	(O) Flexor retinaculum and the hamate (I) Fifth metacarpal	Opposition of digit 5
Lumbricals	Intermediate intrinsic muscles, the only group in the body with no bony attachments!	(O) Tendons of flexor digitorum profundus (I) Extensor expansions of digits 2–5	Flexion and extension of the digits
Palmar interossei	Small intrinsic intermediate muscles, between the bones of the metacarpals on the palmar side	(O) Palmar surfaces of metacarpals (I) Proximal phalanges	Adduction, flexion, and extension of the digits
Dorsal interossei	Small intrinsic intermediate muscles, between the bones of the metacarpals on the dorsal side	(O) Posterior surfaces of metacarpals (I) Proximal phalanges	Abduction, flexion, and extension of the digits

Figure 14.6 **Muscles That Move the Forearm**

The muscles that flex and extend the elbow as well as pronate and supinate the forearm originate in the brachial region. The muscles in the forearm move the wrists, hands, and fingers. (A and B) Anterior and posterior views of the left upper arm, (C and D) palmar and dorsal views of the superficial muscles of the left forearm, and (E and F) palmar and dorsal views of the deep muscles of the left forearm.

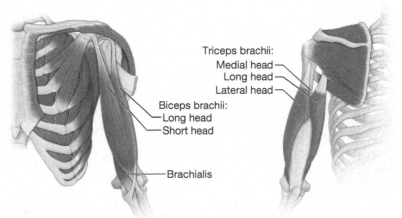

A Anterior view, left upper arm

B Posterior view, left upper arm

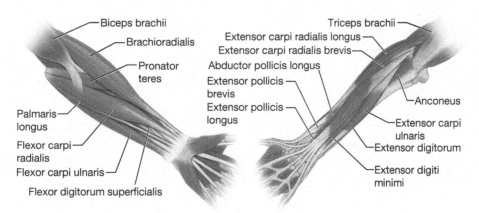

C Palmar view, superficial muscles of the left forearm

D Dorsal view, superficial muscles of the left forearm

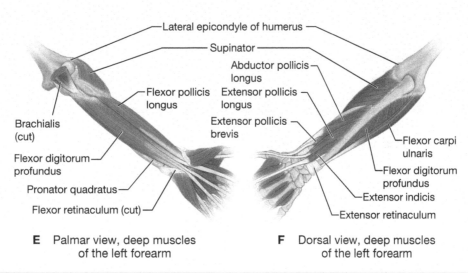

E Palmar view, deep muscles of the left forearm

F Dorsal view, deep muscles of the left forearm

Figure 14.7   **Movements of the Forearm, Wrist, and Fingers**

(A) The elbow is capable of flexion and extension. Muscles on the anterior arm are flexors, and muscles on the posterior are extensors. (B) The radius and ulna are capable of pronation and supination. The movements of the wrist are (C) ulnar and radial deviation, (D) flexion and extension, and (E) pronation and supination. The movements of the fingers are (F) flexion, extension, and hyperextension, (G) abduction and adduction, and (H) circumduction.

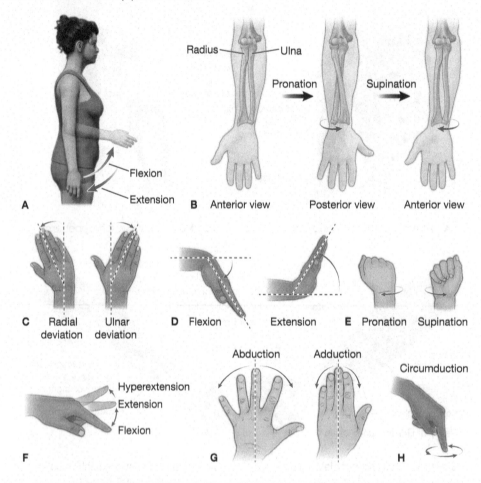

Materials
• Human cadaver or upper limb models or figures

Instructions
1. On the cadaver or model, identify the muscles found in Tables 14.3 and 14.4.

2. Using a mirror or a classmate, fill in Table 14.5.

Figure 14.8 **Intrinsic Muscles of the Hand**

The intrinsic muscles of the hand have origins and insertions within the hand. These muscles provide fine motor control of the fingers. (A and B) Palmar and dorsal views of the superficial muscles of the right hand. (C and D) Palmar and dorsal views of the interossei (interosseous muscles) of the right hand.

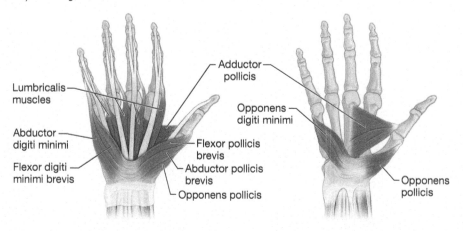

Lumbricalis muscles

Abductor digiti minimi

Flexor digiti minimi brevis

Adductor pollicis

Opponens digiti minimi

Flexor pollicis brevis

Abductor pollicis brevis

Opponens pollicis

Opponens pollicis

A Palmar view, superficial muscles of the right hand

B Palmar view, second superficial muscles of the right hand

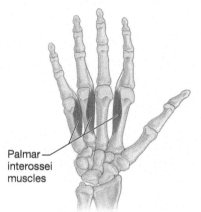

Palmar interossei muscles

Dorsal interossei muscles

C Palmar view, interossei muscles of the right hand

D Dorsal view, interossei muscles of the right hand

Table 14.5 Upper Limb Muscle Movements	
Exercise	**Muscle(s) Involved**
Squeeze your shoulders together toward the midline of your back	
Raise your arms out to the side (abduction)	
Flex your elbow, like a bicep curl	
Extend your fingers, holding them as straight as possible	

(page intentionally left blank)

Activity 14.2 Muscles of the Lower Limb

This activity targets LOs 14.1 and 14.2.

The leg attaches to the trunk at the hip and so many of the muscles that move the leg are found along the lower trunk, crossing the hip and inserting on the femur. These muscles are illustrated in Figure 14.9, and the movements that are possible at the hip are illustrated in Figure 14.10. The muscles that have actions at the hip and knee are listed in Table 14.6.

Figure 14.9	**Hip and Thigh Muscles**

The powerful muscles of the hip originate on the pelvis, cross the hip joint, and insert on the femur. The muscles that flex and extend the knee originate on the femur and insert into the bones of the knee joint or lower leg. Anterior hip muscles provide flexion; posterior muscles provide extension; lateral muscles provide abduction, and medial muscles provide adduction. Knee flexors are on the posterior side, while extensors are on the anterior side. Pelvic and thigh muscles, right leg: (A) anterior view, deep muscles; (B) anterior view of the muscles, and (C) posterior view, superficial muscles.

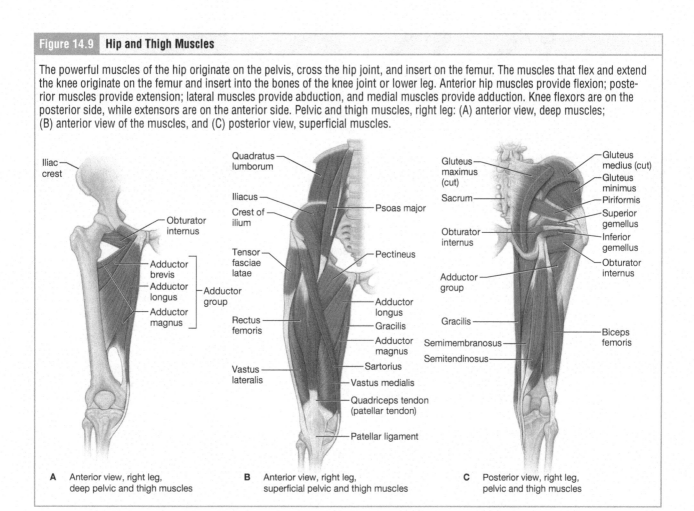

A Anterior view, right leg, deep pelvic and thigh muscles

B Anterior view, right leg, superficial pelvic and thigh muscles

C Posterior view, right leg, pelvic and thigh muscles

Figure 14.10 | Movements of the Hip

The hip joint has a limited range of motion compared to the shoulder. The leg can (A) abduct, (B) adduct, (C) flex, and (D) extend at the hip.

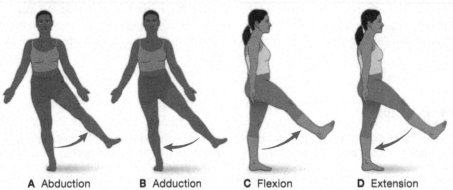

A Abduction **B** Adduction **C** Flexion **D** Extension

Table 14.6 Muscles of the Hip and Thigh

Muscle	Description	Origin (O) and Insertion (I)	Action
Gluteus maximus	Largest and most superficial gluteal muscle	(O) Posterior sacrum (I) Posterior proximal femur, iliotibial tract	Prime mover of hip extension, medial rotation of the hip
Gluteus medius	Intermediate layer of gluteal muscles	(O) Posterior ilium (I) Greater trochanter of the femur	Abduction and medial rotation of the hip
Gluteus minimus	Smallest and deepest gluteal muscle	(O) Posterior ilium (I) Greater trochanter of the femur	Abduction and medial rotation of the hip
Tensor fascia latae	Muscle belly that continues to the iliotibial tract, on the superior lateral aspect of the thigh	(O) Anterior ilium (I) Iliotibial tract	Stabilization of the knee, hip flexion, abduction, and medial rotation
Piriformis	Deep muscle and a short lateral rotator of the hip	(O) Anterior sacrum (I) Greater trochanter of the femur	Lateral rotation of the hip
Obturator internus	Deep posterior muscle that is a part of the short lateral rotator group	(O) Posterior rim of the obturator foramen (I) Greater trochanter of the femur	Lateral rotation of the hip
Obturator externus	Deep anterior muscle that is a part of the short lateral rotator group	(O) Anterior rim of the obturator foramen (I) Greater trochanter of the femur	Lateral rotation of the hip
Superior gemellus	A small and deep gluteal muscle assisting the hip external rotators	(O) Ischial spine (I) Greater trochanter of the femur	Lateral rotation and abduction of the hip
Inferior gemellus	A small and deep gluteal muscle assisting the hip external rotators	(O) Ischial tuberosity (I) Greater trochanter of the femur	Lateral rotation and abduction of the hip
Quadratus femoris	Deep posterior muscle part of the short lateral rotator group	(O) Ischium (I) Posterior proximal femur	Lateral rotation of the hip

(continued)

Table 14.6 Muscles of the Hip and Thigh (*continued*)

Muscle	Description	Origin (O) and Insertion (I)	Action
Adductor longus	Anterior and medial muscle of the thigh, part of the adductor group	(O) Anterior pubis (I) Posterior and medial femur	Adduction and flexion of the hip
Adductor brevis	Anterior and medial muscle of the thigh, part of the adductor group	(O) Inferior pubis (I) Posterior and medial femur	Adduction and flexion of the hip
Adductor magnus	Deepest layer of the medial thigh, this muscle makes up the floor of the posterior thigh and has two heads	(O) Ischium and pubis (I) Posterior medial femur and distal medial femur	Adduction and extension of the hip
Pectineus	Short and deep adductor muscle in the anterior thigh	(O) Superior pubis (I) Proximal and medial femur	Adduction and flexion of the hip
Gracilis	Most medial muscle of the thigh, straplike, and attaching at the knee joint	(O) Pubis (I) Proximal tibia	Adduction of the hip and flexion of the knee
Rectus femoris	Anterior compartment of the thigh, part of the quadriceps group (this is the only muscle of the group that also acts on the hip)	(O) Anterior ilium (I) Tibia via quadriceps tendon	Knee extension and hip flexion
Vastus lateralis	Anterior compartment of the thigh, part of the quadriceps group	(O) Proximal posterior femur and greater trochanter (I) Tibia via quadriceps tendon	Knee extension
Vastus medialis	Anterior compartment of the thigh, part of the quadriceps group	(O) Proximal posterior femur (I) Tibia via quadriceps tendon	Knee extension
Vastus intermedius	Anterior compartment of the thigh, part of the quadriceps group. This muscle is deep to rectus femoris	(O) Anterior and lateral femur (I) Tibia via quadriceps tendon	Knee extension
Sartorius	Anterior and superficial thigh muscle performing similar actions on both the hip and knee	(O) Anterior ilium (I) Proximal medial tibia	Flexion, abduction, and lateral rotation of the hip, flexion and medial rotation of the knee
Biceps femoris	Posterior thigh muscle with two heads, part of the hamstring group	(O) Ischium and posterior femur (I) Lateral surfaces of tibia and fibula	Knee flexion and hip extension
Semitendinosus	Posterior superficial thigh muscle, part of the hamstring group	(O) Ischium (I) Proximal medial tibia	Knee flexion and hip extension
Semimembranosus	Posterior thigh muscle, part of the hamstring group and deep to semitendinosus	(O) Ischium (I) Posterior and medial tibia	Knee flexion and hip extension

Like the forearm and hand, the lower leg houses many extrinsic muscles of the ankle and foot. The foot also has intrinsic muscles, many of which share a name with their counterparts in the hands! For example, both the hands and the feet have muscles termed "interossei." The muscles of the feet are illustrated in Figures 14.11 and 14.12 and are listed in Table 14.7.

Materials
• Human cadaver or lower limb models or images

Instructions
1. On the cadaver or model, identify the muscles found in Tables 14.6 and 14.7.

2. Using a mirror or a classmate, fill in Table 14.8.

Figure 14.11 | Muscles of the Lower Leg

The muscles of the lower leg act on the ankle and foot. (A) The anterior compartment muscles are dorsiflexors, and (B) the muscles of the posterior compartment are plantar flexors. (C) The lateral and medial muscles invert, evert, and rotate the foot.

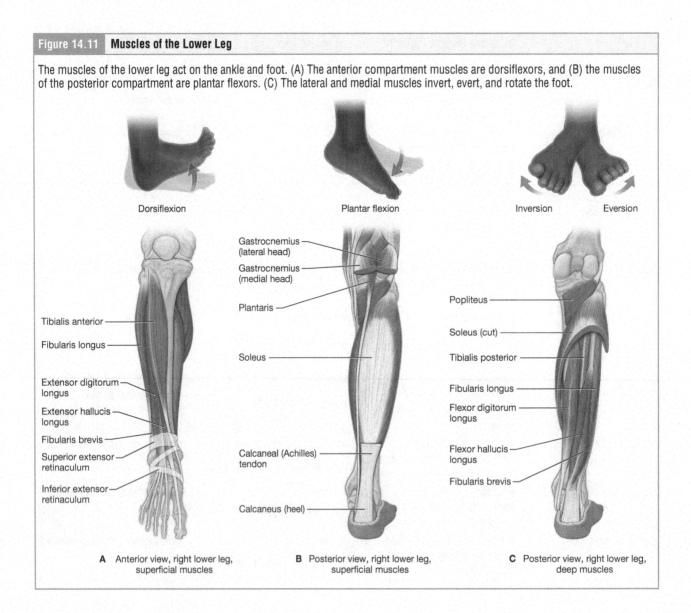

Dorsiflexion — Plantar flexion — Inversion — Eversion

Tibialis anterior
Fibularis longus
Extensor digitorum longus
Extensor hallucis longus
Fibularis brevis
Superior extensor retinaculum
Inferior extensor retinaculum

Gastrocnemius (lateral head)
Gastrocnemius (medial head)
Plantaris
Soleus
Calcaneal (Achilles) tendon
Calcaneus (heel)

Popliteus
Soleus (cut)
Tibialis posterior
Fibularis longus
Flexor digitorum longus
Flexor hallucis longus
Fibularis brevis

A Anterior view, right lower leg, superficial muscles

B Posterior view, right lower leg, superficial muscles

C Posterior view, right lower leg, deep muscles

Figure 14.12 | Intrinsic Muscles of the Foot

The muscles of the sole flex the toes. There are three layers of plantar muscles: (A) superficial, (B) intermediate, and (C) deep. (D) The muscles along the dorsal side of the foot extend the toes.

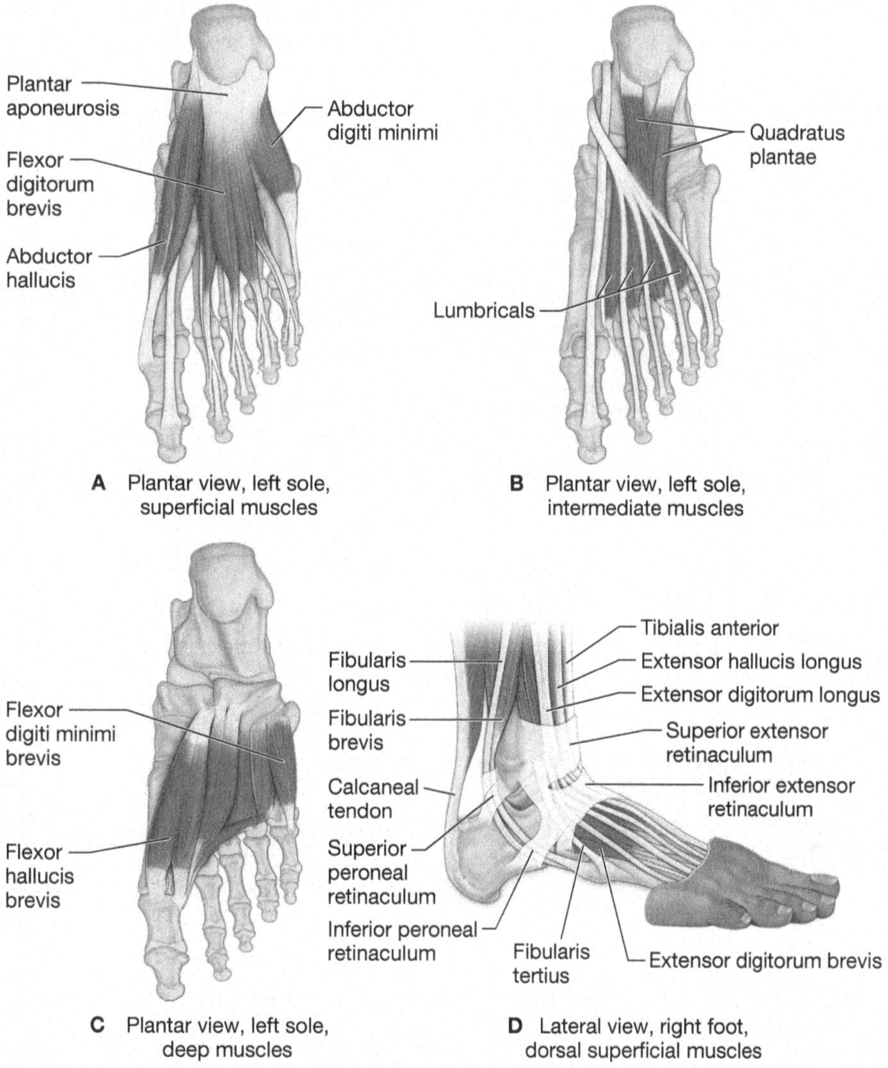

A Plantar view, left sole, superficial muscles

Labels: Plantar aponeurosis; Abductor digiti minimi; Flexor digitorum brevis; Abductor hallucis

B Plantar view, left sole, intermediate muscles

Labels: Quadratus plantae; Lumbricals

C Plantar view, left sole, deep muscles

Labels: Flexor digiti minimi brevis; Flexor hallucis brevis

D Lateral view, right foot, dorsal superficial muscles

Labels: Fibularis longus; Fibularis brevis; Calcaneal tendon; Superior peroneal retinaculum; Inferior peroneal retinaculum; Tibialis anterior; Extensor hallucis longus; Extensor digitorum longus; Superior extensor retinaculum; Inferior extensor retinaculum; Fibularis tertius; Extensor digitorum brevis

Table 14.7 Muscles of the Leg and Foot

Muscle	Description	Origin (O) and Insertion (I)	Action
Tibialis anterior	Anterior lower leg, overlaying the lateral portion of the tibia	(O) Lateral condyle and upper portion of the shaft of the tibia (I) Tendon inserts on the inferior surface of medial cuneiform and first metatarsal	Prime mover for dorsiflexion, inversion of the foot
Extensor hallucis longus	Extrinsic muscle in the anterior compartment of the lower leg	(O) Anterior shaft of the fibula (I) Distal phalanx of the great toe	Extension of the great toe
Extensor digitorum longus	Extrinsic muscle in the anterior compartment of the lateral lower leg	(O) Proximal anterior shaft of the fibula (I) Extensor expansion of toes 2–5	Extension of toes 2–5 and dorsiflexion of the ankle
Fibularis tertius	Small and deep muscle on the lateral lower leg whose tendon is visible but whose muscle belly is not in a superficial view	(O) Distal anterior fibula (I) Base of fifth metatarsal	Eversion of the foot and dorsiflexion of the ankle
Fibularis longus	Superficial muscle of the lateral compartment of the lower leg	(O) Proximal and lateral shaft of the fibula (I) First metatarsal and medial cuneiform	Eversion of the foot and plantarflexion of the ankle
Fibularis brevis	Muscle of the lateral compartment of the lower leg, deep to fibularis longus	(O) Distal and lateral shaft of the fibula (I) Lateral surface of the fifth metatarsal	Eversion of the foot and plantarflexion of the ankle
Soleus	Wide and flat intermediate muscle of the posterior lower leg	(O) Posterior tibia and proximal posterior fibula (I) Calcaneus via the calcaneal tendon	Plantarflexion of the ankle
Gastrocnemius	Large and powerful superficial muscle with two heads, found in the posterior compartment of the lower leg	(O) Medial and lateral epicondyles of the femur (I) Calcaneus via the calcaneal tendon	Prime mover for plantarflexion of the ankle and flexion of the knee
Plantaris	Very thick, tendinous muscle running between the soleus and gastrocnemius	(O) Lateral supracondylar ridge of the femur (I) Posterior calcaneus	Plantarflexion of the ankle and flexion of the knee
Popliteus	Small muscle behind the knee joint in the deep layer of the posterior lower leg	(O) Lateral condyle of the femur (I) Proximal posterior tibia	Medial rotation of the tibia and lateral rotation of the femur to assist in knee flexion
Flexor digitorum longus	Deep and extrinsic muscle of the lower leg, tendons run medially	(O) Posterior tibia (I) Distal phalanges of toes 2–5	Flexion of toes 2–5 and plantarflexion of the ankle
Flexor hallucis longus	Deep and extrinsic muscle of the lower leg, tendons run medially	(O) Distal posterior fibula (I) Distal phalanx of great toe	Flexion of the great toe
Tibialis posterior	Deep muscle in the posterior leg running along the surface of the tibia	(O) Posterior tibia and proximal posterior fibula (I) Plantar surfaces of tarsals and metatarsals	Inversion of the foot and plantarflexion of the ankle
Extensor digitorum brevis	Intrinsic muscle of the foot on the dorsal surface	(O) Anterior and lateral calcaneus (I) Extensor expansion of toes 1–4	Extension of toes 1–4

Table 14.8 Lower Limb Muscle Movements	
Exercise	**Muscle(s) Involved**
Flexing your hips, as if mimicking a sitting position	
Flexing your knee, as if bringing your foot toward your butt	
Plantar flexion of your foot, so that you are standing on your toes	
Dorsiflexion of your foot, bringing your toes toward your shins	
Flexing your toes so that they curl within your shoes	

(page intentionally left blank)

Activity 14.3 Case Study: Fibular Nerve Damage

The common peroneal nerve (also known as the common fibular nerve) is one of the two nerves of the lower leg. It passes along the lateral side of the knee very superficially and winds around the proximal portion of the fibula before branching into the superficial peroneal (fibular) nerve and the deep peroneal (fibular) nerve.

Bess is in a car accident and suffered an injury to their lower leg. Upon arrival at the hospital they report a loss of sensation along the anterior ankle, the top of the foot and numbness in all but their smallest toe. The neurologist that consults on Bess's case asks if they are able to point or flex their toes and Bess is only able to point their toes. The neurologist writes "drop foot" down on Bess's chart, this is a condition in which a patient has trouble lifting their toes.

1. Specifically the neurologist asks Bess to perform the two movements illustrated here. What are the correct names for these movements?

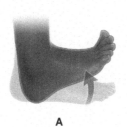
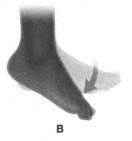

   A   B

   a. Movement A is plantarflexion, movement B is dorsiflexion.
   b. Movement A is dorsiflexion, movement B is plantarflexion.
   c. Movement A is eversion, movement B is inversion.
   d. Movement A is inversion, movement B is eversion.

2. Bess cannot do the motion seen in the image. Which muscles are responsible for this movement?
   a. Tibialis anterior
   b. Fibularis brevis
   c. Soleus

3. The neurologist asks Bess to perform eversion and inversion. Which of the images here shows eversion?
   a. A
   b. B
   c. C

   A   B   C

4. Bess also cannot perform eversion. Which muscle is responsible for this movement?
   a. Fibularis tertius
   b. Soleus
   c. Gastrocnemius
   d. Plantaris

Bess is diagnosed with peroneal nerve damage and is scheduled for surgery to repair the nerve through nerve grafting. They have a long course of physical therapy ahead of them, but the doctors are hopeful that all peroneal nerve function will be restored.

Lab 14 Post-Lab Quiz

1. This muscle is the only muscle that flexes both the hip and the knee (hint: it needs to cross both of these joints).

   a. Adductor magnus
   b. Sartorius
   c. Gracilis
   d. Semitendinosus

2. Which of the muscles shown in the image is capable of pronating the forearm?

   a. A
   b. B
   c. C

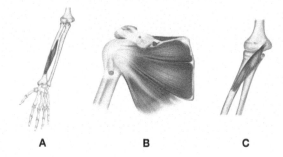

A   B   C

3. Which of the following arm muscles does NOT act at the elbow?

   a. Biceps brachii
   b. Brachialis
   c. Coracobrachialis
   d. Triceps brachii

4. Which of the following produces adduction of the hip/thigh?

   a. Iliopsoas
   b. Gracilis
   c. Gluteus maximus

5. If adductor longus was injured and could not perform its function, would a person still be capable of adduction of the hip/thigh?

   a. Yes, other muscles in the adductor compartment contribute to adduction and could still function.
   b. Yes, other muscles in the gluteal compartment could take over this function.
   c. No, adductor longus is the prime mover and the only adductor, without this muscle, adduction would be impossible.

6. Which of the following muscles is the prime mover of hip flexion?

   a. Iliopsoas
   b. Gracilis
   c. Sartorius
   d. Tensor fascia lata

7. Rebecca has just begun running. After 4 weeks she begins to fell pain in her right foot and ankle. Her partner is a massage therapist and tries to heal her through lower leg massage. The massage therapist notes that her right peroneus (fibularis) muscles are rigidly contracted and refuse to relax. When Rebecca is standing, which of the following might be noticeable about Rebecca's foot?

   a. Inversion
   b. Eversion
   c. Dorsiflexion
   d. Plantarflexion

8. Foot drop is a condition in which the toes of one foot never fully leave the ground during walking. It is caused by a weak or inhibited muscle. Weakening of which of the following muscles could cause foot drop?

   a. Flexor digitorum longus
   b. Tibialis anterior
   c. Tibialis posterior
   d. Gastrocnemius

9. Which of these muscles does not attach to the scapula?

   a. Pectoralis major
   b. Trapezius
   c. Teres major
   d. Serratus anterior

10. Which of the adductor muscles listed is the shortest? (Hint: you should be able to answer this by the names alone)

   a. Adductor magnus
   b. Adductor longus
   c. Adductor brevis

Lab 15 Introduction to the Nervous System

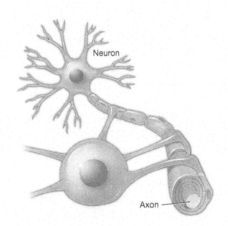

Neuron

Axon

Learning Objectives: After completing this lab, you should be able to:

15.1 Compare and contrast the central nervous system (CNS) and peripheral nervous system (PNS) with respect to structure and function.

15.2 Describe the nervous system as a control system with the following components: sensory receptors, afferent pathways, control (integrating) center, efferent pathways, and effector (target) organs.

15.3* Organize the nervous system based on an innervation approach that differentiates somatic and autonomic subdivision.

15.4 Identify and describe the major components of a typical neuron (e.g., cell body, nucleus, nucleolus, chromatophilic substance [Nissl bodies], axon hillock, dendrites, and axon) and indicate which parts receive input signals and which parts transmit output signals.

15.5 Define neural (neuronal) circuit.

15.6 Describe the structure, location, and function of each of the six types of neuroglia (glial) cells.

15.7 Define myelination and describe its function, including comparing and contrasting how myelination occurs in the CNS and PNS.

15.8 List the major ion channels of neurons and describe them as leak (leakage or passive) or voltage-gated channels, mechanically gated channels, or ligand-gated (chemically gated) channels, and identify where they typically are located on a neuron.

15.9 Define and describe depolarization, repolarization, hyperpolarization, and threshold.

15.10 Describe the physiological process involved in the conduction (propagation) of an action potential, including the types and locations of the ion channels involved.

15.11 Describe the structures involved in a typical chemical synapse (e.g., axon terminal [synaptic knob], voltage-gated calcium channels, synaptic vesicles of presynaptic cell, synaptic cleft, neurotransmitter receptors of the postsynaptic cell).

* Objective is not a HAPS Learning Goal.

The Human Anatomy and Physiology Society includes more than 1,700 educators who work together to promote excellence in the teaching of this subject area. The HAPS A&P Learning Outcomes measure student mastery of the content typically covered in a two-semester Human A&P curriculum at the undergraduate level. The full Learning Outcomes are available at https://www.hapsweb.org.

Introduction

The nervous system is involved in receiving information about the environment (both our internal environment and what is around us) and generating responses to that information. The nervous system is complex in both structure and function. To understand these complexities, scientists have devised several ways of describing and organizing it. We can approach our study of the nervous system through a functional lens, a structural lens, and an innervation lens. From a functional perspective, the nervous system can be divided into regions that provide sensory functions, regions that integrate sensory input with higher cognitive functions, and regions that respond to the sensory input based on that integration. From a structural perspective, the nervous system is made up of the brain, spinal cord, and nerves. Finally, the nervous system can be examined from an innervation perspective, in which it is separated into activities that generate skeletal muscle actions and activities that generate actions in cardiac muscle, smooth muscle, and glands.

All nervous tissue is formed from the same cells: **neurons**. Neurons function by sending messages, called **action potentials**, along their length to their target cells. While the main communication function of the brain, spinal cord, and nerves is carried out by these neurons, they are also supported by a group of smaller cells called **glial (neuroglial) cells**.

One important function served by glial cells is **myelination**, which is the process of insulating axons in the nervous system with **myelin**, a type of specialized cell membrane. Myelination increases the speed of action potentials by ten to twenty times compared with unmyelinated axons, allowing for efficient communication between distant parts of the body.

Lab 15 Pre-Lab Quiz

This quiz will strengthen your background knowledge in preparation for this lab. For help answering the questions, use your resources to deepen your understanding. The best resource for help on the first five questions is your text, and the best resource for help answering the last five questions is to read the introduction section of each lab activity.

1. Anatomy of a neuron. Label the figure with the following terms:

    axon
    axon terminals
    cell body
    dendrites
    hillock
    myelin sheath
    nucleus

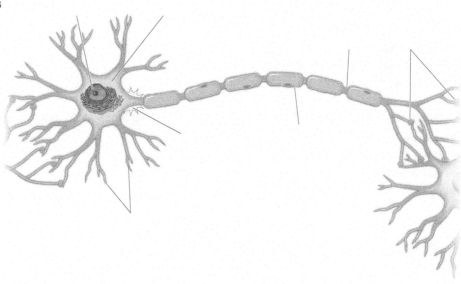

2. Match each of the roots to its definition:

Root	Meaning
	nerve
	small
	self
	body
	membrane
	star
	many

List of Roots

astro
auto
lemma
micro
neuro
oligo
soma

3. Graded potentials occur on the _____ of a neuron, while action potentials occur on the _____ of a neuron.

    a. dendrites and cell body; axon
    b. axon; dendrites and cell body

4. Are all neurotransmitters excitatory?

   a. Yes
   b. No

5. What cell releases neurotransmitter at a synapse?

   a. Presynaptic neuron
   b. Postsynaptic neuron
   c. Glial cell
   d. Ependymal cell

6. What is the primary function of neurons?

   a. Myelinate axons
   b. Support the brain and spinal cord
   c. Send action potentials

7. Which structures make up the central nervous system (CNS)? (choose all correct answers)

   a. Spinal nerves
   b. Spinal cord
   c. Cranial nerves
   d. Brain

8. What is the junction where neurons connect with other cells?

   a. Axon hillocks
   b. Synapses
   c. Myelin

9. What makes up the myelin sheath that surrounds the axons of many of the neurons in the human body?

   a. Phospholipids
   b. Cytoplasm
   c. Both phospholipids and cytoplasm

10. What is contained in the synaptic vesicles of the synaptic terminals of neurons?

   a. Ca^{2+}
   b. Na^+
   c. Neurotransmitter

Activity 15.1 Organization of the Nervous System

This activity targets LOs 15.1, 15.2, and 15.3.

Consider the three different lenses that have been presented for studying the nervous system: a functional approach, a structural approach, and an innervation approach.

The function of the nervous system is to take in sensory information, process that information and integrate it with higher cognitive functions such as memory and learning, and then produce a response that is often called a motor function (whether it results in movement of the body or not). We most often think of the sensation–integration–response cycle through the lens of our conscious perception and voluntary responses such as the example shown in Figure 15.1. But these cycles also function in areas of our nervous system that are unconscious, such as when our body adjusts blood pressure as we change body position or when our body releases digestive enzymes into our stomach as we consume food.

The functions that send sensory information are also called **afferent**, as this directional term means "to" or "toward." Thus, we might describe incoming neural signals to the integrative areas of our brain as afferent messages and the neurons that send these signals can be called afferent neurons. In contrast, the functions that send motor information are called **efferent**, as this directional term means "away" or "exiting." We can describe neural signals leaving the integrative areas of the brain and spinal cord as efferent signals and the neurons that send these signals as efferent neurons. Further, the structures that receive signals from efferent neurons are called **effectors**.

Figure 15.1	The Nervous System Divisions of Sensation, Integration, and Response

The nervous system has sensory components that receive and deliver sensory information to the brain and spinal cord. The central nervous system is the location where information is integrated and processed. A response is sent to an effector along a motor pathway.

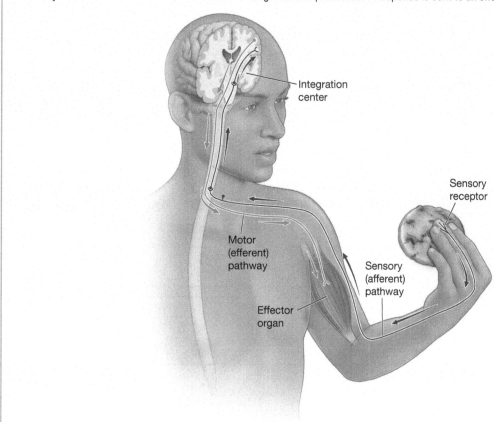

When we examine the nervous system from an anatomical perspective, it may seem simple on the surface. The nervous system is made up of only three structures: the brain, the spinal cord, and many bundles of nerves. Further, we can categorize these three structures into the **central nervous system (CNS)** or the **peripheral nervous system (PNS)**. The central nervous system structures, the brain and spinal cord, are protected within bony cavities and many of the substructures of these two organs perform the integrative function described above. In contrast, the peripheral nervous system consists only of the many nerves that branch from the CNS structures, including both cranial nerves and spinal nerves. The nerves of the body do not serve any integrative function; rather, nerves send sensory input to the integrative areas of the brain and spinal cord or motor output to effectors.

The nervous system can also be divided based on what structures are innervated by the nerves. In such an innervation approach, the **somatic nervous system (SNS)** is responsible for conscious perception and voluntary motor responses. Conscious perception occurs through the senses of touch, taste, smell, vision, and hearing; voluntary motor responses occur through the contraction of skeletal muscle. The **autonomic nervous system (ANS)** is responsible for involuntary control of cardiac muscle, smooth muscle, and glands.

The purpose of this activity is to allow you to clarify your own understanding of possible methods of organizing the nervous system.

Instructions

1. Complete Table 15.1. Use the introduction above and your textbook as resources. Some of the information in the first row has been completed for you to get you started, but you should add more to each of the boxes in this row.

2. Share your completed table with your lab partner. Discuss your ideas, and make additions and alterations as necessary.

Table 15.1 Organizational Approaches to the Nervous System

Approach to Studying the Nervous System	Subdivisions of This Approach	Definition of Each Subdivision	Example or Further Clarification
Functional approach	Afferent	Take in sensory information	Seeing or touching something, food in the stomach
	Integrate		
	Efferent		
Anatomical approach			
Innervation approach			

Activity 15.2 **Anatomy of a Neuron**

This activity targets LOs 15.4 and 15.5.

Neurons are responsible for the communication that the nervous system provides. As with all cells of the body, their structure is uniquely suited to their function. Examine Figure 15.2 and consider how each of the following structural components of a neuron support the ability of the cell to send a message along its length. As you read ahead, note that this communication, in the form of an electrical signal, flows through a neuron from the dendrites, across the cell body, and down the axon.

The part of a neuron where most of the organelles reside is the cell body. Areas of significant rough endoplasmic reticulum in the cell body stain darkly, giving them the alternate name **chromatophilic substance**. Neurons have many extensions from their cell body that allow them to interact with other cells, including cells of sensory structures, effector cells, and other neurons. One of the extensions is an **axon**. A single axon often branches repeatedly at its terminus enabling it to communicate with many target cells, such as neurons or effector cells. The ends of these branches that reach toward neighbor cells are called **axon terminals**. Axons allow information that has entered the neuron to be passed on to target cells. Typically, a neuron will have a single axon departing from the cell body. Where the cell body narrows to become the axon is called the **axon hillock**. Most axons are wrapped in a fatty insulation called **myelin,** to speed the transmission of this information. The myelin wraps the axon in short sections so that there are brief bare patches of axon called **neurofibril nodes** (nodes of Ranvier) between each myelinated section. Emanating from the cell body are other processes of the neuron called **dendrites**, which receive information from other neurons or from sensory structures of the body.

Consider this information as you complete Table 15.2.

| Figure 15.2 | **Anatomy of a Neuron** |

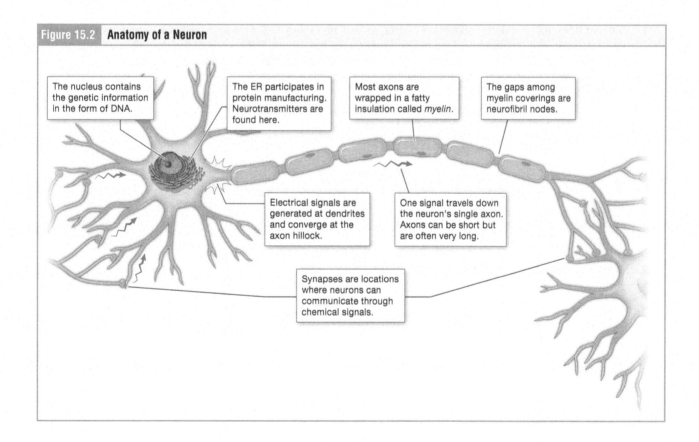

The nucleus contains the genetic information in the form of DNA.

The ER participates in protein manufacturing. Neurotransmitters are found here.

Most axons are wrapped in a fatty insulation called *myelin*.

The gaps among myelin coverings are neurofibril nodes.

Electrical signals are generated at dendrites and converge at the axon hillock.

One signal travels down the neuron's single axon. Axons can be short but are often very long.

Synapses are locations where neurons can communicate through chemical signals.

Table 15.2 Comparing a Neuron's Membrane Projections		
Membrane Projection	Types of Cells That Will Be Reached by This Membrane Extension	Direction of Information Flow (circle one)
Axon		Entering the neuron/exiting the neuron
Dendrites		Entering the neuron/exiting the neuron

As you can see in Figure 15.2, both axons and dendrites form **synapses**, these are the junctions where neurons connect with other cells. Synapses are not physical connections where one cell touches another, but rather locations where the two cells are so close to each other that they can efficiently communicate via secreted chemical signals that have to travel only a very short distance across the synapse.

The purpose of this exercise is to help you to think about neuron anatomy at two levels. First, you will create a neuron, naming all of the structures that make it up. Then you will connect your neuron to others so that you can consider how the anatomy of neurons allows the nervous system to perform its overall function of communication throughout the body.

Materials
• Blank paper
• Colored pencils, if desired
• Models and/or charts of a neuron, if available

Instructions
1. With aid from the provided models and charts, as well as Figure 15.2 and your textbook, use a blank sheet of paper to draw your own neuron. Use the entire page, so that some of the ends of the dendrites of your neuron touch one edge of the page and some of the axon terminals touch another edge of the page. It may look something like one of these sketches (but with more detail!).

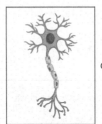 or 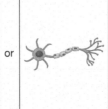 or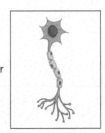

2. Label the following items on your sketch:

   axon
   axon hillock
   cell body
   dendrites
   myelin sheaths
   neurofibril nodes
   nucleus
   rough ER
   synaptic terminals

3. Draw arrows to represent the direction of information flow: where does information enter the neuron? Where does it exit?

4. Form a group of at least four people. Alternately, if you have a group with fewer students, sketch out several more neurons so that you have at least four neurons to work with. Arrange your neurons on the table so that they form a neural circuit (pathway) in which one neuron leads to the next. Think about what structures connect one neuron to another as you arrange your neurons.

5. Find the first neuron in the neural circuit.
   a. If this first neuron is an afferent neuron receiving somatic sensory messages, what structure in the body will house it? Write this location down on the page with this neuron.
   b. What must attach to the dendrites of that neuron? Take another blank piece of paper and write or draw what will be here and place it at the start of the pathway.

6. Find two neurons in the middle of the neural circuit. Consider these neurons the ones that serve an integrative function. What possible locations in the body will contain these neurons? Write down these possible locations on the pages with these neurons.

7. Find the final neuron in the neural circuit.
   a. If this final neuron is an efferent neuron sending somatic motor messages, what structure in the body will house it? Write this location down on the page with this neuron.
   b. What must attach to the axon synapses of that neuron? Take another blank piece of paper and write or draw what will be here and place it at the end of the pathway.

8. Are neural circuits this simple? In other words, do neurons receive messages from only a single neuron before them in the chain? And do neurons send a message to only a single neuron following them in the chain? Do all neural circuits have only one incoming afferent neuron and one outgoing motor neuron? How might you elaborate on the pathway on the table to show the complexity of many neural pathways? Discuss with the group and execute your plan.

9. So that you can recall your work later, take a picture of your completed neural circuit or quickly sketch it out on a single page of blank paper.

(page intentionally left blank)

Activity 15.3 **The Glial Cells**

This activity targets LO 15.6.

Glial (neuroglial) cells do not transmit electrical impulses, but their supportive actions in the tissue of the nervous system are important for ensuring that the actions of the neurons occur. Glial cells are much smaller than the neurons, and are found surrounding neurons in the extracellular matrix of the brain, spinal cord, and nerves. There are six types of glial cells, each with specific supportive functions. Four are found in the CNS and two are found in the PNS.

Each of the glial cells of the brain and spinal cord have specialized shapes and specific functions, as can be seen in Figure 15.3. **Astrocytes**, the most numerous of the CNS glial cells, have the most diverse actions. Generally, these star-shaped cells regulate the extracellular environment around neurons. This includes regulating the

Figure 15.3	Glial Cells of the CNS

(A) Astrocytes make up the blood-brain barrier and regulate the environment around neurons. (B) Microglia serve as resident immune cells in the CNS. (C) Oligodendrocytes myelinate CNS axons. (D) Ependymal cells create the cerebrospinal fluid.

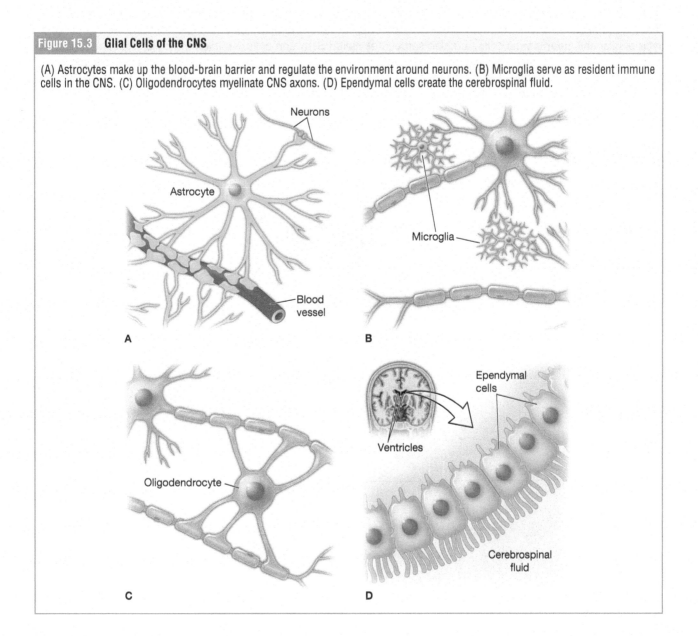

concentration of extracellular chemicals, removing excess neurotransmitter from the synaptic cleft, and either blocking or facilitating the formation of synapses between nearby neurons. Additionally, astrocytes wrap their processes around capillaries in the brain, covering them and creating a barrier that regulates which materials in the blood enter the extracellular space around the neurons. **Microglia** are specialized immune system cells for the CNS since the other white blood cells cannot cross the astrocyte barrier. These tiny cells are excellent phagocytes, ingesting wastes and debris as well as damaged cells and pathogens. **Oligodendrocytes** are the providers of myelin in the CNS. These large octopus-shaped cells reach out processes to wrap around sections of nearby neurons; the wrappings of many oligodendrocytes create a series of insulated wrappings around the length of each axon in the CNS. **Ependymal cells** are shaped like simple cuboidal epithelial cells with cilia. They line spaces in the brain called **ventricles**. The ventricles contain **cerebrospinal fluid (CSF)**, which is produced by the ependymal cells as they carefully select which components of nearby capillaries should be added to the space. The ependymal cells beat their cilia to move the CSF through the connected ventricles, bathing the tissues of the brain in CSF.

There are only two kinds of glial cells found in the peripheral nervous system, shown in Figure 15.4. The **satellite cells**, as their name implies, flatten themselves all around the cell bodies of PNS neurons. Satellite cells help regulate the extracellular environment around these neurons, similar to the activities of the CNS astrocytes. The second type of glial cell, **neurilemma cells**, are also sometimes called Schwann cells. Neurilemma cells, like oligodendrocytes in the CNS, myelinate PNS axons by wrapping around them. The structure of a neurilemma cell is somewhat different from an oligodendrocyte in that each neurilemma cell fully wraps around only a single section of axon, rather than reaching out to myelinate multiple sections as oligodendrocytes do.

Figure 15.4 | **Glial Cells of the PNS**

Satellite cells (A) and neurilemma cells (B) are the types of glial cells present in the peripheral nervous system.

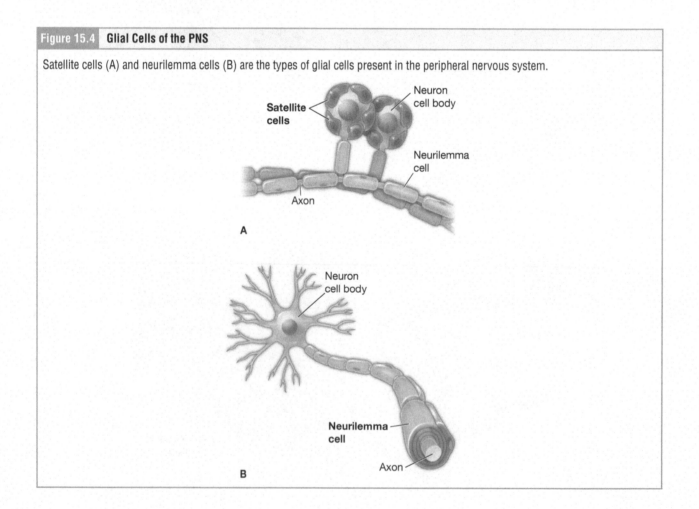

Instructions

1. Glial cells can be distinguished from each other based on their structure and location. For each of the structural descriptions in Table 15.3, identify which glial cell type it describes.

Table 15.3 Types of Glial Cells Based on Structure	
Structure Description	**Glial Cell Type**
Large cells of the CNS that wrap processes around capillaries	
Cube-shaped cells with cilia lining the ventricles of the brain	
Flattened cells surrounding PNS neuron cell bodies	
Large cells of the CNS that wrap processes around neuron axons	
Flattened cells wrapping around PNS neuron axons	
Small cells of the CNS that roam the extracellular environment	

2. Glial cells perform many necessary functions in order to support neuron activity. For each of the functions listed in Table 15.4, identify which cell will perform the task. Some tasks are performed by more than one glial cell type, so some functions will have more than one answer. Some glial cells perform more than one task, so cell types can be used more than once.

Table 15.4 Types of Glial Cells Based on Function	
Function Description	**Glial Cell Type**
Myelinate neuron axons	
Produce CSF	
Regulate the extracellular environment	
Regulate the materials from the blood that enter CNS extracellular spaces	
Remove debris and dead/damaged cells from the extracellular environment	
Circulate the CSF through the ventricles of the brain	

(page intentionally left blank)

Activity 15.4 Anatomy of the Myelin Sheath

This activity targets LO 15.7.

Both oligodendrocytes and neurilemma cells form myelin, the insulation for axons in the nervous system. This insulation is necessary to provide an additional barrier around the axon in addition to the selectively permeable barrier of the phospholipids that make up the cell membrane. The **myelin sheath** is actually many layers of additional barrier as the cell membrane of the oligodendrocyte or neurilemma cells wraps around the axon many times, squishing any cytoplasm in that section of the cell back toward the unwrapped areas. This creates many extra layers of phospholipids between the inside of the neuron and its external environment. In fact, the abundant layers of fatty phospholipids give myelinated areas of the brain and spinal cord a white coloration; these areas are called **white matter** as a result. As the cell membrane wraps around the axon, these phospholipids greatly reduce the ability of ions, which are small charged particles, to leak in or out of the axon where it is covered by the myelin sheath. While the length of the axon is covered in this series of wrappings, the dendrites, cell bodies, and axon terminals are not insulated.

While the function of oligodendrocytes and neurilemma cells is the same, their structures differ in several ways. As can be seen in Figure 15.5, while a neurilemma cell wraps its entire flattened surface around a single section of axon, an oligodendrocyte reaches out many small flattened "arms" that can each wrap a small area of an axon. Neither the final cytoplasm-filled section of the neurilemma cell nor the separate cytoplasm-filled "head" of the oligodendrocyte are part of the myelin sheath that each cell provides. This is because myelination is only the hydrophobic, phospholipid layers of the cell membrane; the cytoplasm of a cell is not a resistor to ion leakage.

The purpose of this activity is to explore the structure and function of myelinating glial cells by creating a large-scale representation of their myelination of a neuron axon.

Figure 15.5 The Process of Myelination

In both the peripheral (A) and central (B) nervous system, glial cells wrap their membranes many times around the axons of neurons, providing electrical insulation to prevent ion leakage.

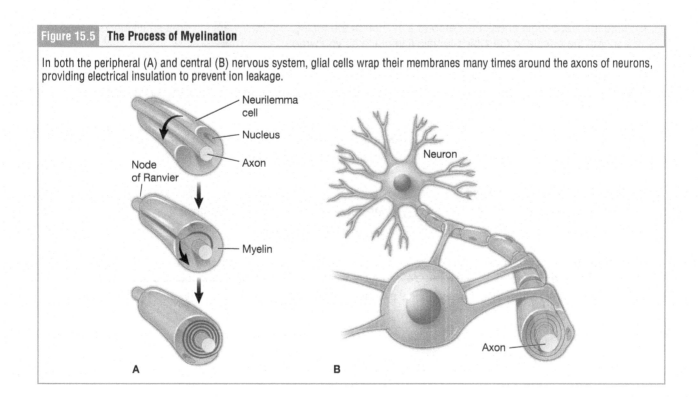

A

B

Materials
- Sandwich-sized zippered baggies, 2
- Tape (clear, if available)
- Tubing, 1-meter-long section
- Water colored with a small amount of dye

Instructions
1. Fill each zippered baggie about ¼ full of dye-colored water. Press the remaining air out and seal each baggie. Do this over a sink or bin to reduce the possibility of spilling.

2. Each baggie represents a cell that can myelinate a section of axon. The section of tubing represents a neuron axon.

3. Answer the following questions:
   a. Given what you know about the structural differences between neurilemma cells and oligodendrocytes, which type of cell is better represented by the water-filled baggie?

   b. What part of the cell does the baggie represent?

   c. What part of the cell does the water inside the baggie represent?

4. Take the first baggie and wrap it around a section of the tubing. You may want to use a small piece of tape to connect the baggie to the tubing to hold it in place as you wrap.

5. As you wrap, think about how this represents the action of the myelinating cell. Wrap the "cell" around the tubing so that it accurately represents the location of the cell membrane and the cytoplasm of the cell. You may have to hold the "cell" in place once it is fully wrapped.

6. Have your lab partner repeat the process with a second baggie. Consider the placement of this second baggie – should it be right next to the first, or should there be space between the two?

7. Answer the following questions:
   a. What is the term to describe the space you left between the two wrapped baggies?

   b. Which portions of the wrapped baggie represent the myelin sheath? In other words, what is the difference between the inner portion of the baggie, empty of water, that wraps the tubing multiple times and the final portion of the baggie, full of water, that is around the outside?

   c. What does the myelin sheath do? How might you represent this on the representation you built here?

8. Once you are done, empty the baggies of water and return the supplies to the location your instructor directs.

Activity 15.5 The Physiology of Nervous Tissue

This activity targets LOs 15.8, 15.9, 15.10, and 15.11.

The mechanism by which a neuron receives a signal, transmits that signal down its length, and then passes that signal on can be thought of as a story. The story is somewhat cyclical, since when a neuron passes its signal on to another neuron the story will start again.

The purpose of this activity is to practice telling this story in ways that support further exploration of how it occurs.

First, examine the visual representation of the story of signal transduction along a neuron shown in Figure 15.6. Then, read through the brief written version of the story of neuron action that follows, referring back to the visual representation as necessary to help you imagine the process. As you examine the visual representation and the written version, you may need to recall what you have already learned about graded potentials and action potentials in this or other courses. Refer to your textbook and other resources as necessary to support your learning. In addition, note that this story assumes the neurotransmitter released by the presynaptic neurons is excitatory and that this creates a graded potential great enough to reach threshold at the trigger zone. This story also assumes that the axon of the neuron is not myelinated.

Visual representation of the story:

Figure 15.6	Transmitting a Neural Signal Across a Synapse

The physiology of neuron transmission begins with (A) a graded potential at the synapse and continues with (B) the initiation of an action potential at the axon hillock. Eventually the action potential will reach the axon terminals.

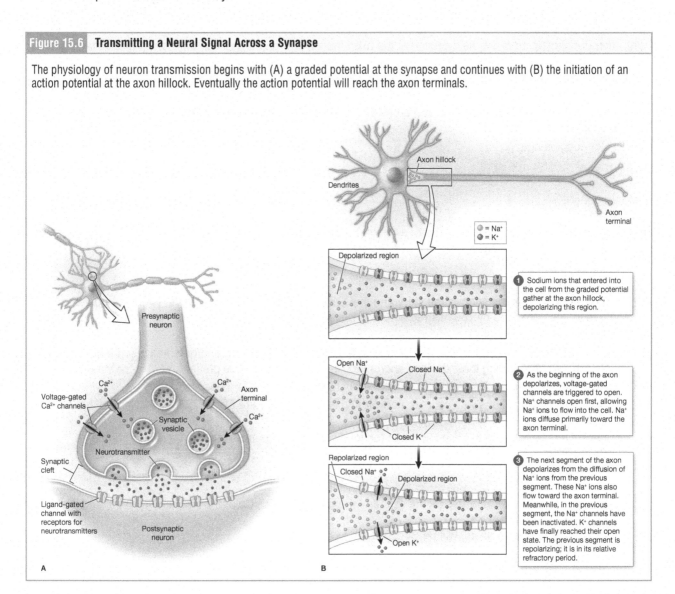

Written version of the story:

1. The action potential reaches the axon terminal of the presynaptic neuron, opening voltage-gated Ca^{2+} channels and allowing Ca^{2+} to enter the membrane at the axon terminals.

2. Ca^{2+} binds to neurotransmitter vesicles, causing them to release neurotransmitter into the synaptic cleft.

3. Neurotransmitter crosses the synaptic cleft and binds to ligand-gated channels with receptors for that neurotransmitter, causing the channels to open.

4. These ligand-gated channels are excitatory and allow Na^+ ions to enter the neuron. Other excitatory ligand-gated receptors embedded in the membrane of the dendrites are triggered to open, allowing more Na^+ ions to enter, depolarizing the membrane to create a graded potential that is excitatory.

5. At the axon hillock, this graded potential depolarizes the membrane potential from -70 mV to at least -55 mV (threshold voltage).

6. -55 mV triggers voltage-gated Na^+ channels at the axon hillock to open, allowing Na^+ to enter the neuron.

7. The entrance of this Na^+ at the axon hillock depolarizes the membrane potential to $+30$ mV.

8. Depolarization of the axon hillock closes voltage-gated Na^+ channels here (no more Na^+ will enter) and opens voltage-gated K^+ channels here (K^+ will leave the cell); these two actions then repolarize the membrane potential in this location to -70 mV.

9. The Na^+ that had entered the axon diffuses toward the synaptic terminals, depolarizing the membrane potential in the next section from -70 mV to -55 mV, triggering the voltage-gated Na^+ channels in this next section to open.

10. After depolarization of this next section of the axon the voltage-gated Na^+ channels close and voltage-gated K^+ channels open; this will repolarize this section of the membrane.

11. The process of entrance of Na^+ causing depolarization to open more voltage-gated Na^+ channels further along the axon continues moving the action potential toward the synaptic terminals with each repeat.

12. After repeated cycles of depolarization and repolarization of the membrane, the action potential reaches the axon terminals.

and the story begins again...

Materials
- Colored pencils
- Envelopes
- Index cards, 3×5, and glue stick or tape (if available)
- Scissors

Instructions

1. Use the visual representation of the story and the written version of the story above, your textbook, and any resources provided by your instructor to add necessary details to each step of the process in the boxes below.

    a. Choose what details you want to add; these details will include drawing and/or labeling some structures, drawing and/or labeling ions, drawing and/or labeling types of channels present, arrows for the direction of movement of ions, adding voltage changes, and anything else you find necessary.

    b. Compare your additions with your lab partner and decide if there are further notes you want to make in each box.

    c. Assign a different color to each ion. Color the ions and their respective ion channels appropriately.
    Ca^{2+} will be _____
    Na^+ will be _____
    K^+ will be _____

    d. Assign a fourth color to represent the ligands (neurotransmitters). Color the neurotransmitters and their ligand-gated ion channels.
    Ligands (neurotransmitters) will be _____

2. Number the steps *on the back of each step* in light pencil. The goal is to have the order recorded once each step is cut out, but not in a location that can be seen as you practice!

3. Cut out each step of the story, removing the numbering that is to the left of each step.

4. If available, glue or tape each step onto a 3 × 5 index card. Glue or tape one step per card, transferring the numbering to the back side of the index card so you don't lose the order.

5. Once you have all of your cards complete, mix them up. Lay the mixed cards image-up in front of you and re-assemble the steps into the correct order. Avoid peeking at the numbers on the back of each card unless absolutely necessary.

6. When you are done with the activity, use an envelope to gather all of your cards so that you can use them later to practice. See Learning Connections: Broken Process, for further suggestions for how to use these cards.

Learning Connection

Broken Process

You discovered one way to practice the story of neuron physiology in Activity 15.5, but can you find other ways to use these cards to deepen your understanding? Try the following activities with your cards:

- What's Missing? Mix up the cards and remove one or two cards randomly from the pack. Then assemble the story in order, using a post-it note to write in the missing steps.

- Talk it Out. Make a blank set of cards that contain the pictures but no labeling or words. Assemble the cards in order and then tell the story out loud using the pictures as clues.

- Draw it Out. Use the written description in the Activity 15.5 introduction to draw out a visual representation of the steps yourself, labelling as you go.

There are many instances in which it is useful to know how active different areas of the brain are. Several approaches to measuring brain activity have been developed. Three of the most common machines used are functional magnetic resonance imaging (fMRI) magnetoencephalography (MEG), and electroencephalography (EEG).

An fMRI machine measures differences in blood flow to various areas of the brain; this approach assumes that blood flow and neural activity coincide with each other. As neurons expend more energy during firing, they must produce more ATP, which should increase their demand for oxygen delivered by the bloodstream. In contrast, both magnetoencephalography (MEG) and electroencephalography (EEG) measure the actual electrical activity of the neurons in the brain, albeit by different methods. While less-expensive EEG machines map out activity in real time on the surface of the scalp (which may obscure where in deep regions of the brain those neural firings began), pricey MEG can identify the specific brain area, deep or superficial, in which the electrical activity is happening. An example of a fixed image from a functional MEG is shown in Figure 15.7.

Figure 15.7	**Functional Magnetoencephalography**

A three-dimensional image of the brain based on magnetic resonance imaging (MRI) and functional magnetoencephalography (fMEG). The green and orange areas are regions of greater activity when the person was asked to repeat a set of words several times as measured by fMEG.

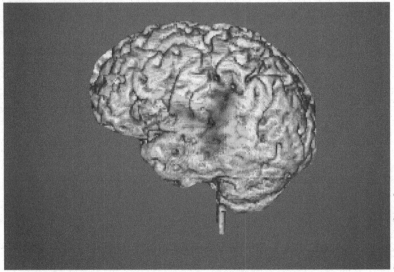

Sim Laboratoire/Science Source

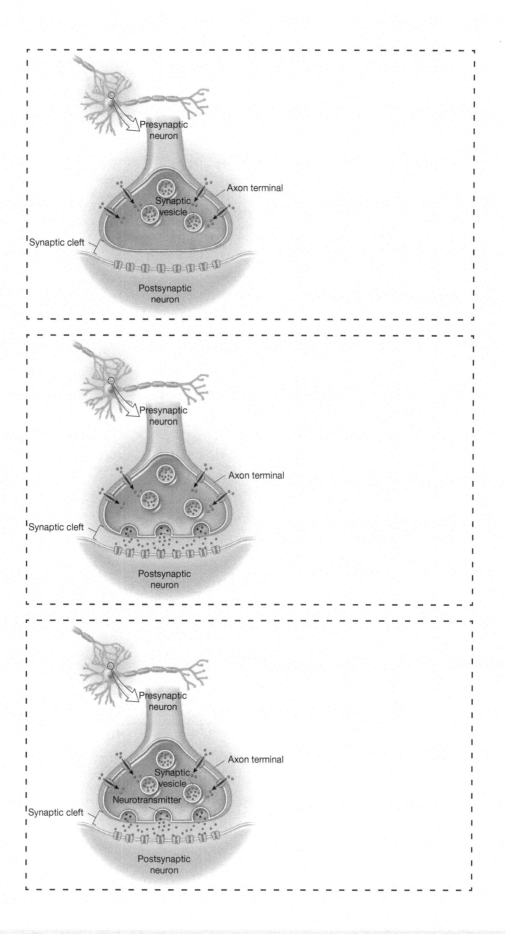

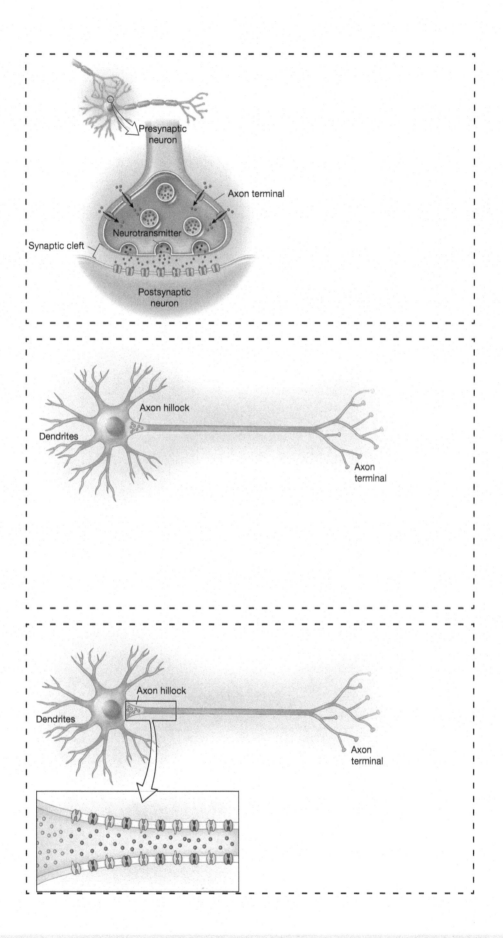

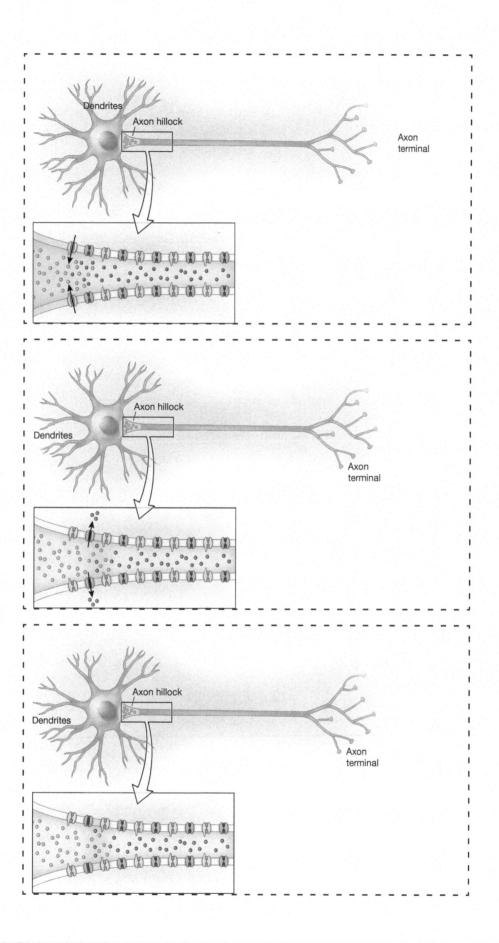

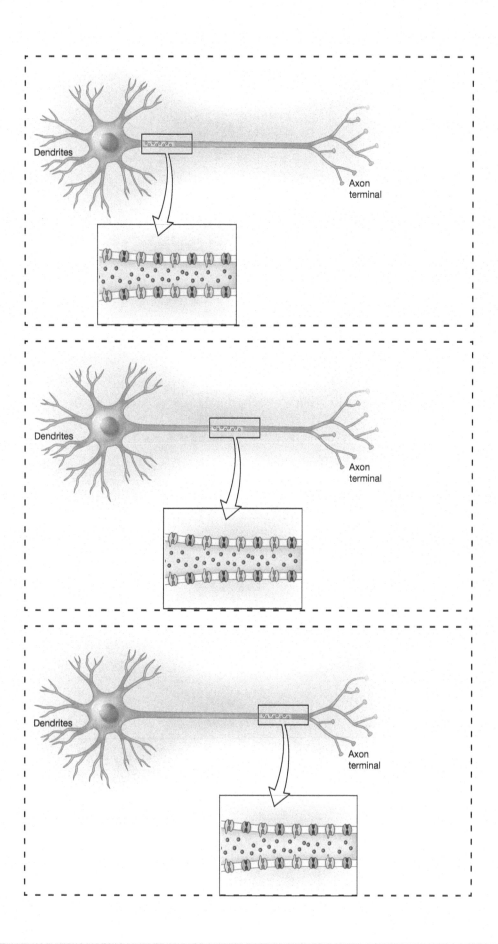

Activity 15.6 Case Study: Having a Stroke

This activity targets LOs 15.2, 15.3, 15.6, and 15.10.

As David is sitting down to dinner, he notices that his speech is slurring, and his eyelid and face are drooping on the left side. His partner calls 911 and an ambulance soon arrives. At the emergency room, the medical staff suspects an ischemic stroke: a loss of blood flow to one or more sections of the brain. David's airway, breathing, and circulation are monitored as blood tests are administered. Functional magnetic resonance imaging (fMRI) is ordered to measure David's brain activity. The fMRI supports the diagnosis of an ischemic (reduced blood flow) stroke and shows reduced blood flow to the right primary motor cortex.

Case Study

1. When blood flow is reduced to an area of the brain this reduces that area's ability to (choose all correct answers)

   a. Produce oxygen

   b. Produce ATP

   c. Produce glucose

2. If an EEG to measure the electrical activity in the brain were also performed, you would expect to see _____ electrical activity in the right primary motor cortex.

   a. less

   b. more

3. The term "electrical activity" refers to _____ occurring on neurons.

   a. action potentials

   b. resting membrane potentials

David is treated for an ischemic stroke and remains in the hospital overnight. The next morning, David and his partner are seen by the physician, Dr. Garcia. Dr. Garcia examines David, who still has left-side issues: his left eyelid is drooping, his mouth is tipped down on the left, and he is having trouble lifting his left arm or closing his left hand. The doctor shares that the lack of blood flow to David's right primary motor cortex has damaged the area of his brain that controls his ability to make his skeletal muscles move on his left side.

4. Which type of neurons have been damaged by the ischemic stroke?

   a. Afferent neurons

   b. Efferent neurons

   c. Integrative neurons

5. Given the symptoms and Dr. Garcia's explanation, were these damaged neurons innervating the somatic nervous system or the autonomic nervous system?

   a. Somatic nervous system

   b. Autonomic nervous system

Dr. Garcia continues to talk with David and his partner. In addition to the damaged neurons, Dr. Garcia shares that recent research shows that some glial cells are also greatly damaged during ischemic stroke events. In the short term, Dr. Garcia is concerned about inflammation, as these damaged glial cells are reduced in their ability to regulate the extracellular environment of the brain. In the long term, Dr. Garcia is concerned about the production of scar tissue that will form between nearby undamaged neurons, preventing them from being able to form synapses.

6. Which type of CNS glial cell is Dr. Garcia referring to?

   a. Ependymal cells

   b. Oligodendrocytes

   c. Astrocytes

   d. Microglia

(page intentionally left blank)

Lab 15 Post-Lab Quiz

1. The projection tracts of the brainstem are columns of white matter that are in the spinal cord and brainstem. Some of the axons in these tracts begin in the spinal cord and terminate at the cerebrum of the brain. These axons serve a(n) _____ function in the nervous system.

   a. integrative

   b. sensory

   c. motor

2. You are walking down a trail when you turn a corner and see a large bear. Immediately, your heart rate quickens. What portion of your nervous system has caused the muscle of your heart to do this? Choose all correct answers.

   a. Somatic nervous system

   b. Autonomic nervous system

   c. Efferent nervous system

   d. Afferent nervous system

3. A damaged neuron of the PNS may be able to repair itself, but only if the cell body is intact. What important structure in the cell body would be used to make these repairs?

   a. Axon hillock

   b. Dendrites

   c. Rough ER (chromatophilic substance)

4. Which of the following cells would be the last cell in a neuronal circuit?

   a. Olfactory neurons that function to give you a sense of smell

   b. Cardiac muscle cells in your heart

   c. Neurons in an afferent nerve

   d. Neurons in an efferent nerve

5. Multiple sclerosis is a disease that causes demyelination in the central nervous system. This disease attacks oligodendrocytes. You can predict that this disease would lead to which of the following outcomes?

   a. An increase in the number of pathogens entering the brain and spinal cord

   b. A decrease in production of cerebrospinal fluid

   c. A reduced ability for CNS axons to send action potentials

6. Areas of the brain and spinal cord that contain the portions of neurons that are myelinated will have a white coloration that is called white matter. Other areas of the brain are called gray matter. What components of neurons would you expect to find in gray matter? Choose all correct answers.

   a. Axons

   b. Dendrites

   c. Cell bodies

   d. Axon terminals

7. Where on a neuron will voltage-gated Ca^{2+} channels be found?

   a. Dendrites

   b. Cell body

   c. Axon

   d. Axon terminals

8. If the voltage-gated K$^+$ channels in neurons were blocked, what will be inhibited?

    a. The production of action potentials

    b. The production of graded potentials

    c. The release of neurotransmitter

9. An action potential starts at what location on a neuron?

    a. Dendrites or cell body

    b. Axon terminals

    c. Neurofibrillary nodes

    d. Axon hillock

10. As astrocytes regulate the extracellular environment of the CNS they clear away neurotransmitter from a chemical synapse after it has been released. What will occur if astrocytes do not perform this function?

    a. Synaptic vesicles will continue to migrate to the edge of the axon terminals and release their contents.

    b. Voltage-gated Ca^{2+} channels will stay open and Ca^{2+} will continue to enter the axon terminals.

    c. Ligand-gated channels will stay open and continue to produce graded potentials.

Lab 16A Chemical Senses

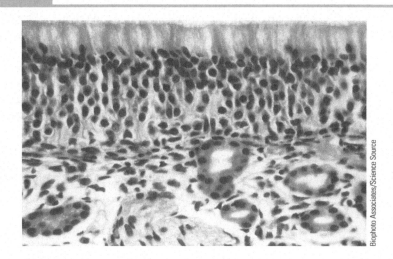

Biophoto Associates/Science Source

Learning Objectives: After completing this lab, you should be able to:

16A.1* Explain the phenomenon of adaptation and the difference between central and peripheral adaptation.

16A.2 Identify and describe the composition and location of the olfactory epithelium.

16A.3 Trace the path of olfaction from the olfactory receptors, to the initiation of an action potential in the olfactory nerves, through the olfactory bulb, the olfactory tract, and to the various parts of the brain.

16A.4 Identify and describe the location and structure of taste buds.

16A.5 Explain the process by which tastants activate gustatory receptors.

16A.6 Trace the path of gustation from gustatory receptors through specific cranial nerves to various parts of the brain.

16A.7 Describe the primary taste sensations.

*Objective is not a HAPS Learning Goal.

Introduction

The peripheral nervous system can be divided into the system of information coming into the brain and spinal cord, also known as the afferent division, about our internal and external environment and the information coming out of the brain and spinal cord, also known as the efferent division, to control our movements and responses. The incoming information is broadly classified as sensory information; the outgoing information is classified as motor information. Our senses can be further divided into **general senses**, which occur all over the body and include information such as touch, pressure, and pain, and **special senses**, which are limited to the head and involve special structures. The eye is the special structure in the head dedicated to the sense of vision, for example.

The special senses are taste (gustation), smell (olfaction), vision, hearing (audition), and equilibrium (balance).

The Human Anatomy and Physiology Society includes more than 1,700 educators who work together to promote excellence in the teaching of this subject area. The HAPS A&P Learning Outcomes measure student mastery of the content typically covered in a two-semester Human A&P curriculum at the undergraduate level. The full Learning Outcomes are available at https://www.hapsweb.org.

In this lab, we will cover taste and smell, which are both chemical senses, as the stimuli that receptors detect in both these senses are chemicals in the food and air, respectively. Vision, hearing, and equilibrium will be covered in Lab 16B. General senses will be covered in Lab 16C.

Sensory information comes into the central nervous system (CNS) and is delivered to a dedicated brain region. The first neuron in the chain is the **sensory receptor,** and this neuron has the unique job of translating a sensory stimulus into an action potential. The sensory receptor must translate the sensory stimulus into an electrochemical signal that can be understood by the nervous system. For example, a receptor in the ear translates the vibration of sound into an action potential.

Have you ever noticed that you sometimes stop sensing something? For example, when you walk into a home for the first time you may initially be aware of the smells of that home. Perhaps you smell cookies that are baking in the kitchen or the scent of the cleaning products used in the home. But when you check in with yourself 10 minutes later, it is likely that not only have you stopped paying attention to the smells; you might not even be able to smell them again if you tried. The phenomenon of **adaptation**—becoming less sensitive to a constant stimulus—enables the nervous system to become aware of a new stimulus or a change in a stimulus (such as the smells you detect when stepping into a new place), but then move on to prioritizing other information. Adaptation can happen at the receptor or centrally, meaning within the CNS.

Lab 16A Pre-Lab Quiz

This quiz will strengthen your background knowledge in preparation for this lab. For help answering the questions, use your resources to deepen your understanding. The best resource for help on the first five questions is your text, and the best resource for help answering the last five questions is to read the introduction section of each lab activity.

1. Anatomy of a sensory pathway. Place the following labels on the diagram:

    gustatory cortex
    secondary neuron
    sensory receptor
    tertiary neuron
    thalamus

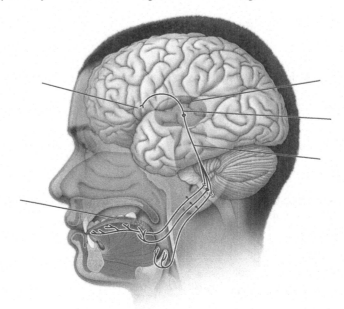

2. Match each of the roots to its definition:

Root	Meaning
	deliciousness
	two
	one
	small protuberance

List of Roots

di
mono
papilla
umami

3. A molecule that binds to a receptor is called its _____.

    a. enzyme
    b. cofactor
    c. ligand
    d. target

4. Which of the following is the correct order of structures from smallest to largest?

    a. Taste receptor cell–taste bud–papilla–tongue
    b. Taste receptor cell–papilla–tongue–taste bud
    c. Taste receptor cell–papilla–taste bud–tongue
    d. Taste bud–taste receptor cell–papilla–tongue

5. Which has a greater variety of receptors?

   a. The tongue
   b. The olfactory epithelium
   c. They have the same diversity of receptors

6. Which cranial nerve has axons that communicate smell information to the brain?

   a. CN1
   b. CN2
   c. CN5
   d. CH7

7. What bone do the axons of olfactory receptors pass through?

   a. Sphenoid bone
   b. Frontal bone
   c. Nasal bone
   d. Vomer
   e. Ethmoid bone

8. The olfactory epithelium and the taste buds each contain cells that are not receptor cells. Identify the pairs where the correct cell type is matched to its location.

   a. Taste bud: supporting cell
   b. Taste bud: basal epithelial cells
   c. Olfactory epithelium: supporting cell
   d. Olfactory epithelium: basal epithelial cells

9. Which of the senses sends some of its axons to the limbic system?

   a. Olfaction
   b. Gustation
   c. Vision
   d. Equilibrium

10. What type of molecule is likely the ligand for sweet taste receptors?

   a. Fatty acids
   b. Amino acids
   c. Na^+ ions
   d. Monosaccharides

Activity 16A.1 Olfaction

This activity targets LOs 16A.1, 16A.2, and 16A.3.

> **Note:** This activity involves exposure to products. Please communicate with your partner and instructor as needed if you have any allergies to the components of this lab.

Olfaction and gustation are both chemical senses. In these senses, small molecules in the air or in our food and drink bind to receptors in our nose or mouth. The receptors interact with neurons that carry the signal of the chemical stimulus to the brain via electrochemical signaling. Scientists estimate that we are capable of detecting approximately 10,000 different odorant molecules. The olfactory receptor cells are located in the superior nasal cavity (Figure 16A.1). The olfactory receptor cells are bipolar neurons with their cell bodies in the nasal epithelium. One extension from their cell body is coated with cilia, which dangle into the nasal cavity and are exposed to odorant molecules. The other extension weaves through the cribriform plate of the ethmoid bone and forms a synapse with olfactory neurons sitting within the cranium. These olfactory neurons and the synapse between the olfactory receptor cells and the neurons are housed within the **olfactory bulbs**. The olfactory bulbs become the olfactory tracts, or cranial nerve one (CNI), posteriorly. In other words, the olfactory bulbs are the terminal end of CNI, but their function is to transmit sensory, or afferent, signals coming from the olfactory epithelium to the CNS.

Figure 16A.1	The Olfactory Epithelium

The olfactory epithelium is found lining the superior region of the nasal cavity.

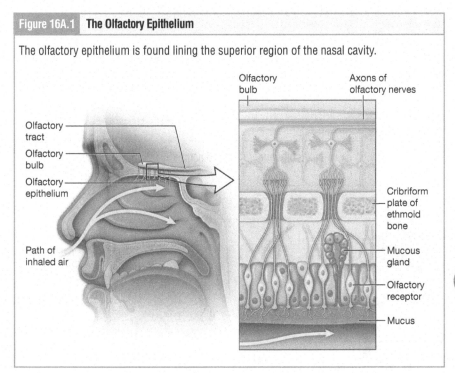

Olfactory bulb

Axons of olfactory nerves

Olfactory tract

Olfactory bulb

Olfactory epithelium

Cribriform plate of ethmoid bone

Mucous gland

Path of inhaled air

Olfactory receptor

Mucus

 Learning Connection

Broken Process

Have you ever noticed that when your nose is congested you can't taste as well? Smell enhances our sense of taste; but why? Think about the anatomical connections between these two locations (hint: the pharynx) and develop a hypothesis as to why this is so.

The axons of olfactory neurons travel through CNI but eventually reach several portions of the brain. Some travel to the primary olfactory cortex, a region of the temporal lobe that decodes smell molecules. Other axons project to structures within the limbic system and hypothalamus, where smells become associated with long-term memory and emotional responses.

Smell is complex and not only is smell important in its own right, it also enhances taste to a significant degree. Smell is a sense that undergoes adaptation readily. In the case of olfaction, central adaptation occurs to allow us to perceive new smells while previously recognized smells are ongoing.

Materials
- Bottle of isopropyl alcohol
- Bottle of oil of wintergreen
- Compound microscope
- Microscope slide of olfactory epithelium
- Models or images of olfactory epithelium
- Stopwatch (or clock with second hand or app on one's phone)

Structures Identified in This Activity
- Olfactory epithelium
- Olfactory bulb
- Olfactory tract
- Cribriform plate of the ethmoid bone
- Olfactory glands

Instructions
1. On the models or images of the olfactory epithelium, identify the structures in the list above.

2. Using your compound microscope, examine the slide of olfactory epithelium at low magnification (40×). Increase magnification to 10× and refocus. Find the supporting cells and olfactory receptors (you might not be able to distinguish between these) of the olfactory epithelium and the olfactory glands. Notice the cilia of the olfactory receptor cells, this is where the odorants bind. Use Figure 16A.2 as a guide as necessary.

| Figure 16A.2 | The Olfactory Epithelium Lines the Nasal Cavity |

There are bipolar olfactory receptors cells interspersed among an epithelium of supporting cells and mucus-producing glands.

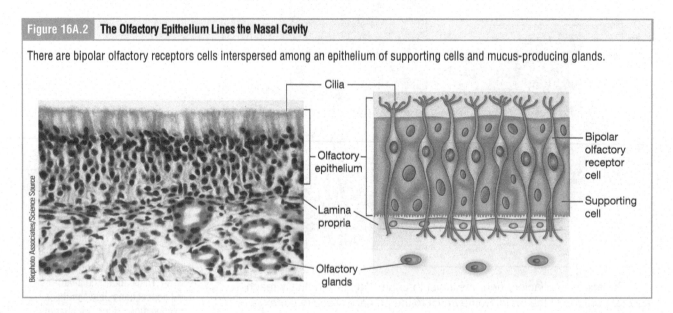

3. To demonstrate olfactory adaption, we will smell two strong odors. First set up your work area with a stopwatch and the bottles of isopropyl alcohol and oil of wintergreen.

4. a. One partner should start the stopwatch as the second partner holds the bottle of wintergreen oil close to their nose.

   b. The partner with the oil should continue breathing through their nose and noticing the smell of the oil.

   c. Continue smelling until they can no longer smell the odor. Have the partner with the stopwatch record the elapsed time in Table 16A.1.

Table 16A.1 Adaptation to Odors		
	Olfactory Adaptation Time for Wintergreen Oil	**Olfactory Adaptation Time for Isopropyl Alcohol**
Trial 1		
Trial 2		
Trial 3		

5. Immediately following the moment they no longer are able to smell the wintergreen oil, this student should uncap the bottle of isopropyl alcohol and determine if they can smell it. The other partner should restart the stopwatch.

6. Continue smelling the isopropyl alcohol with the bottle held close to the nose until you can no longer smell the odor. Have the partner with the stopwatch record the elapsed time in Table 16A.1.

7. Immediately following the moment they are no longer able to smell the isopropyl alcohol, uncap the bottle of wintergreen oil and determine if they can smell it.

8. Repeat steps 4 through 7 two more times, and see if adaptation is affected by repeat exposure.

Questions

1. Were you able to smell the isopropyl alcohol in step 5? Why or why not?

2. Were you able to smell the wintergreen oil in step 7? Why or why not?

(page intentionally left blank)

Activity 16A.2 Gustation

This activity targets LOs 16A.4, 16A.5, 16A.6, and 16A.7.

> **Note:** This activity involves exposure to products. Please communicate with your partner and instructor as needed if you have any allergies to the components of this lab.

Gustation, or taste, is also a chemical sense. Similar to olfaction, in the process of gustation, chemical molecules from our foods bind to cell-surface receptors on the tongue (and cheeks and pharynx). These molecules are known as **tastants**. In response to the bound taste molecules, the receptor cells release neurotransmitters, which activate neurons. The axons of these taste neurons join other axons within the facial nerve (CNVII), glossopharyngeal nerve (CNIX), and vagus nerve (CNX). All three pathways converge on the solitary nucleus in the medulla oblongata (Figure 16A.3). Taste information is carried along second neurons to the thalamus, and third neurons bring this signal to the insula of the cerebrum, which is where the taste cortex is located.

Figure 16A.3	The Pathway of Taste Information

Taste information travels to the CNS via three of the cranial nerves. The first synapse occurs in the medulla oblongata; the second synapse is in the thalamus. Taste information is processed in the insula.

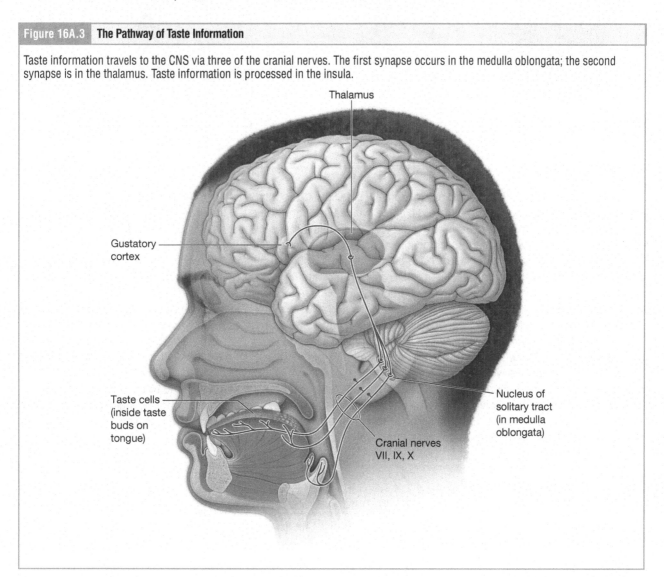

The taste receptors themselves are housed within **taste buds**, small collections of **taste receptors cells** and **basal epithelial cells** (Figure 16A.4). Contrary to popular belief, the raised bumps you can see on your tongue are not taste buds! These are called **papillae**, and each papilla houses one to five taste buds. The receptor cells of the taste bud are sensitive to chemicals contained within the foods we eat. When taste receptor cells are damaged, for example by drinking a hot beverage, they are replaced when the basal epithelial cells reproduce.

Figure 16A.4 **Taste Buds and Papillae**

Taste buds, which are housed within papillae, are composed of taste cells and basal epithelial cells.

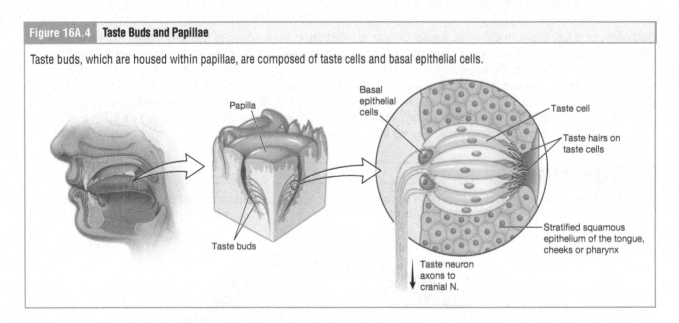

The structure of a taste bud includes a relatively small pore through which taste molecules must pass for sensory reception to occur. When a taste molecule passes through the pore and binds to its receptor on the taste receptor cell surface, an action potential is produced in the taste receptor cell. As a result of the action potential, the taste receptor cell will release a neurotransmitter and activate a primary taste neuron.

The receptors for six distinct tastes have been identified. Sodium ions (Na^+) have the ability to bind to cell surface proteins on taste receptor cells. These cells, when activated by Na^+ will activate neurons that communicate the sense of salty taste to the brain. Likewise, hydrogen ions (H^+), which are more abundant in foods and drinks with a lower pH, can bind to molecules on taste receptor cells and cause neurons to communicate a sour taste to the brain. Sweet receptors are activated by monosaccharides; and fatty flavors occur when fatty acids bind to their taste receptor cells. The taste known as umami, which is a savory taste, is caused by the amino acid L-glutamate binding to its taste receptor cells. There is one taste that is more complex and unique than the other five: bitter taste reception. Unlike the other receptors which bind a single molecule (e.g., Na^+) or a closely related group of molecules (e.g., monosaccharides), there are 23 different bitter receptors that have been identified to date. Which bitter receptors are expressed by one individual may be entirely different than those expressed by another individual. This is the reason why one person may enjoy a bitter food like coffee or broccoli and another person may find it intolerable. It all has to do with what receptors each person expresses!

Digging Deeper:
Picky Eaters

Ever wondered why kids are pickier about their food than grown-ups? The answer lies in the anatomy of taste reception. Just like with bitter receptor expression, the number of taste receptors expressed varies widely from one person to the next. The person who sits next to you in class may have up to 10,000 more taste receptors than you do. The more taste receptors a person has, the more brain signaling they will receive from each bite. It is similar to volume, a person with more taste receptors experiences foods at a higher volume than someone with fewer taste buds and therefore less taste reception. We have the greatest expression of taste receptors around 18 to 24 months of age. From toddlerhood on we are slowly losing taste receptors. So kids experience more sensation from food and therefore might prefer blander foods. As they age, they likely will be able to tolerate more flavor.

Materials

- Blindfold
- Compound microscope
- Cotton swabs
- Models or images of tongue and taste buds
- MSG solution
- Nose clips
- Salt solution
- Slide of the tongue in cross-section
- Sugar solution and/or corn syrup
- Tonic water and/or PTC paper and/or coffee and/or a solution of unsweetened cocoa powder
- Vegetable oil and/or avocado oil
- Vinegar and/or lemon juice

Instructions

1. On the models or images of the tongue and taste buds, identify the following structures: basal epithelial cell, papilla, taste bud, taste receptor cell.

2. Using your compound microscope, examine the slide of the tongue cross section at low magnification (4×). Identify a papilla and increase magnification until you can identify the features of a taste bud. Use Figure 16A.5 as a guide.

Figure 16A.5	Taste Cells

Taste cells, housed within papillae, have microvilli on their apical surface that can come in contact with tastants as they enter through the taste pore.

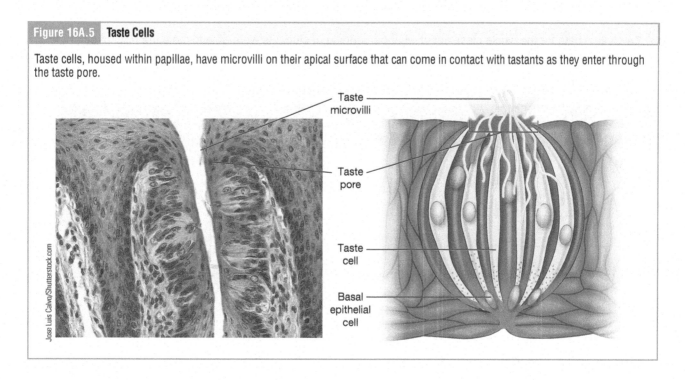

Jose Luis Calvo/Shutterstock.com

- Taste microvilli
- Taste pore
- Taste cell
- Basal epithelial cell

3. Obtain cotton swabs, nose clips, a blindfold, and a tray of the solutions taken from Table 16A.2, each in its own cup or container.

4. One student should wear the nose clips and blindfold (the student does not have to wear a blindfold if their back is turned to the tray of solutions). Dip a cotton swab into one of the solutions listed in Table 16A.2, hand the swab to the student tester, and ask them to touch the swab to their tongue and identify if the solution is bitter, salty, sweet, umami, or fat. Use a different cotton swab for each solution and each student tester. Each used cotton swab should be disposed of immediately, not placed on the lab bench where it could spread pathogens to others.

Table 16A.2 Possible Tastants for Activity 16A.2	
Solutions That Can Be Used	**Type of Tastant**
Tonic water	Bitter
PTC paper	Bitter
Coffee	Bitter
Unsweetened cocoa solution	Bitter
Vegetable oil	Fat
Avocado oil	Fat
Salt solution	Salt
Vinegar solution	Sour
Lemon juice	Sour
Sugar solution	Sweet
Corn syrup	Sweet
Monosodium glutamate solution	Umami

5. Record their responses in Table 16A.3. Allow the student tester at least 1 minute and a few sips of water between the tests.

6. Now remove the taster's nose clips and repeat the tests. Record their responses in Table 16A.3.

7. Collect data for the whole class about what percentage of students were able to taste the PTC paper. How can you explain this?

8. If desired, the two students can change roles so that each has the opportunity to taste.

Table 16A.3 Ability to Identify Tastants			
Material Tested	**Type of Tastant**	**Identified as with Nose Clips on**	**Identified as with Nose Clips off**

Activity 16A.3 Case Study: Chemical Sense Alterations in COVID-19

Case Study

This activity targets LOs 16A.2, 16A.3, and 16A.6.

The symptoms of COVID-19, the disease caused by a novel coronavirus, SARS-CoV-2, discovered in 2019, are varied and diverse. Among many patients (the percentage varies by virus variant) alterations in smell and/or taste are reported (Figure 16A.6). The cause of these changes is not entirely understood. In this case study, we will explore possible mechanisms (some that are hypothesized by scientists and some that are fictional) and you and your partner will need to deduce which make physiological sense, and which do not.

1. If the virus infected the frontal lobe of the brain, would this cause a loss of taste reception?

    a. Infection of the frontal lobe would directly impact taste reception or processing.

    b. Infection of the frontal lobe would probably not impact taste reception or processing.

2. The virus has the ability to cause cranial nerve inflammation. Inflammation of which cranial nerve could result in a decrease in taste sensation?

    a. CN1

    b. CN7

    c. CN8

    d. CN12

3. The virus has the ability to cause cranial nerve inflammation. Inflammation of which cranial nerve could result in a decrease in olfactory sensation?

    a. CN1

    b. CN2

    c. CN4

    d. CN7

Figure 16A.6   **Viral Infections and Our Chemical Senses**

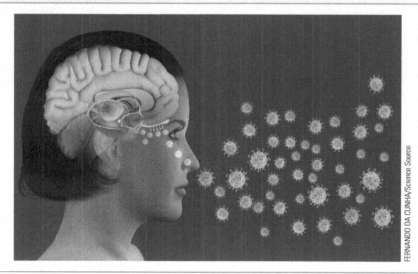

FERNANDO DA CUNHA/Science Source

4. Let's assume that viral infection of the olfactory epithelium causes a change in the olfactory receptor cell cilia. Would damaging the cilia, but not damaging the rest of the olfactory receptor change one's sense of smell?

   a. Damage to the cilia of olfactory receptor cells would directly impact the sense of smell.

   b. Damage to the cilia of olfactory receptor cells would not impact the sense of smell because these cilia do not play a role in olfactory reception.

5. Now let's assume the virus infects the entire olfactory receptor cell. Could the virus ascend into the brain within the axons or cell extensions of these cells?

   a. This is a likely mechanism for the virus to enter the brain.

   b. While viruses can use axons to travel in the body, the axons of olfactory receptor cells do not extend into the brain.

Lab 16A Post-Lab Quiz

1. Your friend sticks their tongue out at you in jest, but you notice that they have a lot of papillae on the surface of their tongue. Is it reasonable to assume that they have a high degree of taste sensation? Which of the statements below is correct?

   a. Each papilla contains one taste bud, the more the friend has, the more she can taste!

   b. Each papilla contains a variable number of taste buds, so it's likely that more papillae means more taste sensation, but it is not necessarily true.

   c. Taste buds are in the regions between papillae, so the more papillae there are, the less room for taste buds.

2. What are the functions of basal epithelial cells?

   a. Mitosis to replace damaged taste receptor cells

   b. Clear away waste products

   c. Bind tastant molecules and hold them in place for reception

3. Sucrose is a disaccharide that tastes sweet to us, but a polysaccharide such as wheat also has glucose in it but it does not taste sweet on the tongue. What is the most likely explanation for this phenomenon?

   a. The carbohydrate components of wheat, like starch, are too large to bind to the sweet receptors. Salivary amylase will begin the digestion of larger carbs, but it won't be completed by the time the food is moved out of the oral cavity.

   b. Wheat contains monosaccharides that we do not have receptors for.

   c. Sucrose can bind our taste receptors because disaccharides are smaller than monosaccharides.

4. The presence of which molecule causes sour sensation?

   a. Na^+

   b. H^+

   c. Amino acids

5. Both taste receptor cells and olfactory receptor cells have extensions of their plasma membranes (cilia on olfactory cells and microvilli on taste cells). What do you think the function of these cell features is?

   a. Expansion of surface area

   b. Sweep molecules off the surface of the cell

   c. Store neurotransmitter

6. The olfactory nerves run along the inferior side of which lobe of the brain?

   a. Insula

   b. Occipital

   c. Temporal

   d. Frontal

7. What is the significance of the fact that many of the neurons from the olfactory bulb project to the limbic system?

a. The limbic system responds to smell information by controlling the secretion of digestive enzymes in anticipation of eating.

b. The limbic system responds to smell information with emotional regulation.

c. The limbic system responds to smell information by activating the visual system to help us locate foods.

8. If a person did not have adaptation for olfaction, which of the following would be true?

a. They would have a difficult time telling one smell from another.

b. Their smell would be fine, but smell would not enhance taste the way we saw in Activity 16A.2.

c. They would experience prolonged smell sensation, instead of getting used to smells.

9. Which label in this diagram is pointing to an olfactory receptor cell?

a. A
b. B
c. C
d. D

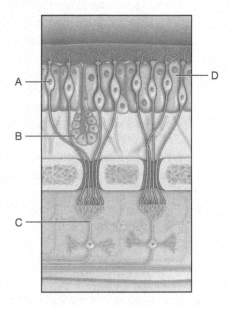

10. Which one of the labels in the image points to a taste bud?

a. A
b. B
c. C

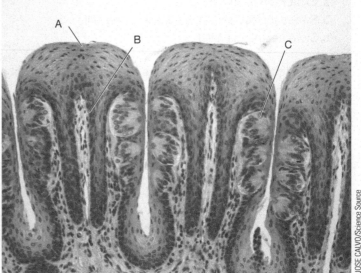

Lab 16B — Vision, Hearing, and Equilibrium

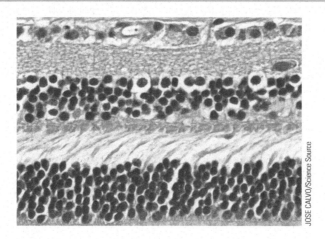

JOSE CALVO/Science Source

Learning Objectives: After completing this lab, you should be able to:

16B.1 Identify the tunics of the eye and their major components (e.g., cornea, sclera, iris, ciliary body), and describe the structure and function of each.

16B.2 Identify and describe the anterior and posterior cavities of the eye and their associated humors.

16B.3 Describe the lens and its role in vision.

16B.4 Identify and describe the accessory eye structures (e.g., conjunctiva and lacrimal apparatus).

16B.5 Trace the path of light as is passes through the eye to the retina, and describe which structures are responsible for refracting the light rays.

16B.6 Compare and contrast the functions and locations of rods and cones.

16B.7* Relate the anatomy of the retina to vision.

16B.8* Describe acuity and give examples in various senses.

16B.9 Identify the macroscopic structures of the outer (external), middle, and inner (internal) ear and their major components (e.g., auditory ossicles, auditory [pharyngotympanic] tube), and describe the structure and function of each.

16B.10 Trace the path of sound from the external ear to the inner ear, including where sound is amplified.

16B.11* Compare and contrast how a single receptor cell, the hair cell, functions in both hearing and equilibrium.

16B.12 Compare and contrast static and dynamic equilibrium.

* Learning Objective not a HAPS Learning Goal

The Human Anatomy and Physiology Society includes more than 1,700 educators who work together to promote excellence in the teaching of this subject area. The HAPS A&P Learning Outcomes measure student mastery of the content typically covered in a two-semester Human A&P curriculum at the undergraduate level. The full Learning Outcomes are available at https://www.hapsweb.org.

Introduction

The portion of the nervous system dedicated to detecting changes in our internal or external environment is our sensory system. The sensory system can be broadly divided into the **general senses**, which occur all over your body and involve no unique structures, and the **special senses**, which are senses that require unique structures and are found in the head. Our special senses are: vision, hearing, equilibrium, taste, and smell. In Lab 16A we explored taste and smell; in this lab, 16B, we will explore hearing, equilibrium, and vision.

These three senses each have their own, unique receptors, which transduce sensory information into a perception. For example, vibrations in the air are captured by the ears and transduced into signals that the brain perceives as hearing. We have dedicated regions of the cerebral cortex which receive and interpret the incoming sensory information.

Lab 16B Pre-Lab Quiz

This quiz will strengthen your background knowledge in preparation for this lab. For help answering the questions, use your resources to deepen your understanding. The best resource for help on the first five questions is your text, and the best resource for help answering the last five questions is to read the introduction section of each lab activity.

1. Anatomy of the pathway of light perception. Place the structures in the order through which a photon of light will pass. Then place the structures through which the electrical signal will pass in their proper order.

Photon of Light Path	**Electrical Signal Path**
Cornea	Photoreceptor cells
Lens	Bipolar cells
Vitreous humor	Ganglion cells
Retina	Optic nerve
	Occipital lobe

2. Anatomy of a hair cell. Label the following structures in the figure:

   hair cell body
   nerve fiber
   nucleus
   stereocilia

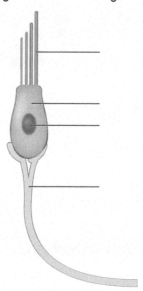

3. Anatomy of the retina. Label the three cell types in the drawing. The cell types are: bipolar cells, ganglion cells, and photoreceptor cells (rods and cones).

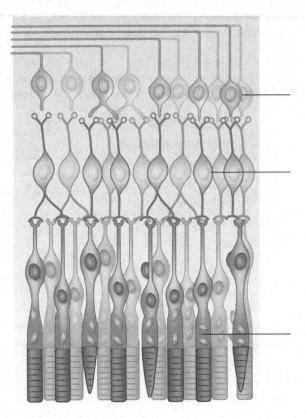

4. Match each of the roots to its definition:

Root	Meaning
	drum
	a fluid
	light
	net

List of Roots

humor
photo
retin-
tympan-

5. The iris is a component of which layer of the eyeball?

a. The fibrous layer
b. The vascular layer
c. The retina

6. A higher density of _____ is associated with higher acuity.

a. rods
b. cones
c. bipolar cells
d. ganglion cells

7. The axons of _____ form the optic nerve.

   a. rods
   b. cones
   c. bipolar cells
   d. ganglion cells

8. The role of the ossicles, the tiny bones inside the ear, is which of the following?

   a. To stabilize the middle ear chamber ceiling
   b. To enable the membranes to stay in place
   c. To amplify vibration
   d. To hold open the auditory tube

9. Which of the following structures of the ear is the most superficial?

   a. Tympanic membrane
   b. Ossicles
   c. Round window
   d. Basilar membrane

10. Choose the option that correctly orders the structures through which sound passes or affects as the sound is transmitted to the brain.

    a. Vestibulocochlear nerve–basilar membrane–tympanic membrane–ossicles
    b. Ossicles–tympanic membrane–basilar membrane–vestibulocochlear nerve
    c. Tympanic membrane–ossicles–basilar membrane–vestibulocochlear nerve

(page intentionally left blank)

Activity 16B.1 Micro and Gross Anatomy of the Eye

This activity targets LOs 16B.1, 16B.2, 16B.3, 16B.4, and 16B.5.

The eye is a small but complex organ with many structures that serve functions for vision or physical protection. The bony orbits surround the eyes, protecting and anchoring them in place. Within the orbit, the eye is able to move due to six extraocular eye muscles (Figure 16B.1). The anterior surface of the eye is coated with a layer of stratified columnar epithelium containing goblet cells, blood vessels, and sensory neurons called the **conjunctiva** (Figure 16B.2). The mucus of the conjunctiva keeps the eye moist along with the tears produced by the **lacrimal glands**. The moisture produced by these two sources sweeps across the eye lateral to medial and drains into the **nasolacrimal duct**, which empties into the nose (Figure 16B.3).

Figure 16B.1	Muscles of the Eye

(A) The extrinsic eye muscles originate on the skull and insert onto the surface of the eye. (B) Muscles surround the eye to enable a wide range of motion.

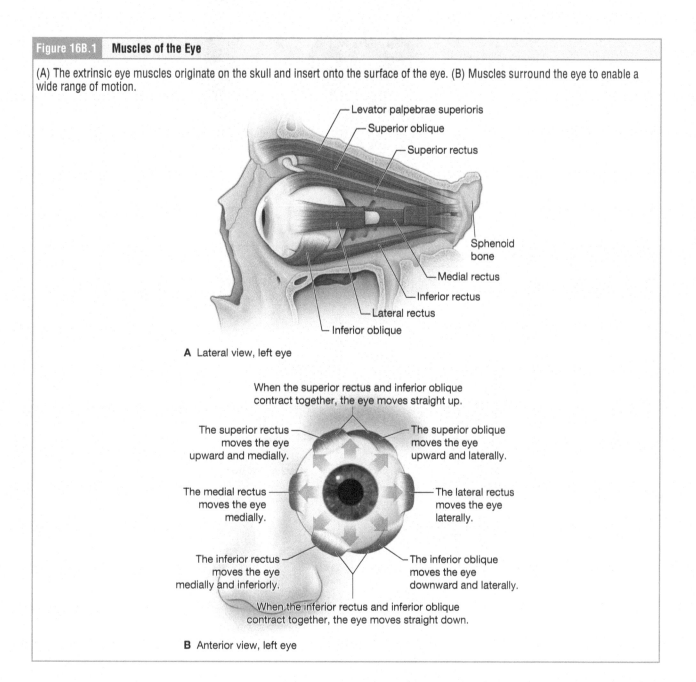

A Lateral view, left eye

B Anterior view, left eye

The eye has hard and soft layers of protection. It sits within the hard, bony shell of the orbit, and is protected by the eyelids, eyebrows, and conjunctiva.

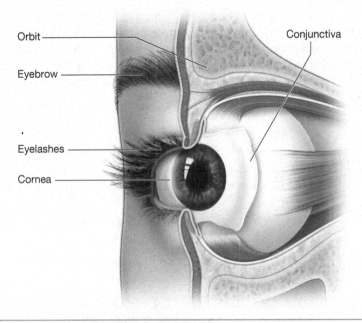

The lacrimal gland produces tears on the lateral side of the eye. The tears sweep across the eye from the lateral to the medial side and are drained into the nose by the nasolacrimal duct.

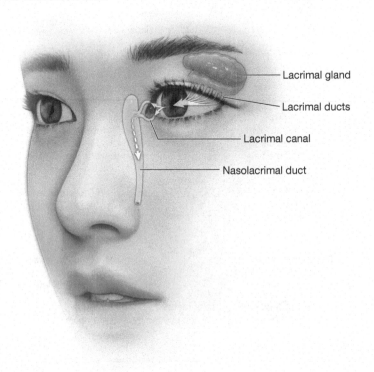

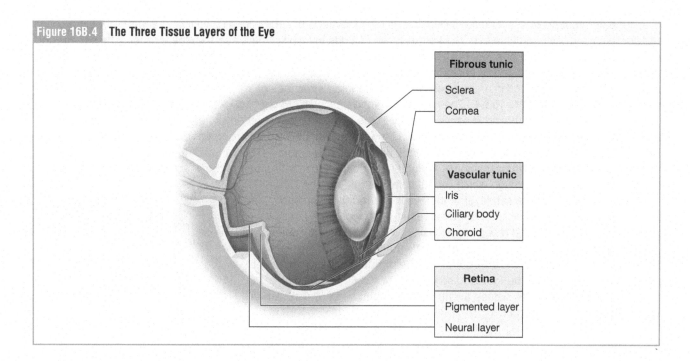

Fibrous tunic

Sclera

Cornea

Vascular tunic

Iris

Ciliary body

Choroid

Retina

Pigmented layer

Neural layer

The eyeball itself is a fluid-filled ball composed of three layers of tissue. The layers from superficial to deep are the fibrous layer, the vascular layer, and the retina (Figure 16B.4). The eye is also divided into two cavities, each is filled with fluid. The **anterior chamber** of the eye is filled with **aqueous humor**, the much larger **posterior chamber** of the eye is filled with **vitreous humor** (see Figure 16B.5).

Figure 16B.5 | Anatomy of the Eye

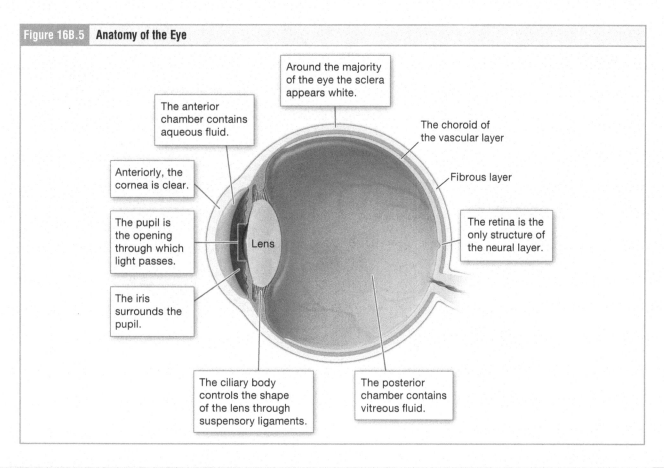

Around the majority of the eye the sclera appears white.

The anterior chamber contains aqueous fluid.

The choroid of the vascular layer

Anteriorly, the cornea is clear.

Fibrous layer

The pupil is the opening through which light passes.

Lens

The retina is the only structure of the neural layer.

The iris surrounds the pupil.

The ciliary body controls the shape of the lens through suspensory ligaments.

The posterior chamber contains vitreous fluid.

The outermost **fibrous layer** (also called fibrous tunic) is made of dense connective tissue and provides protection and scaffolding to the rest of the eye. The fibrous layer includes the white sclera and the clear cornea. The **sclera**—the white of the eye—is continuous with the transparent **cornea** that covers the anterior surface of the eye and allows light to enter (Figure 16B.5). The middle **vascular layer** (also called vascular tunic) is also primarily connective tissue, but is rich with blood vessels and intrinsic muscles. There are three structures of the vascular layer: the choroid, the ciliary body, and the iris (see Figure 16B.4). The **choroid** is highly vascularized connective tissue that provides the blood supply to the eyeball. The choroid lines the majority of the eyeball except anteriorly where the ciliary body and iris are found. The **ciliary body** is a muscle that is attached to the lens by suspensory ligaments. The **lens** is a clear disc through which light passes into the eye. The ciliary body is able to adjust the curvature of the lens via its contraction.

The lens sits just posterior to the iris, a pigmented donut-shaped structure of the vascular layer. The **iris** surrounds the pupil, a hole through which light is able to enter the eye.

The innermost, or deepest, layer of the eye is the **retina** which contains the nervous tissue responsible for photoreception. The photoreceptor cells, the **rods** and **cones**, are the receptors for vision. When photons of light activate these cells, they activate a chain of signals that brings light messages to the brain for perception and interpretation. The **bipolar cells** are the direct recipients of signals from the rods and cones (Figure 16B.6). The bipolar cells signal to the **ganglion cells**, whose axons form the optic nerve which carries visual information to the brain.

 Learning Connection

Try Drawing It

Draw the pathway of light as it enters the eye through the cornea to its ultimate destination.

Materials

- Compound microscope
- Cow eyeball
- Dissection tray
- Eye model
- Gloves
- Prepared slide of retina
- Probe
- Scalpel
- Scissors

Figure 16B.6	Photoreceptor

The tissue of the retina consists of three layers of cells. The most posterior layer consists of the photoreceptor cells themselves. Anterior to this layer are bipolar cells. The most anterior (bordering on the posterior cavity of the eye) are the ganglion cells. Light enters anteriorly through the pupil, and the photons of that light pass between the ganglion cells, then between the cells of the bipolar layer, and then finally to the photoreceptor cells.

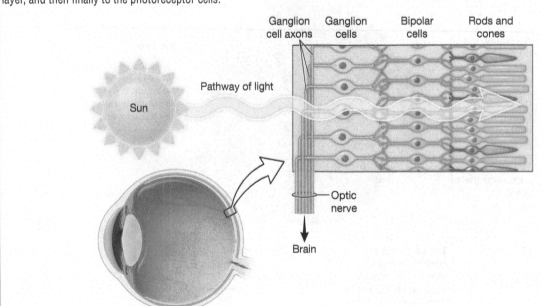

Instructions

1. Cow eyeball dissection.

   a. Obtain a dissection tray, scissors, scalpel, gloves, and an eyeball. Put on your gloves. A lab coat and/or eyewear may be used in addition.

   b. Use scissors to cut away any adipose tissue attached to the eyeball. Identify the optic nerve. What does it feel like? Is it hard or soft? Hollow or solid?

   c. Holding the eyeball firmly with gloved hands, make an incision that cuts the eyeball into anterior and posterior halves while keeping your hand away from the direction of the scalpel. This is a cut in the frontal plane. This is a messy process, the aqueous and vitreous humor will spill out, so do this over a dissecting tray. If your scalpel is dull, it may be helpful to poke the eyeball with the tip of the scalpel to make the initial incision, then cut from there.

   d. Separate the anterior and posterior portions of the eyeball. Note the fragile retina, which is a thin, delicate beige inner layer of tissue. Behind it you will notice the dark choroid, which may have some iridescent regions (refer to Figure 16B.7 for reference).

   e. You can peel back the neural layer using the probe. You will find that it is all attached at one site, where the neural layer exits the eyeball as the optic nerve.

   f. On the anterior portion, find the pupil, cornea, ciliary body, and iris. Find the lens. It may have fallen out of place once the eye was opened, or push it out using the probe.

   g. Dispose of the eyeball and PPE in the locations indicated by your instructor. Wash your tray and instruments using water and soap.

Figure 16B.7	A Dissected Cow Eyeball

This eyeball has been cut in the frontal plane, separating the anterior and posterior sections. The thin, translucent retina and dark choroid are visible.

2. Eye model. On the eye model in your laboratory, locate the structures in Table 16B.1, use Figure 16B.8 as a reference as needed. As you locate them, fill in Table 16B.1 with the function of each structure.

Table 16B.1 Structures of the Eye	
Structure	**Function**
Lacrimal apparatus	
Nasolacrimal duct	
Conjunctiva	
Extrinsic eye muscles (lateral rectus, medial rectus, superior rectus, inferior rectus, superior oblique, inferior oblique).	
Lens	
Anterior cavity	
Aqueous humor	
Posterior cavity	
Vitreous humor	
Fibrous tunic	
Sclera	
Cornea	
Vascular tunic	
Choroid	
Ciliary muscle	
Suspensory ligaments	
Sensory tunic	
Iris	
Pupil	
Retina	
Optic nerve	

The laboratory eye model can be separated into superior and inferior halves. The retina and vascularization are visible lining the posterior cavity.

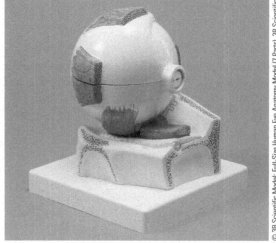

3. Histology of the retina.

Focus on the slide using low magnification and use Figure 16B.9 as a guide. Locate the pigmented and neural parts of the retina and increase magnification to examine these layers more carefully. Identify the three cell layers of the retina.

Figure 16B.9   Histology of the Retina

The neural tissue of the retina is composed of three cell layers: the photoreceptor cells, the bipolar cells, and the ganglion cells.

- Ganglion cell layer
- Bipolar cell layer
- Rods and cones
- Choroid
- Sclera

Questions

1. Which of the structures in Table 16B.1 does light pass *through*?

2. Which humor was more fluid, the aqueous or vitreous humor? Do you imagine that the consistency of the humors affects the light passing through? How so?

Activity 16B.2 Vision Tests

This activity targets LOs 16B.6, 16B.7, and 16B.8.

Vision is the result of photons of light being received by the photoreceptor cells, the rods and cones. Once activated by a photon traveling at a particular wavelength, the rod or cone cell will signal to the bipolar cells and they signal to the ganglion cells. The axons of the ganglion cells weave together to form the optic nerve. You may wonder why we have two types of photoreceptor cells and what the difference is between rods and cones. The answer is that we actually have more than just these two types. Each distinct type of photoreceptor cell is capable of receiving a photon of light traveling within a certain wavelength range (Figure 16B.10). Cones provide color vision as well as visual **acuity**, that is to say, the accuracy of what we see. Rod vision functions in much lower light, but produces only blurry images.

Figure 16B.10	Comparison of Color Sensitivity of Cones

This is a comparison of the wavelength sensitivity of the four photopigments.

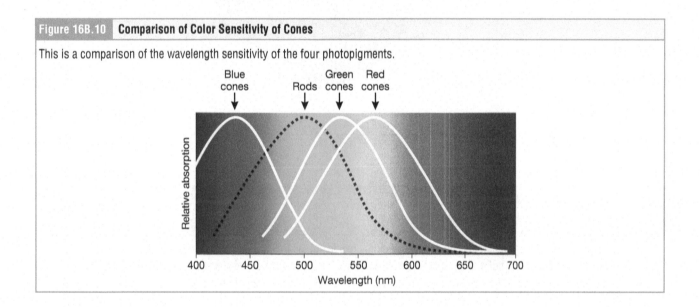

To achieve visual acuity, the structures at the front of the eye, the lens and ciliary body, work to focus the incoming light onto the region of the retina with the highest concentration of cones, the **fovea centralis** (Figure 16B.11). Every aspect of our visual perception is dependent upon the photons of light reaching photoreceptor cells, and the more photoreceptor cells that are activated, the more accurate our visual perception is. In the exercises in this activity, we will explore the factors that contribute to visual acuity.

Figure 16B.11 | **Light Path Through the Eye**

As light travels in through the anterior eye, the lens bends to channel the light toward the region of highest acuity of the retina, the fovea centralis.

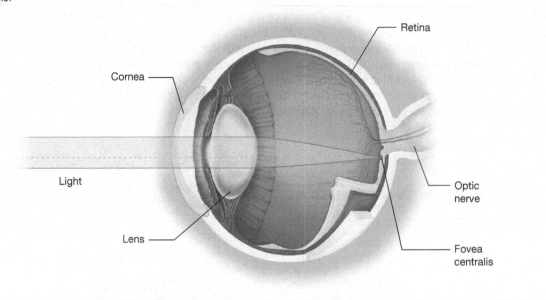

Retina

Cornea

Light

Lens

Optic nerve

Fovea centralis

 Digging Deeper:
Why Do Some People Need Glasses?

Visual acuity is sharpest when incoming light is focused on the regions of the retina with the highest proportion of cones: the fovea centralis. When the focal point of the light is just ahead or just behind the cones, the image appears blurry. Focusing the light in the right plane requires the lens to be able to bend the incoming light. As we age, all tissues, including the lens, lose elasticity. When the lens becomes less able to bend the light, reading glasses are often necessary. But some people require glasses earlier in life, often in late childhood or early puberty. Typically, this happens when the skull grows and changes shape, causing the eye itself to take on a new shape, perhaps more oblong or rounder than it was before. In this new shape, if the lens can no longer focus the light on the right plane, then glasses will be necessary to bend the light to compensate for the lens.

Materials

• Copies of the image in Figure 16B.12

• Paper and pen or pencil

• Ruler

Figure 16B.12 | **Blindspot Test**

+   ●

Instructions

1. On two separate pieces of paper, write a phrase on each one of similar length about any topic of your choice.

2. Hold the first piece of paper about 10–12 inches in front of your lab partner's face; adjust for optimal reading as necessary, and ask them to read the phrase on the paper. If necessary, adjust the distance for optimal reading.

3. Now have your partner continue to sit or stand in the same position, with eyes forward, and hold the second piece of paper the same distance away but to the side (directly lateral) of their face in their peripheral vision. With eyes continuing straight ahead, is your partner able to read the second phrase when relying only on their peripheral vision?

4. Now ask your partner to close their left eye. Hold a copy of the image in Figure 16B.12 approximately 18 inches in front of their right eye so that the + symbol is in line with their right eye.

5. Slowly move the image closer and closer to your lab partner asking them to focus on the + symbol but to report when the black dot is and isn't visible.

6. When the black dot disappears, measure the distance between the image and the eye with a ruler. This is the blind spot.

7. Repeat with the left eye.

8. Change lab partner roles and repeat steps 1–7.

Questions

1. When your subject stared straight ahead to view the first phrase, what type of photoreceptors were predominantly functioning? What about when the subject was viewing the second phrase in their peripheral vision? Where on the retina are these types of photoreceptors in higher relative concentrations?

2. If the subject experiences a difference in their ability to read the two phrases, what does that tell us about the difference in acuity of the two types of photoreceptor cells?

3. Explain the anatomical reason for the blind spot.

(page intentionally left blank)

Activity 16B.3 Anatomy of the Ear

This activity targets LOs 16B.9, 16.B10, and 16B.11.

The ear houses the structures for two different senses, hearing and equilibrium. Interestingly, the same receptor structure, the **hair cell**, is used for both senses, though it is positioned differently to be able to receive different stimuli.

The ear is divided into three regions: the **outer ear**, **middle ear**, and **inner ear** (Figure 16B.13). The outer ear includes the external structure that we can see, the **auricle**, as well as the canal through which sound passes, the **external acoustic canal**. The most internal structure of the outer ear is the tympanic membrane, a sheet of epithelial and connective tissue that vibrates when sound waves hit it.

The middle ear is an air-filled chamber that is bordered laterally and medially by membranes. The tympanic membrane is its lateral border, and its medial border is another membranous sheet, the **oval window**. Between these two membranes, each of which can vibrate like a drum, are the three tiniest bones in the human body: the **malleus**, **incus**, and **stapes**. These three bones are called the **ossicles**, and their job is to receive vibrations from the tympanic membrane and both transmit and amplify those vibrations onto the surface of the oval window. As the stapes strikes the oval window like a drumstick on a drum, the vibrations are sent into the inner ear (Figure 16B.14). The inferior border of the middle ear empties into a long tube that connects the air-filled middle ear to the throat. This tube, the **auditory tube**, allows the pressure within the middle ear to equalize to the outside.

Figure 16B.13   **Structure of the Ear**

The auricle, ear canal, and tympanic membrane are all part of the external ear. The middle ear is an air-filled space that houses the ossicles. The cochlea and the vestibule are responsible for audition and equilibrium, respectively.

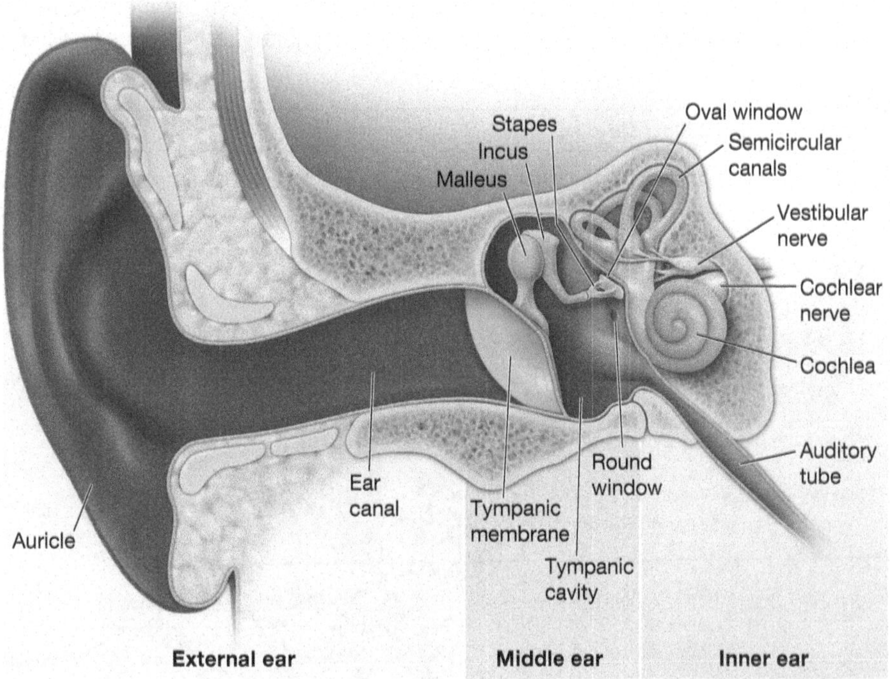

The ossicles amplify the vibrations received by the tympanic membrane. They occupy the air-filled middle ear, but the vibrations must be amplified in order to impact the fluid-filled inner ear.

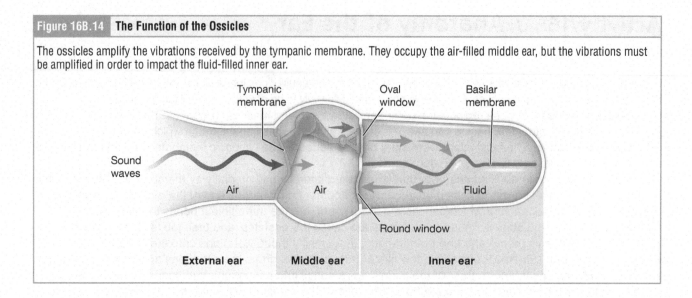

The inner ear has three separate compartments: the **vestibule**, the **semicircular canals**, and the **cochlea**. The cochlea is responsible for our sense of hearing, and the vestibule and semicircular canals are responsible for our sense of balance (Figure 16B.15).

Extending out from the oval window is the long fluid-filled tube of the cochlea. Within the cochlea, the receptor cells of the ear, hair cells, are suspended within the **cochlear duct** between two membranes (Figure 16B.16). As the vibrations are transmitted inwards from the **oval window**, the membrane that the hair cells sit on top of, the **basilar membrane**, will vibrate and bounce. The hair cells will be pushed gently against the membrane that is suspended above them, the **tectorial membrane**, and as they collide with the membrane, the hair cells will depolarize and send

Figure 16B.15 | The Structures of the Inner Ear

The inner ear contains three separate anatomical and functional regions: the vestibule, the semicircular canals, and the cochlea.

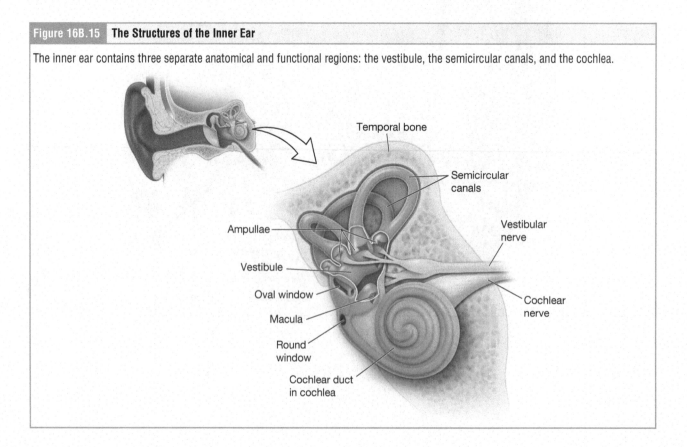

(A) The cochlea is a long tube curled like a snail. Unspooled and cross sectioned, we can see that it actually consists of three anatomically separate tubes rolled together. (B) Within the cochlear duct, hair cells sit on top of the basilar membrane. Their stereocilia project upward toward the unmoving tectorial membrane.

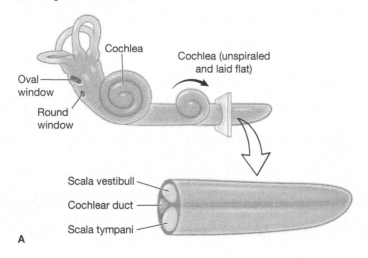

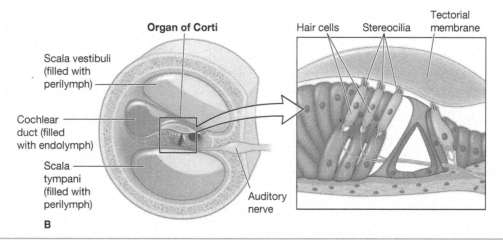

signals to the axons of the **vestibulocochlear nerve** that they connect to. Based on the pattern and intensity of signals sent, the brain can interpret sound. The cochlea has a hole in its lateral wall, the **round window**, which also connects the inner ear to the middle ear and participates in the vibrations of the fluid within the inner ear.

In the vestibule and the semicircular canals, the stereocilia of the hair cells are covered by gelatinous layers. When the head or body moves, the semi-solid gelatinous layers move too, and this motion causes the stereocilia of the hair cells to bend, causing the hair cells to send signals to the brain about the movement. The vestibule conveys information about the anterior/posterior tilt of the head; and the semicircular canals convey information about the rotation of the body. The brain is able to put this information together to understand body position and movement.

Learning Connection

Try Drawing It

Draw the hair cell and its stereocilia and illustrate for yourself how the stereocilia function in three different ways to generate the signals for hearing and equilibrium.

Materials
- Compound microscope
- Ear model
- Slide of cochlea

Instructions

1. Ear model. On the ear model in your laboratory, locate the structures in Table 16B.2, use Figure 16B.17 as a reference as needed. As you locate them, fill in Table 16B.2 with the function of each structure.

2. Histology of the cochlea.

   a. Focus on the slide using low magnification, and use Figure 16B.18 as a guide. At this magnification, you should be able to identify the vestibular membrane, basilar membrane, and the cochlear duct between them. You may also be able to identify the tectorial membrane.

Table 16B.2 Structures of the Ear	
Structure	**Function**
Outer ear	
Middle ear	
Inner ear	
Auricle	
External auditory canal	
Oval window	
Tympanic membrane	
Ossicles (malleus, incus, stapes)	
Auditory tube	
Vestibule	
Semicircular canals	
Cochlea	
Cochlear duct	
Basilar membrane	
Tectorial membrane	
Round window	

A model of a human ear can be useful for learning the anatomy. In most models you should be able to identify the three ear regions, as well as the three components of the inner ear.

© 3B Scientific Model: Organ Of Corti, 3B Scientific GmbH, Germany, 2023, www.3bscientific.com; © Photo by: Cengage Learning.

Figure 16B.18   **Histology of the Cochlea**

A cross section of the cochlear duct reveals the hair cells, tectorial membrane, and basilar membrane.

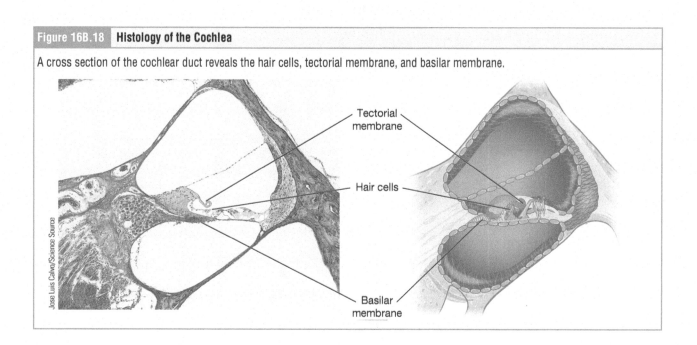

Tectorial membrane

Hair cells

Basilar membrane

Jose Luis Calvo/Science Source

(page intentionally left blank)

Activity 16B.4 Hearing and Equilibrium Tests

This activity targets LO 16B.12.

For the brain to understand a sound, the vibrations caused by that sound must travel through the outer ear, transmit through the middle ear and into the inner ear and travel along axons within the vestibulocochlear nerve to the brain. Changes to any of these environments can lead to a reduction in hearing. When hearing loss is a concern, a clinician may first try to narrow in on the anatomical region in which the reduction is occurring. In conductive hearing loss, sound cannot be conducted through either the middle or outer ear. In sensorineural hearing loss, damage to the inner ear or vestibulocochlear nerve prevents sound reception.

Equilibrium is our sense of balance, and it is established by the hair cells of the vestibule and the semicircular canals. The brain, however, combines the signals from these inner ear structures with visual information to create a composite understanding of where the body is in space. When the equilibrium information from the inner ear does not match with the visual information, for example when you're reading an unmoving book while inside of a moving car, often a sense of dizziness or nausea will result. Typically we call this motion sickness. In a clinical setting, we may want to measure inner ear function by testing to see if a patient can maintain their balance without visual stimuli.

In this activity we will complete two tests: the Rinne test, which uses vibration conducted both through bone and air to examine hearing function, and the Romberg test of equilibrium.

Materials
- Ruler
- Stopwatch or stopwatch app
- Tuning fork
- Whiteboard and markers OR chalkboard and chalk OR wall and painters' tape

Instructions
1. Strike the tuning fork lightly against the side of the table to cause it to vibrate.

2. Place the base of the tuning fork against the subject's mastoid process as shown in Figure 16B.19 and begin a stopwatch.

Figure 16B.19 | The Rinne Test of Hearing

The tuning fork is first placed on the subject's mastoid process, and then in the air very close to the external acoustic canal.

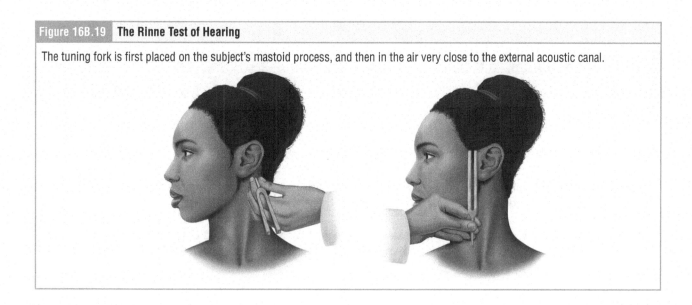

---

3. Ask the subject to tell you when they can no longer hear the ringing of the tuning fork, and record the time in Table 16B.3.

Table 16B.3 Rinne Test Results			
	Trial 1	Trial 2	Trial 3
Time until ringing stopped during bone conduction (seconds)			
Time until ringing stopped during air conduction (seconds)			

4. Once the subject reports no longer being able to hear the ringing, keep the stopwatch going and quickly move the tuning fork to the side of the external acoustic canal (the fork should not be touching the subject, but get it as close to the external acoustic canal as possible).

5. Typically, air-conducted sound can be heard much longer than bone-conducted sound. Ask the subject if they can hear the ringing now, and if yes, ask them to tell you when it stops. Record that time in Table 16B.3.

6. Repeat steps 1–5 two more times, recording the times in Table 16B.3.

7. Now have the subject stand tall and straight next to a whiteboard, chalkboard, or wall. Draw lines (or use painters, tape) to mark the outline of the lateral sides of the subject's body on the board or wall behind them. Note also the position of the subjects' feet.

8. Have your partner stand as straight and unmoving as possible for 60 seconds with their eyes open.

9. Mark on the board or wall where the lateral sides of the subject's body are now. Use a ruler to measure any deviation from their original lines, and note if there was any change in foot position. Record your observations in Table 16B.4.

Table 16B.4 Romberg Test Results		
	Eyes Open	Eyes Closed
Deviation (inches)		

10. Now have the subject return to standing within their marked lines. Ask them to maintain their posture as straight and unmoving as possible for 60 seconds with their eyes closed this time.

11. Mark on the board or wall where the lateral sides of the subject's body are now. Use a ruler to measure any deviation from their original lines, and note if there was any change in foot position. Record your observations in Table 16B.4.

Questions

1. What did you observe in the three trials for sound perception during bone conduction versus air conduction? Why do you think this is?

2. Do you think that visual input affects body positioning? Why or why not?

(page intentionally left blank)

Lab 16B Post-Lab Quiz

1. The rods and cones are part of which layer of the eye?

   a. The fibrous layer

   b. The vascular layer

   c. The retina

2. In Activity 16B.2 you had your subject try to read a phrase written on a paper in their peripheral vision. During this part of the vision test, which photoreceptor cells were being used?

   a. Rods

   b. Cones

   c. An even mix of rods and cones

3. In Activity 16B.2 you had your subject try to read a phrase written on a paper in their peripheral vision. Did you expect that the subject would be equally likely to be able to read the phrase on the paper during this part of the vision test?

   a. Yes, because photoreceptor cells are arranged evenly in all parts of the retina.

   b. No, because the paper was held in the subject's blind spot.

   c. No, because peripheral vision is mostly provided by rods, which provide less accurate vision.

4. In Activity 16B.2 you tested the subject's blind spot, an area of the peripheral vision where no visual stimuli can be detected. Anatomically, what is the blind spot?

   a. A region of the retina where only cones are located.

   b. A region of the retina where only rods are located.

   c. A region of the retina where no rods or cones are located, because the optic nerve is located here.

   d. A region of the retina where rods or cones are located, but covered by blood vessels.

5. As you ascend in an airplane a sense of pressure builds up in your ear. Swallowing, yawning, or chewing gum are all actions that take place in your mouth and throat, yet, they often can relieve this pressure in the ear. What anatomical structure connects the ear to the throat?

   a. The tympanic membrane

   b. The round window

   c. The auditory tube

6. When you have a respiratory infection and the resulting mucus production, you sometimes develop the sensation that your ears feel "full" and at the same time, hearing is diminished or muffled. This sensation can occur because mucus has reached the middle ear. During healthy times, what substance is typically in the middle ear?

   a. Mucus is always in the middle ear but it becomes thicker during a respiratory infection.

   b. A thin fluid usually fills the middle ear.

   c. Air usually fills the middle ear.

7. If the subject in Activity 16B.4 could not hear the vibrations of the tuning fork after moving the fork from the mastoid process, what kind of hearing loss is indicated?

   a. Conductive hearing loss

   b. Sensorineural hearing loss

8. You are administering a vision test using an eyechart such as the one in this image. The person taking the test mistakes an *A* for an *O*. This is an issue with _____.

   a. accommodation
   b. acuity
   c. equilibrium
   d. conduction

Adam Gault/Science Source

9. The hair cell can function in both equilibrium and hearing because in the different regions of the inner ear movement or vibration causes the _____ to bend. The bending of this part of the hair cell allows the cell to signal to nearby neurons.

   a. cell body
   b. nucleus
   c. stereocilia
   d. ganglion cell

10. The _____ humor is found in the posterior cavity of the eye.

   a. aqueous
   b. viscous
   c. vitreous
   d. ganglious

Lab 16C General Senses

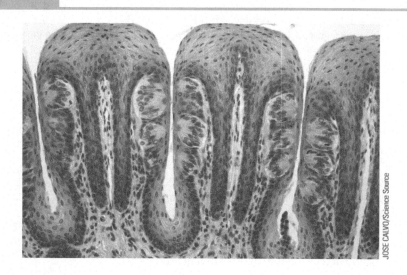

JOSE CALVO/Science Source

Learning Objectives: After completing this lab, you should be able to:

16C.1 Define a sensory receptor.

16C.2* Define and explain the terms receptive field and acuity.

16C.3 Define transduction, perception, sensation, and adaptation.

16C.4 Distinguish between tonic and phasic receptors.

16C.5 Compare and contrast a general sense receptor and a special sense (complex) receptor.

16C.6* Explain how the neurons in a pathway function together to produce a sensation.

16C.7* Define and explain the terms phantom pain and lateral inhibition.

16C.8 Compare and contrast the types of sensory receptors based on the type of stimulus (i.e., thermoreceptor, photoreceptor, chemoreceptor, baroreceptor, nociceptor [pain receptor], mechanoreceptor).

* Objective is not a HAPS Learning Goal.

Introduction

There are two criteria that distinguish general senses from special senses. One, general senses do not have specialized, unique receptor cells, rather they are received and transduced simply by the dendritic endings of neurons embedded within tissues. Two, general senses are received all over the body, whereas special senses are the result of reception only within the head. In this lab we will investigate common characteristics of the general senses.

The **general senses** are typically listed as proprioception, pain, temperature, touch, pressure, and vibration. Sometimes the list includes tickle and itch. But these sensations are complicated. How can tickle be a general sensation if not everyone is ticklish, and no one is ticklish all over the body? How can the list be this limited, when we can perceive the difference between touching a wet surface or a dry one. A soft surface or a hard one? The answer is in the complexity of the brain. While our general senses are anatomically simple, just the dendritic

The Human Anatomy and Physiology Society includes more than 1,700 educators who work together to promote excellence in the teaching of this subject area. The HAPS A&P Learning Outcomes measure student mastery of the content typically covered in a two-semester Human A&P curriculum at the undergraduate level. The full Learning Outcomes are available at https://www.hapsweb.org.

endings of a neuron, the pathway carrying this information can activate multiple regions of the brain. Ultimately the brain combines several different components of sensory information to create one composite understanding.

The brain, however, has a complex and unending job. While it is working to decode and incorporate sensory information from one receptor or region, it is receiving input from all over the body, and from many different receptors. One way that the brain minimizes and focuses its sensory work is through the phenomenon of adaptation. **Adaptation** within the nervous system is a decrease in sensitivity to a stimulus after extended exposure to it. Adaptation can occur in two ways. First, though the receptor continues to receive the stimulus, the receptor stops or slows the signals it sends to the central nervous system. This is called **peripheral adaptation** because it occurs in the peripheral nervous system at the receptor itself. The second form of adaptation occurs in the brain, which has its own ways of diminishing sensory input, and this is called **central adaptation**.

In addition to adapting to their stimulus, neurons also may change the rate at which they fire action potentials due to inhibition by their neighboring neurons. Remember that an action potential travels down a neuronal axon, and changes the voltage at all axon terminals of that neuron. Some of these axon terminals may be inhibitory synapses, inhibiting neighboring neurons (Figure 16C.1). This allows one neuron in a local area to be dominant by reducing the signals sent by nearby neurons. This phenomenon is known as **lateral inhibition**. Undoubtedly you have experienced lateral inhibition before you even knew the physiological mechanism by which it occurs. Have you ever rubbed a site of pain in order to reduce the pain sensation? Or applied heat or a compound like IcyHot™ to a sore muscle and found that the sensation of soreness was reduced. In these examples other neurons, such as touch neurons or thermoreceptors, are firing signals nearby to the nociceptor sending pain information and inhibiting the pain neuron.

Learning Connection

Broken Process

What do you think would happen in an individual who did not have the same level of adaptation as most others? What would they experience?

Figure 16C.1 Lateral Inhibition

This is a phenomenon in which two neighboring neurons have the ability to inhibit each other. In this image, if neuron A was subject to a more intense stimulus, and therefore firing more action potentials, it could inhibit neuron B.

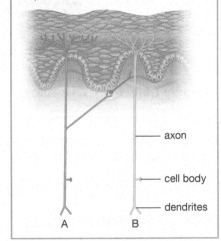

axon

cell body

dendrites

A   B

Digging Deeper:
Are Hot Peppers Actually Hot?

In 2021, David Julius and Ardem Patapoutain won the Nobel Prize in Physiology or Medicine for their discovery of a mechanism that activates thermoreceptors. Thermoreceptors activate at particular temperature ranges in order to give the brain information about the temperature in the environment (Table 16C.1).

Some thermoreceptors, however, can be opened by other stimuli. Capsaicin is a chemical found in spicy peppers and, while it doesn't activate taste receptors and therefore does not produce taste sensation, it does bind to and activate thermoreceptors. Therefore the brain registers an increase in temperature in your mouth when you eat a hot pepper, even though there is no actual temperature change. You could test this out by placing a slice of jalapeño or other spicy pepper in your mouth along with a thermometer (not recommended!).

Table 16C.1 Thermoreceptor Channels

Thermoreceptor Channel	Activation Temperature
TRPV1	>42°C
TRPV2	>52°C
TRPV3	27–34°C
TRPV4	<25°C
TRPV8	34–38°C
TRPA1	<18°C

Lab 16C Pre-Lab Quiz

This quiz will strengthen your background knowledge in preparation for this lab. For help answering the questions, use your resources to deepen your understanding. The best resource for help on the first five questions is your text, and the best resource for help answering the last five questions is to read the introduction section of each lab activity.

1. Anatomy of a sensory pathway. Label the following structures in a sensory pathway:

    Primary somatosensory cortex
    Secondary neuron
    Sensory receptor
    Tertiary neuron
    Thalamus

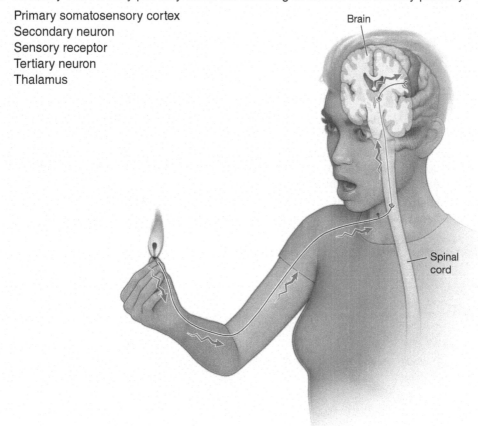

2. Match each of the roots to its definition:

Root	Meaning
	pertaining to a physical change
	sharpen
	pertaining to temperature
	harm
	to fit to a purpose

List of Roots

acu
adapt
mechano
noci
thermo

3. How many neurons are in a somatosensory pathway?

    a. One   c. Three
    b. Two   d. Four

4. Which of the following best describes a receptive field?

    a. The amount of tissue space that a single receptor occupies
    b. The number of receptor neurons in a region of the body
    c. The specific range of a receptor, for example, the temperatures that a thermoreceptor can recognize

5. A neuron that functions as a touch receptor would be classified as which type of receptor?

    a. Thermoreceptor
    b. Nociceptor
    c. Mechanoreceptor
    d. Touch receptors could be any of these

6. The phenomenon of adaptation can be best explained by which of the following statements?

    a. Adaptation is the adjustment by the sensory receptor to increase perception of a stimulus, for example, amplifying the signals caused by a quiet sound.
    b. Adaptation is the adjustment by the sensory system to combine incoming information from multiple receptors and create a composite sensation, for example combining pressure and temperature information to understand that you are touching something wet.
    c. Adaptation is the adjustment by the sensory system to decrease or stop sensation of an ongoing stimulus, for example, no longer being aware of the touch sensation caused by an article of clothing.

7. A receptor that adapts quickly is termed a/an _____ receptor.

    a. tonic
    b. phasic
    c. adaptive
    d. zonal

8. Which types of receptor(s) would be activated if you placed your fingertip on a cold stone? (select all that apply)

    a. Thermoreceptor
    b. Nociceptor
    c. Mechanoreceptor
    d. Photoreceptor

9. There is _____ relationship between acuity and receptive field, meaning that larger receptive fields _____ acuity.

    a. a direct; yield better
    b. an inverse; yield worse
    c. no; do not impact

10. What would you guess about the size of receptive fields in the fingertips compared to the back?

    a. The fingertips have smaller receptive fields than the back.
    b. The fingertips have larger receptive fields than the back.
    c. The fingertips and the back have similar-size receptive fields.

Activity 16C.1 Two-Point Discrimination

This activity targets LOs 16C.1 and 16C.2.

Our sensory system is most useful when it accurately conveys the changes in our internal and external environments. The sharpness or accuracy of our sensory perceptions is known as **acuity**. Let's take visual acuity, for example. The letters *e* and *o* are both round shapes, our visual acuity allows us to distinguish the difference between these shapes. Acuity is helpful, for example if you read the sentence "*please wash the p<u>o</u>t with soap*" and thought that it read "*please wash the p<u>e</u>t with soap*" you might take a very different action. All of our senses have acuity, one aspect of acuity in touch sensation is to localize where a sensation is coming from.

The **receptive field** of a sensory neuron is the area from which its receptor can detect stimuli (Figure 16C.2A). Any location that is stimulated within the receptive field will cause the sensory neuron to send a signal to the CNS (Figure 16C.2B). When multiple stimuli occur within a single receptive field, the neuron cannot communicate any difference among the stimuli to the nervous system, and so no difference is detected. Smaller receptive fields enable more precise communication to the CNS and improve sensory acuity (Figure 16C.2C). In this activity as well as in Activity 16C.3, we will explore touch acuity in the lab.

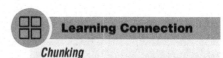

Learning Connection

Chunking

How can you use the term acuity to describe different senses? What would acuity look like for vision? For smell? For touch?

Figure 16C.2 **Receptive Fields**

(A) The receptive field of a sensory neuron is the area from which its receptors can detect stimuli. (B) When a touch stimulus is applied anywhere within the receptive field of a sensory neuron, the neuron sends an electrical stimulus to the CNS. (C) The number of electrical signals sent to the CNS from multiple touch stimuli is determined by the size and density of the receptive fields and sensory neurons present. Touch stimuli activating small, densely packed receptive fields from three adjacent sensory neurons cause three electrical signals to be sent to the CNS. Touch stimuli activating three regions within the same large receptive field of a sensory neuron cause one electrical signal to be sent to the CNS.

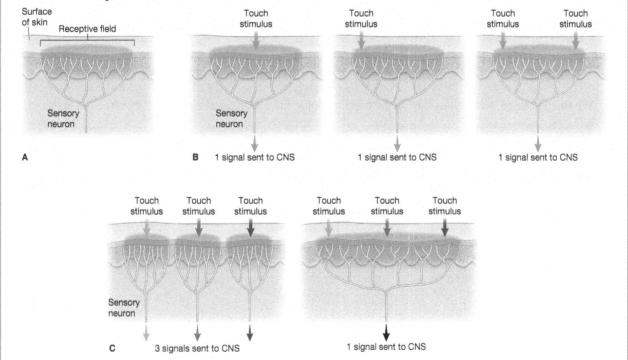

Materials

• Vernier calipers (either digital or dial) and a metric ruler, or a two-point discrimination tool

Instructions

1. Choose one lab partner to be the subject of the test. Have the subject close their eyes. Move the two caliper arms as close as they will go. Beginning with the subject's forearm, touch the calipers to the skin and have the subject report how many points they feel.

2. Gradually move the caliper arms farther and farther apart. Measure the distance at which the subject first feels two points instead of one. Record this distance in Table 16C.2.

Table 16C.2 Touch Discrimination Test Results	
Body Area Tested	**Two-Point Threshold (mm)**
Forearm	
Face	
Fingertip	
Back of neck	
Palm of hand	

3. Repeat steps 1 and 2 on the different locations listed in Table 16C.2, recording each value observed.

4. When completed, swap roles, having the tester become the subject and the subject become the tester.

Questions

1. Did all locations have roughly the same size receptive field?

2. What are the physiological reasons for this observation?

Activity 16C.2 Adaptation

This activity targets LOs 16C.3 and 16C.4.

This activity explores adaptation, the phenomenon of decreased sensation of a stimulus, despite that stimulus remaining constant. Common examples of adaptation are no longer smelling a smell after a while, no longer sensing the feeling of your clothes on your skin after you've been wearing them all day, no longer hearing a consistent sound. Adaptation allows your brain to focus on a stimulus only when it changes. Adaptation can occur in the brain or in the sensory receptor. Receptors can be classified as **phasic receptors**, which adapt to constant stimuli by slowing down or stopping the signals being sent from those receptors to the brain. In contrast, **tonic receptors** are receptors that continue to send signals toward the CNS throughout a consistent stimulus.

Materials
• Three coins of the same size (i.e. all pennies or all nickels, could also be flat washers or similar)

Instructions
1. Have the subject rest their forearm on the lab bench and close their eyes. Place a coin on the middle of the anterior forearm of your subject (see Figure 16C.3). Start a stopwatch.

Figure 16C.3	Coin Placement for Adaptation Test

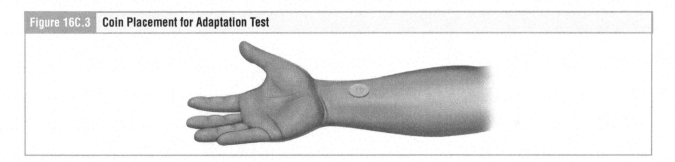

2. At 15-second intervals ask them if they can still feel the coin. Record the time that has elapsed when they can no longer feel the coin in Table 16C.3.

Table 16C.3 Adaptation		
Location	**Time When Subject no Longer Feels Coin**	**Change Perceived When Another Coin Is Added?**
Middle anterior forearm		
Near elbow		
Palm of hand		

3. Add another coin on top of the coin and ask the subject if the sensation changed. If they answer no, add a third coin onto the stack.

4. Move one coin to the forearm just distal to the elbow and repeat steps 1–3.

5. Move one coin to the palm of the hand and repeat steps 1–3.

6. When completed, swap roles, having the tester become the subject and the subject become the tester.

Questions

1. Did all regions you tested have the same adaptation rate?

2. What do you think it would feel like if adaptation did not occur in the skin?

Activity 16C.3 Tactile Localization

This activity targets LOs 16C.1 and 16C.2.

One aspect of somatosensory acuity is precise localization of the stimulus. Imagine if you need to react to a harmful stimulus, such as a mosquito biting you or a thorn that has pierced your skin. If you are unable to locate the stimulus, you cannot save yourself from it. You can imagine that the nervous system has evolved more precise localization in regions where we are more likely to come into contact with stimuli, such as the fingertips.

   The degree to which we can accurately localize a stimulus has a lot to do with the size of the receptive fields of sensory neurons. The receptive field of a sensory neuron is the area from which its receptor can detect stimuli (see Figure 16C.2A). Any location that is stimulated within the receptive field will cause the sensory neuron to send a signal to the CNS (see Figure 16C.2B). When multiple stimuli occur within a single receptive field, the neuron cannot communicate any difference among the stimuli to the nervous system, and so no difference is detected. Smaller receptive fields enable more precise communication to the CNS and improve sensory acuity (see Figure 16C.2C). In an area with very large receptive fields, such as the back, it can be very difficult to precisely locate a given stimulus. Have you ever asked a friend to scratch an itch on your back, only to find that you cannot accurately tell them where the itch is? This is an example of poor acuity. In this lab activity we will explore stimulus localization of different regions.

Materials
- Alcohol wipes
- Blindfold
- Markers, two different colors
- Millimeter ruler

Instructions

1. Have the subject put on a blindfold.

2. Touch the palm of the subject's hand with one marker (it does not have to be right in the center of their palm—Figure 16C.4).

Figure 16C.4	Localization of Sensory Stimuli

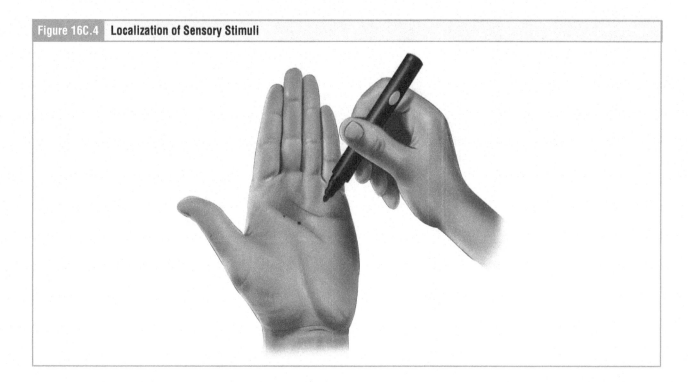

3. Hand the subject a marker of a different color and, with their blindfold still on, ask them to touch their marker to the location where they just felt the other marker tip.

4. Measure the difference between the two marks and record it in Table 16C.4.

Table 16C.4 Tactile Localization Data				
Location	Distance in Test 1 (mm)	Distance in Test 2 (mm)	Distance in Test 3 (mm)	Average Distance (mm)
Palm of hand				

5. Use alcohol wipes to erase the marks from the trial and repeat this process on the palm of the hand two more times. Calculate the average and record.

6. Repeat the process two more times in different locations such as the posterior forearm, the back of the hand, or other accessible places. Record your results and calculate the averages.

7. When completed, swap roles, having the tester become the subject and the subject become the tester.

Questions

1. Did you notice any differences in touch acuity among the three locations measured?

2. Choose two other bodily locations that you have not tested and predict whether they would have greater or less touch acuity than the palm of the hand. Explain your reasoning.

Activity 16C.4 Case Study: Itch

This activity targets LOs 16C.1, 16C.3, 16C.5, 16C.6, and 16C.7.

In your text and in this lab you have learned about several types of neuron sensory receptors.

Case Study

1. Which of the following (select all that apply) are types of neuronal receptors?

   a. Magnetoreceptors

   b. Photoreceptors

   c. Nociceptors

   d. Electroreceptors

   e. Mechanoreceptors

You've also learned that some sensations are detected by the brain due to a combination of stimuli. Wetness, for example, is sensed through a combination of pressure and temperature. Scientists long wondered how the sensation of itch was detected. Does itch result from a combination of signaling from multiple receptors or does it come from its own dedicated receptor?

2. As scientists search for an itch receptor, are they trying to classify a general sense or a special sense?

   a. General sense

   b. Special sense

   c. Could be either of these

Itch receptors were first characterized in 1997. Scientists discovered that when they applied the endogenous chemical histamine an itch sensation was reported in all subjects studied.

3. If an itch receptor fires an action potential when histamine is injected 2 centimeters away, and fires another action potential when histamine is injected 4 centimeters away, but does not fire an action potential when histamine is injected 7 centimeters away, we can say that:

   a. The receptive field is much smaller for itch receptors than other types of receptors.

   b. The receptive field for itch receptors is at least 4 centimeters in size.

   c. The receptive field for itch receptors is less than 2 centimeters in size.

Scientists noted that the conduction speed of itch receptors was very slow. Because of the slow conduction speed, the sensation of itch is slow to build up and also slow to dissipate.

4. How can we explain this anatomically?

   a. Itch receptors must be longer than other types of receptors.

   b. Itch receptors are probably unmyelinated.

   c. Itch receptors have different populations of channels than other types of neurons.

The writer Atul Gawande wrote an article about a patient referred to as "M," who experienced an over-active itch sensation in her scalp. The itch would not go away and the resulting scratching was endangering M.'s health. Through her scratching, M. had damaged most of the neurons in a local area of her scalp. She was unable to detect touch or temperature, but she still itched.

5. Could the lack of other receptors in the area be contributing to the itch?

   a. Yes, with the empty space that used to be occupied by other receptors, itch receptors will multiply and take over.

   b. No, other receptors would enhance itch reception, so we would expect lower levels of itch reception without other receptors around.

   c. Yes, without other types of receptors, the amount of lateral inhibition on the itch receptors has decreased.

One neurosurgeon consulted on M.'s case suggested that they could surgically cut the axon of the itch receptors in the area, and that it would cure M. of the itch sensation. Another neurosurgeon did not think that this would work.

6. Why is it possible that cutting the axon might not cure M.?

   a. It is possible that the over-active neuron(s) in the itch pathway are the secondary or tertiary neurons.

   b. This itch receptor would be quickly replaced by a new itch receptor neuron.

   c. Because of the redundancy of the nervous system, there are receptors all over the body that can trigger the same sensation in the same location.

Lab 16C Post-Lab Quiz

1. Sensory receptors are referred to as transducers. Which of the following statements most accurately describes a receptor's role as a transducer?

   a. A sensory receptor is considered a transducer because it both receives and sends electrical signals.

   b. A sensory receptor is considered a transducer because it connects one place in the nervous system to another.

   c. A sensory receptor is considered a transducer because it receives stimuli in one form and sends stimuli in another.

2. In Activity 16C.2 you explored touch receptor adaptation. Are the specific receptors you were testing tonic receptors or phasic receptors?

   a. Tonic receptors

   b. Phasic receptors

   c. A mix of both kinds

   d. It is not possible to know

3. Your friend is holding your hand while they receive an injection. At first the hand holding is pleasant, but then, as they squeeze harder, it becomes painful. How did a pleasant sensation transform into a painful one?

   a. The touch receptors are firing action potentials faster.

   b. The touch receptors are firing bigger action potentials.

   c. The touch receptors are firing, but now nociceptors are firing too.

4. Menthol is a common chemical ingredient of cough drops, when placed on the tongue or the skin it creates a cold sensation, which can be helpful in soothing a sore throat, for example. What type of receptor is likely the target of menthol?

   a. Nociceptors

   b. Thermoreceptors

   c. Mechanoreceptors

   d. Chemoreceptors

5. In Activity 16C.1, though you administered two points of touch to the subject, they reported feeling only one point. When this occurs, the two points must be

   a. So soft that they do not activate touch receptors

   b. Activating neurons that inhibit each other

   c. So close together that they are within the same receptive field

6. In the image, how many points of touch will the subject be able to feel?

   a. One

   b. Two

   c. Three

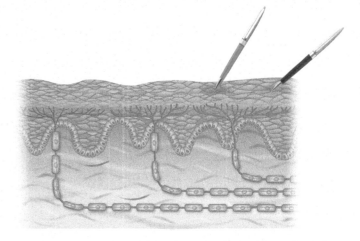

7. Which of the following is a plausible physiological explanation for peripheral adaptation?

   a. The primary neuron in a sensory pathway decreases the amount of neurotransmitter it releases.

   b. The secondary neuron in a sensory pathway dies, eliminating its synapses.

   c. The tertiary neuron in a sensory pathway changes its expression of ion channels, lowering its threshold membrane potential.

8. Endogenous opioids (also called endorphins) are molecules that are released at a synapse within the spinal cord, where they inhibit the axon terminal of nociceptors (see the figure). What is this an example of?

   a. Central adaptation
   b. Peripheral adaptation
   c. Lateral inhibition
   d. Acuity

9. A general sense pathway ends in which region of the brain?

   a. Thalamus
   b. Hypothalamus
   c. Primary somatosensory cortex
   d. Occipital lobe

10. The fingertips have many sensory neurons, each with small receptive fields. The skin of the thigh has fewer sensory neurons but each has larger receptive fields. Which region is likely to have the best touch acuity?

    a. Fingertips
    b. Thighs
    c. They are likely to have similar levels of acuity

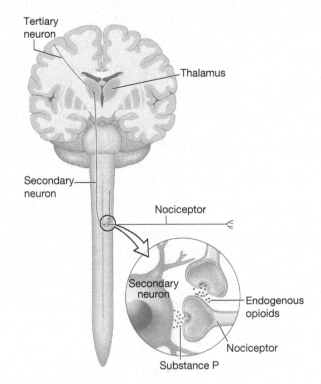

Lab 17 Brain

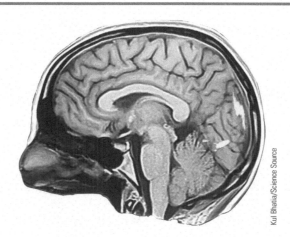

Kul Bhatia/Science Source

Learning Objectives: After completing this lab, you should be able to:

17.1 Identify and define the general terms gyrus, sulcus, and fissure.

17.2 Identify and describe the four major parts of the adult brain (i.e., cerebrum, diencephalon, brainstem, cerebellum).

17.3 Identify and describe the cerebral hemispheres and the five lobes of each (i.e., frontal, parietal, temporal, occipital, insula).

17.4 Identify and describe the major landmarks of the cerebrum (e.g., longitudinal fissure, lateral sulcus [fissure], central sulcus, transverse fissure, precentral gyrus, postcentral gyrus).

17.5 Identify and describe the three major cerebral regions (i.e., cortex, white matter, cerebral nuclei [basal nuclei]).

17.6 Name the major components of the diencephalon.

17.7 Describe the structure, location, and major functions of the thalamus.

17.8 Describe the structure, location, and major functions of the hypothalamus, including its relationship to the autonomic nervous system and the endocrine system.

17.9 Describe and identify the epithalamus, including the pineal gland and its function.

17.10 Name the three subdivisions of the brainstem.

17.11 Describe the structure, location, and major functions of the midbrain (mesencephalon), including the cerebral peduncles, superior colliculi, and inferior colliculi.

17.12 Describe the structure, location, and major functions of the pons.

17.13 Describe the structure, location, and major functions of the cerebellum.

17.14 Identify and describe the ventricular system components.

17.15 Identify the layers of the meninges and describe their anatomical and functional relationships to the CNS (brain and spinal cord).

17.16 Describe the structure and location of the dural venous sinuses, and explain their role in drainage of blood from the brain.

The Human Anatomy and Physiology Society includes more than 1,700 educators who work together to promote excellence in the teaching of this subject area. The HAPS A&P Learning Outcomes measure student mastery of the content typically covered in a two-semester Human A&P curriculum at the undergraduate level. The full Learning Outcomes are available at https://www.hapsweb.org.

17.17 Identify and describe the structure and function of the cranial dural septa.

17.18 List and identify the cranial nerves by name and number.

17.19 Describe the major functions of each cranial nerve and identify each cranial nerve as predominantly sensory, motor, or mixed (i.e., sensory and motor).

17.20* Predict functional changes that could occur with injury to specific locations within the brain

*Objective is not a HAPS Learning Goal.

Introduction

Because brains are not able to be removed and studied until after death, the intricacies of their anatomy remained largely mysterious to scientists for a long time. We learned about functional brain regions mostly through case studies on injuries, connecting the dots between the location of the injury and the functions lost as a result. All of this changed in 1990 with the invention of the functional magnetic resonance imaging machine (fMRI). fMRI imaging can be done on a living, awake person. It measures changes in blood flow that occur with brain activity. As a given brain region is being used more during a task, the blood flow to that region increases (Figure 17.1). With this technology we are able to discern discrete locations that are responsible for different functions within the brain. For example, there are brain regions that are specific for appreciating visual beauty, mathematical beauty, and interpersonal beauty. The first two are located within the frontal lobe, which you will discover in this lab is a region that is involved in decision-making, personal preferences, and higher-order cognitive behaviors. In this lab we will learn how to recognize brain structures as well as the tissues that support and protect the brain.

Most body functions are controlled by dedicated brain regions. A person's conscious experiences, the regulation of homeostasis, muscle movements and sensory input are all functions that can be localized to particular brain regions.

The adult brain has four major regions: the cerebrum, the diencephalon, the brainstem, and the cerebellum. The **cerebrum** is the iconic wrinkled, gray globe of the human brain. The wrinkled superficial portion of the cerebrum is the **cerebral cortex**, while the rest of the structure is deep to the cortex. As the brain develops within the cranium it runs out of space for all of the neurons; therefore the developing brain folds in on itself repeatedly. Each fold is a **gyrus** (plural: gyri) and the grooves between the folds are **sulci** (singular: sulcus). Larger grooves are **fissures** (Figure 17.2).

Figure 17.1

Functional magnetic resonance imaging (fMRI) allows clinicians and scientists to observe areas of the brain that are actively working, which is measured as increased blood flow (yellow/orange regions).

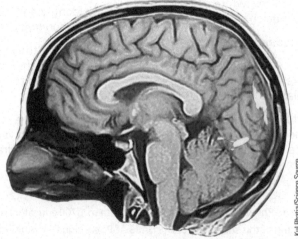

Kul Bhatia/Science Source

Figure 17.2   **The Cerebrum**

The cerebrum is the largest brain structure. Its surface is characterized by many gyri and sulci. It is divided into right and left hemispheres by the longitudinal fissure.

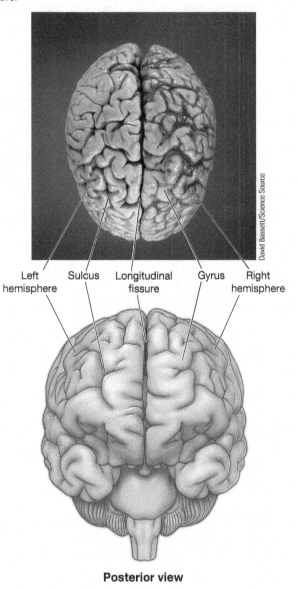

Left hemisphere   Sulcus   Longitudinal fissure   Gyrus   Right hemisphere

David Bassett/Science Source

Posterior view

The Meninges

In the central nervous system (CNS) the brain and spinal cord are shrouded in three layers of connective tissue, collectively called **meninges**. The three meninges are, from superficial to deep, the dura mater, the arachnoid mater, and the pia mater. The **dura mater** is a thick, collagen-rich fibrous layer that acts as a strong protective sheath over the entire brain and spinal cord. It is anchored to the inner surface of the cranium and vertebral cavity, and in some locations it separates into two layers, one just inside the bony structure of the skull and the other over-laying the surface of the brain. When the dura separates into two layers, the more superficial layer—associated with the skull—is called the *periosteal dura mater* and the deeper layer is the *meningeal dura mater*. Beneath the dura is the **arachnoid mater**, a thinner fibrous membrane that forms a loose sac around the CNS. Under the arach-noid, sitting directly on top of the surface of the CNS is the **pia mater**, the thinnest of the three membranes, which hugs all of the bumps, grooves, and indentations of the brain like a thin layer of paint (Figure 17.3).

Figure 17.3   **The Meninges of the Brain**

There are three layers of connective tissue membranes that surround the central nervous system; these are called the *meninges*. (A) The dura, arachnoid, and pia maters are continuous between the brain and spinal cord. (B) At some points within the cranium, the dura splits into two layers, the periosteal and meningeal layers. The space between them accommodates venous blood as it drains from the brain and returns to the right side of the heart.

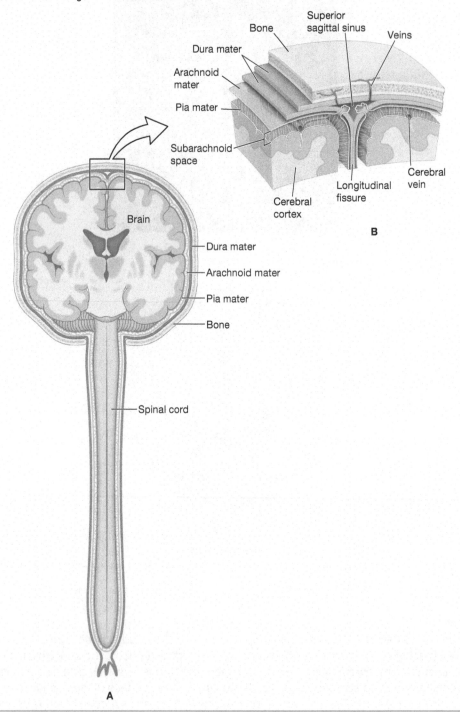

The arachnoid and pia mater cover the brain as flat, continuous sheets, but the dura mater has several folds with notable functions. Some of the spaces where the periosteal and meningeal dura mater separate serve as collecting wells for venous blood. Like many spaces within the body, these are termed sinuses, specifically dural venous sinuses, and are illustrated in Figure 17.4. The blood in the dural venous sinuses is oxygen-poor after passing throughout the brain and collects here before returning to the right side of the heart.

In addition to the blood-filled sinuses, the dura mater also invaginates to form stabilizing walls within the cranium called septa. There are four of these cranial dural septa, which subdivide the cranial space. The four structures are: the **falx cerebri**, **tentorium cerebelli**, **falx cerebelli**, and **diaphragma sellae** (Figure 17.5).

The Ventricles

The neurons of the brain and spinal cord are bathed in a clear, nourishing fluid that circulates throughout the CNS. This fluid, **cerebrospinal fluid** or **CSF**, is derived from blood and functions to both provide nutrients to neurons as well as physically protect the brain and spinal cord from injury. CSF is produced in the **choroid plexus**, a tangle of capillaries and glial cells that are found within fluid filled in the brain known as **ventricles**. The ventricles are physically connected to each other and CSF can flow among the chambers. The two **lateral ventricles** are within each cerebral hemisphere (Figure 17.6). On their posterior ends they each connect to a small drain called the **interventricular foramen** which opens into the **third ventricle**. The third ventricle is connected to the **fourth ventricle** through the **cerebral aqueduct**. Each of the four ventricles has a choroid plexus and produces CSF. CSF can drain from the fourth ventricle into two locations. It can seep upwards into the subarachnoid space (see Figure 17.3) and it can flow downwards into the **central canal**, a hollow space within the spinal cord.

Figure 17.4	Dural Venous Sinuses

Blood that is leaving the brain drains first through spaces within the dura mater called *dural venous sinuses* before joining the veins that lead to the heart.

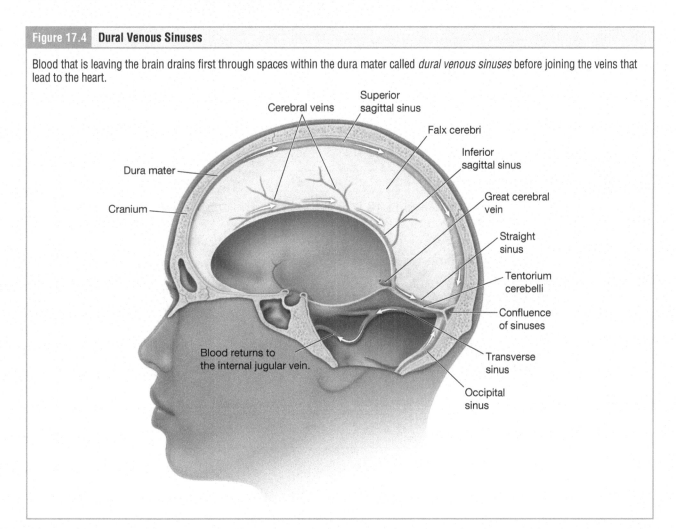

Learning Connection

Explain a Process

Explain or draw the path of CSF formation and recycling.

Figure 17.5 | **Cranial Dural Septa**

At some locations, the dura mater invaginates deeply into the cranium, separating and stabilizing the parts of the brain.

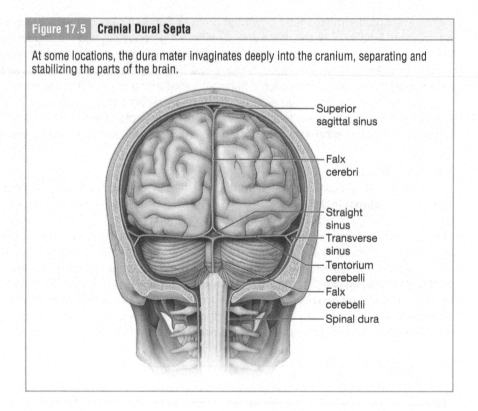

- Superior sagittal sinus
- Falx cerebri
- Straight sinus
- Transverse sinus
- Tentorium cerebelli
- Falx cerebelli
- Spinal dura

Figure 17.6 | **The Ventricles of the Brain**

There are four ventricle structures deep within the brain. Each is a fluid-filled space with a choroid plexus. The choroid plexus generates cerebrospinal fluid from blood plasma.

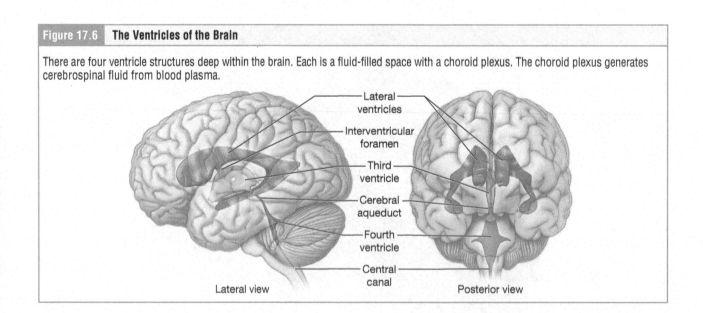

- Lateral ventricles
- Interventricular foramen
- Third ventricle
- Cerebral aqueduct
- Fourth ventricle
- Central canal

Lateral view   Posterior view

The Cerebral Lobes

Using the largest sulci as landmarks, the cerebrum can be divided into four major lobes: the frontal, parietal, temporal, and occipital lobes. Notably, there is one frontal lobe across the anterior brain, one occipital lobe across the posterior brain, but paired temporal and parietal lobes (a right and left of each). Figure 17.7 illustrates that the central sulcus separates the frontal and parietal lobes. We can also see that the lateral sulcus separates the temporal lobe from the parietal and frontal lobes. If the frontal and temporal lobes are pulled apart at the lateral sulcus, another small lobe is evident beneath. This is the insula.

The **frontal lobe** contains areas for planning and initiation of muscle contraction, including the contractions that allow for speech formation and eye movements. The frontal lobe is also involved in a lot of decision-making and other higher-order cognitive behaviors. The **parietal lobes** are mainly dedicated to interpreting the general sensations associated with the body such as stretch of muscles, touch, or pain on the skin, for example. The paired **temporal lobes** are dedicated to the tasks of hearing and smelling. The paired **insula** lobes are involved in taste association, meaning that they function to help you connect a new taste to taste memories. The single **occipital lobe** spans the posterior of the brain and is dedicated entirely to processing visual information and visual memories.

Deep to the gray matter of the cerebral cortex are central white matter tracts. A tract is a bundle of axons within the CNS. These tracts connect regions of the brain to each other and the brain to the spinal cord (Figure 17.8). Beneath the cerebral cortex and among the white matter tracts are sets of nuclei, **basal nuclei**, that contribute to brain function. Here the term *nucleus* means a cluster of cell bodies within the CNS. It can be confusing that a cluster of cell bodies, involving multiple cells, is described with the same word that we use to describe a part of a cell, however the root of the word nucleus means a cluster or a little nut, so early anatomists found this description to be useful for both the nucleus of the cell and the nuclei of the brain.

Figure 17.7	Lobes of the Cerebrum

The cerebral cortex is divided into five functional regions. The frontal lobe, parietal lobes, occipital lobes, and temporal lobes are visible here; the insula is the fifth region, which lies deep to the temporal lobe; part of the insula is visible in this figure.

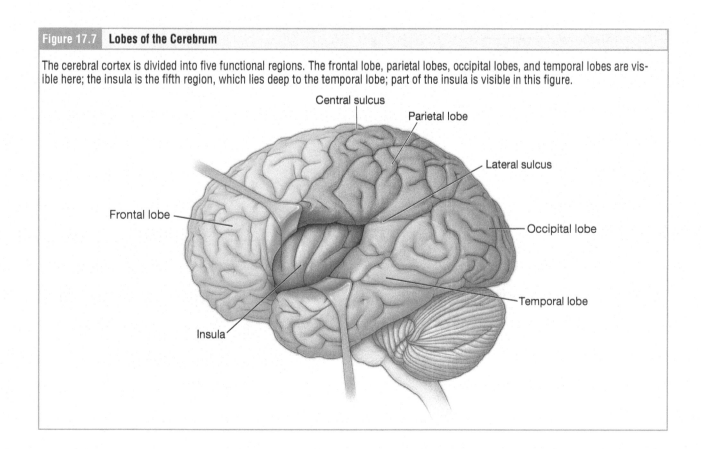

Figure 17.8 | White Matter Tracts of the Brain

Different functional areas of the brain are connected through tracts of axons that make a notable pattern in the brain. (A) Frontal sections demonstrate the long tracts of descending motor and ascending sensory axons that are continuous from the brain to the spinal cord. (B) Midsagittal sections show the connections from the cortex to underlying structures throughout the brain.

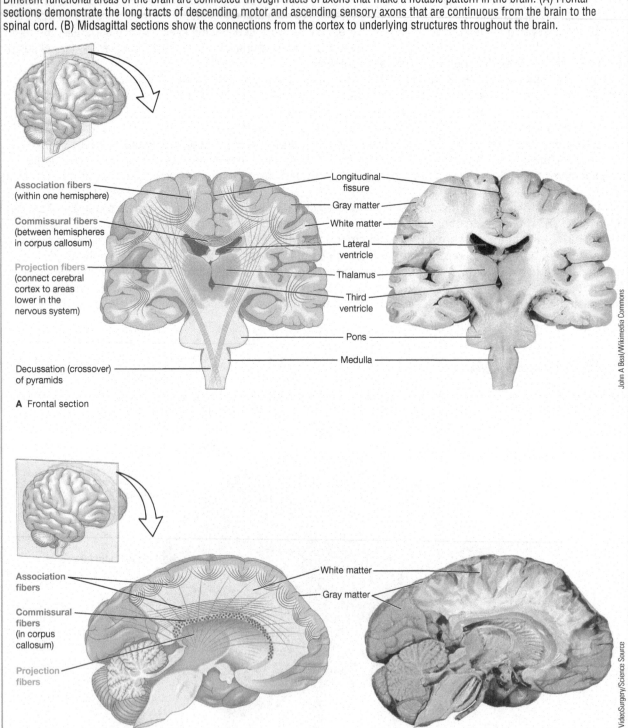

A Frontal section

Association fibers (within one hemisphere)

Commissural fibers (between hemispheres in corpus callosum)

Projection fibers (connect cerebral cortex to areas lower in the nervous system)

Decussation (crossover) of pyramids

Longitudinal fissure

Gray matter

White matter

Lateral ventricle

Thalamus

Third ventricle

Pons

Medulla

John A Beal/Wikimedia Commons

B Midsagittal section

Association fibers

Commissural fibers (in corpus callosum)

Projection fibers

White matter

Gray matter

VideoSurgery/Science Source

These nuclei contribute to the functions of the cortex, particularly filtering and metering responses. For example, basal nuclei control the intensity of a motion such as how hard to swing a racket at a tennis ball.

The Diencephalon

The diencephalon is the connection between the cerebrum and the rest of the nervous system with one exception. The rest of the brain, the spinal cord, and the PNS all send information to the cerebrum through the diencephalon (Figure 17.9). Output from the cerebrum passes through the diencephalon. The single exception is the system associated with olfaction (smell) which connects directly with the cerebrum. The diencephalon is deep beneath the cerebrum surrounding the third ventricle. The major regions of the diencephalon are the thalamus, epithalamus, and hypothalamus.

The thalamus consists of two egg-shaped masses through which almost all incoming sensory information passes before being processed by the cortex (some incoming smell information does not pass through the thalamus, but this is the only exception). The thalamus has the ability to edit, amplify, or diminish sensory information before sending it along to the primary regions of the cerebrum.

Inferior and slightly anterior to the thalamus is the hypothalamus, another major region of the diencephalon. The hypothalamus is a collection of nuclei that are largely involved in regulating homeostasis. The hypothalamus sits directly superior to and is physically connected to the **pituitary gland**. Through its own hypothalamic hormones as well as its regulation of pituitary gland function, the hypothalamus is considered to be the central regulator of the entire endocrine system. The hypothalamus works with the pineal gland to regulate sleep–wake cycles.

The Cerebellum

The **cerebellum** is covered in gyri and sulci like the cerebrum. Like the cerebrum, the cerebellum also has an outer cortex of gray matter and inner white matter tracts (Figure 17.10). These tracts form a treelike shape when the cerebellum is cut in midsagittal section, and are called the **arbor vitae**, which translates to "tree of life." The cerebellum functions to fine tune motor activities that are initiated by the cerebrum.

Figure 17.9	The Diencephalon

The diencephalon includes the hypothalamus, the thalamus, and the pineal gland.

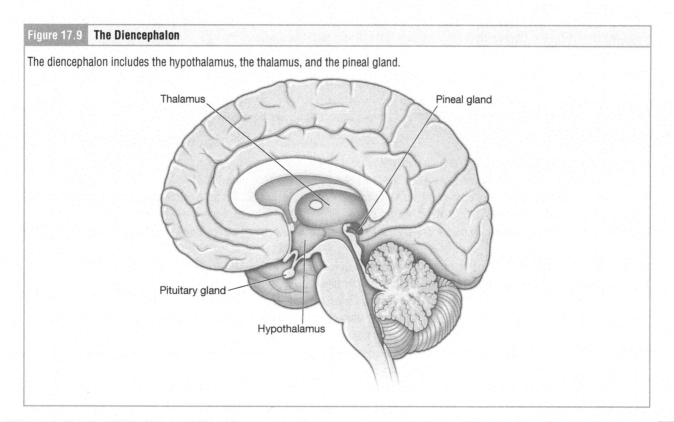

Learning Connection

Chunking

Try making a list of all of the brain structures discussed in this lab's intro. Now separate them into different categories. Which brain structures are involved in movement? In learning? In our senses? If you drew a transverse line through the brain from a single point (for example from the eyes to the back of the head), which brain structures are superior to that line? Which structures are inferior?

Figure 17.10 **Brainstem**

The three regions of the brainstem are the midbrain, the pons, and the medulla.

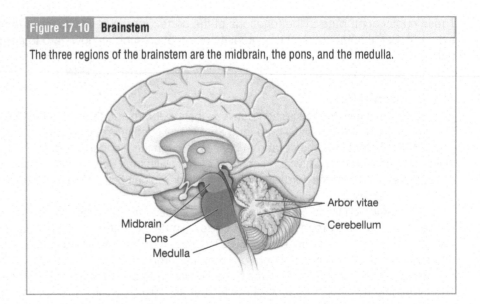

Midbrain
Pons
Medulla
Arbor vitae
Cerebellum

The Brainstem

The **brainstem** consists of three regions: the midbrain, pons, and medulla oblongata (Figure 17.10). The inferior edge of the medulla is continuous with the spinal cord.

The **midbrain** is a small region between the thalamus and the pons. The cerebral aqueduct, which connects the third and fourth ventricles (see Figure 17.6), passes through the center of the midbrain. On the posterior surface of the midbrain are four noticeable bumps, the corpora quadrigemina. Each of these four bumps is known as a colliculus. The two most inferior of these, the **inferior colliculi**, are part of the auditory pathway. The two superior bumps, the **superior colliculi**, function to combine sensory input and motor output, for example, turning to look when you notice movement in your peripheral vision.

The pons is a visible bulge on the anterior surface of the brainstem. The medulla oblongata, or simply the medulla, serves as the connection point between the brain and the spinal cord.

Digging Deeper:
Neurodiversity

Prior to the early 2000's the majority of scientific knowledge about the anatomical and functional specificity of the brain came from case studies of individual injuries or diseases. Scientists gathered evidence over time that individuals who had strokes or injuries in particular areas ended up with the same functional deficits. In the early 2000s the functional magnetic resonance imagery (fMRI) machine was invented. With this, we gained the ability to measure blood flow and neurological activity changes in living, healthy humans. Suddenly the field of neurobiology exploded with studies about the anatomical and functional specificity of the brain. But fMRI technology is both time-consuming and costly. Many of these studies involve a handful or few dozen participants. We are just starting to understand now that in the brain as well as in all other facets of anatomy and physiology, we need to understand diversity by widening the lens of our studies to include many different participants. For example, a 20XX study discovered a higher degree of connectivity between the thalamus and the primary somatosensory cortex in individuals with Autism Spectrum Disorders (ASD) than in neurotypical individuals. This heightened degree of connection between the thalamus and the primary somatosensory cortex can lead to increased somatosensory awareness. But simply in looking around a given college classroom we can appreciate that there is a tremendous amount of diversity in the ways that our brains function, thus, increasing the diversity in neurological studies will shed light on a wider spectrum of brain function.

The Cranial Nerves

All of the structures we have discussed so far in this lab are parts of the central nervous system. The central nervous system is defined as the brain and the spinal cord. Off of these two structures, bundles of axons enter and exit from the spinal cord. These bundles are called nerves and are all part of the peripheral nervous system. The nerves that emerge from the surface of the brain are called **cranial nerves**; the nerves that emerge from the spinal cord are called **spinal nerves**. Both cranial nerves and spinal nerves are part of the peripheral nervous system. Even though the cranial nerves are part of the PNS, we are investigating them here as the cell bodies of the neurons making up the cranial nerves originate in the brain and brainstem

The cranial nerves are primarily responsible for the sensor and motor functions of the head and neck (though one of these nerves targets organs in the thoracic and abdominal cavities). There are 12 cranial nerves and they have both names and numbers. Typically, the number is expressed in roman numerals, for example cranial nerve 10 is abbreviated CNX. The cranial nerves are listed in **Table 17.1**, along with a brief description of their location and function. Where the cranial nerves exit the brain can be seen in Figure 17.11.

Table 17.1 Cranial Nerves			
Nerve	**Function**	**Source**	**Target**
Olfactory I	Sensory	Cerebrum	Nose
Optic II	Sensory	Cerebrum	Eyes
Oculomotor III	Motor	Brainstem: Midbrain-pontine junction	Muscles of the eyes
Trochlear IV	Motor	Brainstem: Midbrain	Superior oblique muscle of the eye
Trigeminal V	Sensory and Motor	Brainstem: Pons	Face, sinuses, teeth (sensory), and muscles of mastication (motor)
Abducens VI	Motor	Brainstem: Pontine-medulla junction	Lateral rectus muscle of the eye
Facial VII	Sensory and Motor	Brainstem: Pons	Muscles of the face and sensory for a portion of the tongue.
Vestibulocochlear VIII	Sensory	Brainstem: Pontine-medulla junction	Inner ear
Glossopharyngeal IX	Sensory and Motor	Brainstem: Medulla oblongata	Posterior part of the tongue, tonsils, pharynx (sensory), and pharyngeal musculature (motor)
Vagus X	Sensory and Motor	Brainstem: Medulla oblongata	Heart, lungs, bronchi, gastrointestinal tract
Accessory XI	Motor	Brainstem: Medulla oblongata	Sternocleidomastoid and trapezius muscles
Hypoglossal XII	Motor	Brainstem: Medulla oblongata	Muscles of the tongue

Figure 17.11 The Cranial Nerves

The cranial nerves are numbered for the order in which they attach to the brain.

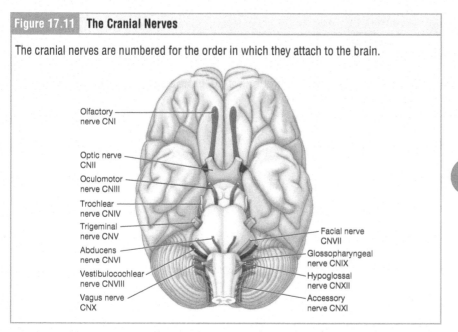

Olfactory nerve CNI
Optic nerve CNII
Oculomotor nerve CNIII
Trochlear nerve CNIV
Trigeminal nerve CNV
Abducens nerve CNVI
Vestibulocochlear nerve CNVIII
Vagus nerve CNX
Facial nerve CNVII
Glossopharyngeal nerve CNIX
Hypoglossal nerve CNXII
Accessory nerve CNXI

Learning Connection

Chunking

One of the ways that anatomists "chunk" the cranial nerves is by function. Make a list of which cranial nerves transmit sensory information, which ones transmit motor information and which ones transmit both.

Lab 17 Pre-Lab Quiz

This quiz will strengthen your background knowledge in preparation for this lab. For help answering the questions, use your resources to deepen your understanding. The best resource for help on the first five questions is your text, and the best resource for help answering the last five questions is to read the introduction section of each lab activity.

1. Anatomy of the brain. Apply the following labels to the figure: fissure, gray matter, gyrus, white matter

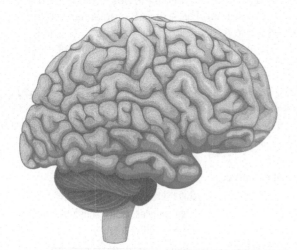 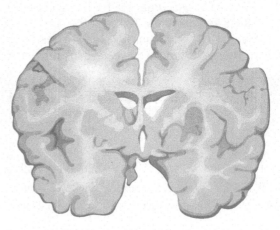

2. Match each of the following roots to its definition:

Root	Meaning		List of Roots
	a little nut, a cluster		arach-
	long lasting, tough		cereb-
	brain		corp
	body		dura
	spider		gemini
	twin		mater
	mother		nucle

3. The vagus nerve exits the brain at the medulla and travels throughout the thorax and abdomen innervating many muscles and nerves. How would we classify this nerve?

   a. Cranial
   b. Spinal
   c. Neither

4. The occipital lobe occupies much of the _____ region of the brain.

   a. anterior
   b. lateral left
   c. lateral right
   d. posterior

5. A bundle of axons is called a _____ when it's within the central nervous system and a _____ when it's within the PNS.

   a. nerve, tract
   b. tract, nerve
   c. ganglion, nerve
   d. neuron, ganglion

6. The cerebellum is located _____ to the pons.

   a. superior
   b. inferior
   c. lateral
   d. posterior
   e. anterior

7. Which of the following would be a reasonable way to test the function of CNI?

   a. Ask a patient if they can see the letters on an eye chart.
   b. Ask a patient if they can smell a notable scent.
   c. Ask a patient if they can turn their head to the side.
   d. Ask a patient if they can close and open their eyes.

8. On the inferior side of the brain, an x-shaped structure can be seen where the black arrow points. What is the correct name of this structure?

   a. Optic chiasm
   b. Suprachiasmatic cross
   c. Crossed ventricle

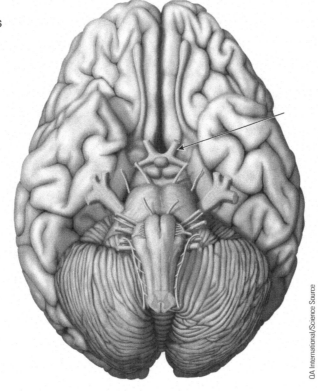

QA International/Science Source

9. Which of these is the correct order of the meninges from superficial to deep?

    a. Pia, arachnoid, dura

    b. Dura, pia, arachnoid

    c. Arachnoid, dura, pia

    d. Dura, arachnoid, pia

10. What is the function of CNIII?

    a. It carries information about sight.

    b. It carries auditory information.

    c. It provides motor control to the eyes.

    d. It provides motor control to the tiny muscles in the nose.

Activity 17.1 Gross Anatomy of the Brain

This activity targets LOs 17.1, 17.2, 17.3, 17.4, 17.5, 17.6, 17.7, 17.8, 17.9, 17.10, 17.11, 17.12, and 17.13.

The human brain with its 86 billion neurons is relatively uniform in appearance. While we use colorful diagrams and models to learn brain anatomy, the regions of a real brain are difficult to discern from one another. In this lab we will use brain models to discern the different functional regions of the brain from one another.

Materials
- Lab tape for labeling
- Model of a human brain

Instructions
1. Using your model of a human brain, identify the following:

   a. The inferior, superior, lateral, and midsagittal views.

   b. The structures listed in Table 17.2, labeling each with tape as you go.

Table 17.2 Structures to identify in Activity 17.1			
Brain Structure		**Found in Image**	**Best Viewed from**
Brainstem	Parietal lobe	Figure 17.7	Lateral
Central sulcus	Pons		
Cerebellum	Postcentral gyrus		
Frontal lobe	Precentral gyrus		
Lateral sulcus	Temporal lobe		
Medulla oblongata	Transverse fissure		
Occipital lobe			
Left hemisphere		Figure 17.12	Superior
Longitudinal fissure			
Right hemisphere			
Infundibulum	Optic chiasm	Figure 17.13	Inferior
Mammillary bodies	Optic nerves		
Olfactory bulbs	Optic tracts		
Olfactory tracts			
Fornix	Septum pellucidum	Figure 17.9	Midsagittal
Hypothalamus	Superior colliculus		
Inferior colliculus	Tectal plate		
Midbrain	Thalamus		

| Figure 17.12 | Superior View of the Sheep Brain | | Figure 17.13 | Inferior View of the Sheep Brain |

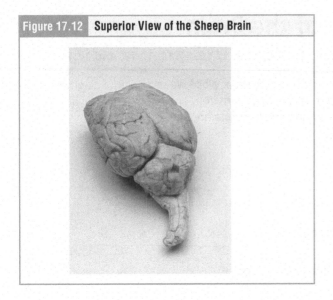

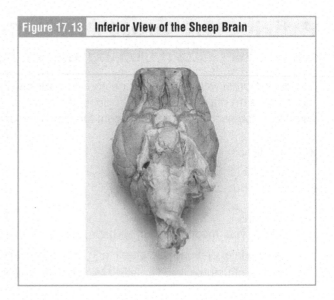

2. Using your brain model as a reference, stretch your skills by labeling the two views of the brain and on the next page. Use the lab's introduction and Table 17.2 to identify as many structures as you can.

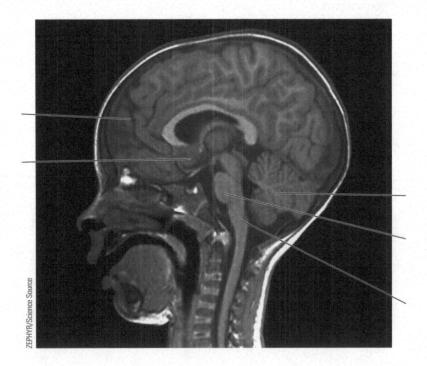

ZEPHYR/Science Source

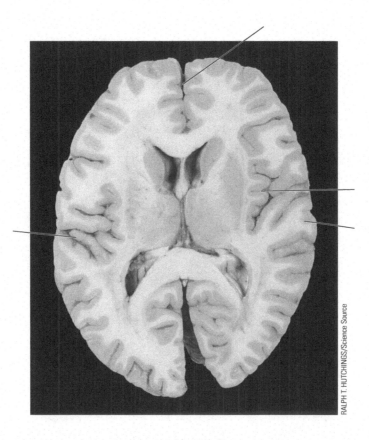

RALPH T. HUTCHINGS/Science Source

Digging Deeper:
The Growing (or Shrinking!) Brain

Do you think you have more or fewer neurons in your brain compared to a baby? It's often surprising to learn that our brains steadily lose neurons throughout our lifespans, and not in an unhealthy way. Most neurons are non-mitotic, meaning that they are incapable of replication, so when neurons are damaged, they are typically not replaced. But it's not just through accumulated damage that we decrease neuronal number. As our neural circuits become more mature, established, and efficient, neurons are removed, a process called neural pruning. While neuron number decreases steadily, the number of synapses, or connections among neurons, increases with age and especially with learning.

(page intentionally left blank)

Activity 17.2 **Sheep Brain Dissection**

This activity targets LOs 17.1, 17.2, 17.3, 17.4, 17.5, 17.6, 17.7, 17.8, 17.9, 17.10, 17.11, 17.12, 17.13, 17.14, 17.15, 17.16, and 17.17.

Introduction
The sheep brain is quite similar to the human brain, though notably smaller. Because the brain is relatively uniform in appearance, we will use landmarks to help identify individual structures. When inside the body, the brain and spinal cord are shrouded with three layers of meninges, which are described in the introduction to this lab. When removed from the skull for dissection, the brain is sometimes removed with all three meningeal layers intact, or the dura and/ or arachnoid may be removed. Remembering that the pia mater hugs the grooves and fissures of the cerebrum, this layer will still be in place on your dissection specimen. When you first obtain your brain specimen you will need to determine which meningeal layers are intact and remove the dura if present.

Materials
- Dissecting tray
- Forceps
- Gloves and other PPE if instructed
- Knife or scalpel
- Scissors
- Sheep brain, ideally with the dura mater intact

Instructions
1. Obtain a sheep brain from your instructor and place on your dissecting tray. Determine with your lab partner whether the brain has the dura mater intact or not. If the dura mater has been removed proceed to step 8.

2. With your gloved hands, feel the dura. What words would you use to describe it? Can you see through it?

3. On the superficial surface of the dura, locate the confluence of sinuses, transverse sinuses, and superior sagittal sinus using Figure 17.4 as your guide.

4. Turn the brain over so that it rests on its superior surface. Find the pituitary gland. Around the pituitary gland you should be able to see capillary tufts and the trigeminal nerves (CN V) all surrounding the pituitary gland. The dura mater that sits between the pituitary gland and the brain is the diaphragma sellae.

5. The dura mater is attached to all these structures, so to remove it you will need to cut a circle lateral to this area, as shown in Figure 17.14. Use your scissors to slowly and gently make this circular cut.

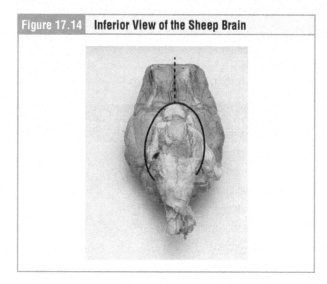

Figure 17.14	Inferior View of the Sheep Brain

6. Now identify the olfactory bulbs and olfactory tracts. Using your scissors, make a cut along the midsagittal plane between these olfactory structures. Connect your midsagittal cut to your circular cut around the diaphragma sellae.

7. You should now be able to remove the dura mater by grasping at the posterior superior region, and pulling gently and slowly, allowing the falx cerebri and tentorium cerebellum to wiggle free from the deep folds of nervous tissue.

8. Identify the structures listed in Table 17.3.

9. Gently bend the cerebellum away from the cerebrum to reveal the single, superior pineal gland and paired inferior corpora quadrigemina of the midbrain.

10. Using a knife or scalpel cut the sheep brain along the midline. Take your time with this step, go slowly and carefully.

11. Locate the structures listed in Table 17.4.

12. Dispose of the sheep brain as directed by your lab instructor.

Table 17.3 Structures Visible on the Intact Sheep Brain		
Inferior View	**Superior View**	**Lateral View**
Longitudinal fissure	Cerebral hemispheres	Parietal lobe
Olfactory bulbs	Gyri	Occipital lobe
Optic nerves	Sulci	Frontal lobe
Optic chiasma	Longitudinal fissure	Temporal lobe
Optic tract	Frontal lobe	Optic nerve
Mammillary body	Parietal lobe	Oculomotor nerve
Pituitary stalk	Temporal lobe	Trochlear nerve
Midbrain	Occipital lobe	Trigeminal nerve
Pons	Cerebellum	Cerebellum
	Medulla oblongata	Facial nerve
	Spinal cord	Vestibulocochlear nerve
		Glossopharyngeal nerve
		Vagus nerve
		Accessory nerve
		Hypoglossal nerve
		Medulla
		Pons
		Midbrain

Table 17.4 Structures Visible in the Midsagittal View
Midsagittal View
Cerebrum
Corpus callosum
Cerebellum white matter
Cerebellum gray matter
Lateral ventricle
Third ventricle
Fourth ventricle

Activity 17.3 Cranial Nerve Function

This activity targets LOs 17.18 and 17.19.

In this activity we will imagine an action (or you can ask someone to perform the action) like the ones in Table 17.5.

Instructions

1. Match each activity to the cranial nerve responsible.

2. List whether that cranial nerve carries sensory, motor, or both kinds of information.

3. Use reference materials to figure out what structure is innervated by this cranial nerve.

4. When you have completed the table, compare your answers with your lab partner. Discuss your reasoning and where your answers differ.

Note that the functions of several cranial nerves are also tested in the sensory lab (Lab 16B).

Table 17.5 Actions Controlled by Cranial Nerves			
Action	**Cranial Nerve Responsible**	**Sensory/Motor or Both?**	**Innervated Structure**
Hear the voice of the instructor	CNIII Vestibulocochlear	sensory	cochlea
Smile			
Open and close the jaw			
Elevate and depress the shoulders			
Follow the path of a ball as it flies through the air with your eyes			
Pupils constrict when you step outside into bright sunlight			
Swallow			
Stick tongue out			

(page intentionally left blank)

Activity 17.4 Case Study: Stroke

This activity targets LO 17.20.

Strokes are types of brain injuries that occur when the brain does not get enough blood. They often have serious or life-threatening outcomes. Approximately 795,000 strokes occur in the United States per year.

    You are spending time looking at graphs of your data with your research advisor, Dr. Smith, when you notice that Dr. Smith's speech begins to slur a little. Looking up at their face, you notice that one side of their face is drooping like the person's in the image (Figure 17.15). You are very concerned and wonder if Dr. Smith may be having a stroke. You remember the acronym FAST (Figure 17.16).

Case Study

Figure 17.15 During a Stroke, It Is Common to Observe Drooping or Paralysis on one Side of the Face

SpeedKingz/Shutterstock.com

1. If Dr. Smith is indeed having a stroke, which arm would they be unable to raise?

    a. The ipsilateral arm (the arm on the same side as the drooping face)

    b. The contralateral arm (the arm on the opposite side as the drooping face)

You call 911. Dr. Smith is transported to the Emergency Department and images are taken of their brain.

Figure 17.16 Remember the Acronym F.A.S.T. for Recognizing the Signs of Stroke

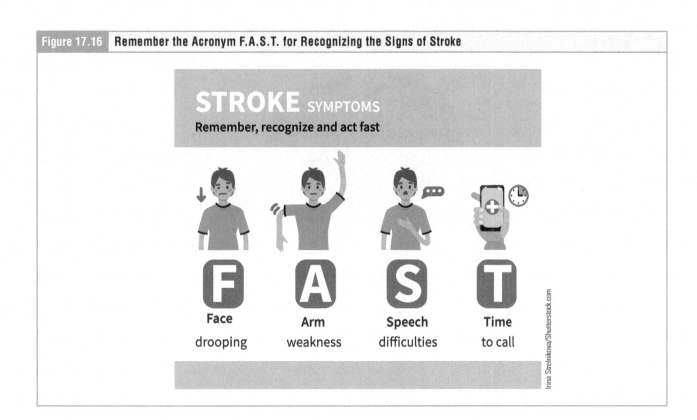

Irina Strelnikova/Shutterstock.com

2. Which of these images of a stroke (a highlighted part of the image) could explain your observations?

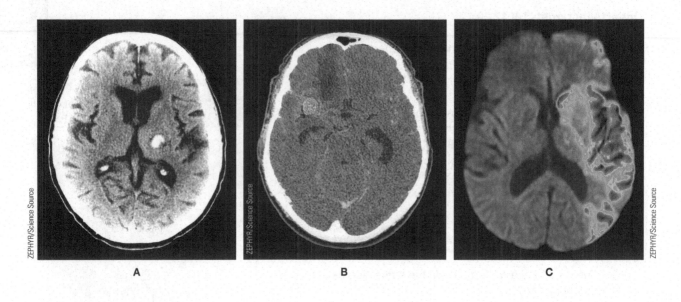

A   B   C

a. A

b. B

c. C

There are several types of stroke. Ischemic strokes occur when there is a clot or other blockage that prevent or limits blood flow through a blood vessel. Hemorrhagic strokes occur when a blood vessel in the brain bursts allowing blood to escape from the vessel (Figure 17.17).

Figure 17.17

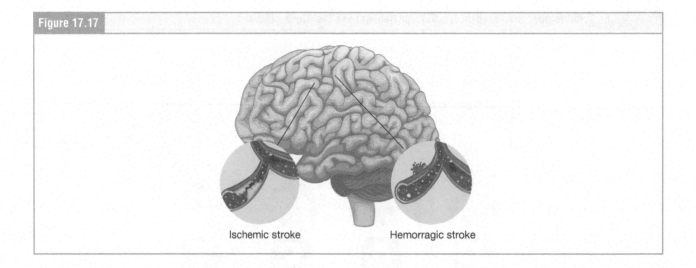

Ischemic stroke   Hemorragic stroke

3. Hemorrhagic strokes are described based on the location of the burst vessel. The figure shows three types of hemorrhagic strokes. Label them as either subdural, intraventricular, or deep cerebral.

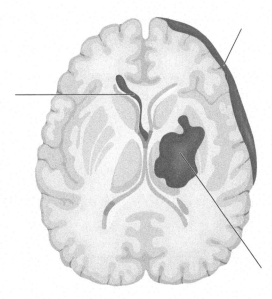

Dr. Smith is having a subdural hemorrhage. The doctors tell you that the blood is escaping from the blood vessel and building up under the dura, pushing both out on the dura as well as down onto the brain.

4. Do you think Dr. Smith is in pain?

   a. Yes, because the brain is sensitive to pain.

   b. No, because neither of these structures are innervated.

   c. Yes, because the dura is sensitive to pain.

5. Sometimes stroke patients are treated with medications that help dissolve clots and/or prevent clots from forming. Would this therapy be helpful for Dr. Smith? Explain why or why not.

(page intentionally left blank)

Lab 17 Post-Lab Quiz

1. Which structure is out of place in this image?

    a. Pituitary gland
    b. Thalamus
    c. Cerebellum
    d. Pineal gland

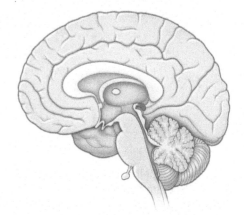

2. Which of the following functions occurs in the insula?

    a. Sight
    b. Muscle coordination
    c. Taste
    d. vision

3. Left neglect is a condition in which there is damage to a portion of the occipital lobe. In left neglect, the patient is unable to interpret anything on the left side of their field of vision, even though both of their eyes are functioning normally. For example, looking at a clock would appear as shown. Which portion of the occipital lobe would be damaged to result in left neglect?

    a. Deep anterior portion
    b. Superficial posterior portion
    c. Right hemisphere
    d. Left hemisphere

4. Phineas Gage was a famous case of brain anatomy because he suffered a dramatic brain injury (see the figure) and yet most of his functions were unaffected. His injury was mostly limited to the frontal lobe. With damage to this region, what might be a notable change to Phineas Gage?

    a. Loss of vision
    b. Loss of taste
    c. Change in personality
    d. Difficulty controlling body movements

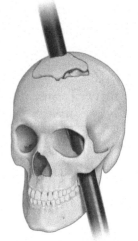

5. What is found within the ventricles?

    a. Blood
    b. Axons
    c. Cerebrospinal fluid
    d. Cell bodies

6. In this image, which meninge is the red arrow pointing to?

   a. Dura
   b. Arachnoid
   c. Pia

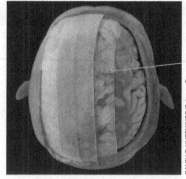

RALPH T. HUTCHINGS/Science Source

7. Which of the following terms best describes the material in the structure that the arrow is pointing to?

   a. Cell bodies
   b. Axons
   c. Nerves

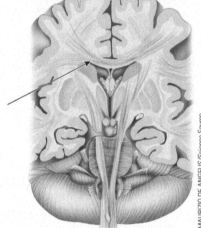

MAURIZIO DE ANGELIS/Science Source

8. Damage to the medulla oblongata would result in _____.

   a. changes in personality
   b. challenges in processing sound
   c. vision changes
   d. disruptions to heart rate and breathing

9. The pituitary gland is part of the _____ system.

   a. digestive
   b. respiratory
   c. endocrine

10. What cranial nerve is the arrow pointing to in this image?

   a. Optic
   b. Oculomotor
   c. Trigeminal
   d. Vagus

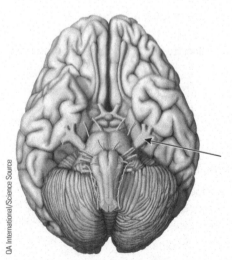

QA International/Science Source

Lab 18 Spinal Cord

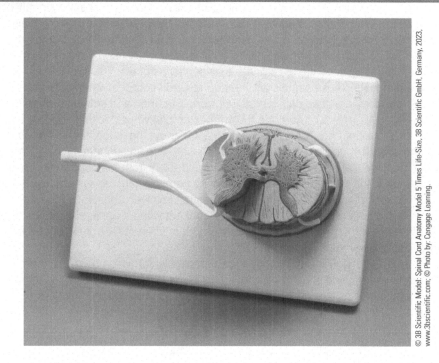

© 3B Scientific Model: Spinal Cord Anatomy Model 5 Times Life-Size, 3B Scientific GmbH, Germany, 2023. www.3bscientific.com; © Photo by: Cengage Learning.

Learning Objectives: After completing this lab, you should be able to:

18.1 Identify and describe the anatomical features seen in a cross-sectional view of the spinal cord (e.g., anterior horn, lateral horn, posterior horn, gray commissure, central canal, anterior funiculus, lateral funiculus, posterior funiculus).

18.2 Describe the structure, location, and function of ascending and descending spinal cord tracts.

18.3 Identify and describe the formation, structure, and branches of a typical spinal nerve, including the roots and the rami (e.g., anterior [ventral], posterior [dorsal]).

18.4 Identify and describe the gross anatomy of the spinal cord, including its enlargements (i.e., cervical and lumbar), conus medullaris, cauda equina, and filum terminale.

18.5* Predict alterations in neuronal function after injury, disease, or damage to the spinal cord or spinal nerves.

18.6 Compare and contrast the location, composition, and function of the anterior (ventral) roots, posterior (dorsal) roots, and posterior (dorsal) root ganglion with respect to the spinal cord.

18.7* Define and classify afferent and efferent signals.

18.8 Describe reflex responses in terms of the major structural and functional components of a reflex arc.

18.9 Identify the layers of the meninges and describe their anatomical and functional relationships to the CNS (brain and spinal cord).

*Objective is not a HAPS Learning Goal.

The Human Anatomy and Physiology Society includes more than 1,700 educators who work together to promote excellence in the teaching of this subject area. The HAPS A&P Learning Outcomes measure student mastery of the content typically covered in a two-semester Human A&P curriculum at the undergraduate level. The full Learning Outcomes are available at https://www.hapsweb.org.

Introduction

The **central nervous system (CNS)** consists of the brain and spinal cord. These structures are composed of **neurons**, the cells that send the signals of the nervous system, and **glial cells**, which cannot send signals but support and protect the neurons. Neurons and glial cells are also found outside the brain and spinal cord; all nervous tissue outside the brain and spinal cord is grouped as the **peripheral nervous system (PNS)**. Signals sent along neurons toward the brain and spinal cord are **afferent signals**. Afferent signals carry information about the internal or external environment, so another way to think of afferent signals is as sensory signals. The signals sent out from the CNS to the periphery are **efferent signals**. These signals are headed toward internal or external muscles in order to cause a change in movement, and they are often called motor signals. The brain and spinal cord are the connection point between afferent and efferent signals. Consider the following example: while cooking you feel and see that your hand has touched the side of a hot pan. The sensory information about the touch, that is the visual input, the heat, and the pain, is heading toward your brain, which compiles these pieces of afferent information and sends out efferent signals to contract the necessary muscles to move your hand away from the hot pan. In this example the touch of the hot pan is the **stimulus**, the thing that evokes a response from the nervous system (Figure 18.1A).

Sometimes the integration of afferent and efferent information involves conscious thought, but there are some anatomical pathways in which afferent and efferent signals connect without conscious thought. These are called reflexes. In a **reflex**, afferent information can produce an automatic efferent response, often before the brain receives information about the stimulus (Figure 18.1B).

The spinal cord is not only a highly organized highway of tracts of information coming to and from the brain but it is also the location where afferent and efferent information connects in most reflexes. In this lab we will examine the anatomy of the spinal cord as well as explore functional consequences of spinal cord injury. Lab 19 examines reflexes.

Figure 18.1 | **Spinal Cord Functions**

The spinal cord is both a traffic-filled highway of information connecting the brain to the periphery as seen in (A), as well as a hub where incoming and outgoing signals can connect in reflexes as shown in (B).

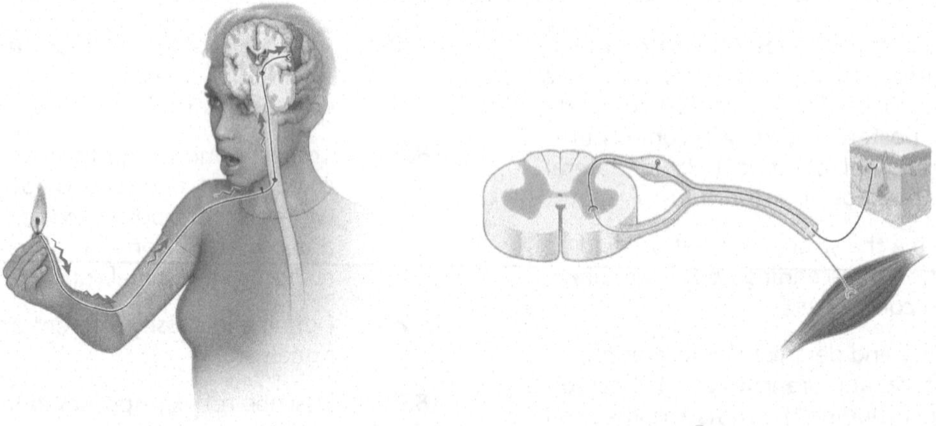

A   B

Digging Deeper:
Spinal Cord Injury

There are only a few structures of the human nervous system in which adult neurogenesis, the generation of new neurons through mitosis, is known to be possible. For the vast majority of the central nervous system, an injury that destroys neurons will not fully heal. In cases of spinal cord injury, signals will be prevented from traveling up or down across the injury site, depending on which structures are damaged. This blockage in communication to and from the brain can prevent both sensory information from being able to be perceived and motor information from being able to be sent from the brain to the muscles. While currently there is no way to regain these functions when lost, scientists are researching ways to heal spinal cord injuries and these methods may be available in the future.

Lab 18 Pre-Lab Quiz

This quiz will strengthen your background knowledge in preparation for this lab. For help answering the questions, use your resources to deepen your understanding. The best resource for help on the first five questions is your text, and the best resource for help answering the last five questions is to read the introduction section of each lab activity.

1. Anatomy of a spinal cord cross section. Apply the following labels to the figure below:

   afferent region
   anterior horn
   efferent region
   gray matter
   lateral horn
   posterior horn
   white matter

   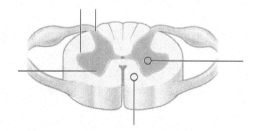

2. Match each of the roots to its definition:

Root	Meaning	List of Roots
	horse	af-
	toward	cauda-
	tough	dura
	tail	ef-
	out, away from	equina
	branch	ganglion
	a swelling	mater
	mother	ramus

3. Which of the following describes the difference between the cauda equina and the spinal cord?

   a. The cauda equina is part of the central nervous system and the spinal cord is part of the peripheral nervous system.
   b. The cauda equina is a bundle of nerves, whereas the spinal cord is a solid structure composed of axons, cell bodies, and synapses.
   c. The cauda equina has no ganglia, but the spinal cord has many ganglia.

4. What is the indicated structure (arrow) in the image?

   a. Root
   b. Ramus
   c. Plexus

   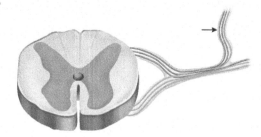

5. A spinal nerve carries which of the following structures? (Select all that apply)

   a. Motor neurons
   b. Sensory neurons
   c. Cell bodies
   d. Ganglia

6. In this image, what would you find at the indicated location (arrow)?

   a. Sensory cell bodies
   b. Motor cell bodies
   c. Sensory axons
   d. Motor axons

   Posterior

   Anterior

7. The filum terminale is an extension of which of the meninges?

   a. Dura mater
   b. Arachnoid mater
   c. Pia mater

8. Which of the following is the correct order of the meninges from superficial to deep?

   a. Dura—arachnoid—pia
   b. Pia—arachnoid—dura
   c. Dura—pia—arachnoid
   d. Arachnoid—pia—dura

9. Which of the following structures contains both motor and sensory neurons?

   a. Rootlet
   b. Root
   c. Ramus
   d. None of these

10. Compared with most regions of the spinal cord, there are more axons coming and going from the spinal cord at the _____.

    a. cervical enlargement
    b. thoracic region
    c. lumbar region
    d. cauda equina

Activity 18.1 Microanatomy of the Spinal Cord

This activity targets LOs 18.1, 18.2, and 18.3.

The surface of the spinal cord is marked by two deep indentations. The **anterior median fissure** is a deep longitudinal groove located between the anterior horns. The **posterior median sulcus** is a similar longitudinal groove located on the posterior surface.

Both the gray matter and the white matter are organized anatomically by function. The white matter is organized into columns. Remember that the white matter is white because it is largely composed of myelinated axons, so the white matter of the spinal cord is tracts of bundled myelinated axons traversing the spinal cord to carry signals to and from the brain. Ascending tracts are carrying sensory information up toward the brain, descending tracts are carrying motor information down from the brain. The **posterior columns**, located between the two posterior horns, is composed of ascending tracts. The **anterior columns**, located between the anterior horns and the **lateral columns** located on the lateral sides of the spinal cord, carry many different groups of axons of both ascending and descending tracts (Figure 18.2).

In cross section, the gray matter of the spinal cord resembles a capital letter H or a butterfly (Figure 18.2). Each of the projections of gray matter—what we would say are wings if we are comparing the gray matter to a butterfly—are called *horns*. Each **posterior horn** contains the ends of axons of sensory neurons. These short sensory axons enter the spinal cord in a bundle called the **posterior root**. Along this root, the cell bodies of these sensory neurons are bundled together in the **posterior root ganglion**. Each **anterior horn** contains the cell bodies of somatic motor neurons that are sending out motor signals to skeletal muscles. In the thoracic and upper lumbar regions of the spinal cord there is another set of horns, the **lateral horns**, which contain autonomic motor neuron cell bodies that send out motor signals to the smooth muscle of the organs. The motor axons from the anterior horn and lateral horn (in the regions of the spinal cord where there is a lateral horn) bundle together and emerge from the spinal cord as the **anterior root**. Each posterior or anterior root is actually composed of smaller bundles called **rootlets** (Figure 18.3). The anterior and posterior roots merge together to form the **spinal nerve**. This spinal nerve carries both motor and sensory neurons. As we trace the spinal nerve further laterally from the spinal cord, it soon splits into two branches, each referred to by the Latin word for branch, ramus. The **posterior ramus** carries motor neurons that will innervate the muscles of the back and

| Figure 18.2 | Cross Section of the Spinal Cord |

Cross sections of the spinal cord reveal the central gray matter to have a butterfly shape. (A) The anterior horns contain the cell bodies of motor neurons, and the posterior horns contain the axons of afferent sensory neurons. In the thoracic region, the lateral horns contain the cell bodies of autonomic neurons. (B) A stained cross section of the spinal cord reveals the gray and white matter when viewed under the microscope.

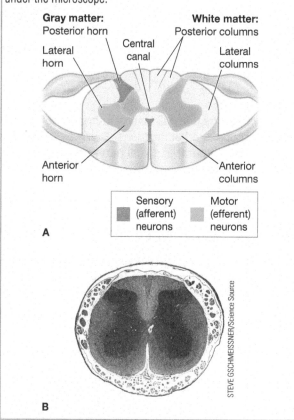

STEVE GSCHMEISSNER/Science Source

Figure 18.3 | Nerves Associated with the Spinal Cord

The neurons emanating from the spinal cord bundle first into small rootlets, which are bundled into larger roots.

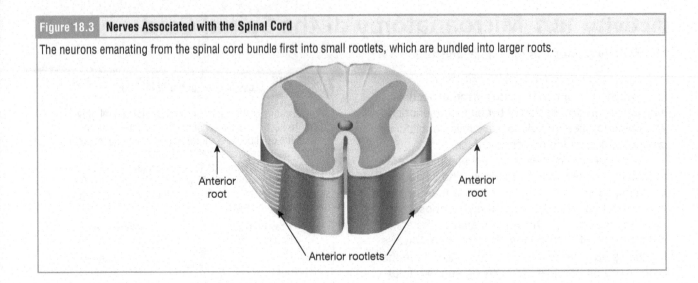

sensory neurons that will innervate the skin of the back. The **anterior ramus** will carry all of the sensory and motor neurons to other locations (Figure 18.4). In Figure 18.4 we draw these structures on one side of the spinal cord, but remember that spinal nerves emerge out of each side, to innervate each side of the body. There are 31 pairs of spinal nerves which are part of the PNS.

Figure 18.4 | The Spinal Nerve Is Composed of Posterior and Anterior Roots

The spinal nerve quickly splits into two branches, the posterior and anterior ramus. Think of this as a single tree trunk (the spinal nerve) that forms from many roots. The tree trunk splits into two main branches (the rami).

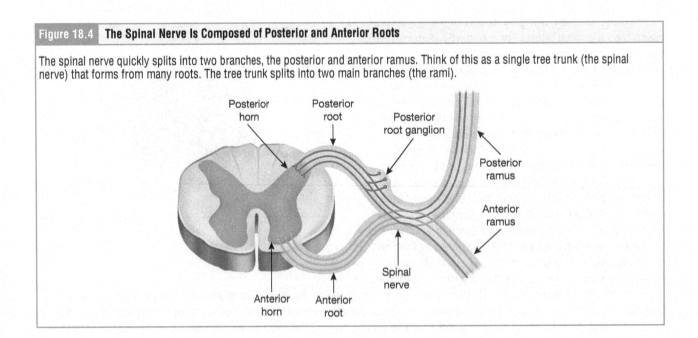

Materials
- Compound microscope
- Model of the spinal cord in cross section
- Slide of a spinal cord cross section

Instructions

1. On the model of the spinal cord in cross section, identify the structures listed in Table 18.1. Draw them in the space provided on the next page.

2. Place the slide of the spinal cord cross section on the microscope stage and bring the image into focus. Using Figure 18.2 and your spinal cord section model for reference, try to identify as many of the structures listed in Table 18.1 as you can. As you identify them, also identify them on the model and place a check mark in the column in Table 18.1.

Learning Connection

Broken Process

Ask yourself what would happen to function if each of these spinal cord structures was damaged? This is a fantastic way to quiz yourself on functions.

Table 18.1 Spinal Cord Structures Seen in Cross Section	
Structure on Spinal Cord Model	**Also Able to Identify on Slide?**
Anterior horn	
Anterior median fissure	
Anterior root	
Anterior rootlets	
Central canal	
Lateral horn	
Posterior horn	
Posterior median sulcus	
Posterior root ganglion	
Posterior rootlets	
Spinal nerve	

Drawing of the spinal cord cross section:

Activity 18.2 Gross Anatomy of the Spinal Cord

This activity targets LO 18.4.

The length of the spinal cord is divided into regions that correspond to the regions of the vertebral column. The name of each spinal cord region corresponds to the level at which spinal nerves that exit it pass through into the intervertebral foramina. From superior to inferior the regions are: cervical, thoracic, lumbar, and sacral (Figure 18.5). The spinal cord is shorter than the full length of the vertebral column. The solid spinal cord ends in the superior lumbar region, and from its end dangles a long bundle of nerves encased within the inferior spine. This bundle of nerves is the **cauda equina**. The solid spinal cord tapers to an end called the conus medullaris (Figure 18.6).

In the CNS, the brain and spinal cord are both shrouded in protective membranes of connective tissue. These three layers are called **meninges**. Each of these layers has a first name that describes its appearance and its second name is mater, the Latin word for 'mother.'[1] The dura mater is the opaque, tough outer covering of the CNS.

Figure 18.5	Longitudinal View of the Spinal Cord

The spinal cord generally tapers from superior to inferior but has two swellings, the cervical and lumbar enlargements.

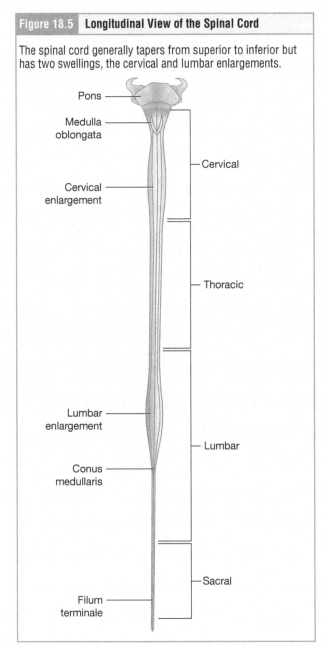

Figure 18.6	The Inferior End of the Spinal Cord Tapers Like a Cone

This tapering end is known as the conus medullaris. It is enmeshed in the long dangling nerves of the cauda equina and covered in meninges as shown.

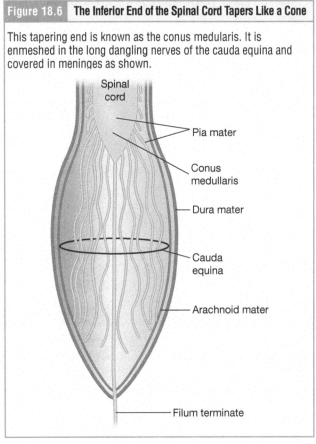

---

[1]Why "mother"? The meninges were seen as protective mothers that hovered over the brain as if the brain were a child.

Beneath it is found the arachnoid mater. The term arachnoid comes from the same root as arachnid, or spider, because the internal surface of the arachnoid is marked by thin fibrous legs that hold this membrane out a bit, creating a space between the arachnoid mater and the meninge deep to it. Through this space flows cerebrospinal fluid (CSF). The deepest meninge is the pia mater. At the conus medullaris, the pia mater comes to a thin string that continues inferiorly even after the dura and arachnoid maters end, this is called the **filum terminale** (Figure 18.6).

The spinal cord generally gets slightly narrower inferiorly, but there are two enlargements where more axons are coming and going from the cord. The **cervical enlargement** and the **lumbar enlargement** are both locations where there are increased numbers of axons entering and exiting the cord from the upper and lower limbs, respectively (see Figure 18.5).

Materials
- Disposable gloves and any other PPE recommended by your instructor
- Dissecting tray
- Preserved spinal cords from cows or cadavers with meninges intact
- Probe
- Scalpel
- Scissors

Instructions
1. If the dura mater is intact, use scissors to gently cut it open vertically.

2. Identify the structures listed in Table 18.2.

3. Using the scalpel, cut a cross section of the cord and identify the anterior and posterior horns of gray matter. Which are larger?

4. On your cross section, identify as many of the structures listed in Table 18.1 as possible (see Activity 18.1).

5. Dispose of the spinal cord as instructed. Wash the tray and tools in the lab sink.

Learning Connection

Chunking

Which of the structures in this lab are technically made of nervous tissue, and which are made of connective tissue?

Table 18.2 Spinal Cord Structures: Gross Anatomy	
Anterior roots	Lumbosacral enlargement
Cauda equina	Pia mater
Cervical enlargement	Posterior median sulcus
Conus medullaris	Posterior root ganglion
Dura mater	Posterior roots
Filum terminale	

Digging Deeper:
Damage to the Spinal Cord

Spinal cord damage often leads to permanent disability or loss of function. Damage to the central nervous system typically does not heal completely, in part because most neurons are not capable of mitosis. Since the spinal cord is a massive collection of axons, the site of injury often severs these axons and damages nearby glial cells. Recently, researchers from Northeastern University published a study in which a unique injection of structural proteins and growth factors was able to stimulate axon regeneration and glial cell growth in mice, offering hope that someday we may be able to do more to help healing in humans with spinal cord injuries.

Activity 18.3 Case Study: Spinal Injury

This activity targets LO 18.5.

A 4-year-old child, Robin, was rushed to the hospital one morning. Her parents observed that she had difficulty swallowing and she seemed to not be using her left arm. Two weeks ago, Robin developed symptoms of a respiratory illness; she has just finished recovering from it. After running tests, the doctor suspects Robin may have acute flaccid myelitis (AFM). This is a rare but serious condition that affects the nervous system of children. The virus is thought to enter the body through the lungs, but breaches the epithelium and moves into the body, targeting the gray matter of the spinal cord. Medical professionals do not know exactly what causes the disease, but they believe environmental pollutants make the respiratory membranes more vulnerable to viral entry. In most cases a mild respiratory illness precedes paralysis symptoms.

1. Based on the figure, identify where the disease affects the spinal cord.

   a. A
   b. B
   c. C

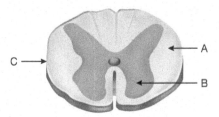

2. Using the answer from Question 1, which of the following is likely affected if the infection is targeting this area of the spinal cord?

   a. Axon of the motor neuron
   b. Synapse in posterior root ganglion
   c. Synapse within the anterior horn
   d. Cell body in the posterior horn

3. There is a test for the diagnosis of AFM. Which body fluid is most likely to be used in this test?

   a. Urine
   b. Blood
   c. Cerebrospinal fluid
   d. Expectorant (lung mucus)

4. Which pathway most likely describes the route the virus used to get into the bloodstream?

   a. Pulmonary (lung) capillaries–mucus–bloodstream–CSF
   b. Mucus–pulmonary (lung) capillaries–CSF–bloodstream
   c. Pulmonary (lung) capillaries–mucus–CSF–bloodstream
   d. Mucus–pulmonary (lung) capillaries–bloodstream–CSF

5. How did the virus get from the bloodstream into the spinal fluid?

   a. Enters the choroid plexus of the brain
   b. Passes through the blood–brain barrier via diffusion at the spinal cord
   c. At the dural venous sinus
   d. Passes through from the blood at the arachnoid mater

6. Which neural pathways are affected by this disease?

   a. Both afferent and efferent pathways
   b. Neither afferent nor efferent
   c. Afferent pathway
   d. Efferent pathway

(page intentionally left blank)

Lab 18 Post-Lab Quiz

1. In this figure, which arrow points to the lateral columns?

   a. A
   b. B
   c. C
   d. D

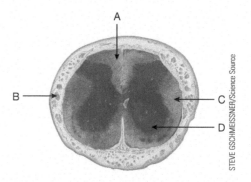

2. Some reflexes can be observed in patients with spinal cord injury. Which of the following correctly explains why?

   a. Because reflexes occur entirely in the peripheral nervous system, central nervous system injury does not affect them.

   b. Because reflexes do not require the brain, they can still occur below the point of injury.

   c. While technically the spinal cord cannot recover from injury, the reflex pathways alone are capable of recovery.

3. What is the best word to describe the indicated structure (arrow)?

   a. Root
   b. Ramus
   c. Nerve

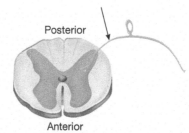

4. In what direction are the action potentials traveling among the neurons of the indicated structure (arrow)?

   a. Afferent
   b. Efferent
   c. A mix of both afferent and efferent signals

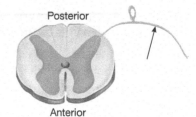

5. In a case of spinal **nerve** injury, would you expect both paralysis and numbness or only one of these?

   a. Paralysis only
   b. Numbness only
   c. Both paralysis and numbness

6. Which of the following is the correct order progressing away from the spinal cord?

   a. Root—rootlet—ramus—nerve
   b. Rootlet—root—ramus—nerve
   c. Root nerve—rootlet—ramus
   d. Rootlet—root—nerve—ramus

7. In this image, the directions of the tracts are labeled. Is the star on the anterior, posterior, or lateral side of the spinal cord?

   a. Anterior
   b. Posterior
   c. Lateral

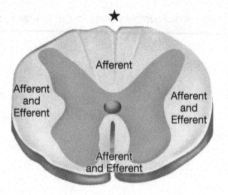

8. What type of neuron or neurons can be found in the posterior ramus?

   a. Sensory only
   b. Motor only
   c. Both sensory and motor

9. Is the reflex in the figure a monosynaptic or polysynaptic reflex arc?

   a. Monosynaptic
   b. Polysynaptic

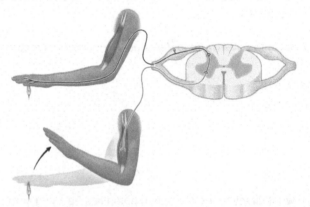

10. While meningitis is often associated with brain damage, neck aches and back aches are common symptoms. How could a disease that affects the membranes around the brain also cause pain in the neck and back?

   a. The neck and back are fine, but the pressure on the brain causes widespread pain perception.
   b. The meninges are continuous from the brain and along the spinal cord.
   c. The meninges only surround the brain, but compress the spinal cord where it exits the skull, causing radiating pain.

Lab 19 | Reflex Arcs and Reflexes

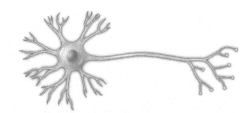

Learning Objectives: After completing this lab, you should be able to:

19.1 Describe reflex responses in terms of the major structural and functional components of a reflex arc.

19.2 Distinguish between each of the following pairs of reflexes: intrinsic versus learned, somatic versus visceral, monosynaptic versus polysynaptic, and ipsilateral versus contralateral.

19.3 Describe the following reflexes and name all components of each reflex arc: stretch reflex, (Golgi) tendon reflex, flexor (withdrawal) reflex, and crossed-extensor reflex.

19.4* Describe and provide examples of reciprocal activation and reciprocal inhibition associated with reflex arcs.

19.5* Perform stretch reflex tests, superficial reflex tests, a somatic reflex test, and an autonomic reflex test.

19.6* Predict the possible location of neurological damage based on the results of reflex tests.

19.7 Define the term reflex.

* Learning Objective not a HAPS Learning Goal

Introduction

Anatomical **reflexes** are quick, involuntary motor responses to stimuli that do not involve the higher brain centers or include conscious, voluntary aspects of movement. Reflexes are rapid and predictable so that the body can respond quickly to a stimulus. Reflexes take advantage of neural pathways called reflex arcs. A **reflex arc** is a pathway of neurons that receives a stimulus at a receptor located at one end of the path and outputs to an effector (a muscle or gland) at the other end of the path. The pathway of a reflex arc passes through the central nervous system: at least one synapse in a reflex arc occurs in the spinal cord or brain stem. Because of this, reflex tests can provide very useful diagnostic information about nervous system functioning, allowing a practitioner to target the location of a spinal cord or brain stem injury.

An exploration of reflex arcs requires you to remember what you have studied regarding the basic anatomy of a neuron and the physiology of how a neuron is stimulated to send an electrical signal from dendrites to the end of its axon. The content in this lab will also refer to specific skeletal muscles and specific bones and bone markings. Review the appropriate sections of your lab manual and textbook if you need a refresher.

The Human Anatomy and Physiology Society includes more than 1,700 educators who work together to promote excellence in the teaching of this subject area. The HAPS A&P Learning Outcomes measure student mastery of the content typically covered in a two-semester Human A&P curriculum at the undergraduate level. The full Learning Outcomes are available at https://www.hapsweb.org.

Lab 19 Pre-Lab Quiz

This quiz will strengthen your background knowledge in preparation for this lab. For help answering the questions, use your resources to deepen your understanding. The best resource for help on the first five questions is your text, and the best resource for help answering the last five questions is to read the introduction section of each lab activity.

1. Anatomy of a neuron. Label the figure with the following:

   axon
   dendrites
   input end of neuron
   output end of neuron

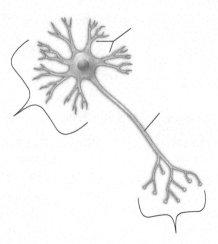

2. Match each of the roots to its definition:

Root	Meaning
	many
	opposite
	same
	one

List of Roots

contra
ipsi
mono
poly

3. If a stimulus to the sensory receptor of a neuron creates a great enough graded potential, what will happen on the sensory neuron?

   a. The dendrites of the neuron will secrete neurotransmitter.

   b. It will synapse with a skeletal muscle to produce contractions.

   c. An action potential will be relayed along the axon to the synaptic knobs.

4. Motor neurons synapse with effectors. What structures of the body can be effectors? Choose all correct answers.

   a. Glands

   b. Skeletal muscle

   c. Smooth muscle

   d. Cardiac muscle

5. What term is used to describe neurons that are entirely within the central nervous system?

    a. Afferent neurons
    b. Efferent neurons
    c. Interneurons

6. Where is the integrating center of a reflex arc?

    a. At the effector
    b. At the site of sensory stimulation
    c. In the central nervous system

7. What kind of reflex stimulates contraction of the biceps brachii muscle of the arm?

    a. A visceral reflex
    b. A somatic reflex

8. What is the purpose of a stretch reflex?

    a. To move the body away from a painful stimulus
    b. To protect a muscle from over-contraction
    c. To help the body maintain contractions for balance and posture

9. Superficial reflexes are a subset of what kind of functional reflex?

    a. Withdrawal
    b. Stretch
    c. Crossed-extensor

10. What is it called when a skeletal muscle contracts repeatedly?

    a. Jerk
    b. Tetany
    c. Clonus

(page intentionally left blank)

Name: _____ Section: _____

Activity 19.1 Anatomy of a Reflex Arc

This activity targets LOs 19.1, 19.2, 19.3, and 19.4.

There are five basic components to all reflex arcs (Figure 19.1): a sensory receptor, a sensory neuron, an integrating center, a motor neuron, and an effector. The **sensory receptor** is the site where a stimulus is received, such as touch or heat on the skin. If the stimulus is great enough, an action potential occurs on the **sensory neuron**. This neuron will convey the action potential along its axon all the way to the spinal cord or brain stem. Here, at the central nervous system, the sensory neuron will synapse. It may synapse directly with a motor neuron, or it may synapse with one or more interneurons. Either way, this portion of the reflex arc is the **integrating center**. Most integrating centers are in the spinal cord; these reflexes are called **spinal reflexes** as a result. A few integrating centers are in the brain stem; these reflexes are called **cranial reflexes**. The integrating center will prompt an action potential that will travel down the axon of an outgoing **motor neuron**. Finally, this motor neuron synapses on an **effector** (skeletal muscle, smooth muscle, cardiac muscle, or gland). The action of the effector at the end of a reflex arc is the reflex itself.

Like other systems in the human body, reflexes and reflex arcs can be described either structurally or functionally. From a structural perspective, there are four ways to group reflexes:

Learning Connection

Try Drawing It

Whether you are considering a stretch reflex, withdrawal reflex, or one of the other types of reflex arcs, the five components of the neural pathway remain the same. Take a few moments to close your lab manual and textbook and sketch, from memory, these five components, labeling them as you go. This short activity will help you quickly double-check that you remember the terminology and the order in which the components of the reflex arc work together.

• Reflexes can be separated based on what types of motor output are generated. **Somatic reflexes** are those in which the effector is skeletal muscle. **Visceral reflexes** (also called autonomic reflexes) are those in which the effector is smooth muscle, cardiac muscle, or a gland.

• Reflexes are sometimes described as being **intrinsic** (genetically programmed and present at birth) versus **learned**. Generally, learned reflexes have more complex reflex arc pathways that develop after birth, but the distinctions and usefulness in separating and defining reflex arcs in this manner are somewhat unclear.

Figure 19.1 | **The Five Basic Components of a Reflex Arc**

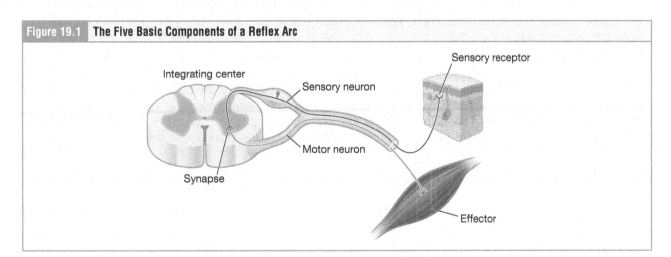

Figure 19.2 | **Monosynaptic Versus Polysynaptic Reflex Arcs**

(A) Polysynaptic reflex of the biceps brachii muscle and (B) monosynaptic reflex of the quadriceps femoris muscle.

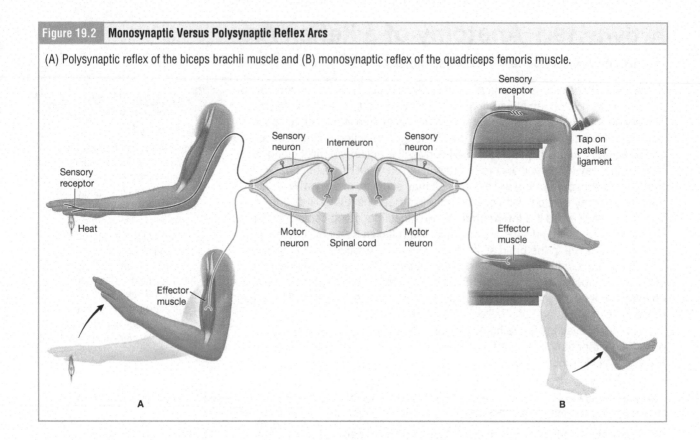

- The structure of the integrating center is considered when describing reflex arcs as either monosynaptic or polysynaptic (Figure 19.2). A **monosynaptic reflex arc** has a single synapse in the integrating center: the synapse between the afferent sensory neuron and the efferent motor neuron. In contrast, a **polysynaptic reflex arc** has more than one synapse in the integrating center, where one or more interneurons exists between the sensory and motor neurons in the arc.

- Finally, a reflex arc may initiate a reflex on either the same side of the body as the stimulus or on the opposite side of the body (Figure 19.3). **Ipsilateral reflexes** begin and end on the same side of the body. In other words, the reflex occurs on the side of the body where the stimulus occurred. In contrast, **contralateral reflexes** are those in which the stimulus causes a response on the other side of the body. For example, when you step on something sharp, a reflex arc through the spinal cord causes the muscles in the opposite leg to engage so that you can shift your weight to the non-affected leg.

Reflexes and reflex arcs can also be categorized functionally into four different categories. This approach considers why the effector action is needed:

- A **withdrawal reflex** (also called a flexor reflex) is used to pull a body part away from a painful stimulus, such as the example in Figure 19.4 in which the biceps brachii of your arm will contract to pull your arm away from something painful. Nociceptors receive the stimulus and sensory neurons relay the action potential to the CNS where motor neurons are activated to prompt skeletal muscle in the area to contract and move the body part away from harm. Another withdrawal reflex occurs if you step on something painful: the hamstrings of your leg will contract to lift and pull the leg away. As you may have noted, withdrawal reflexes are ipsilateral.

Figure 19.3 **Ipsilateral versus Contralateral Reflexes**

(A) In ipsilateral reflexes, the sensory stimulation and motor output occur on the same side of the body. (B) In contralateral reflexes, the sensory stimulation and motor output occur on opposite sides of the body.

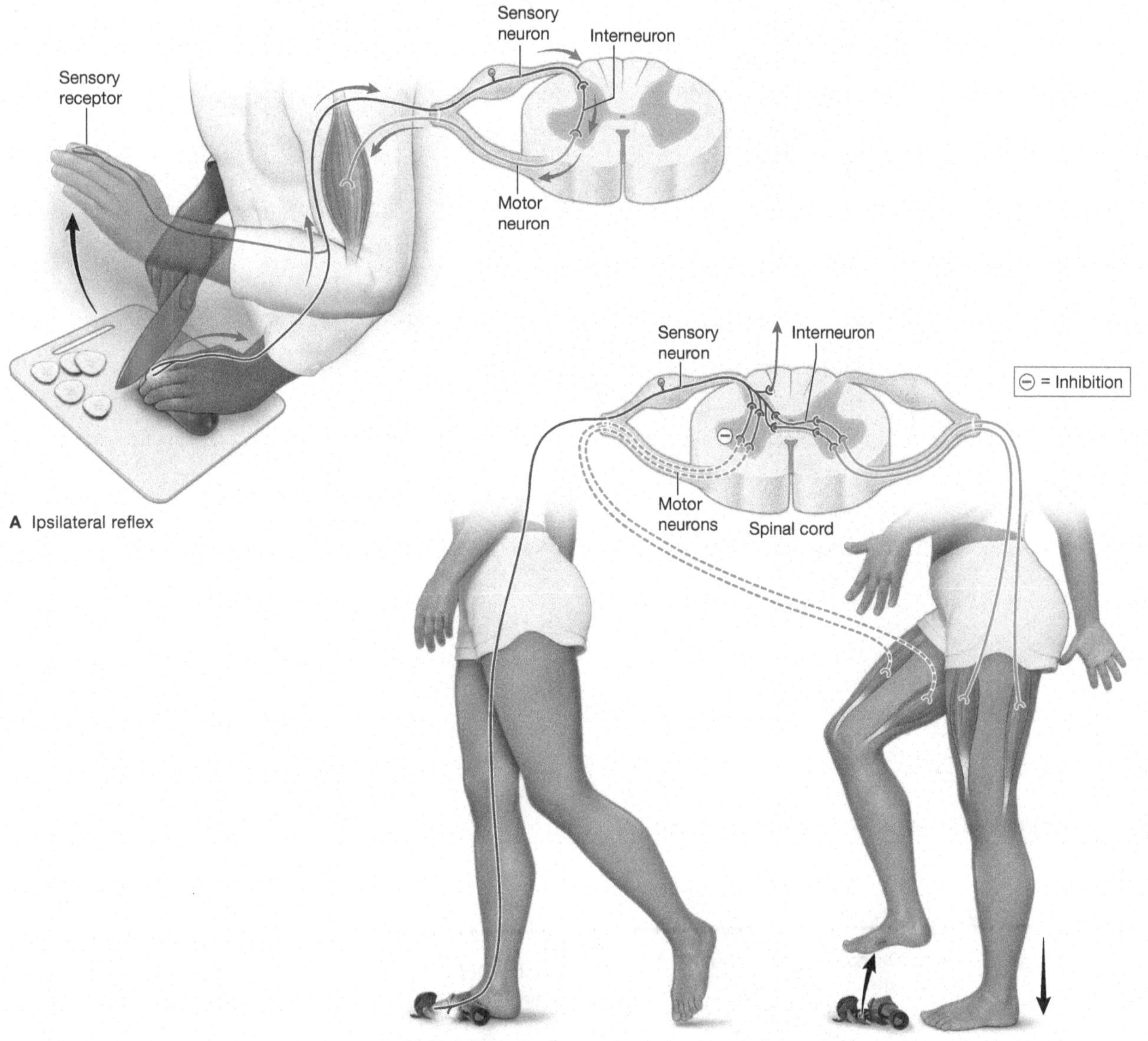

A Ipsilateral reflex

B Contralateral reflex

- A **crossed-extensor reflex** is another way to describe a contralateral reflex, shown in Figure 19.3. Like a withdrawal reflex, a crossed-extensor reflex helps the body move away from something painful or damaging. In contrast to a withdrawal reflex, a crossed-extensor reflex causes skeletal muscle action on the opposite side of the body from where the pain occurred. Sometimes a crossed-extensor reflex and a withdrawal reflex occur together, as shown in Figure 19.3.

- **Stretch reflexes**, shown in Figure 19.5, occur in response to a skeletal muscle being stretched. The lengthening of the muscle activates sensory neurons called a **muscle spindle** that are wrapped around the muscle cells, The action potential on these sensory neurons reaches the CNS to stimulate motor neurons that extend back to the same muscle. These stretch reflexes allow us to maintain muscle tone and balance without conscious thought. The quadriceps femoris, biceps brachii, triceps brachii, and gastrocnemius muscles all exhibit strong stretch reflexes.

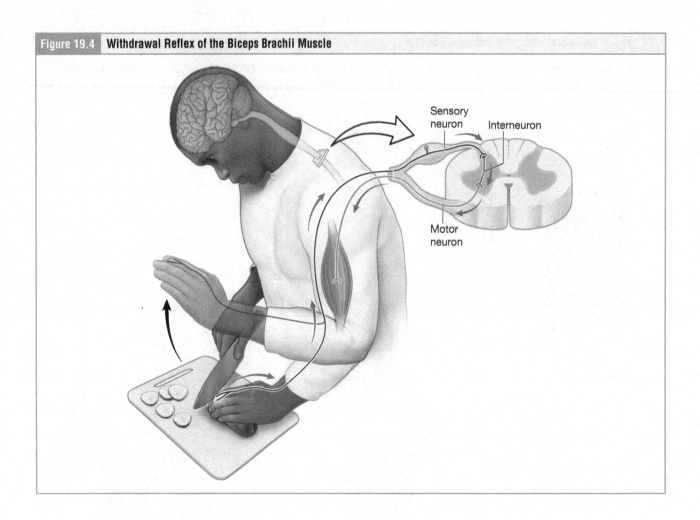

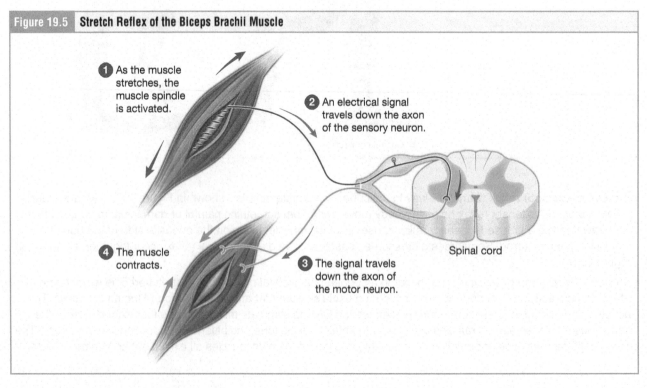

1 As the muscle stretches, the muscle spindle is activated.

2 An electrical signal travels down the axon of the sensory neuron.

4 The muscle contracts.

3 The signal travels down the axon of the motor neuron.

Spinal cord

Figure 19.6 Tendon Reflex of the Quadriceps Femoris Muscle

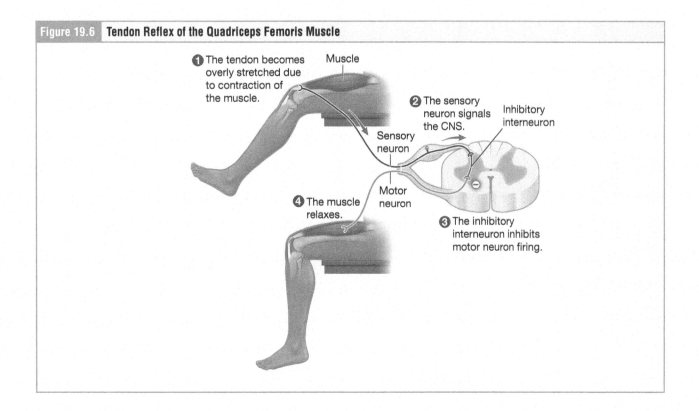

Figure 19.6 Tendon Reflex of the Quadriceps Femoris Muscle

1 The tendon becomes overly stretched due to contraction of the muscle.

Muscle

2 The sensory neuron signals the CNS.

Inhibitory interneuron

Sensory neuron

4 The muscle relaxes.

Motor neuron

3 The inhibitory interneuron inhibits motor neuron firing.

- While the stretch reflex keeps muscles from stretching too far, a **tendon reflex** (also called a Golgi reflex), shown in Figure 19.6, prevents a muscle from over-contracting. Sensory structures in the tendon associated with a muscle are stimulated by the increased tension that is placed on the tendon as the muscle contracts. The action potentials on these sensory neurons integrate in the CNS to prompt motor neurons extending back to the same muscle. These motor neurons inhibit muscle contraction, thereby allowing the muscle to relax somewhat so that it does not become damaged due to over-contraction.

Reflex arcs may also indirectly influence antagonistic muscles on the other side of the joint so that they do not resist the actions of the effectors of the reflex arc. Withdrawal reflexes and stretch reflexes initiate **reciprocal inhibition** in which antagonistic muscles are inhibited from contracting so that the reflex muscular contraction can occur effectively. Conversely, tendon reflexes initiate **reciprocal activation** in which antagonistic muscles are stimulated to contract so that the reflex muscular relaxation will lead to muscle lengthening.

It is important to note that a single muscle in the body may be affected by more than one reflex arc. Each reflex arc that affects a muscle is made of a different grouping of sensory and motor neurons. For example, the biceps brachii muscle is the effector in both a withdrawal reflex arc and a stretch reflex arc (shown in Figures 19.4 and 19.5, respectively). These reflex arcs are separate pathways of neurons.

The purpose of this activity is to help you familiarize yourself with the components of a reflex arc, the ways of categorizing reflex arcs, and the terminology associated with reflex arcs.

Instructions

1. Use the introduction and figures to fill in the information in Table 19.1. Each of the reflexes listed in the table is shown in one of the figures above.

2. The first reflex has been done for you.

3. After completing Table 19.1, compare your table with your lab partner's and discuss any discrepancies.

Table 19.1 Classifications of Typical Reflexes

Reflex	Structural Classifications			Functional Classification: Withdrawal, Crossed-Extensor, Stretch, or Tendon Reflex?
	Somatic or Visceral?	Monosynaptic or Polysynaptic?	Ipsilateral or Contralateral?	
Stretch reflex of the biceps brachii muscle	Somatic	Monosynaptic	Ipsilateral	Stretch reflex
Withdrawal reflex of the biceps brachii muscle				
Withdrawal reflex of the hamstring muscles				
Quadriceps stretch reflex (knee-jerk reflex)				

Name: _____ Section: _____

Activity 19.2 Exploring an Autonomic Reflex

This activity targets LO 19.5.

Autonomic reflexes are those that cause a response in smooth muscle, cardiac muscle, or glands. Autonomic reflex arcs differ from somatic reflex arcs in the output portion of the neural pathway. Because autonomic reflexes use the autonomic motor pathways to reach the effector, two neurons are involved in the output: the preganglionic neuron and the postganglionic neuron. This difference in the neural pathways is shown in Figure 19.7.

| Figure 19.7 | Comparison of the Output of Autonomic and Somatic Reflexes |

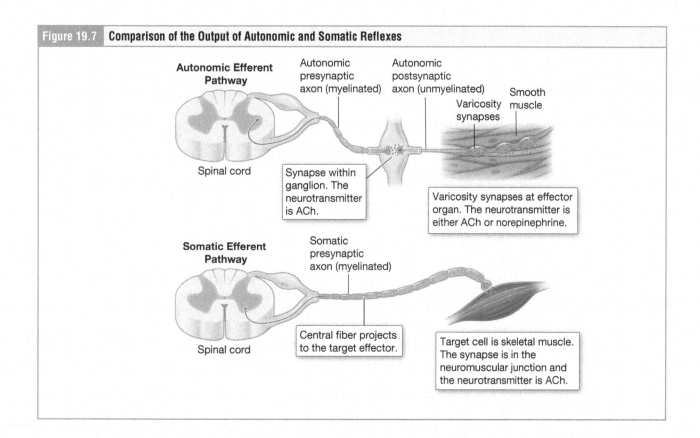

One autonomic reflex that is easy to examine in a laboratory setting is the salivary reflex. The purpose of this activity is to initiate the salivary reflex and examine the body's response.

Materials
- Forceps
- Graduated cylinder, one per group
- Lemon juice
- pH paper
- Sterile dropper

Instructions
1. For this activity, work with your lab partner. Choose one of you to be the subject; the other will be the examiner.

2. Ask your subject to resist the temptation to swallow for 2 minutes. If they need to, they can spit/drool into the graduated cylinder during this time.

3. At the end of the 2 minutes, ask your subject to spit all of the saliva that remains in their mouth into the same graduated cylinder, adding to the volume accumulated over the 2 minutes.

4. Measure the volume and record it in the top row of Table 19.2.

Table 19.2 The Salivary Reflex		
	Volume	pH
Saliva before lemon juice		
pH of lemon juice		
Saliva after lemon juice		

5. The examiner should hold the pH paper with the forceps, then dip it into the graduated cylinder. Immediately remove the pH paper and compare it to the pH key to determine the pH of the saliva. Record the pH of the "saliva before lemon juice" in Table 19.2.

6. Empty the graduated cylinder of the saliva as directed by your instructor. Rinse the cylinder and empty it of any water.

7. Using a sterile dropper, drop one drop of lemon juice onto a fresh piece of pH paper. Immediately compare the pH paper to the key to determine the pH of the lemon juice. Record the pH of the lemon juice in Table 19.2.

8. Using a sterile dropper, drop two to three drops of lemon juice onto the subject's tongue. Have the subject close their mouth.

9. Ask your subject to resist the temptation to swallow for 2 minutes. If they need to, they can spit/drool into the graduated cylinder during this time.

10. At the end of the 2 minutes, ask your subject to spit all of the saliva that remains in their mouth into the graduated cylinder, adding to the volume accumulated over the 2 minutes.

11. Measure the volume and record it in the bottom row of Table 19.2.

12. The examiner should hold the pH paper with the forceps, then dip it into the graduated cylinder. Immediately remove the pH paper and compare it to the key to determine the pH of the saliva. Record the pH of the "saliva after lemon juice" in Table 19.2.

13. Empty the graduated cylinder of the saliva as directed by your instructor.

14. Answer the questions below.

Questions

1. What was the effect of the lemon juice, which is an acid, on the volume of saliva production?

2. What was the effect of the lemon juice on the pH of the saliva that was produced?

3. Why do you think this reflex arc initiates this effector response?

(page intentionally left blank)

Activity 19.3 Exploring Stretch Reflexes

This activity targets LO 19.5.

Stretch reflexes help the body maintain the correct amount of muscle tone in muscles. Normally we are unaware of these reflexes as they are occurring. They operate to minutely adjust how much tension a muscle provides as it contracts, which helps maintain our posture and balance our body as we move. However, we can initiate a strong stretch reflex if we "trick" a muscle into acting as if it has been stretched. We can do this by tapping sharply on the ligament associated with that muscle or the tendon that leads to that muscle, this temporarily stretches the ligament and therefore the attached muscle, creating many sensory stimuli at the muscle spindles (sensory neurons) that initiate the reflex arc.

If there is a problem in a portion of the reflex arc, it is possible for the effector muscle to exhibit **clonus.** Clonus is repetitive muscle contraction, rather than the single muscle contraction (often called a jerk) that is expected in response to a stretch reflex arc. Instead, with clonus the contraction in response to the stimulus happens several times before ceasing.

Four stretch reflexes are easy to examine in the laboratory setting.

1. The patellar reflex (also called the knee-jerk reflex) is familiar to many students. In this reflex, stretch to the patellar tendon initiates a reflex arc that terminates with contraction of the quadriceps femoris muscle as shown in Figure 19.6. The integration center for the patellar reflex is in the L2–L4 section of the spinal cord.

2. The calcaneal reflex (also known as the Achilles reflex or ankle-jerk reflex) stimulates stretch to the calcaneal tendon that extends from the gastrocnemius muscle of the calf. The resulting contraction of the gastrocnemius jerks the foot toward its plantar surface. The integration center for this reflex is in the S1 part of the spinal cord.

3. The biceps brachii reflex stimulates stretch to the tendon that connects the biceps brachii to the anterior surface of the proximal ulna and radius. Stretch to this tendon initiates a reflex arc that terminates with the contraction of the biceps brachii muscle, causing flexion at the elbow as shown in Figure 19.4. The integration center for this reflex arc is the C5 and C6 section of the spinal cord.

4. The triceps brachii reflex stimulates stretch to the tendon that connects the triceps brachii to the olecranon of the ulna. Stretch to this tendon initiates a reflex arc that terminates with the contraction of the triceps brachii muscle, causing extension at the elbow. The integration center for this reflex arc is the C6 and C7 region of the spinal cord.

Materials
- Disinfectant wipes
- Reflex hammer

Instructions
1. Work with your lab partner. One of you will be the subject and the other will be the examiner. After completing the stretch reflex tests, switch roles and repeat these steps. Use the disinfectant wipes to disinfect the reflex hammer between subjects.

2. If the subject is not able to expose the skin of the knee because their pant leg is too tight try to complete the patellar reflex test over the pants. It is possible to complete the patellar reflex test with pants made of thin material.

3. The stretch reflex response to each test is measured somewhat subjectively using the following scale:
   - 0 = no response
   - +1 = little response
   - +2 = a brisk normal response
   - +3 = a very brisk response
   - +4 = exaggerated or repeating (clonus) response

4. For each test:

   a. The examiner will read the description of how to perform the test in Table 19.3.

   b. The subject will position themselves as described or shown in Table 19.3.

   c. Have the subject close their eyes so that they don't anticipate the tap and quietly count down from 100 to distract themselves.

   d. The examiner taps the tendon, observes the response, and grades the response using the scale above. Write this information in Table 19.3 in the subject's lab manual.

   e. Repeat the test with the subject's eyes closed but without a counting distraction; the subject should actively anticipate the tap but not resist the reflexive action. The examiner grades the response and writes the information in the subject's lab manual.

   f. Repeat the test a third time with the subject's eyes open so that they can see the tap coming. The subject should attempt to resist the reflex. The examiner grades the response and writes the information in the subject's lab manual.

5. After both lab partners have played the role of the subject, work with your lab partner to examine your data and answer the following questions:

   a. Are the reflexes you tested here contralateral or ipsilateral? How do you know?

   b. Generally, what happened when you were anticipating the tap of the reflex hammer?

   c. Generally, what happened when you could see the reflex hammer coming and were able to actively resist it?

   d. Explain what you think was happening in your answers to (b) and (c) above.

Table 19.3 Data from Stretch Tendon Tests

Stretch Reflex	How to Perform the Test	Subject Position during Test	Subject Is Distracted (Eyes Closed and Counting)	Subject Is Ready (Eyes Closed but Anticipating)	Subject Is Resisting (Eyes Open and Resisting Reflex)
Patellar reflex (knee jerk reflex)	Palpate (feel for) the patella and the tibial tuberosity; the patellar ligament is between these two structures. With the broad side of the hammer, gently but firmly tap the patellar ligament.	Sit on edge of table or tall chair so thigh is supported and leg is dangling, with the skin of the knee exposed, if possible. wavebreakmedia/Shutterstock.com			
Calcaneal reflex (Achilles tendon reflex or ankle jerk reflex)	Palpate the calcaneal tendon. With the broad side of the hammer, gently but firmly tap the calcaneal tendon.	Sit in same position as for patellar reflex, above. Shoe is removed and foot is slightly dorsiflexed.			
Biceps brachii reflex (biceps jerk reflex)	While the subject contracts the biceps brachii, palpate the biceps brachii tendon in the cubital fossa (the inside crook of the elbow). Place your thumb over the tendon and keep it there. Once the subject has relaxed the muscle, gently but firmly tap your thumb with the tapered end of the reflex hammer.	Stand or sit with arm relaxed and hanging at side. Contract biceps brachii muscle so examiner can find the tendon, then relax again for the test.			
Triceps brachii reflex (triceps jerk reflex)	While the subject contracts the triceps brachii, palpate the triceps brachii tendon just proximal to the olecranon. Place your thumb over the tendon and keep it there. Once the subject has relaxed the muscle and bent their arm, remove your thumb and gently but firmly tap the tendon with the tapered end of the reflex hammer.	Stand or sit with arm relaxed and hanging at side. Contract triceps brachii muscle so examiner can find the tendon, then relax the muscle and bend your arm across the front of the body, as shown here. Antonia Reeve/Science Source			

(page intentionally left blank)

Activity 19.4 Exploring a Superficial Reflex

This activity targets LO 19.5.

There are several withdrawal reflexes that are identified as superficial reflexes because they are triggered by stroking or brushing certain areas of the skin (or a mucous membrane) rather than by an obviously painful stimulus. Like with other withdrawal reflexes, superficial reflexes assist in moving the area of the body away from a stimulus that might cause damage.

Table 19.4 describes some superficial reflexes. Like the other reflex arcs that have already been examined, these reflex arcs consist of five components (see Figure 19.1). For each superficial reflex, its reflex arc is outlined as you read across a row from the site of the stimulus to the reflex arc pathway (sensory neuron path to the CNS → location of the integrating center within CNS → motor neuron path from the CNS to effector) to the effector response.

While most of these superficial reflexes are not easily tested in the laboratory setting, the plantar reflex is. As can be seen in Table 19.4, the plantar reflex arc relies on the posterior tibial nerve for both the afferent sensory neurons and the efferent motor neurons to travel. This reflex arc uses levels L5 and S1 of the spinal cord for an integration center. Further afferent signals travel up to integrate with motor centers in the brain, which then respond with motor output via the corticospinal tract. The corticospinal tract helps control and coordinate movements, especially movements of the limbs. Thus, a test of the plantar reflex not only indicates that the L5 and S1 spinal cord is intact and functioning correctly, but also that the corticospinal tract is intact.

The absence of a superficial reflex response is not always pathological as the lack of a response may be due to weak muscles in the area or because the individual is a young child in whom the portion of the reflex that travels to and from the brain has not yet fully developed. For example, the plantar reflex that occurs when the bottom of the foot is stroked will not occur the same way in infants. An opposing muscular action occurs, called *Babinski's sign*, in which digit 1 extends and the rest of the digits fan out. This indicates that while the reflex arc occurred and was integrated at the L5 and S1 spinal cord to prompt an effector response, the corticospinal tract is not influencing and coordinating those muscular actions.

The purpose of this exercise is to examine the plantar reflex as an example of a superficial reflex.

Reflex	Stimulus	Reflex Arc Pathway	Effector Response	Purpose of the Response
		Table 19.4 Superficial Reflexes		
Corneal reflex	Touch on the cornea of the eye	CN V → pons → CN VII	Blinking of both eyelids	*Prevents foreign material from damaging the eye.*
Conjunctival reflex	Touch on the conjunctiva of the eye	CN V → pons → CN VII	Blinking of both eyelids	
Pharyngeal reflex (gag reflex)	Touch to the oral mucosa lateral to the uvula	CN IX → medulla oblongata → CN X	Contraction and elevation of oropharynx	
Scapular reflex	Stroking the skin medial to the scapulae	C4 & C5 nerves → C4 & C5 spinal cord → dorsal scapular nerve	Contraction of scapular muscles	
Abdominal reflex	Light stroke from lateral to medial along lateral surface of abdomen	T8–T12 nerves → T8–T12 spinal cord → T8–T12 nerves (specific level depends on where on the abdomen the skin is stroked)	Contraction of abdominal muscles	
Cremasteric reflex	Light stroke on superomedial thigh	Femoral nerve → L1 & L2 spinal cord → femoral nerve	Elevation of testis ipsilaterally	
Anal reflex	Light stroke of perianal skin	Pudendal nerve → S4 & S5 spinal cord	Contraction of external anal sphincter	
Plantar reflex	Light stroke from heel along lateral foot and across base of toes to digit 1	Posterior tibial nerve → L5 & S1 → posterior tibial nerve	Flexion and abduction of toes on same foot	

Materials
• Reflex hammer

Instructions

1. With your lab partner, consider Table 19.4 and determine what the purpose of each reflex is Complete the table. The first reflex has been completed for you.

2. Work with your lab partner. One of you will be the subject and the other will be the examiner. After completing the plantar reflex test on one of you, you will switch roles and repeat these steps.

3. Superficial reflex tests are usually assessed by whether the effector action is not present, present, or an alternate muscular action (such as Babinski's sign) occurred.

4. The subject will be seated.

    a. Have the subject remove their sock and shoe.

    b. The subject will rest the posterolateral surface of their foot on a second chair, exposing the sole of their foot.

5. The examiner will use the handle of the reflex hammer to firmly draw a line along the lateral edge of the sole of the foot from the heel and then across the base of the toes, as shown by the arrow in Figure 19.8.

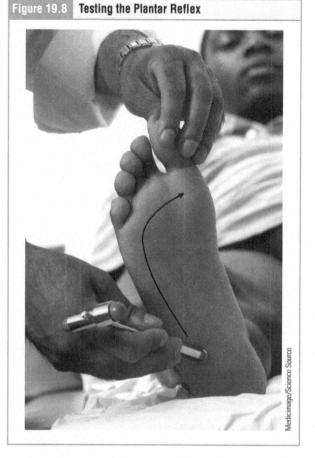

Figure 19.8 Testing the Plantar Reflex

6. The examiner will observe the subject to determine whether an effector response occurs. Write this information in the <u>subject's</u> lab manual here:

    Plantar reflex test result (not present, present, or Babinski's sign): _____

7. Switch roles and repeat the test.

8. After both lab partners have played the role of the subject, work with your lab partner to answer the following questions:

    a. What is the meaning of a positive Babinski's sign in an adult? How is this different than the meaning of a plantar reflex test that shows no reflex present?

    b. Which of the superficial reflexes listed in Table 19.4 are considered cranial reflexes rather than spinal reflexes? How do you know?

(page intentionally left blank)

Activity 19.5 Case Study: Using Reflexes to Measure the Progression of Multiple Sclerosis

Case Study

This activity targets LOs 19.1, 19.2, and 19.6.

Matthew has been recently diagnosed with multiple sclerosis (MS), a disease that affects the brain and spinal cord. Matthew's doctor shares that MS occurs as the body's immune system attacks the myelin sheath around neuron axons, interrupting the transmission of action potentials along neurons in the central nervous system. The doctor would like to establish a baseline for measuring the progression of the disease and so she performs a neurological exam on Matthew. In addition to other tests that are part of the exam, the doctor tests Matthew's reflexes.

1. What portion of the reflex arc is being tested in a reflex test on a person with MS?

    a. The receptor

    b. The sensory neuron

    c. The integrating center

    d. The motor neuron

    e. The effector

2. While performing the stretch reflexes, the doctor asks Matthew about family and work. Why is the doctor distracting Matthew during the stretch reflexes?

    a. To prevent somatic control over effector action

    b. To prevent Matthew pulling away from pain sensations

    c. To trick the integrating center into initiating a reflex arc

The doctor records the following results for the reflex tests.

Test	Response
Biceps test	+2
Triceps test	+2
Abdominal reflexes	present
Cremasteric reflex	present
Patellar reflex	+4 with clonus
Achilles reflex	+4 with clonus
Plantar reflex	Babinski's sign

3. What does the phrase "with clonus" mean for the response to the patellar and Achilles reflexes?

    a. That neither the quadriceps femoris muscle nor the gastrocnemius contracted at all.

    b. That both the quadriceps femoris muscle and the gastrocnemius muscle contracted normally.

    c. That both the quadriceps femoris muscle and the gastrocnemius muscle contracted many times in response to a single tap of the reflex hammer.

4. What does it mean that Matthew exhibits Babinski's sign?

    a. That there is damage to the posterior tibial nerve.

    b. That there is damage to the L5 and S1 spinal cord.

    c. That there is damage to the corticospinal tract.

5. The +2 results for both the biceps test and the triceps test indicate what?
   a. That neither of these muscles contracted in response to the reflex arc test.
   b. That both of these muscles exhibited a normal response to the reflex arc test.
   c. That both of these muscles exhibited a hyperactive response to the reflex test.

Digging Deeper:
The Reflex Hammer

The neurological exam was developed in the late 1800s to assist in diagnosing problems with the nervous system. Some reflex tests were a component of these early neurological exams and, like today, the tests relied on the ability to stimulate stretch reflexes by tapping on a tendon. The initial tool used as a reflex hammer was a tool used by wine makers to hit the side of a wine barrel to determine how much wine was left inside. Later in the 1800s, practitioners developed smaller, gentler reflex hammers designed specifically for testing a person's reflexes, such as the Bennet percussor shown in Figure 19.9.

| Figure 19.9 | The Bennet Percussor, Circa 1860 |

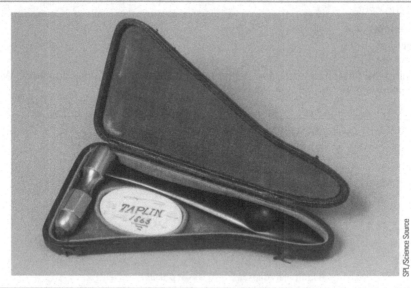

SPL/Science Source

Lab 19 Post-Lab Quiz

1. What is a reflex?

   a. A response to a stimulus that does not rely on higher brain centers to execute.

   b. A neural pathway that includes a stimulus, sensory neuron, integrating center, motor neuron, and effector.

   c. A sensory neuron that picks up information from the environment around the body.

2. Are all the synapses in a reflex arc found within the central nervous system?

   a. Yes

   b. No

3. Unlike the reflexes tested in this lab, the salivary reflex initiates saliva production from the salivary glands when something acidic, such as lemon juice, is placed in the mouth. What kind of reflex arc would the salivary reflex be categorized as?

   a. A somatic reflex arc

   b. A visceral reflex arc

4. How many neurons are in a monosynaptic reflex arc?

   a. One

   b. Two

   c. Three

   d. Four or more

5. Both stretch reflex arcs and tendon reflex arcs are always:

   a. Polysynaptic

   b. Ipsilateral

   c. Spinal

6. When a tendon reflex arc is initiated, what will happen to the skeletal muscle contraction?

   a. Increased contraction of the skeletal muscle

   b. Decreased contraction of the skeletal muscle

7. A crossed-extensor reflex often happens in conjunction with a withdrawal reflex. If a person's right forearm is aggressively grabbed, a withdrawal reflex will cause the triceps brachii and other muscles of the right arm to contract to draw the arm and forearm away posteriorly. Which of the following would be the crossed-extensor reflex that would occur at the same time?

   a. Contraction of the right biceps brachii to counteract the right triceps brachii

   b. Relaxation of the tendons of the triceps brachii to assist the muscle in lengthening

   c. Contraction of the left biceps brachii and other muscles of the left arm to push the person away

8. A stretch reflex arc will also initiate what muscular action?

   a. Reciprocal inhibition

   b. Reciprocal activation

9. If the bicep reflex test is negative, what is the possible location of neural damage?

   a. Corticospinal tract

   b. C5 and C6 spinal cord

   c. Lower thoracic spinal cord

   d. L2–L4 spinal cord

10. If the plantar reflex test of an adult demonstrates Babinski's sign, what is the possible location of neural damage?

    a. Corticospinal tracts

    b. C7 and C8 spinal cord

    c. L1 and L2 spinal cord

    d. L5 and S1 spinal cord

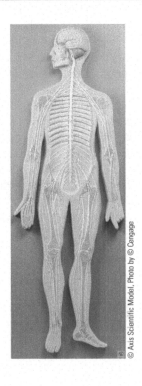

© Axis Scientific Model. Photo by © Cengage

HAPS
LEARNING GOALS
AND OUTCOMES

Learning Objectives: After completing this lab, you should be able to:

20.1 Name the two main divisions of the ANS and compare and contrast the major functions of each division, their neurotransmitters, the origination of the division in the CNS, the locations of their preganglionic and postganglionic cell bodies, and the length of the preganglionic versus postganglionic axons.

20.2 Describe the different anatomical pathways through which sympathetic and parasympathetic neurons reach target effectors.

20.3 Explain the relationship between chromaffin cells in the adrenal medulla and the sympathetic division of the autonomic nervous system.

20.4 Compare and contrast the effects (or lack thereof) of sympathetic and parasympathetic innervation on various effectors.

20.5 Compare and contrast cholinergic and adrenergic receptors with respect to neurotransmitters that bind to them, receptor subtypes, receptor locations, target cell response, and examples of drugs, hormones, and other substances that interact with these receptors.

20.6 Compare and contrast the ANS to the SNS with respect to the site of origination, number of neurons involved in the pathway, effector, receptors, and neurotransmitters.

HAPS
LEARNING GOALS
AND OUTCOMES

The Human Anatomy and Physiology Society includes more than 1,700 educators who work together to promote excellence in the teaching of this subject area. The HAPS A&P Learning Outcomes measure student mastery of the content typically covered in a two-semester Human A&P curriculum at the undergraduate level. The full Learning Outcomes are available at https://www.hapsweb.org.

Introduction

The nervous system can be divided into two functional parts: the somatic nervous system (SNS) and the autonomic nervous system (ANS). The two systems differ from each other in the structures they innervate, the anatomy of that innervation, and the ensuing responses. While the SNS causes contraction of skeletal muscle, the ANS controls cardiac muscle, smooth muscle, and glands. The SNS uses a single neuron to reach from the central nervous system to these skeletal muscle effectors to allow us to voluntarily contract those skeletal muscles. In contrast, the ANS uses a pair of neurons to reach from the CNS to effectors, requiring synapses within **ganglia** in the peripheral nervous system. The ANS neural signals traveling these two-neuron pathways cause involuntary responses that maintain homeostasis. Thus, the ANS works in concert with the endocrine system to enact homeostatic actions in the body.

The ANS uses two aspects, or divisions, to homeostatically regulate the organs: the **sympathetic division** and the **parasympathetic division**. The sympathetic division of the ANS functions to prepare the body to act during emergency situations and is often described as the "fight, flight, and fright" division. The parasympathetic division of the ANS allows the body to recover and repair between emergencies and is often described as the "rest and digest" division. The body maintains homeostasis as it balances its activities between these two divisions depending on what the body is doing in the moment. Almost all effectors (organs) are innervated by both sympathetic and parasympathetic neurons, described as **dual innervation**. Generally, the neurons of one of the two divisions will stimulate the effector while the neurons of the other division inhibit or reduce the activity of the effector.

Lab 20 Pre-Lab Quiz

This quiz will strengthen your background knowledge in preparation for this lab. For help answering the questions, use your resources to deepen your understanding. The best resource for help on the first five questions is your text, and the best resource for help answering the last five questions is to read the introduction section of each lab activity.

1. Anatomy of the brain stem and spinal cord. Label the sections of the brain stem and spinal cord:

   Lumbar spinal cord
   Medulla oblongata
   Midbrain
   Pons
   Sacral spinal cord
   Thoracic spinal cord

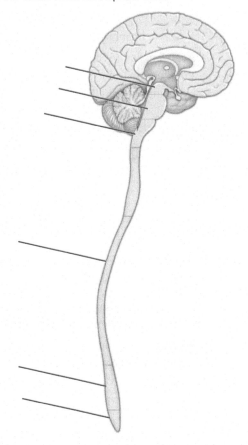

2. Match each of the following roots to its definition:

Root	Meaning	List of Roots
	before	auto
	after	medulla
	self	post
	middle, innermost part	pre

3. Neurotransmitter release onto effectors of the autonomic nervous system occurs at _____ on the postganglionic axons.

   a. varicosities at many places

   b. axon terminals

4. Alpha and beta receptors receive which neurotransmitter?

   a. Acetylcholine (ACh)

   b. Norepinephrine (NE)

5. The term "cholinergic" refers to structures that secrete or interact with:

   a. Acetylcholine (ACh)

   b. Epinephrine (epi)

   c. Norepinephrine (NE)

6. How many total neurons are used to send motor messages to effectors in the autonomic nervous system?

   a. One

   b. Two

   c. Three

7. The neurotransmitter released at the synapse in the ganglia in the ANS is:

   a. Acetylcholine (ACh)

   b. Epinephrine (epi)

   c. Norepinephrine (NE)

8. The parasympathetic division of the ANS will use _____ ganglia as locations for synapses between preganglionic neurons and postganglionic neurons. Select all correct answers.

   a. sympathetic chain

   b. prevertebral (collateral)

   c. intramural

   d. terminal

9. What neurotransmitter will act on muscarinic receptors?

   a. Acetylcholine (ACh)

   b. Epinephrine (epi)

   c. Norepinephrine (NE)

10. The overall goal of the _____ division of the ANS is to maintain homeostasis when the body is active.

   a. parasympathetic

   b. sympathetic

Activity 20.1 Anatomy of an ANS Neural Pathway

This activity targets LO 20.1.

While the outcome of the actions of the sympathetic and parasympathetic divisions of the ANS are usually antagonistic to each other, the two divisions share some anatomical features in the neural pathways that reach to target organs. Both divisions communicate with effectors using a two-neuron system, shown in a generalized manner in Figure 20.1. The first neuron has its cell body in the brain stem or spinal cord and its axon projects out of the CNS to reach a ganglion where it synapses on a second neuron. Because of this arrangement, the first of the two neurons in the pathway is called the **preganglionic neuron** and the second neuron is called the **postganglionic neuron**. The synapse between these two neurons at the ganglion is a cholinergic synapse: the preganglionic neuron releases the neurotransmitter **acetylcholine (ACh)** onto ACh receptors called **nicotinic receptors**. The nicotinic receptors are stimulatory so that a release of ACh by the preganglionic neuron's synaptic knobs will prompt an action potential on the postganglionic neuron. The axon of the postganglionic neuron reaches to the effector where it synapses on the cardiac muscle cells, smooth muscle cells, or glandular cells of that organ. The neurotransmitter released here will either be acetylcholine or **norepinephrine (NE)** and can either excite or inhibit the cells of the effector depending on the type of receptors embedded in them.

While the sympathetic and parasympathetic divisions both use a two-neuron system organized as just described, the arrangement of these two neurons within each division has some distinct anatomic features. These features include the location of the preganglionic neuronal cell bodies within the CNS, the length of the preganglionic and postganglionic axons, the types of ganglia where the synapse occurs between the two neurons, the neurotransmitter secreted by the postganglionic neurons onto the effector's cells, and the types of neurotransmitter receptors found on the effector's cells. Examine Figure 20.2, which illustrates these differences between the two divisions.

This exercise will explore the structure of the two-neuron system and begin to differentiate between the anatomy of the two divisions of the ANS.

Materials
• Colored pencils
• Textbook or other resources showing the two-neuron pathways of the ANS divisions

Instructions
1. Use Figure 20.2, your textbook, and other resources provided by your instructor to complete Table 20.1.

2. Which features of the two-neuron system are the same for both divisions and which differ?
   a. Choose a colored pencil to represent the similarities. Color in the box in the Table 20.1 key and the rows that are the same for both divisions.
   b. Chose a different colored pencil to represent the differences. Color in the other box in the key and the rows that differ between the two divisions.

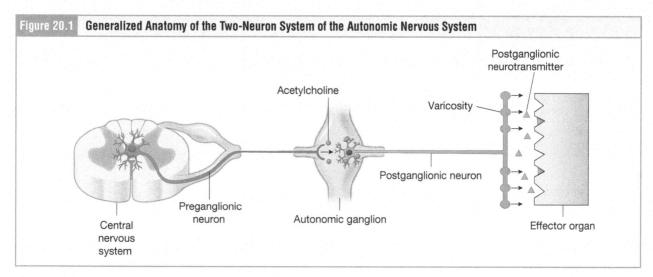

Figure 20.1 **Generalized Anatomy of the Two-Neuron System of the Autonomic Nervous System**

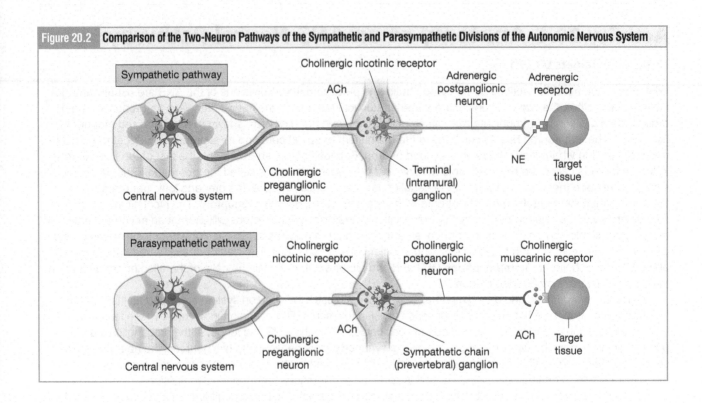

Figure 20.2 Comparison of the Two-Neuron Pathways of the Sympathetic and Parasympathetic Divisions of the Autonomic Nervous System

Table 20.1 Anatomical Features of the Two-Neuron System of the Two Divisions of the Autonomic Nervous System

	Parasympathetic Division	Sympathetic Division
Location of the preganglionic neuron cell body within the CNS		
Length of preganglionic axon (long or short?)		
Types of ganglia where preganglionic neuron and postganglionic neuron synapse		
Neurotransmitter released by preganglionic neuron		
Neurotransmitter receptor type on postganglionic neuron cell body		
Length of postganglionic axon (long or short?)		
Neurotransmitter released by postganglionic neuron		
Neurotransmitter receptor types on effector cells		

Table Key

❑ Both the parasympathetic and sympathetic divisions share the same feature.

❑ The parasympathetic and sympathetic divisions differ in this feature.

Activity 20.2 Anatomy of the Parasympathetic Division

This activity targets LO 20.2.

The parasympathetic division of the ANS can also be referred to as the **craniosacral system** because the cell bodies of the preganglionic neurons are located in the brain stem ("cranio-") and the lateral gray horn of the sacral spinal cord ("-sacral"). Once the axons of these preganglionic neurons leave the central nervous system, they will extend within various nerves until they reach ganglia located either near (**collateral ganglia**) or within the wall of (**intramural ganglia**) the target organ. The ganglion is always the site of the synapse between the preganglionic neuron and the postganglionic neuron. The postganglionic neuron axon extends only a short distance to innervate the cells of the target organ, releasing acetylcholine onto either stimulatory or inhibitory cholinergic receptors called **muscarinic receptors**.

Examine Figure 20.3 as you read about these pathways of parasympathetic innervation to the body's organs.

• The **midbrain** houses parasympathetic neuron cell bodies that depart the CNS through the **oculomotor nerve** (CN III). These preganglionic neurons synapse at collateral ganglia within the orbit. The postganglionic neurons extend to the sphincter pupillae (a smooth muscle in the iris of the eye); when this muscle contracts it will narrow the pupil diameter to reduce the amount of light entering the eye.

Figure 20.3	Parasympathetic Division Pathways to the Organs of the Body

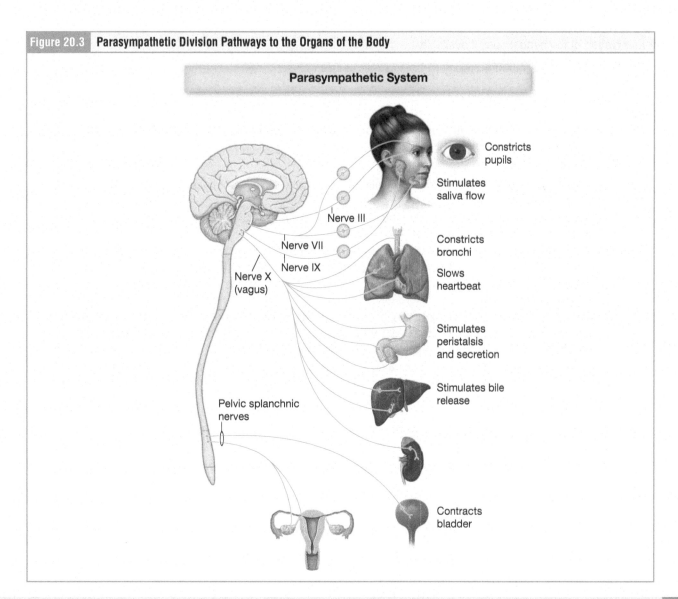

- The **pons** houses parasympathetic neuron cell bodies that depart the CNS through the **facial nerve** (CN VII).
  - Some of these preganglionic neurons synapse at collateral ganglia within the orbit. The postganglionic neurons extend to the lacrimal glands and mucus glands of the nasal cavity; when these glands are stimulated, tears will wash the eyes and mucus will be produced.
  - Other preganglionic neurons in CN VII will synapse at collateral ganglia inferior to the tongue. The postganglionic neurons extend to the salivary glands here; when these glands are stimulated, saliva is produced.
- The **medulla oblongata** houses parasympathetic neuron cell bodies that depart the CNS through the **glossopharyngeal nerve** (CN IX). These preganglionic neurons synapse at collateral ganglia in the cheek. The postganglionic neurons extend to another pair of salivary glands located here; they also produce saliva when stimulated.
- The medulla oblongata also houses parasympathetic neuron cell bodies that depart the CNS through the **vagus nerve** (CN X). This nerve sends preganglionic neuron axons to intramural ganglia in the bronchi (airways of the lungs), heart, digestive tract organs, gall bladder, and liver.
- The lateral gray horns within S2 to S4 spinal cord sections house parasympathetic neuron cell bodies that depart the CNS through the **pelvic splanchnic nerves**. These preganglionic neurons synapse on intramural ganglia in the bladder and reproductive structures (including the erectile tissue of the penis or clitoris and the wall of the uterus and vagina).

The goal of this activity is to help you learn the terminology associated with these parasympathetic nervous system pathways in preparation for connecting this anatomy to the physiology of initiating parasympathetic actions.

Materials
- Innervation of the head model, if available
- Spinal cord model, if available
- Sticky notes

Instructions
1. Use the description in the introduction and Figure 20.3 to find the listed anatomical structures related to the parasympathetic division of the ANS.

   a. If using the models:
      i. Depending on how many models are available for the whole classroom, you may be working in a larger group of students.
      ii. Divide the listed CNS structures among the group. Write your structure(s) on sticky notes (or labeling tape) and attach them to the model in the correct location.
      iii. Divide the listed nerves among the group. Write your nerve(s) on sticky notes (or labeling tape). Then list the organs of the body that will be reached by the nerve on the sticky note. Attach to the model in the correct location.

   b. If using the photo below:
      i. Draw lines to each listed CNS structure, but don't write the names in yet. Place a sticky note next to each line and write the name of the structure UNDER the sticky note, so that it acts as a cover to hide the label you write in. If your sticky notes are large, cut or tear them in half.
      ii. Draw lines to each listed nerve. Place a sticky note next to each line and write the name of the nerve UNDER the sticky note, then list the organs of the body that will be reached by the nerve under the sticky note, too.

2. After completing the labeling activity, work with your lab partner to quiz each other on the photo below, referring to the anatomical structure list when necessary.

   a. One of you will point to structures on the photo while the other student identifies them.
   b. Switch roles.

3. As a final practice for identifying each structure, work individually to label the photo without the help of the anatomical structure list.

   a. If you used the models for step 1, label each structure on the photo.
   b. If you used the photo already in step 1, label each structure again on top of the sticky note.

Anatomical Structures of the Parasympathetic Division to identify:

CNS Structures
- Midbrain
- Pons
- Medulla oblongata
- S2-S4 spinal cord

Nerves
- Oculomotor nerve (CN III)
- Facial nerve (CN VII)
- Glossopharyngeal nerve (CN IX)
- Vagus nerve (CN X)
- Pelvic splanchnic nerves

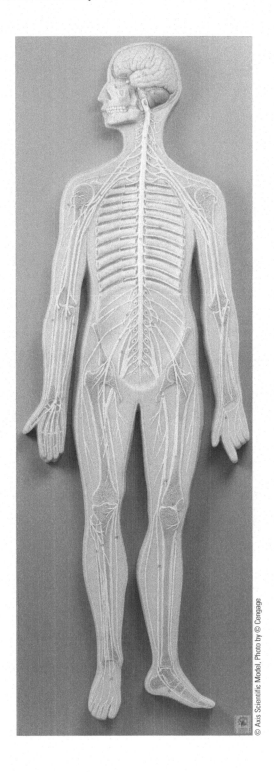

© Axis Scientific Model, Photo by © Cengage

(page intentionally left blank)

Activity 20.3 Anatomy of the Sympathetic Division

This activity targets LOs 20.2 and 20.3.

The sympathetic division of the ANS can also be referred to as the *thoracolumbar system* because the cell bodies of the preganglionic neurons are located in the lateral gray horns of the thoracic spinal cord ("thoraco-") and the superior portion of the lumbar spinal cord ("-lumbar"). As the axons of these preganglionic neurons leave the spinal cord through the anterior roots they will use the white rami communicantes to reach the sympathetic chain ganglia. From here, the short preganglionic axons can follow a few possible pathways, depending on which organs will be innervated by those fibers, to synapse with postganglionic neuron cell bodies. The long postganglionic axons will project to the effector organs, releasing the neurotransmitter norepinephrine (NE) onto a variety of adrenergic receptor types. It is worth noting that this holds true for sympathetic innervation of almost all effector organs of the body. However, for the target cells on sweat glands and the blood vessels serving skeletal muscle, the postganglionic axons will release the neurotransmitter Ach onto muscarinic receptors.

The sympathetic division's anatomy is more complex than the parasympathetic division but is still organized around the location of the effectors that are innervated. Examine Figures 20.4 and 20.5 as you read the description of these pathways of sympathetic innervation.

Figure 20.4	Sympathetic Division Pathways to the Organs of the Body

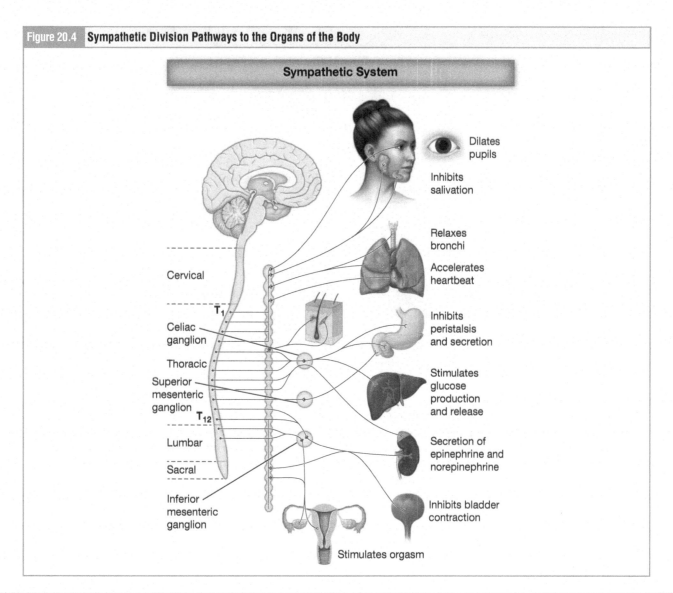

Figure 20.5 | **Sympathetic Pathways**

The preganglionic axon can follow a few pathways. (A) The preganglionic neuron can project out to the ganglion at the same level and synapse on a postganglionic neuron. (B) A preganglionic axon can reach and synapse on a more superior or inferior ganglion in the chain. (C) The axon can project through the white ramus communicans, but not form a synapse within the chain. Instead, it projects through one of the splanchnic nerves to a prevertebral ganglion or the adrenal medulla.

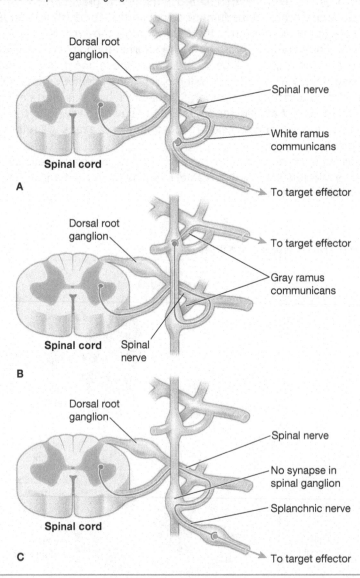

- To reach an effector that is at the same horizontal position in the body as the lateral gray horn from which the preganglionic neurons project, the preganglionic neuron and postganglionic neuron will synapse in the sympathetic chain ganglion. The postganglionic neurons' long axons will then project to the effector organ. Examples of such effectors include much of the skin of the body and the heart.
- To reach an effector that is superior or inferior to the lateral gray horn from which preganglionic neurons project, the preganglionic neurons' axons that have arrived in the sympathetic chain ganglia will ascend to the cervical ganglia or descend to more inferior lumbar ganglia before synapsing with the postganglionic neurons. The postganglionic neurons' long axons will then project to the effector organ. Examples of such effectors include the skin of the head and neck, the effectors of the head, and the lungs.

- To reach the abdominal and pelvic organs, the preganglionic neurons' axons arrive in the sympathetic chain ganglia but do not synapse. Instead, these axons continue to project through a splanchnic nerve leaving the ganglia, and reach a prevertebral ganglion (also called a collateral ganglion) where they finally synapse with the postganglionic neurons. The postganglionic neurons' long axons will then project to the effector organ.

The coordination of responses during sympathetic nervous system activation is supplemented by a hormonal response from the **adrenal medulla**, which is itself innervated by sympathetic neurons. These neurons have their cell bodies in the lateral gray horns of the T8 to T12 spinal cord. Their axons follow the path described above, with the sympathetic preganglionic neurons projecting through anterior roots to the white rami communicantes to reach the sympathetic chain ganglia of T8 to T12. The preganglionic neuron axons continue out of the ganglia, through prevertebral ganglia, and finally synapse on specialized cells within the adrenal medulla. These specialized cells, are called **chromaffin cells**. They are **neurosecretory cells,** meaning they secrete their neurotransmitter into the bloodstream rather than directly onto an effector tissue. When the sympathetic neurons reaching to the chromaffin cells release ACh, nicotinic receptors on the chromaffin cells are stimulated to release a mix of epinephrine and norepinephrine into the bloodstream. These two chemicals, now more accurately called hormones rather than neurotransmitters, are able to reach the adrenergic receptors on all the organs of the body that were also individually innervated by the three pathways described above.

The goal of this activity is to help you learn the terminology associated with these sympathetic nervous system pathways in preparation for connecting this anatomy to the physiology of initiating sympathetic actions.

Materials
- Spinal cord cross-section model, if available
- Spinal cord model, if available
- Sticky notes

Instructions
1. Use the description in the introduction and Figures 20.4 and 20.5 to find the listed anatomical structures related to the sympathetic division of the ANS.

    a. If using the models:

    i. Depending on how many models are available for the whole classroom, you may be working in a larger group of students.

    ii. Divide the listed structures among the group.

    iii. Each student will write their structures on sticky notes (or labeling tape) and attach them to the model in the correct locations.

    b. If using the photos below:

    i. Draw lines to each listed anatomical structure, but don't write the names in yet.

    ii. Place a sticky note next to each line and write the name of the structure UNDER the sticky note, so that it acts as a cover to hide the label you write in. If your sticky notes are large, cut or tear them in half.

2. After completing the labeling activity, work with your lab partner to quiz each other on the photos below, referring to the anatomical structure list when necessary.

    a. One of you will point to structures on the photo while the other student identifies them.

    b. Switch roles.

3. As a final practice for identifying each structure, work individually to label the photos without the help of the anatomical structure list.

    a. If you used the models for step 1, label each structure on the photos.

    b. If you used the photos already in step 1, label each structure again on top of the sticky note.

Anatomical Structures of the Sympathetic Division to identify:

- Lateral gray horn
- Anterior root
- White ramus communicans
- Gray ramus communicans
- Sympathetic chain ganglia
  - Cervical sympathetic chain ganglia
  - T1 to L2 sympathetic chain ganglia
  - L3 to S1 Sympathetic chain ganglia

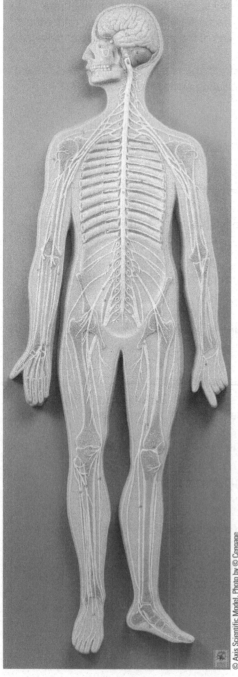

© Axis Scientific Model, Photo by © Cengage

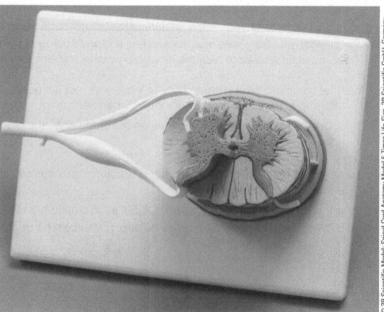

© 3B Scientific Model: Spinal Cord Anatomy Model 5 Times Life-Size. 3B Scientific GmbH, Germany. 2023. www.3bscientific.com; © Photo by: Cengage Learning.

Activity 20.4 Physiology of Antagonistic Dual Innervation

This activity targets LOs 20.1, 20.4, and 20.5.

Organ function is held in balance between the input from the sympathetic and parasympathetic divisions. For example, parasympathetic activation will stimulate smooth muscle in the lining of the stomach to contract to promote food digestion, while sympathetic activation will inhibit those same muscles, slowing digestion so that the body's efforts can be re-directed to skeletal muscles and other sympathetically active structures. While this dual innervation is what typically results in the two divisions of the ANS having opposite effects on any one organ, there are a few effector organs that receive innervation from the sympathetic division only. Most of the blood vessels (other than those to the digestive tract organs), the sweat glands, and the arrector pili muscles to the skin have no parasympathetic innervation. Rather, their responses are based solely on the level of sympathetic activation they receive. When the body is sympathetically active, this innervation will release significant amounts of norepinephrine onto these structures. When the body is in parasympathetic dominance, less NE will be received, and their sympathetic actions will wane.

Remember, the overall goal of the parasympathetic division of the ANS is to maintain homeostasis when the body is at rest. When the body is in parasympathetic dominance, it will prioritize activities that conserve energy, replenish nutrient stores, and perform maintenance functions. In contrast, the overall goal of the sympathetic division of the ANS is to maintain homeostasis during times of activity. When the body is in sympathetic dominance, it will prioritize activities that support skeletal muscle action, release nutrients from storage, and prepare the body to respond to stress or an emergency.

Can you predict how various organs and structures of the body will act under sympathetic or parasympathetic dominance? This activity will allow you to think critically about what sorts of body actions support homeostasis under various circumstances. Then you will be able to apply the terminology and anatomy of the two divisions to those circumstances. Work with your lab partner for all of the following directions.

Instructions

1. Begin with a brief review comparing the neurotransmitters of the two divisions of the ANS. Fill in the first two rows of the table to identify:

   a. What neurotransmitter is released on effectors in each division?

   b. What kinds of receptors are on the effectors to receive that neurotransmitter?

2. Consider the circumstances below. Which ones will stimulate sympathetic dominance? Which ones will stimulate parasympathetic dominance? Choose the division that is likely to be dominant in each of the circumstances, then write them in the correct box in the "Circumstances" row of Table 20.2. The first one has been done for you.

   a. Encountering a large aggressively barking dog.

   b. Sitting on the couch watching your favorite comedy.

   c. Playing soccer with some friends.

   d. Giving a speech in front of a large audience.

   e. Eating dinner with a loved one.

3. Think of another example of a circumstance in which sympathetic dominance will be likely. Add it to the table in the row labeled My Examples.

4. Think of another example of a circumstance in which parasympathetic dominance will be likely. Add it to the table.

5. For each of the body organs/structures listed in the table, predict:

   a. How it will behave under sympathetic and parasympathetic dominance.

   b. Why it will behave in this manner. Two have been done for you.

6. Several of the listed body organs/structures are not innervated by parasympathetic neurons. Put a star next to these structures.

Table 20.2 Physiology of the Sympathetic and Parasympathetic Divisions of the Autonomic Nervous System		
	Sympathetic Dominance	**Parasympathetic Dominance**
Neurotransmitter used		
Receptor(s) on the effectors to receive that neurotransmitter		
Circumstances	*Encountering a large aggressively barking dog*	
My Examples		
Pupil of the eye	*Wide diameter to allow more light for increased ability to navigate surroundings*	*Narrow diameter to allow for sharper close-up vision*
Lacrimal gland (tear production)		
Salivary glands		
Bronchi of the lungs (airway diameter)		
Pacemaker of the heart (heart rate)		
Smooth muscle surrounding blood vessels to the heart		
Smooth muscle in the walls of digestive tract organs		

(*continued*)

Table 20.2 Physiology of the Sympathetic and Parasympathetic Divisions of the Autonomic Nervous System (*continued*)		
	Sympathetic Dominance	**Parasympathetic Dominance**
Smooth muscle surrounding blood vessels to the digestive tract organs	Stimulates smooth muscle to narrow the diameter of these blood vessels; this re-directs blood flow to skeletal and cardiac muscle	Inhibits smooth muscle to widen the diameter of these blood vessels; this increases blood flow to these organs while they are active
Glands of the digestive tract		
Liver		
Smooth muscle in the wall of the bladder		
Sweat glands		
Smooth muscle surrounding blood vessels to the skin		
Arrector pili muscles of the skin		
Smooth muscle surrounding blood vessels to skeletal muscle		

(page intentionally left blank)

Activity 20.5 Case Study: Gastroparesis

This activity targets LOs 20.1 and 20.2.

Layla has recently been experiencing heartburn and nausea after every meal. She seems to feel full almost immediately when she begins to eat and her abdomen is bloated even though she has been losing weight. Concerned, Layla visits her doctor and explains her symptoms. After reviewing Layla's medical history her doctor orders a gastric emptying study to determine the time that it takes for a meal to move through Layla's digestive tract to reach the small intestine. The results of the gastric emptying study show that food is not exiting Layla's stomach in a typical amount of time.

1. Which division of the ANS is dominant when Layla is eating and digesting a meal?

   a. Parasympathetic division

   b. Sympathetic division

2. What nerve carries the ANS signals to Layla's stomach while she is digesting a meal?

   a. T5 to T8 spinal nerves

   b. Pelvic splanchnic nerves

   c. Facial nerve (CN VII)

   d. Vagus nerve (CN X)

3. In addition to innervation of the smooth muscle of the wall of the stomach, the ANS also innervates the pyloric sphincter which, when contracted, closes off the passageway from the stomach to the small intestine. When a person is digesting a meal, the pyloric sphincter should be regularly _____ by the ANS to open the passageway and allow the meal to progress to the next steps of digestion in the small intestine.

   a. relaxed/inhibited

   b. contracted/stimulated

Layla's doctor examines the results of the gastric emptying study and determines that Layla is suffering from gastroparesis. The doctor explains that the term gastroparesis means "stomach paralysis" and that her stomach is not emptying food as it should. It is caused by damage to the vagus nerve, which has reduced the ability for the autonomic nervous system to prompt Layla's rest-and-digest actions when she is eating.

4. Not ALL of Layla's rest-and-digest actions will be reduced due to the gastroparesis. Which of the following actions will still occur? Choose all correct answers.

   a. Contraction of the bladder wall

   b. Narrowing of the bronchi of the lungs

   c. Saliva production

The doctor explains that gastroparesis is a chronic condition; there is no cure for it, but it can be managed to keep the symptoms under control and make Layla's life more comfortable. The doctor prescribes a medication that Layla can take each time she eats to cause contraction of the muscles of the stomach wall, which bypasses the need for the ANS to stimulate these contractions. At a follow-up visit, Layla shares that she still has heartburn, nausea, and has been vomiting even though she is taking the medication. The doctor is concerned that even though the stomach is contracting, material is still not moving forward as it should and is refluxing into the esophagus to cause these symptoms. The doctor recommends a new treatment called a per oral pyloromyotomy (POP) that uses a narrow endoscope to reach into the digestive tract to sever some of the smooth muscle that is not receiving ANS signals.

5. Given the name, you can deduce that the smooth muscle cut by the per oral pyloromyotomy (POP) procedure is:
   a. In the wall of the stomach
   b. In the blood vessels bringing blood to the stomach
   c. In the sphincter between the stomach and small intestine

Digging Deeper:
Beta-blockers

Beta-blockers, also called beta-adrenergic blocking agents, are medications that are prescribed for patients with high blood pressure when other avenues of treatment have not been effective. The medications work to stop the attachment of NE and epi to the beta-adrenergic receptors on blood vessels and the heart. When these receptors are blocked, NE and epi released by sympathetic ANS pathways will not attach to the receptors. This will allow for:

• Relaxation of the smooth muscle surrounding many of the body's blood vessels, reducing blood pressure.
• Reduced force of contraction by the heart muscle, which also reduces blood pressure.

A common side effect of people taking beta-blockers is cold hands or feet because less blood is being transported to these peripheral structures.

Lab 20 Post-Lab Quiz

1. The autonomic nervous system innervates all of the following structures except:

   a. Skeletal muscle cells

   b. Smooth muscle surrounding blood vessels to skeletal muscle

   c. Cardiac muscle cells

   d. Smooth muscle surrounding blood vessels to cardiac muscle

2. The preganglionic cells bodies located in the brain stem are part of the _____ division of the ANS.

   a. parasympathetic

   b. sympathetic

3. While performing a dissection of an unknown nerve in the body, you find some neurons with long axons that terminate somewhere around the heart, but you cannot tell with certainty what structure is innervated by these neurons. When you test these neurons, you find that they release acetylcholine. Therefore, you know that these neurons are:

   a. Parasympathetic preganglionic neurons

   b. Parasympathetic postganglionic neurons

   c. Sympathetic preganglionic neurons

   d. Sympathetic postganglionic neurons

4. The pelvic splanchnic nerves provide parasympathetic innervation to:

   a. The stomach

   b. The salivary glands

   c. The bladder

5. How do the pathways of sympathetic innervation to the skin of the head differ from the pathways of sympathetic innervation to the skin of the thorax and abdomen?

   a. While no pathways of sympathetic innervation to the skin of the head exist, there are pathways of sympathetic innervation to the skin of the thorax and abdomen.

   b. Preganglionic neurons in the pathways of sympathetic innervation to the skin of the head will travel in more than one sympathetic chain ganglion.

   c. The preganglionic neurons in the pathways of sympathetic innervation to the skin of the head will be long, while those in the pathways of sympathetic innervation to the skin of the thorax and abdomen will be short.

6. The chromaffin cells of the adrenal medulla are neurosecretory cells that secrete epinephrine and norepinephrine into the blood to reach:

   a. Nicotinic receptors on postganglionic neurons

   b. Muscarinic receptors on effector organs

   c. Adrenergic receptors on effector organs

7. To reduce the production of sweat by the body's sweat glands, the ANS will:

   a. Increase stimulation to parasympathetic two-neuron pathways to the sweat glands

   b. Decrease stimulation to parasympathetic two-neuron pathways to the sweat glands

   c. Increase stimulation to sympathetic two-neuron pathways to the sweat glands

   d. Decrease stimulation to sympathetic two-neuron pathways to the sweat glands

8. Sympathetic innervation to the urinary system includes not only innervation of the smooth muscle of the wall of the bladder, but also the internal urethral sphincter that is at the base of the bladder. During urination, sympathetic signals decrease so that urine can be eliminated from the body. This decrease in sympathetic signals to the internal urethral sphincter means that there is less NE causing less _____ to the internal urethral sphincter.

    a. stimulation

    b. inhibition

9. Where are nicotinic receptors found? Choose all correct answers.

    a. In intramural and terminal ganglia

    b. In sympathetic and prevertebral ganglia

    c. On effector organs

10. People with asthma generally will not be prescribed a beta-blocker drug because it can trigger an asthma attack. In this context, you can determine that blocked beta-adrenergic receptors in the tissue of the lungs cause the asthma attack as they allow:

    a. Smooth muscle in the walls of the airways to contract

    b. Smooth muscle in the walls of the airways to relax

Lab 21 | The Endocrine System

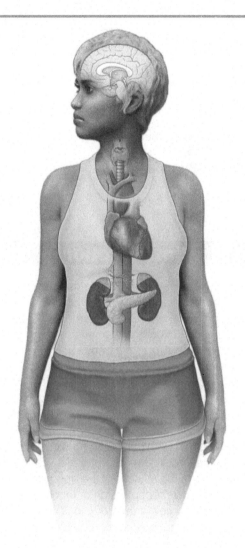

Learning Objectives: After completing this lab, you should be able to:

21.1 Describe the locations and the anatomical relationships of the hypothalamus, anterior pituitary, and posterior pituitary, including the hypothalamic–hypophyseal portal system.

21.2 Describe the anatomy of the thyroid gland, its location, the major hormones secreted, the control pathway(s) for hormone

secretion, and the hormones' primary targets and effects.

21.3 Describe the anatomy of the parathyroid glands, their location, the major hormone secreted, the control pathway(s) for hormone secretion, and the hormone's primary targets and effects.

The Human Anatomy and Physiology Society includes more than 1,700 educators who work together to promote excellence in the teaching of this subject area. The HAPS A&P Learning Outcomes measure student mastery of the content typically covered in a two-semester Human A&P curriculum at the undergraduate level. The full Learning Outcomes are available at https://www.hapsweb.org.

21.4 Describe the anatomy of the adrenal cortex, its location, the major hormones secreted, the control pathway(s) for hormone secretion, and the hormones' primary targets and effects.

21.5 Describe the anatomy of the pancreas, its location, the major hormones secreted, the control pathway(s) for hormone secretion, and the hormones' primary targets and effects.

21.6 Describe major hormones secreted by the anterior pituitary, their control pathways, and their primary target(s) and effects.

21.7* Describe the location and function of the pineal gland.

21.8 Given a disruption in the structure or function of the endocrine system (e.g., hypothyroidism), predict the possible factors or situations that might have caused that disruption (i.e., given an effect, predict possible causes).

*Objective is not a HAPS Learning Goal.

Introduction

The **endocrine** system consists of numerous separate structures that each secrete hormones into the blood or lymph as a means of communication among distant areas of the body. Some of these structures are entirely endocrine; meaning, their only job is to produce and release hormones. The parathyroid glands, adrenal cortex, and the pituitary glands all fall in this category. In contrast, other structures such as the pancreas, heart, and liver secrete hormones in addition to their other functions in the body.

Hormones are signaling molecules that travel through the bloodstream to attach to hormone receptors located in or on cells. By attaching to these receptors, the hormones can alter the actions of these target cells. Because the hormone will then cause an effect by the target cells, the structure to which those cells belong is often called an effector. What kinds of effects are prompted on the target cells? Hormones, generally, regulate the functions of target cells by changing the profile of proteins the target cells produce, altering the cells' metabolic rate, or changing the proteins embedded in the cells' membrane.

For cells to be targets of a hormone they must have receptors for that hormone. Thus, once hormones are released into the bloodstream and therefore reach all cells of the body, only the cells expressing receptors for that hormone will respond. Some hormones have receptors on or in many cells of the body. For example, insulin, secreted by the pancreas, targets almost every cell of the body to take up glucose from the bloodstream as a nutrient source. In contrast, other hormones target only a few structures. The adrenal glands secrete aldosterone, which targets certain cells within the kidney tubules to reabsorb sodium that would otherwise be lost in the urine.

Lab 21 Pre-Lab Quiz

This quiz will strengthen your background knowledge in preparation for this lab. For help answering the questions, use your resources to deepen your understanding. The best resource for help on the first five questions is your text, and the best resource for help answering the last five questions is to read the introduction section of each lab activity.

1. Anatomy of the endocrine system. Label the following structures of the endocrine system:

adrenal glands   pineal gland
pancreas   pituitary gland
parathyroid glands   thyroid gland

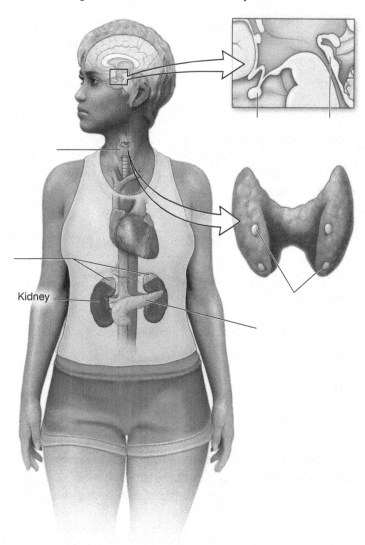

Kidney

2. Match each of the following roots to its definition:

Root	Meaning	List of Roots
	below	adeno-
	of or relating to a gland	adreno-
	within	cortico-
	pertaining to the nervous system	crine-
	to secrete	endo-
	alongside of	hypo-
	outer region	neuro-
	of or relating to the adrenal glands	para-

3. An endocrine structure secretes:

   a. Hormones into the bloodstream
   b. Substances into ducts
   c. Hormone receptors
   d. All of the above

4. Where are hormone receptors found on a target cell?

   a. Within the cell
   b. Embedded in the cell membrane
   c. Both a and b

5. The neuroendocrine structure in the body that controls many of the hormone pathways is the:

   a. Pineal gland
   b. Thyroid gland
   c. Hypothalamus
   d. Pancreas

6. What actions can hormones cause on a target cell? Mark all correct answers.

   a. A hormone can change the polarity of a cell membrane.
   b. A hormone can change a cell's metabolic activity.
   c. A hormone can alter the pattern of protein synthesis within a cell.

7. In addition to the endocrine system, the other main control system in the human body is the:

   a. Digestive system
   b. Circulatory system
   c. Lymphatic system
   d. Nervous system

8. The general tissue type that makes up endocrine glands is:

   a. Epithelial tissue
   b. Muscle tissue
   c. Connective tissue
   d. Nervous tissue

9. Which of the following structures has only an endocrine function?

   a. Liver
   b. Heart
   c. Adrenal cortex
   d. Testes

10. Which of the following hormones is a tropic hormone?

   a. Thyroid-stimulating hormone
   b. Prolactin
   c. Antidiuretic hormone
   d. Oxytocin

(page intentionally left blank)

Activity 21.1 Gross Anatomy of the Endocrine System

This activity targets LOs 21.1, 21.2, 21.3, 21.4, 21.5, 21.6, and 21.7.

The endocrine system is made up of a number of distinct structures and organs located throughout the body (Figure 21.1). Within the endocrine system, there are 10 structures in the body whose primary function is to secrete hormones. As you read this introduction, look at Figure 21.1 as a guide for where each endocrine structure is located.

- The **hypothalamus** and **pituitary gland** are often thought of as the "command center" of the endocrine system because together they play a central role in the release of many hormones. As can be seen in Figure 21.2, the hypothalamus is in the diencephalon, anterior to the thalamus. It is connected to the pituitary gland through a narrow section of tissue called the **infundibulum**. The tissue of the infundibulum provides a pair of distinctly different communication routes between the hypothalamus and two glands: the anterior pituitary gland and the posterior pituitary gland.

Figure 21.1	The 10 Primary Organs with Endocrine Function

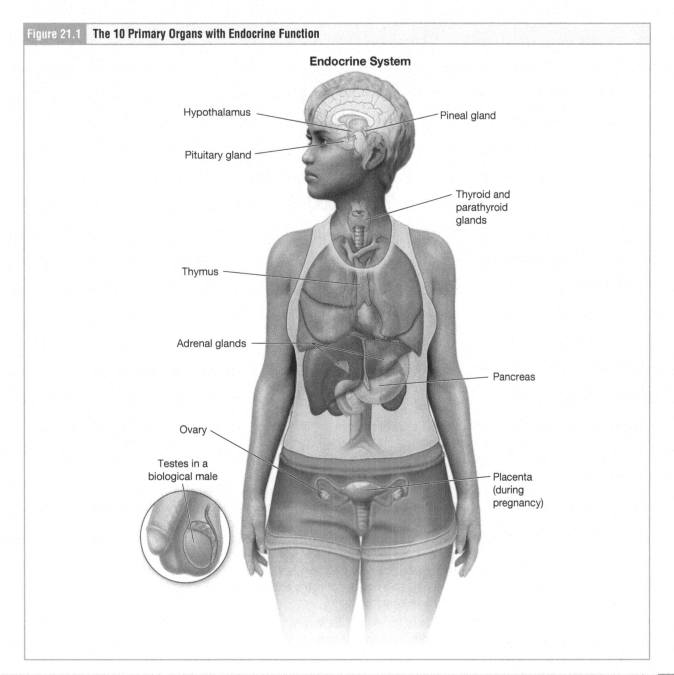

Endocrine System

Hypothalamus

Pineal gland

Pituitary gland

Thyroid and parathyroid glands

Thymus

Adrenal glands

Pancreas

Ovary

Testes in a biological male

Placenta (during pregnancy)

Figure 21.2 **The Hypothalamus and the Pituitary Gland**

The hypothalamus is part of the diencephalon. The pituitary gland, which is suspended beneath it, really consists of two glands. The anterior pituitary is made of glandular tissue and secretes six different hormones. The posterior pituitary is an extension of the hypothalamus and contains neurons that originate in the hypothalamus.

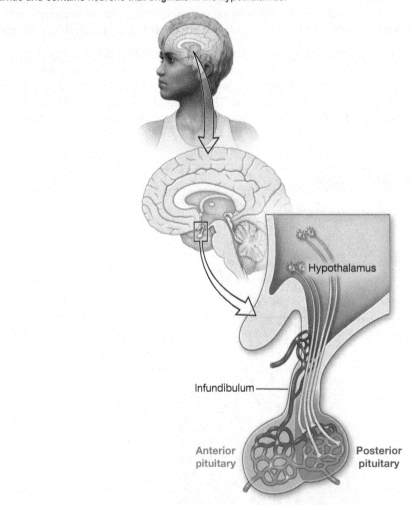

- The hypothalamus communicates with the anterior pituitary gland by secreting hormones into the **hypophyseal portal veins**. These hormones are received by cells of the anterior pituitary gland, which responds by either secreting or ceasing to secrete its own hormones. Six hormones are secreted by the anterior pituitary gland; these will be addressed in a separate lab exercise.
- The hypothalamus communicates with the posterior pituitary gland via the **hypothalamic neurons**. These neurons, which have cell bodies in the hypothalamus and synaptic terminals in the posterior pituitary gland, secrete hormones into the general bloodstream when the hypothalamus prompts them to fire. One group of hypothalamic neurons secretes **oxytocin (OT)** from the posterior pituitary gland; oxytocin stimulates contraction of the smooth muscle of the uterus and the release of milk from the breasts of a lactating female. A second group of hypothalamic neurons secretes **antidiuretic hormone (ADH)** from the posterior pituitary gland; ADH raises blood pressure by simultaneously targeting the kidneys to reduce urine output, the thirst center of the brain to increase water consumption, and the smooth muscle of the walls of blood vessels to vasoconstrict.
- The **pineal gland** is located in the anterior region of the diencephalon, superior to the cerebellum. It secretes the hormone **melatonin** in response to low-light conditions perceived by the eye. The target cells of melatonin are other areas of the brain that induce sleep.

- The **thyroid gland** is located on the anterior surface of larynx within the neck. It secretes **thyroid hormone (TH)** from its **follicular cells** in response to a hormone pathway initiated in the hypothalamus, which is triggered when circulating thyroid hormone levels are low. TH targets all cells of the body, prompting them to increase their metabolic rate and take up amino acids to support the building of more proteins. Throughout the body, specific tissues will respond to TH to provide both the glucose and O_2 that the cells will need to fuel an increase in metabolism: the liver and adipose tissue will release energy stores into the blood while heart rate, force of heart contractions, and respiration rate will all increase. The **parafollicular cells** of the thyroid gland act in an unrelated hormone pathway, producing and secreting **calcitonin** in response to high levels of calcium in the blood to prompt bone building.

- The **parathyroid glands** are small round glands embedded on the posterior surface of the thyroid gland. The parathyroid glands secrete **parathyroid hormone (PTH)** when blood calcium levels are low. PTH signals to osteoclasts to break down bone extracellular matrix, releasing calcium into the blood. PTH also targets the kidneys, prompting them to produce a partner hormone, calcitriol, which stimulates calcium absorption from digesting food in the small intestine.

- The **thymus**, an organ found on the superior aspect of the heart, produces a group of hormones called **thymosins** that prompt immune system development of T cells in young children. Little is known about this group of hormones.

- The **adrenal glands**, shown in Figure 21.3, are found on the superior aspect of each kidney and are more accurately described as two glands: the deeper **adrenal medulla** and the more superficial **adrenal cortex**.
  - The adrenal medulla secretes **epinephrine** and **norepinephrine** in response to neural stimulation by the sympathetic division of the autonomic nervous system (ANS). Epinephrine and norepinephrine target smooth muscle, cardiac muscle, and glands that are also innervated by the sympathetic division of the ANS.
  - The adrenal cortex produces a group of hormones collectively known as corticoids. The corticoid **aldosterone** is released in response to low blood pressure and targets the kidneys to retain Na^+, which in turn increases blood volume and pressure. The corticoid **cortisol** is secreted in response to a hormone pathway initiated in the hypothalamus in response to long-term stress. Cortisol inhibits tissue building while releasing stored nutrients from the liver and adipose tissue.

- The **pancreas** is located in the upper left quadrant of the abdomen, posterior to the stomach. It contains small groups of cells, the **pancreatic islets**, that secrete hormones. The **alpha cells** of the pancreatic islets secrete the hormone **glucagon** in response to low blood glucose levels between meals. Glucagon stimulates the liver and adipose tissue to release stored nutrients to raise blood glucose levels. The **beta cells** of the pancreatic islets secrete the hormone **insulin** in response to elevated blood glucose levels during digestion of a meal. Insulin stimulates the cells of the body to take up circulating glucose for their own metabolic needs; insulin also stimulates the liver and adipose tissue to store extra nutrients for later needs.

- The **ovaries** of a biological female are located in the pelvic cavity. They secrete **estrogen** and **progesterone** in response to a cyclical hormone pathway initiated in the hypothalamus. Generally, this pair of hormones functions in the development of oocytes and the maintenance of secondary sex characteristics of biological females.

- The **testes** of a biological male are located in the scrotum. They secrete **testosterone** in response to a hormone pathway initiated in the hypothalamus. Testosterone functions in the development of sperm and the maintenance of secondary sex characteristics of biological males.

The goal of this activity is to practice learning the anatomical terms associated with the endocrine system and locating these glands and organs within the body.

Materials
- Anatomical chart of the structures of the endocrine system
- Model of the digestive system, if available
- Model of the human brain, if available
- Model of the human torso, if available
- Model of the kidney, if available
- Model of the thyroid gland, if available

Figure 21.3	Structure of the Adrenal Gland

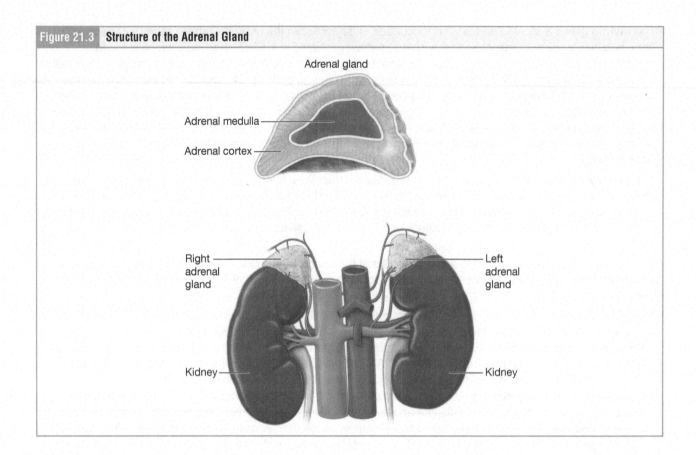

Adrenal gland

Adrenal medulla

Adrenal cortex

Right adrenal gland

Left adrenal gland

Kidney

Kidney

Structures to identify in this activity:

Hypothalamus	Thymus
Pituitary gland	Adrenal glands
Pineal gland	Pancreas
Thyroid gland	Ovaries
Parathyroid glands	Testes

Instructions

1. Your instructor will have stations with the available models. At each station:

   a. Determine which of the structures listed are on that model.

   b. Locate the structures on the model.

   c. Add to Figure 21.1 by writing in the hormone(s) produced by each structure (do not list the hormones for the hypothalamus or pituitary gland, those will be addressed in a separate activity).

2. For any listed structures that are not present on the models available to you, find those structures on the anatomical charts provided. Again, add to Figure 21.1 by writing in the hormone(s) produced by these remaining structures.

 Learning Connection

It can be difficult to distinguish between the actions prompted by the hormone cortisol and actions prompted by thyroid hormone. Both hormones will stimulate the liver and adipose tissue to release stored nutrients into the body. However, while thyroid hormone stimulates the cells of the body to increase their metabolism and *build more functional proteins*, cortisol will stimulate the cells of the body to *break down their own proteins* in order to add the resulting amino acids to the nutrient supply in the blood. Contrast the overall goal of thyroid hormone with the overall goal of cortisol; why does thyroid hormone release result in protein construction while cortisol results in protein breakdown?

Activity 21.2 Micro Anatomy of Endocrine System Structures

This activity targets LOs 21.2, 21.4, and 21.5.

The thyroid gland consists of small round structures called follicles that contain a substance called **colloid**. The colloid, produced by the follicular cells that ring each follicle, contains a precursor molecule to thyroid hormone. When prompted, the follicular cells take back this precursor molecule from the colloid, complete its construction into thyroid hormone, and release it into the bloodstream. The other hormone-producing cells of the thyroid gland, the parafollicular cells, are located in small wedges in between the follicles. As noted, parafollicular cells make calcitonin.

The adrenal gland is made up of two different areas of tissue. The adrenal cortex is the outer region of the gland and the adrenal medulla is the center region. As noted, the adrenal cortex produces aldosterone and cortisol (in addition to other corticoid hormones) and the adrenal medulla produces epinephrine and norepinephrine. If you are having trouble orienting your view to determine which section of tissue is superficial and which is deep, note the wrapping around the outside of the adrenal gland, the capsule, which looks like a flaky layer next to the open space on the slide.

The pancreas has many non-endocrine functions, but small round groups of cells called pancreatic islets have endocrine function, secreting the hormones insulin and glucagon. These groups of cells are called "islets" because they look like small islands in the sea of non-endocrine tissue of the pancreas. The pancreatic islets may stain darker or lighter than the rest of the tissue depending on the staining technique used, so search for "islands" rather than a specific color.

The purpose of this exercise is to develop a greater understanding of the function of the thyroid gland, adrenal glands, and pancreas through an exploration of the microanatomy of these structures.

Materials
- Colored pencils
- Compound microscope
- Slide or histological image of the adrenal gland
- Slide or histological image of the pancreas
- Slide or histological image of the thyroid gland

Instructions

1. For each slide, examine the slide first on scanning power (4×) and use the coarse focus knob to locate the tissue. Then increase the magnification by switching to the low power (10×) objective lens and use the fine focus knob. If necessary, switch to the high power (40×) objective lens and again use the fine focus knob until you can clearly see the details.

2. If you are provided with histological images rather than slides, continue to step 3.

3. Draw the microscopic view of each tissue, identifying the requested structures. In a blank area on the drawing, note the magnification used.

Slide 1: Thyroid Gland

• On your drawing, identify: thyroid follicle, follicular cells, colloid, parafollicular cells.

• What hormone is produced by the follicular cells and stored (in its unfinished form) in the follicular lumen? Write this hormone in under your label of the follicular cells.

• What hormone is produced by the parafollicular cells? Write this hormone in under your label of the parafollicular cells.

Slide 2: Adrenal Gland

• On your drawing, identify: adrenal cortex, adrenal medulla.

• What hormones are produced by the adrenal cortex? Write these hormones under your label of the adrenal cortex.

• What hormones are produced by the adrenal medulla? Write these hormones under your label of the adrenal medulla.

Slide 3: Pancreas

• On your drawing, identify: pancreatic islets.

• What hormones are produced by the pancreatic islets? Write these hormones under your label of the pancreatic islets (note that you will NOT be able to visually distinguish between the alpha cells and beta cells within the islets, but you should know which cells produce which hormone).

(page intentionally left blank)

Activity 21.3 Tropic Hormone Pathway Physiology

This activity targets LOs 21.6 and 21.8.

A **tropic hormone** is a hormone whose target is another endocrine gland, and whose effect is to prompt the target gland to release its own hormone. The seven hormones released by the hypothalamus into the hypophyseal portal system are tropic hormones, as all of them target the anterior pituitary gland to regulate the release of six pituitary hormones. Some of the hormones released by the anterior pituitary gland are themselves tropic hormones, targeting other structures to release yet more hormones. In this way, several specific hormone pathways emerge.

Examine Figure 21.4 as we consider the hormone pathway that regulates the release of **growth hormone (GH)** by the anterior pituitary gland. The hypothalamus regulates the release of GH through the release of either **growth**

Figure 21.4	The Growth Hormone Pathway

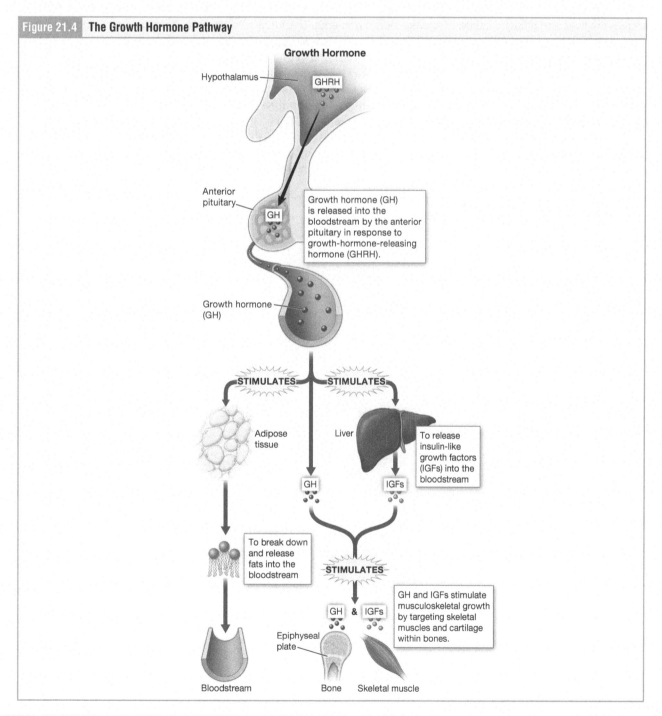

hormone-releasing hormone (GHRH) or **growth hormone-inhibiting hormone (GHIH)** into the hypophyseal portal veins. GH stimulates both adipose tissue to release nutrients into the blood (to support tissue growth) and the liver to release **insulin-like growth factors (IGFs)**, hormones that stimulate cell division in skeletal muscles and bone, increasing growth in these tissues. The hypothalamus secretes GHRH in response to stress and exercise, consumption of dietary protein, and the onset of sleep (so growth and repair can occur while we sleep). Increased levels of GH and IGFs in the blood feed back to inhibit the release of GHRH and prompt the release of GHIH from the hypothalamus, thus regulating levels of circulating GH and IGFs.

Figure 21.5 traces the hormone pathway that regulates the release of thyroid hormone. In response to low levels of circulating thyroid hormone in the blood, the hypothalamus will release **thyrotropin-releasing hormone (TRH)** into the hypophyseal portal veins. The anterior pituitary gland responds to TRH by secreting **thyroid-stimulating hormone (TSH)**. The TSH enters the bloodstream and targets the follicular cells of the thyroid gland. The follicular cells release thyroid hormone in response. Thyroid hormone targets all cells to increase their metabolism and targets the liver and adipose tissue to release stored nutrients. The lungs and heart respond by increasing the respiratory rate and heart rate. As thyroid hormone levels rise, they will negatively feed back on both the anterior pituitary gland and the hypothalamus, inhibiting the release of TSH and TRH, respectively.

In a third example, let's trace the hormone pathway that regulates the release of cortisol from the adrenal cortex (Figure 21.6). In response to low levels of cortisol and high levels of stress (both emotional stress and physical stress), the hypothalamus will release **corticotropin-releasing hormone (CRH)** into the hypophyseal portal veins. CRH stimulates the release of **adrenocorticotropic hormone (ACTH)** by the anterior pituitary gland into the bloodstream. ACTH targets the adrenal cortex, prompting the release of cortisol. Cortisol will target the liver and adipose tissue to release stored nutrients and prompt all cells of the body to break down their own proteins to release amino acids into the bloodstream. Cortisol also inhibits the immune system, reducing inflammation.

Figure 21.5 | **The Thyroid Hormone Pathway**

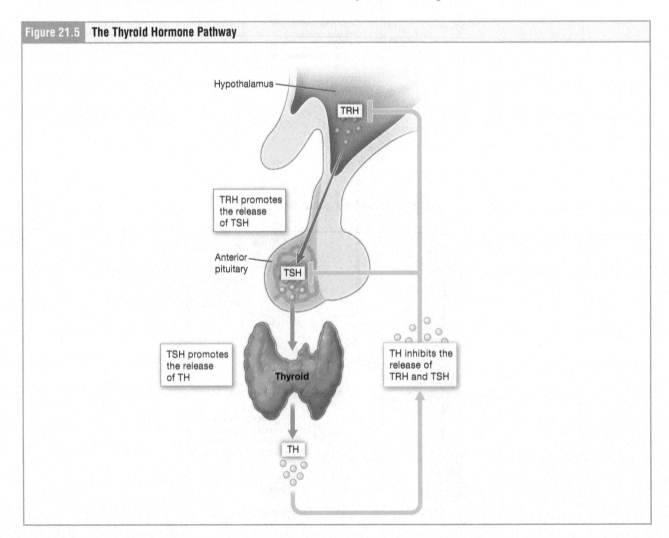

Figure 21.6   **The Cortisol Hormone Pathway**

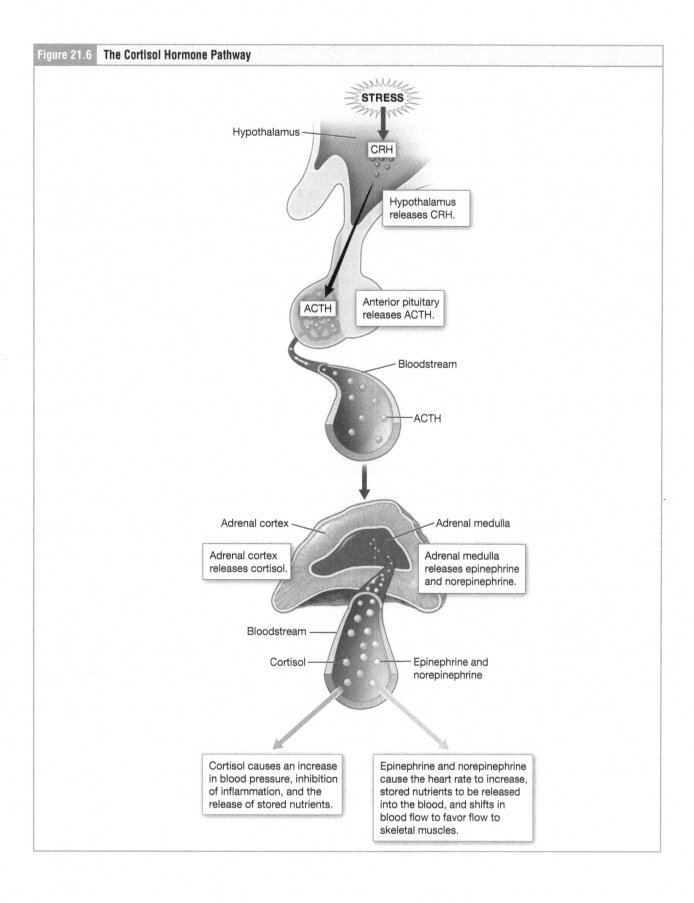

The hypothalamus regulates the release of both **luteinizing hormone (LH)** and **follicle-stimulating hormone (FSH)** by the anterior pituitary gland through the release of **gonadotropin-releasing hormone (GRH)** into the hypophyseal portal system. Together LH and FSH target the ovaries in biological females (to release estrogen and progesterone) and the testes in biological males (to release testosterone). These hormone pathways are more complicated than described here and will be investigated further in the reproductive system chapter.

The hypothalamus regulates the release of **prolactin (PRL)** by the anterior pituitary gland through the release of either **prolactin-releasing hormone (PRH)** or **prolactin-inhibiting hormone (PIH)** into the hypophyseal portal veins. PRL stimulates the production of milk by the breasts of a lactating female.

The purpose of this exercise is to allow you to practice two of the hormone pathways described above, the growth hormone pathway and the thyroid hormone pathway, by acting them out with your classmates. You will also have a chance to examine what happens when these hormone pathways are disrupted. Acting out these pathways may feel clumsy or awkward at first. Take the time to work through each step with your classmates, troubleshooting the next steps if you get stuck.

Materials
- Index cards
- Paper or plastic cups
- Paper plates
- Rectangular blank stickers such as name tag stickers
- Sharpie markers

PART A
Instructions for acting out the thyroid hormone pathway:

1. You will need a total of 10–12 students for this pathway. While one group of students in the class is working on this activity, the rest of the class can work on the other activities in this lab.

2. Assign the following roles; seven students will be a specific structure or group of cells, two to four students will be blood vessels, and one student will be the director. Each student should write who they are on a blank sticker and stick it to their shirt.

   a. Structures/cells
      i. Hypothalamus
      ii. Anterior pituitary gland
      iii. Follicular cells of the thyroid gland
      iv. All cells of the body
      v. Liver and adipose tissue
      vi. Heart
      vii. Lungs
   b. Blood vessels
      i. Hypophyseal portal veins
      ii. One to three students to be the blood vessels of the rest of the body
   c. Director

3. Some of these structures/cells produce hormones.

   a. Students who are structures that produce hormones should write their hormone on five separate index cards.
   b. For example, the follicular cells of the thyroid gland make TH, so the person who is the follicular cells should write "thyroid hormone" on five separate index cards to represent five separate releases of TH.

4. All of the structures/cells receive hormones.

   a. Students who are structures should use a paper or plastic cup to represent the receptor for the hormone they receive.
   b. Write the name of the hormone receptor on the cup. For example, the hypothalamus has receptors for thyroid hormone (since high levels of the hormone will inhibit further release of TRH). The person who is the hypothalamus will write "Thyroid hormone receptor" on their cup.

5. The bloodstream carries hormones from place to place.

    a. The students who are the blood vessels should take a paper plate so they can carry hormones from place to place.

    b. Determine which hormones will be carried by the hypophyseal portal veins and which hormones will be carried by the blood vessels of the rest of the body.

    c. As the blood arrives at a structure with the correct hormone receptor, the blood can put that hormone (index card) in the cup to stimulate the structure.

6. The final effectors of the pathway (cells of the body, liver and adipose tissue, heart, and lungs) all have specific actions they will perform when they receive TH.

    a. Students who are final effectors should determine what actions they will take when they receive TH.

    b. Effectors should yell out what they do each time they receive TH in their receptors.

7. The director will now help get the story started.

    a. Together, arrange yourselves in a section of the room to represent general locations of the structures and blood vessels and begin the hormone pathway, using Figure 21.5 to get started.

    b. The director will tell the blood vessels to arrive at the hypothalamus with no hormone on their plates. How will the hypothalamus respond to this low level of thyroid hormone? Once the blood vessels receive hormone from the hypothalamus, where will they go next to drop it off?

    c. Continue with the blood vessels picking up and dropping off hormones in the correct order, considering how the arrival of a hormone at a structure prompts that structure to do something.

    d. If the group gets stuck, the director should help get things back on track. Remember, acting out this pathway may feel awkward and the group may run into questions or be unclear about what to do next. Troubleshoot as a group, with the director taking the lead.

    e. Don't forget to represent the negative-feedback loop that will stop the hypothalamus and anterior pituitary gland from releasing any more hormone.

8. What happens when the follicular cells of the thyroid gland are unable to make enough TH? Primary hypothyroidism (Hashimoto's disease) is a condition in which the follicular cells are unable to make very much TH because the cells are damaged. Take all but one card away from the follicular cells and re-play the story.

9. When the group is finished acting out the revised pathways to represent Hashimoto's disease, gather up the supplies and return them to the appropriate location so that other groups can use them.

PART B
With your lab partner, address these questions regarding the thyroid hormone pathway:

1. Why do only certain structures/cells of the body respond to any one hormone? For example, why didn't the bloodstream give TSH to the liver and adipose tissue, the heart, etc.?

2. A person with Hashimoto's disease will still make some of the hormones associated with the thyroid hormone pathway. Which structures will continue to produce their hormones?

3. Will the negative feedback loop occur in a person with Hashimoto's disease? Explain.

4. The main way to diagnose Hashimoto's disease is to test for the level of thyroid-stimulating hormone. For a patient with Hashimoto's disease, would you predict their TSH to be higher or lower than expected? Explain.

PART C
Instructions for acting out the cortisol hormone pathway:

1. Your instructor will advise you as to whether to act out the cortisol pathway as a group or to skip steps 2 and 3 to move on to PART D.

2. If time allows, follow steps 1 through 6 of the directions above, with necessary alterations, to act out the cortisol hormone pathway.

3. Let's look at what happens when an exogenous hormone is added to the pathway.

   a. The hormones described in the introduction are endogenous hormones. The term endogenous describes that these hormones were made by the body. In contrast, hormones can also drive pathways or portions of pathways when they are introduced to the body through consumption or injection—these are called exogenous hormones.

   b. A person who is taking corticosteroid medication to treat inflammatory diseases such as rheumatoid arthritis is introducing exogenous cortisol to their body, which can result in Cushing syndrome. Write out 10 more index cards with "exogenous corticosteroid" on them. As you act out the cortisol steroid, the director will add these cards regularly to the cortisol receptors.

PART D
With your lab partner, address these questions regarding the cortisol hormone pathway:

1. How are the results of cortisol different from the results of thyroid hormone? In other words, compare and contrast the actions of the effectors in these two pathways.

2. How did the addition of exogenous cortisol influence the effectors?

3. Is the consumption of exogenous glucocorticoids influenced by any negative feedback? Explain.

Lab 21 Post-Lab Quiz

1. The hormone oxytocin reaches the posterior pituitary gland through:
   a. The hypophyseal portal veins
   b. Hypothalamic neurons
   c. The general bloodstream of the body

2. The hypophyseal portal veins transport hormones from the hypothalamus to target cells in the:
   a. Anterior pituitary gland
   b. Posterior pituitary gland
   c. Infundibulum

3. Anterior pituitary gland hormones that have receptors on the liver and adipose tissue include: (choose all correct answers)
   a. Growth hormone
   b. Thyroid hormone
   c. Cortisol

4. When thyroid-stimulating hormone attaches to receptors on the follicular cells of the thyroid, these cells:
   a. Make colloid
   b. Convert colloid to TH
   c. Make calcitonin

5. The body will produce more thyroid hormone when the _____ senses low levels of thyroid hormone:
   a. follicular cells of the thyroid gland
   b. anterior pituitary gland
   c. hypothalamus

6. The parathyroid glands are located:
   a. Adjacent to the stomach
   b. Superior to the kidneys
   c. On the posterior aspect of the thyroid gland

7. Damage to which of the following structures would inhibit aldosterone production?
   a. The adrenal medulla
   b. The adrenal cortex
   c. The posterior pituitary gland

8. You have just eaten a delicious dinner. As your body digests the meal, you test your blood hormone levels. Which of the following hormones would you expect to increase as you absorb the nutrients in your meal?
   a. Luteinizing hormone
   b. Follicle-stimulating hormone
   c. Glucagon
   d. Insulin

9. In between meals, nutrient levels in your blood will begin to drop. Which cells would you expect to increase their activity as a direct result of lower blood glucose levels?

   a. Beta cells of the pancreas

   b. Alpha cells of the pancreas

   c. Follicular cells of the thyroid gland

   d. Parafollicular cells of the thyroid gland

10. A person with hyperthyroidism, who's body makes too much thyroid hormone, will have which of the following symptoms?

   a. Weight gain

   b. Weight loss

   c. Feeling cold

Lab 22 Blood

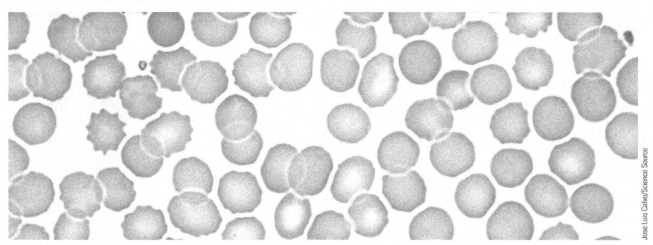

Jose Luis Calvo/Science Source

Learning Objectives: After completing this lab, you should be able to:

22.1 Describe the general composition of blood (e.g., plasma, formed elements).

22.2 Compare and contrast the morphological features and general functions of the formed elements (i.e., erythrocytes, leukocytes, platelets).

22.3* Identify the formed elements of the blood in a blood smear slide.

22.4 List the five types of leukocytes in order of their relative prevalence in normal blood, and describe their major functions.

22.5* Perform a differential white blood cell count and interpret the results considering possible pathologies.

22.6* Perform a hemoglobin and hematocrit blood analysis.

22.7* Interpret a hematocrit and determine whether its values fall within the normal range.

22.8 Explain the role of surface antigens on erythrocytes in determining blood groups.

22.9 Predict which blood types are compatible and what happens when the incorrect ABO or Rh blood type is transfused.

22.10 Describe the development and clinical significance of anti-Rh antibodies.

*Objective is not a HAPS Learning Goal.

Introduction

Blood, as a connective tissue, is made up of cells, proteins, and extracellular matrix. The extracellular matrix of blood is a liquid, **plasma**, in which cells are suspended. Suspended in the plasma are many different proteins; some of these proteins are integral to the functioning of the blood itself, while others are present to be delivered to various areas of the body. Dissolved in the blood are gasses (O_2 and CO_2), nutrients, and wastes.

The term **formed elements** describes both the whole cells and cell fragments in the blood; the formed elements include **erythrocytes** (red blood cells), **leukocytes** (white blood cells), and **platelets**.

The Human Anatomy and Physiology Society includes more than 1,700 educators who work together to promote excellence in the teaching of this subject area. The HAPS A&P Learning Outcomes measure student mastery of the content typically covered in a two-semester Human A&P curriculum at the undergraduate level. The full Learning Outcomes are available at https://www.hapsweb.org.

Lab 22 Pre-Lab Quiz

This quiz will strengthen your background knowledge in preparation for this lab. For help answering the questions, use your resources to deepen your understanding. The best resource for help on the first five questions is your text, and the best resource for help answering the last five questions is to read the introduction section of each lab activity.

1. Identify the layers of a centrifuged blood sample:

Plasma
Leukocytes and platelets
Erythrocytes

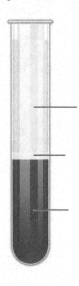
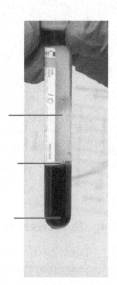

2. Match the root next to its definition:

Root	Meaning
	eating
	cell
	red
	white
	water

List of Roots

cyto-

erythro-

leuko-

lympho-

phago-

3. Which of the following materials are subcomponents of the blood plasma? Choose all correct answers.
   a. Water
   b. Proteins
   c. Gasses (O_2 and CO_2)
   d. Platelets

4. What is the function of the leukocytes found in the bloodstream?
   a. They carry O_2 and CO_2.
   b. They help the blood form clots when necessary.
   c. They are part of the immune response.

5. Where does hematopoiesis, the formation of the formed elements of the blood, occur?
   a. In the spleen
   b. In the kidney
   c. In the red bone marrow

6. What is the main function of the protein hemoglobin?
   a. It carries oxygen.
   b. It carries nutrients.
   c. It breaks down pathogens.

7. Which of the following substances make up the formed elements of the blood? Choose all correct answers.
   a. Leukocytes
   b. Erythrocytes
   c. Plasma
   d. Platelets

8. Which of the following items is/are found inside erythrocytes? Choose all correct answers.
   a. Granules
   b. A nucleus
   c. Hemoglobin

9. Which leukocyte is most common in the bloodstream?
   a. Basophils
   b. Neutrophils
   c. Eosinophils

10. What is the blood analysis that is used to measure the relative amounts of leukocytes in the blood?
    a. Differential white blood cell count
    b. Hemoglobin
    c. Hematocrit

(page intentionally left blank)

Activity 22.1 **Blood Smear Microscopy**

This activity targets LOs 22.1, 22.2, 22.3, and 22.4.

The components within the blood plasma include both dissolved substances and the formed elements. While the dissolved substances are too small to see even under magnification through a microscope, the formed elements are large enough to be seen with the aid of a light microscope. The formed elements of the blood include erythrocytes, platelets, and five different kinds of leukocytes: neutrophils, eosinophils, basophils, lymphocytes, and monocytes.

Erythrocytes, commonly known as red blood cells or RBCs, make up most of the formed elements present in the blood. Structurally, erythrocytes appear very simple. These cells contain no nucleus, and no other organelles will be visible. Rather, erythrocytes are simple membrane-bound sacks of the protein **hemoglobin** that carry O_2 (and CO_2) in the blood. The shape of erythrocytes, as shown in Figure 22.1, is a biconcave disk, somewhat like a circular cushion with a depression on each side.

Platelets are not cells but rather fragments of larger cells called **megakaryocytes** that reside in the bone marrow. As shown in Figure 22.2, megakaryocytes release small membrane-bound cell fragments into the bloodstream; each megakaryocyte can release 2000–3000 platelets during its life span. These platelets are in the blood to assist with hemostasis, the stoppage of blood flow following blood vessel damage.

Leukocytes, commonly known as white blood cells or WBCs, are present in the bloodstream as part of the body's immune system. These cells protect the body against invading organisms, eliminate cells with mutated DNA, and clean up debris. Generally, leukocytes are larger than the erythrocytes that they share the bloodstream with; they are also distinguishable from the erythrocytes due to many visible internal structures including nuclei and, in some types of leukocytes, specialized vesicles called **granules** that help the leukocytes in their fight against pathogens such as bacteria and parasites. In fact, the five kinds of leukocytes are sometimes categorized based on whether they contain visible granules or not. Examine Figure 22.3 to discover the visible differences between the five different kinds of leukocytes that are present in the blood.

| Figure 22.1 | **Red Blood Cell Structure** |

Red blood cells can be described as biconcave discs, meaning they are relatively flat with a curve inward on each side.

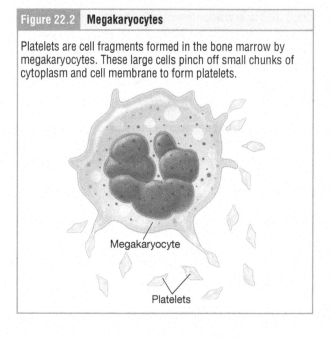

Cross section

Superior view

| Figure 22.2 | **Megakaryocytes** |

Platelets are cell fragments formed in the bone marrow by megakaryocytes. These large cells pinch off small chunks of cytoplasm and cell membrane to form platelets.

Megakaryocyte

Platelets

| Figure 22.3 | **Leukocytes** |

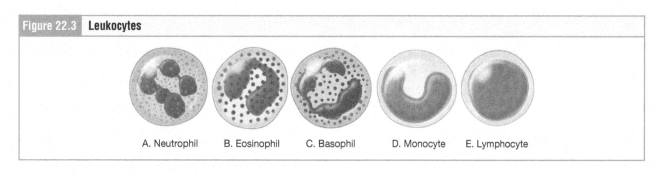

A. Neutrophil B. Eosinophil C. Basophil D. Monocyte E. Lymphocyte

Only a small fraction of the body's leukocyte population is found in the blood at any one time, as these cells spend most of their time in other structures and tissues of the body. You can imagine the bloodstream as a highway system for leukocytes to travel as they transit from one structure to another, so that while the numbers of leukocytes on this "highway" are relatively consistent from moment to moment, these are not the same leukocytes in the blood all the time but a constantly rotating group of them as they go from one location to another. **Neutrophils** are the most abundant leukocytes in the blood. These cells are the rapid responders to the site of infection and will quickly engulf and destroy bacteria they encounter. **Lymphocytes**, the next most abundant WBCs in the blood, are quite small compared to the other leukocytes, typically similar in size or smaller than the RBCs around them. Lymphocytes include three different specialized cell types: natural killer (NK) cells, B lymphocytes, and T lymphocytes. **Monocytes,** the third most abundant leukocytes, differentiate into the phagocytes called **macrophages** once they leave the bloodstream. Less abundant still are the **eosinophils** that are critical to fighting parasitic infections and are also present in large amounts when the body experiences an allergic reaction. Finally, the least prevalent WBCs in the blood are the **basophils**, key players in inflammatory responses in the body. The relative numbers of different leukocytes found in the blood are more representative of the amount of time each type of leukocyte spends travelling in the bloodstream than the total numbers of each kind of leukocyte that are present in the body, though some correlations can be drawn between the two.

The purpose of this activity is to practice visually distinguishing between the various formed elements of the blood and to consider the significance of the relative abundance of each type of leukocyte in a typical blood sample. The prepared and fixed blood smear slides provided for this activity have been preserved and do not pose the safety hazards that are present in samples of fresh, unpreserved blood. As such, personal protective equipment (PPE) is not necessary for this exercise.

Before continuing with the microscope activity below, use the information above to answer the following questions:

1. Considering the source of platelets, would you expect them to be the same size as erythrocytes, smaller than erythrocytes, or larger than erythrocytes?

2. Which leukocyte is likely to be the easiest to find on a blood smear slide? Why?

3. Which leukocyte is likely to be the most difficult to find on a blood smear slide? Why?

4. Use the introduction above and Figure 22.3 to complete Table 22.1, which compares the visible distinguishing features of the leukocytes of the blood.

Table 22.1 Visible Features of the Leukocytes of the Blood				
Leukocyte	Are Visible Granules Present?	Shape of the Nucleus	Color of the Stain Used	Relative Size Compared to RBCs (Smaller, Same Size, or Larger)
Neutrophil				
Eosinophil				
Basophil				
Lymphocyte				
Monocyte				

Materials
- Colored pencils
- Compound microscope
- Prepared blood smear slide

Instructions

1. Place a prepared blood slide on the microscope and examine on scanning power (4×) objective. Use the coarse focus knob to focus on the blood cells present, which will appear very small at this magnification.

2. Switch to the lower power (10×) objective and use the fine focus knob to clearly view the cells present in your view. Switch to the high power (40×) objective and use the fine focus knob as necessary to view more detail.

3. Draw several erythrocytes and platelets in the first space provided below.

   a. Pay attention to the relative size of these two formed elements as you draw them together.
   b. Record the total magnification at which you viewed and sketched near the space.

4. Draw each of the leukocytes in the provided spaces below.

   a. Use pencils and colored pencils to draw in the details of each leukocyte. Make sure that you can represent the information you assembled in Table 22.1 in each of your sketches.
   b. Add one or more erythrocytes into the space to represent the relative size of the leukocyte compared to the erythrocytes. Make a note to assist you in distinguishing the drawn leukocyte from other types of leukocytes.
   c. Write the magnification at which you viewed and sketched each leukocyte near the provided spaces.
   d. You will have to scroll around the slide to find all the formed elements. It will be helpful to switch back to the lower power (10×) objective and scroll methodically either left to right or top to bottom along the slide in order to be organized in your approach to viewing new areas of the slide.

Learning Connection

Never Let Monkeys Eat Bananas

What? This silly phrase, "never let monkeys eat bananas," is a great way to remember the relative abundance of the various types of leukocytes within the bloodstream. It lists the order of abundance by using the first letter of each leukocyte: N – L – M – E – B. In other words, neutrophils are more abundant than lymphocytes, which are more abundant than monocytes, which are more abundant than eosinophils, and the basophils are the least abundant in the blood. Remember that these leukocytes are using the bloodstream to transit from place to place in the body, so this acronym is describing their relative abundance in the blood and not their overall relative abundance in the body as a whole.

Erythrocytes & Platelets

Neutrophil

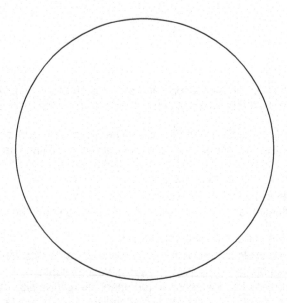

Eosinophil

Basophil

Lymphocyte

Monocyte

(page intentionally left blank)

Activity 22.2 Differential White Blood Cell Count

This activity targets LOs 22.3 and 22.5.

Each of the white blood cell types present in the blood performs specific immune system tasks. These tasks are outlined in Table 22.2, which shows the expected percentages of each WBC type in a blood sample. As you read the descriptions of the roles of each WBC below, consider how the elevated or low levels of each indicate pathologies connected to their functions.

Table 22.2 Differential White Blood Cell Count			
Leukocyte Type	Normal Value (as a Percent of Total Leukocytes)	Elevated Levels May Indicate	Low Levels May Indicate
Neutrophil	40–60%	Bacterial infection Inflammation Burns or other tissue damage Stress	Radiation exposure (chemotherapy) Some anemias Sepsis (severe overwhelming bacterial infection)
Lymphocyte	20–40%	Viral infection Mononucleosis infection Some leukemias	HIV infection Autoimmune disorders (e.g., lupus, rheumatoid arthritis)
Monocyte	2–8%	Fungal infection Tuberculosis Some leukemias	Bone marrow damage
Eosinophil	1–4%	Parasitic infection Allergic reactions Asthma	**Difficult to determine due to normal low numbers in the blood
Basophil	0.5–1%	Rare allergic reactions Inflammation Some leukemias	**Difficult to determine due to normal low numbers in the blood

Neutrophils are present in the blood as the rapid responders to the site of an infection and are effective phagocytes against bacterial infections. Neutrophils use the bloodstream to travel to sites of infection where they engulf large quantities of bacteria and then break them down internally. Lymphocytes include three different specialized cell types: natural killer (NK) cells, B lymphocytes, and T lymphocytes. Both B and T lymphocytes are active in specific immune responses, including responses to viruses. The human immunodeficiency virus (HIV) infects and destroys T lymphocytes. Monocytes, which differentiate into macrophages once they leave the bloodstream to address infections in the tissues of the body, are phagocytes like the neutrophils. Eosinophils are specialized to target parasites such as worms, but also become activated in large numbers when the body initiates an allergic reaction through other immune system cells. Basophils are important in initiating inflammatory responses in damaged or infected tissues of the body; these inflammatory responses are meant to clear debris and pathogens from the site.

A differential white blood cell count is a common blood analysis method used to measure the percentage of each of the different types of leukocytes in a blood sample against the expected values presented in Table 22.2. The results of a differential white blood cell count are valuable for determining if one or more of the leukocyte types have abnormal populations in that person's bloodstream; different conditions and diseases can stimulate changes in the expected amounts of each leukocyte type. In the recent past, differential white blood cell counts were performed by technicians in a lab in a manner like what you will be doing in this activity. Now, these counts are typically completed by a machine, with a lab technician performing a secondary examination if any abnormalities are indicated.

The prepared and fixed blood smear slides provided for this activity have been preserved and do not pose the safety hazards that are present in samples of fresh, unpreserved blood. As such, personal protective equipment (PPE) is not necessary for this exercise.

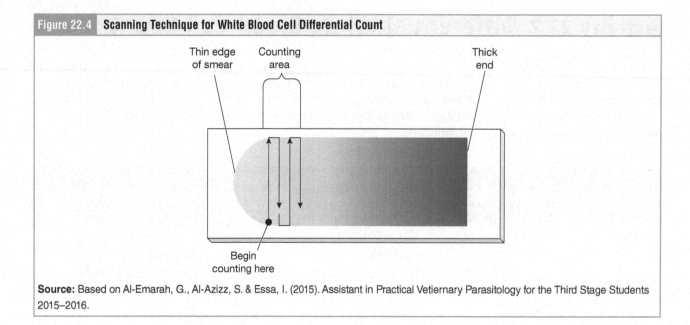

Source: Based on Al-Emarah, G., Al-Azizz, S. & Essa, I. (2015). Assistant in Practical Vetiernary Parasitology for the Third Stage Students 2015–2016.

Materials
- Compound microscope
- Hand clicker counter (also called a hand tally counter)
- Prepared blood smear slide with a normal blood sample
- Prepared blood smear slide with an abnormal blood sample, if available

Instructions
1. Place a prepared blood slide of a normal blood sample on the microscope and examine on scanning power (4×) objective. Use the coarse focus knob to focus on the blood cells present, which will appear very small at this magnification.

2. Switch to the lower power (10×) objective and use the fine focus knob to clearly view the cells present in your view, then switch to the high power (40×) objective and use the fine focus knob as necessary to view more detail.

3. Move the slide so that you are viewing one end of the area of the slide containing blood.

4. Scan the slide slowly, in the pattern shown in Figure 22.4.

   a. Keep a running tally of the types of leukocytes encountered in Table 22.3.

   b. If available, use the hand clicker counter to keep the tally of the abundant neutrophils while making tallies of the other four types of cells on the table.

   c. If you encounter a leukocyte that you cannot identify, skip it and move on.

5. Continue until you have found and identified 100 total leukocytes.

6. Since your total number of leukocytes was 100, the percentage of each type is the number of that type that you counted; for example, if you counted 53 neutrophils out of a total of 100 leukocytes, that represents 53% of the total.

7. If your instructor has provided abnormal blood sample smears, repeat steps 1–6 for up to two provided abnormal slides.
   a. Add these data to Table 22.3.
   b. Identify, on the table, which results are abnormal and if they represent values that are higher or lower than expected.
   c. Identify, on the table, the possible pathologies that these results may indicate (refer to Table 22.2).

Table 22.3 Differential White Blood Cell Count Results				
Leukocyte Type	Normal Value as a Percent of Whole	Normal Blood Sample	Abnormal Blood Sample #1 (If Available)	Abnormal Sample #2 (If Available)
Neutrophil	40–60%			
Lymphocyte	20–40%			
Monocyte	2–8%			
Eosinophil	1–4%			
Basophil	0.5–1%			
Possible pathologies indicated by these results:				

(page intentionally left blank)

Activity 22.3 **Hemoglobin Content and Hematocrit**

This activity targets LOs 22.6 and 22.7.

Using Blood Samples in the Laboratory

Because of the possible risks associated with working with blood samples in the laboratory, some classrooms will not be using samples of your own blood in today's laboratory activities. If alternative sources of blood samples are provided, they will either be virus-free sterile blood samples or simulated blood. No matter the source of the blood, you must follow universal precautions in handling this body fluid. Follow all precautions and instructions given by your lab instructor. Wear the personal protective equipment (PPE) provided for the duration of the lab.

Erythrocytes are 97% full of hemoglobin protein, which carries respiratory gasses, predominantly oxygen, through the blood to the tissues of the body. While the percentage of the blood that is comprised of erythrocytes varies from individual to individual, the normal range is considered to be between 37% and 54% of the total volume of the blood. Since the amount of hemoglobin contained in those erythrocytes can vary, it is also useful to describe the normal range of the hemoglobin itself, which is 12–18g/100 mL of blood.

Changes in the levels of erythrocytes or the amount of hemoglobin that they carry can have significant effects on the body's ability to deliver oxygen to the tissues. Any condition that results in either insufficient numbers of erythrocytes, insufficient amount of hemoglobin, or some other reduced ability to carry oxygen in the blood is called **anemia**.

This activity will allow you to execute a pair of common assessments of a blood sample that can provide insight into the oxygen-carrying capacity of the blood.

Materials
- Centrifuge
- Disposable pipette
- Heparinized microhematocrit capillary tubes
- Personal protective equipment (PPE): safety glasses, gloves, disposable lab apron or protective gown
- Sealing clay tray
- Tallquist paper
- Tallquist scale
- Virus-free sterile blood or simulated blood (if using your own blood, follow all directions provided by your instructor for obtaining the blood sample)

Instructions
1. Put on all of your safety equipment: safety glasses, gloves, and disposable lab apron or protective gown.

2. Determine the hemoglobin content of your blood sample:
   a. Place a drop of blood on a piece of Tallquist paper.
   b. Observe the change in color of the paper quickly and proceed while the paper is damp.
   c. Compare the color of the paper to the Tallquist scale and choose the appropriate estimated hemoglobin value.
   d. Record the estimated hemoglobin value in Table 22.4. Include units.

3. Determine the hematocrit content of your blood sample:
   a. Examine the heparinized microhematocrit capillary tube, one end will be marked red to indicate the inside of the tube is coated with heparin to prevent coagulation.

Table 22.4 Hemoglobin and Hematocrit Values		
Test	**Results**	**Comparison to Expected**
Hemoglobin content		
Hematocrit (PCV)		

Figure 22.5   Sealing a Microhematocrit Tube in Sealing Clay

b. Touch the non-marked end to the blood and hold the tube at a 45 degree angle, immersed in the blood; blood should flow up the tube.

c. When blood reaches about three-fourths of the way up the tube, place a finger over the top end of the tube so blood can't drain; lift and place the blood end of the tube into the sealing clay to plug it as shown in Figure 22.5.

d. Twist the tube gently to adhere the clay within the end of the tube.

e. Provide your sealed tube to your instructor as directed, who will centrifuge your tube along with the tubes of the rest of the class.

f. Once you receive your centrifuged tube, note the three layers: the erythrocytes are packed at the bottom, a thin layer of white- to tan-colored leukocytes and platelets are next, and the liquid plasma is on top (see the figure in the pre-lab quiz for comparison).

g. Use a microhematocrit reader or a millimeter ruler to measure the length of the whole column and the length of the packed erythrocytes.

h. Calculate the packed cell volume (PCV) as a percent as follows: (length of erythrocyte column divided by the length of the whole column) $\times$ 100.

i. Record this value in Table 22.4.

j. How do your values compare to the expected value provided in the introduction? Record in Table 22.4.

4. Continue to the next lab activity, or if you are finished with all lab activities using blood samples, dispose of the material in the provided hazardous materials disposal locations specified by your instructor.

5. Clean all surfaces in your work area with disinfectant provided by your instructor.

Digging Deeper:
H&H and CBCdiff

A hemoglobin and hematocrit, often called an "H&H" for short in medical settings, is a common blood analysis order that reports back on the individual's hemoglobin measurement and the volume of red blood cells in the sample as a percentage.

A complete blood count with differential, often called a "CBCdiff" for short in medical settings, is another common blood analysis order by doctors who suspect an infection or particular disease in their patient. This blood test includes a differential white blood cells count plus a red blood cell count and a hematocrit.

Activity 22.4 **ABO and Rh Blood Typing**

This activity targets LO 22.8.

Using Blood Samples in the Laboratory

Because of the possible risks associated with working with blood samples in the laboratory, some classrooms will not be using samples of your own blood in today's laboratory activities. If alternative sources of blood samples are provided, they will either be virus-free sterile blood samples or simulated blood. No matter the source of the blood, you must follow universal precautions in handling this body fluid. Follow all precautions and instructions given by your lab instructor. Wear the personal protective equipment (PPE) provided for the duration of the lab.

Blood typing refers to a method for describing the presence or absence of certain surface markers called **antigens** on a person's erythrocytes. These antigens are important because they can trigger immune reactions if the body perceives them as a foreign substance, such as when a blood transfusion is given containing the wrong antigen type. While more than 50 different antigens have been identified on the surface of erythrocytes, only three are strongly antigenic, meaning that they are the most likely to provoke an immune response. An immune response results in the production of **antibodies**, which are small proteins that adhere to the antigen that prompted the response. Figure 22.6 shows antibodies reacting to the presence of surface antigens on a cell.

Figure 22.6   **Antigens and Antibodies**

An antibody reacts to an antigen that it recognizes by attaching to the antigen.

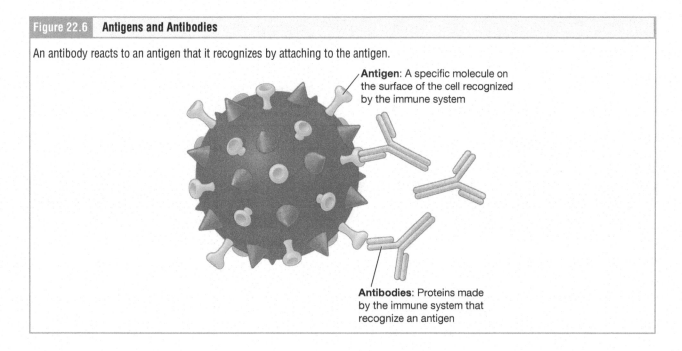

Antigen: A specific molecule on the surface of the cell recognized by the immune system

Antibodies: Proteins made by the immune system that recognize an antigen

---

The first two highly antigenic erythrocyte antigens, A and B, are part of a group collectively known as the ABO blood group. People who have only the A antigens on the surface of their erythrocytes are called "type A" and those with only the B antigens on the surface of their erythrocytes are called "type B." If a person has both A and B antigens on their erythrocytes, they are called "type AB." It is also possible to have neither of these antigens on the surface of one's erythrocytes; such a person is "type O."

The third antigen is often described as antigen D, while this component of blood typing is called the Rh group or Rh factor. If antigen D is present on a person's erythrocytes, they are called Rh+ and if antigen D is not present, they are called Rh−.

These two blood typing groups, ABO and Rh factor, can be combined to fully describe the presence or absence of all three antigens. Hence, a person who is A+ has both A antigens and D antigens on the surface of their erythrocytes. A visual representation of the antigens on the erythrocytes of these blood types is shown in Figure 22.7.

Figure 22.7 | **Erythrocyte Antigens**

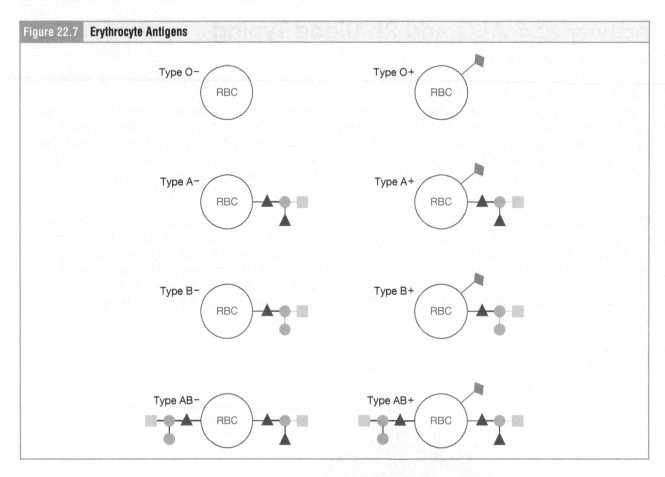

Clarify your understanding of the information above by completing Table 22.5 (the first one has been completed for you):

A Person with This Blood Type:	Has These Antigens on the Surface of Their Erythrocytes:
Table 22.5 Antigens Present on the Erythrocytes of Each Blood Type	
A+	A antigens and D (Rh) antigens
A−	
B+	
B−	
AB+	
AB−	
O+	
O−	

In order to determine which blood type a person is, we allow their blood to come into contact with the antibodies that react against the three antigens. Anti-A antibodies react against A antigens, anti-B antibodies react against B antigens, and anti-D antibodies react against D antigens. The reaction results in **agglutination**, which occurs as antibodies attach to the antigens on many erythrocytes at once, forming clumps of cells. Figure 22.8 shows how agglutination of a blood sample will look.

Figure 22.8   Agglutination of a Blood Sample

The purpose of this exercise is to see how the use of only three different antibody solutions allows one to distinguish between eight different blood types. Use the information above to make your predictions. Circle the correct answer or answers to see how it works:

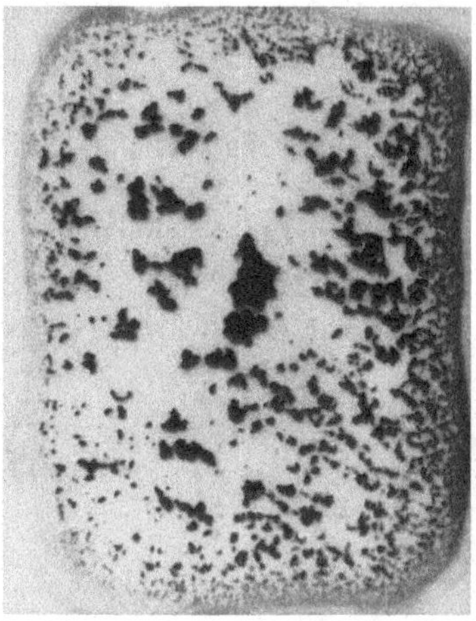

- A person with blood type A (either A+ or A−) will agglutinate when exposed to:

  Anti-A   Anti-B   Anti-D (anti-Rh)   none

- A person with blood type B (either B+ or B−) will agglutinate when exposed to:

  Anti-A   Anti-B   Anti-D (anti-Rh)   none

- A person with blood type AB (either AB+ or AB−) will agglutinate when exposed to:

  Anti-A   Anti-B   Anti-D (anti-Rh)   none

- A person with blood type O (either O+ or O−) will agglutinate when exposed to:

  Anti-A   Anti-B   Anti-D (anti-Rh)   none

- A person with a positive blood type (A+, B+, AB+, or O+) will agglutinate when exposed to:

  Anti-A   Anti-B   Anti-D (anti-Rh)   none

- A person with a negative blood type (A−, B−, AB−, or O−) will agglutinate when exposed to:

  Anti-A   Anti-B   Anti-D (anti-Rh)   none

Materials
- Anti-A serum
- Anti-B serum
- Anti-D (anti-Rh) serum
- Clean glass slides OR a blood typing test card
- Personal protective equipment (PPE): safety glasses, gloves, disposable lab apron or protective gown
- Toothpicks
- Virus-free sterile blood or simulated blood (if using your own blood, follow all directions provided by your instructor for obtaining the blood sample)

Instructions
1. Put on all of your safety equipment: safety glasses, gloves, and disposable lab apron or protective gown.

2. Obtain three clean glass slides or one blood typing test card.

3. If using glass slides, mark one slide as anti-A, one as anti-B, and one as anti-D.

4. Obtain three toothpicks; keep these toothpicks separate as you use them to avoid cross-contamination.

5. Place a drop of blood on each slide or in each circle on the test card.

6. Place the appropriate serum on each drop of blood.

7. Using separate toothpicks for each drop of blood, mix each sample and watch for agglutination. See Figure 22.8 for an example of agglutination.

8. Record your results below and determine which blood type it is.

9. Dispose of the material in the provided hazardous materials disposal locations specified by your instructor.

10. Clean all surfaces in your work area with disinfectant provided by your instructor.

Results of ABO Blood Typing

Did the blood sample agglutinate when exposed to:

Anti-A serum _____

Anti-B serum _____

Anti-D (Rh) serum _____

The blood type of the sample is _____

Activity 22.5 **ABO and Rh Blood Type Matching**

This activity targets LO 22.9.

The possible presence of the ABO and Rh antigens on the surface of erythrocytes makes some forms of blood incompatible with others. This possible incompatibility is why we identify an individual's blood type (both the ABO and Rh groups) in order to determine which blood types they can receive a blood transfusion from and which blood types they can donate blood to.

Antigens, the molecules that are identified in blood typing, are the molecules on the surface of that individual's erythrocytes. Thus, a person's blood type states what is marking their RBCs. But, in addition to what that individual's RBCs look like, their blood type implies what antibodies they have to non-self blood antigens. Antibodies to blood antigens will cause agglutination that clumps the erythrocytes together in large masses. The clumps of erythrocytes block small blood vessels (Figure 22.9) throughout the body, depriving tissues of oxygen and nutrients. As the erythrocyte clumps are degraded their hemoglobin is released into the bloodstream and travels to the kidneys for filtration. If the agglutination is significant, the load of hemoglobin will overwhelm the kidney's filtration system and can lead to kidney failure.

| Figure 22.9 | **Agglutination** |

(A) Antibodies against antigens expressed on red blood cells will bind to red blood cells at each of their two binding sites. (B) This can cause clumping of the red blood cells, which can clog small vessels.

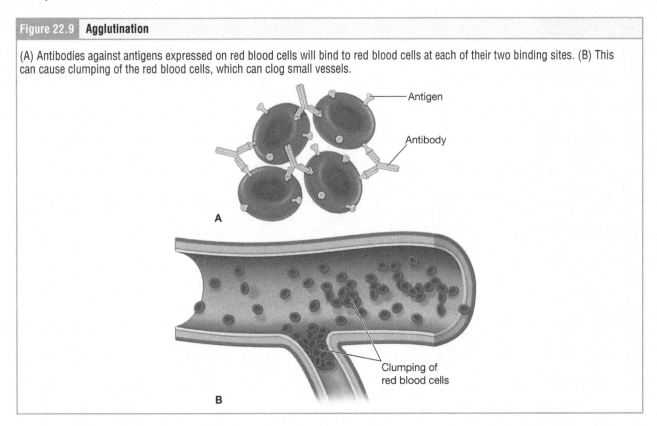

Antibodies to A and B antigens appear in an infant's blood plasma shortly after birth even without prior exposure to those antigens (it should be noted that this contrasts with normal immune function in which the body must be exposed to an antigen before developing antibodies). But these antibodies only appear if that baby's erythrocytes are not marked with those antigens. In other words, the infant develops antibodies to any non-self antigens of the ABO group. Remember, antibodies are proteins that can attach to a specific antigen. The nomenclature of these antibodies is to describe the antigen to which they attach, so that the antibodies that attach to A antigens on the surface of an individual's erythrocytes would be called "anti-A antibodies."

In the previous exercise you determined which antigens are present on the surface of erythrocytes for each of the blood types. Transfer that information to column 1 of Table 22.6. Then predict the ABO antibodies that will be present in the blood plasma for each of the blood groups listed in Table 22.6 and write that information in column 2 (you will complete the other columns as you proceed through this introduction). If no antibodies will be present in the plasma, write "none." The first blood type has been done for you as an example.

Table 22.6 Prediction of Antigens, Antibodies, and Donation Possibilities for the Various Blood Types

A Person with This Blood Type:	1 Has These Antigens on the Surface of Their Erythrocytes	2 Has These ABO Antibodies in Their Blood Plasma	3 May Have These Rh Antibodies in Their Blood Plasma (If There Has Been Prior Exposure)	4 Can Safely Donate Blood (as Packed RBCs) to These Blood Types	5 Can Safely Receive Blood (as Packed RBCs) from These Blood Types
A+	A antigens and D (Rh) antigens	Anti-B antibodies	No	A+ AB+	A+ A− O+ O−
A−					
B+					
B−					
AB+					
AB−					
O+					
O−					

In contrast to the ABO blood group antibodies that are preformed, antibodies to the Rh antigen (anti-D antibodies) are only present in the plasma of a Rh-negative blood type individual if they have been exposed to the D antigen on the erythrocytes of Rh positive blood at some point. This process is called sensitization and can occur following mixing of one person's blood with the open wound of another, a mismatched transfusion (extremely rare), or the birth of a Rh+ baby to a Rh– mother. Use column 3 to predict the blood types that will have anti-D antibodies if sensitization has occurred. If no antibodies will be present, write "none."

When donated blood is used in a clinical setting in order to restore blood volume or to replace lost blood, it can be typical for only the erythrocytes of the donated blood to be transfused rather than whole blood. This is possible because after blood is donated it is centrifuged to separate the erythrocytes, plasma, and platelets for different donation purposes. For the purposes of this exercise, you can consider that the blood transfusions that are discussed are transfusions of erythrocytes only, which is often called a transfusion of packed RBCs. Because you have already assembled the information in columns 2 and 3, it is now possible to predict which blood types can donate packed RBCs to which other blood types. Complete columns 4 and 5 with these predictions.

Materials
- Completed Table 22.6
- Two envelopes of blood type cards per lab group (each envelope contains eight cards, one of each blood type listed in Table 22.6)

Instructions
1. Examine your completed Table 22.6.
   a. Which of the eight blood types can *receive* blood from all of the blood types? This blood type is called the universal recipient. Circle the universal recipient in column 1 of the table and write "universal recipient" next to it.
   b. Which of the eight blood types can *donate* blood to all of the blood types? This blood type is called the universal donor. Circle the universal donor in column 1 of the table and write "universal donor" next to it.

2. Before beginning with the blood transfusion game below, consider why it is important to note that these simulated blood donations are in the form of packed RBCs only. What is found in the blood plasma of most individuals (see columns 2 and 3 of Table 22.6)? If we had to consider the plasma donation as well, would the universal recipient and donor you noted above still apply? Explain why.

3. Work with your lab partner to play the blood transfusion game.
   a. Choose your roles: one partner is the blood donor and one partner is the recipient.
   b. Each lab partner draws a random blood type card from their envelope.
   c. Write in the blood types in Table 22.7.
   d. Work with your lab partner to determine if this transfusion will result in a match or a mismatch? Fill in the table.
   e. If the transfusion is a mismatch, explain specifically what happened in the recipient and why in the designated column (see the example provided).
   f. Return the cards to the envelope and repeat with a new pair of cards. If you draw the same pairing as you have already made (in the same order, since a swapped donor and recipient blood type are not the same), return the cards and draw again.
   g. Make matches (and mismatches!) until you have completed all of the rows in the table.

Table 22.7 ABO and Rh Blood Transfusion Game

Round	Donor Blood Type	Recipient Blood Type	Match or Mismatch?	If a Mismatch, What Happened and Why?
1	A+	B−	Mismatch	The anti-A antibodies in the B− recipient's blood plasma will attach to the A antigens on the surface of the donated RBCs and cause agglutination. If the recipient has been sensitized to D antigens, the anti-D antibodies in the recipient's blood plasma will attach to the D antigens on the surface of the donated RBCs and cause further agglutination
2				
3				
4				
5				
6				
7				
8				

Activity 22.6 Case Study: Fetal Hemolytic Newborn Disease

This activity targets LOs 22.8 and 22.10.

Case Study

Pari is 5 months pregnant with her first pregnancy and is visiting her midwife for a regular check-up. At the visit her midwife tells her she will need to schedule a RhoGAM shot because Pari is a Rh– blood type. Pari asks why she needs the shot, and her midwife explains that the shot will prevent Pari's body from producing antibodies against her developing fetus in case the fetus is a Rh+ blood type.

1. What specific type of antibodies is the RhoGAM shot preventing?
   a. Anti-A antibodies
   b. Anti-B antibodies
   c. Anti-D antibodies

2. If the antibodies did develop in Pari's body, where would they be?
   a. On the surface of Pari's erythrocytes
   b. In Pari's blood plasma

Pari is nervous about the shot, so she goes home to research some information. She reads that mothers with Rh– blood types don't have any Rh antigens, also called D antigens, marking the surface of their red blood cells. Because of this, if the mother is exposed to a blood type that is Rh+, which has D antigens on the RBCs, their own immune system becomes sensitized to D antigens as it identifies the D antigens as foreign and begins making anti-D antibodies. For Rh– women, this sensitization often comes during the birth of their first Rh+ baby when the Rh+ blood of the newborn initially mixes with the mother's during birth. Since this is Pari's first pregnancy, at her next appointment she asks her midwife if she really needs the RhoGAM shot. The midwife explains that it is also possible for a Rh– individual to be exposed to Rh+ blood in other circumstances, so all Rh– pregnancies warrant the protection of the RhoGAM shot.

Pari is also wondering about other blood antibodies she might have that could harm the fetus. Pari is blood type O–.

3. What ABO blood group antigens are on Pari's erythrocytes? Choose all correct answers.
   a. A antigens
   b. B antigens
   c. O antigens
   d. No antigens

4. What ABO blood group antibodies will be in Pari's blood plasma? Choose all correct answers.
   a. A antibodies
   b. B antibodies
   c. O antibodies
   d. No antibodies

Pari asks the midwife if there is concern about these other antibodies. The midwife shares that the antibodies to the ABO blood group antigens are much larger than the antibodies to Rh factor. So while anti-D antibodies can cross the placenta to enter the fetus's blood, the anti-A and anti-B antibodies cannot cross and therefore do no damage to the fetal blood.

5. What kind of damage will anti-D antibodies do to the fetal blood.
   a. They will cause coagulation.
   b. They will cause agglutination.

Digging Deeper:
Developing the RhoGAM Shot

Although the Rh antigen was discovered and explained in the 1930s, development of a method to prevent hemolytic newborn disease took several decades. The Rho(D) immune globulin (also called the RhIG or RhoGAM shot) was developed in the 1950s and 1960s. The first clinical use of the shot was administered on May 29, 1968, to a woman in New Jersey. Prior to this time, Rh– mothers who were pregnant with a Rh+ fetus often lost the fetus or newborn to hemolytic newborn disease. For a sense of how significant this complication can be, consider that in countries without protocols to administer Rh immune globulins as many as 14% of affected fetuses are stillborn and 50% of affected live births have severe brain deficiencies that may lead to death.

Name: _____ Section: _____

1. Blood can be centrifuged to separate it into individual components. What will be absent from the packed RBC portion of centrifuged blood?

   a. Oxygen
   b. Erythrocytes
   c. Platelets

2. Which are the smallest of the formed elements?

   a. Platelets
   b. Lymphocytes
   c. Erythrocytes
   d. Proteins

3. Examine the following blood slide. The marked cell is a:

   a. Neutrophil
   b. Monocyte
   c. Eosinophil

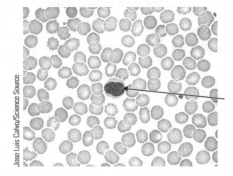

4. What is the main function of the most prevalent leukocyte in the bloodstream?

   a. To engulf and destroy bacteria
   b. To attack virus-containing cells
   c. To attack parasites such as worms
   d. To stimulate inflammation

5. You perform a differential white blood cell count and find the following results. Which type of infection is the most likely?

   a. Fungal infection
   b. Bacterial infection
   c. Viral infection

Neutrophils	40%
Lymphocytes	50%
Monocytes	7%
Eosinophils	2.5%
Basophils	0.5%

6. You examine a hematocrit that shows a PCV of 60%. What interpretation can be made of this result?

   a. This person may be dehydrated.
   b. This person's hematocrit is normal.
   c. This person may be over-hydrated.

7. What ABO blood group antigens are on the surface of a person's erythrocytes who is blood type AB–?

   a. AB– antigens
   b. AB antigens
   c. A antigens and B antigens

8. Your friend tells you that they are type O negative blood. How can you explain to your friend what the surface of their red blood cells looks like?

   a. Their red blood cells have two different antigenic markers: O antigens and – antigens.
   b. Their red blood cells have a single antigenic marker: O– antigens.
   c. Their red blood cells have no antigenic markers.

9. Raul has arrived at the emergency room with a wound and has lost significant blood. There is no time to find out Raul's blood type, but the attending physician requests a transfusion of blood (packed RBCs). Raul is given a transfusion of type O+ blood. The wound is treated. Raul is also blood-typed and found to be blood type AB+. Later while Raul is waiting to be discharged, what can you expect to happen?

   a. Nothing, Raul has received blood from the universal donor blood type.
   b. Nothing, although this is a mismatch, it will not cause any problems in Raul, who is a universal recipient.
   c. Raul's blood will coagulate and he might have serious complications.
   d. Raul's blood will agglutinate and he might have serious complications.

10. Isla and Marta are friends who are 12 years old. Isla is blood type A– and Marta is blood type B+. While out playing, both children fall and scrape their hands. While they walk back home to get help, they hold hands and some of each of their blood enters the wounds of the other. Which of the following best explains what antibodies each child will develop as a result of this?

   a. Isla will develop anti-B and anti-D antibodies; Marta will develop anti-A antibodies.
   b. Isla already has anti-B antibodies but will now develop anti-D antibodies; Marta already has anti-A antibodies and will not develop any new antibodies.
   c. Both children already have both anti-A and anti-B antibodies, they both will now develop anti-D antibodies.

Lab 23 Heart

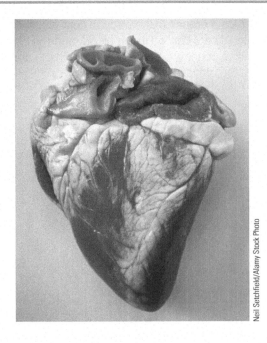

Neil Setchfield/Alamy Stock Photo

Learning Objectives: After completing this lab, you should be able to:

23.1 Describe the position of the heart in the thoracic cavity.

23.2 Identify and describe the location, structure, and function of the fibrous pericardium, parietal and visceral layers of the serous pericardium, serous fluid, and the pericardial cavity.

23.3 Explain the structural and functional differences between atria and ventricles.

23.4 On the external surface of the heart identify the four chambers, the coronary (atrioventricular) sulcus, anterior interventricular sulcus, posterior interventricular sulcus, apex, and base.

23.5 Identify and describe the structure and function of the primary internal structures of the heart, including chambers, septa, valves, papillary muscles, chordae tendineae, fibrous skeleton, and venous and arterial openings.

23.6 Describe the structure and functions of each layer of the heart wall (i.e., epicardium, myocardium, endocardium).

23.7 Describe the blood flow to and from the heart wall, including the location of the openings for the left and right coronary arteries, left coronary artery and its major branches, right coronary artery and its major branches, cardiac veins, and coronary sinus.

23.8 Trace the path of blood through the right and left sides of the heart, including its passage through the heart valves, and indicate whether the blood is oxygen-rich or oxygen-poor.

23.9 Describe the changes in major fetal cardiovascular structures (i.e., umbilical vessels, ductus venosus, ductus arteriosus, foramen ovale) that typically occur beginning at birth, and the ultimate postnatal remnants (fates) of these structures.

The Human Anatomy and Physiology Society includes more than 1,700 educators who work together to promote excellence in the teaching of this subject area. The HAPS A&P Learning Outcomes measure student mastery of the content typically covered in a two-semester Human A&P curriculum at the undergraduate level. The full Learning Outcomes are available at https://www.hapsweb.org.

Introduction

The cardiovascular system functions, like any transportation system, to distribute materials around a vast space. The materials in this case are nutrients like amino acids, glucose, and fats, gasses such as oxygen and carbon dioxide, and hormones that allow disparate organs in the body to communicate and coordinate function. In order to achieve this distribution, these materials are carried in the blood, and the blood flows through a series of connected vessels. Like the road system in a country, nearly every place in the body is accessible through the system of blood vessels. The heart generates the pressure to drive blood flow. In this lab, you will examine the gross anatomy of the heart and investigate the relationship between structure and function.

Lab 23 Pre-Lab Quiz

This quiz will strengthen your background knowledge in preparation for this lab. For help answering the questions, use your resources to deepen your understanding. The best resource for help on the first five questions is your text, and the best resource for help answering the last five questions is to read the introduction section of each lab activity.

1. Anatomy of the heart. Assign the following labels to the correct locations on the diagram:

   aorta

   inferior vena cava

   left atrium

   right ventricle

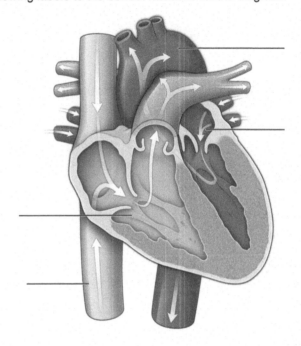

2. Match the root next to its definition:

Root	Meaning
	a wrinkle
	an entry way
	crown
	heart
	pertaining to the lungs
	small bump

List of Roots

atrium

cardia

corona

papilla

pulmo

sulcus

3. Which of these structures is the pacemaker of the heart?

   a. Purkinje fibers

   b. SA node

   c. AV node

   d. Bundle branches

4. What component of the conduction system is located within the interventricular septum?

a. Purkinje fibers
b. SA node
c. AV node
d. Bundle branches

5. The right side of the heart fills with _____ blood.

a. oxygenated
b. deoxygenated
c. pulmonary

6. Which circuit is highlighted in color in the figure to the right?

a. Pulmonary circuit
b. Systemic circuit

7. Which of the following vessels carries oxygen-rich blood to the heart muscle tissue?

a. Thoracic aorta
b. Circumflex artery
c. Pulmonary artery

8. In a serous membrane there are _____ layers of tissue and the _____ is the deeper/deepest layer.

a. two, parietal
b. two, visceral
c. three, fibrous
d. three, visceral

9. The _____ is the layer of simple squamous epithelium that lines the heart.

a. epicardium
b. pericardium
c. myocardium
d. endocardium

10. In this lab you will examine a sheep heart in _____ -section

a. transverse
b. frontal
c. oblique

Activity 23.1 Position of the Heart and Pericardium

This activity targets LOs 23.1 and 23.2.

The heart is located within the thoracic cavity, in a space between the lungs known as the **mediastinum** (Figure 23.1). Within the mediastinum the heart is held within a tough case known as the **pericardium**, which envelopes the heart. The pericardium consists of two distinct layers: the sturdy outer **fibrous pericardium** and the more delicate **serous pericardium**. The serous pericardium has two layers: the outer **parietal pericardium** and the inner **visceral pericardium**. The serous pericardia, like other serous membranes, are separated by a fluid-filled cavity, called the **pericardial cavity** (Figure 23.2).

Figure 23.1	Position of the Heart in the Thorax

The heart is located within the thoracic cavity within the mediastinum. It is wide at the top and tapered inferiorly.

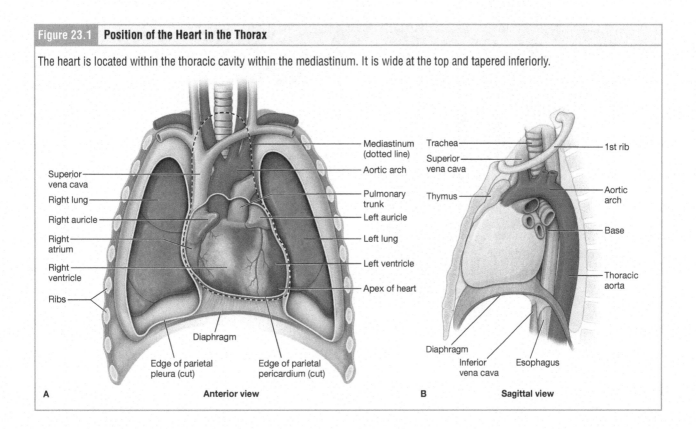

A   **Anterior view**   B   **Sagittal view**

Figure 23.2 **Frontal Section of the Heart Within the Pericardium**

The pericardium is composed of three layers: the fibrous pericardium and the two layers of the serous pericardium, (the parietal and visceral layers, inset).

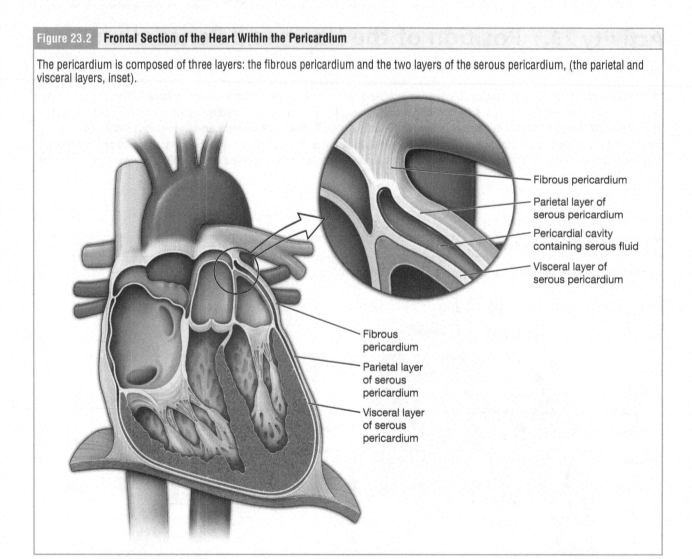

- Fibrous pericardium
- Parietal layer of serous pericardium
- Pericardial cavity containing serous fluid
- Visceral layer of serous pericardium

- Fibrous pericardium
- Parietal layer of serous pericardium
- Visceral layer of serous pericardium

Materials
- Heart model
- Torso model

Instructions
1. Examine the available models in your laboratory (and Figure 23.2, if necessary) and locate the following structures:

   - Mediastinum
   - Fibrous pericardium
   - Serous pericardium visceral layer
   - Serous pericardium parietal layer

Learning Connection

Broken Process

Imagine an individual with inflammation of their pericardium. What do you think the effects on the heart, if any, would be?

Activity 23.2 Heart Dissection

This activity targets LOs 23.2, 23.3, 23.4, 23.5, 23.6, and 23.7.

In this lab you will dissect the heart of a sheep. Sheep are good organisms for our heart study in A&P because their hearts are similar in both size and structure to human hearts. Both sheep and human hearts are divided by two walls into four distinct chambers. Each wall is called a **septum**. The top two chambers are the **right atrium** and **left atrium**. The bottom two chambers are the **right ventricle** and **left ventricle**. Right and left here describes the heart in anatomical position.

This activity is divided into two parts. First, we examine the superficial structures of the heart. Second, we look at the heart's internal structures.

Superficial Structures of the Heart

Each atrium has an **auricle**, a superficial, pouch-like extension that lends the atrium extra filling capacity (Figure 23.3). Also prominent on the surface of the heart is a series of grooves, each of which is known as a **sulcus**. The coronary sulcus is located between the atria and the ventricles. Located between the left and right ventricles are two additional sulci: the **anterior interventricular sulcus** on the anterior surface of the heart and the **posterior interventricular sulcus** on the posterior surface (Figure 23.4). Blood vessels that transport blood to and from the wall of the heart, collectively called **coronary vessels**, are nestled within these sulci.

The **left coronary artery** distributes blood to the left side of the heart and the interventricular septum and is nestled within the coronary sulcus. The left coronary artery splits into two branches: the circumflex artery and the anterior interventricular artery. The **circumflex artery** supplies the left atrium and posterior walls of the left ventricle. The **anterior interventricular artery** is nestled within the anterior interventricular sulcus and supplies the anterior-inferior heart wall and the interventricular septum. The **right coronary artery** is nestled within the coronary sulcus and distributes blood to the right atrium, portions of both ventricles, and the heart's conduction system. The right coronary artery gives rise to **marginal arteries** (the number of marginal arteries can vary from one person to another). The marginal arteries supply blood to the superficial right ventricle. On the posterior surface of the heart, the right coronary artery gives rise to the **posterior interventricular artery**, supplying both ventricle walls with oxygenated blood.

Figure 23.3	Anterior View of the Heart

On the anterior view, the following structures are prominent: the large ventricles, the smaller auricles (which form the outer walls of the atria), the aorta, which emerges from the left ventricle, and the pulmonary trunk, which emerges from the right ventricle.

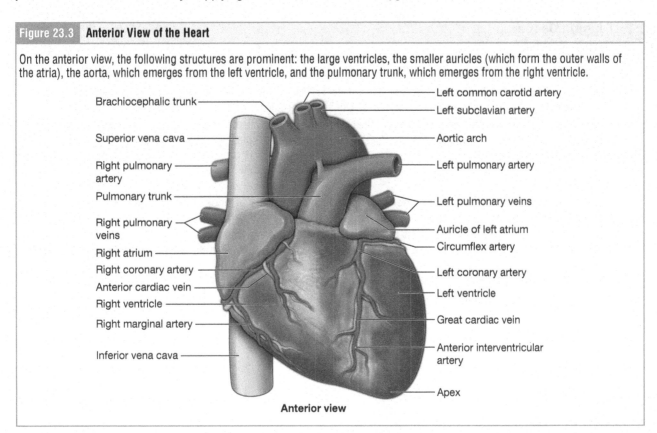

Brachiocephalic trunk

Superior vena cava

Right pulmonary artery

Pulmonary trunk

Right pulmonary veins

Right atrium

Right coronary artery

Anterior cardiac vein

Right ventricle

Right marginal artery

Inferior vena cava

Left common carotid artery

Left subclavian artery

Aortic arch

Left pulmonary artery

Left pulmonary veins

Auricle of left atrium

Circumflex artery

Left coronary artery

Left ventricle

Great cardiac vein

Anterior interventricular artery

Apex

Anterior view

Figure 23.4 | Posterior View of the Heart

In the posterior view, the prominent structures include the pulmonary veins, which enter the left atrium, and the superior and inferior venae cavae, which enter the right atrium. The coronary blood vessels are visible in both views.

Aorta

Left pulmonary artery

Left pulmonary veins

Auricle of left atrium

Left atrium

Circumflex branch
of left coronary artery

Great cardiac vein

Posterior vein of
left ventricle

Left ventricle

Superior vena cava

Right pulmonary artery

Right pulmonary veins

Right atrium

Inferior vena cava

Base of heart

Coronary sinus

Right coronary artery

Posterior interventricular
artery

Middle cardiac vein

Right ventricle

Posterior view

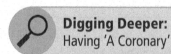

Digging Deeper:
Having 'A Coronary'

Myocardial infarctions are also called heart attacks or coronaries. These are small or large areas of dead cardiac muscle tissue. The death of the tissue occurs because of an interruption in blood supply, usually because of a blockage in a coronary artery. Myocardial infarctions can happen to anyone due to a blood clot becoming lodged in one of these vessels, but are more likely to happen in individuals who have regions of narrowing of the blood vessels due to atherosclerotic plaques. Atherosclerotic plaques are build-ups of cholesterol and inflammatory tissue. Risk factors for developing these plaques are high blood pressure, genetics, and diets high in saturated fats.

Coronary veins drain the heart wall and generally parallel the coronary arteries in location (see Figures 23.3. and 23.4). Companion vessels of the coronary circulation are listed in Table 23.1. The **great cardiac vein** can be seen on the anterior surface of the heart nestled within the interventricular sulcus along with the anterior interventricular artery. As it ascends, it traces the coronary sulcus and merges with the **coronary sinus** posteriorly. The **posterior cardiac vein**, **middle cardiac vein** and **small cardiac vein** all drain into the great cardiac vein.

The two ventricles narrow inferiorly, forming a pointed end of the heart known as the **apex**. On the posterior surface of the heart the pulmonary veins converge to make a broad flat superior surface known as the **base**.

Materials
• Dissecting tray
• Gloves
• Lab coat, if instructed
• Preserved sheep heart
• Probe
• Scalpel or Scissors

Table 23.1 Coronary Vessels		
Artery	**Vein**	**Sulcus**
Left coronary artery		Coronary sulcus
Circumflex artery	Posterior cardiac vein	
Anterior interventricular artery	Great cardiac vein	Anterior interventricular sulcus
Right coronary artery	Small cardiac vein	Coronary sulcus
Marginal arteries	Anterior cardiac veins	
Posterior interventricular artery	Middle cardiac vein	Posterior interventricular sulcus
	Coronary sinus	Coronary sulcus

Instructions for the Superficial Structures of the Heart

1. Obtain the preserved sheep heart and line up your instruments next to your tray. Put on gloves.

2. Typically, your sheep heart will still have the pericardium intact. Is that true of the heart in front of you? If so, touch the fibrous pericardium and observe its texture. How would you describe it? Record your observation in Table 23.2 using the describing words in the table or your own words.

3. Cut open the fibrous pericardium (CAUTION: cut away from your hand), using your scalpel or scissors (cut gently if using the scalpel). On the inside of the fibrous pericardium you can find the parietal serous pericardium. How would you describe it? Record your observation in Table 23.2 using the describing words in the table or your own words.

4. Now you can see the epicardium and the external surface of the heart. You'll note adipose tissue that is tightly adhered to the surface of the heart. If instructed, gently scrape or pull away what you can of the adipose tissue.

5. Identify the base, the apex, and the auricles. How would you describe the auricles? The ventricular walls? Record your observation in Table 23.2 using the describing words in the table or your own words.

6. Identify the sulci, arteries, and veins listed in Table 23.1.

7. Hold the heart in anatomical position, how can you tell the anterior from the posterior? Describe the landmarks you use here:

Table 23.2 Describing the External Features of the Heart	
Describing words: tough, silky, slippery, thick, thin, bumpy, glistening, wrinkled, solid, papery	
Fibrous pericardium	Serous pericardium
Ventricular walls	Auricles

Table 23.3 Major Vessels of the Heart	
Vessel	**How to Identify**
Pulmonary veins	Between two and four vessels that converge to form the base, viewable posteriorly
Aorta	Extends superiorly in the midline and curves into the aortic arch (may or may not be present in your specimen). Posterior to the pulmonary trunk. Walls are thicker than any other vessel
Pulmonary trunk	Extends superiorly in the midline, anterior to the aorta, splits into right and left pulmonary arteries (may or may not be present in your specimen)
Superior vena cava	Located superior to and entering the right atrium, walls are thinner than pulmonary trunk or aorta
Inferior vena cava	Located posteriorly and entering the right atrium, walls are thinner than pulmonary trunk or aorta

8. Examine the vessels associated with the superior heart. Identify the vessels in Table 23.3, using Figure 23.5 as your guide. Label the diagram below with the structures you identified.

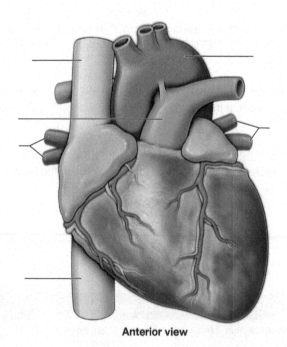

Anterior view

9. Make a longitudinal cut into the aorta until you can see the aortic semilunar valve (note that this valve is discussed in more detail in the next section).

10. Identify the two openings to the coronary arteries just superior to the aortic valve. See if you can insert a probe into either of the openings to trace the connection to the coronary artery.

11. Turning the heart over to view it posteriorly, place your probe into the vessels on this side, and trace the path of blood flow into the heart from the pulmonary veins and the inferior and superior venae cavae.

Figure 23.5 The Great Vessels of the Heart

The great vessels of the heart leave from the heart's superior side.

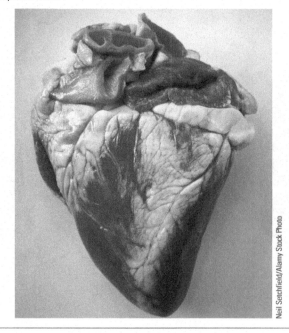

Neil Setchfield/Alamy Stock Photo

The Internal Structures of the Heart

The heart wall is composed of three layers. From superficial to deep these are the **epicardium**, which you viewed in the preceding section, the **myocardium**, and the **endocardium**.

Located between the two atria is the **interatrial septum**. In an adult heart, the interatrial septum bears an oval-shaped depression known as the **fossa ovalis**, a remnant of an opening in the fetal heart known as the **foramen ovale**. Between the two ventricles is the **interventricular septum**. Between the atria and the ventricles is a fibrous divider, the **fibrous skeleton of the heart**, which is not composed of muscle and does not conduct electrical signals. Embedded into this fibrous layer are two valves that serve as doorways allowing blood to pass from the atria to the ventricles. Each of these is an atrioventricular valve. Between the right atrium and the right ventricle is the **right atrioventricular valve**. Between the left atrium and the left ventricle is the **left atrioventricular valve**.

The interior walls of the right atrium are smooth with a few exceptions. The fossa ovalis is a notable depression on the interatrial wall. The anterior atrial walls are lined with rough ridges of muscle called **pectinate muscles** (Figure 23.6). The right atrium is continuous with the right auricle, which also has pectinate muscles. If you slip your gloved finger into the auricle you'll find a cavernous space. The auricles provide extended volume capacity for the atrium during times when a large amount of blood is returning through the veins.

Blood reaches the right ventricle by passing through the right atrioventricular valve. The three flaps of this valve are anchored to the walls of the ventricle by fibrous strings that prevent the flaps from flipping up into the atrium during contraction. These strings, the **chordae tendineae**, are each attached to a **papillary muscle** that extends from the floor of the ventricle (Figure 23.7). The walls of the ventricle are rough, with ridges of cardiac muscle called **trabeculae carneae**. When the ventricles contract, blood within the right ventricle is pushed up and out of the ventricle through the **pulmonary semilunar valve** and into the pulmonary trunk.

Figure 23.6 Frontal Section of the Heart

One feature of the right atrium are its pectinate muscles.

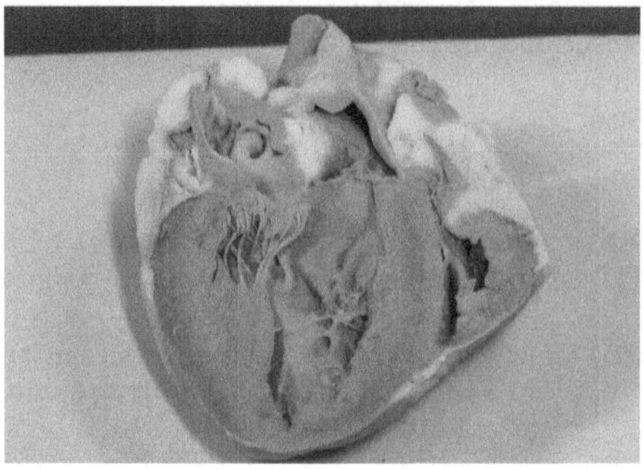

Figure 23.7 | Internal Structures of the Heart

The chordae tendineae attach to papillary muscles, anchoring the flaps of the atrioventricular valves.

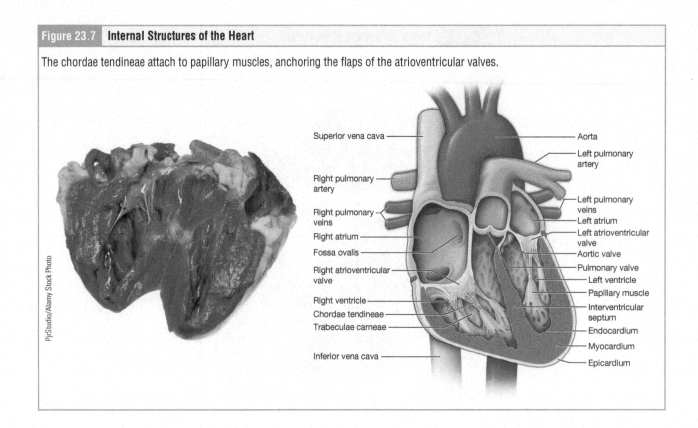

Superior vena cava

Right pulmonary artery

Right pulmonary veins

Right atrium

Fossa ovalis

Right atrioventricular valve

Right ventricle

Chordae tendineae

Trabeculae carneae

Inferior vena cava

Aorta

Left pulmonary artery

Left pulmonary veins

Left atrium

Left atrioventricular valve

Aortic valve

Pulmonary valve

Left ventricle

Papillary muscle

Interventricular septum

Endocardium

Myocardium

Epicardium

PjrStudio/Alamy Stock Photo

The interior walls of the left atrium are completely smooth. The left atrium has fewer and thinner pectinate muscles than are in the right atrium. Blood reaches the left ventricle by passing through the left atrioventricular valve, the flaps of which are also anchored by chordae tendineae to papillary muscles. The walls of the left ventricle also have trabeculae carneae. When the ventricles contract, blood within the left ventricle is pushed up and out of the ventricle through the **aortic semilunar valve** and into the aorta.

 Learning Connection

Chunking

Try making a list of all the internal AND external structures you can name associated with each chamber of the heart.

Materials
- Dissecting tray
- Gloves
- Lab coat, if instructed to wear one
- Preserved sheep heart
- Probe
- Ruler
- Scalpel or Scissors

Instructions for the Internal Structures of the Heart
1. Obtain a preserved heart and line up your instruments next to your tray. Put on gloves.

2. Turn the heart in the dissecting pan so that it rests on its base with its apex pointing toward you. Make a frontal section through the heart cutting all the way to the apex, but not completely separating its anterior and posterior halves (Figure 23.8).

3. Within the right atrium, identify the pectinate muscles. Gently rubbing the interatrial septum between your gloved thumb and forefinger, find the fossa ovalis.

4. Within each ventricle and using Figure 23.9 as a guide, identify the papillary muscles, chordae tendineae, and trabeculae carneae.

5. Using a small ruler, measure the thickness of the wall of the right ventricle and record this number in Table 23.4.

6. Now measure the thickness of the wall of the left ventricle and record this number in Table 23.4. Approximately how much thicker is the left than the right? What is the functional significance of this difference in wall thickness?

7. At the end of your dissection and observations, clean the scalpel as directed by your lab instructor. Dispose of your heart according to the directions of your lab instructor and clean all your instruments and dissecting tray with soapy water.

| Figure 23.8 | How to Make a Frontal Section of the Heart |

Using a scalpel or knife, make a frontal section through the sheep heart (as shown by the dashed line) to examine the internal structures.

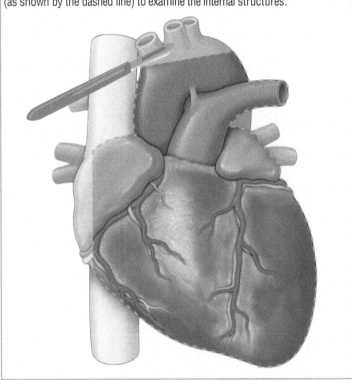

| Figure 23.9 | Internal Structures of the Heart |

The internal structures of the ventricles include papillary muscles, chordae tendineae and trabeculae carne.

Jasmin Merdan/Moment/Getty Images

Table 23.4 Ventricle Wall Thicknesses		
Thickness of the Right Ventricle Wall (cm)	Thickness of the Left Ventricle Wall (cm)	Approximate Difference

(page intentionally left blank)

Activity 23.3 Understanding Blood Flow Through the Heart

This activity targets LO 23.8.

Materials
• Set of index cards with anatomical structures written on them OR blank index cards

Instructions

1. Either collect a pre-made set of cards from your instructor or take blank note cards and write the names of the following structures on them, using one card per structure.

   aorta
   aortic semilunar valve
   left atrioventricular valve
   left atrium
   left ventricle
   pulmonary semilunar valve
   pulmonary trunk
   pulmonary veins
   pulmonary circulation
   right atrioventricular valve
   right atrium
   right ventricle
   systemic circulation
   venae cavae

2. Blood enters and leaves the heart's right and left sides simultaneously, but for the purpose of this exercise place the note cards in order following a single drop of blood from the venae cavae to the aorta.

(page intentionally left blank)

Activity 23.4 Case Study: Congenital Heart Defects

This activity targets LO 23.9.

During fetal development the heart initially develops as a tube, and then folds in on itself twice in the initial months of pregnancy. As the external structure is forming, the separations among the heart chambers and vessels begin to form. Notably, for much of development there is one large vessel, the truncus arteriosus, that carries blood away from the heart. This vessel eventually divides into the pulmonary trunk and aorta.

Congenital heart defects (CHDs) are structural issues that arise when developmental movements and separations do not follow the typical patterns. Some CHDs are caused by the truncus arteriosus not dividing fully into the two adult vessels.

In truncus arteriosus, there is no division of the two arteries, only one artery leaves the ventricles. In patent ductus arteriosus, the two vessels separate, but incompletely, a small duct joins them.

1. Which of the diagrams represents truncus arteriosus?

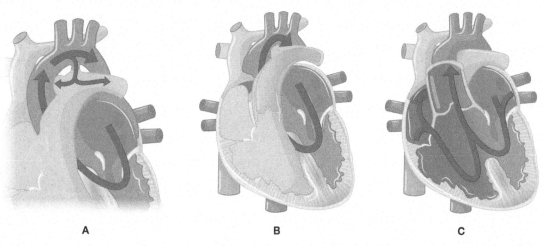

a. A
b. B
c. C

2. Which of the diagrams represents patent ductus arteriosus?

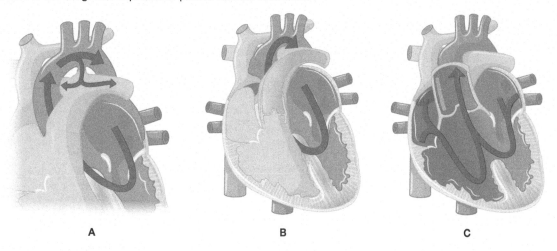

a. A
b. B
c. C

3. In both truncus arteriosus and patent ductus arteriosus, which of the following is true?

    a. A lower hematocrit (red blood cell volume) would be observed.

    b. A lower oxygen saturation in the systemic circulation would be observed.

    c. A lower oxygen saturation in the pulmonary circulation would be observed.

    d. A lower white blood cell volume would be observed.

4. CHDs can often be pinpointed to a moment in developmental time with a set of events that typically either does not occur or occurs in an unusual way. Which developmental moment do you think would come first, the moment that leads to truncus arteriosus or the moment that leads to patent ductus arteriosus? Explain why you think this is so.

Lab 23 Post-Lab Quiz

1. In the diagram, which of the following structures is incorrectly labeled?

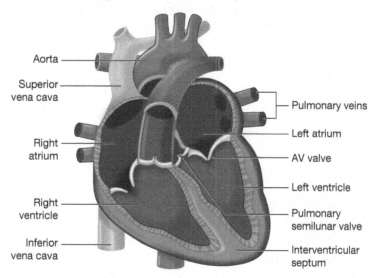

Aorta

Superior vena cava

Right atrium

Right ventricle

Inferior vena cava

Pulmonary veins

Left atrium

AV valve

Left ventricle

Pulmonary semilunar valve

Interventricular septum

2. Which of the images indicates the correct blood flow for the right ventricle?

   a. A
   b. B
   c. Neither of these

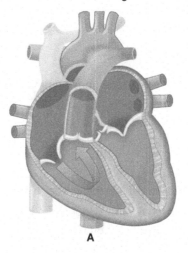

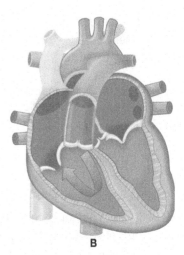

A

B

3. Which valves have flaps that are anchored by chordae tendineae?

   a. Semilunar valves
   b. Atrioventricular valves
   c. Neither of these

4. What is the function of auricles?

   a. To expand the volume of the ventricles
   b. To expand the volume of the atria
   c. To lend extra power to ventricular contraction
   d. To lend extra power to atrial contraction

5. If the fossa ovalis never sealed, which of the following would be true?

   a. A lower hematocrit (red blood cell volume) would be observed.

   b. A lower oxygen saturation in the systemic circulation would be observed.

   c. A lower oxygen saturation in the pulmonary circulation would be observed.

   d. A lower white blood cell volume would be observed.

6. A murmur is diagnosed based on an alteration in heart sounds. Which of the following structure(s) contribute to heart sounds? (select all that apply)

   a. Pectinate muscles

   b. Atrioventricular valves

   c. Semilunar valves

   d. Fossa ovalis

7. Blood in the _____ is headed toward the _____.

   a. posterior interventricular artery; right atrial wall

   b. middle cardiac vein; coronary sinus

   c. middle cardiac vein; left ventricle wall

   d. right marginal artery; right atrial wall

8. Stenosis is a condition in which a valve can stiffen and be less flexible. Stenosis of the left atrioventricular valve would lead to blood backing up in the _____.

   a. right atrium

   b. right ventricle

   c. left atrium

   d. left ventricle

9. The structures that prevent the AV valve flaps from everting into the atria during contraction are the _____ muscles.

   a. pectinate

   b. cardiac

   c. papillary

10. In which of the following are the structures in the correct order of blood flow?

    a. Right atrium → pulmonary semilunar valve → right ventricle

    b. Right atrium → atrioventricular valve → left atrium

    c. Superior vena cava → left atrium → atrioventricular valve

    d. Pulmonary vein → left atrium → atrioventricular valve

Lab 24 | Blood Vessels

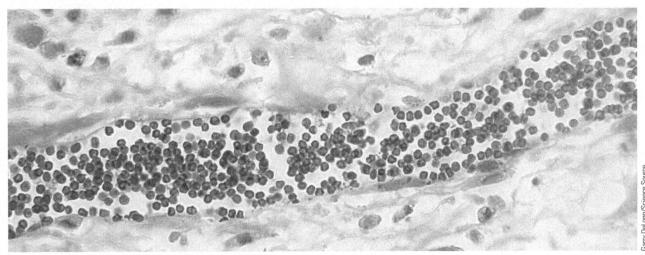

Garry DeLong/Science Source

Learning Objectives: After completing this lab, you should be able to:

24.1 Describe the systemic and pulmonary circuits (circulations) and explain the functional significance of each.

24.2 Identify the major arteries and veins of the pulmonary circuit.

24.3 Identify the major arteries and veins of the systemic circuit.

24.4 Define the terms artery, capillary, and vein.

24.5 List the three tunics associated with most blood vessels and describe the composition of each tunic.

24.6 Compare and contrast tunic thickness, composition, and lumen diameter among arteries, capillaries, and veins.

24.7 Define vasoconstriction and vasodilation.

24.8 Apply an understanding of blood flow to a pathological situation.

Introduction

Blood travels through the body within blood vessels. **Arteries** carry blood away from the heart and branch into ever-smaller vessels. As the vessels branch, they cover more surface area, but become smaller in diameter. Eventually the smallest arteries—vessels called **arterioles**—further branch into tiny **capillaries**, where the walls are so thin that nutrients and wastes can be exchanged between the inside of the blood vessel, and the surrounding tissues. Capillaries merge to form **venules**, small blood vessels that carry blood to a **vein**, a larger blood vessel that returns blood to the heart. The words artery and arteriole describe blood vessels in which blood is moving away from the heart, and the words venule and vein describe blood vessels in which blood is moving toward the heart. All over the body, but especially in the limbs, we can observe that veins are usually closer to the skin and arteries are deeper. This arrangement offers protection to the body because, should injury occur, much less blood would be lost if a vein was injured than if an artery was.

The Human Anatomy and Physiology Society includes more than 1,700 educators who work together to promote excellence in the teaching of this subject area. The HAPS A&P Learning Outcomes measure student mastery of the content typically covered in a two-semester Human A&P curriculum at the undergraduate level. The full Learning Outcomes are available at https://www.hapsweb.org.

Arteries and veins carry blood in two distinct circuits: **the systemic circuit** and **the pulmonary circuit** (Figure 24.1). One of the central functions of blood is to distribute oxygen, which enters the body at the lungs, to the cells throughout the body. The systemic circuit carries oxygen-rich blood throughout the body. This blood is pumped from the left side of the heart. Once it has distributed oxygen to the tissues, the blood is now oxygen-poor but carbon dioxide-rich, the blood returns to the right side of the heart. The right side ejects this oxygen-poor blood into the pulmonary circuit, which carries blood to the lungs, where it releases CO_2 into the lungs and becomes reoxygenated with fresh O_2 from the air. The pulmonary circuit returns the oxygen-rich blood back to the left side of the heart and the cycle begins again.

Figure 24.1	Cardiovascular Circulation Pathways

The body's vast system of blood vessels is grossly divided into two main circuits. The pulmonary circulation transports blood traveling from the right side of the heart toward the lungs for oxygenation; the systemic circuit transports blood from the left side of the heart toward the tissues of the body.

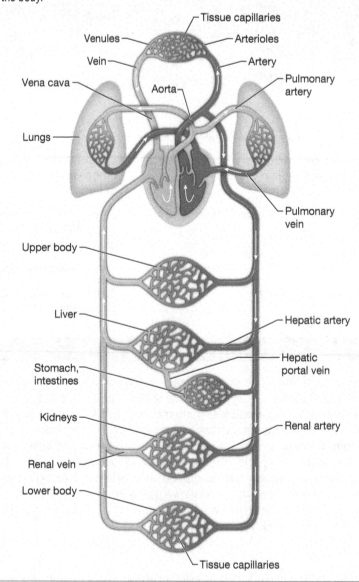

Lab 24 Pre-Lab Quiz

This quiz will strengthen your background knowledge in preparation for this lab. For help answering the questions, use your resources to deepen your understanding. The best resource for help on the first five questions is your text, and the best resource for help answering the last five questions is to read the introduction section of each lab activity.

1. Anatomy of an artery. Place the labels on the structures in the image. Use the following labels:

    lumen
    tunica externa
    tunica intima
    tunica media

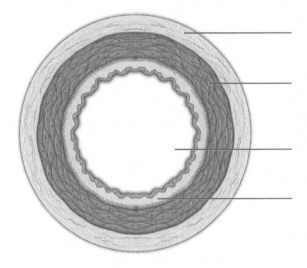

2. Match the root next to its definition:

Root	Meaning	List of Roots
	innermost	artery
	layer	externa
	light	intima
	middle	lumen
	outward	media
	raise	tunic
	vessel or channel of water	vein

The table's "List of Roots" column spans the following roots: artery, externa, intima, lumen, media, tunic, vein.

3. Choose the option that correctly matches the fiber with its description:

    a. Collagen is stretchy and bendable; elastin is tough and fibrous.
    b. Collagen is rigid and solid; elastin is tough and fibrous.
    c. Collagen is tough and fibrous; elastin is rigid and solid.
    d. Collagen is tough and fibrous; elastin is stretchy and bendable.

4. Choose the option that correctly matches the region name and description.

    a. Popliteal—back of the knee
    b. Brachium—lower/forearm
    c. Mental—back of the head

5. Choose the option that correctly matches the region name and description.

a. Femoral—lower leg between knee and ankle

b. Inguinal—front of the hips/lower abdomen

c. Gluteal—back of the shoulders

6. What material can be found in the tunica intima of most vessels?

a. Connective tissue that is rich in collagen

b. Many bundled elastin fibers

c. Simple squamous epithelium

d. Smooth muscle

7. The popliteal artery is located _____.

a. at the back of the thigh

b. traversing the armpit/axilla

c. at the back of the knee

d. at the elbow

8. The superior vena cava collects blood from the _____.

a. head and arms

b. head, arms, and thorax

c. entire systemic circulation

d. head

9. The great saphenous vein is often used in surgeries where a vessel graft is needed. Where is this vein located?

a. In the heart

b. In the abdomen

c. In the leg

d. In the arm

10. The great saphenous vein is often used in surgeries where a vessel graft is needed. When this vessel is removed, more blood will need to travel in the _____ vein to accommodate for the removal of the saphenous vein.

a. femoral

b. basilic

c. cephalic

Activity 24.1 **Gross Anatomy of the Circulatory System**

This activity targets LOs 24.1, 24.2, and 24.3.

The Aorta and Its Branches

The **aorta** is the largest artery in the body (Figure 24.2). It carries the blood ejected from the left ventricle upward as the **ascending aorta**, arches over the top of the heart as the **aortic arch**, and plunges downward through the thorax and toward the abdominal region as the **descending aorta**.

There are three major branches off the aortic arch: the **brachiocephalic trunk**, the **left common carotid artery**, and the **left subclavian artery**. The brachiocephalic trunk is the first branch off the aortic arch. It is a short vessel that splits into two arteries: the **right subclavian artery** and the **right common carotid artery**. The left subclavian artery and the left common carotid arteries arise independently from the aortic arch. The subclavian arteries carry blood to each arm, they are discussed in the upper limb subsection. The carotid arteries carry blood to the brain. They are discussed in the head and brain subsection.

The thoracic aorta has paired branches at each level of the vertebra known as the **intercostal arteries**. As the aorta passes through the diaphragm, it is now called the **descending aorta** (Figure 24.3). Within the abdomen, the aorta branches many times. The abdominal aorta eventually splits into the two **common iliac arteries**. Each of these arteries carries oxygenated blood to the legs. Their branches are discussed further in the lower leg subsection.

Arteries and Veins of the Upper Limbs

Each subclavian artery (Figure 24.4) ascends to the shoulder and passes beneath the clavicle, which protects it from damage. Each subclavian artery then gives rise to three major branches: the internal thoracic artery, the vertebral artery, and the thyrocervical artery which provide blood flow to other regions and are discussed later. As the subclavian artery enters the axillary region, it is renamed the **axillary artery**. This vessel continues into the brachium and becomes the **brachial artery** (Figure 24.4). The brachial artery divides into several smaller branches including the **deep brachial arteries**, which provide blood to the posterior surface of the arm, and the **ulnar collateral arteries**, which supply blood to the region of the elbow. As the brachial artery approaches the elbow it splits into the **radial artery** and **ulnar artery** which continue into the forearm. In the proximal hand, the radial and ulnar arteries fuse to form the **palmar arches** and give rise to the **digital artery** branches that serve each finger.

Figure 24.2 The Aorta

The ascending aorta emerges from the middle of the heart, beginning at the aortic semilunar valve, which separates the aorta from the left ventricle. The aorta arches, curving around the top of the heart, before descending through the thorax posterior to the heart.

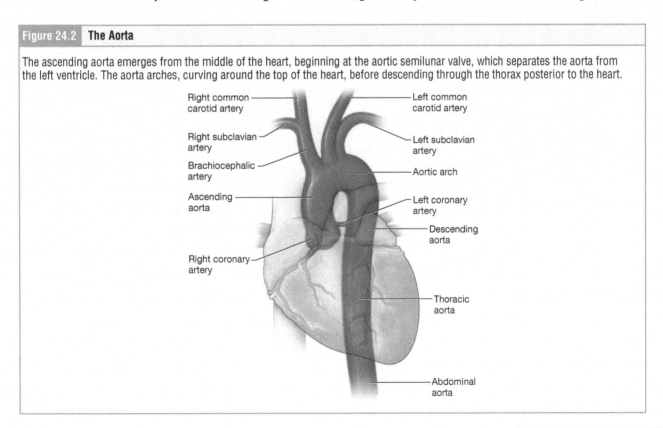

- Right common carotid artery
- Right subclavian artery
- Brachiocephalic artery
- Ascending aorta
- Right coronary artery
- Left common carotid artery
- Left subclavian artery
- Aortic arch
- Left coronary artery
- Descending aorta
- Thoracic aorta
- Abdominal aorta

Figure 24.3 **The Thoracic and Abdominal Aorta**

As the aorta descends through the thorax, it carries oxygenated blood to the lower body. The descending thoracic aorta is renamed the abdominal aorta when it crosses the diaphragm.

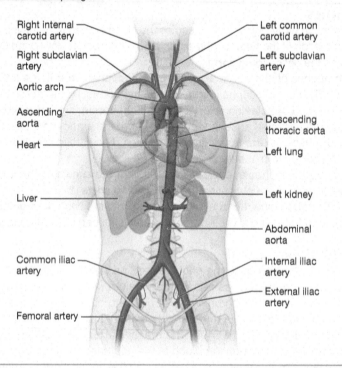

Right internal carotid artery

Right subclavian artery

Aortic arch

Ascending aorta

Heart

Liver

Common iliac artery

Femoral artery

Left common carotid artery

Left subclavian artery

Descending thoracic aorta

Left lung

Left kidney

Abdominal aorta

Internal iliac artery

External iliac artery

Figure 24.4 **Arterial Supply to the Upper Limb**

The subclavian artery, which branches off the aorta on the left and the brachiocephalic trunk on the right, carries oxygenated blood to the upper limbs.

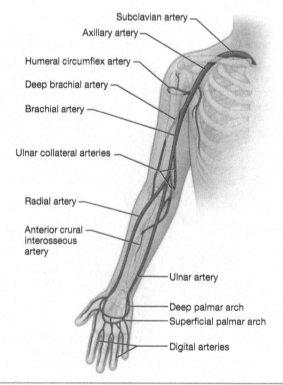

Subclavian artery

Axillary artery

Humeral circumflex artery

Deep brachial artery

Brachial artery

Ulnar collateral arteries

Radial artery

Anterior crural interosseous artery

Ulnar artery

Deep palmar arch

Superficial palmar arch

Digital arteries

The **digital veins** in the fingers come together in the hand to form the **palmar venous arches** (Figure 24.5). From here, the veins merge to form the **radial vein**, the **ulnar vein**, and the **median antebrachial vein**. The ulnar and radial veins merge to form the **brachial vein**, a deep vein that flows into the **axillary vein**. The **median antebrachial vein** joins the **basilic vein** in the forearm. Between the basilic vein and the **cephalic vein**, the **median cubital vein** crosses the elbow to join these two superficial veins.

The **subscapular vein** drains blood from the subscapular region and joins the cephalic vein and the axillary vein, forming the **subclavian vein**. The subclavian vein eventually joins veins descending from the head to form the **superior vena cava**.

The blood vessels of the upper limbs are listed in Table 24.1.

Arteries and Veins of the Head and Neck

The **vertebral arteries** and common carotid arteries are the major arteries that ascend through the neck to bring blood to the face and head (Figure 24.6). The common carotid artery branches into the **internal carotid artery** and the **external carotid artery**. The external carotid brings blood to the face. The vertebral artery and internal carotid artery both enter the cranial cavity and bring their blood to the **cerebral arterial circle** (Figure 24.7) at the base of the brain. The vertebral arteries combine to form the **basilar artery**, which joins the cerebral arterial circle posteriorly. The two common carotid arteries join the circle anterolaterally. Branches off the cerebral arterial circle include the **anterior cerebral artery**, the **anterior communicating artery**, the **middle cerebral arteries**, the **posterior communicating artery**, and the **posterior cerebral arteries**.

Deoxygenated blood from the brain and the face drain into the **internal jugular vein**. Blood from the scalp and cranium flow into the **external jugular vein**. Both these veins join the **subclavian veins**, which eventually drain blood into the superior vena cava (Figure 24.8).

The blood vessels of the head and neck are listed in Table 24.2.

Arteries and Veins of the Thorax and Abdomen

The thoracic aorta descends through the thorax, giving off several branches before it reaches the diaphragm. Each **bronchial artery** supplies blood to the lungs and visceral pleura. Each **pericardial artery** supplies blood to the pericardium. The **esophageal artery** provides blood to the esophagus. The **mediastinal artery** provides blood to the mediastinum. The paired **intercostal arteries** supply blood to the chest wall and the **superior phrenic artery** provides blood to the superior surface of the diaphragm (Figure 24.9). As the thoracic aorta passes through a gap in the diaphragm known as the aortic hiatus, it is then called the **abdominal aorta**.

Figure 24.5 Veins of the Upper Limb

The structures of the upper limb are drained through either superficial or deep veins. The deep veins share names with companion arteries, which are located nearby.

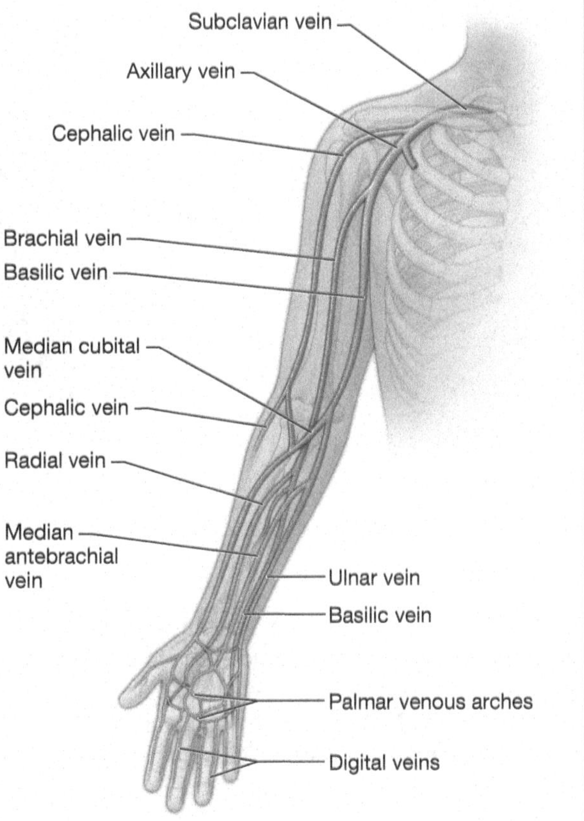

- Subclavian vein
- Axillary vein
- Cephalic vein
- Brachial vein
- Basilic vein
- Median cubital vein
- Cephalic vein
- Radial vein
- Median antebrachial vein
- Ulnar vein
- Basilic vein
- Palmar venous arches
- Digital veins

Table 24.1 Blood Vessels of the Upper Limb

Arteries	Veins
Axillary	Axillary
Brachial	Basilic
Deep brachial	Brachial
Digital	Cephalic
Palmar arch	Digital
Radial	Left subclavian
Ulnar	Median antebrachial
Ulnar collateral	Palmer venous arch
	Radial
	Right subclavian
	Ulnar

Figure 24.6 **Arteries Supplying the Head and Neck**

The common carotid arteries branch off the brachiocephalic trunk or aorta and carry oxygenated blood to the neck, scalp, face, and brain.

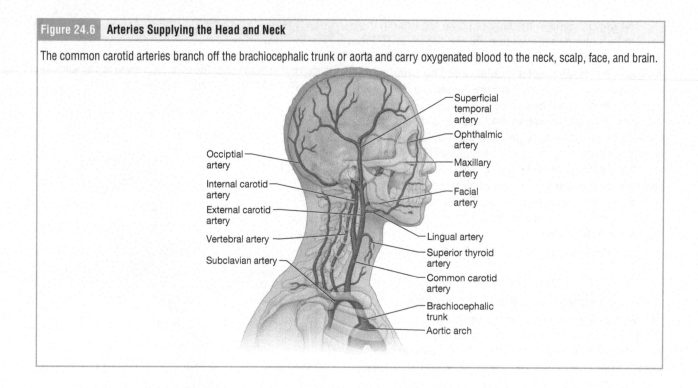

The branches of the abdominal aorta supply oxygenated blood to the abdominal organs. A single **celiac trunk** emerges from the aorta and divides into three arteries: the **left gastric artery**, which supplies blood to the stomach and esophagus; the **splenic artery**, which supplies blood to the spleen; and the **common hepatic artery**, which supplies blood to the liver. Branches off the common hepatic artery include the **hepatic artery** proper, which supplies blood to the liver, the **right gastric artery**, which supplies blood to the stomach, and the **cystic artery**, which supplies blood to the gallbladder.

Figure 24.7 **(A) Arteries of the Brain (B) Graphical Representation of the Cerebral Arterial Circle**

(A) The internal carotid arteries and the vertebral arteries carry blood upward toward the brain. The vertebral arteries join at the brainstem to form the basilar artery. The two internal carotid (and one basilar) arteries contribute to an arterial circle—the cerebral arterial circle—at the base of the brain. (B) In this representation of the cerebral arterial circle, it is evident that blood flow to the brain is ensured even if blood flow is interrupted in one of the supplying vessels.

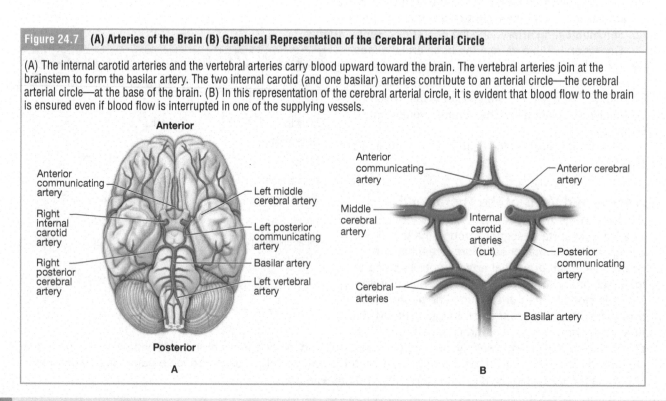

Figure 24.8 **The Venous Drainage of the Head and Neck**

Within the cranium, venous blood collects in dural venous sinuses and then drains into the internal jugular vein. All blood returns to the heart through the internal jugular, external jugular, or vertebral veins.

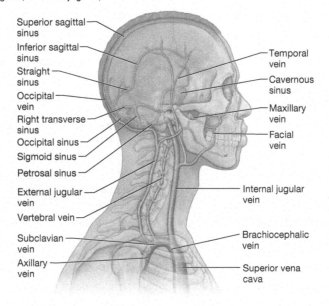

- Superior sagittal sinus
- Inferior sagittal sinus
- Straight sinus
- Occipital vein
- Right transverse sinus
- Occipital sinus
- Sigmoid sinus
- Petrosal sinus
- External jugular vein
- Vertebral vein
- Subclavian vein
- Axillary vein
- Temporal vein
- Cavernous sinus
- Maxillary vein
- Facial vein
- Internal jugular vein
- Brachiocephalic vein
- Superior vena cava

Figure 24.9 **The Arterial Supply to the Thorax and Abdomen**

(A) All of the arteries in the thorax and abdomen can be traced back to branches off the thoracic and abdominal aortas. (B) The branches of the celiac trunk.

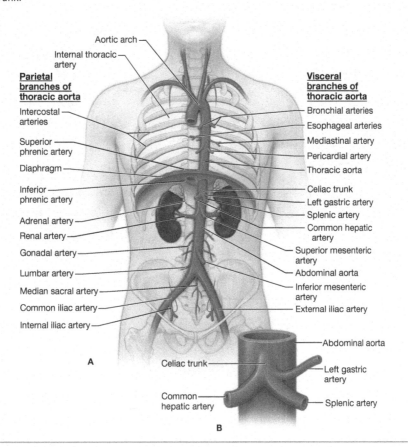

- Aortic arch
- Internal thoracic artery

Parietal branches of thoracic aorta
- Intercostal arteries
- Superior phrenic artery
- Diaphragm
- Inferior phrenic artery
- Adrenal artery
- Renal artery
- Gonadal artery
- Lumbar artery
- Median sacral artery
- Common iliac artery
- Internal iliac artery

Visceral branches of thoracic aorta
- Bronchial arteries
- Esophageal arteries
- Mediastinal artery
- Pericardial artery
- Thoracic aorta
- Celiac trunk
- Left gastric artery
- Splenic artery
- Common hepatic artery
- Superior mesenteric artery
- Abdominal aorta
- Inferior mesenteric artery
- External iliac artery

A

- Celiac trunk
- Common hepatic artery
- Abdominal aorta
- Left gastric artery
- Splenic artery

B

Table 24.2 Blood Vessels of the Head and Neck	
Arteries	**Veins**
Anterior cerebral	External jugular
Anterior communicating	Internal jugular
Basilar	
Cerebral arterial circle	
Internal carotid	
Middle cerebral	
Posterior cerebral	
Posterior communicating	
Vertebral	

The next two vessels that arise from the abdominal aorta are the **superior mesenteric artery** and **inferior mesenteric artery**. The superior mesenteric artery branches into several major vessels that supply blood to the small intestine, pancreas, and a majority of the large intestine. The inferior mesenteric artery supplies blood to the distal large intestine.

The abdominal aorta also gives rise to the **inferior phrenic arteries**, which supply blood to the inferior surface of the diaphragm, the **adrenal arteries**, which supplies blood to the adrenal gland, and the **renal arteries** which supply the kidneys. The **gonadal arteries** branch off the abdominal aorta to supply blood to the gonads, and the four, paired **lumbar arteries** supply blood to the posterior abdominal wall.

Each **intercostal vein** (Figure 24.10) drains the chest wall and contribute their venous blood to the azygos or hemiazygos veins. The blood of the esophagus is drained into the **esophageal vein**. The **bronchial vein** drains the lungs. Each of these will contribute their venous blood to the azygos vein which brings blood to the superior vena cava.

Most of the blood inferior to the diaphragm will drain into the **inferior vena cava** before it is returned to the heart (Figures 24.10 and 24.11A). The posterior body wall is drained into **lumbar veins** which drain into the inferior vena cava or the azygous and hemiazygos veins. Blood supply from the kidneys flows into each **renal vein**, which are the largest veins that contribute blood to the inferior vena cava. Each **adrenal vein** drains the adrenal glands.

Figure 24.10	The Veins of the Thorax and Abdomen

The veins of the thorax and abdomen drain into the inferior vena cava.

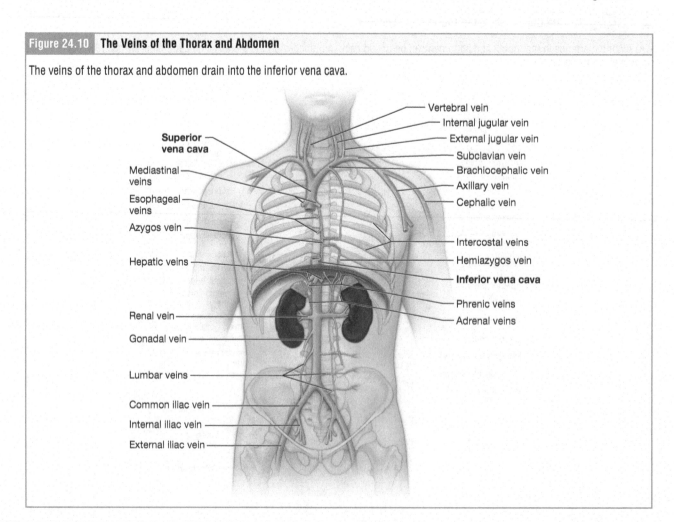

Figure 24.11 | The Hepatic Portal System and Inferior Vena Cava

(A) The inferior vena cava drains the abdominal cavity. Several pairs of lumbar veins contribute blood from the body wall. (B) The venous blood from the spleen, stomach, and intestines drains into the hepatic portal vein, which brings blood to the liver. The renal veins that drain the kidneys connect directly to the inferior vena cava. (C) The drainage of the abdominal organs into the hepatic portal system is represented graphically.

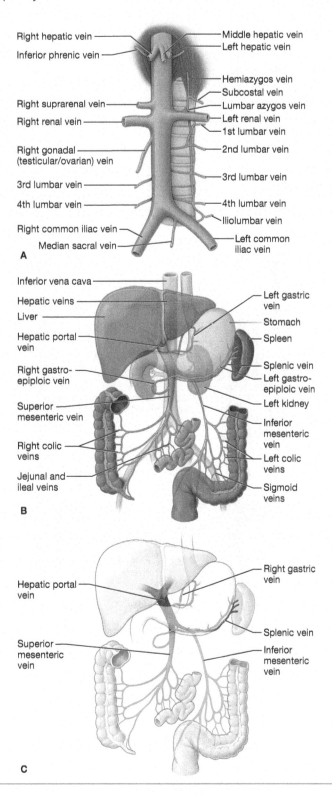

The gonads are drained by **gonadal veins**; each side of the diaphragm drains into a **phrenic vein**. Blood supply from the liver drains into a **hepatic vein**. Each of these veins eventually contributes its blood to the inferior vena cava.

Together the **inferior mesenteric vein** and the **superior mesenteric vein** drain the stomach, and the small and large intestine. The **gastric veins** also drain some of the stomach, the **cystic veins** drain the gallbladder, and the **splenic veins** carries blood coming from the spleen, pancreas, and portions of the stomach. These veins combine to form the **hepatic portal vein**, a large vein that brings blood from the gastrointestinal organs to the liver for filtering before being drained into the inferior vena cava (Figure 24.11B and C).

The blood vessels of the thorax and abdomen are listed in Table 24.3.

Arteries and Veins of the Lower Limbs

The aorta divides at approximately the level of the L4 vertebra into a **left common iliac artery** and **right common iliac artery** (see Figure 24.9). The common iliac arteries split into the **internal iliac artery** that sends branches to the urinary bladder, the walls of the pelvis, and the external genitalia, and the **external iliac artery** supplies blood to each of the lower limbs (Figure 24.12).

Figure 24.12	Arteries in the Lower Limb

The common iliac arteries branch from the abdominal aorta and supply the pelvic organs, genitals and lower limb with oxygenated blood.

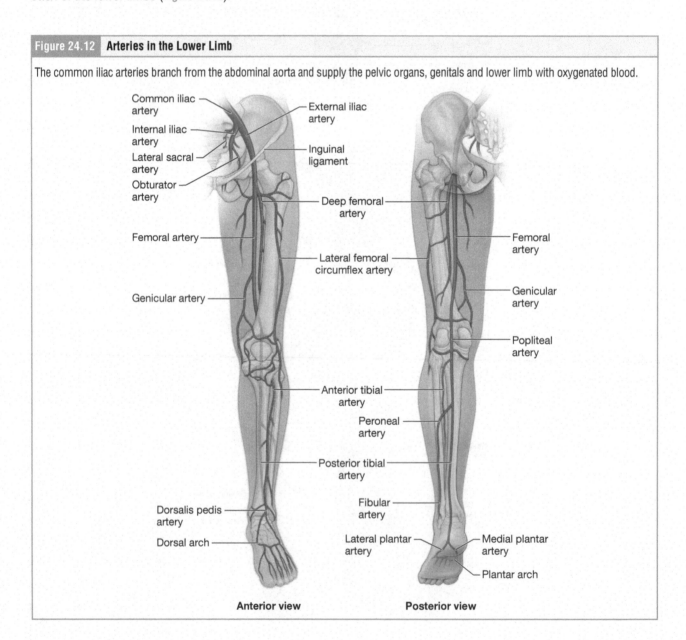

Common iliac artery
External iliac artery
Internal iliac artery
Lateral sacral artery
Inguinal ligament
Obturator artery
Deep femoral artery
Femoral artery
Femoral artery
Lateral femoral circumflex artery
Genicular artery
Genicular artery
Popliteal artery
Anterior tibial artery
Peroneal artery
Posterior tibial artery
Fibular artery
Dorsalis pedis artery
Lateral plantar artery
Medial plantar artery
Dorsal arch
Plantar arch

Anterior view　　**Posterior view**

Table 24.3 Blood Vessels of the Thorax and Abdomen	
Arteries	**Veins**
Aorta	Adrenal
Descending aorta	Azygos
Adrenal	Gastric
Aortic arch	Gonadal
Brachiocephalic trunk	Hemiazygos
Bronchial	Hepatic
Celiac trunk	Hepatic portal
Common hepatic	Inferior mesenteric
Cystic	Inferior vena cava
Esophageal	Intercostal
External carotid	Internal iliac
Gonadal	Lumbar
Hepatic	Phrenic
Inferior mesenteric	Renal
Inferior phrenic	Splenic
Intercostal	Subscapular
Internal iliac	Superior mesenteric
Left common carotid	Superior vena cava
Left gastric	
Left subclavian	
Lumbar	
Mediastinal	
Pericardial	
Renal	
Right common carotid	
Right common iliac	
Right gastric	
Right subclavian	
Splenic	
Superior mesenteric	
Superior phrenic	

Table 24.4 Blood Vessels of the Lower Limb	
Arteries	**Veins**
Anterior tibial	Deep femoral
Common iliac	Digital (of the foot)
Deep femoral	External iliac
Dorsal arch	Femoral
Dorsalis pedis	Great saphenous
External iliac	Plantar venous arch
Femoral	Popliteal
Genicular	Posterior tibial
Lateral circumflex	Small saphenous
Lateral plantar	
Left common iliac	
Medial plantar	
Plantar arch	
Popliteal	
Posterior tibial	

As the external iliac artery passes into the leg it is renamed the **femoral artery**. It gives off several smaller branches as well as a larger branch, the **deep femoral artery**, which in turn gives rise to a **lateral circumflex artery**. The femoral artery also gives rise to the **genicular artery**, which provides blood to the anterior knee. Just above the knee, the femoral artery passes toward the posterior leg, where it becomes the **popliteal artery**. The popliteal artery branches into the anterior and posterior tibial arteries.

The **anterior tibial artery** becomes the **dorsalis pedis artery** on the anterior ankle surface. The **posterior tibial artery** splits and becomes the **medial plantar artery** and the **lateral plantar artery**, both of these arteries

provide blood to the bottom or plantar surface of the foot. The dorsalis pedis, medial, and lateral plantar arteries all contribute to the **dorsal arch** and **plantar arch**, which provide blood to the remainder of the foot and toes.

The superior surface of the foot drains into the **digital veins** and the inferior surface drains into the **plantar veins** (Figure 24.13). Two arches then collect venous blood from these veins: the **dorsal venous arch** on the dorsal side of the foot and the **plantar venous arch** on the plantar side of the foot. From here, blood drains into the **anterior tibial vein** and **posterior tibial vein**. These two veins combine, along with the **fibular vein** to form the **popliteal vein**. The **small saphenous vein** drains blood from the superficial regions of the lower leg and foot and flows into the popliteal vein. The **popliteal vein** passes behind the knee and becomes the **femoral vein**. The **great saphenous vein** is a large superficial vein located on the medial surface of the leg that collects blood from superficial tissues. The **deep femoral vein** drains deeper structures. Both of these veins eventually drain into the femoral vein, which is renamed the **external iliac vein** as it passes into the pelvic region.

The blood vessels of the lower limbs are listed in Table 24.4.

 Learning Connection

Chunking

Circulation patterns are fairly similar in the arms and legs. Create a chart where you fill in the analogous vessel in each limb. For example, the brachial artery and the femoral artery could be considered analogous.

Figure 24.13 **Venous Drainage of the Lower Limb**

The structures of the lower limb are drained by superficial and deep veins. The deep veins share names with companion arteries, which are located nearby.

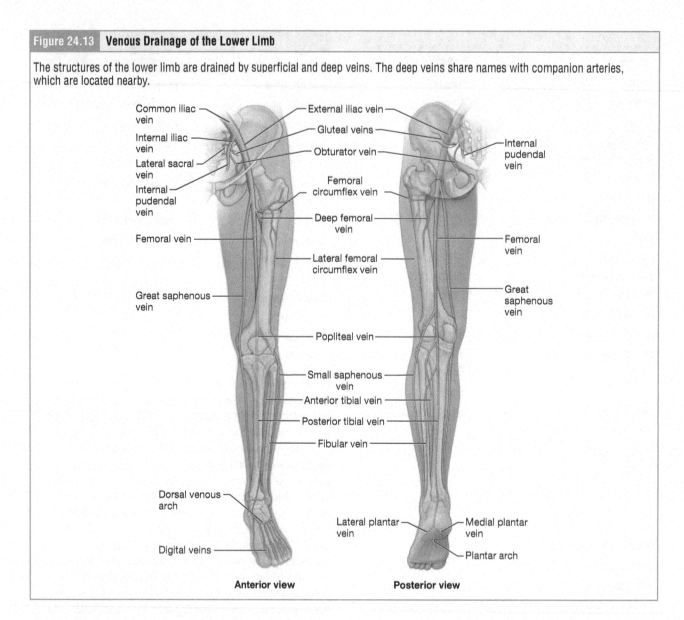

Anterior view   Posterior view

The Pulmonary Circuit

Remember that all veins return deoxygenated blood to the right atrium via the superior and inferior venae cavae. The pulmonary circuit is the path of blood flow that carries oxygen-poor blood from the right side of the heart to the lungs where it can pick up oxygen (Figure 24.14). Blood leaves the right ventricle in the **pulmonary trunk**, which branches into the **left pulmonary artery** and the **right pulmonary artery**. Just like arteries in the systemic circuit, the pulmonary arteries branch into smaller and smaller arteries and arterioles. The pulmonary capillaries encase the alveoli like a net, creating a structure of maximum surface area for gas exchange between the alveoli and the blood. Here the blood becomes oxygenated as well as gives off some of its carbon dioxide. The oxygenated blood returns to the left side of the heart in **pulmonary veins**.

Materials

• Models of human blood vessels (systemic or regional models)

Instructions

1. Using the available models, identify the structures found in Tables 24.1–24.4, using lab tape to label each as you identify it.

2. Write out the path of vessels that a drop of blood would take to get to and from the locations in Table 24.5.

Figure 24.14 **The Pulmonary Circuit**

The pulmonary arteries carry blood away from the right side of the heart and toward the lungs, where the blood gives off accumulated carbon dioxide and picks up oxygen before returning, via the pulmonary veins, to the left side of the heart.

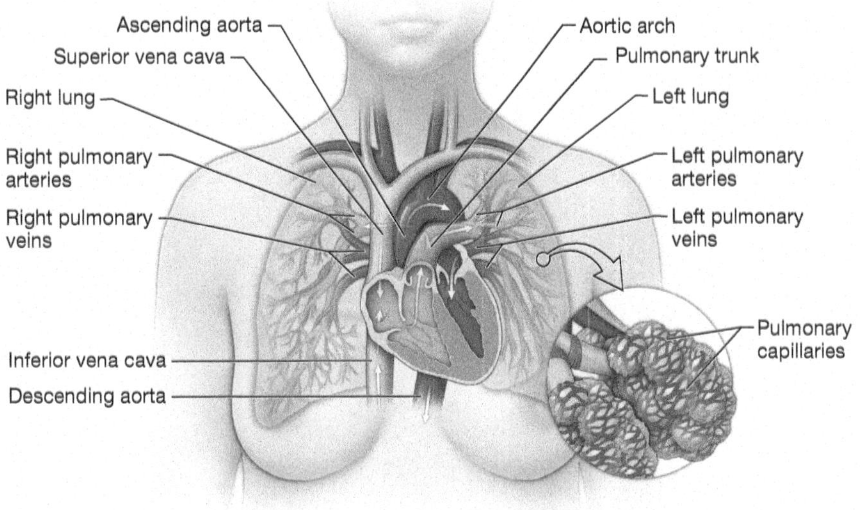

Ascending aorta
Superior vena cava
Right lung
Right pulmonary arteries
Right pulmonary veins
Inferior vena cava
Descending aorta
Aortic arch
Pulmonary trunk
Left lung
Left pulmonary arteries
Left pulmonary veins
Pulmonary capillaries

To	Table 24.5 Path of Blood
	Pathway
Fingertips	
Right atrium	
Stomach wall	

(*continued*)

Table 24.5 Path of Blood (*Continued*)	
From	**Pathway**
Left ventricle	
Capillary in toes	
Right ventricle	

Activity 24.2 Histology of Blood Vessels

This activity targets LOs 24.4, 24.5, 24.6, and 24.7.

Though the different types of blood vessels vary in their structures, they share the same general features. All vessels have a lumen, the hollow passageway through which blood flows. The wall of a blood vessel surrounds the lumen, but varies in thickness and composition in different types of vessels.

Arteries and arterioles have thicker walls than do veins and venules because they receive blood that is at a much higher pressure. As the blood moves further from the heart through arteries, arterioles, and capillaries, it loses pressure, so the venules and veins are subjected to much less pressure and therefore have much thinner walls. The walls of arteries and arterioles not only have to withstand a higher pressure, but they also have to withstand fluctuations in that pressure between when the heart contracts (systole) and when it relaxes (diastole). The walls of arteries and arterioles therefore have greater elasticity than venules and veins due to a higher concentration of elastin fibers. The walls of arteries and arterioles have more smooth muscle in their walls than do veins and venules to direct the flow of blood (Figure 24.15).

Figure 24.15	A Comparison of Arteries and Veins under the Microscope

Arteries carry blood away from the heart, and therefore are exposed to blood traveling at a higher velocity and pressure than veins. Their walls must be able to accommodate that pressure and their lumens are smaller as the blood spends less time there. Veins, by comparison, do not have wall structure that tolerates high pressure, and their walls are thinner, their structure baggier and less rigid. Because blood moves more slowly through veins, they have larger lumens.

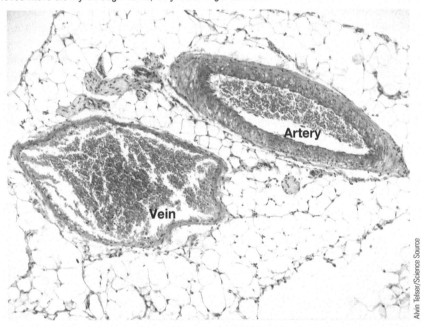

Alvin Telser/Science Source

Digging Deeper:
Stretchy Vessels

As we age, we lose elasticity in our tissues. In our skin, the loss of elasticity causes wrinkles. In our blood vessels, the loss of elasticity causes stiffening. Vascular stiffening is deleterious to health in two ways. One, the stiffer walls offer more resistance to the blood being ejected by the heart, in turn, the heart must work harder to eject blood. Two, without the moderation in pressure provided by the wave-like expansion and contraction of the artery walls, the blood maintains a significant pressure difference between systole and diastole as it reaches smaller vessels. These smaller vessels are at an increased risk of bursting. Not everyone's vessels stiffen at the same rate. Exercise plays a huge role in preventing arterial stiffening and active adults have much more compliant blood vessels then sedentary adults.

Venules and veins have much thinner walls and much larger lumens than arteries and arterioles. In addition, many veins of the body, particularly those of the limbs, contain valves. The valves assure one-way flow toward the heart; once blood has passed a valve, the valve closes and prevents backflow.

Both arteries and veins have the same three layers, called tunics. From the most interior layer to the outer layer, these tunics are the tunica intima, tunica media, and tunica externa (Figure 24.16).

The **tunica intima**, the innermost layer of blood vessels, is composed of endothelium, the simple squamous epithelium, and its basement membrane that lines the entire inner surface of the cardiovascular system. The middle layer of blood vessel walls is the **tunica media**. In larger arteries, there is a layer of elastic fibers, the **internal elastic membrane**, between the tunica intima and the tunica media.

The tunica media is composed largely of smooth muscle. When this smooth muscle contracts it narrows the lumen of blood vessels and reduces blood flow through them—this is called **vasoconstriction**. The opposite, relaxation of the smooth muscle and therefore an increase in lumen size and blood flow is **vasodilation**. In addition to smooth muscle, the tunica media has many collagen fibers. In larger arteries there is a thick **external elastic membrane** that contributes stretchiness. The elastic fibers of this membrane appear wavy in histological preparations (Figure 24.17).

The tunica externa is the outermost tunic, composed primarily of collagen fibers. The outer layers of this tunic may not be distinctive, but may instead blend with the surrounding connective tissue.

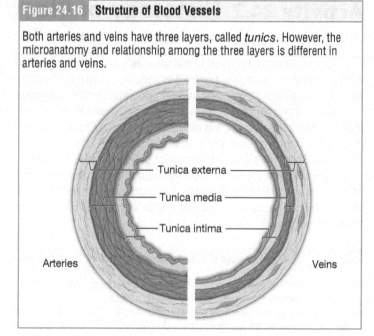

Figure 24.16 **Structure of Blood Vessels**

Both arteries and veins have three layers, called *tunics*. However, the microanatomy and relationship among the three layers is different in arteries and veins.

Tunica externa

Tunica media

Tunica intima

Arteries

Veins

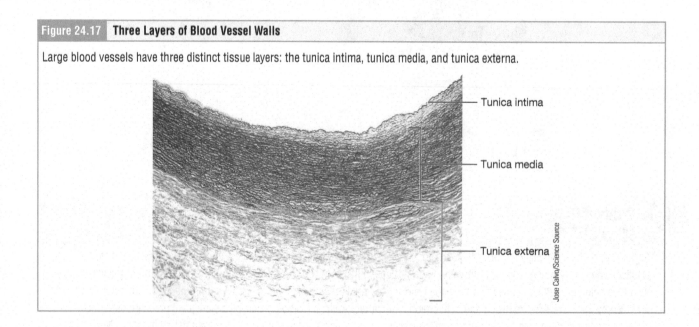

Figure 24.17 **Three Layers of Blood Vessel Walls**

Large blood vessels have three distinct tissue layers: the tunica intima, tunica media, and tunica externa.

Tunica intima

Tunica media

Tunica externa

Jose Calvo/Science Source

Materials

- Compound microscope
- Slide that features cross-sectional view of arteries and veins

Instructions

1. Using the low power objective, focus and scan the slide until you can identify an artery and a vein.

2. Using Figure 24.18 as a guide, compare the lumen size, wall thickness, and general shape of the artery and vein. Draw what you see in the space provided.

3. Increase magnification, refocus, and identify the tunica intima, media, and externa of each vessel.

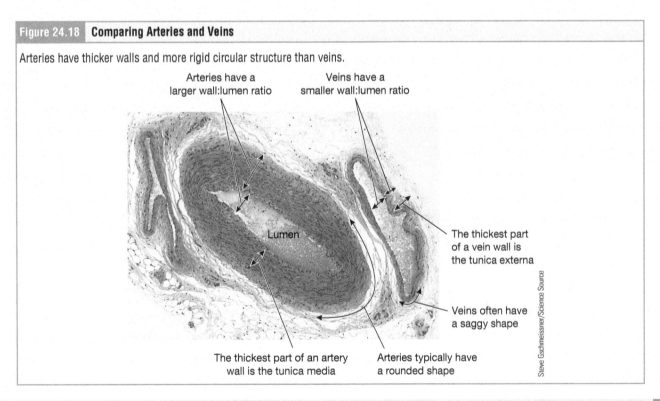

Figure 24.18 Comparing Arteries and Veins

Arteries have thicker walls and more rigid circular structure than veins.

Arteries have a larger wall:lumen ratio

Veins have a smaller wall:lumen ratio

Lumen

The thickest part of a vein wall is the tunica externa

Veins often have a saggy shape

The thickest part of an artery wall is the tunica media

Arteries typically have a rounded shape

Steve Gschmeissner/Science Source

(page intentionally left blank)

Activity 24.3 Case Study: Pulmonary Embolism

This activity targets LO 24.8.

Ben, a 59-year-old, collapses in the baggage claim area of Logan International Airport. EMTs transport them to Massachusetts General Hospital, noting that their breath rate and pulse are elevated. Ben feels severely short of breath and complains of dizziness. Upon arrival in the ER echocardiography—ultrasound imaging of the heart (Figure 24.19)—is performed, and it is noted that his right ventricle appears slightly larger than his left. The echocardiogram also reveals that his right atrioventricular valve is leaking fluid. The doctors note that this size increase in a heart chamber typically appears when that chamber is working harder than usual. The ER doctors perform blood tests for proteins typically found in the blood during heart attacks and found that these tests were negative.

Figure 24.19	Ultrasound Imaging of the Heart

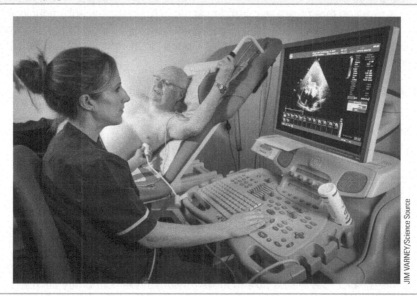

JIM VARNEY/Science Source

A CT scan of his chest was performed and bilateral pulmonary emboli (blood clots, PEs) were found (Figure 24.20).

Figure 24.20	Thoracic CT Scan

CT scan of the chest shows bilateral pulmonary emboli (blood clots).

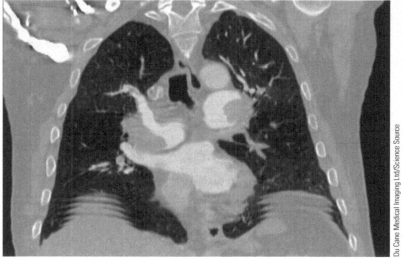

Du Cane Medical Imaging Ltd/Science Source

1. If there are blood clots in the lungs, what vessel did they travel in to get to where they are now lodged?

    a. Pulmonary veins

    b. Pulmonary arteries

    c. Intervertebral arteries

    d. Intervertebral veins

2. Considering your answer from Question #1, which side of the heart did these blood clots most recently pass through?

    a. The right side

    b. The left side

To figure out how to treat the emboli, the doctors perform further tests to investigate their identity and origin. Ultra-sounds of the man's legs revealed three deep vein thromboses (DVTs) in the patient's right lower extremity. Both of the words thrombus and embolus mean blood clot. Thromboses are blood clots at the site where they develop. When the clots or pieces of the clots come loose, they travel in the blood vessels. If they get stuck in a smaller vessel, they are then known as emboli.

3. Which of the following scenarios now seems most likely about Ben's pulmonary blood clots?

    a. The blood clots developed in the lungs and stayed there.

    b. The blood clots developed in the lungs and traveled to the legs.

    c. The blood clots developed in the legs and traveled to the lungs.

4. Why did the blood clots not get stuck elsewhere in the venous circulatory system?

    a. The veins get progressively larger as they approach the heart.

    b. The blood clots developed very close to where they became stuck.

    c. The oxygen in the pulmonary circulation caused the blood clots to become larger and stickier.

For immediate treatment, the patient is put on medications known as blood thinners that dissolve blood clots. As the clots dissolve and become smaller they will be able to move forward in the pulmonary circulation.

5. Will they pass through any heart structures as they dissolve away? If so, which ones?

    a. No, the dissolving clots will not pass through the heart.

    b. Yes, the dissolving clots will pass through the right atrium and right ventricle.

    c. Yes, the dissolving clots will pass through the left atrium and left ventricle.

DVTs are linked to genetics and certain lifestyle factors. Thirty percent of patients who have experienced a DVT before will get one again. To avoid further PEs, the doctors implant a filter into the bloodstream to catch any emboli moving from the extremities or pelvis before reaching the heart.

6. Would the doctors be able to place **one** filter that would catch emboli originating in both the upper and lower extremities?

    a. Yes, one filter will catch emboli from both locations.

    b. No, more than one filter would be needed.

7. To catch any emboli from the lower extremities or pelvis, which vein would be the best location for the filter?

    a. Superior vena cava

    b. Inferior vena cava

    c. Hepatic portal vein

    d. Femoral vein

    e. Subclavian vein

Lab 24 Post-Lab Quiz

1. Blood in the popliteal vein will next be in the _____ vein.

   a. femoral
   b. great saphenous
   c. small saphenous
   d. tibial

2. The next time blood that is in the brachial artery reaches the heart, which chamber will it enter first?

   a. Right atrium
   b. Right ventricle
   c. Left atrium
   d. Left ventricle

3. What type of blood vessel is shown here?

   a. Vein
   b. Artery
   c. Capillary

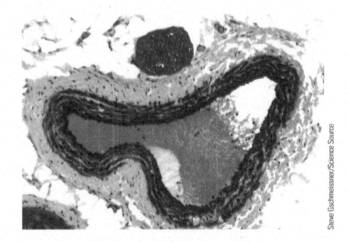

4. What type of blood vessel is shown here?

   a. Vein
   b. Artery
   c. Capillary

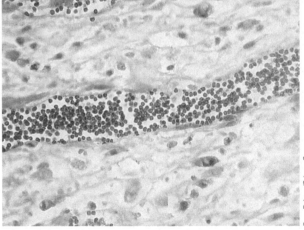

5. Blood in the median cubital vein will next be in the _____ vein.

   a. basilic
   b. great saphenous
   c. jugular
   d. superior vena cava

6. Blushing is an emotional response to embarrassment. During blushing an increase in blood flow to the skin of the face can sometimes cause a color change, especially in the cheeks. Blushing is an example of _____ caused by a change in the smooth muscle of the _____.

    a. vasoconstriction; tunica intima

    b. vasoconstriction; tunica media

    c. vasodilation; tunica intima

    d. vasodilation; tunica media

7. The vessel marked with an X is called a(n) _____.

    a. vein

    b. artery

    c. capillary

    d. arteriole

    e. venule

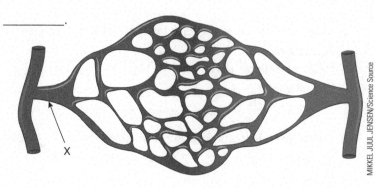

MIKKEL JUUL JENSEN/Science Source

8. The arrow is pointing to what vessel?

    a. Radial artery

    b. Ulnar artery

    c. Basilic artery

    d. Brachial artery

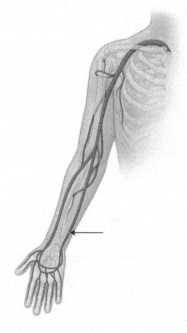

9. The arrow is pointing to what vessel?

    a. Superior vena cava

    b. Inferior vena cava

    c. Internal jugular vein

    d. Subclavian vein

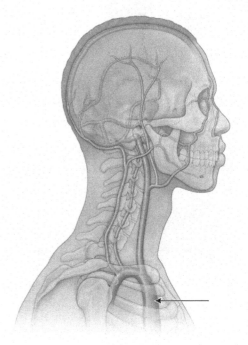

10. Compliance of blood vessels is defined as the vessel's ability to stretch in response to increasing pressure and recoil or decrease in volume in times of lower pressure. Which vessel type do you think will be the most compliant?

    a. Vein

    b. Artery

    c. Capillary

    d. Arteriole

    e. Venule

Lab 25 Cardiovascular Physiology

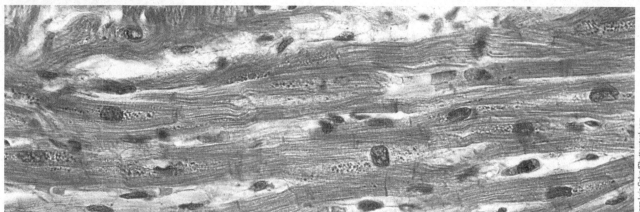

Jose Luis Calvo/Shutterstock.com

Learning Objectives: After completing this lab, you should be able to:

25.1 Trace the path of blood through the right and left sides of the heart, including its passage through the heart valves, and indicate whether the blood is oxygen-rich or oxygen-poor.

25.2 Describe the phases of the cardiac cycle including ventricular filling, isovolumic (isovolumetric) contraction, ventricular ejection, and isovolumic (isovolumetric) relaxation.

25.3 Relate the opening and closing of specific heart valves in each phase of the cardiac cycle to pressure changes in the heart chambers and the great vessels (i.e., blood vessels entering and leaving the heart).

25.4 Relate the heart sounds to the events of the cardiac cycle.

25.5 Define cardiac cycle, systole, and diastole.

25.6 Define the terms artery, capillary, and vein.

25.7 Name the waveforms in a normal electrocardiogram (ECG or EKG) and explain the electrical events represented by each waveform.

25.8 Relate the electrical events represented on an electrocardiogram (ECG or EKG) to the normal mechanical events of the cardiac cycle.

25.9 Explain how atrial systole is related to ventricular filling.

25.10 List the parts of the electrical conduction system of the heart in the correct sequence for one contraction and explain how the electrical conduction system functions.

25.11 Explain how the cardiac conduction system produces coordinated heart chamber contractions.

25.12* Describe the clinical relevance of blood pressure readings compared to cardiovascular physiology.

25.13 Identify the major arteries and veins of the systemic circuit.

25.14* Define the two numbers in blood pressure readings and relate each to cardiac activity.

* Objective is not a HAPS Learning Goal.

The Human Anatomy and Physiology Society includes more than 1,700 educators who work together to promote excellence in the teaching of this subject area. The HAPS A&P Learning Outcomes measure student mastery of the content typically covered in a two-semester Human A&P curriculum at the undergraduate level. The full Learning Outcomes are available at https://www.hapsweb.org.

Introduction

The heart is a pump that runs by electrical signaling. Therefore, when we examine heart function or heart physiology, we can focus on the plumbing system (the pumping of blood), the electrical system (the electrical signals that trigger contractions), or how the two work together. When we examine the plumbing of the heart, we can study how blood moves through the heart to be pumped to the lungs or the body, or how blood moves along the vessels of the heart, the **coronary vessels**, to supply oxygen and nutrients to the heart muscle itself. Ultrasound technology is useful to image and measure how effectively the plumbing system is working (Figure 25.1A), an ultrasound of the heart is called an **echocardiogram**. When we study the electrical conduction system of the heart, we can look at the cycle of electrical signals that move through the heart to understand more about the rhythm of heart contractions and relaxations. An **electrocardiogram** (**ECG** or **EKG**) is the best tool to visualize and understand the electrical signals within the heart. During an ECG, electrodes are placed on the skin (Figure 25.1B) where they detect disruptions in the concentration of ions in the interstitial fluid that are caused when the heart contracts. Essentially, when the heart contracts it creates a wave of electrical changes in the fluid surrounding the heart that ripples outwards and can be detected in the skin. The ECG machine creates a graph (Figure 25.1C). When the line on the graph is flat there are no electrical changes occurring in the fluid beneath the skin. When electrical changes do occur, the ECG line will move up or down, creating a waveform on the graph. **Electrocardiologists**, the doctors who specialize in the electrical system of the heart, can read the waves and the spaces between them on the ECG graph to understand how the heart's conduction system is performing. The simplest tool that clinicians use to evaluate heart function is **auscultation**, or listening! With auscultation a clinician can listen to the heart using a stethoscope and hear each of the heart valves closing. This gives the clinician some information about the structure of the heart.

Figure 25.1 | **Echocardiogram**

(A) An echocardiogram is used to visualize and measure blood flow through the heart. It is a form of ultrasound imaging. (B) An electrocardiogram (ECG) is used to detect the rhythm of cardiac contractions. ECG measures electrical changes caused by the flow of ions. (C) An ECG produces a tracing of electrical changes that can be read and interpreted to understand cardiac function.

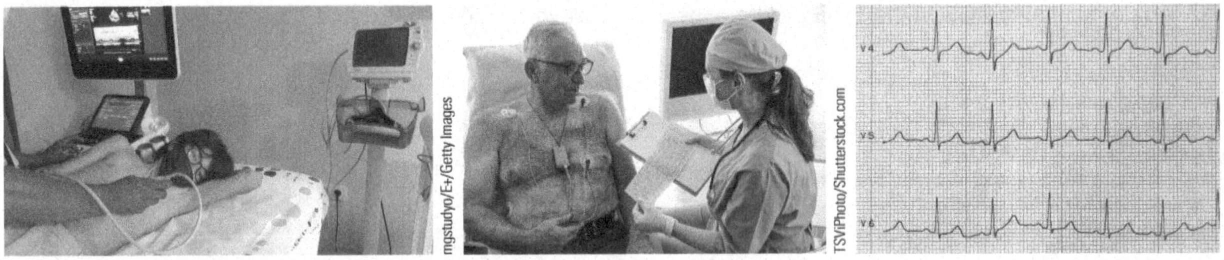

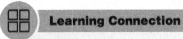

A   **B**   **C**

Two tests that evaluate blood flow in the blood vessels beyond the heart are blood pressure and pulse. Blood pressure testing uses an inflatable cuff and valve to determine how much pressure the blood is exerting against the walls of the arteries after it leaves the heart. Pulse is also determinable because of the pressure of blood against the walls of the vessels, however, the only physiological factor that can be determined when taking the pulse is how frequently the heart is beating.

In this lab we will learn how to take and interpret some of these measurements as well as explore two clinical case studies relating to cardiac physiology.

Learning Connection

Chunking

Which parts of the heart could be "chunked" in a group as having electrical functions, and which parts could be "chunked" has having plumbing functions?

Lab 25 Pre-Lab Quiz

This quiz will strengthen your background knowledge in preparation for this lab. For help answering the questions, use your resources to deepen your understanding. The best resource for help on the first five questions is your text, and the best resource for help answering the last five questions is to read the introduction section of each lab activity.

1. Anatomy of the conduction system of the heart. Label the figure with the following terms:

   atrioventricular node
   bundle branches
   bundle of His
   internodal pathway
   Purkinje fibers
   sinoatrial node

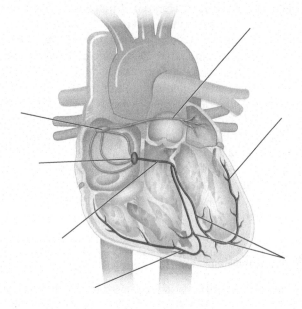

2. Anatomy of an ECG. On the ECG tracing, label one P wave, QRS complex, and T wave.

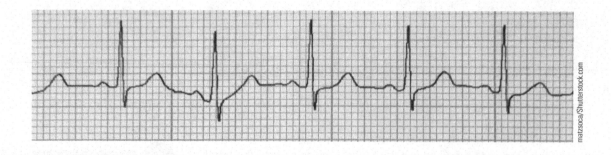

matzsoca/Shutterstock.com

3. Anatomy of Wiggers diagram. Label the figure with the following terms:

AV valves close
AV valves open
diastole
semilunar valves close
semilunar valves open
systole

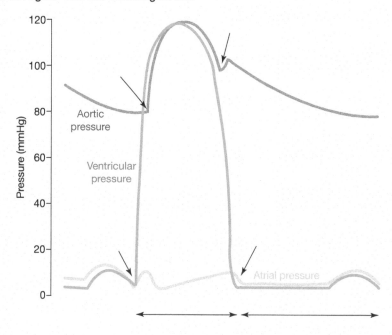

4. Match each of the roots to its definition:

Root	Meaning
	contract
	heart
	rapid
	relax
	without

List of Roots

a
cardi
diastol
systol
tachy

5. Match each of the roots to its definition:

Root	Meaning
	a period of time in which either the atria or ventricles are contracting
	a period of time in which either the atria or ventricles are relaxing
	the pressure exerted by a fluid onto the walls of its container

List of Roots

diastole
hydrostatic pressure
systole

6. Which of the following is commonly measured at the radial artery?

   a. Electrocardiogram
   b. Blood pressure
   c. Pulse
   d. Echocardiogram

7. If a doctor suspects a blockage is disrupting blood flow through the heart, which type of test are they most likely to order?

   a. Electrocardiogram
   b. Blood pressure
   c. Pulse
   d. Echocardiogram

8. The heart sounds are caused by what event?

   a. The smack of blood against the walls of the atria as it enters the heart.
   b. The expansion of the walls of the elastic arteries as blood is forced into them.
   c. The closure of the flaps of the heart valves against each other.

9. When measuring blood pressure, which of the following is the correct location to place your stethoscope?

   a. On the proximal region of the brachial artery
   b. On the cubital fossa
   c. On the radial artery at the distal part of the antebrachium
   d. On the chest to the lower left of the sternum

10. If a doctor suspects an arrhythmia, a condition in which the heart does not follow its typical pattern, which type of test are they most likely to order?

    a. Electrocardiogram
    b. Blood pressure
    c. Pulse
    d. Echocardiogram

Activity 25.1 Heart Sounds

This activity targets LOs 25.1, 25.2, 25.3, and 25.4.

The sounds associated with the heart pumping—heart beats—are the sounds of the valves closing. The valves closing is a mechanical event that occurs due to the pressure of blood flow on the underside of the cusps (Figure 25.2). Notice, as in Figure 25.3, that the opening and closing of the atrioventricular (AV) valves and the semilunar valves are in

Figure 25.2	The Structure of the Heart Valves Is Simple, as Is Their Function

As blood pushes on the tops of the atrioventricular valve flaps, it opens them, allowing blood to flow from the atria to the ventricles. As the ventricles contract, blood is pushed against the bottom of the atrioventricular flaps, forcing them closed. Similarly, as blood pushes against the bottom of the semilunar valves, the flaps are forced open and blood can flow into the leaving vessels. As the blood flows backwards, it catches on the upside-down cusps of the semilunar valves and forces them closed.

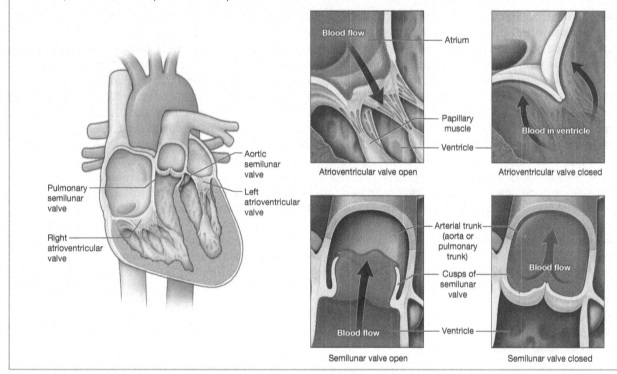

Figure 25.3	Transverse Section of the Heart

A transverse section through the heart, viewed from above, shows that the atrioventricular valves have the opposite orientation from the semilunar valves.

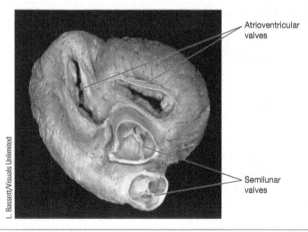

Heart murmurs are atypical sounds detected during auscultation. There are three types, described by *when* they are heard. Diastolic murmurs are heard when the heart is in diastole. These murmurs can occur when there is stenosis (narrowing) of the atrioventricular valves or regurgitation (backflow) at the semilunar valves. Systolic murmurs are heard when the heart is in systole. They typically occur when there are anatomical issues at the aortic semilunar valve that interferes with blood ejection. Continuous murmurs are heard all through the cardiac cycle. They occur when there is a physical shunt between high- and low-pressure vessels such as a connection between an artery and a vein. Because blood is always flowing in one of these, the sound is heard continuously.

opposite orientations. Recall from readings in your textbook that the AV valves close as the ventricles begin to contract, and blood is squeezed upwards onto the underside of their cusps (Figure 25.2). As the ventricles continue to contract, the flowing blood opens the semilunar valves and blood is ejected into the pulmonary trunk and aorta. As the ventricles relax, their pressure drops below the pressure in the pulmonary trunk and the aorta, so the blood sloshes backwards from the vessels toward the ventricles, but catches on the underside of the cusps, forcing them closed (Figure 25.2).

We typically describe the heart sounds as a "lub-dub," a two-part sound. For students familiar with music, this is often described in music as a syncopated rhythm. The first sound is quicker and the second sound is longer and more forceful. The first heart sound, the "lub" is the sound of the AV valves closing at the beginning of ventricular **systole** (contraction). The second heart sound, the "dub" is the sound of the semilunar valves closing as the ventricles enter **diastole** (relaxation). Because the closures of the valves are due to pressure changes, and because the left side of the heart has a greater capacity for generating pressure, the heart sounds from the right and left sides can be heard at slightly different times if you listen carefully and at the optimal location.

In this activity we will listen to heart sounds. The word that clinicians use to describe listening to the heart is *auscultate*. You can choose to auscultate your own heart or your lab partner's. It is easier to hear with less fabric between the stethoscope and the skin, so dress in light layers, or wear a loose sweater that allows you to operate the stethoscope beneath it. If you are auscultating a partner, be sure to ask them if what you are doing is ok and check-in to make sure they are comfortable.

Materials
- Alcohol swabs
- Stethoscope
- Watch/clock/timer that displays seconds

Instructions
1. Use an alcohol swab to clean the earpieces of a stethoscope. Notice the angle of the earpieces and hold the stethoscope so that the earpieces are angled forward before putting them in your ears (it's more comfortable this way, see Figure 25.4).

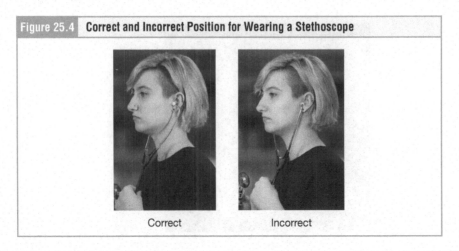

| Figure 25.4 | Correct and Incorrect Position for Wearing a Stethoscope |

Correct   Incorrect

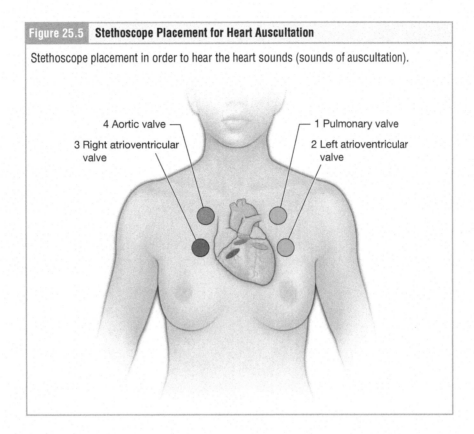

Figure 25.5 Stethoscope Placement for Heart Auscultation

Stethoscope placement in order to hear the heart sounds (sounds of auscultation).

4 Aortic valve

3 Right atrioventricular valve

1 Pulmonary valve

2 Left atrioventricular valve

2. Follow the diagram in Figure 25.5 and listen to your or your partner's heart at the location marked 1. You can also have your partner hold the stethoscope themselves if they are more comfortable doing so.

3. After listening for a few heart beats, try to time the pause between the second sound of one heartbeat and the first sound of the next heartbeat.

4. Record that interval here: _____ seconds.

5. Now try listening at positions 2 and 3. At position 2 you should be able to hear the closure of the left AV valve just before and slightly more clearly than you hear the right AV valve. At position 3 you should be able to hear the right AV valve more clearly than the left AV valve.

6. Now compare the sounds heard at positions 4 and 1. At position 1 the valve closure you can hear most clearly is the pulmonary semilunar valve, and at position 4 you can most clearly hear the aortic semilunar valve.

(page intentionally left blank)

Activity 25.2 Blood Pressure and Pulse

This activity targets LOs 25.1, 25.5, 25.6, 25.7, 25.8, and 25.9.

In this activity we will determine the blood pressure in the brachial artery and measure the pulse at the radial artery. **Blood pressure** is the force that the blood exerts against the walls of the artery. This pressure is generated by the heart when it squeezes onto the blood and propels it through the vessels. Blood pressure is measured using a blood pressure cuff or **sphygmomanometer** and the units of measurement are millimeters of mercury (for more on how pressure is measured please see the section on Units of Measurement in the beginning of this manual). This cuff is placed proximal to the elbow on the brachium (see Figure 25.6) and a stethoscope is placed just distal to the cuff over the interior of the elbow, the region known as the cubital fossa. The cuff is inflated, putting pressure on the structures below it until blood flow through the brachial artery is obstructed. Then, the pressure of the cuff is slowly released until it is entirely deflated. As the cuff deflates, at some point the pressure of the blood in the brachial artery is sufficient to overcome and flow past the cuff. The blood will begin to flow past the cuff when it is at its highest pressure, its systolic pressure.

When the ventricles contract, they squeeze blood at a high pressure into the arteries. The artery walls expand at this moment, but arterial walls are highly elastic, and so they snap back after expansion, like a rubber band. This exerts some force onto the blood, contributing to its driving pressure. After ventricular contraction (systole) they relax (diastole). When they are relaxed, no new blood is pushed into the arteries and their walls relax as well. Blood pressure can therefore be measured at systole and at diastole, and these two

---

Figure 25.6	Measuring Blood Pressure

Blood pressure is always measured at the brachial artery, just above the elbow. Blood pressure is measured using a cuff with a pressure valve, called a sphygmomanometer and a stethoscope to listen for the sounds of blood flow.

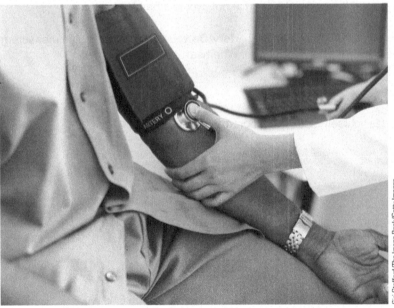

Siri Stafford/The Image Bank/Getty Images

Table 25.1 The American Heart Association's Guidelines for When Elevated Blood Pressure Introduces Risk of Heart Disease

Blood Pressure Category	Systolic mm Hg (upper number)		Diastolic mm Hg (lower number)
Normal	Less than 120	and	Less than 80
Elevated	120–129	and	Less than 80
High blood pressure (hypertension) stage 1	130–139	or	80–89
High blood pressure (hypertension) stage 2	140 or higher	or	90 or higher
Hypertensive crisis (consult doctor immediately)	Higher than 180	and/or	Higher than 120

numbers may be quite different from each other. Blood pressure readings are expressed as a ratio with the systolic pressure expressed over the diastolic number. For example, 120/80mmHg. Table 25.1 indicates some standard ranges for resting blood pressure.

Pulse, which we will be measuring in the radial artery, is a measure of the number of times that the ventricles contract in one minute, expressed in beats per minute or bpm. Pulse is measured by counting the number of times that the artery walls expand during systole. Pulse can only be measured at arteries that are close to the skin.

Using Figure 25.7, find the two vessels we are using to measure blood pressure and pulse. If you measured the blood pressure further from the heart, for example, in the posterior tibial artery, do you think you would obtain the same pressure reading as at the brachial artery? If not, would it be higher or lower?

Pulse can also be measured in the popliteal artery. Do you think that the pulse measurement would change if you measured it at the popliteal artery, farther from the heart? Why or why not?

Learning Connection

Broken process

What do you think happens to blood flow and blood pressure when the arteries lose elasticity?

Materials
- Alcohol wipes
- Another human besides yourself (usually a laboratory partner)
- Sphygmomanometer
- Stethoscope

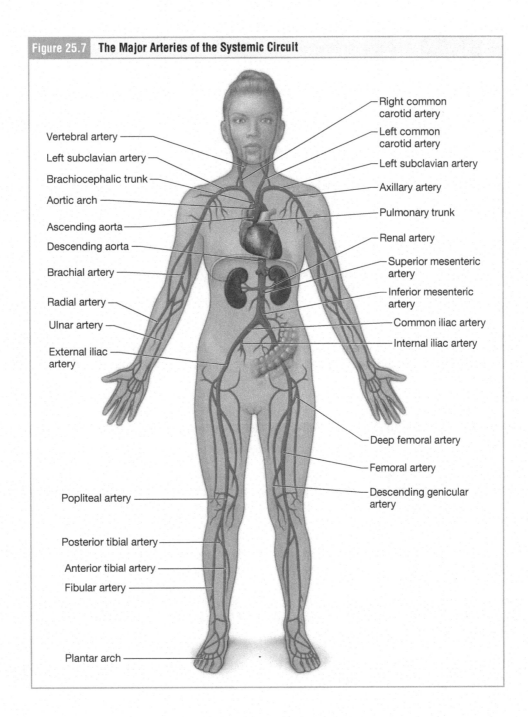

Figure 25.7 **The Major Arteries of the Systemic Circuit**

Right common carotid artery

Left common carotid artery

Left subclavian artery

Axillary artery

Pulmonary trunk

Renal artery

Superior mesenteric artery

Inferior mesenteric artery

Common iliac artery

Internal iliac artery

Deep femoral artery

Femoral artery

Descending genicular artery

Vertebral artery

Left subclavian artery

Brachiocephalic trunk

Aortic arch

Ascending aorta

Descending aorta

Brachial artery

Radial artery

Ulnar artery

External iliac artery

Popliteal artery

Posterior tibial artery

Anterior tibial artery

Fibular artery

Plantar arch

Instructions

Measuring Blood Pressure

1. Have your lab partner sit comfortably for a few minutes. Ask them to roll up the sleeve of one arm past the elbow.

2. While your partner is resting, identify the following parts of your sphygmomanometer using Figure 25.8 for reference: cuff, pressure gauge, rubber bulb, release valve.

Figure 25.8 | A Sphygmomanometer Is Used to Measure Blood Pressure

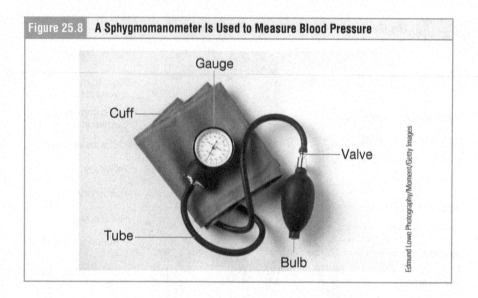

Gauge

Cuff

Valve

Tube

Bulb

Edmund Lowe Photography/Moment/Getty Images

3. Use an alcohol swab to clean the earpieces of a stethoscope. Notice the angle of the earpieces and hold the stethoscope so that the earpieces are angled forward before putting them in your ears (it's more comfortable this way, see Figure 25.4).

4. Loosely wrap the cuff around your partner's arm, then tighten the cuff so that it is snug, but not uncomfortable for your partner. Place the bell of the stethoscope on the distal edge of the cuff, just over the cubital fossa.

5. Check in with your partner and ask if they are comfortable and ready to proceed. Once you begin to inflate the cuff, you need to proceed quickly to avoid discomfort for your partner, so read through the rest of the directions now if you haven't already. When you and your partner are both ready, proceed to step 6.

6. Close the valve on the cuff and squeeze the rubber bulb to inflate the cuff to approximately 160mmHg. Do not inflate higher than 160mmHg as this would be very uncomfortable for your partner!

7. Carefully open the valve of the sphygmomanometer, and slowly deflate the cuff while listening for the sounds of blood once again flowing through the brachial artery. When the first sound is heard, take a note of the pressure at this moment, this is the systolic pressure.

8. Continue to release the valve slowly, the moment when you can no longer hear sounds, note the pressure gauge reading. This is the diastolic pressure.

9. Quickly deflate the cuff and remove it from your lab partner's arm.

10. Record your partner's values here:

11. You might want to repeat this process with your partner in different conditions. Choose your experiments and fill out Table 25.2 with your measurements. You can design your own condition and describe it in the last row. Be sure to remove the sphygmomanometer before moving around the room or building and replace it when you're ready to measure again.

Table 25.2 Measuring Blood Pressure in Various Positions

Condition	Description	Prediction (higher or lower than seated resting BP)	Blood Pressure (systolic/diastolic)	Pulse (bpm)
Lying supine	have your partner lie on a cot or on your lab table for approximately 5 min. Measure BP while still in lying position.			
Standing	Have your partner stand still for approximately 5 min. Measure BP while still in standing position.			
After walking down a hallway	Have your partner walk back and forth down a hallway or across a classroom for approximately 5 min. Measure BP immediately upon stopping.			
After quickly climbing stairs	Have your partner climb stairs as quickly as they are comfortable for approximately 5 min. Measure BP immediately upon stopping.			

Measuring Pulse

12. Have your partner sit down and relax for several minutes, alternatively you can perform a pulse measurement on yourself.

13. Find the pulse of the radial artery as shown in Figure 25.9. Use either your index finger or index and middle fingers but do not use your thumb (it has its own pulse which may confuse your measurement). Applying light but comfortable pressure, record the number of pulses in a 15 second period.

Figure 25.9 Proper Location for Measuring the Radial Artery Pulse

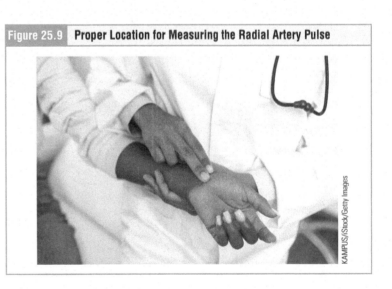

KAMPUS/iStock/Getty Images

14. Record that number here:

_____

Now multiply by 4 to get bpm:

_____

15. If you would like, repeat the conditions in Step 11 and measure pulse at the end of each. Record in Table 25.2.

Activity 25.3 **Electrocardiogram**

This activity targets LOs 25.10 and 25.11.

Materials
- Alcohol swabs
- Cot or table for lying down
- Electrocardiograph
- Electrode paste or gel
- Electrodes and cables

Instructions
Note: your instructor may have specific instructions for operating the type of ECG you have in your lab.

1. Apply electrode gel to four electrodes. Have your partner prepare their skin at the locations in Figure 25.10 by wiping the skin surface with alcohol swabs. Your partner should then apply the prepared electrodes. You can help them if they are comfortable with you doing so.

2. Attach the electrodes to the corresponding cables of the ECG system.

3. Choose the position for your first measurement (seated, standing, lying supine).

4. Turn on the recording instrument and make any adjustments necessary.

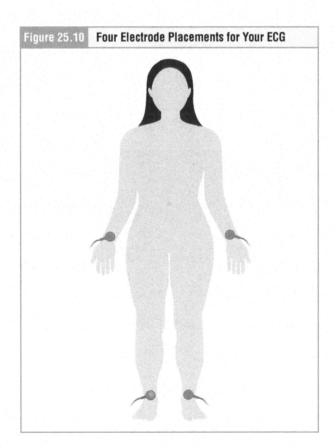

Figure 25.10 **Four Electrode Placements for Your ECG**

5. Once you have a few minutes of recording, your partner may move to a different position and you can record again.

6. On your recordings identify the P wave, QRS complex, and T wave.

7. Measure the PQ interval, the ST interval, and the interval between the T wave and the subsequent P wave.

Activity 25.4 Case Study: Rheumatic Heart Disease

This activity targets LOs 25.1, 25.3, 25.4, 25.6, 25.7, 25.8, and 25.12.

A 34-year-old nurse has been referred to a cardiologist's office. Their chart notes persistent **tachycardia** (increased heart rate) and light-headedness. A complete medical history is taken and it turns out that this adult had rheumatic fever at age 8. Rheumatic fever occurs when strep throat is not treated with antibiotics and the bacteria are able to move past the throat and infect other parts of the body, including the heart. The cardiologist wonders if the bacterial infection in the patient's heart during childhood may be causing the symptoms occurring now.

    The cardiologist orders blood tests, chest x-rays, an echocardiogram, and an ECG.

Case Study

1. What is the difference between an echocardiogram and an electrocardiogram?

2. The image below resulted from which of the tests, echocardiogram or electrocardiogram?

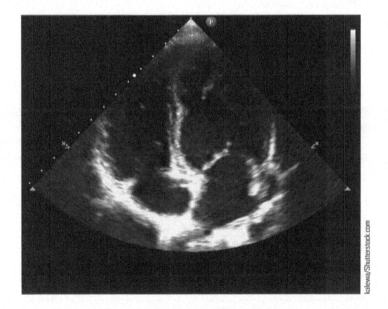

kalewa/Shutterstock.com

During auscultation, the cardiologist notes a gurgling sound at the area marked with the number 2 in Figure 25.5. The sounds at the other locations were more typical. Based on this sound, the cardiologist hypothesizes that the patient may have a leaky valve.

3. Which valve do they suspect a leak in?

   a. Right AV valve

   b. Pulmonary semilunar valve

   c. Left AV valve

   d. Aortic semilunar valve

The chest x-rays showed an enlarged left atrium as well as some fluid in the lungs. Based on these findings, the cardiologist suspects that the patient has regurgitation, or backflow at this valve. In conditions of regurgitation, some of the blood flows backwards through the valve when it closes. Examine the frontal section of the heart in the figure and circle the valve you answered in question 3. Discuss the consequences of backflow at this valve with your lab partner.

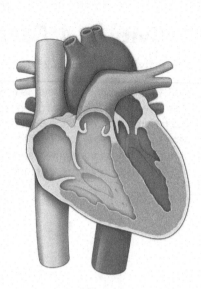

To confirm the diagnosis of regurgitation, the cardiologist turned to the results of the echocardiogram. Thinking through the consequences of backflow at the valve you suggested, fill out Table 25.3 with your predictions of what the cardiologist will find in this patient's lab results.

4. The patient's chart indicates that they have been experiencing tachycardia. Which of the following resting pulse values is consistent with tachycardia?

a. 50 bpm

b. 75 bpm

c. 100 bpm

5. Why is this patient experiencing tachycardia?

Table 25.3 Predicted Lab Values for This Patient			
	Typical Values	Your Prediction for the Patient's Values	Why You Made This Prediction
Blood pressure (measured at the brachial artery)	120/80 mmHg	Higher, lower, or typical?	
Cardiac output	4.9 L/min	Higher, lower, or typical?	
Left atrial pressure	8 mmHg	Higher, lower, or typical?	

Activity 25.5 **Case Study: Arrhythmias**

This activity targets LOs 25.10, 25.11, 25.12, 25.13, and 25.14.

Case Study

The term **arrhythmia** refers to a condition in which the heart rhythm does not follow the typical pattern of lub-dub. Arrythmias are diagnosed using an ECG similar to the one performed in Activity 25.3.

Examine the ECG tracing below. In the boxes in cycle A, write the labels for the P wave, QRS complex, and T wave. Label these waveforms on other cardiac cycles of the ECG.

Measure the RR interval between cycle A and cycle B using the guide in Figure 25.11.

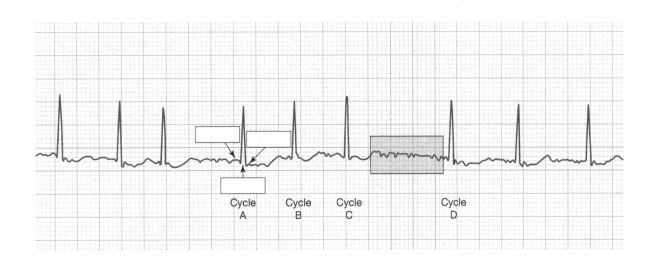

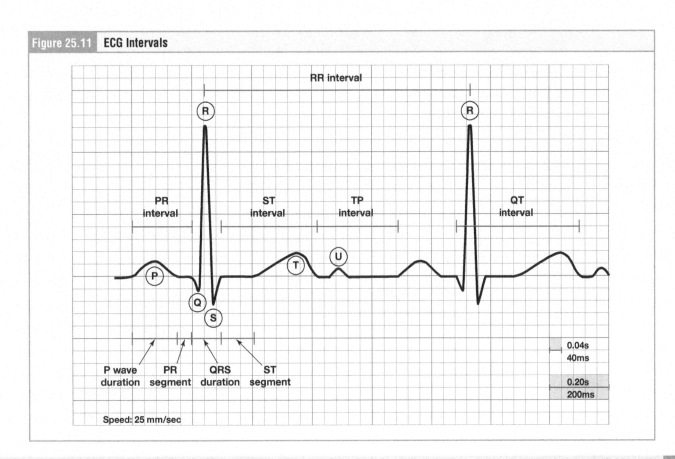

Figure 25.11 | **ECG Intervals**

1. What is the patient's heart rate in bpm?

2. Examine the region shaded in yellow between cycles C and D. It should be clear that there is extra electrical activity there, so which of the following seems the most plausible?

    a. The extra waveforms indicate that the ventricles are depolarizing additional times between cycle C and cycle D.

    b. The extra waveforms indicate that the atria are depolarizing additional times between cycle C and D.

3. Why did you choose your answer choice in Question 2?

This type of arrhythmia is called fibrillation, which involves fast and irregular contractions of either the atria (in which case it's called atrial fibrillation) or ventricles (in which case it's called ventricular fibrillation).

4. Match the muscular event with the blood flow event.

    Contraction   Chamber fills with blood
    Relaxation   Blood ejection

5. In a given fibrillation contraction, do you think the chamber will eject as much blood as during a normal contraction?

    a. Yes, there will be more contractions, but each one will be as powerful as a normal contraction.

    b. No, the fibrillation contractions will not eject as much blood as the normal contractions because the chamber did not have enough time to fill with blood.

6. If this was atrial fibrillation, do you think that the ventricles would fill with as much blood as they would if the patient was not experiencing fibrillation?

    a. Yes, there would be as much blood filling the ventricles as during times when the patient is not experiencing fibrillation because the atria are still contracting.

    b. No, due to inadequate atrial filling time or incomplete atrial contraction, there would be very little blood that reached the ventricles.

    c. Because the majority of blood reaches the ventricles through gravity alone, atrial fibrillation may reduce, but only slightly reduce, the amount of blood reaching the ventricles.

7. One treatment for fibrillation is cardiac ablation, which uses a laser placed inside the interior of the heart, to reshape the conduction pathway. Which portion of the conduction pathway illustrated in the figure would be the target of ablation in this case?

    a. The portion of the conduction system shown by the dotted line

    b. The portion of the conduction system shown by the solid line

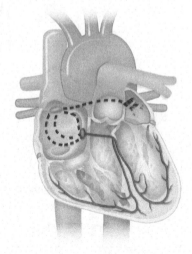

Lab 25 Post-Lab Quiz

1. The left AV valve closes slightly before the right AV valve. Which of the following statements correctly explains why?

   a. The ventricular walls do not depolarize at the same time, as evidenced by the different waveforms for each ventricle on the ECG.

   b. The pressure is greater on the left side.

   c. The left AV valve has four cusps, making it easier to close.

2. In Activity 25.1, heart sounds, you heard the left AV valve on the left side of the heart, and the right AV valve was clearer on the right side of the heart. But the pulmonary semilunar valve, which is connected to the right ventricle, was more clearly auscultated on the *left*. Why do you think that may be? (hint: reviewing the heart anatomy may help!)

   a. Because the vessels cross at the midline as they ascend

   b. Because the sternum prevents auscultation of the pulmonary valve

   c. Because the pulmonary semilunar valve is located far into the pulmonary arteries

3. On the diagram of pressure changes, each arrow indicates the time of a valve opening or closing. Write the name of the valve and if it is opening or closing next to each arrow. Then on the line labeled "Sound," found below the graph, indicate the timing of each heart sound by drawing a squiggle over the line when a heart sound is heard.

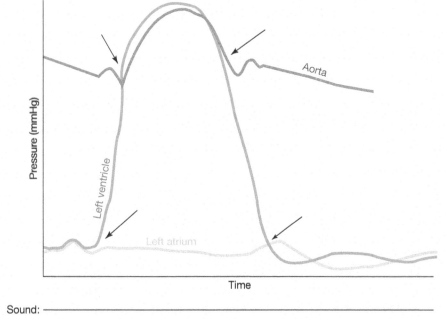

Sound: ———————————————————————————————————————

At the moment when you expect to hear the "hub" and the "dub" draw a squiggle over the sound line like this: —————~~~~~~~~~~—————

4. The arteries of young people and athletic individuals tend to be quite elastic, but as we age (and especially in sedentary individuals) elasticity decreases over time. Below are two blood pressure measurements. One is from a 38-year-old athlete and the other is from a 62-year-old sedentary individual. Which one do you think is from the athlete?

   a. 125/85

   b. 120/55

5. AV node reentry tachycardia (AVNRT) is a type of arrhythmia which can cause sudden (and potentially fatal) increases in heart rate. AVNRT is caused by miscommunication between the sinoatrial node and the atrioventricular node. In a healthy heart, which of these nodes fires first?

   a. Sinoatrial node

   b. Atrioventricular node

   c. They fire at the same time

6. AV node reentry tachycardia (AVNRT) is a type of arrhythmia which can cause sudden (and potentially fatal) increases in heart rate. AVNRT is caused by miscommunication between the sinoatrial node and the atrioventricular node. AVNRT can be cured through a surgical procedure in which the cells causing this miscommunication are burned with a laser. Which type of cells are the target of this procedure?

   a. Autorhythmic cells

   b. Contractile cells

   c. Neither of these types

7. Cardiac muscle cells contract as a single unit because the cells share ions for depolarization via _____ cell junctions.

   a. gap

   b. desmosomes

   c. tight junctions

8. In the ECG of a healthy individual, the __ wave corresponds to the depolarization of the atria.

   a. T

   b. Q

   c. R

   d. S

   e. P

9. Imagine the cuff of a sphygmomanometer that is occluding (blocking) blood flow through the brachial artery, as occurs during a blood pressure measurement. As the cuff loosens, flow is restored through the artery. In what artery is blood flow restored immediately after deflating the cuff?

   a. Radial artery

   b. Axillary artery

   c. Palmar artery

   d. Proximal brachial artery

10. Why is blood pressure measured in arteries and not veins?

    a. Arteries have thicker walls and can better withstand the pressure of the sphygmomanometer.

    b. Blood is ejected from the heart into arteries, whereas veins receive blood from capillary beds, therefore the pressure generated by the heart can be measured more easily.

    c. Arteries have more elastic fibers in their walls and so they expand and retract with the systole and diastole phases more than veins, which have fewer elastic fibers.

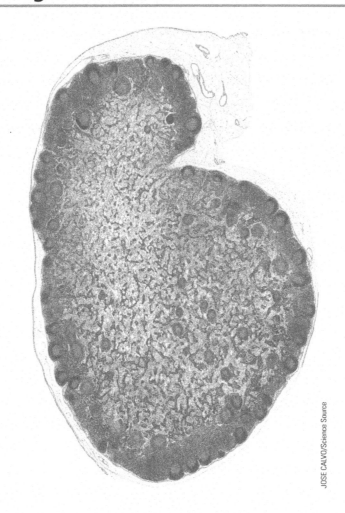

JOSE CALVO/Science Source

Learning Objectives: After completing this lab, you should be able to:

26.1 Describe the structure, functions, and major locations of the following lymphatic organs: lymph nodes, thymus, and spleen.

26.2 Describe the structure, function, and major locations of lymphatic nodules (e.g., mucosa-associated lymphoid tissue [MALT], tonsils).

26.3 Describe the path of lymph circulation.

26.4 Describe the mechanisms of lymph formation and circulation.

26.5* Describe the path of lymphatic fluid into, through, and out of a lymph node.

26.6* Predict the likely area of infection given a swollen lymph node in an area of the body.

* Objective is not a HAPS Learning Goal.

The Human Anatomy and Physiology Society includes more than 1,700 educators who work together to promote excellence in the teaching of this subject area. The HAPS A&P Learning Outcomes measure student mastery of the content typically covered in a two-semester Human A&P curriculum at the undergraduate level. The full Learning Outcomes are available at https://www.hapsweb.org.

Introduction

The lymphatic system is made up of the vessels, cells, and organs that carry excess fluids from the tissues to the bloodstream. During transport it filters pathogens from that fluid and from the blood. The lymphatic system acts as a partner to the circulation of the blood in the blood vessels as it collects excess interstitial fluid from the tissues to return it to the blood. Structures of the lymphatic system are called lymphoid structures and include the **red bone marrow**, **thymus**, **spleen**, **lymph nodes**, **tonsils**, and **diffuse lymphatic tissue**.

The immune system is the complex collection of cells and organs that destroys invaders that could cause disease or death. The immune system depends on the lymphatic system for its function, so we typically discuss the two systems together. The immune system uses the flow of lymphatic fluid through the lymph vessels to monitor the tissues of the body; as lymph fluid leaves each tissue it passes through the lymph nodes, where waiting immune cells can check the fluid for pathogens. The immune system uses similar lymphatic structures in other areas of the body, as well, to monitor for pathogens that may have made their way into the tissues.

Lab 26 Pre-Lab Quiz

This quiz will strengthen your background knowledge in preparation for this lab. For help answering the questions, use your resources to deepen your understanding. The best resource for help on the first five questions is your text, and the best resource for help answering the last five questions is to read the introduction section of each lab activity.

1. Anatomy of the organs of the lymphatic and immune systems. Label the following structures on the image:
   Red bone marrow
   Spleen
   Thymus
   Tonsils

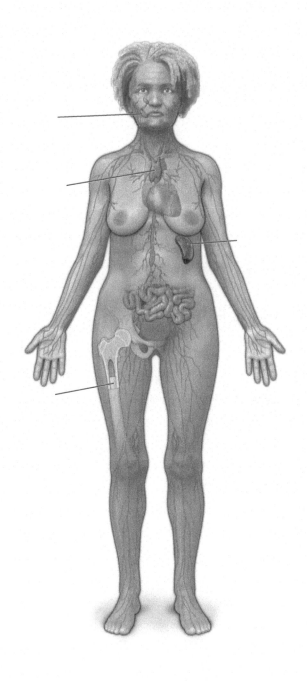

2. Match the root next to its definition:

Root	Meaning
	center
	hollow
	little thing
	rind

List of Roots

cortex
hilus
medulla
sinus

3. What fluid of the body becomes lymphatic fluid?
   a. Urine
   b. Interstitial fluid
   c. Saliva

4. Is there a pump in the body to move lymphatic fluid through the vessels?
   a. Yes
   b. No

5. What structure of the body monitors the lymph for pathogens?
   a. Spleen
   b. Thymus
   c. Lymph nodes

6. Where is the mucosa-associated lymphatic tissue (MALT) found in the body?
   a. In the lining of the digestive tract
   b. In the lining of the respiratory tract
   c. In the lining of the oral cavity

7. Which of these structures is NOT a primary lymphatic organ?
   a. Spleen
   b. Red bone marrow
   c. Thymus

8. Which type of structure directly returns lymphatic fluid into the subclavian veins?
   a. Trunks
   b. Ducts
   c. Vessels

9. What is the outer layer of a lymph node called?
   a. Hilus
   b. Cortex
   c. Medulla

10. Where are the structures called Peyer's patches found?
    a. The lymph nodes
    b. The spleen
    c. The small intestine

Activity 26.1 Overview of the Lymphatic System

This activity targets LOs 26.1 and 26.2.

Lymphoid structures can be categorized based on the basic anatomical difference between the lymphoid organs and lymphoid tissues that are found scattered throughout the body. But what is the difference between a lymphoid organ and a lymphoid tissue? The term "organ" is reserved for those structures that are wrapped in a capsule of dense irregular connective tissue so that it is structurally separated from other tissues around it. Lymphoid organs include the red bone marrow, the thymus, the spleen, and lymph nodes.

In contrast, lymphoid tissue is not wrapped and can be found inside other structures; typically the lymphoid tissue is found within structures that are being monitored for pathogens. Lymphoid tissue includes the tonsils, mucosa-associated lymphatic tissue, and bronchus-associated lymphatic tissue. The tonsils include one **pharyngeal tonsil** (which is sometimes also called the adenoids), a pair of **palatine tonsils**, and an aggregation of **lingual tonsils**, all shown in Figure 26.1. **Mucosa-associated lymphatic tissue (MALT)** is lymphoid tissue that is scattered through the mucosal layer of the digestive tract. **Bronchus-associated lymphatic tissue (BALT)** is lymphoid tissue that is scattered within the wall of the bronchi of the lungs.

Figure 26.1 | Tonsils

(A) Tonsils are specialized lymphatic tissue in the pharynx. (B) The surface of tonsils is deeply cracked or invaginated to capture food particles and expose the lymphocytes to them. These tissues are thought to be particularly important for the building and maintenance of immune tolerance.

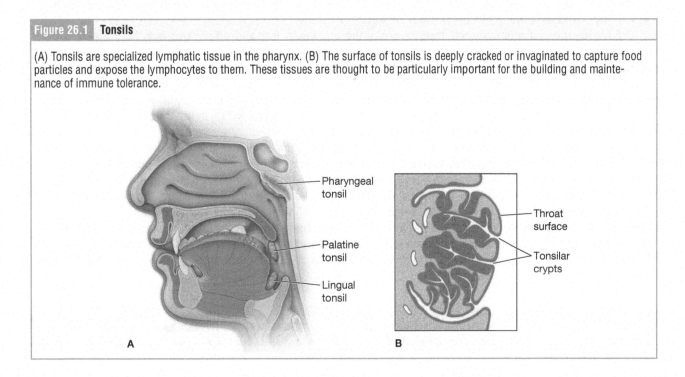

A

B

In addition to distinguishing between lymphoid organs and tissues due to their structural differences, as described in the introduction, it is also possible to categorize the structures of the lymphatic system based on a functional approach. Two organs are called primary lymphatic organs because they are the locations where lymphatic cells that operate in the adaptive immune responses are trained to recognize pathogens: B lymphocytes are trained in the red bone marrow while T lymphocytes are trained in the thymus. Red bone marrow is located in the medullary cavities of longs bones and within the small spaces in spongy bone. The thymus is located superior to the heart within the **mediastinum**. Both of these structures provide for the training of the lymphocytes within a person's first few years of life. As the person ages and the training is complete, these primary lymphatic structures shrink as their job shrinks: some red bone marrow is replaced with adipose tissue to form yellow bone marrow, and the thymus becomes smaller and less active above the heart.

The other structures of the lymphatic system are considered secondary lymphatic structures because they do not train the lymphocytes, but rather serve other necessary immune functions. This category includes the spleen, the lymph nodes, and all the lymphoid tissues listed above.

Whether you are considering the lymphatic system from a functional approach or an anatomical approach, it is helpful to think about how different structures have developed to monitor the body for pathogens that may be present in the various "fluids" of the body: the air that enters the body, the food and liquids that enter the body, the blood that flows through the body, and the lymph fluid that flows through the body. While different lymphatic tissues filter each of these fluids, all these tissues share a common structure within them to facilitate this filtering process: the **lymphoid nodule**. Lymphoid nodules are dense clusters of lymphocytes and other immune cells, such as macrophages, that the fluid flows past. As it flows near the lymphocytes and macrophages in the nodule, these cells can monitor the fluid for pathogens and engage in immune responses as necessary. Which lymphatic system structures monitor which types of fluid? Read the descriptions below and then complete Table 26.1 to analyze and organize the information.

Air that enters the body often contains pathogens. In fact, you can probably remember a time that you got sick with a cold or the flu after breathing the same air as a person who was already sick. The three different kinds of tonsils in your body represented in Figure 26.1 are in areas of high air flow. As air flows past the tonsils, some of the pathogens in it will become trapped in the **tonsillar crypts**. Lining the edges of these crypts are vast quantities of lymphoid nodules (sometimes called lymphoid follicles) that can monitor what has arrived in the deep crevices of the crypts. If pathogens remain in the air flow beyond the tonsils and make it to the **bronchi** (large air passages) of the lungs, BALT in the walls of the bronchi will further filter and hopefully remove any pathogens that try to enter the body here.

The tonsils monitor more than air, though. Since the oral cavity is a shared passageway for both air and food, the tonsils monitor food as well as air. Food particles collect on the tonsils, where the immune cells in the lymphoid nodules can check for potential invaders. Of course, most of the food is not getting checked in the tonsillar crypts, so much of the immune monitoring of what we ingest falls to the MALT within the walls of the digestive tract organs. The lymphoid nodules here will assess any pathogens that attempt to leave the lumen of the digestive tract to pass into the tissues of the body. Large clumps of MALT in the small intestine called Peyer's patches monitor especially vulnerable areas of the tract.

While it is less likely for pathogens to thrive in the bloodstream, the body monitors this fluid as well. The spleen, located left lateral to the stomach, uses specialized versions of lymphoid nodules called **white pulp**, which contain lymphocytes and macrophages, to assess the blood. The spleen then directs the blood leaving the white pulp to pass to nearby tissue called **red pulp**, where old and damaged erythrocytes can be removed and recycled.

Finally, the lymphatic fluid of the body, which is a product of excess interstitial fluid, is assessed by the lymph nodes. These tiny structures are scattered at intervals along the vessels of the lymphatic system so that the lymphatic fluid flowing through these vessels is checked by many successive lymph nodes before it is eventually returned to the bloodstream. The internal structure of each lymph node is arranged so that the fluid flows past the lymphoid nodules within.

Table 26.1 Function and Structure of Lymphatic System Tissues		
Lymphatic Tissue	**Fluid That Is Monitored**	**Characteristics of the Lymphoid Nodules in This Location**
Tonsils and BALT		
Tonsils and MALT		
Spleen		
Lymph nodes		

The structures and tissues of the lymphatic system can be difficult to categorize. The goal of this activity is to practice thinking about the structure and function of each lymphoid organ and tissue to help you organize your thinking.

Materials

- Envelope with nine cards, each listing a lymphatic system organ or tissues. Cards contain the following terms:

  red bone marrow
  thymus
  pharyngeal tonsil
  palatine tonsils
  lingual tonsils
  mucosa-associated lymphatic tissue (MALT)
  bronchus-associated lymphatic tissue (BALT)
  spleen
  lymph nodes

Instructions

1. Working with your lab partner, sort the cards into two piles: primary lymphatic organs and secondary lymphatic organs and tissues.

   a. List the structures that fall into each of the categories in Table 26.2.

   b. Discuss with your lab partner what distinguishes each of these categories and add that information to the table.

2. Return all the cards to a single pile. Then sort all the cards into two new piles: lymphoid organs and lymphoid tissues.

   a. List the structures that fall into each of the categories in Table 26.2.

   b. Discuss with your lab partner what distinguishes each of these categories and add that information to the table.

3. Return all the cards to a single pile. Then sort all the cards into two new piles: structures that contain lymphoid nodules and structures that do not.

   a. List the structures that fall into each of the categories in Table 26.2.

   b. Discuss with your lab partner what distinguishes each of these categories and add that information to the table.

4. Return all the cards to the envelope.

Table 26.2 Organizational Approaches to the Structures of the Lymphatic System

Organizational Approach	Functional Approach		Structural Approach			Lymphoid Nodule Presence	
Categories	Primary lymphatic organs	Secondary lymphatic organs	Lymphoid organs	Lymphoid tissues	Structures that contain lymphoid nodules	Structures that do not contain lymphoid nodules
Characteristics that distinguish these categories from each other						
Structures present within each category						

Activity 26.2 Anatomy of the Lymphatic System

This activity targets LOs 26.3 and 26.4.

Excess interstitial fluid is always accumulating in the spaces between cells because the blood pressure within the blood vessels causes some plasma from the bloodstream to leak into the interstitial spaces. This fluid must be returned to the bloodstream so that blood volume is maintained; it is the function of the lymphatic system to do so. The smallest lymphatic vessels of the body, the **lymphatic capillaries**, receive excess interstitial fluid from between the cells. As the lymphatic capillaries travel away from individual tissues they join to form larger lymphatic vessels. You can see this pattern of merging lymphatic capillaries in Figure 26.2.

Figure 26.2	Lymphatic Capillaries

The smallest lymphatic capillaries leaving a tissue join to allow lymph to flow into larger lymphatic vessels.

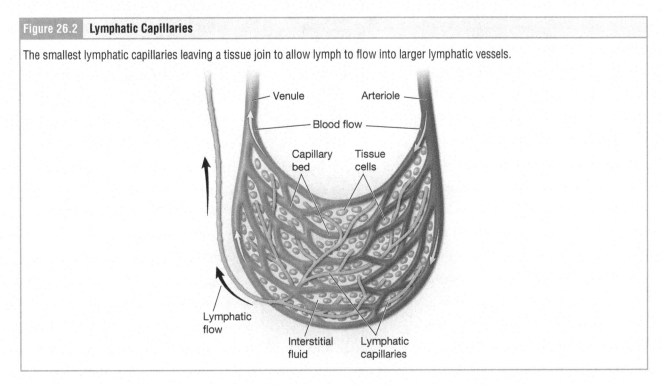

As this excess fluid continues through the lymphatic vessels on its return to the blood, the fluid is monitored for pathogens each time it reaches a lymph node. Lymphatic vessels will enter a lymph node where the fluid is checked as it passes by lymphoid nodules before exiting again to continue through more lymphatic vessels and lymph nodes. If pathogens are detected by the immune system cells in the lymphoid nodules within a lymph node, the immune cells of the body will rapidly proliferate to make a large army of cells that is prepared to fight the invading pathogen. These lymph nodes will swell with the increasing numbers of immune cells.

Eventually, as the lymphatic vessels merge they form even larger structures. These large, merged passageways are called **lymphatic trunks**. The *right and left lumbar trunks* drain the lymph of the right and left legs, respectively. A single *intestinal trunk* drains the lymph of the abdomen. This lymph in the intestinal trunk contains more than just interstitial fluid from the tissues of the abdomen. In addition, it contains the dietary fats that have been absorbed from the digestive tract. These fat molecules must first travel through the lymph before they enter the bloodstream on their way to supply the cells of the body with nutrients. The *right bronchomediastinal trunk* drains the right thorax, the *right subclavian trunk* drains the right arm, and the *right jugular trunk* drains the right neck and head. Similarly, the *left bronchomediastinal trunk*, *left subclavian trunk*, and *left jugular trunk* drain these same structures on the left side of the body. The lymphatic trunks are shown in Figure 26.3.

These nine lymphatic trunks drain the lymph from the entire body, but where is this lymph going? Remember that it is being returned to the bloodstream from which it originated to maintain the fluid balance of the blood. The lymphatic fluid is returned to the bloodstream at the right and left subclavian veins (Figure 26.3). But this lymphatic fluid return is not balanced equally between the two veins. A very short **right lymphatic duct** gathers lymph from the right bronchomediastinal trunk, the right subclavian trunk, and the right jugular trunk to add back to the right sub-clavian vein. The rest of the lymph is gathered along a long **thoracic duct** to be returned to the left subclavian vein.

Figure 26.3 | Anatomy of the Lymphatic Vessels

Lymph flows from the body to be returned at the right and left subclavian veins.

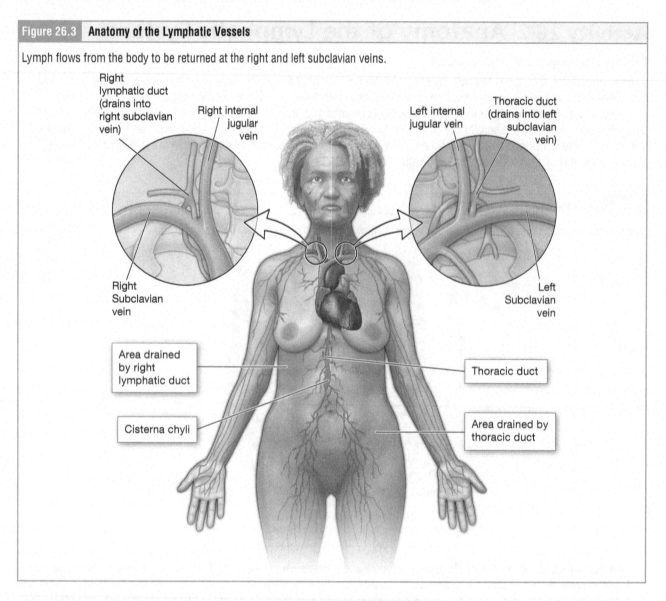

Right lymphatic duct (drains into right subclavian vein)

Right internal jugular vein

Left internal jugular vein

Thoracic duct (drains into left subclavian vein)

Right Subclavian vein

Left Subclavian vein

Area drained by right lymphatic duct

Thoracic duct

Cisterna chyli

Area drained by thoracic duct

The thoracic duct begins mid-abdomen, where the two lumbar trunks and the intestinal trunk join to form a small receptacle called the **cisterna chyli.**

Ascending from the cisterna chyli is the thoracic duct, which extends superiorly to the thorax where the left bronchomediastinal trunk, left subclavian trunk, and left jugular trunk all empty into it before it reaches the left subclavian vein. Examine Figure 26.3 and re-read the description of which trunks empty into which duct to visualize how the shading in the figure represents what area of the body is served by each of the two ducts.

Organize this content in your mind with the following brief exercise. Write in the following terms to the flow chart below to place these structures in the correct order of lymph flow.
• Lymphatic capillaries
• Lymphatic ducts
• Lymphatic trunks
• Lymphatic vessels and lymph nodes

Interstital spaces of the body tissues

Subclavian veins

How does the lymph flow from the tissues of the body to the subclavian veins through these passageways? Remember, for drainage of most structures of the body, the direction of flow is against gravity, and there is no integrated pump for this circulation like the heart of the cardiovascular system. Rather, these vessels have small one-way valves within them, so fluid can only travel in the direction toward the subclavian veins. The regular contraction of the skeletal muscles of the body pushes on the lymphatic vessels sandwiched between them, forcing the fluid forward; it can't flow back the other direction due to those one-way valves.

The purpose of this exercise is to practice tracing the flow of lymph from various tissues of the body to its return to the bloodstream. Use the introduction above, Figures 26.2 and 26.3, your text, and any other resources provided by your instructor to complete this exercise.

Instructions

1. Working with your lab partner, choose one column in Table 26.3, so that you have either the right or left side of the body and your lab partner has the other side.

2. For each structure, trace the path of lymph flow from the structure to its return to the blood.
   a. The first structure has been completed for you.
   b. Work on one structure (i.e., one row) at a time, completing the paths of flow on the left and right sides of the body. That is, while one lab partner is completing the path from the right side of the body, the other lab partner is completing the path from the left side of the body.
   c. After you are both done with the right and left paths for a structure, compare your paths. How similar or different are they? Why? Add this information to the last column of the table.

Learning Connection

Differentiate It

There are several pairs of similar terms in this section of material that are easy to confuse. For each of the pairs of terms below, think about what distinguishes them from each other:

- What is the difference between a lymphoid nodule and a lymph node?

- Which type of lymphocyte is trained in the red bone marrow and which type of lymphocyte is trained in the thymus? (note that the B in "bone marrow" and T in "thymus" can be a memory aid!)

- What function is served by the white pulp and what function is served by the red pulp? (note that the terms "white" and "red" can refer to which type of cells, WBCs or RBCs, are involved!)

- What is the difference between a lymphatic trunk and a lymphatic duct?

Table 26.3 Lymph Flow Pathways

Structure	Right Side of the Body	Left Side of the Body	Explanation of Similarities and Differences
From the foot	Tissue of the R foot → lymphatic capillaries of the R foot → lymphatic vessels of the R leg → lymph nodes of the R leg → R lumbar trunk → cisterna chyli → thoracic duct → L subclavian vein	Tissue of the L foot → lymphatic capillaries of the L foot → lymphatic vessels of the L leg → lymph nodes of the L leg → L lumbar trunk → cisterna chyli → thoracic duct → L subclavian vein	Because both the right and left lumbar trunks enter the cisterna chyli, the path is the same once the lymph reaches this location; both of these paths end at the left subclavian vein (even the path from the right leg!)
From the arm			
From the abdominal structures			
From the thoracic structures (lungs, mediastinum, etc.)			
From the head and neck			

Activity 26.3 **Anatomy and Physiology of a Lymph Node**

This activity targets LO 26.5.

Lymph nodes are specialized for monitoring the lymph. Consider how the following anatomy prioritizes as much exposure as possible between the arriving lymph fluid and the lymphatic nodules containing the immune system cells. Figure 26.4 shows the structure of a single lymph node, wrapped in a layer of dense irregular connective tissue called the **capsule**. The node consists of an outer layer called the **cortex** that contains the lymphatic nodules and a central layer called the **medulla** that gathers the filtered lymph so that it can leave at the **hilus**.

Figure 26.4	A Lymph Node

Cross section of a lymph node shows its internal organization.

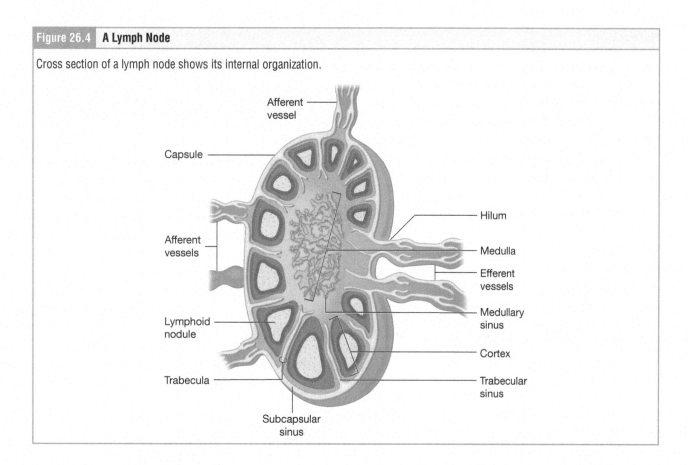

Follow the flow of lymph as it moves through the node. Lymphatic fluid arrives through several **afferent lymphatic vessels**. These vessels are bringing the lymphatic fluid that was gathered from lymphatic capillaries within a tissue and now this fluid has arrived at the node. As the fluid enters through the afferent vessels, it flows under the capsule wrapping the node to pool in the first of many spaces: the **subcapsular sinus**. From this space, the fluid begins to flow toward the medulla of the node. It can head this direction because there are small channels, each called a **trabecular sinus**, that pass around the lymphatic nodules: the columns of tissue that enclose the trabecular sinuses are called **trabeculae** (singular = trabecula). As the fluid flows through the trabecular sinuses it bathes the lymphatic nodules of the cortex, allowing the lymphocytes that are present to check the fluid for the presence of pathogens. The fluid flows from the trabecular sinuses to converge in the **medullary sinuses** of the medulla, where immune cells are present, especially those that have bound a pathogen and begun to divide rapidly. As these cells proliferate, some will remain in the lymph node to encounter more of the same pathogen as it arrives in the node and some will hitch a ride on the departing lymph fluid to reach the areas of the body that contain the pathogen. The lymphatic fluid leaves the node at the hilus where the medullary sinuses converge to dump the departing lymphatic fluid into a single efferent lymphatic vessel. This vessel will continue toward the waiting lymphatic trunk that drains that area of the body but will pass the lymphatic fluid through a series of several more lymph nodes before this occurs.

The purpose of this exercise is to transfer what you understand about lymph node anatomy from a "tidy" artist's rendering of a node to the much less tidy (but more realistic) histological image.

Materials
- Colored pencils or pens
- Histological image of a lymph node, below

Instructions

1. Use Figure 26.4, the activity introduction, your textbook, and any other resources provided by your instructor to label the histological image of a lymph node on the next page.

2. The afferent and efferent vessels are not captured in this histological image, but you can predict where they would be based on the visible structure. Draw in several afferent vessels and an efferent vessel in the proper locations.

3. Bracket the section of the node that is the cortex and label it. Then bracket the section of the node that is the medulla and label it.

4. Using a different color, label the following structures:
   a. Afferent vessels
   b. Efferent vessels
   c. Capsule
   d. Hilus
   e. Lymphoid nodules
   f. Trabeculae

5. Using a third color, identify the spaces where lymph will flow:
   a. Capsular sinus
   b. Trabecular sinus
   c. Medullary sinus

6. Using a fourth color, draw arrows to show the flow of lymphatic fluid into, through, and out of the node. Place your first arrows within the afferent vessels and your last arrow within the efferent vessel.

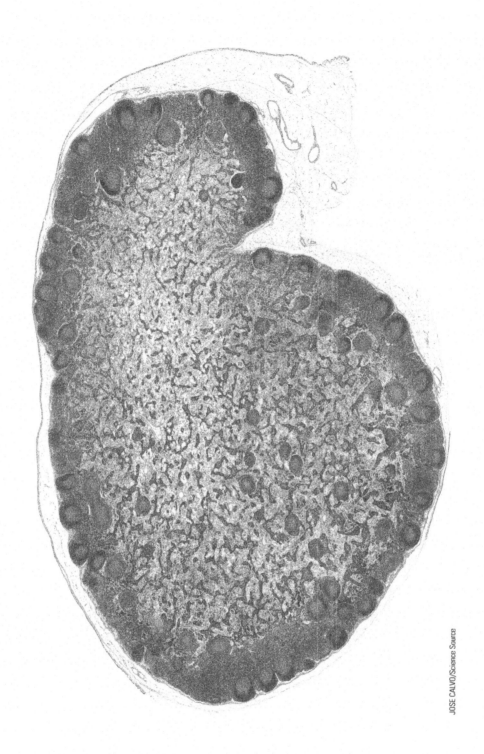

JOSE CALVO/Science Source

Digging Deeper:
Sentinel Lymph Node Biopsy

When an individual is diagnosed with breast cancer, there is concern that the cancer is not localized to the breast tissue but has spread to other areas of the body. One of the first places that breast cancer cells will spread is to lymph nodes in the area. The cancer cells have hitched a ride on the lymph fluid leaving the breast tissue and have gotten caught in the lymph node. Here, the cancer cells may be able to begin growing without interference from the immune cells in the node. Usually, during surgery to remove cancerous breast tissue the oncologist will remove the lymph node in the drainage path that is closest to the cancer, this is called the sentinel lymph node and the procedure to remove it is a sentinel lymph node biopsy (SLNB). The removed node will be sent to a pathologist to determine if it contains any cancer cells. Figure 26.5 shows the many lymph node groups in the area and how each group of nodes drains a different section of the breast tissue.

Figure 26.5 | **Lymph Nodes of the Left Breast**

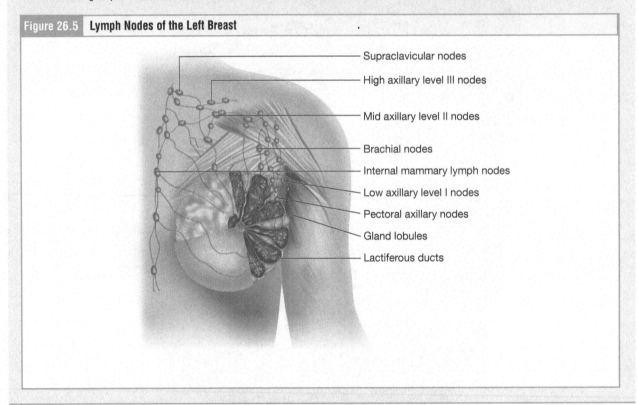

- Supraclavicular nodes
- High axillary level III nodes
- Mid axillary level II nodes
- Brachial nodes
- Internal mammary lymph nodes
- Low axillary level I nodes
- Pectoral axillary nodes
- Gland lobules
- Lactiferous ducts

Activity 26.4 Microscopic Anatomy of Lymphatic Structures

This activity targets LOs 26.1 and 26.2.

The repeating structural feature of lymphoid nodules is often best visualized by examining histological images of the various lymphatic tissues that contain these nodules.

The lymph nodes that filter the body's lymph fluid check for pathogens that may be in the tissues and have now been washed to the nodes. The lymphatic nodules are arranged throughout the cortex to encounter as much of the lymph flowing through the node as possible.

The arrangement of the spleen is somewhat different, as is to be expected because it is filtering the blood rather than the lymph. As shown in Figure 26.6, blood arrives at the spleen via the splenic artery, which branches and gives rise to many small capillaries that spread throughout the tissue of the spleen. The blood first arrives from these capillaries into the white pulp. Don't be deceived by the name of this area of the spleen. Although it is named for the abundance of leukocytes (white blood cells) that are present here, the typical stains used to prepare the tissue on the slide usually render these sections of the tissue a dark purple. The tell-tale clue to distinguish the white pulp is that it will look like small islands in a sea of the rest of the splenic tissue, which is red pulp.

Figure 26.6	Anatomy of the Spleen

Blood arrives at the spleen and first travels through the white pulp before continuing to the red pulp.

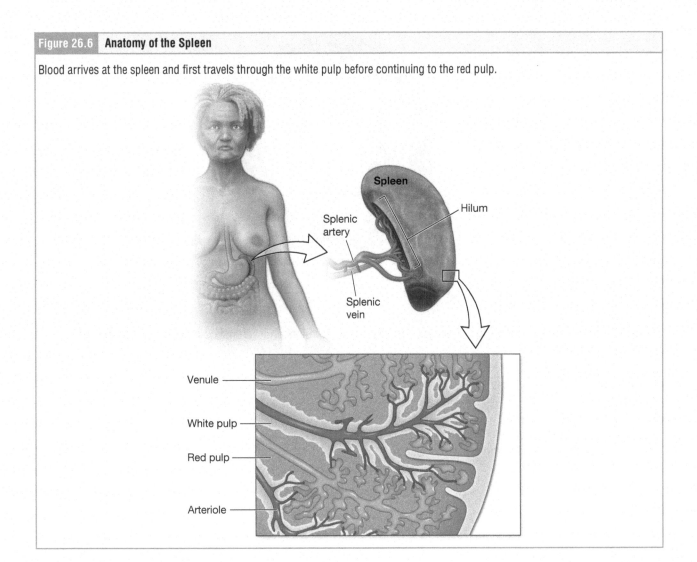

The tonsils of the oral cavity (one pharyngeal tonsil, a pair of palatine tonsils, and an aggregation of lingual tonsils, all shown in Figure 26.1), all share a similar structure. They have deep crevices, the tonsillar crypts, that provide spaces that trap pathogens. Lining the tonsillar crypts are lymphoid follicles, where immune system cells are ready and waiting to encounter those pathogens.

The mucosa-associated lymphatic tissue (MALT) is found in the mucosal layer of the gut lining, as shown in Figure 26.7. While the lymphatic nodules that make up the MALT are scattered throughout the gut lining, high densities of these nodules are clustered in the final section of the small intestine, where they are called Peyer's patches. The MALT is more robust in this location because the bacteria-laden large intestine is nearby. The immune cells of the Peyer's patches monitor the mucosa of this area of the small intestine in case any bacteria from the large intestine attempt to cross the mucosa to enter the body. The lymphatic nodules of the Peyer's patches may be less defined and clear than the nodules you have seen in the other slides. To locate them, find the lumen of the small intestine and look for the finger-like projections of the villi that mark the edge of the tissue. Scroll away from the lumen to look at the base of the villi, you should see spherical clusters that stain more darkly than the surrounding tissue; these are the lymphatic nodules that make up the Peyer's patches.

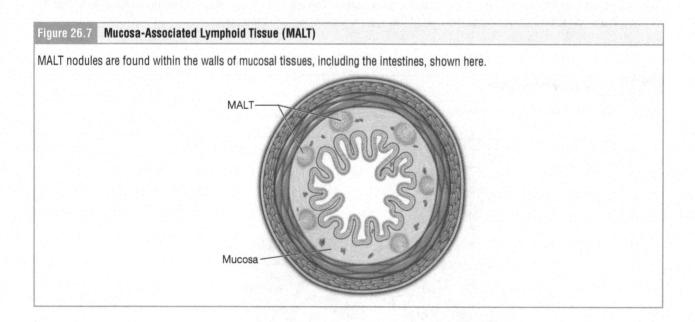

Figure 26.7   **Mucosa-Associated Lymphoid Tissue (MALT)**

MALT nodules are found within the walls of mucosal tissues, including the intestines, shown here.

The purpose of this exercise is to help you identify key anatomical characteristics of some of the lymphatic organs and tissues.

Materials
• Microscope
• Prepared slides of:
  • Lymph node (alternately, use the histological image in Activity 26.3)
  • Peyer's patches or ileum of the small intestine
  • Spleen
  • Tonsil

Instructions

1. Obtain the slides.

2. For each slide, examine the slide first on scanning power (4×) and use the coarse focus knob to locate the tissue. Then increase the magnification by switching to the low power (10×) objective lens and use the fine focus knob. Then switch to the high power (40×) objective lens and again use the fine focus knob until you can clearly see the tissue.

3. Draw and describe each tissue in the spaces below, identifying all requested structures that are listed. In a blank area on the drawing, note the total magnification used.

Lymph Node

On your drawing, identify: capsule, lymphatic nodules, cortex, and medulla.

Spleen

On your drawing, identify: red pulp, white pulp, and blood vessels.

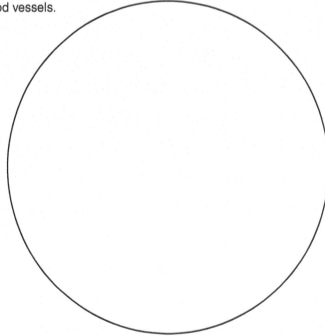

Tonsil

On your drawing, identify: tonsillar crypt and lymphoid nodules.

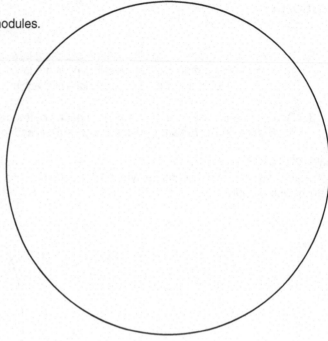

Peyer's Patches

On your drawing, identify: Peyer's patches (these are lymphoid nodules), villi of the small intestine, and lumen of the small intestine.

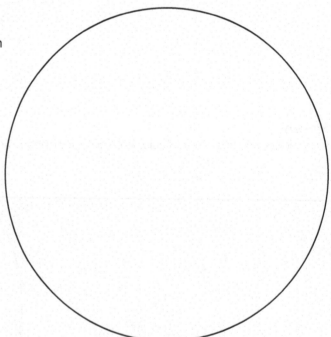

Activity 26.5 Case Study: Lymphadenopathy

This activity targets LO 26.6.

Jordan wakes up the morning of an important exam and isn't feeling well. It hurts to swallow and the muscles in their arms and legs feel sore. They decide to tough it out and take their test, but by the end of the day they feel worse, and they have a significant swelling on either side of their neck. The next day, Jordan goes to the campus health clinic. The nurse examines Jordan and feels the swelling on the sides of their neck. "You have swollen cervical lymph nodes," says the nurse.

1. Assuming that Jordan is infected with a pathogen, why have the lymph nodes swelled?
   a. The pathogen is getting caught in the nodes, as more gathers there the nodes get larger.
   b. The immune system cells of Jordan's body are dividing to make more immune cells to fight the pathogen, as their numbers increase the nodes get larger.
   c. Lymphatic fluid is arriving in the nodes and not moving forward. As more fluid arrives the nodes get larger.

2. Considering the location of the swollen lymph nodes, which location in the body most likely is infected with pathogen?
   a. Arms
   b. Throat
   c. Lungs

3. The lymphatic fluid leaving the cervical lymph nodes will flow through lymphatic vessels. These lymphatic vessels will then join to form which trunk?
   a. Jugular trunk
   b. Subclavian trunk
   c. Bronchomediastinal trunk

4. Jordan's cervical lymph nodes are swollen on both sides of their neck. The lymph leaving these nodes will eventually be returned to the blood at what location(s)?
   a. From the right lymphatic duct into the right subclavian vein
   b. From the thoracic duct into the left subclavian vein
   c. Both a and b

The nurse suspects that Jordan may have strep throat or infectious mononucleosis (mono). Both infections can present with a sore throat, achy body, and swollen lymph nodes. The nurse notices some swelling in Jordan's tonsils and swabs the back of Jordan's throat to send off to the lab (Figure 26.6). The results come back positive for group A *Streptococcus* (group A strep throat). Jordan's nurse prescribes a round of antibiotics to help Jordan fight off the infection.

5. Which tonsils, shown here, was the nurse able to see in Jordan's throat?
   a. Lingual tonsils
   b. Pharyngeal tonsils
   c. Palatine tonsils

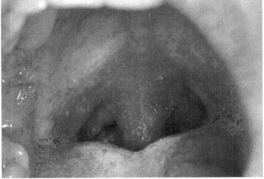

DR P. MARAZZI/SCIENCE Source

6. Jordan's tonsils are swollen because the *Streptococcus* bacteria are growing there. How did the bacteria get there?

    a. Jordan picked up the bacteria in a cut, and the bacteria used the blood to travel to the tonsils.
    b. Jordan inhaled some air that contained the bacteria, and the bacteria were caught in the tonsillar crypts.
    c. The bacteria were in the tissues of Jordan's head and neck, and the bacteria used the lymph fluid to travel to the tonsils.

7. How did the *Streptococcus* bacteria make its way from the tonsils to the lymph nodes?

    a. The bacteria made its way into the bloodstream and then on to the lymph fluid.
    b. The bacteria was already in the lymph nodes before it arrived at the tonsils.
    c. Interstitial fluid in the tonsils drained into the lymphatic capillaries.

Lab 26 Post-Lab Quiz

1. Where is the thymus found?
   a. Superior to the heart
   b. Anterior to the larynx
   c. Left lateral to the stomach

2. If a person's spleen is damaged it may have to be removed. Which complication would be most likely in a person who has had their spleen removed?
   a. Their lymph fluid will be more likely to have pathogens in it.
   b. They are more likely to become infected with pathogens from the food they eat.
   c. They are more likely to develop a blood infection.

3. If a young child has an infection in their thymus and must get it removed, which health concern may arise?
   a. The child may produce less lymphatic fluid.
   b. The child may produce fewer B lymphocytes.
   c. The child may produce fewer T lymphocytes.

4. What is found inside lymphatic nodules?
   a. Lymph fluid
   b. Immune system cells
   c. Blood

5. The tonsils of the body monitor what substance for the presence of pathogens?
   a. The air
   b. The blood
   c. They lymph

6. The lymphatic capillaries and vessels of the abdominal area all lead to the intestinal trunk, which then flows into the cisterna chyli. The lymphatic fluid in these vessels is white rather than clear. Which of the following substances, found in the lymph here but nowhere else, makes it look white?
   a. Pathogens
   b. Dietary fats
   c. Lymphocytes

7. The term edema refers to the swelling of a body area due to excess fluid trapped in those tissues. One thing that can cause edema is when lymph nodes are blocked. Examine this image. The lymph nodes in which area of this person's body are blocked?
   a. The lymph nodes of their right neck
   b. The lymph nodes of their right hand
   c. The lymph nodes of their right axilla (armpit)

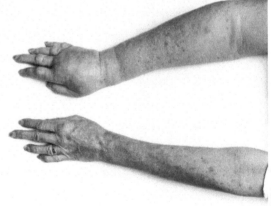

SPL/Science Source

8. Which lymphatic trunk drains lymph from the right lung?

a. Right bronchomediastinal trunk

b. Right jugular trunk

c. Right lumbar trunk

9. Once lymph fluid enters a lymph node, it will first pass into which area?

a. Medullary sinuses

b. Trabecular sinuses

c. Capsular sinuses

10. The inguinal lymph nodes are a group of nodes found in the groin where the leg joins the pelvis. If a person's right inguinal lymph nodes are swollen, you can predict that they may have an infection in what area of the body?

a. Their right abdomen

b. Their right thorax

c. Their right leg

Lab 27 | Anatomy of the Respiratory System

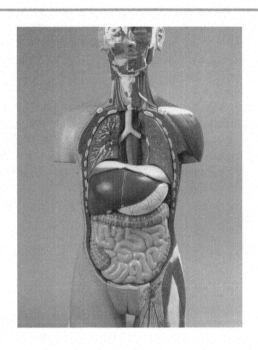

Learning Objectives: After completing this lab, you should be able to:

27.1 Describe the major functions, gross anatomical features, and epithelial lining of the nasal cavity, paranasal sinuses, and pharynx.

27.2 Describe the anatomical features of the larynx, including the laryngeal cartilages.

27.3 Compare and contrast the location, composition, and function of the vestibular folds (false vocal cords) and vocal folds (true vocal cords).

27.4* Explain the position and function of the tissues of the larynx in the process of speech production.

27.5 Describe the gross anatomical features of the trachea, including its positioning with respect to the esophagus.

27.6 Identify and describe the anatomic features of the bronchial tree (e.g., main [primary] bronchi, lobar [secondary] bronchi, segmental [tertiary] bronchi, smaller bronchi, bronchioles, terminal bronchioles, respiratory bronchioles, alveolar ducts, alveolar sacs, and alveoli).

27.7 Compare and contrast the gross anatomic features of the left and right lungs, and explain the reasons for these differences.

27.8 Describe the histological changes that occur along the bronchial tree from larger to smaller air passageways.

* Objective is not a HAPS Learning Goal

The Human Anatomy and Physiology Society includes more than 1,700 educators who work together to promote excellence in the teaching of this subject area. The HAPS A&P Learning Outcomes measure student mastery of the content typically covered in a two-semester Human A&P curriculum at the undergraduate level. The full Learning Outcomes are available at https://www.hapsweb.org.

Introduction

All cells of the body need oxygen to perform aerobic cellular respiration, which generates large amounts of ATP. The process of aerobic cellular respiration also produces carbon dioxide as waste. The respiratory system is critical because it moves these two gasses to and from the very effective O_2-delivery system of the blood. Oxygen is in the air we breathe; this air travels through the respiratory tract to small sacs in the lungs called **alveoli**. Here, oxygen diffuses into the blood and carbon dioxide diffuses out of the blood.

The respiratory tract is a series of specialized structures that enables the air to move rhythmically in and out of the lungs. The lungs are positioned lateral to the heart and occupy most of the thoracic space. The **diaphragm** is a thin sheet of skeletal muscle that lays against the base of each lung. It is the rhythmic contraction of this muscle that changes the dimensions of the thoracic cavity which, in turn, increases or decreases the volume of each alveolus to drive air in and out of the lungs.

Structurally, the respiratory tract is divided into upper and lower regions. The **upper respiratory tract** includes the **nasal cavity**, **pharynx** (also called the throat), and **larynx** (also called the voicebox). The **lower respiratory tract** consists of the **trachea** (also called the windpipe), **bronchial tree**, and alveoli. The respiratory tract can also be described from a functional perspective, comprised of two different zones. The **conducting zone** is the tubes of the tract through which air travels, while the **respiratory zone** is the location where gas exchange occurs between the air in the alveoli and the bloodstream.

Lab 27 Pre-Lab Quiz

This quiz will strengthen your background knowledge in preparation for this lab. For help answering the questions, use your resources to deepen your understanding. The best resource for help on the first five questions is your text, and the best resource for help answering the last five questions is to read the introduction section of each lab activity.

1. Anatomy of the respiratory system. Label the image with the following structures:

    Diaphragm
    Larynx
    Left lung
    Nasal cavity
    Right lung
    Trachea

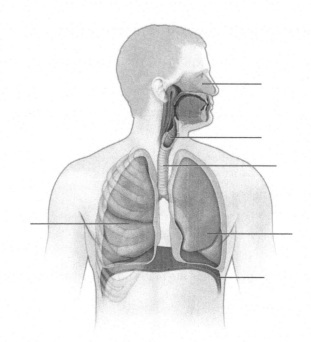

2. Match the root next to its definition:

Root	Meaning
	hollow sac
	mouth
	nose
	space
	wing

List of Roots

ala
alveus
nasus
ori
sinus

3. What structure is positioned directly posterior to the trachea?
    a. The larynx
    b. The lungs
    c. The esophagus

4. Which lung is larger?
   a. The right lung
   b. The left lung
   c. They are of equal size

5. What position are the vocal cords in when sound is produced?
   a. Open
   b. Closed

6. Which of the following structures is commonly called the throat?
   a. The trachea
   b. The larynx
   c. The pharynx

7. The tissue that makes up the nasal septum is rigid due to what tissue?
   a. Cartilage
   b. Bone
   c. Both cartilage and bone

8. Other than the epiglottis, all other cartilage in the airways is made up of what specific tissue type?
   a. Hyaline cartilage
   b. Elastic cartilage
   c. Fibrocartilage

9. The trachea branches into what paired structures?
   a. Bronchioles
   b. Bronchi
   c. Alveolar sacs

10. The walls of the alveoli or air sacs of the lungs are made up of what tissue type?
    a. Hyaline cartilage
    b. Pseudostratified ciliated columnar epithelium
    c. Simple squamous epithelium

Activity 27.1 Anatomy of the Nose

This activity targets LO 27.1.

The nose is the beginning of the conducting zone of the respiratory tract. Examine Figure 27.1 to view the features of the external nose. The **root** of the nose is the area between the eyebrows, while the **bridge** of the nose is the area between the orbits. The **dorsum nasi** is the length of the nose that leads to the **apex** or tip. A pair of rounded flared tissue called **ala** on either side of the apex make space for two openings to the nasal cavity, the **nares** or nostrils. Underneath the thin skin of these external nasal features is a combination of flexible cartilage and hard bone.

Figure 27.1	The External Features of the Nose

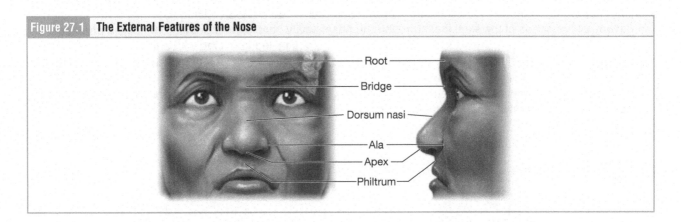

Figure 27.2 shows the internal anatomy of the nose. Each of the nares opens into a separate cavity within the nose, separated by the **nasal septum**. Anteriorly, the nasal septum is made of cartilage. These anterior-most spaces of the nasal cavity, surrounded by cartilage, are called the **nasal vestibules**. The vestibules are filled with short, coarse hairs, **vibrissae**, that catch large particles in the air before the air enters further into the nose. Posterior to

Figure 27.2	Structures of the Nose

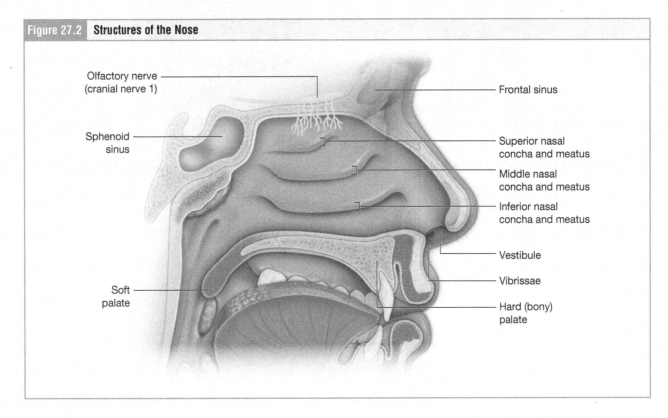

the cartilage of the nasal septum, the bony nasal septum continues the division of the nasal cavity; it is made of the perpendicular plate of the ethmoid bone and the vomer bone. On the lateral wall of each of the nasal cavities are the superior, middle, and inferior **nasal conchae** (also called nasal turbinates). These bony projections, covered with mucous membrane, create turbulent flow as air is inhaled through the nose. The air swirls over and between the spaces between the conchae (called the superior, middle, and inferior **meatuses**), warming and humidifying the incoming air. The floor of the nasal cavity is made of the **hard palate**, this sheet of bone (that is formed by the palatine process of the maxilla and the horizontal process of the palatine bone) separates the nasal cavity from the oral cavity. Posteriorly, where the horizontal process of the palatine bone terminates, the soft tissue covering the hard palate continues as the **soft palate**. The mucosal tissue lining the nasal cavity is pseudostratified ciliated columnar epithelium rich with mucus-producing goblet cells except on the roof of the cavity. Here, the olfactory nerves project through the cribriform plates of the ethmoid bone to create olfactory epithelium used to detect odors.

Several of the bones of the skull that surround the nasal cavity have spaces within them called **paranasal sinuses**, shown in Figure 27.3. These spaces within the bone are air-filled and lined with a thin layer of epithelial tissue, the sinuses open through narrow passageways into the nasal cavity. Each of the sinuses is named for the bone in which it is found: a pair of **frontal sinuses**, a pair of **maxillary sinuses**, a group of small spaces called either the **ethmoid sinuses** or *ethmoid air cells*, and a single **sphenoid sinus**. These spaces are thought to both reduce the weight of the skull and act as resonance chambers to amplify the sound of the voice.

| Figure 27.3 | The Paranasal Sinuses Surrounding the Nasal Cavity |

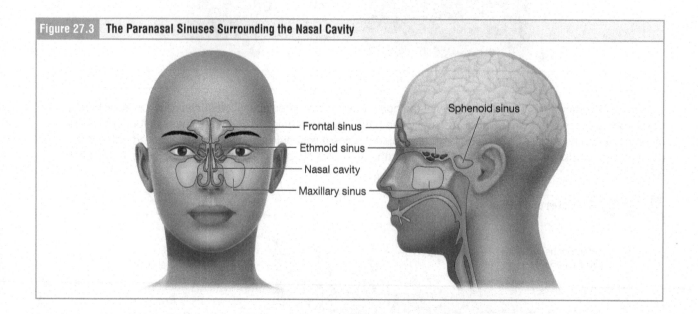

The purpose of this exercise is to help you identify key anatomical structures of the nose and nasal cavity.

Materials
- Colored pencils
- Midsagittal anatomical model of the head and neck, if available
- Scissors
- Skull or skull model with removable cranium
- Sticky notes

Instructions

1. Find the bones and bone structures listed below.

   List of bones and bone structures to find on the skull model and image:
   - i. Frontal bone
   - ii. Nasal bones
   - iii. Maxillary bones
   - iv. Perpendicular plate of the ethmoid bone
   - v. Vomer
   - vi. Superior nasal conchae (may not be visible on your skull model; not visible on the image below)
   - vii. Middle nasal conchae (should be visible on your skull model; not visible on the image below)
   - viii. Inferior nasal conchae
   - ix. Palatine processes of the maxillae (not visible on the image below)
   - x. Horizontal plates of the palatine bones (not visible on the image below)

   a. Locate each bone or bone structure on the skull model and then identify it (when possible) on the image of the skull shown on the next page.
   b. Label and color each bone or structure with colored pencils on the skull image.

2. Using Figure 27.3 as a guide, determine the location of each of the paranasal sinuses on the skull.
   a. For each of the paranasal sinuses, determine the approximate shape of the spaces that are inside the bones in your skull or skull model and sketch the shape onto a sticky note (make sure some of your sketch overlaps with the sticky section so that it will stick to your skull when you cut out the shape).
      - i. For the frontal and maxillary sinuses, which each comes in a pair, cut out a pair of shaped sticky notes. Stick them to the anterior surface of the skull.
      - ii. Repeat this process for the ethmoid sinuses (draw these as a group on a single sticky note) and the single sphenoid sinus. These sinuses may be more difficult to determine where you should adhere your sticky note. There are several different ways to correctly identify their location; discuss with your lab partner how you would like to do so.
   b. Sketch and label the locations of the paranasal sinuses on the skull image below using either a different color or dashed lines for each sinus.

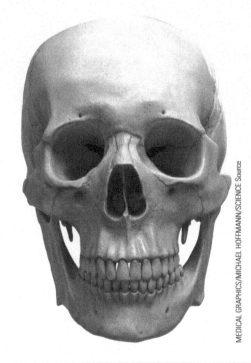

MEDICAL GRAPHICS/MICHAEL HOFFMANN/SCIENCE Source

3. Find the soft tissue structures listed below.

List of soft-tissue structures to find on the model and/or image:

    i. Root of nose
    ii. Bridge of nose
    iii. Dorsum nasi
    iv. Apex of nose
    v. Nasal vestibule
    vi. Superior meatus
    vii. Middle meatus
    viii. Inferior meatus

a. Locate each structure on the midsagittal head and neck model (if available) and then identify them on the midsagittal cadaver dissection shown below.
b. Label each structure on the image.
c. Label the frontal sinus and sphenoid sinus, which are also visible on the image.

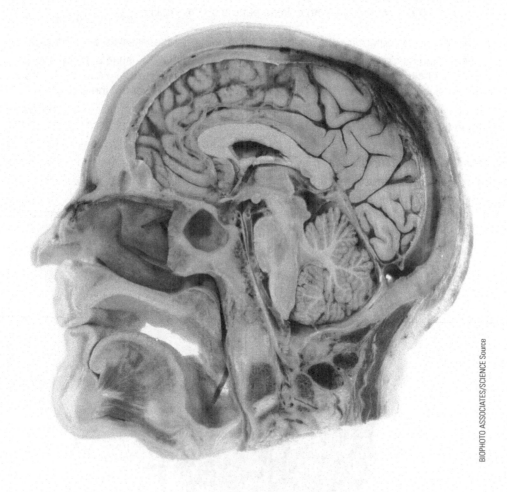

Activity 27.2 Anatomy of the Pharynx

This activity targets LO 27.1.

The pharynx, or throat, is the space posterior to the nasal cavity and oral cavity. It serves as a shared passageway, providing a route for inspired air to reach the larynx and for food to reach the esophagus.

The pharynx is separated into three areas (Figure 27.4). The **nasopharynx** is posterior to, and continuous with, the nasal cavity. The nasopharynx is continuous with the **oropharynx**, which is posterior to the oral cavity. Inferior to the oropharynx, the **laryngopharynx** is posterior to the larynx.

As air is inspired and passes through the nasal cavities it joins together into a single cavity at the nasopharynx; this region is structured with the expectation that only air will pass through it. This area of the pharynx is lined with the same pseudostratified ciliated columnar epithelium as the nasal cavity. Here the air slows as it turns to descend through the nasopharynx. At this superior posterior aspect of the nasopharynx are the **pharyngeal tonsils**. The slowing air enables some of the debris that was not caught by the vibrissae or mucus of the nasal cavity to become caught in the deep folds and pits of the **tonsillar crypts** of the pharyngeal tonsils. The tonsils are aggregations of lymph tissue similar to lymph nodes that function in pathogen destruction. The inferior openings of the **auditory tubes** are located on the lateral walls of the nasopharynx; these tubes enable the air of the middle ear to equalize with the air of the environment. When you feel your ears "pop" as you ascend or descend a mountain or ride in an airplane, you are feeling the changing air pressure of the environment move through the auditory tubes to equalize with the air pressure in the middle ear.

The oropharynx is a shared passageway for both air and food and is lined with stratified squamous epithelium to accommodate the friction of passing food. This area of the pharynx contains two distinct sets of tonsils: the **palatine tonsils** are located laterally in the oropharynx and the **lingual tonsils** are located at the base of the tongue. The tonsils serve to monitor and remove debris from the inspired air.

The laryngopharynx continues the route for both food and air until its distal end, where the digestive system and respiratory system split. As food passes through the laryngopharynx, it presses downward on the flexible **epiglottis**, which closes the trachea, and the food is directed posteriorly into the esophagus. Once the food has passed, the epiglottis springs upward and anteriorly, re-opening the passageway for air to enter and exit the larynx. As is true of the oropharynx, this area is also lined with stratified squamous epithelium that is resistant to abrasive forces from passing food.

The purpose of this brief exercise is to familiarize yourself with the areas of the pharynx and the path of inspired air through them.

Materials

• Midsagittal anatomical model of the head and neck, if available

Instructions

1. Figure 27.4 shows the divisions of the pharyngeal areas but does not identify the structures described in the introduction above. Use a midsagittal anatomical model of the head and neck (if available), your text, and other resources provided by your instructor to label Figure 27.4 with the following structures:

   a. Pharyngeal tonsils
   b. Palatine tonsils
   c. Lingual tonsils
   d. Opening of auditory tube

2. Assuming that air is inspired through the nose rather than the mouth, draw arrows on **Figure 27.4** to show the path of air from the environment to the trachea.

Figure 27.4	**The Parts of the Pharynx**

The pharynx is separated into three areas.

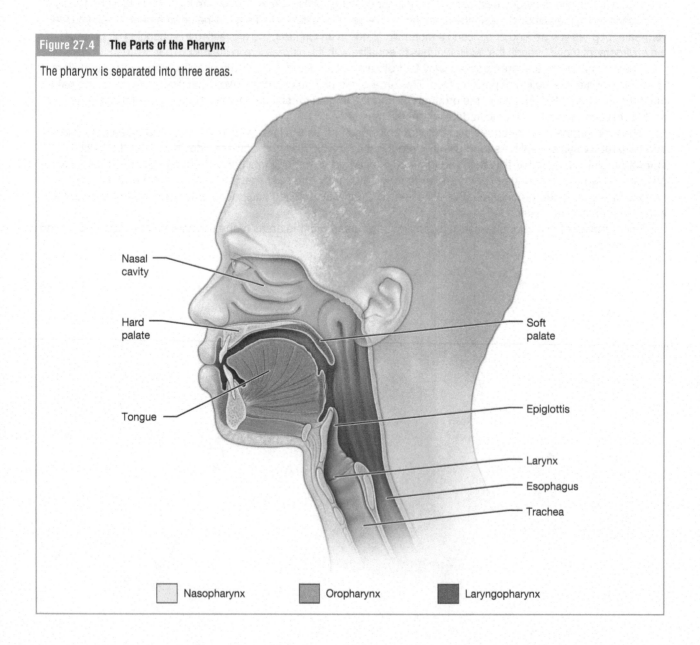

Nasal cavity

Hard palate

Tongue

Soft palate

Epiglottis

Larynx

Esophagus

Trachea

☐ Nasopharynx   ☐ Oropharynx   ◼ Laryngopharynx

Activity 27.3 **Anatomy of the Larynx**

This activity targets LOs 27.2, 27.3, and 27.4.

The larynx is a cartilaginous structure that serves as the entrance to the lower respiratory tract. Because this structure is what enables us to generate sound for language it is sometimes called the voice box. Anatomically, a total of nine separate pieces of cartilage come together to make the larynx. Three large pieces form the greater structure and three smaller pairs of cartilage internal to them create the mechanism for controlling the vocal cords.

Examine Figure 27.5 to see the three large pieces of cartilage that make up the bulk of the larynx: the epiglottis, the thyroid cartilage, and the cricoid cartilage. The epiglottis, attached to the anterior aspect of the larynx, is made of very flexible elastic cartilage. The epiglottis is like a springy trap door that bends inferiorly to close off the airway when food is swallowed, re-directing it posteriorly toward the esophagus. At the same time, the rest of the larynx rises to meet it, creating an effective barrier against swallowed material entering the airway. The epiglottis then springs back open to re-establish the airway. To imagine how it works, bend your ear (which is also made of elastic cartilage) forward and then let it go.

The remaining cartilage pieces of the larynx and trachea are all composed of hyaline cartilage to provide firm but flexible structure to these areas. The **thyroid cartilage** is a large shield-shaped piece of cartilage that makes up the anterior and lateral walls of the larynx and creates an anchor point for other laryngeal structures. On its exterior, anterior surface the thyroid cartilage forms a peak, called the **laryngeal prominence**, that you may be able to feel if you palpate your own larynx. Inferior to the thyroid cartilage is a thick ring that widens posteriorly: the **cricoid cartilage**. The cricoid cartilage acts as a base or platform for the larynx, with the thyroid cartilage balanced on its anterior surface and several of the smaller pairs of cartilage anchored to its posterior surface. The trachea begins inferior to the cricoid cartilages with closely spaced C-shaped pieces of cartilage.

Before describing the three smaller pairs of cartilage at work in the larynx, it is useful to imagine the larynx as it exists in a living body, covered with tissue. Figure 27.6 shows a view downward into the larynx. The bottom of the image is the anterior aspect of the larynx. Deep to the tissue you see in the bottom of the image is the anterior aspect of the thyroid cartilage. Attached to this anterior aspect of the thyroid cartilage are two important tissues. The **vestibular folds** are folded sections of epithelial tissue; these highly innervated folds are easily irritated and will prompt a dramatic coughing reaction if touched by food or liquid entering the larynx. Think of the vestibular folds as aggressive guards at the entrance to the larynx who are under strict orders to only allow air to pass. Inferomedial to the vestibular folds are a second pair of folds: the **vocal folds** (also called the vocal cords). The vocal folds wrap over a pair of fibrous connective tissue **vocal ligaments** that are fixed at their anterior attachment to the thyroid cartilage but are not fixed at their posterior attachment to moveable cartilage. They can pinch closed or swing open (as shown in the photo) to change the dimensions of the **glottis** (the space between the vocal folds).

Figure 27.5 | **The Cartilaginous Structure of the Larynx**

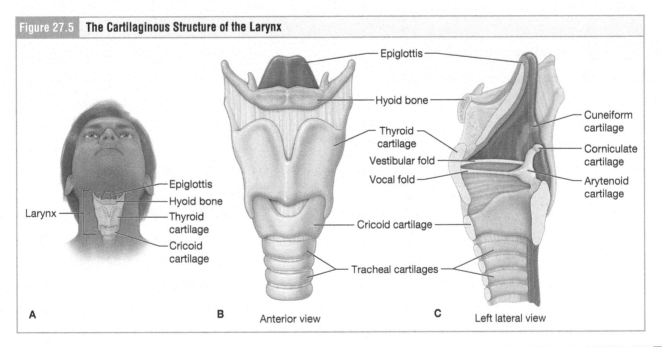

A

B   Anterior view

C   Left lateral view

Figure 27.6 | Structure of the Larynx, in Situ

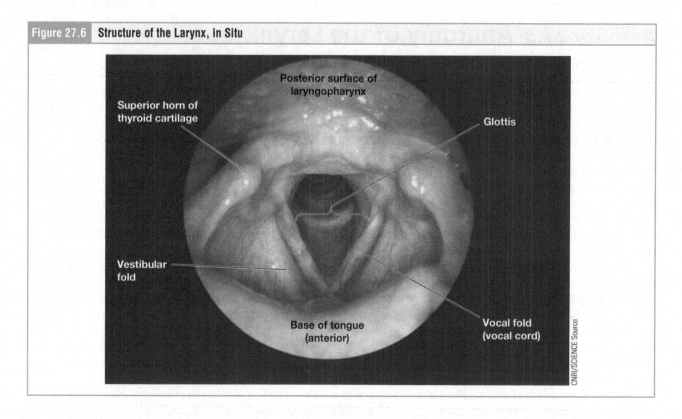

The three small pairs of cartilage create a framework for the muscles and ligaments of voice production (Figure 27.7). The **arytenoid** cartilages and their attached **corniculate cartilages** work together to widen and narrow the glottis. The vocal ligaments, wrapped in the vocal folds, reach from their fixed attachment at the anterior thyroid cartilage to the arytenoid cartilages. The arytenoids can swing when pulled on by muscle. Two pairs of muscle attach to the arytenoid cartilages, indirectly, by attaching to the corniculate cartilages. The laterally positioned **cuneiform cartilages** bracket these muscles to hold them in position. As muscles pull on the corniculate cartilages and the arytenoids swing medially, they close the glottis. As other muscles pull on the corniculate cartilages they cause the arytenoid cartilages to swing laterally and open the glottis. When air moves through the glottis between the closed vocal folds, the folds vibrate and produce sound. Changes in both volume and pitch will occur as the amount of air moving through and the tension of the closed vocal folds change.

Figure 27.7 | **Vocal Ligament Movement**

Movement of the vocal ligaments occurs due to the rotation of the arytenoid cartilages.

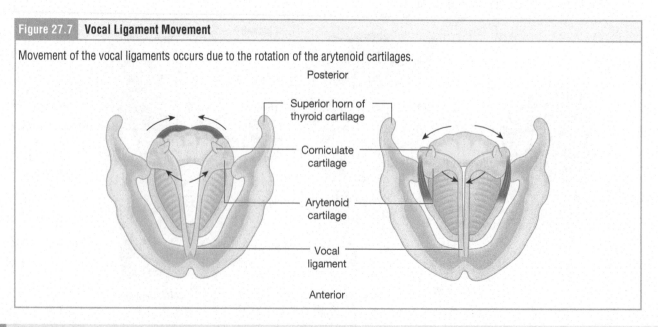

The three-dimensional nature of the larynx can make it difficult to picture. The purpose of this exercise is to connect the images shown here with a three-dimensional construction of a model of this important structure.

Materials
- Empty toilet paper tube
- Midsagittal model of the larynx, if available
- Scissors
- Stapler
- Tape
- Wide rubber band

Instructions

1. Your instructor will direct you to work either individually or in pairs, depending on the supplies available.

2. Use scissors to cut out the pieces of the larynx found on page 697. You will have the following pieces:
   a. One epiglottis
   b. one thyroid cartilage
   c. One cricoid cartilage
   d. Two arytenoid + corniculate cartilages

3. Create each cartilage piece:
   a. Gently fold the thyroid cartilage in half at the dotted line to represent the laryngeal prominence.
   b. Loop the cricoid cartilage into a ring so that letter A and letter B meet and tape it closed.
   c. Cut the rubber band so that it is a single long piece. Fold the rubber band in half to represent the two vocal folds (the fold will represent where the vocal folds attach close together at the thyroid cartilage). Where it is folded it should be attached to the inside of the anterior surface of the thyroid cartilage (use a stapler to do this rather than tape).
   d. Staple each arytenoid + corniculate cartilage to each cut end of the rubber band (consider with your lab partner whether you should attach the rubber band to the arytenoid or corniculate portion of the paper piece to be anatomically accurate).

4. The toilet paper tube will represent the trachea. Draw on the tracheal cartilages, if desired.

5. Use Figures 27.5, 27.6, and 27.7 and your pieces to assemble the larynx. Use tape to attach pieces together. These directions are purposefully vague to give you space to think about how the pieces attach to become the larynx. Work with your lab partner as necessary to determine how the pieces should attach to make the larynx.

6. Note that you are building the larynx with the vocal ligaments in a fixed position (either open or closed). If you want to get creative, consider how you might attach the arytenoid + corniculate cartilages to the cricoid cartilage in a way that enables them to rotate to open and close the glottis.

An emergency cricothyrotomy (also called a cricothyroidotomy or "cric" for short) is an emergency medical procedure done when an individual has a failed airway due to a foreign body in the pharynx or facial trauma that obstructs air flow to and from of the lungs. The procedure, shown in Figure 27.8, involves finding the laryngeal prominence and making an incision just inferior to that landmark. This enables the practitioner to target the space between the thyroid cartilage and the cricoid cartilage, which is spanned by the cricothyroid membrane. Once the incision is made, a narrow tube is inserted into the incision to reach the trachea. Thus, air can once again flow between the environment and the patient's lungs, bypassing the obstruction.

Figure 27.8 Cricothyroidotomy Emergency Airway Puncture Procedure

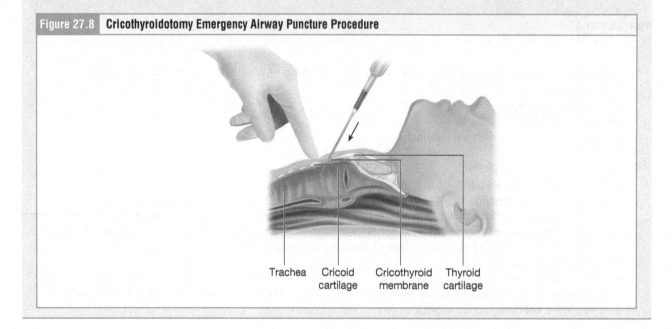

Trachea Cricoid cartilage Cricothyroid membrane Thyroid cartilage

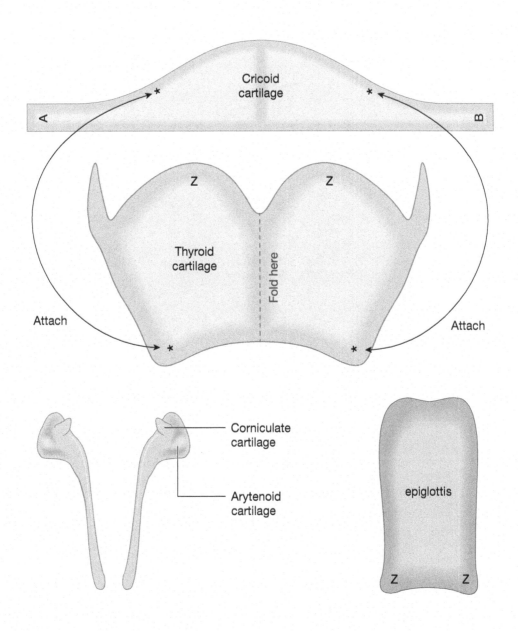

Cricoid
cartilage

A

B

Z　　　　　Z

Thyroid
cartilage

Fold here

Attach　　　　　　　　　　　　　　　　　　　Attach

*　　　　　　　　　　　　　*

Corniculate
cartilage

Arytenoid
cartilage

epiglottis

Z　　　　Z

Activity 27.4 Gross Anatomy of the Trachea, Bronchial Tree, and Lungs

This activity targets LOs 27.5, 27.6, and 27.7.

The trachea, or windpipe, contains hyaline cartilage in its walls. Each of the approximately 16 to 20 small pieces of C-shaped **tracheal cartilage** is separated by connective tissue and smooth muscle. The incomplete shape of each ring of tracheal cartilage allows space for the esophagus to expand as food is swallowed, as shown in Figure 27.9. The final piece of tracheal cartilage, the **carina**, is a saddle shape where the single passageway of the trachea splits into two primary bronchi. The trachea is lined with pseudostratified ciliated columnar epithelium.

The gross anatomy of the lungs can be seen in Figure 27.10. The right lung consists of three lobes, each separated by a visible fissure. The right horizontal fissure separates the superior and middle right lobes. The right oblique fissure separates the middle right lobe from the inferior right lobe. The left lung, on the other hand, is smaller due to the space occupied by the heart. This creates the cardiac notch, an indent on the left lung. There are only two lobes in the left lung, a superior and inferior left lobe separated by a single left oblique fissure.

The **bronchial tree** (Figure 27.11) is the collective term used for the many branches of the bronchi that finally terminate at the small alveoli of the lungs. At about the level of the sternal angle (where the manubrium and the body of the sternum meet) the trachea branches into the right and left **primary bronchi**. The primary bronchi resemble the trachea both in the C-shaped rings of cartilage in their walls and their lining of pseudostratified columnar epithelium. The primary bronchi enter each lung at the hilum, a concave region where pulmonary blood vessels and nerves also enter the lungs. The branching of the bronchi is not symmetrical between each lung. In the right lung, the right primary bronchus diverges into three right **secondary bronchi**, each leading to one of the three lobes of the left lung. The secondary bronchi are sometimes called the lobar bronchi because each enters a single lobe of the lung. These secondary bronchi branch further into tertiary bronchi, smaller bronchi, and terminal bronchioles. The left primary bronchus divides into only two left secondary bronchi to reach each of the two lobes of the left lung. Again, these secondary bronchi branch into many tertiary bronchi, smaller bronchi, and terminal bronchioles. The epithelial tissue lining these airways is specialized for conducting the air and managing any possible debris: pseudostratified ciliated columnar epithelium with many mucus-producing goblet cells transitions to simple ciliated columnar epithelium at the smaller bronchi.

Figure 27.9	Structure of the Trachea

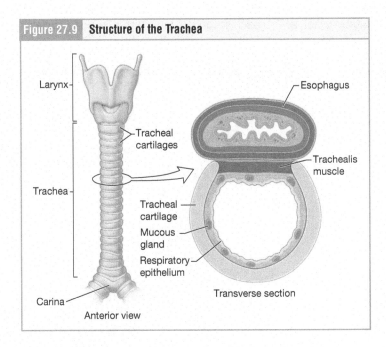

Figure 27.10	The Lobes of the Lungs

Both of the lungs are divided into separate functional lobes. The right lung has three lobes; the left lung, which is smaller due to the presence of the heart, has only two.

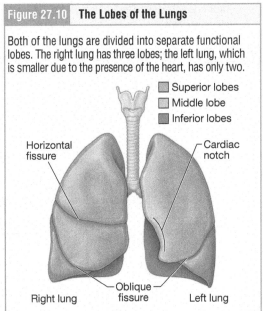

The secondary, tertiary bronchi, and smaller bronchi, like the primary bronchi, contain cartilage in the walls for structure, but these pieces of cartilage get smaller and spaced further apart as the airways become smaller. Unlike the bronchi, the vastly smaller bronchioles do not contain cartilage in their walls. Rather they are surrounded by ribbon-like threads of smooth muscle that can change the diameter of the airways. The tiny terminal bronchioles mark the end of the conducting zone of the airways.

The final respiratory zone portion of the airways, which all may have alveoli budding off, consist of the **respiratory bronchioles**, **alveolar ducts**, and **alveolar sacs**, shown in Figure 27.12. These areas of the airways contain neither cartilage nor smooth muscle in their walls. Mucus-producing goblet cells are also absent from the respiratory zone. In contrast to the conducting zone this respiratory zone consists of the surfaces across which gasses diffuse and are exchanged between the lungs and the blood. Therefore, the epithelial tissue in these areas is thin: simple cuboidal epithelium lining the respiratory bronchioles and simple squamous epithelium lining the alveolar ducts and alveoli. A small net of capillaries extends around each alveolar sac to provide close contact between the alveoli and the blood.

The purpose of this exercise is to connect the anatomy of the bronchial tree to the gross anatomy of the lung.

Materials
- Colored pencils
- Model of the bronchial tree, if available
- Model of the gross anatomy of the lungs

Instructions
1. Work with your lab partner to find as many structures described in the introduction above as possible on the available models.

2. Label the following structures on the figure found on the next page.
   a. All lobes of each lung
   b. All fissures of each lung

3. Using the models and Figure 27.11 as a reference, sketch in the primary and secondary bronchi on the figure on the next page.

4. Sketch in a few tertiary bronchi branching from each secondary bronchus to visualize how the branching will continue.

5. Chose five different colored pencils. Choose a color for each lobe, coloring in the lobe lightly and the secondary bronchus that serves that lobe with the same color but darker, so that each secondary bronchus is linked to its own lobe by color.

Figure 27.11	The Structure of the Trachea, Bronchi, and Lungs

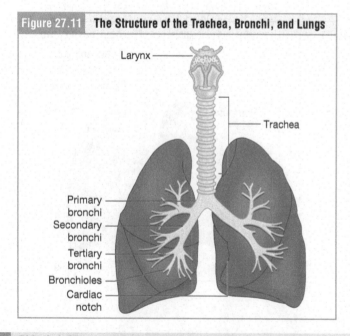

Larynx

Trachea

Primary bronchi
Secondary bronchi
Tertiary bronchi
Bronchioles
Cardiac notch

Figure 27.12	Structure of the Respiratory Zone of the Lungs

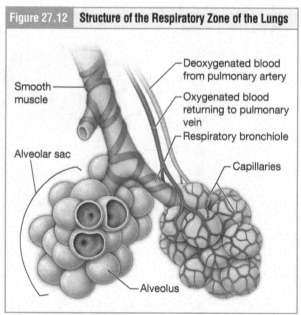

Deoxygenated blood from pulmonary artery

Smooth muscle

Oxygenated blood returning to pulmonary vein

Respiratory bronchiole

Alveolar sac

Capillaries

Alveolus

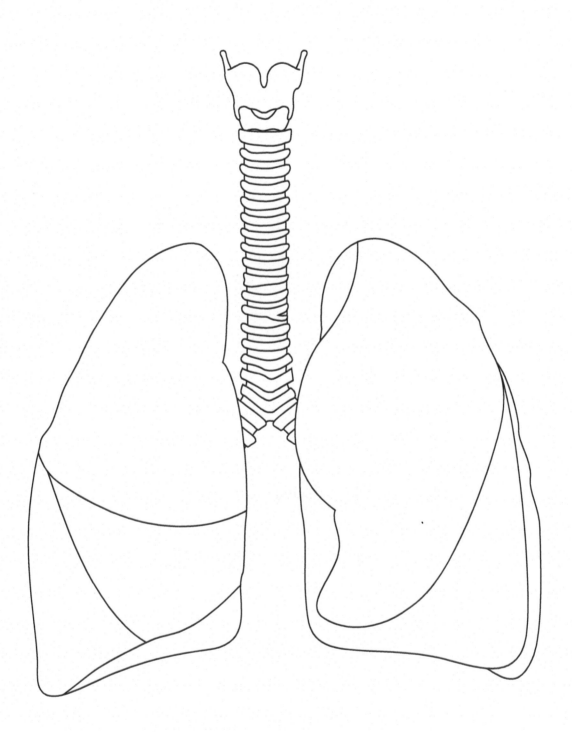

(page intentionally left blank)

Activity 27.5 Microanatomy of the Lungs

This activity targets LO 27.6.

An alveolar sac is a cluster of many individual alveoli that are responsible for gas exchange (see Figure 27.12). An alveolus (Figure 27.13) has elastic walls that enable it to stretch during air intake. Tiny alveolar pores connect each hollow ball-shaped alveolus to its neighbors. The thin wall of each alveolus is made of squamous epithelial cells called **type I alveolar cells**. These cells are flattened so that oxygen and carbon dioxide can easily diffuse across them. Scattered among the type I alveolar cells are **type II alveolar cells**. These cells secrete **pulmonary surfactant**, a slippery liquid that reduces surface tension so that the wet walls of the alveoli do not stick to each other as the alveoli deflate somewhat during exhalation. A third type of cell, **alveolar macrophages**, roams the interior surfaces of the alveoli, scavenging for debris and pathogens.

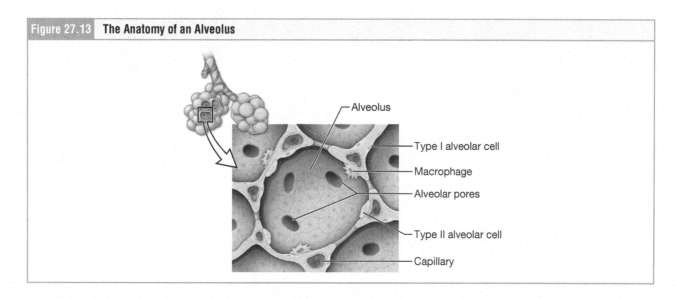

Figure 27.13 **The Anatomy of an Alveolus**

— Alveolus
— Type I alveolar cell
— Macrophage
— Alveolar pores
— Type II alveolar cell
— Capillary

The purpose of this exercise is to connect the drawing above with a histological slide of lung tissue.

Materials
- Colored pencils
- Compound microscope
- Prepared slides of alveoli

Instructions

1. Obtain a prepared slide of alveoli.

2. Examine the slide first on scanning power (4×) and use the coarse focus knob to locate the tissue. Then increase the magnification by switching to the low power (10×) objective lens and use the fine focus knob. Then switch to the high power (40×) objective lens and again use the fine focus knob until you can clearly see the tissue.

3. Check with your instructor if you are unsure about what you are looking at or need assistance before proceeding.

4. Draw and label the tissue using the space on the next page.
   a. Label an alveolus, an alveolar pore, and the type I alveolar cells.
   b. Find a capillary as it travels along the wall of an alveolus, draw and label it.
   c. Find a smooth muscle-lined bronchiole, draw and label it. Note that bronchioles are not visible on all slides or in all locations on a single slide. Work with your lab partner or other groups of students to find a view of a bronchiole before proceeding.

5. Are there other lung structures visible in your view, such as alveolar ducts or alveolar sacs? Draw and label the other visible structures.

Activity 27.6 Structural Differences Along the Respiratory Tract

This activity targets LOs 27.1, 27.6, and 27.8.

As you have observed and studied the airways of the respiratory tract you have seen that there are structural changes along its path as the function of each section of the passageway changes. The type of epithelium lining the airway changes, first depending on whether the space is a shared passageway for food and later as the airways transition into smaller and smaller spaces. The amount and shape of cartilage present also changes. Consider that cartilage provides structure to the walls, but this structure comes at a price: it is less possible to adjust the diameter of an airway when it has a lot of cartilage holding it open. The presence or absence of smooth muscle occurs in relation to the presence of cartilage. Smooth muscle allows for **broncho-constriction** and **bronchodilation** to decrease or increase, respectively, the amount of air passing through the bronchial tree.

Instructions

1. Use this introduction, your textbook, and the other exercises in this lab to organize the information for yourself in Table 27.1. The first row has been completed for you.

2. When your table is complete, compare your work with your lab partner and make additions or adjustments to the content as necessary. This completed summary table will be an excellent tool for study in preparation for a quiz or exam.

Table 27.1 The Structural Changes Along the Airways				
Structure	**Function(S) Served by This Structure**	**Epithelial Tissue Type**	**Cartilage**	**Smooth Muscle (Circle One)**
Nasal cavity	Conditions the air, allows for sense of smell	Pseudostratified ciliated columnar epithelium, olfactory epithelium only in superior roof	In anterior aspect only to provide structure to external nose	present or ⟨absent⟩
Nasopharynx				present or absent
Oropharynx				present or absent
Laryngopharynx				present or absent
Superior larynx				present or absent

(continued)

	Table 27.1 The Structural Changes Along the Airways (*Continued*)			
Structure	**Function(S) Served by This Structure**	**Epithelial Tissue Type**	**Cartilage**	**Smooth Muscle (Circle One)**
Inferior larynx				present or absent
Trachea				present or absent
Primary, secondary, and tertiary bronchi				present or absent
Smaller bronchi				present or absent
Bronchioles				present or absent
Respiratory bronchioles				present or absent
Alveolar ducts and alveoli				present or absent

Learning Connection

Try Drawing It

The many changes to the epithelial lining of the airways can be difficult to keep track of. Try sketching out these changes to visualize how the epithelium adjusts as the needs of the passageways adjust. Don't forget details like the cilia and goblet cells!

Activity 27.7 Case Study: Cigarette Smoke and the Mucociliary Escalator

Case Study

This activity targets LOs 27.1, 27.5, and 27.8.

Chris has been a smoker for the past ten years. When a friend who is a fellow smoker approaches Chris about trying to quit, the two agree to quit together. Chris visits the doctor to review their options for quitting. First, though, Chris wants to know how the daily cigarettes have been affecting their lungs. The doctor tells Chris that cigarette smoke contains both toxic chemicals and particulate matter that damage the airways and alveoli.

1. What structures present at the very start of Chris' airways should be able to assist in removing the debris from air that enters the airways? Choose all correct answers.

   a. Vibrissae
   b. Epiglottis
   c. Mucus

2. Why are upper respiratory structures less able to assist in removing the debris that is present in cigarette smoke as compared to debris that is in the air that enters your nasal vestibules?

   a. These structures are in the nasal cavity, which is bypassed.
   b. These structures are in the esophagus that leads to the stomach.
   c. These structures are obstructed by the epiglottis.

Chris wonders about these toxic chemicals and debris, "I thought I had heard that the tissue lining my lungs is specialized for removing dust and other particles in the air that might make it down into my lungs?"

3. What tissue type lines most of the airways?

   a. Pseudostratified ciliated columnar epithelium
   b. Keratinized stratified squamous epithelium
   c. Transitional epithelium

4. What feature of this tissue type will assist in trapping debris?

   a. Keratin
   b. Mucus
   c. Olfactory nerves

The doctor agrees with Chris that the tissue lining the airways can help. "In fact," says the doctor, "this tissue creates a structure called the mucociliary escalator, which traps debris and then uses cilia to move it out of the airways to be swallowed. Any material that makes it all the way to the alveoli will be engulfed by specialized cells there, which will then crawl out of the alveoli to get a ride on the mucociliary escalator. Unfortunately, cigarette smoke both shortens and paralyzes the cilia. The good news, though, is that it only takes about two weeks after you stop smoking for the mucociliary escalator to recover!"

5. Given the description above, where does the mucociliary escalator end superiorly?

   a. At the nasopharynx
   b. At the laryngopharynx
   c. At the carina

6. Given the description above, where does the mucociliary escalator begin inferiorly?

   a. The smaller bronchi
   b. The alveolar ducts
   c. The bronchioles

7. What cells are present in the alveoli to engulf debris and pathogens?
   a. Type I alveolar cells
   b. Type II alveolar cells
   c. Alveolar macrophages

Lab 27 Post-Lab Quiz

1. If you measure the air in the nasopharynx and compare it to the external air, the air in the nasopharynx will _____ than the external air.
   a. have less debris
   b. be less humid
   c. be colder than

2. If your maxillary sinuses are blocked due to an infection, you would expect to feel the pressure or pain in what area?
   a. At the root of your nose
   b. In your larynx
   c. Lateral to your nares

3. What epithelial tissue type lines the oropharynx?
   a. Simple cuboidal epithelial tissue
   b. Pseudostratified ciliated columnar epithelial tissue
   c. Stratified squamous epithelial tissue

4. When Botox is injected into a skeletal muscle, the nerves can no longer stimulate that muscle to contract, paralyzing it. If you inject Botox into the muscles that attach to the corniculate cartilage of the larynx, what will the result be?
   a. The epiglottis will not be able to close.
   b. Air will not move into the person's airway.
   c. The person will not be able to make sounds.

5. The vestibular folds are _____ to the vocal folds.
   a. superior and lateral
   b. inferior and medial
   c. posterior

6. What is the shape of the tracheal cartilages?
   a. Rings
   b. C-shaped pieces
   c. Irregular plates

7. How many total secondary bronchi are present in an individual?
   a. Two
   b. Three
   c. Five

8. Which of the following are functions of the cilia found in the respiratory tract? Choose all correct answers.
   a. The cilia move mucus out of the airways.
   b. The cilia move air through the airways.
   c. The cilia assist in ridding the airways of pathogens.

9. What important transition occurs at the bronchioles of the lungs?
   a. The transition from stratified squamous epithelium to pseudostratified ciliated columnar epithelium
   b. The transition from the conducting zone to the respiratory zone
   c. The transition from the superior respiratory tract to the inferior respiratory tract

10. A person with asthma may carry an inhaler for helping when they have an asthma attack and have difficulty breathing. Some inhalers contain a bronchodilator that relaxes the smooth muscle of the airways to cause bronchodilation. This means that the bronchodilator medicine in this type of inhaler is most influential on what part of the bronchial tree?

a. The trachea

b. The carina

c. The bronchioles

Lab 28 | Respiratory Physiology

Martyn F. Chillmaid/Science Source

Learning Objectives: After completing this lab, you should be able to:

28.1 Define pulmonary ventilation, inspiration (inhalation), and expiration (exhalation).

28.2 Identify the muscles used during quiet inspiration, deep inspiration, and forced expiration.

28.3 Explain the inverse relationship between gas pressure and volume of the gas (i.e., Boyle's law) and apply this relationship to explain airflow during inspiration and expiration.

28.4 Define, identify, and determine values for the pulmonary volumes (inspiratory reserve volume [IRV], tidal volume [TV], expiratory

reserve volume [ERV], and residual volume [RV]) and the pulmonary capacities (inspiratory capacity [IC], functional residual capacity [FRC], vital capacity [VC], and total lung capacity [TLC]).

28.5 Define and calculate minute ventilation and alveolar ventilation.

28.6 Compare and contrast the solubility of oxygen and carbon dioxide in plasma.

28.7 Describe oxygen and carbon dioxide concentration gradients and net gas movements between the alveoli and the pulmonary capillaries.

The Human Anatomy and Physiology Society includes more than 1,700 educators who work together to promote excellence in the teaching of this subject area. The HAPS A&P Learning Outcomes measure student mastery of the content typically covered in a two-semester Human A&P curriculum at the undergraduate level. The full Learning Outcomes are available at https://www.hapsweb.org.

28.8 Describe oxygen and carbon dioxide concentration gradients and net gas movements between systemic capillaries and the body tissues.

28.9 Analyze how oxygen and carbon dioxide movements are affected by changes in partial pressure gradients (e.g., at high altitude), area of the exchange surface, permeability of the exchange surface, and diffusion distance.

28.10 Explain the influence of cellular respiration on oxygen and carbon dioxide gradients that govern gas exchange between blood and body tissues.

28.11 Describe the ways in which oxygen is transported in blood, and explain the relative importance of each to total oxygen transport.

28.12 Describe the ways in which carbon dioxide is transported in blood and explain the relative importance of each to total carbon dioxide transport.

28.13 Given a factor or situation (e.g., pulmonary fibrosis), predict the changes that could occur in the respiratory system and the consequences of those changes (i.e. given a cause, state a possible effect).

28.14 Describe the processes associated with the respiratory system (i.e., ventilation, pulmonary gas exchange [gas exchange between alveoli and blood], transport of gases in blood, tissue gas exchange [gas exchange between blood and body tissues]).

Introduction

The term "respiration" refers to a group of physiological processes that work together to provide oxygen to the cells of your body. Since this term can be somewhat vague, more specific terminology will be used to describe each individual process.

- **Ventilation** is the movement of air in and out of the lungs, and the terms **inhalation** (or *inspiration*) and **exhalation** (or *expiration*) can then be used to describe the direction of air movement. The process of pulmonary ventilation occurs due to the body's ability to change the dimensions of the thorax to either increase or decrease the volume of the cavity within.
- **Gas exchange across the respiratory membrane** is the simultaneous diffusion of oxygen and carbon dioxide that occurs between the alveoli and the capillaries that surround them. Each of the gasses will diffuse down their own partial pressure gradients.
- **Gas transport** describes the movement of oxygen and carbon dioxide within the blood. Due to their differing solubilities in the blood, oxygen and carbon dioxide are carried differently in the blood.
- **Gas exchange between the blood and tissues** is the diffusion of oxygen and carbon dioxide that occurs between systemic capillaries and the individual cells of your body. Just as for gas exchange across the respiratory membrane at the alveoli of the lungs, gas exchange at the tissues occurs as each gas diffuses down its own partial pressure gradient.
- **Aerobic cellular respiration** is the process of cells using glucose and oxygen to generate ATP energy. This process produces carbon dioxide as waste. Note that cells carrying out aerobic cellular respiration are continually using up the oxygen that enters the cell and producing more carbon dioxide at the same time.

Lab 28 **Pre-Lab Quiz**

This quiz will strengthen your background knowledge in preparation for this lab. For help answering the questions, use your resources to deepen your understanding. The best resource for help on the first five questions is your text, and the best resource for help answering the last five questions is to read the introduction section of each lab activity.

1. Anatomy of the thoracic cavity. Identify the following structures on the image:

   Diaphragm   Pleural cavity
   Lung   Ribs and intercostal muscles of the thoracic wall
   Parietal pleura   Visceral pleura

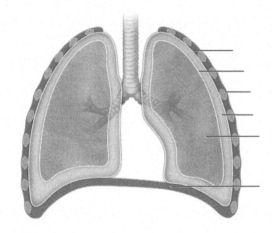

2. Match the root next to its definition:

Root	Meaning
	channel
	lung
	rib
	side of something

List of Roots

alveo
costa
pleura
pulmo

3. What is the air pressure within the alveoli of the lungs called?

   a. Intrapulmonary pressure
   b. Atmospheric pressure
   c. Intrapleural pressure

4. During exhalation, the volume of the lungs decreases. What will happen to the pressure of the gas within the lungs as a result?

   a. It will also decrease.
   b. It will increase.
   c. It will stay the same.

5. Which cells of the blood transport most of the oxygen and some of the carbon dioxide within the blood?

   a. Platelets
   b. Leukocytes
   c. Erythrocytes

6. What is the name of the process that most of the cells of your body go through that consumes oxygen and gives off carbon dioxide as waste?

   a. Ventilation
   b. Aerobic cellular respiration
   c. Gas exchange

7. What causes the bulk flow of air in and out of the lungs?

   a. Pressure differences
   b. The partial pressures of oxygen and carbon dioxide
   c. The beating of the cilia lining the airways

8. Which of the following muscles will reduce the volume of the lungs when they contract? Choose all correct answers.

   a. Diaphragm
   b. External intercostals
   c. Internal intercostals
   d. Abdominal muscles

9. What information is needed to calculate respiratory capacities?

   a. Minute ventilation
   b. Breaths per minute
   c. Respiratory volumes

10. What is the name of the tool used to measure respiratory volumes and capacities?

    a. Spirometer
    b. Ventilator
    c. Bag valve mask

Activity 28.1 Volume, Pressure, and Pulmonary Ventilation

This activity targets LOs 28.1, 28.2, and 28.3.

Pulmonary ventilation moves air in and out of the lungs. This is sometimes described as *bulk flow*, as the movement of the air is not dependent on the partial pressures of the gasses in the air. In other words, the composition of this air may have more or less oxygen, carbon dioxide, water vapor, or other gasses, but the content of the air is not what is driving its movement either in or out of the lungs. What is driving the movement of the air is a difference in pressure. Air will flow from an area of higher pressure to an area of lower pressure. Figure 28.1 shows this principle.

Figure 28.1	Air Pressure and Air Flow in the Lungs

Air will flow from an area of higher pressure to an area of lower pressure.

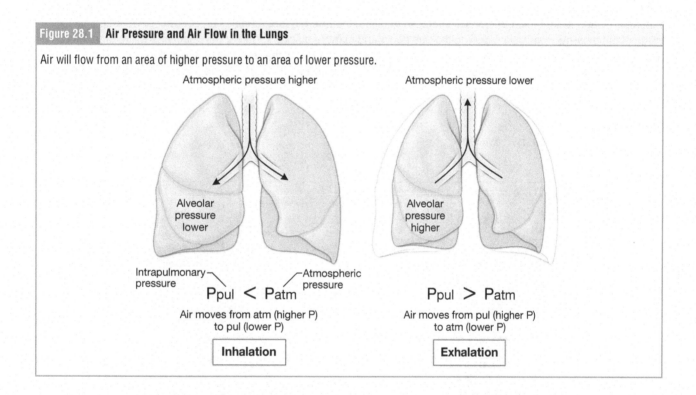

In pulmonary ventilation, air flows between the atmosphere and the alveoli. Find these two locations in Figure 28.1 and note how the relative pressures change upon inhalation and exhalation. Our bodies cannot change the atmospheric pressure in order to adjust these relative pressures, so we change the pressure within the alveoli—the **intrapulmonary pressure**—to either be greater than the atmospheric pressure or less than the atmospheric pressure.

But how does our body change the intrapulmonary pressure? When a structure increases in volume, the pressure within that structure decreases. Likewise, when a structure decreases in volume, the pressure within that structure increases. This means that:

To inhale we must _____ the pressure in the alveoli by _____ the volume of the lungs.
And
To exhale we must _____ the pressure in the alveoli by _____ the volume of the lungs.

These two competing volume changes are achieved by using four groups of skeletal muscles. As these skeletal muscles contract and change the volume of the thoracic cavity, the stretchy tissue surrounding each alveolus allows the alveoli to follow suit, changing their own volume, and thereby changing the volume of the entire lung. Examine Figure 28.2, which shows two of the four muscles of inspiration (and how they can act to cause both inhalation and exhalation).

The muscles of inspiration increase thoracic volume during inhalation. These muscles include the **external intercostal muscles** and the **diaphragm.** When the external intercostal muscles contract they pull the ribs upward and laterally, increasing the volume of the thoracic cavity both vertically and side-to-side. When the diaphragm contracts it flattens out inferiorly into the abdomen, increasing the volume of the thorax vertically, as well. These two muscle groups cause the regular breaths you take, called *quiet inspiration*. If you want to take a more forceful breath you contract the diaphragm even more, flattening it further and increasing the volume of the thoracic cavity further. This is called *forced inspiration*.

Quiet expiration happens passively as these two muscles relax. The ribs drop and the vertical dimensions reduce, reducing the overall volume of the thoracic cavity. If a more forceful exhalation (*forced expiration*) is required, muscles of the abdomen will contract, forcing abdominal organs up against the relaxing diaphragm to cause it to move upwards more quickly and further into the thoracic space. In addition, the **internal intercostals** can contract to work against the action of the external intercostals, pulling the ribs down and in more quickly. Together these actions will reduce the volume of the thorax more quickly than would occur only due to relaxation of the muscles of inspiration.

The purpose of this activity is to create connections between the ideas of volume, pressure, and the movement of air.

Materials
• Bell-jar model of the lungs, if available

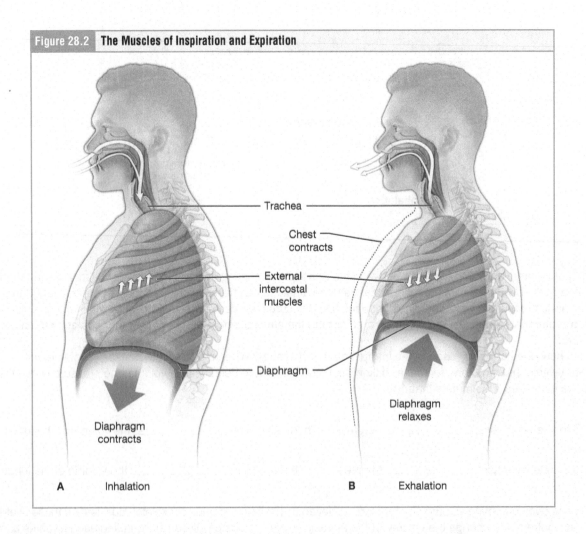

| Figure 28.2 | The Muscles of Inspiration and Expiration |

Trachea

Chest contracts

External intercostal muscles

Diaphragm

Diaphragm contracts

Diaphragm relaxes

A   Inhalation

B   Exhalation

Instructions

1. With your lab partner, compare your answers to the two prompts in the introduction—did you get the same answers? Discuss if necessary to understand these concepts.

2. Depending on how many bell-jar models are available, you will either be given a bell-jar model to use with your lab partner or you will watch a demonstration by your instructor. The model should look similar to the image shown in Figure 28.3.

3. Identify what parts of the model represent the following structures by labeling Figure 28.3.

   a. Trachea
   b. Primary bronchi
   c. Lungs
   d. Thoracic cavity
   e. Diaphragm

4. Before demonstrating any action with the model, consider the two balloons that represent the lungs. A lung is not an empty sac, is it? Consider what you know about lung anatomy. Rather than representing a whole lung, the balloon is really representing what is happening in each _____ of each lung.

Figure 28.3 Bell-Jar Model of the Lungs, Showing the Initiation of Inspiration

Martyn F. Chillmaid/Science Source

5. Press up on the membrane that represents the diaphragm (or watch as your instructor does this). This is what the diaphragm muscle looks like when it is relaxed.

   a. Did the pressure inside the thoracic cavity increase or decrease?

   b. What happened to the volume of air in the lungs?

6. Gently pull inferiorly on the membrane so that it flattens (or watch as your instructor does this). This represents the contraction of the diaphragm.

   a. Did the pressure inside the thoracic cavity increase or decrease?

   b. What happened to the volume of air in the lungs?

7. The internal and external intercostal muscles are not represented by this model, but you can consider where they would be if they were present.

   a. Where would these muscles be located in the bell-jar model?

b. How would contraction of the external intercostals change the volume of the thoracic cavity? What action would follow?

c. How would contraction of the internal intercostals change the volume of the thoracic cavity? What action would follow?

8. The abdominal muscles and organs of the abdominal cavity are not represented by this model, but you can consider where they would be if they were present.

   a. Where would the abdominal muscles and organs of the abdominal cavity be located in the bell-jar model?

   b. How would contraction of the abdominal muscles change the volume of the thoracic cavity? What action would follow?

Digging Deeper:
Bag Valve Masks

A ventilator is used to deliver air to a person who is unable to breathe or is taking breaths that are insufficient for their oxygen needs. A ventilator works by actively moving air in and out of the lungs. While ventilators are often a computerized machine, a hand-held ventilator (also called a bag valve mask, shown in Figure 28.4) can be used for short-term needs. How does a ventilator move air? Consider the need for a difference in pressure between the atmosphere and the alveoli to move air in and out of the lungs. While we breathe by adjusting the pressure in our lungs to be either greater or lower than atmospheric pressure, the ventilator moves air by adjusting the pressure in the atmosphere inside the ventilator itself. When the bag valve mask is squeezed, the volume is reduced, and the pressure is increased within the bag. Which way will air flow when the bag is squeezed?

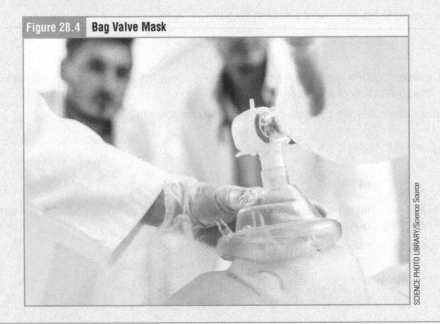

Figure 28.4   Bag Valve Mask

SCIENCE PHOTO LIBRARY/Science Source

Activity 28.2 Pulmonary Volumes and Capacities

This activity targets LOs 28.4 and 28.5.

The volume of air we exchange with each breath is variable. Try taking a deep breath and then exhaling all of the air you just inhaled. That probably felt different than the normal quiet breaths you were taking prior to that. We can identify and measure the various pulmonary volumes of our lungs using a **spirometer**. These volumes, shown in Figure 28.5, include:

- **Tidal Volume (TV)** is the amount of air that normally enters and exits the lungs during quiet breathing.
- **Expiratory Reserve Volume (ERV)** is the amount of air you can forcefully exhale past the normal tidal expiration.
- **Inspiratory Reserve Volume (IRV)** is the amount of air you inhale after you have already performed a tidal inspiration.
- **Residual Volume (RV)** is not a volume of air that you can move (and cannot be measured with a spirometer). It is the amount of air that remains in your lungs after maximum forced expiration. It is typically about 1,100 to 1,200 mL of air.

While these volumes are interesting on their own, we can also use them to calculate respiratory capacities:

- **Total Lung Capacity (TLC)** is all the air that could possibly be held in the lungs. It is the sum of all four of the volumes together.
- **Vital Capacity (VC)** is all the air that can possibly be moved by the lungs. It is the sum of the tidal volume, expiratory reserve volume, and inspiratory reserve volume.
- **Inspiratory Capacity (IC)** is the maximum amount of air you can inspire. This is the sum of tidal volume plus inspiratory reserve volume.
- **Functional Residual Capacity (FRC)** is the amount of air that remains in the lungs after a normal tidal expiration. It is the sum of expiratory reserve volume and residual volume.

Figure 28.5	The Lung Volumes and Capacities

Various volumes of the lungs are measured using a spirometer, a machine that measures the volume of air inhaled and exhaled in various circumstances.

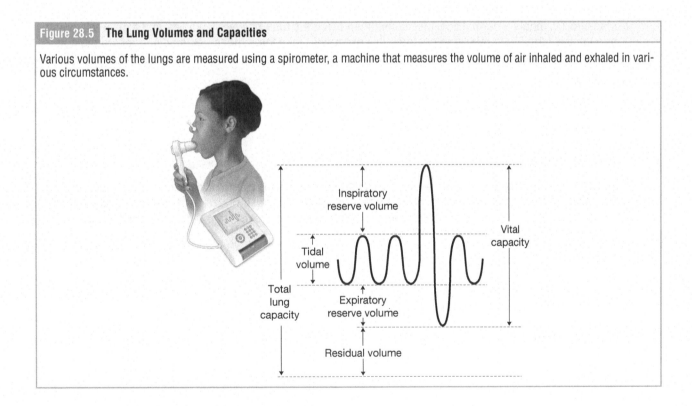

We use a spirometer to measure respiratory volumes and capacities because this allows us to assess an individual's pulmonary function. Lung volumes and capacities vary from person to person depending on factors such as age and height. Due to these individual variations, measurements that fall within 80% of the expected measurements are considered normal. Reductions in respiratory volumes and/or capacities below 80% may indicate a respiratory disorder.

A final important measurement, **minute ventilation**, is the total volume of air in liters that enters the lungs per minute. At rest, it is calculated as tidal volume multiplied by the respiratory rate (breaths per minute, BPM):

$$\text{Minute ventilation} = TV \times BPM$$

The purpose of this exercise is to use a spirometer to measure your own volumes and capacities. You will directly measure some of them and then use the formulas above to calculate others. There are three common types of spirometers available in anatomy and physiology labs. Your instructor will demonstrate the method for using the type of spirometer available to you.

Materials
• Alcohol swabs
• Disposable mouthpieces for spirometer
• Spirometer (wet spirometer, computerized spirometer, or handheld spirometer)

Instructions
1. Disinfect the nozzle of the spirometer with an alcohol swab.

2. Place a clean disposable mouthpiece on the nozzle.

3. Following the instructions provided by your instructor and those below, measure the tidal volume (TV), expiratory reserve volume (ERV), and vital capacity (VC). Measure all three while sitting in a chair with your back straight and body relaxed. Practice each volume first before recording any measurements. Your lab partner will record the volumes in your table.

   a. Measure TV. Inhale a normal tidal inspiration, then place the tube between your lips and exhale this breath through the tube. Try to relax and not force this expiration. Take three measurements and record them here. Then average the three measurements and record this in Table 28.1.

   Measurement 1: _____ mL

   Measurement 2: _____ mL

   Measurement 3: _____ mL

   b. Measure ERV. Inhale and exhale several tidal volumes, finish with a normal tidal expiration. Place the tube between your lips and exhale as forcibly as possible (pinch your nose or use the nose clips to prevent any air escaping through your nose). Take three measurements and record them here. Then average the three measurements and record this in Table 28.1.

   Measurement 1: _____ mL

   Measurement 2: _____ mL

   Measurement 3: _____ mL

Table 28.1 Respiratory Volumes, Capacities, and Minute Ventilation at Rest

Measurement	Formula	My Spirometer Results or Calculation
Tidal volume	N/A	
Expiratory reserve volume	N/A	
Inspiratory reserve volume	$IRV = VC - (TV + ERV)$	
Vital capacity	N/A	
Inspiratory reserve capacity		
Functional residual capacity		
Total lung capacity		
Minute volume		

c. Measure VC. Inhale and exhale several tidal volumes, finish with as deep an inspiration as possible. Place the tube between your lips and exhale all the air out of your lungs for as long as you possibly can (pinch your nose or use the nose clips to prevent any air escaping through your nose). Take three measurements, but rest in between each one. Record them here. Then average the three measurements and record this in Table 28.1.

Measurement 1: _____ mL

Measurement 2: _____ mL

Measurement 3: _____ mL

4. Discard the used mouthpiece.

5. Disinfect the nozzle of the spirometer with an alcohol swab.

6. Place a clean disposable mouthpiece on the nozzle.

7. Switch roles and complete your lab partner's table while they use the spirometer.

8. Discard the used mouthpiece.

9. Disinfect the nozzle of the spirometer with an alcohol swab.

10. Each lab partner should determine resting breaths per minute.
   a. Sit in a chair with your back straight and body relaxed. Sit quietly for a moment while you inhale and exhale tidal volumes (remember, this is quiet breathing).
   b. Your lab partner will watch the clock and tell you when to begin counting and stop counting each tidal exchange (one inhale + one exhale). Count for one minute.
   c. Record that information here.
   My resting breaths per minute: _____

11. Use the information in the introduction and the data you gathered in these steps to calculate the remaining information in Table 28.1.

a. Write the formula you will need to determine these additional measurements in the provided column. Inspiratory reserve volume has been done for you.

b. Calculate each of these measurements.

12. Work with your lab partner to answer the following questions.

a. How do your results and calculations compare to each other?

b. Hypothesize why your results are different or the same.

Activity 28.3 Gas Transportation within the Blood and Gas Exchange at the Alveolus and Tissues

This activity targets LOs 28.6, 28.7, 28.8, 28.9, 28.10, 28.11, and 28.12.

The purpose of the respiratory system is to perform gas exchange, first at the respiratory membrane and then again at the tissues of the body. Pulmonary ventilation provides air to the alveoli for the gas exchange process at the alveoli. At the respiratory membrane where the alveolar and capillary walls meet, gasses move across the membranes. Oxygen enters the bloodstream and carbon dioxide exits to the alveolus. Each of these gasses is moving down its own partial pressure gradient, as shown in Figure 28.6.

| Figure 28.6 | Gas Exchange at the Alveoli |

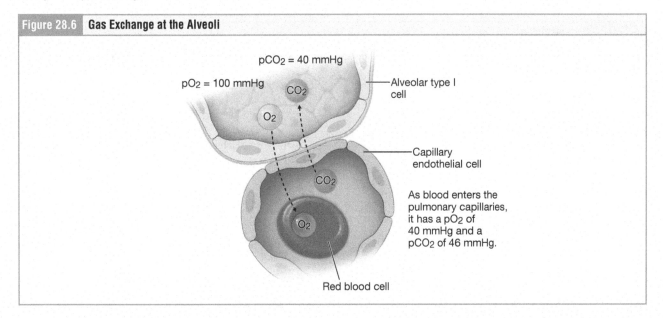

At the tissues of the body, the partial pressure of O_2 is greater in the blood than in the tissues, while the partial pressure of CO_2 is greater in the tissues than in the blood. Here O_2 moves into the tissues and CO_2 moves into the blood, as shown in Figure 28.7. At the tissues, gas exchange is driven by the process of aerobic cellular respiration. This process continually uses up oxygen (essentially removing it from the inside of the cell and therefore decreasing its partial pressure) and continually produces carbon dioxide (increasing its partial pressure inside the cell).

| Figure 28.7 | Gas Exchange at the Tissues |

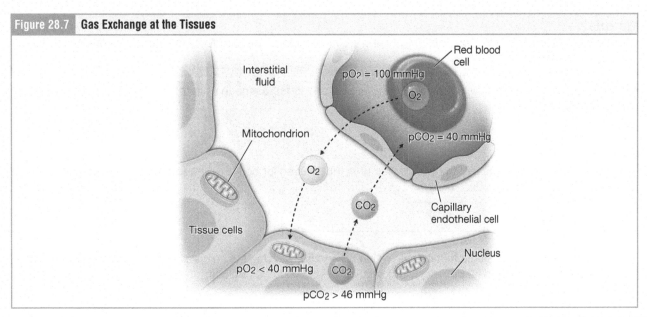

As you can see from Figures 28.6 and 28.7, the two gasses are traveling in the blood to move between the alveoli and the body's other tissues. How, exactly, are these gasses being carried by the blood? Almost all of the oxygen that is carried in the blood (98%) is transported within erythrocytes by attaching to the iron-containing heme groups on hemoglobin molecules (Figure 28.8). The remaining 2% of the oxygen is dissolved in the blood plasma. Some of the carbon dioxide also travels attached to the hemoglobin molecules, but it is not occupying the heme groups. Instead, about 23% of the carbon dioxide molecules are attracted to the amine groups of the hemoglobin proteins; you can imagine a few CO_2 molecules "stuck" to the protein chains of some of the hemoglobin molecules within each erythrocyte. Carbon dioxide is more soluble in water; 7% will be dissolved in the plasma. The final 70% of the carbon dioxide will combine with water to form bicarbonate, which will also travel in the plasma:

$$CO_2 + H_2O \rightarrow HCO^{3-} + H^+$$

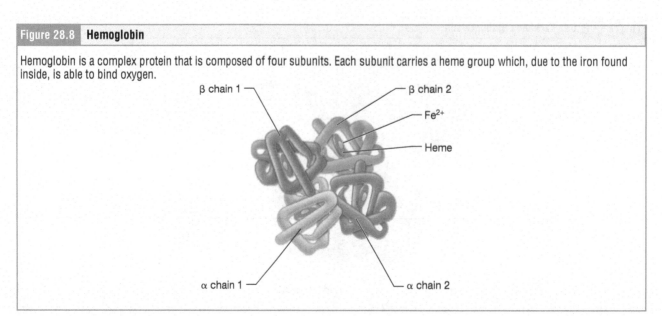

Figure 28.8   Hemoglobin

Hemoglobin is a complex protein that is composed of four subunits. Each subunit carries a heme group which, due to the iron found inside, is able to bind oxygen.

β chain 1   β chain 2 Fe^{2+} Heme   α chain 1   α chain 2

In this activity, you will act out the processes of gas exchange and gas transportation. This activity can take a bit of time to get comfortable with. Take your time and don't rush the process; repeating the steps a few times will make it clearer. In addition, the directions are vague as to how many of each bead (representing oxygen molecules and carbon dioxide molecules) you should use. Remember to consider the relative amounts of each at different moments in the process as you choose how many to use; you can always adjust as you go.

Materials
- Colored beads, one color to represent oxygen and a second color to represent carbon dioxide. Approximately 15 to 20 beads of each color per group
- Colored pencils in the same two colors as the beads
- Petri dishes, four of them: one to represent the alveolus, one to represent an erythrocyte, one to represent the blood plasma, and one to represent a cell of the body
- Sticky notes

Instructions
1. Choose colors to represent the two gasses (this will be used for both the beads and colored pencils)

   Oxygen is _____

   Carbon dioxide is_____

2. Figure 28.8 shows a single hemoglobin within an erythrocyte. Use the colored pencils to draw in where the oxygen molecules and carbon dioxide molecules will attach to hemoglobin.

3. Set up the "map" found on page 726:
   a. Place one Petri dish over the alveolus and a second Petri dish over the cell representing the tissues of the body.
   b. Label the last two Petri dishes using the sticky notes, one as erythrocyte and one as blood plasma.
   c. Place the latter two Petri dishes over the bloodstream (start with the blood going TO the tissues).

4. Distribute the beads among the alveolus and the cell of the body. You won't be representing the actual partial pressures of each gas in the two locations, but you do want to represent the relative amounts of each gas at each place. Work with your lab mate to develop a plan.

5. In the Petri dishes representing the erythrocyte and blood plasma going TO the tissues:
   a. Move some oxygen molecules from the alveolus to these dishes (how many will go into the erythrocyte and how many into the plasma?).
   b. Some carbon dioxide was still left in this blood, so add some carbon dioxide molecules (how many will go into the erythrocyte and how many into the plasma?).

6. Slide the erythrocyte and plasma dishes toward the cell of the body. What will happen when this blood arrives? Move the beads to represent this.

7. Slide the erythrocyte and plasma dishes over to the bloodstream returning TO the alveolus and follow them to their return at the alveolus.
   a. Before you exchange gasses again, remember that another breath of air has occurred while you were away at the tissues. Change the beads in the alveolus to represent this.
   b. Exchange gasses between alveolus and blood (both erythrocytes and plasma).

8. Return the erythrocyte and plasma once again to the tissues via the bloodstream.
   a. Before you exchange gasses again, remember that more aerobic cellular respiration has occurred while you were away at the alveolus. Change the beads in the cell of the body to represent this.
   b. Exchange gasses between cell and blood (both erythrocytes and plasma).

9. Repeat this process a few more times until you and your lab partner are comfortable with all the events that are occurring.

10. When you ascend in altitude, like climbing a high mountain, the air pressure reduces, reducing the partial pressures of oxygen and carbon dioxide as a result. What will happen if the partial pressure of both oxygen and carbon dioxide in the alveolus are reduced? Make this happen in your process.
    Take notes here:

11. When you exercise, your skeletal muscle cells go through more aerobic cellular respiration. What will happen at these cells as a result? Make this happen in your process and predict what might change at the alveolus to help support greater rates of aerobic cellular respiration.
    Take notes here:

12. If desired, snap a few pictures of the steps to review later.

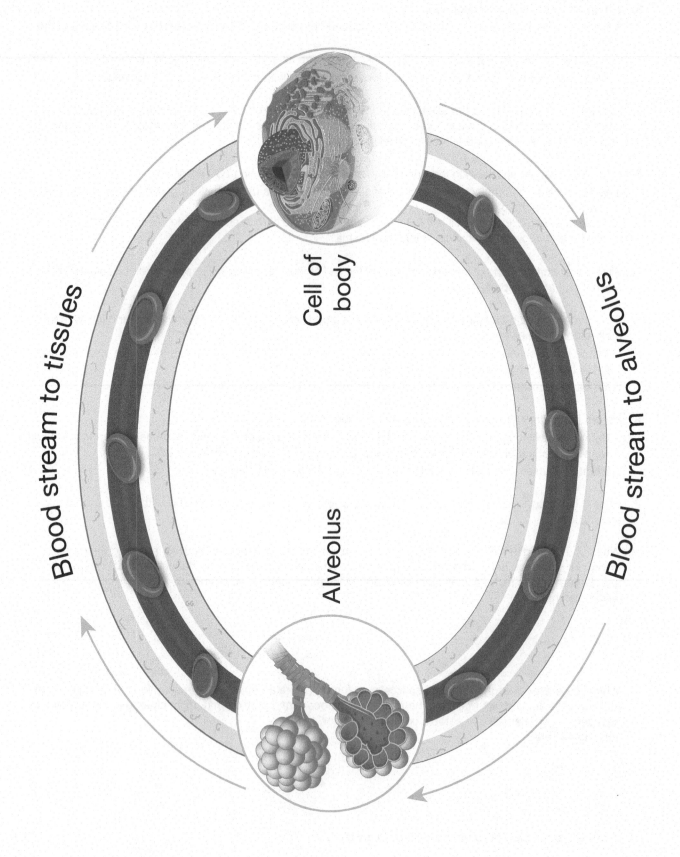

Cell of
body

Alveolus

Blood stream to tissues

Blood stream to alveolus

Activity 28.4 Case Study: Tuberculosis

This activity targets LOs 28.2, 28.4, 28.9, and 28.13.

Jordie has a fever and is very fatigued. Their appetite has reduced, and they have been losing weight. Jordie makes an appointment to see a doctor and explains these symptoms. Since the symptoms are vague, the doctor asks Jordie some follow-up questions to assist in a diagnosis. During this questioning, Jordie shares that they are a smoker and enjoy traveling around the world. Upon further investigation, the doctor determines that outbreaks of tuberculosis are common in the country that Jordie has recently visited. Jordie, who has not been vaccinated against tuberculosis, returns to the doctor's office for some tests, including a chest X-ray and sputum cultures. The results indicate that is it likely that Jordie is experiencing an active tuberculosis infection. While the doctor waits on the definitive test results that can take up to six weeks, Jordie starts on the recommended treatment regimen for tuberculosis that includes several different antibiotics. The doctor tells Jordie that being a smoker increases a person's risk of being infected by tuberculosis two-fold. Jordie wants to know more about the tuberculosis infection and begins to do some of their own research.

1. The tuberculosis bacterium infects a person when it reaches the alveoli of the lungs. Here it can multiply and fill the alveoli. When alveoli are full of bacteria, what important component of respiration cannot occur as efficiently?

   a. Ventilation

   b. Gas exchange across the respiratory membrane

   c. Gas transport

2. As the disease progresses, the damaged alveoli will be replaced by scar tissue. This scar tissue is less stretchy compared to healthy lungs. Because the tissue surrounding the remaining healthy alveoli is less stretchy, the alveoli will be less likely to:

   a. Increase in pressure upon inhalation

   b. Increase in volume upon inhalation

3. Which of Jordie's muscles must contract more forcefully to cause inhalation? Choose all correct answers.

   a. The diaphragm

   b. The internal intercostals

   c. The external intercostals

After several weeks of treatment with antibiotics, Jordie returns to the doctor for a check-up to see if their infection is under control. A chest X-ray and sputum cultures are promising; the active infection appears to be greatly reduced. The doctor is concerned about the damage that has occurred to Jordie's lungs before diagnosis and treatment and orders a spirometry assessment. The doctor tells Jordie that tuberculosis is sometimes called a *restrictive disease*, as it restricts the patient's ability to inspire air but has less of an effect on expiration.

4. Which of the following spirometry results would you expect to see most reduced in Jordie's assessment if there has been significant scar tissue development in their lungs?

   a. Inspiratory reserve volume

   b. Residual volume

   c. Expiratory reserve volume

5. Because of the change in volume that you noted in your previous question, which capacities will also be reduced? Choose all the correct answers.

   a. Vital capacity

   b. Inspiratory capacity

   c. Total lung capacity

Jordie, a fan of historical novels, is surprised to learn that tuberculosis used to be called "consumption" in the nineteenth century. In fact, about one-fourth of all deaths in the 1800s were caused by tuberculosis before the development of antibiotics in the early 1900s. Prior to antibiotics, people with active tuberculosis cases were often sent to live in a place called a sanitorium so that they would not infect family members at home. These separated living spaces, often resembling prisons for those who could not afford more expensive resort-type lodging, offered little in the way of treatment and over 50% of the people who entered them died within five years.

Lab 28 Post-Lab Quiz

1. What is pulmonary ventilation?

   a. The movement of gasses in the blood

   b. The movement of air in and out of the lungs

   c. The movement of gasses across the respiratory membrane

2. What is another term that means "inhalation"?

   a. Inspiration

   b. Contraction

   c. Bulk flow

3. Which muscles relax to cause quiet expiration?

   a. The internal intercostals and external intercostals

   b. The internal intercostals and abdominal muscles

   c. The diaphragm and external intercostals

4. Air moves into the lungs when:

   a. Intrapulmonary pressure is greater than atmospheric pressure.

   b. Intrapulmonary pressure is less than atmospheric pressure.

   c. The two pressures are equal.

5. If you consume botulism toxin it can paralyze your diaphragm. Partial paralysis of the diaphragm will cause which of these respiratory volumes to decrease?

   a. Tidal volume

   b. Expiratory reserve volume

   c. Residual volume

6. If a person's tidal volume is 3000 mL and their respiratory rate is 10 breaths per minute, what is their minute ventilation?

   a. 30 L/minute

   b. 300 mL/minute

   c. 3 minutes/L

7. Which is more soluble in plasma, oxygen or carbon dioxide?

   a. Oxygen

   b. Carbon dioxide

   c. They are equally soluble in plasma.

8. In comparing the relative concentration of oxygen between an alveolus and the bloodstream, the oxygen is _____ in the alveolus as compared to the blood that is arriving there.

   a. higher

   b. lower

9. The cells of the brain are highly metabolically active and must generate a lot of ATP to function. Because of this, the relative concentration of oxygen within the brain tissue will be _____ than the concentration of oxygen in the blood arriving at the brain.

a. lower

b. higher

10. Carbon monoxide poisoning is deadly because the carbon monoxide molecules attach to the heme groups of hemoglobin molecules and do not detach. What percent of which gas will be affected by carbon monoxide poisoning?

a. 23% of the carbon dioxide

b. 77% of the carbon dioxide

c. 2% of the oxygen

d. 98% of the oxygen

Lab 29

Anatomy of the Digestive System

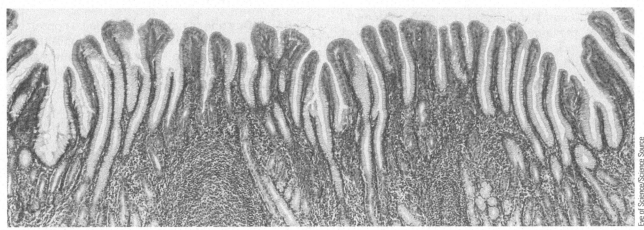

Eye of Science/Science Source

Learning Objectives: After completing this lab, you should be able to:

29.1 Compare and contrast mechanical digestion and chemical digestion, including where they occur in the digestive system.

29.2 Identify and describe the boundaries of the oral cavity.

29.3 Compare and contrast the composition and functions of the hard palate, soft palate, and uvula.

29.4 Describe the structure and function of teeth.

29.5 Describe the structure and function of the salivary glands.

29.6 Describe the major functions of the digestive system.

29.7 Explain the differences between the gastrointestinal (GI) tract (alimentary canal) and the accessory digestive organs.

29.8 Trace the pathway of ingested substances through the gastrointestinal (GI) tract.

29.9 List and identify the organs that compose the gastrointestinal (GI) tract.

29.10 Identify and describe the different regions of the pharynx with respect to the passage of air and/or food.

29.11 Identify and describe the gross anatomy of the esophagus, including its location relative to other body structures.

29.12 Describe the general functions of the esophagus.

29.13 Identify and describe the gross anatomy of the stomach, including its location relative to other body structures.

29.14 Describe the general functions of the stomach.

29.15 Describe the compositions, locations, and functions of the inferior esophageal (cardiac, lower esophageal) sphincter and the pyloric sphincter.

29.16 Identify gastric folds (rugae) and discuss their functional significance.

29.17 Identify and describe the gross anatomy of the small intestine, including its location relative to other body structures.

The Human Anatomy and Physiology Society includes more than 1,700 educators who work together to promote excellence in the teaching of this subject area. The HAPS A&P Learning Outcomes measure student mastery of the content typically covered in a two-semester Human A&P curriculum at the undergraduate level. The full Learning Outcomes are available at https://www.hapsweb.org.

29.18 Describe the general functions of the small intestine.

29.19 Identify the specific segments of the small intestine (i.e., duodenum, jejunum, ileum), including their relative length.

29.20 Identify and describe the gross anatomy of the large intestine, rectum, and anal canal, including their location relative to other body structures.

29.21 Identify the specific segments and related flexures of the large intestine.

29.22 Describe the general functions of the large intestine, rectum, and anal canal.

29.23 Describe the general functions of the liver.

29.24 Identify and describe the structure of the liver, including the individual lobes, ligaments (e.g., coronary ligament, falciform ligament, round ligament [ligamentum teres]), and the porta hepatis.

29.25 Describe the location of the liver relative to other body structures.

29.26 Identify and describe the structure and functions of the gallbladder.

29.27 Describe the location of the gallbladder relative to other body structures.

29.28 Identify and describe the structure and functions of the pancreas.

29.29 Describe the location of the pancreas relative to other body structures.

29.30 Identify and describe the major histological components of the pancreas (pancreatic acini and pancreatic islets [islets of Langerhans]) and discuss their major functions.

29.31 Describe the major functions of the biliary apparatus.

29.32 Identify and describe the biliary apparatus components (i.e., left and right hepatic ducts, common hepatic duct, cystic duct, common bile duct, main pancreatic duct, hepatopancreatic ampulla [ampulla of Vater], hepatopancreatic sphincter [sphincter of Oddi], major duodenal papilla).

29.33 Trace the path of bile and pancreatic juice through the biliary apparatus.

29.34 Identify and describe the gross anatomic and microscopic structure and function of each of the gastrointestinal (GI) tract tunics (layers): mucosa, submucosa, muscularis (muscularis externa), and serosa or adventitia.

29.35 Describe the anatomic specializations of the esophageal tunics (e.g., composition of the mucosa and muscularis [muscularis externa]) compared to the tunics of the rest of the GI tract.

29.36 Describe the anatomic specializations of the stomach tunics compared to the tunics of the rest of the GI tract.

29.37 Relate the anatomic specializations of the stomach tunics (e.g., number of layers of muscle in the muscularis [muscularis externa]) to the organ's functions.

29.38 Identify and describe the gastric glands, including their cells (e.g., parietal cells, chief cells).

29.39 Describe the anatomic specializations of the small intestine tunics (e.g., circular folds [plicae circulares], villi, microvilli) compared to the tunics of the rest of the GI tract.

29.40 Relate the anatomic specializations of the small intestine tunics (e.g., circular folds [plicae circulares], villi, microvilli) to the organ's functions.

29.41 Identify and describe the function of the following small intestine structures: duodenal glands (Brunner glands), intestinal glands (crypts of Lieberkuhn), and Peyer patches (lymphoid [lymphatic] nodules).

29.42 Describe the specializations of the large intestine tunics (e.g., composition of the muscularis [muscularis externa]) compared to the tunics of the rest of the GI tract.

29.43 Relate the specializations of the large intestine tunics (e.g., composition of the muscularis [muscularis externa]) to the organ's functions.

29.44 Identify and describe the histological components of the classic hepatic lobule.

29.45* Predict disruptions in function and/or symptoms if various anatomical components are altered.

* Learning Objective not a HAPS Learning Goal

The functions of the digestive system are to break down the foods you eat, release their nutrients, and absorb those nutrients into the body. The digestive system also generates, stores, and excretes some of our waste and is responsible for absorbing water. When you take a bite of food, it enters a long and convoluted tract of connected organs. Each of the organs in the **gastrointestinal (GI) tract** contributes its own function to breakdown and absorption of nutrients. In addition to those organs within the tract, there are organs that are outside the tract, that never have food pass through them, but that contribute to the process of digestion. These **accessory digestive organs** are the liver, gall bladder, salivary glands, and pancreas. The organs of the digestive system are illustrated in Figure 29.1.

Figure 29.1	Organs of the Digestive System

The digestive system is composed of the GI tract and the accessory digestive organs (liver, pancreas, gallbladder, salivary glands).

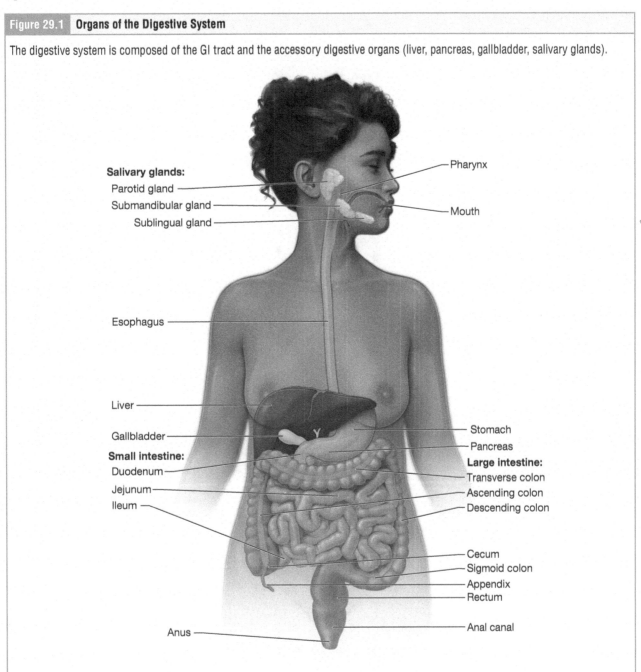

Salivary glands:
Parotid gland
Submandibular gland
Sublingual gland

Esophagus

Liver

Gallbladder

Small intestine:
Duodenum
Jejunum
Ileum

Anus

Pharynx

Mouth

Stomach
Pancreas
Large intestine:
Transverse colon
Ascending colon
Descending colon

Cecum
Sigmoid colon
Appendix
Rectum

Anal canal

Though we eat complex foods such as burritos, dumplings, chicken, and rice, our cells actually can utilize only the **monomers**, the smallest building blocks of these foods. The process of digestion breaks down the foods we eat into these monomers. You may remember that all of the biological molecules in our cells can be categorized as lipids, proteins, carbohydrates, or nucleic acids. The monomers of these molecules are **fatty acids**, **amino acids**, **monosaccharides**, and **nucleotides**.

As is the case with all body systems, the digestive system does not work in isolation. Where do the nutrients go after they have been absorbed by GI organs? The answer is the blood! The nutrients are absorbed into the circulating blood and distributed all around the body. From the blood, cells can pick up the nutrients they need, such as amino acids for generating new cellular proteins or glucose molecules for performing cellular respiration.

Lab 29 Pre-Lab Quiz

This quiz will strengthen your background knowledge in preparation for this lab. For help answering the questions, use your resources to deepen your understanding. The best resource for help on the first five questions is your text, and the best resource for help answering the last five questions is to read the introduction section of each lab activity.

1. Anatomy of digestive system organs. Using the figure, label the following organs:

Esophagus	Mouth	Small intestine
Gall bladder	Pancreas	Stomach
Large intestine	Parotid gland	Sublingual gland
Liver	Pharynx	Submandibular gland

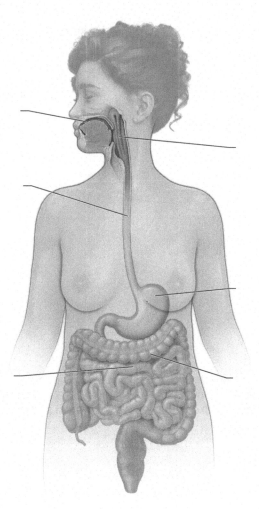
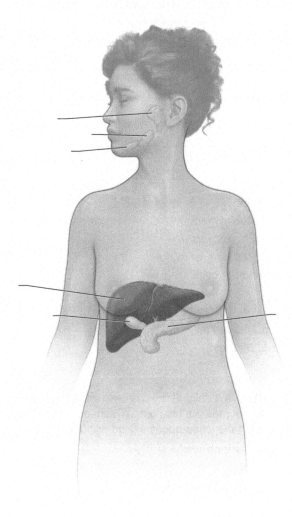

Label each panel in the figure with one of the following:
GI tract organs   Digestive accessory organs

2. Match the root next to its definition:

Root	Meaning
	a bag
	shaggy hair
	sheet
	the liver
	under

List of Roots

asci
hepa-
lamin-
sub-
villi

3. Match the organ next to its type of epithelium:

Organ	Type of Epithelium
	Stratified squamous epithelium
	Simple columnar epithelium
	Simple columnar epithelium
	Stratified squamous epithelium
	Simple columnar epithelium

List of Organs

Mouth
Esophagus
Stomach
Small intestine
Large intestine

4. What type of cells generate the mucus that coats an epithelium?

   a. Goblet cells
   b. Enterocytes
   c. Hepatocytes
   d. Acinar cells

5. What is the shortest region of the small intestine?

   a. Duodenum
   b. Jejunum
   c. Ileum

6. Harry is a child at the cafeteria table drinking milk. His friend tells a funny joke and Harry laughs so hard that milk comes out his nose! Which region of the pharynx is not typically exposed to the things we eat and drink, but in this case became filled with milk?

   a. Nasopharynx
   b. Oropharynx
   c. Laryngopharynx

7. How many layers of muscle are found in the muscularis of the small intestine?

   a. Zero
   b. One
   c. Two
   d. Three

8. In which of the following organs are clusters of endocrine cells, called islets, visible under the microscope?

   a. Stomach
   b. Small intestine
   c. Large intestine
   d. Liver
   e. Pancreas

9. Which of the following are gross anatomical structures (i.e. visible without a microscope)?

   a. Intestinal villi
   b. Rugae
   c. Pancreatic acini
   d. Liver lobules

10. Where would you need to look to find cementum?

   a. Inside the liver
   b. Under the gingiva
   c. Behind the uvula
   d. In the lumen of the esophagus

(page intentionally left blank)

Activity 29.1 The Mouth

This activity targets LOs 29.1, 29.2, 29.3, 29.4, and 29.5.

The mouth, also known as the oral cavity, is the official entrance to the digestive system. Food is ingested in the mouth, where both the processes of mechanical digestion and chemical digestion begin. **Mechanical digestion** is the physical breaking up of the food, which begins in the mouth through the act of chewing and continues in the stomach and duodenum through the contraction of smooth muscle in the walls of these organs. **Chemical digestion**, as the name implies, is the breaking of chemical bonds within the food to release the monomers of the nutrients. Once the food leaves the mouth, during the act of swallowing, both kinds of digestion have already begun.

The lips, or **labia**, are the anterior border of the mouth. The posterior border of the mouth is defined by the **fauces**, the two arched openings to the throat, and the **uvula**, which is suspended between them (Figure 29.2). The lateral walls are the **cheeks**. The superior border is defined at the palate, which has an anterior **hard palate**, made of bone, and a posterior **soft palate**. The inferior border of the mouth is the **tongue**, which is anchored to the floor of the mouth. The **lingual frenulum**, a fold of tissue in the midline, is visible when the tongue is raised.

Chemical digestion is made possible in the mouth because of saliva, which is a watery substance that is impregnated with digestive enzymes. Three pairs of salivary glands produce saliva. The **submandibular glands** are found inferior and medial to the mandible. The **sublingual salivary glands** are found under the tongue. The **parotid salivary glands** are found in the cheeks, just at the curve of the jaw (Figure 29.3).

Mechanical digestion is made possible in the mouth through the act of chewing, which is dependent upon the teeth. Each tooth forms a joint with a socket of the maxilla or mandible. The **gingivae** (commonly called the gums) line the sockets and surround the necks of the teeth. Teeth are held in their sockets by short bands of dense connective tissue called **periodontal ligaments** (Figure 29.4). The two main parts of the tooth are the **crown**, which is the portion projecting above the gums, and the **root**, which is embedded within the maxilla or mandible. Both parts contain an inner **pulp cavity**, which contains the nerves and blood vessels enmeshed in loose connective tissue.

Figure 29.2	The Mouth

The mouth is the entrance to the GI tract. Its structures enable mechanical digestion of ingested food.

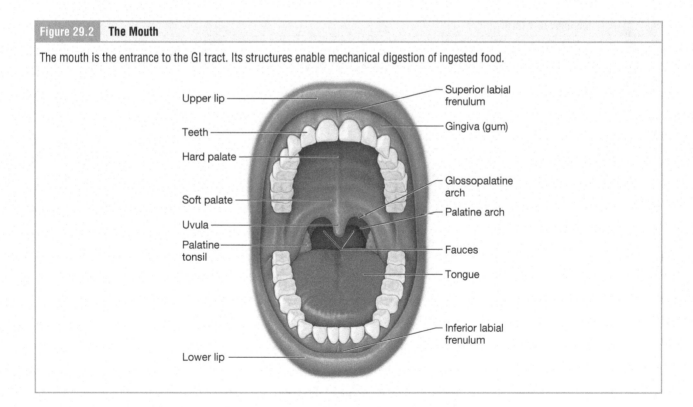

Figure 29.3 **The Salivary Glands**

There are three salivary glands on each side of the mouth. The parotid gland is under the skin in the posterior cheek. The submandibular gland sits below the mandible. The sublingual salivary gland is found under the tongue.

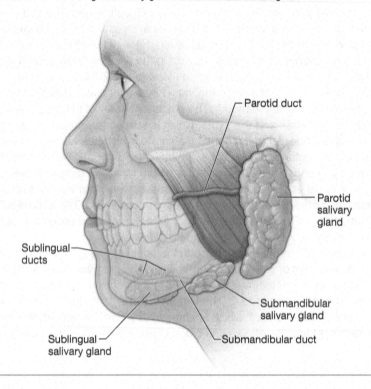

Figure 29.4 **Anatomy of a Tooth**

An individual tooth has a root, which is housed within a socket in the bone, and through that root runs a canal containing blood vessels and nerves. The top of the tooth, which is visible outside the gums, consists of enamel and dentin; both substances are hard and durable materials.

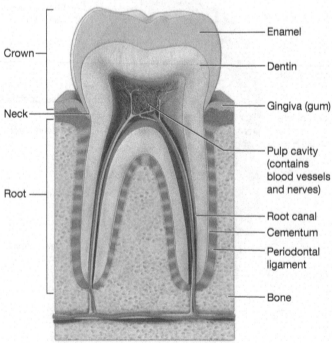

The blood vessels and nerves within the pulp cavity run through the root of the tooth and into the bone. This passage is called the **root canal**. Surrounding the pulp is **dentin**, a bonelike tissue. In the root of each tooth, the dentin is covered by an even harder bonelike layer called **cementum**. In the crown of each tooth, the dentin is covered by an outer layer of **enamel**, the hardest substance in the human body.

Materials
- Hand mirror
- Model or figure of a longitudinally sectioned tooth
- Model or figure of a sagittal section of the head

Instructions

1. Using the head model or figure and Figure 29.5 as a guide, identify the following structures:

   - Labia
   - Uvula
   - Hard palate
   - Soft palate
   - Tongue

   - Lingual frenulum
   - Submandibular gland
   - Sublingual gland
   - Parotid gland

2. Label the figure of a sagittal section of a head found on the next page.

3. Using the hand mirror, open your own mouth and identify your fauces, uvula, tongue, and lingual frenulum.

4. Using the model of the tooth or figure and Figure 29.6 as a guide, identify the following structures:

   - Gingivae
   - Periodontal ligament
   - Root
   - Pulp cavity

   - Root canal
   - Dentin
   - Cementum
   - Enamel

5. Label the figure of a sectioned tooth found on the next page.

| Figure 29.5 | Sagittal Model of the Head |

© Axis Scientific Model, Photo by © Cengage

| Figure 29.6 | Model of a Tooth |

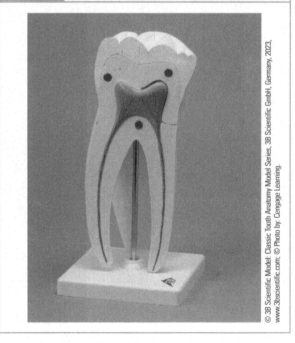

© 3B Scientific Model: Classic Tooth Anatomy Model Series, 3B Scientific GmbH, Germany, 2023. www.3bscientific.com; © Photo by: Cengage Learning.

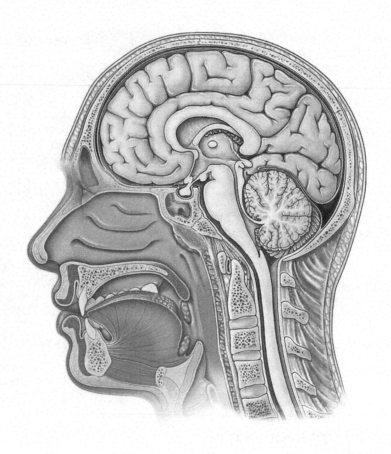

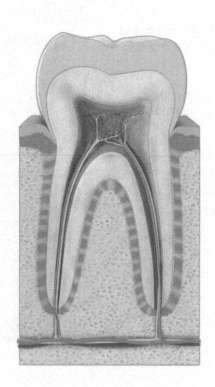

Activity 29.2 Gross Anatomy of the Digestive System

This activity targets LOs 29.1, 29.6, 29.7, 29.8, 29.9, 29.10, 29.11, 29.12, 29.13, 29.14, 29.15, 29.16, 29.17, 29.18, 29.19, 29.20, 29.21, 29.22, 29.23, 29.24, 29.25, 29.26, 29.27, 29.28, 29.29, 29.30, 29.31, 29.32, and 29.33.

The gastrointestinal tract is a convoluted tube that begins at the back of the mouth and ends at the anus. Within the organs of the GI tract, our food is broken down into the components that are able to be absorbed. Nutrients are absorbed across the wall of the GI tract organs and into the bloodstream or lymph vessels. Any material that cannot be broken down into pieces small enough to absorb, or material that is unusable to us, is compacted into feces and eliminated from the end of the GI tract.

If you were dissecting a human cadaver, you would cut into the skin of the abdomen and peel back both the skin and muscle layers. The first structure visible to you once inside the abdominal cavity would be the **greater omentum**, one of the structures of the **peritoneum**. The peritoneum and the mesenteries are broad serous membranes that hold the abdominal organs in place. These membranes may or may not be represented on the models in your lab but can be seen in Figure 29.7.

When you swallow, the food from your mouth enters the **pharynx**, a short tube of skeletal muscle lined with a mucous membrane. The pharynx runs from the posterior oral and nasal cavities to the opening of the esophagus. The pharynx has three regions, the most superior, the **nasopharynx**, is involved only in breathing and speech, only air passes through it. The **oropharynx** is the region below the nasopharynx, posterior to the mouth. The **laryngopharynx** is inferior to the oropharynx and at its inferior edge, it splits, like a fork in the road, to lead to two different structures. Anteriorly, the laryngopharynx becomes the **trachea**, the passageway for air traveling to and from the lungs. Posteriorly, the laryngopharynx becomes the esophagus, which allows food and drink to pass through the thorax and toward the stomach (Figure 29.8).

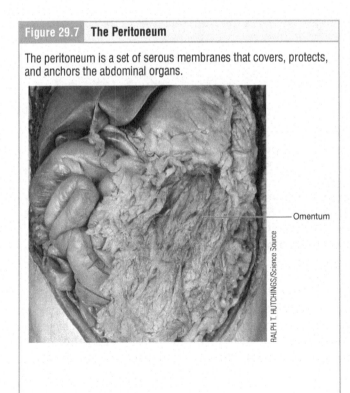

Figure 29.7	**The Peritoneum**

The peritoneum is a set of serous membranes that covers, protects, and anchors the abdominal organs.

— Omentum

RALPH T. HUTCHINGS/Science Source

Figure 29.8	**The Pharynx**

The pharynx, or throat, is an open cavity posterior to the nasal cavity and mouth. It has three zones—the nasopharynx, the oropharynx, and the laryngopharynx—all of which are continuous with one another.

Nasal cavity

Hard palate

Tongue

Soft palate

Epiglottis

Larynx

Esophagus

Trachea

☐ Nasopharynx   ☐ Oropharynx   ■ Laryngopharynx

Figure 29.9 | The Esophagus

(A) The esophagus is a long elastic tube that runs from the mouth to the stomach. It passes through a gap in the diaphragm called the esophageal hiatus. (B) A cross-section of the esophagus. Like all organs of the GI tract, the esophagus has a mucosa that lines the lumen, a submucosa, and a muscularis surrounded by a serosa, a sheath of connective tissue.

The **esophagus** is a muscular tube that connects the pharynx to the stomach (Figure 29.9), passing through an opening in the diaphragm, the **esophageal hiatus**, to reach the abdomen. At the inferior edge of the esophagus, where it meets the stomach, is a ring of smooth muscle called the **lower esophageal sphincter**. This sphincter opens to allow food to enter the stomach and closes behind it. It functions to keep digesting food and acidic gastric fluids in the stomach, so that they don't harm the esophageal lining.

The stomach continues the jobs of mechanical and chemical digestion. Most of the chemical digestion of food takes place in the next organ along the GI tract, the **small intestine**. The enzymes for digestion are added in a small initial stretch of the small intestine, the **duodenum**. Because the duodenum is small, the stomach's larger volume allows it to function as a holding tank for food. This allows us to eat a large meal but digest it in the small intestine slowly over the hours that follow. While ingested food is in the stomach, it is exposed to gastric juice, a combination of acid and digestive enzymes that breaks down proteins. The stomach also has an enormous capacity to grind the food, contributing to the mechanical digestion of our meals. There are four main regions of the stomach: the **cardia**, the **fundus**, the **body**, and the **pylorus** (Figure 29.10). The pylorus is divided from the duodenum by another sphincter, the **pyloric sphincter**. The opening of this sphincter allows food to pass into the duodenum. The opening and closing of the pyloric sphincter are highly controlled to regulate the speed of digestion. On the model, or in Figure 29.10, notice that the stomach is curved like the letter J. The superior curve of the stomach is called the **lesser curvature**; the broader inferior curve is the **greater curvature**. When the stomach is empty the mucosa is highly folded. The stomach stretches when it fills with food and these folds, **rugae**, will not be apparent in a full stomach.

Partially digested food passes through the pyloric sphincter and enters into the **duodenum**, the first segment of the small intestine. The small intestine has two other, much longer, segments called the **jejunum** and the **ileum**. The duodenum makes a C-shaped curve around the head of the pancreas before joining the jejunum. The duodenum is the site of the majority of chemical digestion, using secretions from the **gallbladder** and the **pancreas**. Therefore, all three of these organs are anatomically connected and their network of ducts is collectively referred to as the **biliary**

Figure 29.10 | **The Regions of the Stomach**

The stomach has five regions, all of which are continuous with each other. The food enters the stomach from the esophagus at the cardia, named for its proximity to the heart. The fundus is an expandable dome-shaped superior region. The body—the largest component of the stomach—comprises the middle of the organ. The antrum is the most inferior, with the pylorus containing the food that is about to pass into the small intestine.

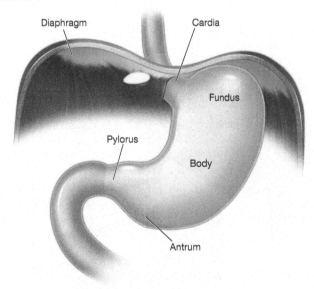

apparatus (Figure 29.11). On your model you may be able to find the common bile duct, which carries bile from its site of production, the liver, toward the duodenum. The common bile duct may carry bile directly from the liver to the duodenum, or the bile may head through the cystic duct to be stored in the gall bladder until it is needed in digestion.

The pancreas generates the majority of our digestive enzymes. These enzymes are carried to the duodenum by the **pancreatic duct**. Both the common bile duct and the pancreatic duct join the duodenum at the **hepatopancreatic ampulla**.

The duodenum ends at its connection to the jejunum, the middle segment of the small intestine. No clear anatomical border separates the jejunum from the duodenum or the ileum. The ileum joins the **cecum**, the first part of the **large intestine**, at the **ileocecal sphincter**.

The large intestine finishes the absorption of nutrients and water and forms feces. Its length is divided into four main regions: the **cecum**, the **colon**, the **rectum**, and the **anus**. The colon is further subdivided into the **ascending colon**, **transverse colon**, **descending colon**, and **sigmoid colon**.

The **liver** is the largest abdominal organ, occupying the upper right quadrant of the abdomen. The liver has many functions including recycling blood components, detoxifying nutrients and storing glucose. It is considered an organ of the digestive system because it produces bile, an important digestive compound. It can be divided into four unequal lobes. The **right lobe** and **left lobe** are separated by the **falciform ligament**. From an inferior view, the **quadrate lobe** can be identified on the inferior surface of the right lobe. The **caudate lobe** is just posterior to the quadrate lobe, with the **hepatic portal vein**, **hepatic artery**, and **common bile duct** in between (Figure 29.12).

The pancreas is found deep to the stomach. Its widest part, the head of the pancreas, is nestled into the C-shaped curve of the duodenum. The **body of the pancreas** tapers to the left and ends in the **tail of the pancreas**.

Learning Connection

Explain a Process

Try writing each of the components of the large intestine on pieces of paper. Mix the papers up and put them into the order that digested material passes through from the ileum to the anus.

Learning Connection

Broken Process

For each of the digestive system organs, imagine what the consequences might be if they were damaged, removed, or not functioning correctly.

Figure 29.11 | The Accessory Organs of Digestion

The liver and gallbladder contribute bile to the duodenum. The head of the pancreas is nestled into the curve of the duodenum, and the exocrine secretions of the pancreas are added to the duodenum. Secretions from all three accessory organs enter the duodenum at the ampulla.

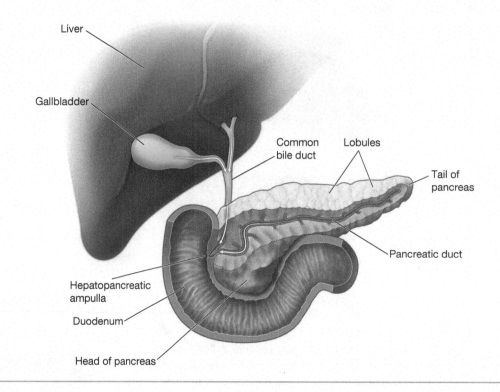

Figure 29.12 | The Liver

In the anterior perspective (A) the liver is divided into two lobes by the falciform ligament. From the posterior perspective (B) the smaller caudate and quadrate lobes are visible.

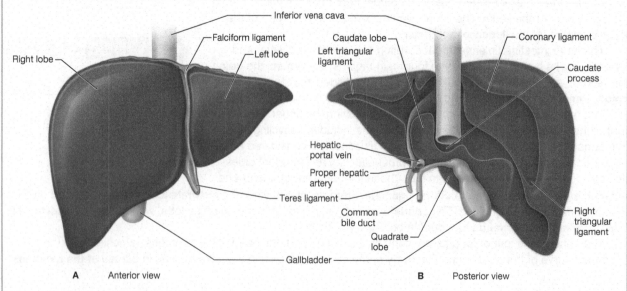

Materials
- Preserved animal stomach (optional)
- Torso model

Structures Identified in This Activity

- pharynx
- nasopharynx
- oropharynx
- laryngopharynx
- trachea
- esophagus
- esophageal hiatus
- lower esophageal sphincter
- cardia
- fundus

- body (of the stomach)
- pylorus
- pyloric sphincter
- greater curvature of the stomach
- lesser curvature of the stomach
- rugae
- duodenum
- jejunum
- ileum

- liver
- gallbladder
- pancreas
- head of pancreas
- body of pancreas
- tail of pancreas
- biliary apparatus
- pancreatic duct
- hepatopancreatic ampulla
- cecum

- large intestine
- ileocecal sphincter
- colon
- rectum
- anus
- ascending colon
- transverse colon
- descending colon
- sigmoid colon

Instructions

1. Identify the gross anatomical structures of the GI tract on the torso model and/or figures using Figure 29.13 as a guide.

2. Write each of the structure names above on a sticky note and use them to label your torso model.

3. If a preserved animal stomach is available, examine it. Locate the regions of the stomach (cardia, fundus, body, pylorus) and the features (rugae, lower esophageal sphincter, pyloric sphincter).

Learning Connection

Chunking

Using the structure list in Activuty 29.2, categorize every structure as either part of the GI tract or part of the accessory digestive organs.

Figure 29.13 **Human Torso Model**

The gastrointestinal tract organs are visible in the human torso model.

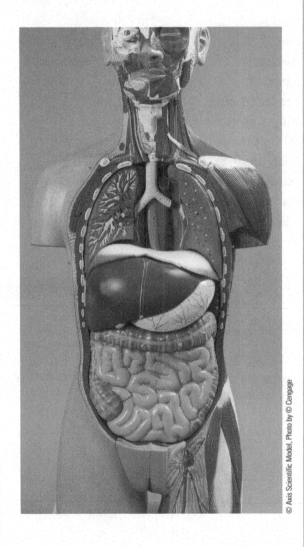

© Axis Scientific Model. Photo by © Cengage

Digging Deeper:
Weight Loss Surgery

Several forms of surgery are used to treat obesity and type 2 diabetes mellitus. The most common is gastric bypass, a surgery that cuts the stomach, leaving only the cardia region to hold food, and connects this abrogated stomach to the jejunum, bypassing the duodenum. The surgery is effective in two ways, first, because the stomach can hold much less food, the patient has a dramatically decreased food intake after surgery, leading to weight loss. Second, because the ingested food is not subjected to the digestive enzymes of the duodenum, food is only partially broken down, and far less nutrients are absorbed. Major complications of the surgery include diarrhea, because a lot of nutrients are left in the lumen of the GI tract, where they draw water, and malnutrition, because many of the nutrients cannot be liberated in order to be absorbed.

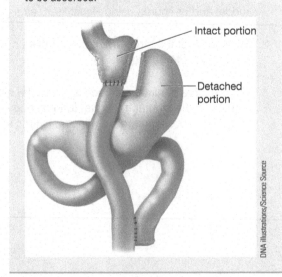

Intact portion

Detached portion

DNA illustrations/Science Source

Activity 29.3 Microanatomy of the Digestive System: The Four Tissue Layers

This activity targets LOs 29.15, 29.16, 29.17, 29.30, 29.34, 29.35, 29.36, 29.37, 29.38, 29.39, 29.40, 29.41, 29.42, 29.43, and 29.44.

Throughout its length the GI tract is composed of the same four tissue layers but the composition and structure of these layers varies with the function of each organ. Remember that the inside of a hollow structure is called the lumen. The four layers of the wall of GI tract organs, from lumen to the outside are: mucosa, submucosa, muscularis, and serosa (Figure 29.14).

Figure 29.14	Layers of the Wall of the GI Tract

(A) Every organ in the GI tract has four layers of different tissues that make up its walls. The hollow center of the organ is the lumen. From the lumen outward, the layers are the mucosa, the submucosa, the muscularis, and the serosa. While all organs have these four layers, their composition and structure vary (B) The four layers of the esophagus can be seen in this cross-section.

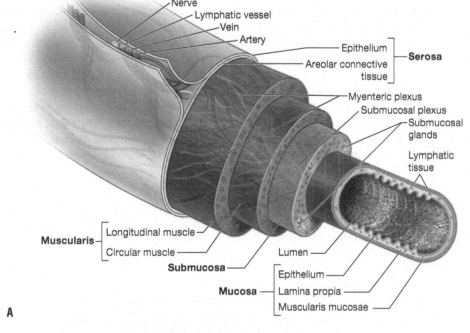

A

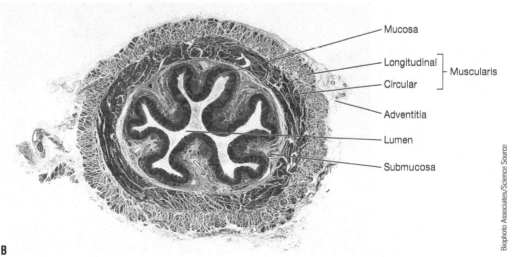

B

Biophoto Associates/Science Source

The **mucosa** is a layer of mucous membrane that is similar to the lining of the respiratory system. The mucosa has a very thin layer of muscle in it, called the muscularis mucosa. The contractions of the muscularis mucosa control the folds of the mucosa, similar to a drawstring, but are not strong enough to propel food down the length of the GI tract.

The epithelium of the GI tract will vary from one organ to another based on the conditions within that organ. In the mouth, pharynx, esophagus, and anal canal, the epithelium is primarily a non-keratinized, stratified squamous epithelium. These are regions that are exposed to much more friction by the contents of their lumens, so a thicker epithelium is required. By the time the food reaches the stomach and intestines it has been mechanically ground to a relatively soupy consistency, and so, in the stomach and intestines, a simple columnar epithelium can be found (Figure 29.15). Moreover, in the small intestine in particular, nutrients must cross the mucosa to be absorbed, so a thicker epithelium is not desirable. Punctuated along the epithelium are **goblet cells** which secrete mucus onto the epithelium surface, lubricating the passage of digested food. Enteroendocrine cells are features of the epithelium of the intestines and function to secrete hormones that regulate some digestive processes.

Recall that one of the characteristics of epithelium is that it is avascular (lacking blood vessels). All epithelia are supported by a vascularized connective tissue and nutrients, and oxygen will diffuse from the connective tissue to the epithelium, nourishing it. The epithelium of the GI mucosa is supported by a sheet of connective tissue, the lamina propria. In addition to blood vessels the lamina propria also contains clusters of lymphocytes, making up the **mucosa-associated lymphoid tissue (MALT)**. Lymphocytes will stain a deep purple in hematoxylin and eosin staining, and the MALT patches can be easily noted as dark clusters (Figure 29.16).

The **submucosa** lies immediately under the mucosa. This is a dense layer of connective tissue that anchors the mucosa to the muscularis. The submucosa contains glands that release digestive secretions into the lumen of the GI tract organs. It also serves as a conduit for a dense branching network of nerves called the submucosal plexus.

The third layer of the GI tract is the **muscularis**. The muscularis is made up of two layers of smooth muscle in most GI organs (a third layer is present in the stomach). The inner layer has smooth muscle fibers that wrap around the circumference of the organ in rings. The outer layer has smooth muscle fibers that stretch along the organ longitudinally. The contractions of the muscularis are responsible for propelling food along the GI tract. If you compare the muscularis from one organ to another, you can see that it varies in thickness according to the functions of the organ.

Figure 29.15	Histology of the Gastroesophageal Junction

At the gastroesophageal junction, the transition between stratified squamous epithelium and simple columnar epithelium can be seen.

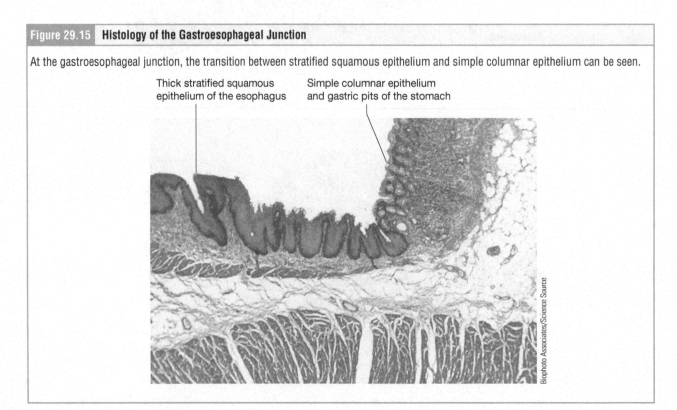

Thick stratified squamous epithelium of the esophagus

Simple columnar epithelium and gastric pits of the stomach

Biophoto Associates/Science Source

Figure 29.16 | Mucosa-Associated Lymphoid Tissue in the GI Lamina Propria

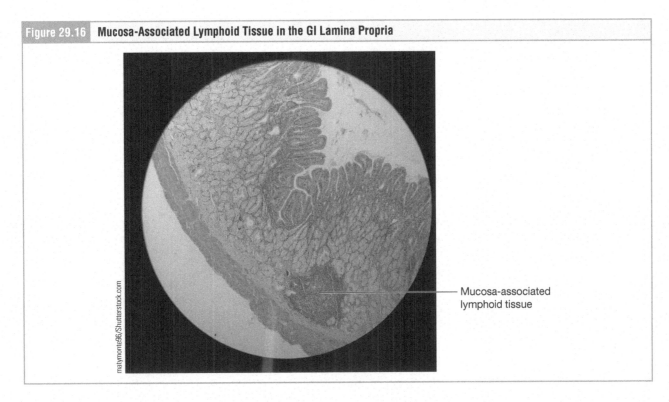

Mucosa-associated lymphoid tissue

All of the organs are wrapped with an outer layer of connective tissue. This layer is called the **serosa** in most of the organs you will examine today. The exception is the esophagus. The esophagus is not found within the abdominal cavity, and so its outer layer has a different name, it is called the **adventitia**.

Materials
• Model or images of the layers of the GI tract

Instructions
1. Identify the four tissue layers of the GI tract on the model and/or figures. Figure 29.17 shows the tissue layers that surround the stomach.

Figure 29.17 | **Human Stomach Model**

Many laboratory models of the stomach can be opened to see the rugae and organ wall layers.

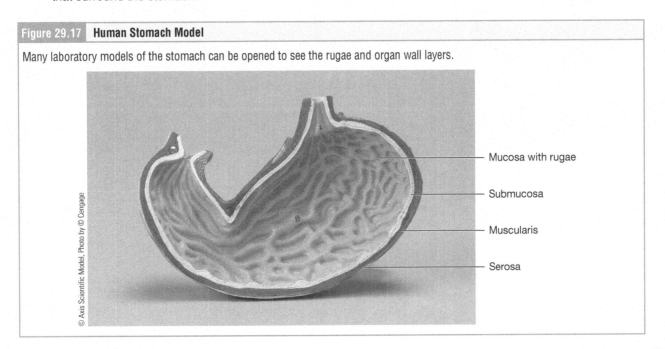

Mucosa with rugae

Submucosa

Muscularis

Serosa

(page intentionally left blank)

Activity 29.4 Histology of GI Organs

This activity targets LOs 29.15, 29.30, 29.34, 29.35, 29.36, 29.37, 29.39, 29.41, 29.42, and 29.44.

Histology of the Esophagus

The mucosa of the esophagus has a thick stratified squamous epithelial lining. This thick epithelium protects the underlying tissue from the friction of the only partially digested food. The mucosa's lamina propria contains mucus-secreting glands.

Histology of the Stomach

The wall of the stomach is made of the same four layers as most of the rest of the GI tract, but with adaptations to the mucosa and muscularis for the unique functions of this organ. In addition to the typical two layers of muscle in the organ wall, the stomach has a third layer which brings extra grinding power to assist in mechanical digestion. The three layers are the **circular layer**, the **longitudinal layer**, and the **oblique layer**. The mucosa is dotted with deep pits called gastric pits, which are long tunnels leading to gastric glands (Figure 29.18). The surface of the stomach and the shallow regions of the gastric pits are lined with specialized simple columnar epithelial cells that secrete a thick mucus that protects them from the acidic interior of the stomach. Cells that secrete mucus typically appear vacuous or partially transparent in histological preparations because the mucus repels the stains (Figure 29.19). Along the gastric glands, primarily in the middle region, are parietal cells which secrete the stomach's acid. Deeper in the pits are chief cells, which secrete pepsinogen, the inactive precursor to the protein-digesting enzyme pepsin.

Figure 29.18	Gastric Pits

The histology of the stomach is notable for the deep gastric pits.

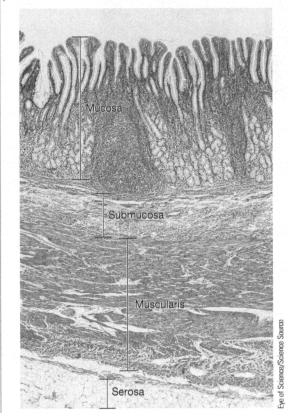

Eye of Science/Science Source

Figure 29.19	Cells of the Stomach

Histologically we can identify cells that secrete mucus because the mucus repels histological stains, making them appear empty.

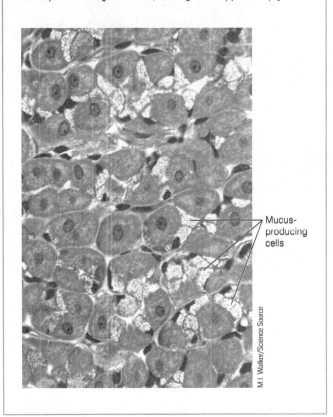

Mucus-producing cells

M.I. Walker/Science Source

Histology of the Small Intestine

The majority of the length of the small intestine functions in the absorption of nutrients across the gut wall and into the underlying blood vessels. Therefore, the unique histological features of the small intestine all maximize surface area, providing as much space for absorption as possible. Collectively, these features increase the absorptive surface area of the small intestine more than 600-fold.

- The **circular folds** are deep ridges in the mucosa and submucosa. The folds are found in the duodenum and the first half of the ileum. They expand the surface area, but their large size also slows the flow of material through the small intestine, allowing for more time for digestion (Figure 29.20).

- The surface of the folds is not smooth, rather, they are covered in tiny **villi** (singular = villus) that greatly expand the available surface area (Figure 29.21). Each villus is a projection of both mucosa and submucosa. Each villus is lined with columnar epithelium. In the core of the villus, a capillary bed and a lymphatic vessel, called a lacteal, can be found.

| Figure 29.20 | Lumen and Walls of the Small Intestine |

The internal surface of the small intestine is marked by circular folds which are seen (A) by endoscope and (B) in a drawing made during dissection.

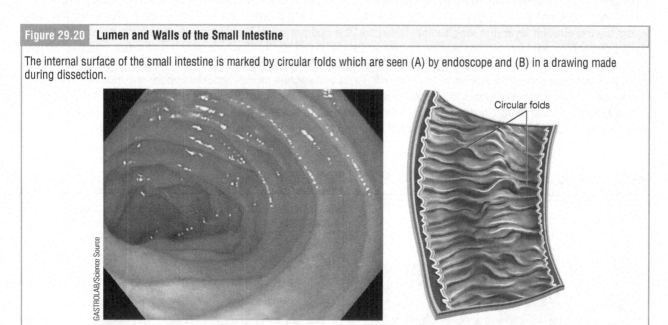

| Figure 29.21 | Histology of the Small Intestine |

The small intestine is marked by villi, which are projections of the mucosa and submucosa.

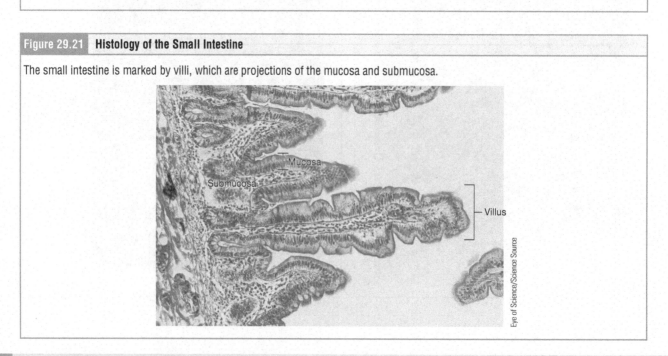

- The apical surface of the villus epithelial cells is rough, with many, many tiny projections of the cell membrane. These **microvilli** constitute the **brush border**, the expansive surface area of the intestinal epithelium. These microvilli are too small to be seen with your microscope, but you can see the fuzziness at the apical surface that they cause.

- Deep crevices may be present where one villus is separated from another. These **intestinal glands** produce intestinal juice, a mixture of mucus and water.

The length of the small intestine is heavily dotted with MALT, as this is a location in which the immune system must both be active to protect against bacterial invasion as well as developing important immunity around our foods. The clusters of MALT in the small intestine are called **Peyer's patches**, and you may, or might not, be able to see some of these in your slide.

Histology of the Large Intestine

As water is removed the digested material becomes thicker and lends more friction to the walls of the large intestine. The large intestine therefore has several adaptions to ease friction. The epithelium contains many goblet cells, which secrete mucus for lubrication (Figure 29.22). Another adaptation is that the muscularis of the large intestine is very thick in order to move the increasingly dehydrated contents. As water is absorbed along the length of the large intestine, the muscle layers become thicker and more powerful closer to the rectum.

Figure 29.22 **The Histology of the Large Intestine**

Compared to the small intestine, the large intestine has a relatively smooth surface, with many mucus-secreting goblet cells.

Garry DeLong/Science Source

Histology of the Liver

The liver is a solid organ composed almost entirely of **hepatocytes**. These cells are capable of filtering the blood to improve it in a number of ways. The hepatocytes are equipped with smooth endoplasmic reticulum and peroxisomes that enable them to filter toxins (like alcohol) and wastes (like urea) from the blood. Hepatocytes are also able to remove glucose and fat-soluble vitamins from the blood and store these nutrients for later use. They also produce and secrete bile. Hepatocytes are arranged in **lobules** (Figure 29.23), which are roughly hexagonal in shape. At the periphery of the hexagon, clusters of vessels can be found grouped in threes. In each of these triads is a branch of the **hepatic portal vein**, a branch of the **hepatic artery**, and a **bile duct**. The hepatic portal vein connects the capillaries of the small intestine, into which

Figure 29.23 **Liver Histology**

The tissue of the liver is organized into lobules, each of which has a central vein and a triad of vessels on the corners.

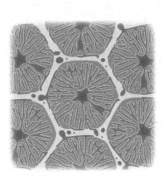
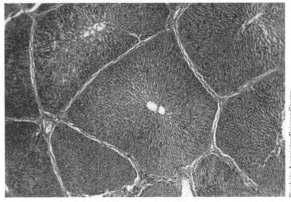

Biophoto Associates/Science Source

nutrients are absorbed, to the liver, where this blood can be filtered. The hepatic artery carries oxygenated blood. The blood from these two vessels mixes together and percolates through the hepatocytes of the lobule. In the center of each lobule is a central vein, which drains all of the blood and empties into the inferior vena cava. The bile ducts drain bile produced by the hepatocytes, and carry it to the central bile duct. Together these three vessels are considered a **portal triad**.

Histology of the Pancreas

The pancreas is both an endocrine and exocrine organ, and the organization of the tissues differs based on these two functions. The exocrine part of the pancreas resembles grapelike clusters, each called an **acinus** (plural = acini), located at the ends of the pancreatic ducts (Figure 29.24). The cells of the acini produce pancreatic juice, a mix of digestive enzymes. Ducts carry these secreted enzymes toward the duodenum. Scattered throughout the sea of exocrine acini are small islands of endocrine cells called **pancreatic islets** which produce a number of hormones involved in nutrient regulation.

Figure 29.24	**Histology of the Pancreas**

The exocrine pancreas tissue is organized into ascini which empty into ducts. The endocrine pancreas is organized into islets.

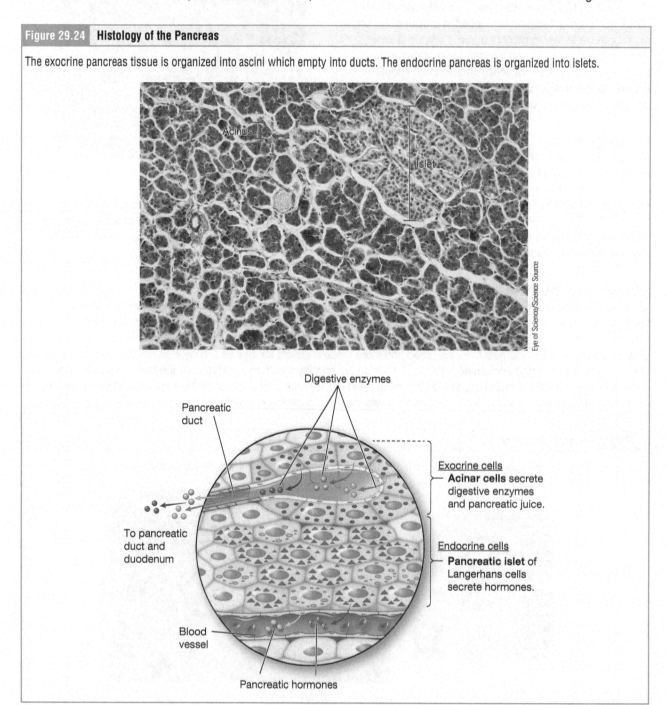

Materials
• Prepared slides of: esophagus, stomach, small intestine, large intestine, liver, pancreas.

Instructions
1. Examine the cross-section of the esophagus. Using Figure 29.14B as your guide, identify the four tissue layers (mucosa, submucosa, muscularis, adventitia). In the space below, draw what you see, label your drawing.

2. On the slide of the stomach, identify the epithelium, the gastric pits, and the muscular layers. They are arranged in this order from the lumen: oblique, circular, longitudinal. The outermost layer of the stomach is the serosa. When examining the gastric pits, can you differentiate between the mucus-secreting cells and the others? In the space below, draw what you see, label your drawing.

3. Using the scanning power objective to examine the slide of the small intestine, identify circular folds. Increase magnification and identify the four tissue layers (mucosa, submucosa, muscularis, serosa). Identify villi and intestinal glands. Scan around your slide using the slide adjustment knobs until you find Peyer's patches. In the space below, draw what you see, label your drawing.

4. Examine the slide of the large intestine. Identify the four tissue layers (mucosa, submucosa, muscularis, serosa). In the space below, draw what you see, label your drawing.

5. Examine the liver slide. Identify the lobules, central vein, and portal triads. In the space below, draw what you see, label your drawing.

6. Examine the pancreas slide. Identify acini and islets. In the space below, draw what you see, label your drawing.

(page intentionally left blank)

Lab 29 Post-Lab Quiz

1. What is the difference between the terms *colon* and *large intestine*?

   a. These two terms mean the same thing, they are synonyms.

   b. The term *large intestine* includes the colon and the cecum, rectum, and anus.

   c. The term *colon* includes the rectum and anus but the term *large intestine* includes the cecum.

   d. The term *colon* includes the colon and the cecum, rectum, and anus.

2. Where is bile produced?

   a. The pancreas

   b. The gallbladder

   c. The intestinal glands

   d. The liver

3. Imagine a disease that involves a dysfunction of the lower esophageal sphincter. In sufferers of this disease, the lower esophageal sphincter becomes leaky. What would the symptoms of this disease be?

   a. Constipation (hard feces)

   b. Irritation or burning of the lining of the esophagus

   c. Flooding of the pancreas with digestive enzymes

   d. Lack of blood flow to the liver

4. Susan has recently had surgery to remove her gall bladder. Her doctor told her that this was necessary because a gall stone (a calcified deposit of bile) became lodged in the duct that leads to her gall bladder. Her doctor assured her, however, that bile would still be able to reach her duodenum from the cells that secrete bile. After Susan's surgery, which duct remains intact?

   a. Cystic duct

   b. Common bile duct

   c. Both of these remain

   d. Neither of these remain

5. Celiac disease is an immune disease in which the immune system is activated by the presence of gluten, a protein found in wheat products. The activation of the immune system causes the destruction of intestinal villi. Which of the following would be a consequence of celiac disease?

   a. Frequent vomiting and the presence of food in the esophagus

   b. Decrease in the production of digestive enzymes

   c. Decrease in the absorption of nutrients

6. Pyloric stenosis is a condition in which the pyloric sphincter does not open as much as it typically does. Individuals with pyloric stenosis would complain of which of the following symptoms?

   a. Constant hunger, even after eating

   b. Feeling full long after a meal

   c. Rapid digestion and diarrhea

   d. Weight gain

7. In some types of weight-loss surgery, an inflatable band is placed around the stomach as shown in the figure. Which region of the stomach is most restricted by the band?

   a. Pyloris
   b. Cardia
   c. Fundus
   d. Body

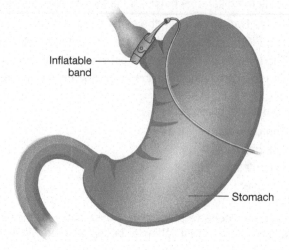

Inflatable band

Stomach

8. In the process of digestion, food must pass through the thoracic cavity to the abdominal cavity. In which organ is the food when it moves from the thoracic to the abdominal cavity?

   a. Pharynx
   b. Esophagus
   c. Stomach
   d. Small Intestine

9. The singer Elvis Presley suffered from a condition called megacolon. In this disease, the colon has an unusually large diameter and the forward movement of digested material is slow or may stop completely. Which layer of the colon wall is not performing its function?

   a. Mucosa
   b. Submucosa
   c. Muscularis
   d. Serosa

10. What type of tissue is the submucosa made of?

   a. Epithelium
   b. Muscle
   c. Connective

Lab 30 Digestive Physiology

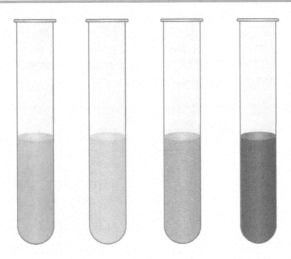

Learning Objectives: After completing this lab, you should be able to:

30.1 List the enzymes, their sources, their substrates, and their products of chemical digestion (enzymatic hydrolysis).

30.2 Identify the locations of chemical digestion of macromolecules (e.g., carbohydrates, proteins, lipids). Define zymogen and describe its importance in chemical digestion.

30.3 Describe the functions, production, and regulation of secretion of hydrochloric acid (HCl).

30.4 Define emulsification, and explain how and where bile salts facilitate fat digestion.

30.5 Given a disruption in the structure or function of the digestive system (e.g., diarrhea), predict the possible factors or situations that might have created that disruption (i.e., given an effect, predict possible causes).

30.6* Describe the uses of lab reagents to determine the presence or absence of monomers and polymers.

30.7 Describe the role of bacteria (microbiome) in digestion.

*Objective is not a HAPS Learning Goal.

Introduction

When you take a bite of food like a delicious juicy hamburger, or a fresh, cool salad you may be considering your food in terms of cheese, tomatoes, or meat, but your cells require your food to be broken down and transported as its smallest and most basic components such as amino acids, vitamins, and glucose molecules (Figure 30.1). The process of digestion breaks food down both physically and chemically to release these tiny components so that they can cross the cells lining your digestive tract and be used by all the cells of the body.

The molecules in our food fit into the categories of nucleic acids, lipids, proteins, and carbohydrates. Each of these is large molecules composed of subunits, similar to how a brick wall is composed of individual bricks. The individual subunits are called **monomers** and the larger molecule build by linking monomers is a **polymer**. Each of

The Human Anatomy and Physiology Society includes more than 1,700 educators who work together to promote excellence in the teaching of this subject area. The HAPS A&P Learning Outcomes measure student mastery of the content typically covered in a two-semester Human A&P curriculum at the undergraduate level. The full Learning Outcomes are available at https://www.hapsweb.org.

the monomers will be connected to another by a covalent bond. These bonds need to be broken in order for the molecule to come apart and the liberated monomers to be absorbed by cells. Covalent bonds are strong bonds and they do not come apart easily, instead they are broken by the action of enzymes. The digestive organs make many different kinds of enzymes that are capable of breaking a wide variety of covalent bonds.

There are many different organs that contribute to digestion. The hollow organs through which food passes are the organs of the gastrointestinal (GI) tract. They are aided by three accessory organs, which do not contact food directly, but manufacture or store chemicals that assist in the digestion processes. These accessory organs are the liver, gall bladder, and pancreas. The GI tract organs that we will focus on in this lab are the stomach and the duodenum, which is the first segment of the small intestine.

 Learning Connection

Quiz Yourself

What are the monomers of each type of food molecule, lipids, proteins, and carbohydrates?

| Figure 30.1 | Food We Eat Is Digested Into Smaller Molecules |

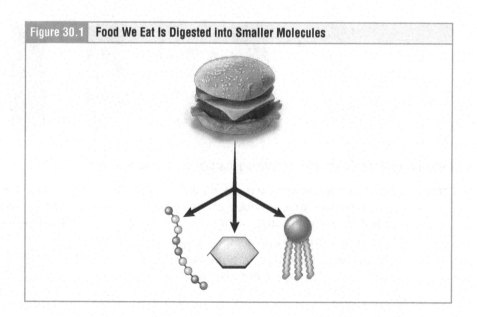

Lab 30 Pre-Lab Quiz

This quiz will strengthen your background knowledge in preparation for this lab. For help answering the questions, use your resources to deepen your understanding. The best resource for help on the first five questions is your text, and the best resource for help answering the last five questions is to read the introduction section of each lab activity.

1. a. Anatomy of digestion. Label the following organs in the figure:

duodenum
gallbladder
liver
pancreas
stomach

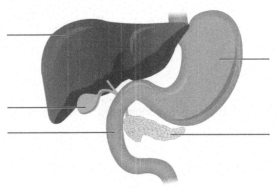

b. Digestive enzymes. Fill in the table below by indicating in which organ (1) each of the digestive compounds is made and (2) where it is utilized. The organ choices are the duodenum, gallbladder, liver, pancreas, stomach.

Digestive compound	Made in	Acts in
amylases		
bile		
HCl		
lipases		
pepsinogen/pepsin		

2. Match the root next to its definition:

Root	Meaning
	origin, birth
	many
	sugar
	enzyme
	one
	two

List of Roots

-ase
-ose
mono-
di-
poly-
-gen (also gen-)

3. Fill in the blanks with the following terms: acidic, basic, high, low.

A solution at low pH is _____ and has a _____ concentration of H⁺ ions. And a solution at high pH is _____ and has a _____ concentration of H⁺ ions.

4. This figure is an illustration of a triglyceride. Lipases are the class of enzyme that digests triglycerides. Where on the triglyceride molecule do lipases act?

a. A
b. B
c. C

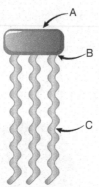

5. What type of bond joins two monosaccharides, such as glucose and galactose, together?

a. Peptide bond
b. Glycosidic bond
c. Hydrogen bond

6. What type of molecule is capable of denaturation?

a. Carbohydrates
b. Proteins
c. Lipids

7. What does Biuret's solution test for the presence of?

a. Amino acids
b. Monosaccharides
c. Starches
d. Fatty acids

8. What is the function of gastric mucus?

a. To get rid of waste products from the body by adding them to digested food to be eliminated
b. To protect the epithelium of the stomach
c. To add hydrochloric acid to the lumen of the stomach
d. To emulsify fats

9. What enzyme is found in the lumen of the stomach?

a. Lipase
b. Pepsin
c. Amylase

10. If you added phenol red to a solution with a pH of 3, what color would the solution become?

a. Dark pink
b. Light pink
c. Bright red
d. Yellow

Activity 30.1 Carbohydrate Digestion

This activity targets LOs 30.1 and 30.2.

Carbohydrates are built from monosaccharides. There are many types of monosaccharides in nature, but the three that are most common in our foods are glucose, fructose, and galactose. The term *sugar* refers to disaccharides, which are molecules made of two monosaccharides joined by a **glycosidic bond**, a specific type of covalent bond that unites two monosaccharides. A common disaccharide is table sugar (sucrose), the ingredient you use in cooking. Disaccharides are too large to be transported across the cell membrane, so the cells of the body can only utilize monosaccharides. Polysaccharides are carbohydrate molecules made of more than two monosaccharides. Starches are particularly large polysaccharides such as those found in wheat or potatoes. Disaccharides and starches are common in our foods, but in order for our cells to utilize the energy stored in these molecules, digestive enzymes must break the bonds that hold the monosaccharides together, so that the monosaccharides can be transported into the cells. For example, **amylases** are secreted from the salivary glands of the mouth and the pancreas. They break complex starches into smaller chains of saccharides. **Brush border enzymes** are anchored to the membrane of duodenal epithelial cells so that they face into the lumen, where they are able to break glycosidic bonds of the disaccharides and oligosaccharides so that the resulting monosaccharides are available to cross the cell membrane (Figure 30.2).

In this lab we will examine the digestion of carbohydrates using two test solutions, Lugol's iodine and Benedict's solution. Lugol's iodine will color a solution yellow in the absence of starch, but in the presence of starch Lugol's iodine will turn the solution a dark purple or even black (Figure 30.3). We can also test for the presence of monosaccharides using Benedict's solution. This solution is a pale blue in the absence of monosaccharides, but turns a yellow, orange, or red color in the presence of monosaccharides (the progression from yellow to red indicates more available monosaccharides; Figure 30.4).

Figure 30.2	Carbohydrate Digestion

Carbohydrates are first broken down into small chains by amylases, then broken into monomers by brush border enzymes.

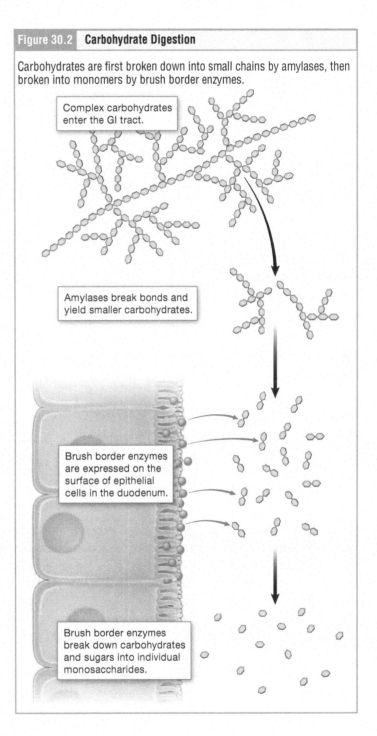

Complex carbohydrates enter the GI tract.

Amylases break bonds and yield smaller carbohydrates.

Brush border enzymes are expressed on the surface of epithelial cells in the duodenum.

Brush border enzymes break down carbohydrates and sugars into individual monosaccharides.

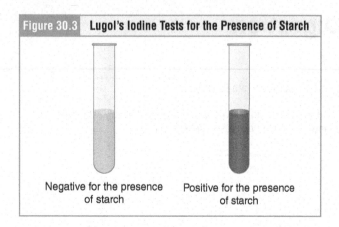

Figure 30.3 | Lugol's Iodine Tests for the Presence of Starch

Negative for the presence of starch

Positive for the presence of starch

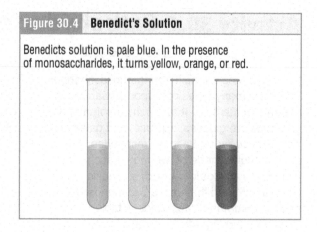

Figure 30.4 | Benedict's Solution

Benedicts solution is pale blue. In the presence of monosaccharides, it turns yellow, orange, or red.

Materials
- Amylase
- Benedict's solution
- Distilled water (dH₂O)
- Glass stirring rods, 8
- Glass test tubes, 8
- Glucose solution, 4%
- Lugol's iodine
- Starch solution
- Test tube holder
- Test tube rack
- Water bath, 37°C
- Water bath, 100°C (or boiling water on a hot plate)

Instructions
1. Label your test tubes 1 through 8.

Table 30.1 Experimental Setup for Activity 30.1								
Experimental Step								
	Tube 1	**Tube 2**	**Tube 3**	**Tube 4**	**Tube 5**	**Tube 6**	**Tube 7**	**Tube 8**
Solution	Starch	Starch	Starch	Starch	Glucose	Glucose	Glucose	Glucose
Enzyme	Amylase	dH₂O	Amylase	dH₂O	Amylase	dH₂O	Amylase	dH₂O
Test	Lugol's	Lugol's	Benedict's	Benedict's	Lugol's	Lugol's	Benedict's	Benedict's
Prediction								

2. Stir solutions. Using Table 30.1 as your guide, distribute 3 mL of each solution (starch solution or glucose solution) into the tubes as directed.

3. Using Table 30.1 as your guide, distribute 3 mL of enzyme (amylase or dH_2O) into the tubes as directed.

4. Set the tubes into the 37°C water bath and leave them for 30 minutes.

5. While you are waiting for the reactions to take place, predict whether you will see a positive or negative reaction in each test tube. Record your predictions in Table 30.1.

6. Remove the tubes from the water bath and stir the solutions with a stirring rod, be sure to use a clean stirring rod for each tube so that you don't contaminate your solutions before testing.

7. Add 4 drops of Lugol's iodine to tubes 1, 2, 5, and 6.

8. Add 20 drops of Benedict's reagent to tubes 3, 4, 7, and 8.

9. Put on protective eyewear. Place the tubes with Benedict's solution (tubes 3, 4, 7, and 8) into the 100°C water bath for 3 minutes.

10. Remove the tubes using test tube holders and place them in a test tube rack. Be careful as the tubes will be very hot. Record your results for both the Lugol's and Benedict's tests in Table 30.2.

11. Dispose of the reagents and wash the test tubes as directed by your instructor.

Table 30.2 Results of Lugol's Iodine and Benedict's Reagent Tests								
	Tube 1	Tube 2	Tube 3	Tube 4	Tube 5	Tube 6	Tube 7	Tube 8
Did you test for monosaccharides (M) or starch (S)?								
Was the result positive (+) or negative (−)?								

Digging Deeper:
Fiber

Fiber is the term used for a collection of food components that are indigestible. But what does it mean to say we cannot digest something? Digestion is breaking a complex food down into its monomers, because only monomers are small enough to cross the gut wall. If we lack the enzymes for a particular bond within a food, we lack the ability to break it into its monomers. Food that remains too big to be digested will stay within the lumen of the GI tract, eventually becoming feces. As humans, we share a common set of enzymes. For example, all humans lack cellulase enzymes, which renders us unable to digest cellulose, the complex carbohydrate found in plant cell walls. This means that many of the calories in plants are unavailable to us and most of the ingested plant mass becomes feces. However, though humans do not have the genes for cellulases, many humans carry species of bacteria in their guts that are able to break down cellulose and liberate the monomers. A person with these bacteria would be better able to digest and gain nutrition from plant matter. The collection of bacteria and other microbes in your gut is called your microbiome, and everyone's microbiome is unique to them. For most of us the number of bacteria in our microbiome is 30–40 trillion cells (that's more bacterial cells in your gut than you have human cells in your whole body!) and those bacteria belong to 500–1,000 different bacterial species. What species you carry determines a lot about what you can and cannot digest.

(page intentionally left blank)

Activity 30.2 Protein Digestion

This activity targets LOs 30.1, 30.2, and 30.3.

The building blocks of proteins are amino acids. The 20 different amino acids in our diet can be combined by peptide bonds to form long chains. Proteins rarely stay in a long chain form, although, they are typically highly folded molecules (Figure 30.5). Just like glycosidic bonds are broken by amylases and brush border enzymes, peptide bonds are broken by enzymes. There are many kinds of enzymes that break peptide bonds, including pepsin, peptidases, trypsin, and chymotrypsin. These enzymes can only break the bonds that are accessible, therefore the work of digesting proteins relies on unfolding these molecules to reveal all of their peptide bonds.

Protein digestion begins in the stomach, where hydrochloric acid (HCl) functions to disrupt many of the bonds that hold the proteins in their folded shape, thus **denaturing**, or unfolding the molecule. The stomach wall also secretes a molecule called pepsinogen. **Pepsinogen** is an inactive precursor to pepsin, an enzyme that cleaves peptide bonds. The digested material that passes from the stomach to the duodenum contains unfolded and partially digested protein fragments. Many protein-digesting enzymes are produced by and secreted by the pancreas into the duodenum. Brush border enzymes further break down peptide bonds, releasing amino acids. The amino acids are small enough to be transported into the membranes of intestinal epithelial cells.

In this lab we will use one of the most common biological proteins, albumin. Albumin is produced by the human liver and found in your blood, but it also makes up the "whites" of chicken eggs, which is what we will use in these experiments. We can test for the presence of free amino acids using a reagent, Biuret's solution. In the presence of free amino acids, Biuret's solution will color a solution purple. If amino acids are absent, the solution will be a very light blue (Figure 30.6).

Figure 30.5	**Protein Digestion**

In order for their bonds to be cleaved by enzymes, proteins must first be unfolded/denatured so that the bonds are exposed. Then enzymes can cut the amino acids into shorter segments and finally liberate individual amino acids for absorption across the gut wall.

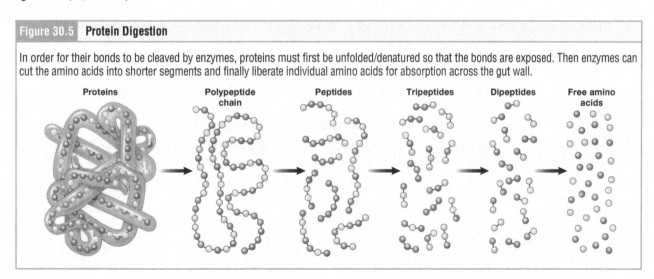

Materials

- Biuret's solution
- Disposable gloves
- Distilled water (dH$_2$O)
- Egg whites (albumen)
- HCl
- Pepsin
- Safety glasses
- Test tube rack
- Test tubes, 8
- Water bath, 20°C
- Water bath, 37°C

Figure 30.6	**Biuret's Solution**

Biuret's solution is sea blue, but turns lavender/purple in the presence of amino acids.

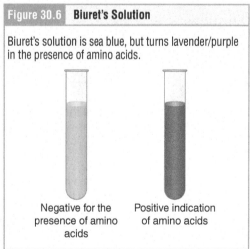

Negative for the presence of amino acids

Positive indication of amino acids

Instructions

1. Start by putting on safety gear, including safety glasses and gloves.

2. Label your test tubes 1 through 8.

	Table 30.3 Experimental Setup for Activity 30.2							
Tube 1	**Tube 2**	**Tube 3**	**Tube 4**	**Tube 5**	**Tube 6**	**Tube 7**	**Tube 8**
Albumin	Albumin	Albumin	Albumin	Albumin	Albumin	Albumin	Albumin
+	+	+	+	+	+	+	+
dH_2O	Pepsin	Pepsin	Amylase	dH_2O	Pepsin	Pepsin	Amylase
+	+	+	+	+	+	+	+
dH_2O	dH_2O	HCl	dH_2O	dH_2O	dH_2O	HCl	dH_2O
Cold **Water Bath**				Warm **Water Bath**			

3. Using Table 30.3 as your guide, add about 3 mL of egg white to each tube (note, it may be difficult to pipet egg white; do not worry if your volume is a little under or over 3 mL).

4. Add 3 mL pepsin to tubes 2, 3, 6, and 7, dH_2O to tubes 1 and 5, and amylase to tubes 4 and 8.

5. Add 1 mL HCl to tubes 3 and 7 and dH_2O to the other tubes.

6. Place tubes 1–4 in the 20°C water bath and tubes 5–8 in the 37°C water bath and incubate for 1 hour.

7. Record your predictions in Table 30.4.

8. Remove the tubes and add five drops of Biuret solution to each tube. Record your observations in Table 30.4.

9. Dispose of the reagents and wash the test tubes as directed by your instructor.

Table 30.4 Results of the Biuret Solution Tests								
	Tube 1	**Tube 2**	**Tube 3**	**Tube 4**	**Tube 5**	**Tube 6**	**Tube 7**	**Tube 8**
Prediction								
Was the result positive (+) or negative (−)?								

Questions

1. Which tube did you think would have the most free amino acids (the most significant degree of digestion)?

2. Why did you hypothesize that the most digestion would occur in that tube?

3. Did you notice a difference in the degree of digestion that occurred at a warm temperature versus a cooler temperature? Why do you think temperature can impact digestion?

Activity 30.3 Lipid Digestion

This activity targets LOs 30.1, 30.2, and 30.4.

Most dietary lipids are triglycerides, which are made up of a glycerol molecule bound to three fatty acid chains (Figure 30.7). In order to absorb and use the energy in lipid molecules, the triglycerides need to be broken down. Two of the fatty acid chains are cleaved off the glycerol molecule by enzymes called lipases. In the human body, lipases are generated by the pancreas and added to the duodenum. As with enzymes that digest proteins, lipases can act only on the bonds that they can access, and as in proteins, these bonds are often inaccessible to enzymes. The fatty acid chains are non-polar and therefore hydrophobic, and in the watery solution inside the digestive tract, lipids tend to cluster together so that their glycerol heads are facing the watery solution and their fatty acid chains are tucked together, protected from the aqueous environment (Figure 30.8). When the lipids are in this arrangement, lipases would not be very effective in accessing the bonds between glycerol and fatty acid chains. Just as acids are used to denature proteins, bile is used to **emulsify**, or break apart, fatty acid droplets (Figure 30.9).

Figure 30.7	Lipid Digestion

The most common form of lipid in the diet is a triglyceride. During digestion, two of the fatty acid chains are freed by lipases, and one fatty acid chain remains bound to a glycerol molecule, producing a monoglyceride.

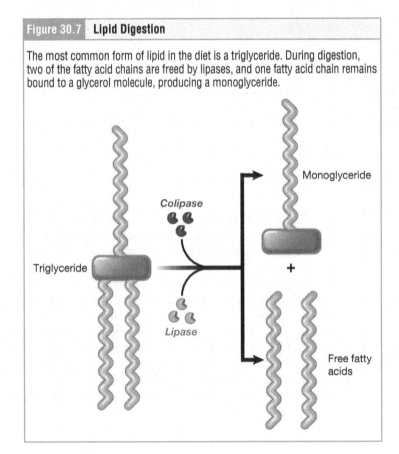

Figure 30.8	Lipid Molecules in Water

In aqueous solutions, lipid molecules, like the triglycerides shown here, will orient themselves so that their non-polar fatty acid tails are protected from water.

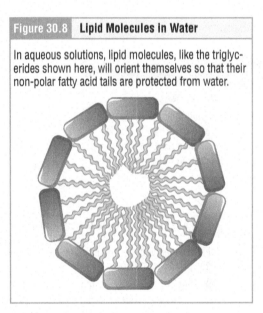

Figure 30.9 | Fat Emulsification

Because lipids (fats) are hydrophobic, they gather into spheres that exclude water. The function of bile is to interrupt their hydrophobic interactions and force the lipids into smaller particles.

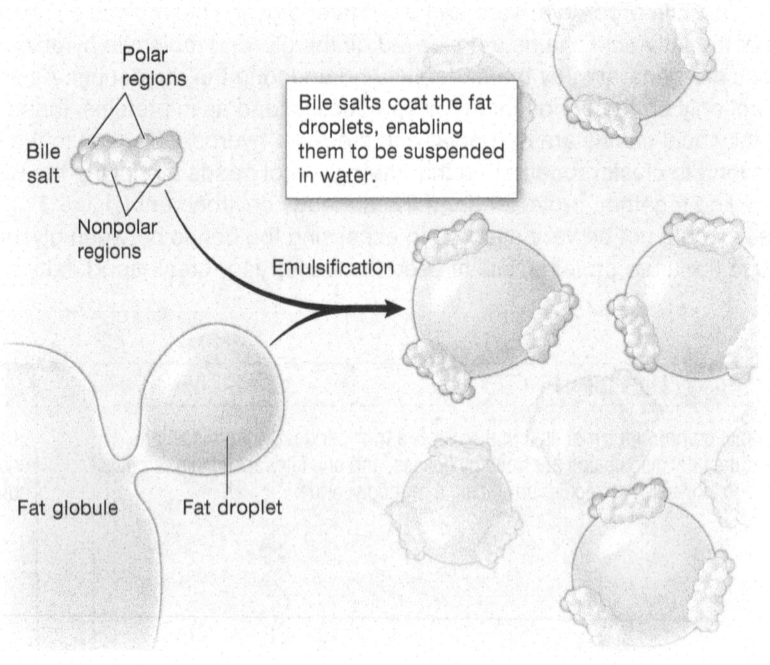

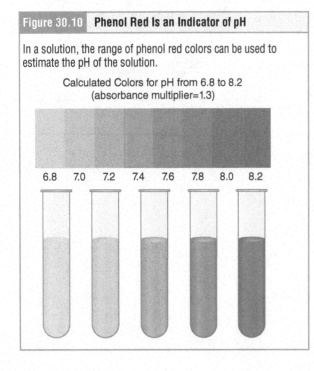

Polar regions

Bile salt

Nonpolar regions

Bile salts coat the fat droplets, enabling them to be suspended in water.

Emulsification

Fat globule

Fat droplet

Learning Connection

Broken Process

What would happen in an individual who produced less bile, or wasn't able to store bile to release during fatty meals?

Figure 30.10 | Phenol Red Is an Indicator of pH

In a solution, the range of phenol red colors can be used to estimate the pH of the solution.

Calculated Colors for pH from 6.8 to 8.2
(absorbance multiplier=1.3)

6.8   7.0   7.2   7.4   7.6   7.8   8.0   8.2

Bile is made by the liver and stored in the gall bladder until digestion begins at mealtime. The gall bladder will contract to eject the bile through a duct and into the duodenum. Bile will emulsify the fats from the meal, allowing the lipases to act on the triglycerides.

In this lab we will use a common dietary lipid, vegetable oil, and test for the presence of fatty acids using a lab reagent called phenol red. Phenol red is an indicator of the pH of a solution, its color ranges from yellow in acids, to red in basic solutions (Figure 30.10). When lipid digestion occurs, fatty acids are released from triglycerides and the presence of these free fatty acids reduces the pH of the solution.

Materials
• Amylase
• Distilled water (dH$_2$O)
• Lipase
• NaOH, 0.1M, in a dropper bottle
• Phenol red
• Test tube rack
• Test tubes, 4
• Vegetable oil
• Water bath, 37°C

Instructions

1. Start by putting on safety gear, including safety glasses and gloves.

2. Label your test tubes 1 through 5.

Table 30.5 Experimental Setup for Activity 30.3				
Tube 1	Tube 2	Tube 3	Tube 4	Tube 5
Oil	Oil	Oil	Oil	Oil
+	+	+	+	+
Lipase	Lipase	Amylase	Amylase	dH_2O
+	+	+	+	+
Bile	dH_2O	Bile	dH_2O	dH_2O

3. Using Table 30.5 as your guide, add 3 mL of vegetable oil to each tube.

4. Add 10 drops of phenol red to each test tube. The solution in each test tube should now be a dark pink. If the solution is not pink, add a drop or two of 0.1 M NaOH to each tube.

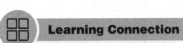

Learning Connection

Chunking

Try making three lists, one called "carbohydrates," one called "proteins," and one called "lipids" and write every term down from this lab that is specific to each one. For example, you'd put phenol red on the list for lipids, and pepsin on the list for proteins.

5. Add 1 mL of lipase to tubes 1 and 2, 1 mL of amylase to tubes 3 and 4, and 1 mL of dH_2O to tube 5.

6. Add 1 mL of bile to tubes 1 and 3, and 1 mL of dH_2O to tubes 2, 4, and 5.

7. Place the tubes in a 37°C water bath for 30 minutes.

8. Record your predictions in Table 30.6.

9. Remove the tubes and record your observations in Table 30.6.

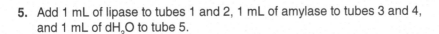

Table 30.6 Results of the Phenol Red Tests					
	Tube 1	Tube 2	Tube 3	Tube 4	Tube 5
Prediction					
Was the result positive (+) or negative (−)?					

(page intentionally left blank)

Activity 30.4 Case Study: Gastric Ulcer

This activity targets LOs 30.3 and 30.5.

The gastric epithelium is endangered by its environment. The lumen of the stomach is one of the most acidic environments on the planet, with an average pH between 1 and 2. The stomach lumen also contains pepsin, an enzyme that cleaves peptide bonds. Together these environmental factors can be very threatening to the physical integrity of the cells. The stomach epithelium is protected by the mucosal barrier (Figure 30.11). This barrier has several components. First, the stomach wall is covered by a thick coating of bicarbonate-rich mucus. The mucus forms both a physical and chemical barrier. Physically, the mucus repels the liquid stomach contents, and its bicarbonate ions neutralize the acid. Second, the stomach epithelial cells are joined by tight junctions, which prevent gastric juice from seeping between the epithelial cells and damaging the tissues underneath. Finally, stem cells quickly replace any damaged epithelial cells.

Figure 30.11	A Mucosal Barrier Protects Cells of the Gastric Epithelium

The gastric epithelium is invaginated with gastric pits, where HCl and pepsinogen are secreted. Along the surface of the gastric epithelium, a protective mucus is secreted.

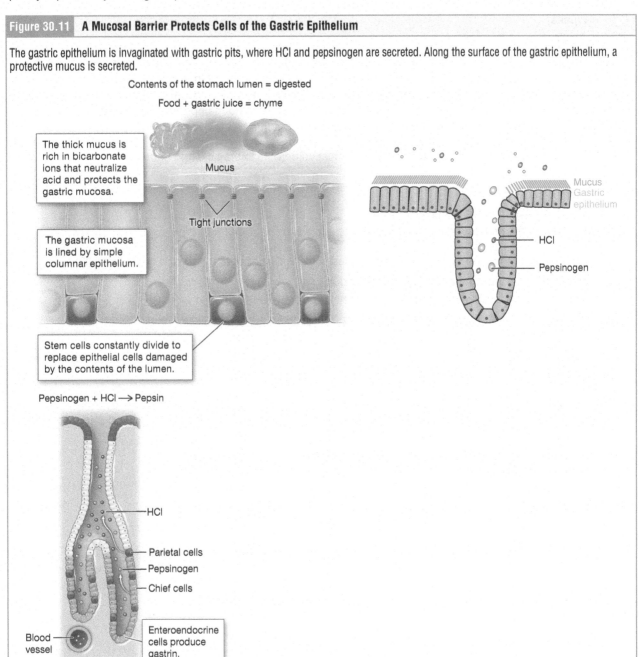

Contents of the stomach lumen = digested

Food + gastric juice = chyme

The thick mucus is rich in bicarbonate ions that neutralize acid and protects the gastric mucosa.

Mucus

Tight junctions

The gastric mucosa is lined by simple columnar epithelium.

Stem cells constantly divide to replace epithelial cells damaged by the contents of the lumen.

Mucus
Gastric epithelium

HCl

Pepsinogen

Pepsinogen + HCl ⟶ Pepsin

HCl

Parietal cells

Pepsinogen

Chief cells

Blood vessel

Enteroendocrine cells produce gastrin.

Several drugs, including caffeine and non-steroidal anti-inflammatory drugs such as aspirin and ibuprofen, inhibit mucus production in gastric epithelial cells.

Julia just had a baby. Julia has always loved coffee, but dutifully avoided any caffeine-containing beverages, including coffee, throughout her pregnancy. Now that the pregnancy is over and Julia is up throughout the night feeding the baby, she's drinking more coffee than she ever has before. As her baby turns 6 months old, Julia begins to notice a gnawing pain in her abdomen from time-to-time. She even notices a new burning sensation after certain meals. She tells her doctor about these new symptoms and the doctor tells Julia that she may have a gastric ulcer. Her doctor refers Julia to a gastroenterologist.

1. Examine the formula. What compound should be where the **X** is?

$$\text{Pepsinogen} + \textbf{X} \rightarrow \text{Pepsin}$$

   a. Hydrochloric acid
   b. Sodium hydroxide
   c. Lipase
   d. Amylase

The gastroenterologist confirms that Julia has a gastric ulcer and recommends that she cut back or cut out caffeinated beverages. The gastroenterologist shows Julia a diagram (Figure 30.11) of the wall of the stomach and explains that the caffeine Julia is drinking is inhibiting the stomach's ability to make mucus, leaving the gastric epithelial cells exposed. Julia thinks back to her college biology class and realizes that the membranes of the gastric epithelial cells will be the parts of the cells in contact with the acidic stomach contents.

2. Which component of the membrane will be the most vulnerable to the acid and pepsin?
   a. Phospholipids
   b. Membrane proteins
   c. Cholesterol embedded in the membrane
   d. Carbohydrates embedded in the membrane

3. Once the cell membrane integrity is compromised, what will happen to the gastric epithelial cell?
   a. Water will rush into the cell and the cell will burst.
   b. The cell will get smaller and smaller, leaving holes in the membrane.
   c. The cell will quickly manufacture replacement components.

4. How many layers thick is the gastric epithelium?
   a. One
   b. Two
   c. Three
   d. Four

5. What type of tissue is directly under the gastric epithelium?
   a. Connective tissue
   b. Muscle tissue
   c. Nervous tissue

6. Thinking about your answer from question 5, what component of this tissue would be most vulnerable to the acid and pepsin of the stomach contents?

    a. Myofibrils

    b. Myelin

    c. Water

    d. Collagen fibers

Julia's gastroenterologist prescribes a drug called a **proton pump inhibitor (PPI).**

7. *Proton* is another term for which kind of ion?

    a. Cl^-

    b. Na^+

    c. OH^-

    d. H^+

8. Preventing the movement of protons across the gastric epithelium will prevent _____ from forming in the lumen.

    a. HCl

    b. NaOH

    c. NaCl

    d. H_2O

9. After Julia starts taking the PPI, will the amount of pepsin in the stomach lumen change?

    a. There will be a decrease in pepsin.

    b. There will be an increase in pepsin.

    c. The pepsin levels will not change.

10. Which of the following components of Julia's diet will become more difficult to digest?

    a. Lipids

    b. Carbohydrates

    c. Fiber

    d. Protein

(page intentionally left blank)

Lab 30 Post-Lab Quiz

1. You are working for a laundry detergent company. The company has noticed that a lot of their customers are complaining that they can't get greasy stains like oil or butter out of their clothes. You're looking at the list of enzymes that the detergent has in it. Which of the following enzymes could the company increase the concentration of to try to improve the grease-fighting ability of the detergent?

   a. Amylases

   b. Lipases

   c. Proteases

2. Human infants are born with the ability to express lactase, the enzyme that breaks the bond between the two monomers in the milk sugar lactose. Many humans (and most mammals) stop expressing this enzyme in childhood. In an adult that didn't express lactase, what would the result be?

   a. Another enzyme, such as lipases, will break the bond allowing the digestion of lactose.

   b. Lactose will not be broken down but will be transported as a disaccharide into epithelial cells.

   c. Lactose will not be broken down but will remain in the gut lumen.

3. Examine the tubes in the figure, which have been used to test for the presence of lipids. Which tube has the higher concentration of free fatty acids?

   a. Tube A

   b. Tube B

   c. Neither tube contains fatty acids.

A   B

4. Examine the tubes in the figure. Both tubes contain all of the same ingredients, but one tube was left sitting on the lab bench while the other was incubated at 37°C. Which one was incubated in the warm water bath?

   a. Tube A

   b. Tube B

   c. It is impossible to tell based on this information.

A   B

5. Examine the figure. Lipase was added to one of the tubes, but you cannot remember which one. Which tube most likely contains lipase?

   a. Tube A
   b. Tube B
   c. Tube C
   d. Tube D

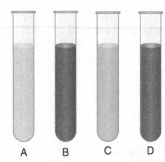

A   B   C   D

6. An individual consumes a meal containing lipids, carbohydrates, and proteins. If you examined the contents of their stomach 30 minutes after the meal, when much of the ingested food is still within the stomach, which of the following would you find?

   a. Free fatty acids
   b. Denatured, partially digested proteins
   c. Free monosaccharides

7. The function of the gall bladder is to store bile until it is needed in digestion. Which of the following is likely to be a trigger for the gall bladder to contract, releasing bile into the bile duct?

   a. Stimulation of fat receptors on the tongue
   b. A sudden increase in blood glucose level
   c. Stimulation of umami (amino acid) receptors on the tongue
   d. Stimulation of stretch receptors in the stomach indicating a large meal has been ingested

8. Imagine the spoonful of table sugar you put into your coffee this morning. Table sugar, or sucrose, is a disaccharide composed of glucose and fructose. In which location are these two monosaccharides separated from each other?

   a. At the membrane of duodenal epithelial cells
   b. In the lysosome of duodenal epithelial cells
   c. In the oral cavity (mouth)
   d. In the lumen of the stomach

9. In the following chemical equation, what should replace the question mark?

$$\text{Proteins} + \text{Water} \xrightarrow{\text{peptidase}} ? + \text{Energy}$$

   a. Fatty acids
   b. Amino acids
   c. Monosaccharides
   d. Glycerol + fatty acids

10. Brady was fond of dairy products as a child, but when they arrived in college they decided to try out being a vegan. After a full semester without any dairy, they went back home for the school break and indulged in their mother's homemade ice cream. Brady felt immediately sick and experienced diarrhea. Which of the following is the most likely explanation?

   a. Brady used to express the enzyme lactase, but their expression of the lactase gene changed in the university setting.
   b. Brady used to have a population of bacteria in their microbiome that digested lactose, but this population died off during the vegan period, and now there is no lactase expression in Brady or their microbiome.
   c. Brady's brain is no longer used to the sight and smell of dairy, so the brain triggered a GI reaction to get rid of the ice cream.

Lab 31

Anatomy of the Urinary System

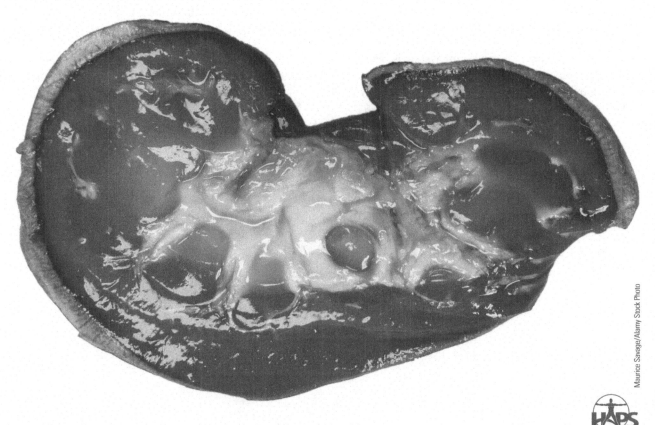

Maurice Savage/Alamy Stock Photo

Learning Objectives: After completing this lab, you should be able to:

31.1* Locate and identify the gross anatomy structures of the urinary system.

31.2 Identify and describe the anatomic structure of the kidney, including its coverings.

31.3 Identify and describe the structure of a typical nephron, including the components of the renal corpuscle and the components of the renal tubule.

31.4 Trace the flow of filtrate from the renal corpuscle through the collecting duct.

31.5 Trace the path of blood flow through the kidney, from the renal artery to the renal vein.

31.6 Identify and describe the microscopic anatomy of the ureters, urinary bladder, and urethra.

31.7 Compare and contrast the anatomy of the male urethra versus the female urethra.

31.8 Compare and contrast the locations, innervation, and functions of the internal urethral sphincter and external urethral sphincter.

* Objective is not a HAPS Learning Goal.

The Human Anatomy and Physiology Society includes more than 1,700 educators who work together to promote excellence in the teaching of this subject area. The HAPS A&P Learning Outcomes measure student mastery of the content typically covered in a two-semester Human A&P curriculum at the undergraduate level. The full Learning Outcomes are available at https://www.hapsweb.org.

Introduction

The urinary system, also called the renal system, contributes to homeostasis by removing metabolic waste products from the blood, regulating the pH of the blood, regulating blood pressure, and regulating the concentration of many solutes in the blood. In the process, the urinary system produces, stores, and excretes urine, a byproduct of these homeostatic actions. The urinary system mechanisms that regulate homeostasis are closely associated with the cardiovascular system, so it should be no surprise that approximately 25% of your cardiac output is directed to your kidneys.

The gross anatomy of the urinary system is relatively simple: two kidneys, two ureters, one urinary bladder, and one urethra. However, that simplicity belies the complex microanatomy at work in the kidneys. As you delve into the gross anatomy and microanatomy of the urinary system, consider how the structure of each component of this system serves the overall functions of homeostasis presented here.

Lab 31 Pre-Lab Quiz

This quiz will strengthen your background knowledge in preparation for this lab. For help answering the questions, use your resources to deepen your understanding. The best resource for help on the first five questions is your text, and the best resource for help answering the last five questions is to read the introduction section of each lab activity.

1. Anatomy of the urinary system. Identify the following structures on the image:

   left kidney

   left ureter

   right kidney

   right ureter

   urethra

   urinary bladder

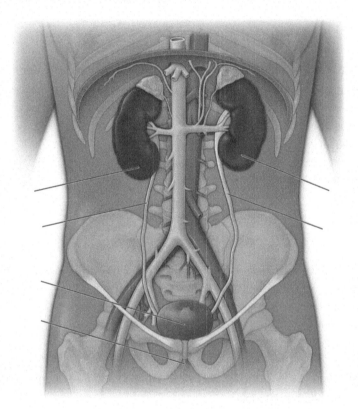

2. Match the root next to its definition:

Root	Meaning
	all around or enclosing
	beside or next to
	between
	kidneys

List of Roots

inter-

juxta-

nephro-

peri-

3. The kidney structure that acts as the entry and exit site for blood vessels, nerves, lymphatics, and the ureter is the:

   a. Renal papilla

   b. Renal pelvis

   c. Hilum

4. The renal corpuscle is made up of the (choose all correct answers):
   a. Distal convoluted tubule
   b. Proximal convoluted tubule
   c. Bowman's capsule
   d. Glomerulus

5. The kidneys contain _____ juxtamedullary nephrons than cortical nephrons.
   a. fewer
   b. more

6. Place these layers of protection around the kidneys in the correct order from deepest to most superficial.
   Renal fascia
   Ribs
   Fibrous capsule
   Perinephritic fat

7. What is the outer region of the kidney?
   a. Medulla
   b. Cortex
   c. Pyramid

8. Place these kidney structures in the order in which fluid flows, beginning with its exit from the renal pyramids.
   Renal pelvis
   Renal papillae
   Minor calyces
   Major calyces

9. The pair of blood vessels that directly deliver blood to and away from the kidneys are the:
   a. Afferent and efferent arterioles
   b. Renal arteries and renal veins
   c. Common iliac artery and common iliac vein

10. The tissue type lining the inside of the ureters, bladder, and urethra is:
    a. Epithelial tissue
    b. Connective tissue
    c. Smooth muscle

Activity 31.1 Gross Anatomy of the Kidney

This activity targets LO 31.2.

Make a fist with your hand: this is the approximate size of one of your two kidneys, though they are likely a bit longer than your fist. Your kidneys are well protected deep to your last few pair of ribs. They are anchored to the posterior wall of your abdominal cavity by renal fascia, holding them in place posterior to the peritoneum surrounding many other abdominal organs. Deep to the renal fascia is a cushioning layer of **perinephritic fat**. A final protective layer of dense irregular connective tissue called a **fibrous capsule** wraps the tissue of the kidney into its familiar bean-shaped form.

    The internal gross anatomy of the kidney looks like a series of wedges (Figure 31.1). Like several other organs, the tissue of the kidney is organized into an outer renal **cortex** and an inner renal **medulla**. Within the medulla are repeating wedges of tissue: the **renal pyramids**. A human kidney contains approximately 8 to 18 renal pyramids. Each renal pyramid is shaped somewhat like a cone, with the tip of each pyramid pointing toward the indented curve of the kidney called the **hilum**. The renal pyramids are separated from each other by connective tissue called **renal columns**. As fluid is processed through the renal pyramids it travels toward the central tips, the **renal papillae**, where it flows into a series of larger and larger passageways called **calyces** (singular = calyx) until it leaves the kidney at the **renal pelvis**, the final passageway out of the kidney that becomes the ureter leading to the bladder.

Figure 31.1	Gross Anatomy of the Human Kidney

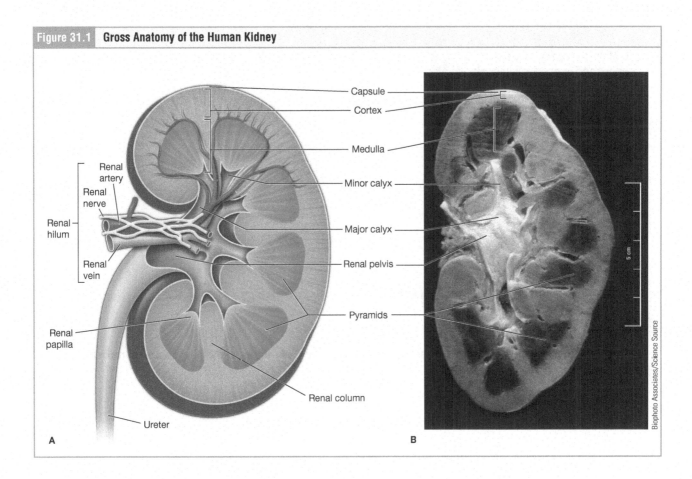

Materials

- Disposable gloves
- Dissecting instruments and tray
- Fresh or preserved kidney specimen
- Lab apron or lab coat
- Safety glasses
- Three-dimensional model or labeled diagram of a kidney sectioned along the frontal plane

Structures Identified in This Activity

Fibrous capsule	Renal cortex	Renal column
Perinephritic fat	Renal medulla	Minor calyx
Renal fascia	Renal pyramid	Major calyx
Hilum	Renal papilla	Renal pelvis
Kidney lobe		

Instructions

1. Don a lab coat or apron, disposable gloves, and safety glasses.

2. Obtain a fresh or preserved kidney specimen, dissecting instruments, and tray. Alternatively, if a kidney specimen is not available, skip to step 5 using a three-dimensional model of a cut kidney.

3. Examine the connective tissue that surrounds the kidney.

   a. Is there adipose tissue? Describe:

   b. What function does the adipose tissue surrounding the kidney serve?

   c. Describe the fibrous capsule that surrounds the kidney (texture, thickness, etc.):

   d. What function does the fibrous capsule serve?

4. Cut through the fibrous capsule, if present, and remove it and any other connective tissue surrounding the kidney.

5. Locate the hilum of the kidney.

   a. What directional term(s) describe the position of the hilum on the kidney?

   b. List the three structures that are entering/exiting at the hilum and describe the characteristics of each that allow you to differentiate between them.

6. Make a frontal section of the kidney as shown in Figure 31.2. Sketch your view of your sectioned kidney below, labeling all structures that you can locate.

7. When you have finished, dispose of the kidney specimen and clean your tray and dissecting instruments as directed by your instructor.

Figure 31.2   Frontal Section of a Fresh Pig Kidney

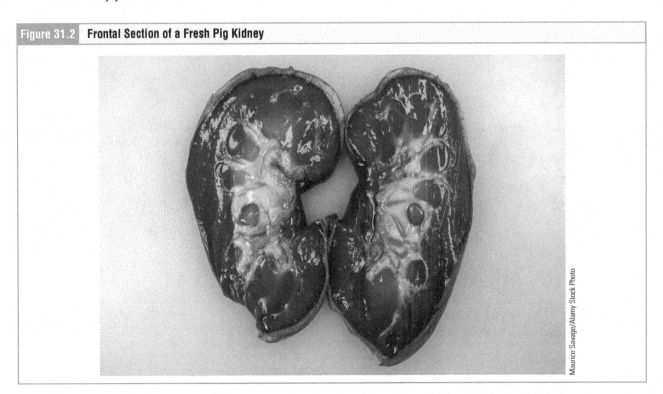

Maurice Savage/Alamy Stock Photo

(page intentionally left blank)

Activity 31.2 Anatomy of the Nephron

This activity targets LOs 31.3 and 31.4.

Each renal pyramid contains approximately 100,000 nephrons, for a total of approximately 1.25 million nephrons per kidney. Each nephron is, in essence, a blood processing unit. As blood travels through a small capillary bed at the start of the nephron, a portion of blood plasma is removed from the blood vessels to become filtrate. This **filtrate** can then be processed by the rest of the nephron—a series of small tubules—to reabsorb some material that was "lost" into the filtrate back into the blood and secrete even more substances that remained in the blood into the filtrate. Eventually the filtrate becomes the urine you void from your body.

The nephron begins with the **renal corpuscle**, as seen in Figure 31.3. This bulging structure contains both the capillary bed called the **glomerulus** and the swollen start to the tubing system, **Bowman's capsule**, wrapped around the glomerulus, creating a small cupped space called the **capsular space**. On the inside of Bowman's capsule, tightly hugging the capillaries of the glomerulus are **podocytes**, which extend finger-like **pedicels** to cover the surface of the capillaries, making thin **filtration slits** through which blood plasma can pass from the capillaries to become filtrate in the capsular space. The podocytes are considered the visceral, or inner, layer of Bowman's capsule, while the outer parietal layer is made of a thin wrapping of simple squamous epithelium.

Once the fluid of the filtrate has entered the capsular space, it flows on through the tubing, which leads from the **proximal convoluted tubule** to the **descending limb** and **ascending limb** of the **nephron loop**. The final portion of the nephron, the **distal convoluted tubule** twists back toward the renal corpuscle, carrying the fluid past the **juxtaglomerular apparatus** (JGA) that is close to the start of that nephron. This allows the **macula densa cells** of the JGA to assess the finished product of the nephron that is now in the distal convoluted tubule (the filtrate) and adjust the process as necessary.

At the end of the distal convoluted tubule the filtrate leaves the nephron and enters one of many collecting ducts. **Collecting ducts** gather the filtrate leaving many nephrons and descend through the pyramids toward their tips, the **renal papillae**. Further reabsorption and secretion occur here to finalize the contents of the urine. The urine is then gathered by the minor and then major calyces, passed on to the renal pelvis, and exits the kidneys to the ureters.

Much of each nephron is found in the renal cortex, including the renal corpuscle, proximal convoluted tubule, the top portions of the nephron loop, and the distal convoluted tubule. But a distinction between nephrons can be made based on the position of the nephron loops. The majority of nephrons in the kidney, about 85%, are **cortical nephrons**. Cortical nephrons have short nephron loops that just barely reach the renal medulla. In contrast, the remaining 15% of nephrons are **juxtamedullary nephrons** that have long nephron loops that extend deep into the medulla.

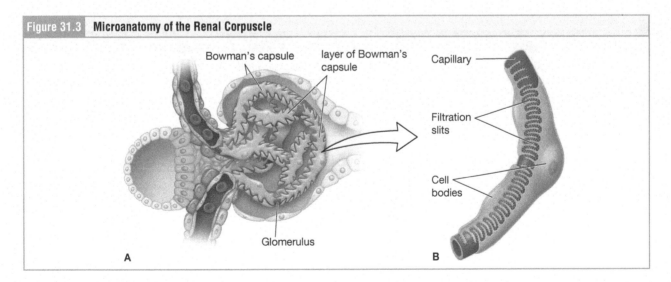

Figure 31.3 **Microanatomy of the Renal Corpuscle**

Bowman's capsule

layer of Bowman's capsule

Capillary

Filtration slits

Cell bodies

Glomerulus

A

B

Materials

- Three-dimensional model or anatomical chart of a nephron with both cortical and juxtamedullary nephrons represented

Structures Identified in This Activity

Nephron

Renal corpuscle

Glomerulus

Bowman's capsule (glomerular capsule)

Parietal layer

Visceral layer

Podocytes

Pedicels

Filtration slits

Capsular space

Renal tubule

Proximal convoluted tubule

Nephron loop (loop of Henle)

Descending limb

Ascending limb

Distal convoluted tubule

Collecting ducts

Juxtaglomerular apparatus

Macula densa

Juxtaglomerular cells (granular cells)

Cortical nephron

Juxtamedullary nephron

Instructions

1. Using the description in the activity introduction, label the figure on the facing page with as many of the structures in the above list as you can locate.

2. Find all structures listed in bold in the activity introduction, on the provided models and/or charts.

3. Not all models and charts show an accurate representation of the position of the distal convoluted tubule. Do the models and charts you were provided accurately represent the distal convoluted tubule? Explain. Why do you think some models and charts do not?

Learning Connection

There are two unrelated words in the urinary system that both begin with the prefix juxta-: the juxtaglomerular apparatus and juxtamedullary nephrons. Review the meaning of the prefix and the remaining portions of the words to keep these two terms separate from each other.

Juxta = _____ glomerular = _____ apparatus = _____

Juxta = _____ medullary = _____ nephron = _____

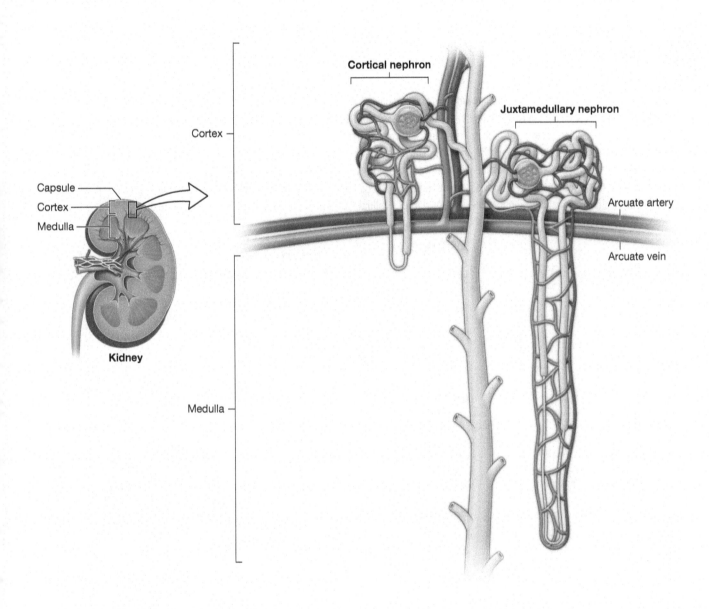

Cortical nephron

Juxtamedullary nephron

Cortex

Medulla

Capsule

Cortex

Medulla

Kidney

Arcuate artery

Arcuate vein

Digging Deeper:
Dialysis

Dialysis is a medical treatment for people who have kidney failure. The process of hemodialysis uses a dialyzer machine as an artificial kidney to perform the work of the nephrons. The person undergoing dialysis will be connected to a line that removes blood from their body, passes it through the machine, and returns the filtered blood back to their body in a continual flow that takes about 3 to 5 hours. In a manner like what happens at the renal corpuscle, the machine passes the blood through mesh tubes that replicate the filtration membrane that exists between the glomerulus and Bowman's capsule. Tiny holes in the tubes allow fluid to leave the blood; this fluid will have a similar composition of solutes to the filtrate produced by the filtration process at the renal corpuscles. The dialyzer monitors the composition of the fluid that is produced and the blood returning to the patient so that adequate water and salts are returned to the patient's blood.

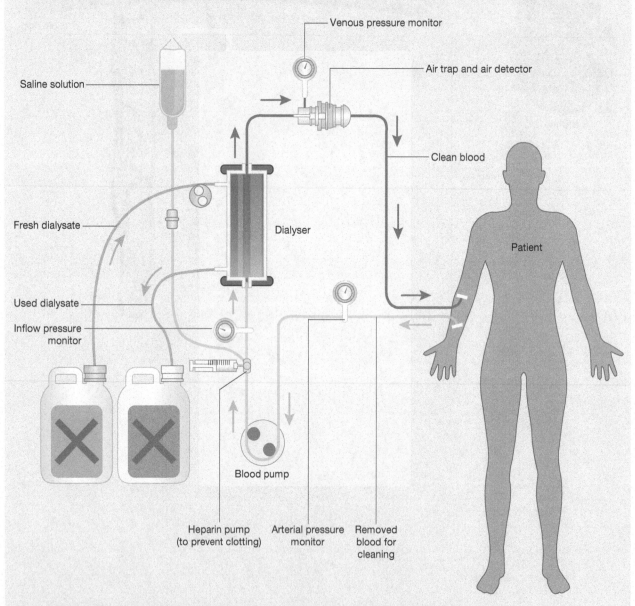

Source: Based on YassineMrabet, Wikimedia Commons. https://commons.wikimedia.org/wiki/File:Hemodialysis-en.svg
Fig32.01 – Based on Neuendorf, J. (2020). Description of Urinary Sediment Constituents. In: Urine Sediment. Springer, Cham.

Activity 31.3 Blood Supply to the Kidney

This activity targets LO 31.5.

Approximately 25% of your cardiac output flows to the kidneys, with upwards of 140 to 190 liters (38 to 50 gallons) of blood filtered through your kidneys every day. This blood needs to reach each of the 1 million nephrons in each kidney, so there is an elaborate vascular network to do so.

Follow the path of blood described below in Figure 31.4 (see page 797). Note that this figure is not labeled, because you will be doing that as part of this activity. Branching from the ascending aorta are two large, thick-walled **renal arteries**, one leading to each kidney. At the hilum, the renal artery branches into **segmental arteries**. The segmental arteries branch further to create **interlobar arteries** that pass between the lobes of the kidney using the renal columns to reach the cortex of the kidney. The interlobar arteries, in turn, branch into **arcuate arteries** that travel along the border between the cortex and medulla. To radiate into the cortex the arcuate arteries branch into small **interlobular arteries** that finally give off the **afferent arterioles** that individually supply each nephron.

The afferent arterioles form a ball of high-pressure capillaries called the **glomerulus**, which is tucked inside Bowman's capsule. After passing through the glomerulus and being filtered the blood leaves the capsules through the **efferent arterioles**. The efferent arterioles lead to the **peritubular capillaries** and **vasa recta** that run along the tubing of each nephron in order to receive back any filtered material that is selected for return to the blood.

As blood passes through the peritubular capillaries and vasa recta, substances are exchanged. In addition, as expected these capillaries will transfer gases, nutrients, and wastes between the blood and the cells of the kidneys. The blood next enters the venous system, traveling from the **interlobular veins** to **arcuate veins** to **interlobar veins**, and finally to the pair of **renal veins**, one departing from each kidney.

Materials
- Human torso model
- Three-dimensional model of a kidney sectioned along the frontal plane
- Three-dimensional model of a nephron

Structures Identified in This Activity

Descending aorta	Afferent arteriole	Arcuate veins
Renal artery	Glomerulus	Lobar veins
Segmental artery	Efferent arteriole	Renal vein
Lobar artery	Peritubular capillaries	Inferior vena cava
Arcuate artery	Vasa recta	
Lobular artery	Lobular veins	

Instructions

1. Each of the three provided models shows a portion of the vascular pathway from the heart to the kidneys and back again. List which vessels, bolded in the activity introduction, are visible on each of the provided models.

    a. Human torso model:

    b. Kidney model:

    c. Nephron model:

2. Use Figure 31.4 to trace this vascular pathway in full.

    a. Starting at the renal artery, use arrows to indicate the directional flow of blood as it travels through the kidney and through a single nephron. The first arrow entering at the renal artery and last arrow departing at the renal vein have been drawn for you.

    b. How did blood arrive at the renal artery from the heart? Trace that path here:

    Left ventricle of heart → _____ → _____ →

    _____ → renal artery

    c. How will the blood return to the heart from the renal vein? Trace that path here:

    Renal vein → _____ → right atrium of heart

    d. Re-read the activity's introductory paragraphs and use the description of blood flow through the kidneys to label the rest of the vessels you listed in step 1 onto Figure 31.4.

    e. Compare your completed figure with that of your lab partner to confirm that you have labeled all the vessels named in this activity.

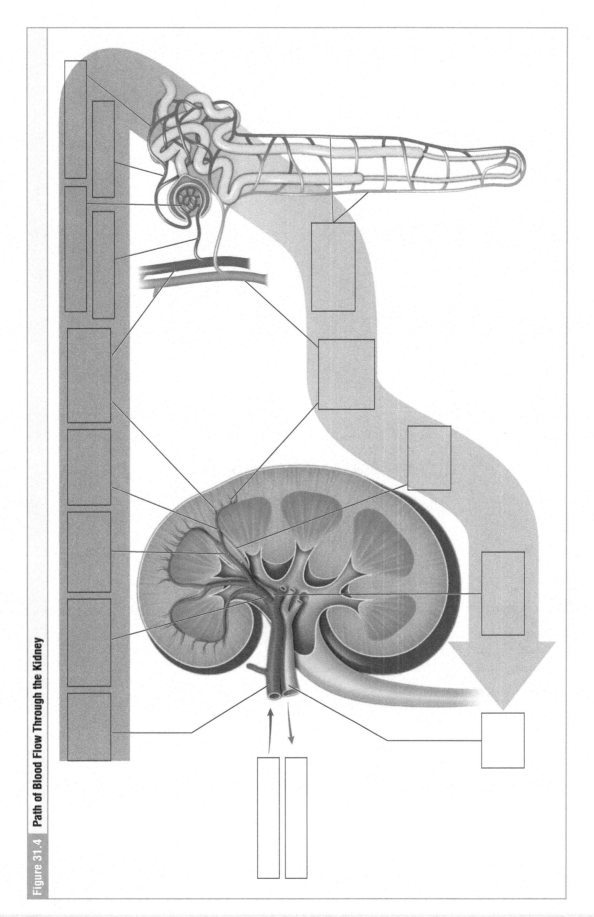

Figure 31.4 Path of Blood Flow Through the Kidney

(page intentionally left blank)

Activity 31.4 Histology of the Ureters, Urinary Bladder, and Urethra

This activity targets LO 31.6.

The ureters, urinary bladder, and urethra manage the urine that has recently been produced by the kidneys. These three accessory structures do so in preparation for the process of **micturition**, or urination.

Which accessory structure of the urinary system performs which function?

The _____ carry the urine away from the kidneys.

The _____ temporarily stores the urine until micturition can occur.

The _____ eliminates the urine from the body through micturition.

These three urinary structures are similar to other hollow organs of the body in that their walls are made up of several layers of tissue. The inner **mucosa** is made of a lining of epithelium with an underlying lamina propria. Surrounding the mucosa is the **muscularis** layer; the muscle is wrapped in an outer layer of connective tissue called **adventitia**. Consider how these three layers serve the specific functions of the ureters, bladder, and urethra: these structures must be able to stretch to accommodate volumes of urine, create rhythmic contractions to propel the urine forward, and remain firmly attached to the dorsal wall of the abdominopelvic cavity. Can you determine which specific tissue type makes up each layer and which function is served by each? Use your textbook and other resources to complete Table 31.1. Then examine these layers on histological slides of the ureters, urinary bladder, and urethra.

Table 31.1 Accessory Structures of the Urinary System		
Layer of the Wall	**Specific Tissue Type**	**Function Served**
Epithelium of the mucosa		
Muscularis		
Adventitia		

A few notes about the specialized anatomy of the bladder, some of which can be seen in Figure 31.5. In addition to the microanatomical structure of the epithelium of the mucosa, the mucosa of the bladder is somewhat folded into **rugae**, which provides another level of ability for the bladder to change in size as it fills or empties of urine. The bladder also contains a visible submucosal layer between the lamina propria of the mucosal layer and the muscularis layer. The extensive submucosal layer provides space for the glands here that secrete a watery mucus to protect the epithelium. Finally, the muscularis layer of the bladder is the detrusor muscle, which contracts during micturition to aid in emptying the bladder of urine.

Figure 31.5	Anatomy of the Bladder

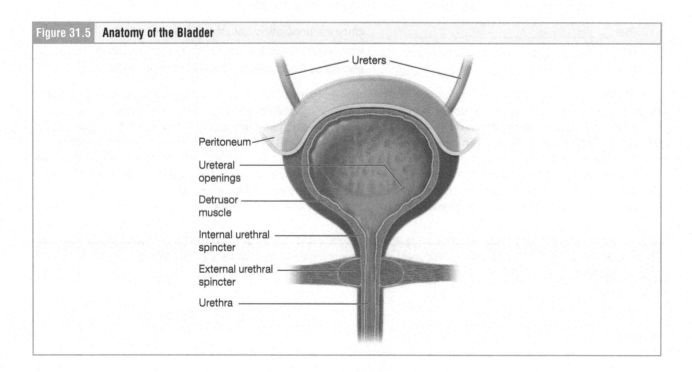

Materials
- Histological slide of bladder wall
- Histological slide of ureters
- Histological slide of urethra

Structures Identified in This Activity

Ureters	Rugae
Urinary bladder	Urethra
Detrusor muscle	

Instructions
1. For each slide, examine the slide first on scanning power (4×) and use the coarse focus knob to locate the lumen. Then increase the magnification by switching to the low-power (10×) objective lens and use the fine focus knob. If necessary, switch to the high-power (40×) objective lens and again use the fine focus knob until you can clearly see the details.

2. Draw the microscopic view of each structure, identifying each of the three layers in your drawings. In a blank area on the drawing, note the magnification used.

Slide 1: Ureters

• On your drawing, identify: lumen, epithelium of the mucosa, lamina propria, smooth muscle, and adventitia.

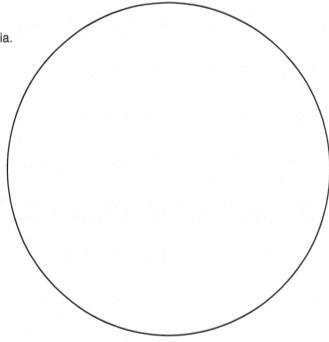

Slide 2: Urethra

• On your drawing, identify: lumen, epithelium of the mucosa, lamina propria, smooth muscle, and adventitia.

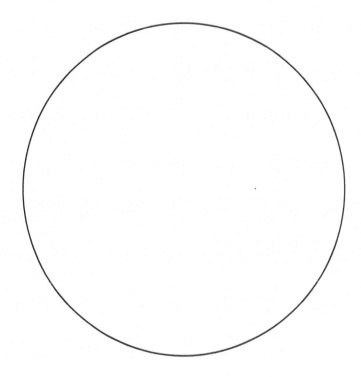

• Only the proximal-most section of the urethra, near the exit from the bladder, is made up of transitional epithelium. The lining transitions to stratified squamous epithelium at the urethral orifice. In biological males, the urethra is much longer and a zone of pseudostratified columnar epithelium lines the urethra between these two tissue types.

Examine the epithelium in your slide and determine which region of urethra you are viewing:

_____.

Slide 3: Urinary Bladder

Many histological slides of the urinary bladder will provide two sections of tissue on a single slide, one with the bladder distended (stretched) and one with the bladder empty. Which section is which?

The thinner section is distended/empty (circle one)

The thicker section is distended/empty (circle one)

Create your drawing based on the empty section.

On your drawing, identify: lumen, epithelium of the mucosa, lamina propria, submucosa, smooth muscle, adventitia, and rugae.

The smooth muscle of the bladder wall is given a specific name; label this on your drawing.

Compare the empty section you just examined to the distended section by carefully examining the apical surface of the epithelium on each section. How do they differ? Explain what is occurring here to cause this difference.

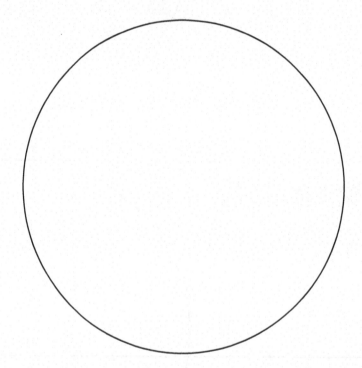

Activity 31.5 Case Study: Urinary Incontinence

This activity targets LO 31.8.

Anna has recently given birth. She delivered vaginally, and her midwife explains to her that she might experience stress incontinence for a time following the delivery. The midwife shares that Anna actually has two separate sphincters that contract to prevent urine from leaking from the bladder. The internal urethral sphincter is an extension of the muscle that surrounds the bladder and the external urethral sphincter is a small group of muscles that surround both the external urethral orifice and the vagina.

1. Which sphincter is associated with the detrusor muscle?

   a. Internal urethral sphincter
   b. External urethral sphincter

2. The sphincter in the previous question is _____ muscle under _____ control.

   a. smooth; voluntary
   b. smooth; involuntary
   c. skeletal; voluntary
   d. skeletal; involuntary

3. Is the sphincter addressed in questions 1 and 2 likely to contribute to Ana's stress incontinence? Explain.

The midwife explains that the external urethral sphincter is a skeletal muscle under voluntary control and that under most circumstances we can control when we urinate, even if stretch receptors in the bladder have sensed a full bladder and have initiated the micturition reflex, by using the external urethral sphincter. But if the external urethral sphincter has been damaged then the exit from the urethra is not fully closed and some urine will begin to leak out. The midwife suggests that Anna perform Kegel exercises to help increase her ability to prevent these leaks. Kegel exercises target the muscles of the pelvic floor by practicing repeated contractions. This activity strengthens these muscles just like other skeletal muscles of the body become stronger with regular use. Anna asks, "so when I practice the Kegel exercises, I am contracting my external urethral sphincter?" But the midwife answers, "no."

4. Use what you know about the muscles of the pelvic floor to explain how strengthening the pelvic floor works to reduce stress incontinence even though the Kegel exercises do not target the external urethral sphincter directly.

(page intentionally left blank)

Lab 31 Post-Lab Quiz

1. If you needed to perform an operation to remove a portion of kidney, the order of tissues that you would need to cut through, from start to finish, would be:

2. The section of the renal tubule found in the renal medulla is the:
   a. Proximal convoluted tubule
   b. Distal convoluted tubule
   c. Nephron loop

3. Through which structure do the collecting ducts pass after they gather filtrate from the nephrons?
   a. Renal medulla
   b. Renal cortex
   c. Renal pelvis

4. Place the following structures in the correct order of filtrate flow through a nephron:

   capsular space
   distal convoluted tubule
   proximal convoluted tubule
   descending limb of the nephron loop
   ascending limb of the nephron loop

5. Blood flow departing from the kidneys will travel from arcuate veins to:
   a. Interlobular veins
   b. Interlobar veins
   c. Renal veins

6. Where does the blood go next after it leaves the glomerulus of the nephron?
   a. It enters the glomerular capsule to go through further reabsorption and secretion.
   b. It flows from efferent arterioles to interlobular veins to be sent to the heart.
   c. It runs into peritubular capillaries to travel next to the nephron tubules.

7. Approximately how many afferent arterioles will be found in a single kidney?
   a. 1
   b. 10
   c. 100 thousand
   d. 1 million

8. When the detrusor muscle of the bladder contracts to void the bladder of urine, the rugae of the bladder will become _____ pronounced (visibly deep).
   a. more
   b. less

9. You and your lab partner are examining histological images of the ureters, bladder, and urethra. Your lab partner wonders why the bladder has a significant submucosal layer with glands, while the ureters and urethra do not. You can explain to them that this is because the bladder is the only location of the three that:

a. Contracts to void the body of urine

b. Always contains some urine

c. Is damaged by stress incontinence

10. You are examining the histological slide shown here, labeled as "wall of the prostatic urethra." From this information and what you see on the slide, you can determine that the histological sample was taken from which of the following? Choose all that apply.

a. The urethra of a biological male

b. The urethra of a biological female

c. The external urethral orifice

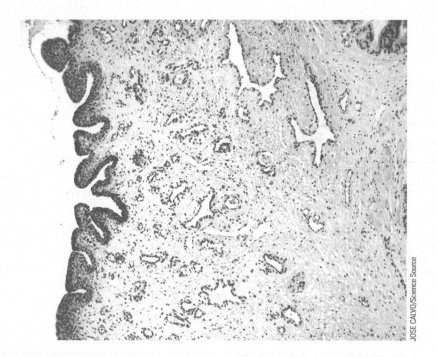

JOSE CALVO/Science Source

Lab 32 | Urinalysis

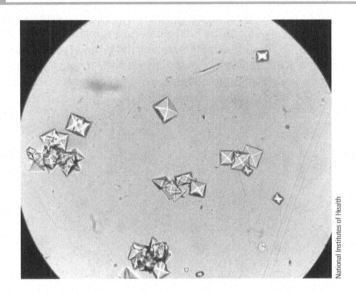

National Institutes of Health

Learning Objectives: After completing this lab, you should be able to:

32.1* Perform a microscopic sediment analysis of a urine sample.

32.2* Evaluate the characteristics of a urine sample.

32.3* Summarize and interpret the results of urinalysis tests.

32.4 Describe the composition of normal urine.

32.5 Given a factor or situation, predict the changes in the urinary system and the consequences of those changes

32.6 Given a disruption in the structure or function of the urinary system, predict the possible factors or situations that might have caused that disruption

*Objective is not a HAPS Learning Goal.

Introduction

The average amount of urine produced by a person depends greatly on age, gender, and fluid consumption, with most adults urinating approximately 500 to 2000 milliliters per day. The production of urine requires three processes at the nephron: filtration at the renal corpuscle and both reabsorption and secretion at the renal tubules. These actions maintain fluid and electrolyte homeostasis in the body and remove wastes from the blood. Because of the dynamic nature of these processes, both the volume and composition of urine range widely from day to day and from person to person. Urine composition will vary based on what that person is consuming, their physical activity, their rate of respiration, and the temperature and humidity of the environment.

The Human Anatomy and Physiology Society includes more than 1,700 educators who work together to promote excellence in the teaching of this subject area. The HAPS A&P Learning Outcomes measure student mastery of the content typically covered in a two-semester Human A&P curriculum at the undergraduate level. The full Learning Outcomes are available at https://www.hapsweb.org.

Even with such variation, evaluation of a person's urine can provide significant information about their kidney function, as well as the physiological actions of some other organs of the body. Urinalysis is a powerful tool of the clinician in evaluating general health and investigating possible pathologies.

The kidneys must be able to consistently make urine even in the face of blood pressure changes due to variations in daily activities. Each nephron does this by carefully monitoring its own production of filtrate. This task of monitoring and adjusting the filtration rate falls to the **juxtaglomerular apparatus**, which consists of **macula densa** cells in the distal convoluted tubule (DCT) and **juxtaglomerular cells** in the afferent arteriole. As blood pressure rises, the macula densa cells sense an increase in Na^+ concentration and fluid flow through the DCT. These cells respond to these changes by stimulating the juxtaglomerular cells to contract, narrowing the afferent arteriole, reducing the blood pressure in the glomerulus, and reducing the glomerular filtration rate back to acceptable levels.

Lab 32 Pre-Lab Quiz

This quiz will strengthen your background knowledge in preparation for this lab. For help answering the questions, use your resources to deepen your understanding. The best resource for help on the first five questions is your text, and the best resource for help answering the last five questions is to read the introduction section of each lab activity.

1. Anatomy of Bowman's capsule and the juxtaglomerular apparatus. Use the following terms to correctly label the figure:

   Filtration slits   Parietal layer of Bowman's capsule
   Glomerulus   Podocytes
   Juxtaglomerular cells   Visceral layer of Bowman's
   Macula densa cells

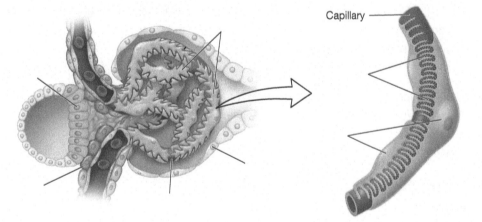

2. Match the root next to its definition:

Root	Meaning
	measure
	calcium
	of the urine

List of Roots

calc

ometer

uria

3. While little protein should be found in the initial filtrate of the capsular space, any small proteins that do make it through the filtration membrane are reabsorbed in the:

   a. Nephron loop
   b. Proximal convoluted tubule
   c. Distal convoluted tubule

4. The yellow color of urine is due to the presence of _____ in the urine.

   a. hemoglobin
   b. bilirubin
   c. urobilin

5. If an adult's urine output is below 500 mL/day it is indicative of:
   a. Diabetes mellitus
   b. An acidic diet
   c. Severe dehydration

6. As the juxtaglomerular cells contract, this will lead to a _____ in glomerular filtration rate.
   a. reduction
   b. increase

7. Is healthy urine clear or cloudy?
   a. Clear
   b. Cloudy

8. What urinalysis procedure allows the clinician to examine the sediment present in urine?
   a. Urinometry
   b. Centrifugation
   c. Dipstick analysis

9. You examine two urine samples. Sample A has a specific gravity of 1.020 and Sample B has a specific gravity of 1.009. Which sample contains more solutes?
   a. Sample A
   b. Sample B

10. Pyuria is the presence of _____ in the urine.
    a. glucose
    b. protein
    c. leukocytes

Obtain a Urine Sample

For use in Activities 32.1–32.4.

The following four activities will allow you to analyze a urine sample. Your will either be instructed to obtain your own urine sample for analysis or artificial urine will be provided to you by your instructor. No matter the source of the urine, you are expected to follow universal precautions in handling this bodily fluid. Follow all precautions and instructions given by your lab instructor. Wear all personal protective equipment (PPE) provided for the duration of the lab.

Materials
• Personal protective equipment: safety glasses, gloves, lab apron or protective gown
• Sterile wipes, two
• Urine sample cup

Instructions

1. Prepare your lab station by gathering the materials listed for Activities 32.1–32.4, below.

2. Obtain a clean catch urine sample as directed in the box below. Alternately, obtain the artificial urine provided by your instructor.

3. Put on your safety equipment: safety glasses, gloves, and lab apron or protective gown.

4. Urine is a bodily fluid; follow the safety precautions given by your lab instructor for the remainder of lab.

Clean Catch Urine Sample

Obtaining a urine sample through the clean catch method helps prevent contaminating the urine sample with material that may be on your genitals but that was not present in the urine or urinary tract.

Directions for collecting a clean catch urine sample if you have labia:
1. Wash your hands with soap and water.
2. While sitting on the toilet, use two fingers to spread your labia open.
3. Use a sterile wipe to clean the inner folds of your labia, wiping from front to back. Use a second sterile wipe to clean over your urethra.
4. Urinate a small amount into the toilet, pause, then move the urine sample cup into position and urinate into the cup.
5. Try to fill the urine sample cup at least half full. If you have more urine than that you can finish urinating into the toilet.

Directions for collecting a clean catch urine sample if you have a penis:
1. Wash your hands with soap and water.
2. Use a sterile wipe to clean the head of your penis. If your penis is uncircumcised, pull the foreskin back to clean.
3. Urinate a small amount into the toilet, pause, then move the urine sample cup into position and urinate into the cup.
4. Try to fill the urine sample cup at least half full (approximately 50 to 60 mL of urine). If you have more urine than that you can finish urinating into the toilet.

(page intentionally left blank)

Activity 32.1 Sediment Analysis of a Urine Sample

This activity targets LO 32.1.

There are a variety of particles suspended in urine that can provide helpful information about the individual, so a complete urinalysis includes examining these particles. The microscopic solid material in urine is normally partially suspended in the fluid. To examine these substances under the microscope without having to scan past vast fields of fluid, it is helpful to concentrate the sediment. This is done by centrifugation, which forces the solid material to the bottom of the tube where it can be recovered, stained, and viewed.

While some sediments in urine are present normally, others are considered abnormal and may indicate disease or infection. Many of the possible sediments are shown in Figure 32.1 and include:

- Epithelial cells   Epithelial cells line the entire urinary tract and small numbers are regularly shed as new layers take their place; therefore, some transitional often, squamous, and a small number of cuboidal epithelial cells are often found in urine.

- Blood cells   The presence of significant numbers of erythrocytes or leukocytes in urine is abnormal. It should be noted that urine samples can contain erythrocytes from menstruation that have been picked up during the process of urination.

- Casts   Casts are cylindrical masses made when denatured proteins and sloughed cells clump together and harden in the distal convoluted tubule or collecting ducts, taking on the tubular shape of the passageway in which they form. This can occur because the urine is acidic or contains a lot of protein or salts.

- Microorganisms   The presence of a small number of microorganisms is considered normal. These are usually picked up at the external urethral orifice during the process of urination. In contrast, large numbers of microorganisms are indicative of an infection of one or more locations along the urinary tract.

- Crystals   Crystals are made of salts, acids, cholesterol, or phosphates that have come out of solution to form small geometric particles. Like the presence of microorganisms, small amounts of crystals, especially calcium oxalate crystals and magnesium ammonium phosphate crystals, are considered normal. Larger amounts of these crystals or the presence of other crystal types such as cystine crystals may be prompted by diet, disease, or infection.

Figure 32.1 | Components of Urine Sediment

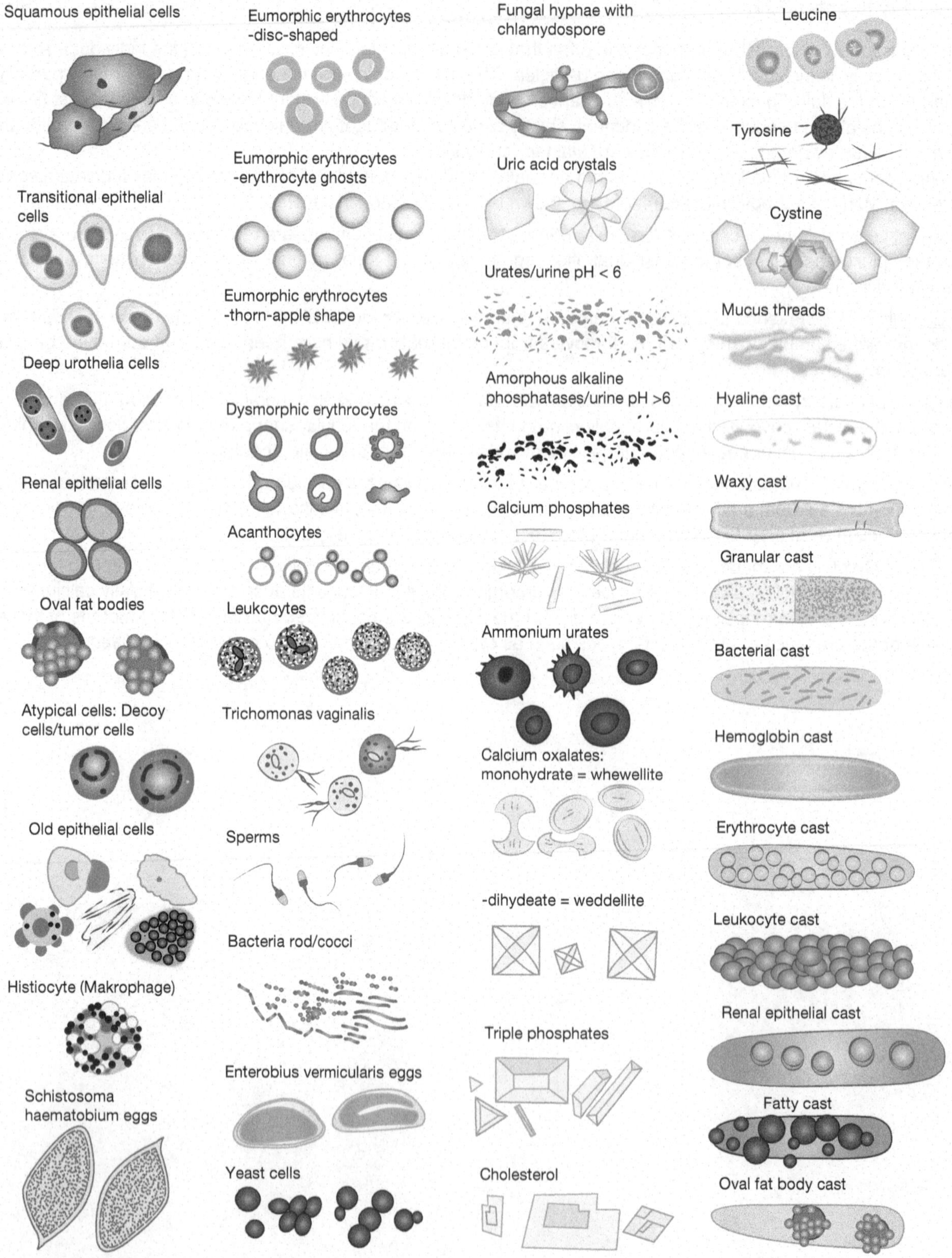

Squamous epithelial cells

Transitional epithelial cells

Deep urothelia cells

Renal epithelial cells

Oval fat bodies

Atypical cells: Decoy cells/tumor cells

Old epithelial cells

Histiocyte (Makrophage)

Schistosoma haematobium eggs

Eumorphic erythrocytes -disc-shaped

Eumorphic erythrocytes -erythrocyte ghosts

Eumorphic erythrocytes -thorn-apple shape

Dysmorphic erythrocytes

Acanthocytes

Leukcoytes

Trichomonas vaginalis

Sperms

Bacteria rod/cocci

Enterobius vermicularis eggs

Yeast cells

Fungal hyphae with chlamydospore

Uric acid crystals

Urates/urine pH < 6

Amorphous alkaline phosphatases/urine pH >6

Calcium phosphates

Ammonium urates

Calcium oxalates: monohydrate = whewellite

-dihydeate = weddellite

Triple phosphates

Cholesterol

Leucine

Tyrosine

Cystine

Mucus threads

Hyaline cast

Waxy cast

Granular cast

Bacterial cast

Hemoglobin cast

Erythrocyte cast

Leukocyte cast

Renal epithelial cast

Fatty cast

Oval fat body cast

Source: Adapted from Neuendorf, J. (2020). Description of Urinary Sediment Constituents. In: Urine Sediment. Springer, Cham.

Materials
- Centrifuge
- Centrifuge tube, 10-mL
- Compound light microscope
- Cover slip
- Dropper or pipette
- Graduated cylinder, 10-mL (if using unmarked centrifuge tubes)
- Hazardous material sharps box or other disposal location for used slides
- Microscope slide
- Sedi-stain
- Urine sample
- Urine sediment charts
- Waste beaker for decanted urine

Instructions

1. Label your centrifuge tube as directed by your instructor.

2. Stir or shake your urine sample to re-suspend any heavier sediment that has settled.

3. Pour 10 mL of urine into the centrifuge tube, using a graduated cylinder if there are no markings on the provided centrifuge tubes. Keep the remaining urine in the sample cup and set it aside for use in other activities in this lab.

4. Give your centrifuge tube to your instructor; your tube will be centrifuged on slow speed (1,500 rpm) for 5 to 10 minutes.

Note: Your instructor will gather the urine samples from the class and centrifuge them; this process takes some time. As you wait for your instructor to return your centrifuged sample to you, proceed to Activities 32.2–32.4. You may return to this activity once your sample is ready and continue with the rest of the procedure.

---

5. Examine your centrifuged sample. It should resemble Figure 32.2.

6. Decant approximately 9 mL of supernatant into the provided waste beaker, taking care to not disturb the sediment.

7. Gently re-suspend the sediment in the remaining 1 mL of supernatant using the dropper or pipette.

8. Use the dropper or pipette to obtain some of the sediment mixture; place one drop on the microscope slide.

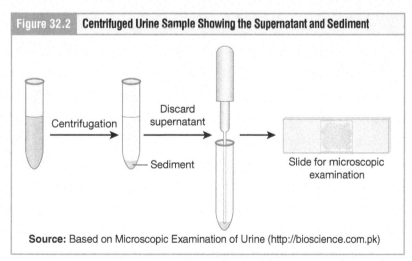

| Figure 32.2 | Centrifuged Urine Sample Showing the Supernatant and Sediment |

Source: Based on Microscopic Examination of Urine (http://bioscience.com.pk)

9. Add a small drop of sedi-stain, taking care not to over-fill the slide.

10. Add a cover slip.

11. View the slide on scanning power (4×). Use the coarse focus and then fine focus to locate stained sediment on the slide. It should appear as darker blue material on the slide.

12. As you locate some sediment, observe it more clearly by switching to the 10× and then the 40× objective lenses, adjusting only the fine focus and the light as necessary.

13. Use the space below to draw, label, and describe the sediments you observe, using Figure 32.1, and any other resources such as urine sediment charts provided by your instructor as references to identify what you see.

14. When you have finished observing your sample, dispose of the slide in the provided hazardous material sharps box or the location provided by your instructor.

Activity 32.2 Evaluate the Color and Transparency of a Urine Sample

This activity targets LO 32.2.

You have probably noticed that the color of fresh urine ranges from clear to deep yellow. The yellow color comes from urobilin (also called urochrome), a pigment that occurs as a waste product from the breakdown of the heme in hemoglobin. As the proportion of water to solutes in urine increases, the color of the resulting urine becomes lighter.

Urine is normally transparent, but it can become cloudy due to the presence of larger suspended substances such as mucus, microbes, dead cells, crystals, or casts.

Materials
• Blank sheet of white paper or a light-colored surface
• Sheet of paper with writing on it
• Urine sample

Instructions
1. Examine the color of the urine sample against a white or light-colored background.

   a. Is your urine sample clear or does it have a yellow color? Define the color as clear, light yellow, yellow, dark yellow, or amber.

   b. If your urine color is something other than yellow, describe that color. Other likely colors include yellow-brown, brown, red, orange, green, or even blue. While some of these colors are considered abnormal results, other possible sources include drugs, vitamins, or certain foods.

   c. Record your results in Table 32.2 in Activity 32.5.

2. Examine the transparency of the urine sample against a sheet of paper with writing on it, such as this lab manual.

   a. Choose from the following descriptions of the transparency of your sample: clear, slightly cloudy, very cloudy.

   b. Record your results in Table 32.2 in Activity 32.5.

(page intentionally left blank)

Activity 32.3 Specific Gravity Analysis of a Urine Sample

This activity targets LO 32.2.

While analysis of urine color can provide a qualitative method for assessing solute concentration, a more quantitative approach is using **specific gravity**. Specific gravity is a comparison between the weight of the substance you are measuring and the weight of the same volume of distilled water, which has a specific gravity of 1.000. A substance that weighs more than water will have a specific gravity greater than 1.000 and a substance that weighs less than water will have a number below 1.000. Urine is mostly water, but contains some solutes, so it will have a specific gravity slightly greater than 1.000.

The range of the specific gravity of urine is usually between 1.003 to 1.032. Note that there are no units for a measurement of specific gravity. High urine specific gravity measurements can be cause for concern, as this indicates that the kidneys are reabsorbing a significant amount of water from the filtrate.

Materials
- Cleaning supplies suck as Kimwipes, isopropyl alcohol, and distilled water
- Urine sample
- Urinometer

Instructions
1. Make a prediction based on your analysis of the color of your urine sample.

    a. Do you think your urine will have: (choose one option)
    - A low specific gravity (approximately 1.003 to 1.015)
    - A moderate specific gravity (approximately 1.016 to 1.025)
    - A high specific gravity (approximately 1.026 to 1.032)
    - A very high specific gravity (above 1.032)

    b. Support your prediction. What evidence do you have?

    c. What does this mean about the solute concentration of your urine sample?

    d. Identify some behaviors that would lead to a solute concentration at either the high or low end of the range.

2. Use the urinometer to measure the specific gravity of the urine sample using Figure 32.3 as a guide.

   a. Examine the markings on the long stem of the hydrometer (the flotation device that fits inside the urinometer cylinder). The markings indicate specific gravity. What range of specific gravity is provided?

   From _____ to _____

   b. Gently swirl the urine sample cup to suspend any particulates that have settled and pour the urine into the clean cylinder of the urinometer until it is about three-quarters full.

   c. Gently place the hydrometer into the urine with the long stem facing up and allow it to freely float in the urine. If it sinks and touches the bottom of the cylinder, add more urine to the cylinder until it is floating.

   d. Read the marking on the stem of the hydrometer that aligns with the lower part of the curved meniscus in the cylinder. This is the specific gravity of the sample.

3. Record your results in Table 32.2 in Activity 32.5.

4. Do not dispose of the urine in the cylinder; pour it back into your sample cup to use for Activity 32.4.

5. Clean the urinometer with kimwipes and isopropyl alcohol. Then rinse it with distilled water.

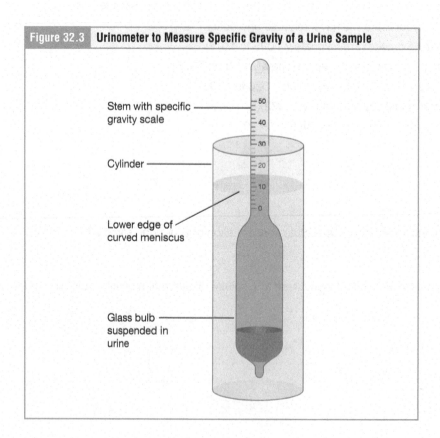

Figure 32.3   Urinometer to Measure Specific Gravity of a Urine Sample

Stem with specific gravity scale

Cylinder

Lower edge of curved meniscus

Glass bulb suspended in urine

Activity 32.4 Dipstick Analysis of a Urine Sample

This activity targets LO 32.2.

A dipstick analysis of a urine sample provides a quick and accurate method for quantitatively assessing the presence of certain substances dissolved in a urine sample. Each test pad on a dipstick strip contains a reagent that will change color when it interacts with a substance that is possibly found in urine. While urine normally contains a range of substances such as ions, small molecules, and nitrogenous wastes, some of these substances (and the presence of other substances that should not be found in normal urine) may be indicative of disease or infection. Table 32.1 includes possible factors that you may be measuring using a dipstick analysis that represent urine abnormalities.

Table 32.1 Possible Abnormal Urine Dipstick Results

Abnormal Urine Component	Abnormal Value	Clinical Term	Description	Reason for Abnormality
Glucose	Present	Glucosuria	While glucose is freely filtered at the glomerulus, all glucose should be reabsorbed at the PCT	Can be present temporarily due to a very high carbohydrate meal Is chronically present in untreated diabetes mellitus
Bilirubin	1 mg/dL or greater	Bilirubinuria	Bilirubin is produced as hemoglobin is recycled. A small amount (less than 1 mg/dL) is normal	Liver damage or a block to the bile duct will increase bilirubin in the blood which will increase its presence in the urine
Ketone	Present	Ketonuria	Ketones are produced as the body metabolizes fat as an energy source	A low carbohydrate diet or starvation, uncontrolled diabetes mellitus
Blood (hemoglobin)	Present	Hematuria	Erythrocytes or fragments of erythrocytes should not be able to cross the filtration membrane	Can be present due to a contaminated urine sample due to menstruation. Damage to the filtration membrane will allow formed elements such as erythrocytes to cross. Irritation of the ureters, bladder, or urethra can cause bleeding
Protein	Greater than trace	Proteinuria (albuminuria)	Most blood proteins are too large to cross the filtration membrane	Can be present temporarily due to physical exercise, high-protein meals, or pregnancy. Damage to the filtration membrane will allow large proteins to cross
Urobilinogen	1 mg/dL or greater	Urobilinogenuria	Bilirubin from the bile is converted to urobilinogen in the digestive tract. Some of this urobilinogen is reabsorbed into the blood and converted to urobilin before being filtered by the kidneys	A small amount of urobilinogen in the urine is normal, greater than 1 mg/dL is indicative of liver damage
Nitrites	Present	Nitriuria	Bacteria can convert the nitrates normally present in urine to nitrites	The presence of nitrites indicates a bacterial infection in the urinary tract
Leukocytes	15 leu/µL or greater	Pyuria	Leukocytes travel to sites of infection or damage	Leukocytes in the urinary tract are present due to inflammation and/or infection

Materials

- Biohazard disposal containers for used dipstick test strips and emptied urine sample cups
- Individual dipsticks or a combination dipstick test strip (such as URS-10)
- Laminated color copies of standard dipstick test strip color chart
- Urine sample

Instructions

1. Identify which factors in Table 32.1 will be measured using the dipstick strip(s) provided by your instructor. Draw a line through all factors in the table for which you do not have a dipstick test to reduce any possible confusion.

2. Depending on the factors on your dipstick strip(s) you should read the results immediately or after a brief period of time, but never beyond 2 minutes. Take a moment to read the directions on the dipstick container(s) to confirm the timing of reading the results of the tests.

3. Locate the standard color scale(s) provided with the dipstick container(s). Your instructor may provide you with laminated color copies. Examine the scale and compare it to the dipstick strip(s) to make sure you understand how to read the results before you start.

4. Dip the strip into the urine sample and wait the length of time indicated in the directions.

5. Compare the color of the test pads on your strip(s) to the colors on the standard color scale(s). Record the value or amount indicated by the matching color in each relevant row in Table 32.2 in Activity 32.5.

6. Dispose of the dipstick in the biohazard disposal containers provided by your instructor.

7. Dispose of all remaining urine into the toilet or into the container provided by your instructor. Dispose of the now-empty urine sample cup in the biohazard disposal container provided by your instructor.

Activity 32.5 Interpretation of Results from Urinalysis Tests

This activity targets LOs 32.3, 32.4, 32.5, and 32.6.

As already mentioned, evaluation of a person's urine can provide significant information about the body's physiology. Evaluate the results of the urine sample you examined today.

Materials
• Table 32.2 (on page 825) with your completed results column

Instructions
1. For each urine characteristic, compare your result to the normal value.

2. Analyze your results.

   a. If you consider the result to be normal, explain why.

   b. If you consider the result to be abnormal, explain possible causes of the abnormal result(s).

   c. Are all of your test results consistent with each other? For example, if you found blood cells in your sediment analysis, did you also get positive results for blood and leukocytes on the dipstick analysis?

3. Answer the questions below about possible urinalysis results and pathologies.

Questions
1. You and a friend eat two very different diets the day before a urinalysis test. Your friend eats a very high-fat, low-carbohydrate diet, while you consume the opposite: high carbohydrates and very little fat.

   a. Which two dipstick tests should you examine to see if your temporary diets have affected your urine composition?

   b. What possible abnormal results from these tests would you expect to see for the two of you?

   c. Should these results be of concern? Explain.

2. What abnormal urinalysis result(s) would you expect from a patient with uncontrolled diabetes mellitus? How do those results differ from the abnormal results that you and your friend may have found in question #1, above?

3. Your patient has the following urinalysis results. Comparing these urinalysis results to the dipstick results in Table 32.2, what is the most likely diagnosis?

Color = light yellow
Transparency = very cloudy

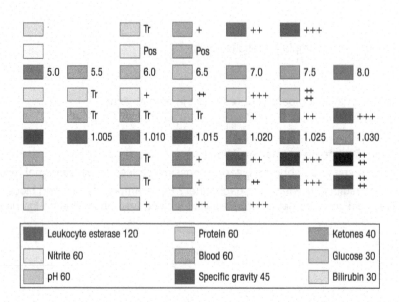

| | | Tr | | + | | ++ | | +++ |

Leukocyte esterase 120 Protein 60 Ketones 40
Nitrite 60 Blood 60 Glucose 30
pH 60 Specific gravity 45 Bilirubin 30

4. What abnormal urinalysis results would you expect from a patient with severe dehydration? Explain why you would expect these results.

Urine Characteristic	Results	Normal Value	Analysis
Color		Light yellow to amber	
Transparency		Clear	
Specific Gravity (measured with urinometer)		1.003–1.025	
Glucose		Negative	
Bilirubin		Negative	
Ketone		Negative	
Specific gravity (measured with dipstick)		1.0–1.3	
Blood (hemoglobin)		Negative	
pH		4.6–8.0	
Protein		Negative to Trace	
Urobilinogen		Normal (0.2 mg/dL)	
Nitrite		Negative	
Leukocytes		Negative to Trace	

Table 32.2 Physical and Chemical Characteristics of a Urine Sample

(page intentionally left blank)

Activity 32.6 Case Study: Chronic Kidney Disease

This activity targets LOs 32.3, 32.5, and 32.6.

Muhammed has had high blood pressure for many years and has been struggling to manage it. With this in mind, Muhammed's doctor has been carefully watching his health. At a routine visit, his doctor requests urinalysis to examine the composition of Muhammed's urine. The doctor explains that chronic high blood pressure can cause problems at the nephrons of his kidneys.

1. What particular structure of the kidneys will directly experience this high blood pressure?

    a. Glomerulus

    b. Renal tubule

    c. Collecting ducts

2. What cells at the entrance of the afferent arteriole to the glomerulus will respond to high blood pressure by contracting to attempt to reduce the effects of this chronic high blood pressure?

    a. Podocytes

    b. Juxtaglomerular cells

    c. Macula densa cells

3. These cells contract to reduce the effects of chronic high blood pressure because high blood pressure will increase the _____ at the filtration slits.

    a. glomerular filtration rate

    b. specific gravity

    c. caliculi

The doctor examines his most recent urinalysis results and finds two results that are considered abnormal. The doctor thinks that these results are indicative of glomerulonephritis, which can occur when untreated chronic high blood pressure begins to create scar tissue at the glomerulus.

4. Examine Muhammed's dipstick results. Which two are abnormal?

    a. pH

    b. Protein

    c. Blood

    d. Specific gravity

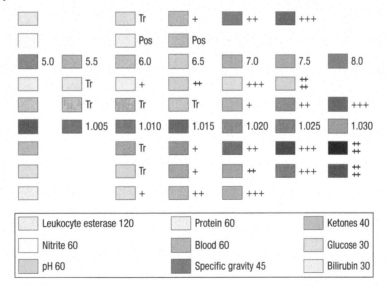

5. Considering the abnormal results, what can you conclude has changed as scarring of the glomerulus occurs?

   a. The scarring blocks many of the filtration slits.

   b. The scarring creates filtration slits that are too wide.

   c. The scarring creates breaks in the parietal layer of Bowman's capsule.

The doctor orders more tests, including a test to determine the glomerular filtration rate (GFR). The results show a reduced GFR from normal. Combined with Muhammed's blood pressure and urinalysis results, the doctor tells Muhammed that his glomerulonephritis is due to undiagnosed chronic kidney disease (CKD). Luckily, his CKD is at a mild stage. The doctor asks to work with Muhammed to get his high blood pressure under control, high pressure at the glomerulus will continue to damage the kidneys and reduce kidney function. The doctor recommends several approaches to reduce Muhammed's proteinuria and lower his blood pressure, including an ACE-inhibitor medicine and a diuretic.

6. An ACE-inhibitor is used to target the renin–angiotensin system, inhibiting the production of angiotensin II. Consult your textbook or another source. Lowering the production of angiotensin II will directly lower blood pressure because it will:

   a. reduce vasoconstriction

   b. reduce the glomerular filtration rate

   c. reduce Na^+ excretion

7. A diuretic will inhibit water reabsorption at the nephrons. What will be the result of reducing water reabsorption in Muhammed's body?

   a. This will increase the amount of water lost in the urine.

   b. This will decrease the amount of filtrate produced.

   c. This will decrease the amount of water that moves from kidney tubules to blood.

Activity 32.7 Case Study: Kidney Stones

This activity targets LOs 32.3, 32.5, and 32.6.

Case Study

Gary has just arrived at work when he feels a sharp pain in his side and back, just below his ribs. The pain is severe and feels like it is washing across his back, sides, and abdomen in waves. He feels nauseous and as he heads to the bathroom he also feels an intense need to urinate. His urine is pink and cloudy. Since the pain is so severe he heads to the emergency room to seek treatment. At the emergency room they suspect a kidney stone, called a renal calculus (pl. renal calculi), and so perform a blood test, urinalysis, and an ultrasound.

1. Pink, cloudy urine likely indicates:
   a. Normal urine
   b. Blood in the urine
   c. High specific gravity of the urine

2. Which of the following sediments might indicate the presence of a renal calculus?
   a. Epithelial cells
   b. Uric acid crystals
   c. Bacteria

The results of Gary's ultrasound show a small calculus located in his left renal pelvis. The results of the sediment analysis of his urine are shown in the figure below.

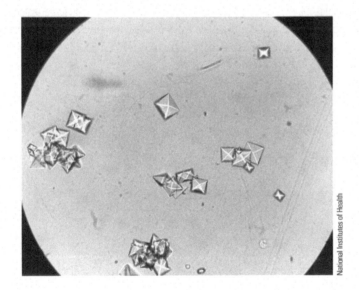

National Institutes of Health

3. What sediment is present in his urine?
   a. Calcium oxalate crystals
   b. Hyaline casts
   c. Simple squamous epithelial cells

Gary is advised to drink a lot of water and to wait for the calculus to pass. He is asked to try to catch the stone with a filter so that its composition can be confirmed. Gary is also prescribed medication to manage the pain.

4. Place these structures in the correct order to describe the path the stone must take to exit Gary's body:

bladder   membranous urethra
external urethral orifice   prostatic urethra
left renal pelvis   spongy urethra
left ureter

Digging Deeper:
Lithotripsy

Lithotripsy is a procedure that uses shock waves produced by a machine called a lithotripter to break up renal calculi (kidney stones). Breaking up the calculi into smaller pieces allows them to pass more easily through the urinary tract to leave the body in the urine. A clinician will focus the lithotripter to send shock waves through the skin and tissue of the body to reach the calculus.

While patients benefit from a procedure that does not require entering the body, not all patients respond well to the approach. The size and composition of the calculus may dictate if the clinician chooses this approach, as larger calculi are less likely to break apart under lithotripsy and stones composed of cystine and certain types of calcium are resistant to fragmentation by lithotripsy.

Lab 32 Post-Lab Quiz

1. Urine that is dark in color indicates that a _____ proportion of the water in the original filtrate was reabsorbed.

   a. greater
   b. lesser

2. In performing a urinalysis you note that the sample is very cloudy. Which dipstick analysis tests would you expect might be abnormal because of the cloudy urine? Choose all correct answers.

   a. Leukocytes
   b. Glucose
   c. pH

3. If the filtration membrane of Bowman's capsule is healthy and intact, how can erythrocytes still be found in the urine?

   a. Erythrocytes can be excreted at the distal convoluted tubules.
   b. Erythrocytes are both reabsorbed and excreted at the nephron loop.
   c. Damage to the ureters, bladder, or urethra can introduce erythrocytes.

4. Compare the specific gravity of the following three urine samples. Which sample contains a greater proportion of water compared to the other two?

   Sample A = specific gravity of 1.033
   Sample B = specific gravity of 1.008
   Sample C = specific gravity of 1.019
   a. Sample A
   b. Sample B
   c. Sample C

5. You are examining a slide of urine sediment. Order the following particles from smallest to largest: erythrocytes, bacteria, epithelial cast

6. The presence of squamous epithelial cells in urine sediment occurs as:

   a. Wide filtration slits allow formed elements through
   b. The cells lining the ureters slough off
   c. The filtrate moves through the distal convoluted tubule

7. The dipstick analysis below produces positive results for protein and ketones. In addition, you might also expect to find which of the following sediments in the urine?

   a. Uric acid crystals
   b. Bacteria
   c. Leukocytes

Glucose	neg
Bilirubin	neg
Ketone	40 mg/dL
SG	1.10
Blood	neg
pH	5.0
protein	100 mg/dL
urobilinogen	0.2 mg/dL
leukocytes	neg

8. Which of the following characteristics from a clean catch urine sample are considered normal? Assume the individual is not menstruating at the time of the sample. Choose all correct answers.

   a. Small numbers of leukocytes

   b. Small numbers of epithelial cells

   c. Small numbers of erythrocytes

9. What abnormal urinalysis result(s) would you expect from a patient with liver damage due to hepatitis?

   a. Bilirubinuria and urobilinogenuria

   b. Proteinuria (albuminuria) and ketonuria

   c. Hematuria and pyuria

10. Abnormal urinalysis results showing 500 mg/dL of glucose, 80 mg/dL of ketones, and a pH of 5.0 indicate a possible disease of:

    a. Kidney damage due to kidney disease

    b. An infection of the urinary tract

    c. Uncontrolled diabetes mellitus

Lab 33 Reproduction

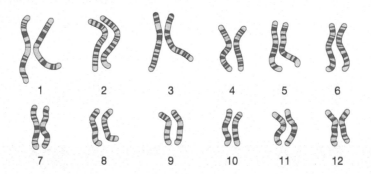

1 2 3 4 5 6

7 8 9 10 11 12

Learning Objectives: After completing this lab, you should be able to:

33.1 Compare and contrast the major anatomy of the male and female reproductive systems.

33.2 Identify male and female homologues of various reproductive system structures (e.g., ovary is homologous to the testis).

33.3 Identify and describe the gross anatomy, microscopic anatomy, and functions of the testes.

33.4 Describe the pathway of sperm from the seminiferous tubules to the external urethral orifice of the penis.

33.5 Identify and describe the gross anatomy, microscopic anatomy, and functions of the ovaries.

33.6 Identify and describe the gross anatomy, microscopic anatomy, and functions of the uterus and uterine (fallopian) tubes.

33.7 Identify and describe the structure and function of the vagina.

33.8 Identify and describe the structure and functions of the external genitalia (e.g., mons pubis, labia majora, labia minora, clitoris, greater vestibular glands).

33.9 Describe the pathway of the oocyte from the ovary to the uterus.

33.10 Define the processes of spermatogenesis and spermiogenesis.

33.11 Describe the stages of spermatogenesis in the seminiferous tubule, including the roles of the nurse (sustentacular, Sertoli) cells and interstitial cells (of Leydig).

33.12 Define the process of oogenesis (oocyte development).

33.13 Compare and contrast somatic cell division (mitosis) and reproductive cell division (meiosis).

33.14 Describe the functions of the hormones involved in the regulation of the reproductive processes (e.g., gonadotropin-releasing hormone [GnRH], follicle-stimulating hormone [FSH], luteinizing hormone [LH], androgens, inhibin, estrogens, progesterone).

33.15 Define ovulation, and explain the role of luteinizing hormone (LH) in ovulation.

33.16 Describe endocrine regulation of oogenesis, folliculogenesis, and the ovarian cycle.

33.17 Describe the functional changes in the woman's body during pregnancy.

The Human Anatomy and Physiology Society includes more than 1,700 educators who work together to promote excellence in the teaching of this subject area. The HAPS A&P Learning Outcomes measure student mastery of the content typically covered in a two-semester Human A&P curriculum at the undergraduate level. The full Learning Outcomes are available at https://www.hapsweb.org.

33.18 Given a factor or situation (e.g., ectopic pregnancy), predict the changes you could see in the reproductive system and the consequences of those changes (i.e., given a cause, state a possible effect).

33.19 Identify and describe the structure and functions of the male external genitalia (e.g., scrotum, penis).

33.20 Identify and describe the structure and functions of accessory glands (i.e., seminal glands [seminal vesicles], prostate gland, bulbourethral [Cowper] glands).

33.21 Describe the correlation between the uterine and ovarian cycles.

33.22 Given a disruption in the structure or function of the reproductive system (e.g., vasectomy), predict the possible factors or situations that might have caused that disruption (i.e., given an effect, predict possible causes).

Introduction

Reading through the list of learning objectives from the Human Anatomy and Physiology Society that open this chapter you may notice two words threaded throughout: female and male. These terms describe the sex of an individual. **Sex**, sometimes called anatomical or biological sex, can include considerations of a person's anatomy, chromosomes, or hormones. Sex may be difficult to put into absolute terms, given the wide diversity of anatomical and hormonal arrangements among humans. Biological sex is very different from gender or gender expression. A person's gender (sometimes called gender identity) reflects who a person knows themselves to be. **Gender** may or may not match biological sex, and it also may not fit neatly our two biological categories. In most cases, sex is assigned at birth by the attending medical team, based entirely on external genitalia. The development of these anatomical structures is based largely on hormone signaling; that is, both the levels of pertinent hormones as well as the expression of their receptors. The influence of hormone signaling on anatomy does not stop at birth, rather, hormones and their receptors continue to influence anatomical structures both within and outside of the reproductive tract throughout the lifespan.

In this lab we will explore a typical arrangement of anatomical structures for each biological sex, as well as the development of gametes, the influence of hormones and the functional changes that occur during pregnancy.

Lab 33 Pre-Lab Quiz

This quiz will strengthen your background knowledge in preparation for this lab. For help answering the questions, use your resources to deepen your understanding. The best resource for help on the first five questions is your text, and the best resource for help answering the last five questions is to read the introduction section of each lab.

1. Anatomy of a mature follicle. Label the follicle drawing with the following terms:

   antrum
   granulosa cells
   oocyte
   thecal cells
   zona pellucida

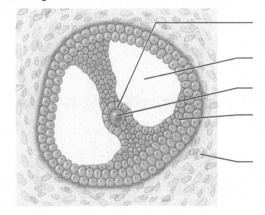

2. Match the root next to its definition:

Root	Meaning
	egg
	little bag
	lips
	same

List of Roots

follicle
homo-
labia
oo-

3. Which of the following is true of mitosis and meiosis?
   a. Both processes begin with diploid cells.
   b. Meiosis is the only process that can produce eggs and sperm.
   c. Mitosis is essential for growth and repair, but meiosis is not.
   d. All of the above are true.

4. The target tissue of a hormone is defined as a tissue that expresses:
   a. The receptor for the hormone
   b. The amino acids required to make the hormone
   c. The organelles required to assemble the hormone

5. Which of the following is the best definition for the term tropic hormone?
   a. A tropic hormone is a hormone that causes a response in another endocrine gland.
   b. A tropic hormone is a hormone that causes a response within the reproductive system.
   c. A tropic hormone is a hormone that inhibits its target tissue from making or secreting any cellular products.

6. Which of the following hormones is released by the anterior pituitary?
   a. Gonadotropin-releasing hormone (GnRH)
   b. Follicle-stimulating hormone (FSH)
   c. Estrogen
   d. Testosterone

7. Which of the following directional terms would most accurately describe the position of the penis relative to the urinary bladder?
   a. Anteroinferior
   b. Anterosuperior
   c. Distal

8. Sperm are produced through the process of _____ which occurs in the _____.
   a. mitosis; epididymis
   b. mitosis; seminiferous tubules
   c. meiosis; epididymis
   d. meiosis; seminiferous tubules

9. Match each of the following structures with its homologue found in the opposite sex:

Matching Structure	Structure Found in One Sex
	uterine tube
	testes
	penis

Strucutre Found in Opposite Sex
clitoris
ductus deferens
ovaries

10. As a pregnancy progresses and the uterus increases in size, it pushes superiorly more than inferiorly. This is due to:
    a. The pubic symphysis prevents the uterus from expanding inferiorly
    b. The abdominal muscles are stronger inferiorly than superiorly
    c. The embryo always implants on the fundus of the uterus

Activity 33.1 Identifying Reproductive Structures on Anatomical Models

This activity targets LOs 33.1, 33.2, 33.3, 33.4, 33.5, 33.6, and 33.7.

In all embryos, the structures known as **gonads** develop in the abdominopelvic cavity. Gonads are structures that will eventually be capable of **gametogenesis**, the production of **gametes**. The **testes** are gonads that generate sperm (Figure 33.1). The **ovaries** are gonads that generate eggs (also called oocytes). During development, usually shortly before or after birth, the testes emerge through small portals in the anterior body, each known as an **inguinal canal**. From this point on, the testes are housed in a pouch of skin known as the **scrotum** and remain outside the body cavity. In contrast, the ovaries remain within the body cavity throughout life. The ovaries and testes are examples of **homologous structures**, structures that have the same function but have different anatomies.

The internal structures of ovaries and testes are examined in more detail in Activity 33.2.

Figure 33.1   **The Testes are Gonads, Housed within the Scrotum**

The gametes (sperm) are released from the testes and travel within the ductus deferens to exit the body at the tip of the penis. The structures involved in spermatogenesis, spermiogenesis and semen development can be seen both in a drawing (A) and laboratory model (B) viewed from a midsagittal perspective.

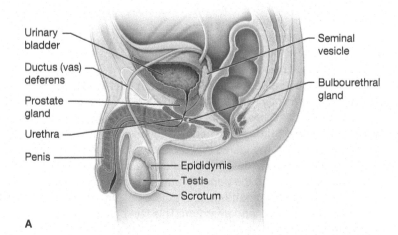

A

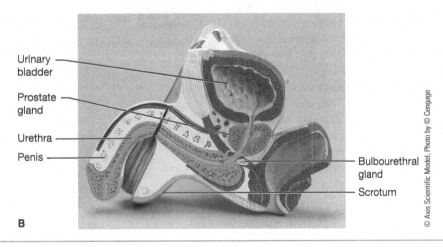

B

It connects to the vagina by way of the cervix. The uterine wall has three layers, the endometrium, myometrium, and perimetrium. On either side of the uterus are the two ovaries. The uterine tubes allow for travel of the egg from ovary to uterus.

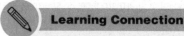
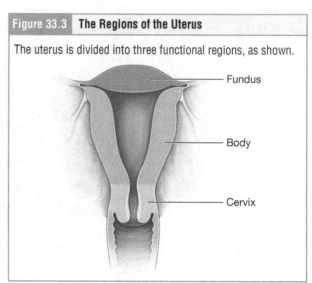

Uterine tube
Ovary
Uterus
Lumen of uterus
Endometrium
Myometrium
Perimetrium
Cervix
Vagina

ericsphotography/E+/Getty Images

In all bodies, the gametes must have a path to get to the site of fertilization and outside the body. Eggs are released from the ovaries and into **uterine tubes** which span the distance between the ovaries and the **uterus**, which is in the midline of the body. The uterus houses developing embryos or fetuses when pregnancy occurs (Figure 33.2). Sperm travel through two tubes, each known as a **ductus deferens**, which merge to join the **urethra** (Figure 33.1). The uterine tubes and ductus deferens are examples of homologous structures.

The uterus is a muscular organ located in the pelvic cavity posterior to the bladder. While superiorly it connects to the ovaries, inferiorly it connects to the vagina by way of the cervix.

The internal structure of the uterus includes three regions and three layers. The region of the uterus superior to the opening of the uterine tubes is called the **fundus**. The middle section of the uterus is called the **body** of the uterus. The cervix is the narrow inferior portion of the uterus that projects into the vagina (Figure 33.3). The **cervix** is a narrow channel through which material can pass between the uterus and the vagina. The wall of the uterus is made up of three layers. The most superficial layer is the **perimetrium**, a serous membrane that covers the exterior uterus (Figure 33.2). The middle layer is the **myometrium**, a thick layer of smooth muscle which can contract rhythmically. The innermost layer of the uterus is the **endometrium**. This layer grows through proliferation and then is shed in cyclical fashion throughout the reproductive years.

In biological males, the testes are housed within the scrotum instead of within the pelvic cavity where ovaries are found. The reason for this is that the optimal sperm production temperature is actually 35°C (93.2°F), two degrees lower than the core body temperature of 37°C. In order to maintain this ideal temperature, the testes are raised closer to the pelvis or lowered further from the pelvis depending on body and external temperatures. The two muscles that do this raising and lowering for temperature regulation are the **dartos muscle** and the **cremaster muscle** (the dartos is external and the cremaster is internal). These muscles may be visible on your lab model and can also be viewed in Figure 33.4B.

Deep to these muscles, the surface of the testes can be seen. Along their posterior surface is a structure known as the **epididymis**, where sperm mature. Inferiorly, the epididymis releases sperm into the ductus deferens, discussed earlier (Figure 33.4A).

Learning Connection

Try Drawing It

Try drawing the path that sperm take from their development in the testes to their exit out of the body. Label all the structures as you go.

Figure 33.3 | The Regions of the Uterus

The uterus is divided into three functional regions, as shown.

Fundus
Body
Cervix

Figure 33.4 | The Reproductive Anatomy of a Biological Male

The testes are housed within the scrotum. When sperm are released from the testes they travel within the ductus deferens to exit the body. (A) Midsagittal view of biological male sexual system. (B) Anterior view of external male sexual organs. (C) Cross-section through the shaft of the penis.

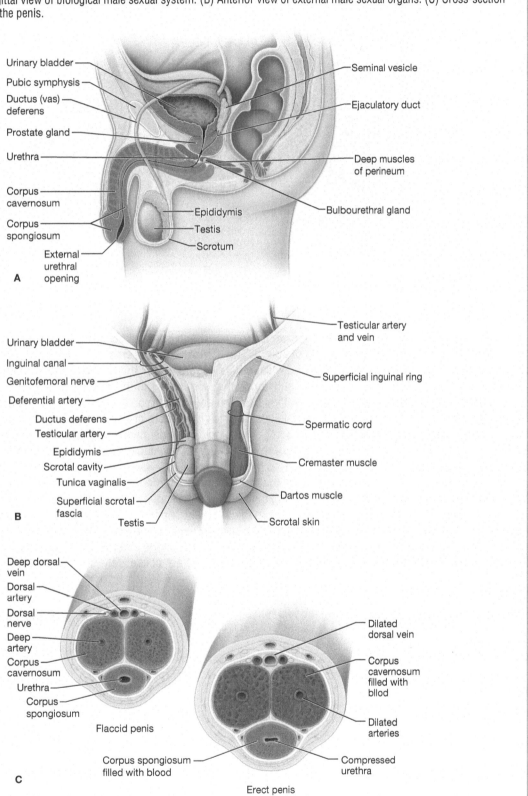

A
- Urinary bladder
- Pubic symphysis
- Ductus (vas) deferens
- Prostate gland
- Urethra
- Corpus cavernosum
- Corpus spongiosum
- External urethral opening
- Epididymis
- Testis
- Scrotum
- Seminal vesicle
- Ejaculatory duct
- Deep muscles of perineum
- Bulbourethral gland

B
- Urinary bladder
- Inguinal canal
- Genitofemoral nerve
- Deferential artery
- Ductus deferens
- Testicular artery
- Epididymis
- Scrotal cavity
- Tunica vaginalis
- Superficial scrotal fascia
- Testis
- Testicular artery and vein
- Superficial inguinal ring
- Spermatic cord
- Cremaster muscle
- Dartos muscle
- Scrotal skin

C
- Deep dorsal vein
- Dorsal artery
- Dorsal nerve
- Deep artery
- Corpus cavernosum
- Urethra
- Corpus spongiosum
- Flaccid penis
- Corpus spongiosum filled with blood
- Dilated dorsal vein
- Corpus cavernosum filled with bllod
- Dilated arteries
- Compressed urethra
- Erect penis

Figure 33.5 | The Vulva

The term vulva encompasses all of the external female genital structures: the clitoris, the labia majora, and the labia minora. (A) Superficial view of the external female reproductive organs. (B) Enlarged view of some of the external female reproductive structures.

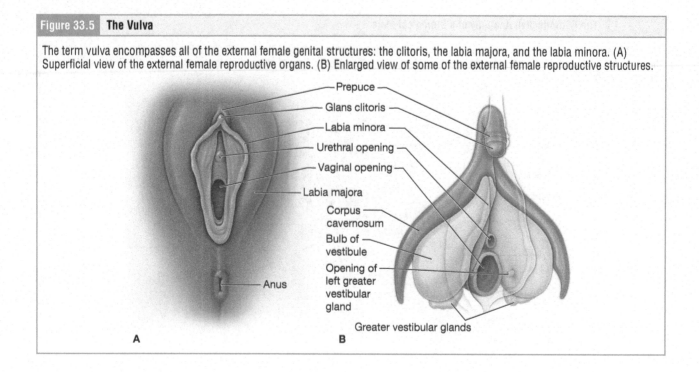

Prepuce
Glans clitoris
Labia minora
Urethral opening
Vaginal opening
Labia majora
Corpus cavernosum
Bulb of vestibule
Opening of left greater vestibular gland
Anus
Greater vestibular glands

A   B

Sperm exit the body within a mixture known as semen, which is a liquid composed of sperm along with additions such as fructose that support the life of sperm. The components of semen are produced by three glands positioned along the reproductive tract of biological males (Figure 33.1). The **prostate gland**, **seminal vesicle**, and **bulbourethral glands** each make a different substance that contributes to the mixture of semen.

The external genitalia in both sexes develops from the same embryonic structures and so can be considered a set of homologues. Both sexes develop erectile tissue which is capable of swelling during arousal. A cross-section of these erectile tissues can be seen in Figure 33.4C. In biological males, this tissue comprises the **penis**, in biological females it is mostly an internal structure, with only a small portion available externally, and is known as the **clitoris**. The clitoris may or may not be visible on your lab model.

Embryos have two external folds of tissue that hug the anterior/inferior curve of the pelvis. In biological males these folds fuse and become the scrotum. Along the midline of the adult scrotum a line remains where the labia are fused (Figure 33.4). In biological females, these folds remain separate, where they are called **labia**. Two sets of labia (minor and major) surround the opening to the muscular vagina. The **vagina** serves as the entrance/exit to the internal reproductive structures. The internal erectile tissue of the clitoris straddles the vaginal opening (Figure 33.5). These external structures (clitoris, labia, and associated glands) are collectively called the **vulva**.

Materials
• Male and female reproduction system models

Instructions
Your lab may be equipped with a variety of model sources for this activity. You also can use the images and information in this lab manual as well as your textbook for references.

1. In Table 33.1, identify each of the structures listed. List the sources (textbook, name of model, etc.) and write a brief description of the location and function of the structure.

Table 33.1 Important Reproductive Structures		
Structure	**Location**	**Function**
Body of uterus	*Lateral to the uterus*	*oogenesis*
Bulbourethral gland		
Cervix		
Clitoris		
Cremaster muscle		
Dartos muscle		
Ductus deferens		
Endometrium		
Epididymis		
Fundus of uterus		
Inguinal canal		
Labia		
Myometrium		
Ovaries		
Penis		
Penis glans		
Penis root		
Penis shaft		
Perimetrium		
Prostate gland		

(continued)

Table 33.1 Important Reproductive Structures (*Continued*)		
Structure	**Location**	**Function**
Raphe		
Scrotum		
Seminal vesicle		
Testes		
Urethra		
Uterine tubes		
Uterus		
Vagina		

2. Let's consider where an oocyte travels following ovulation.
   2a. In Table 33.2, list the structures, in order, through which an unfertilized oocyte will pass after ovulation, beginning at the ovary and ending at the vulva.
   2b. On a blank piece of paper, draw the path that an oocyte takes, based on your list in Table 33.2.

3. Let's consider where sperm travel after leaving the seminiferous tubules.
   3a. In Table 33.2, list the structures, in order, through which sperm pass after they leave the seminiferous tubules, beginning at the lumen of the seminiferous tubule and ending at the external opening on the glans of the penis.
   3b. On a blank piece of paper, draw the path that sperm take, based on your list in Table 33.2.

4. Consider the following as you explore the models and diagrams.
   4a. A vasectomy is a surgical procedure in which each ductus (vas) deferens is cut to prevent pregnancy. Most commonly, the ductus deferens is cut at its most superficial point at either side of the base of the penis shaft. Identify on the models or diagrams where this would be. Mark this location on the drawing you made in step 3. Questions 1 and 2 in your post-lab quiz relate to this exploration.
   4b. A tubal ligation is a surgical procedure in which each of the uterine tubes are cut to prevent pregnancy. Identify the uterine tubes and their surrounding structures. Mark this location on the drawing you made in step 2. Question 3 in your post-lab quiz relates to this.

Table 33.2 Pathways of an Egg and Sperm	
Structures an Egg Passes Through	**Structures Sperm Pass Through**

Digging Deeper:
Pelvic Inflammatory Disease (PID)

The uterine tubes are lined with both smooth muscle and a ciliated epithelium. The smooth muscle contractions and rhythmic beating of the cilia work together to propel eggs in one direction, from ovary to uterus. The ovarian ends of the uterine tubes are open to the abdominal cavity, so the forces that contribute to flow away from the open ends are instrumental in keeping material, such as fertilized eggs, out of the abdominal cavity. Like all tissues, the lining of the uterine tubes can become inflamed in response to infection. This inflammation is called pelvic inflammatory disease (PID). Any introduction of bacteria into the reproductive tract could lead to PID, but the sexually transmitted bacterial infections chlamydia and gonorrhea cause 85% of PID cases. When treated promptly, there are rarely long-lasting side effects from chlamydia or gonorrhea infection, but if left untreated for a period of time, the effects of inflammation can be long-lasting. After the inflammation, scar tissue is often left behind. This tissue does not have cilia and is often thicker than the healthy epithelium, reducing the impact of the smooth muscle contractions. Together these effects limit the motility of eggs within the uterine tubes, allowing eggs to become stuck in the tube, or flow backwards into the abdominal cavity. If eggs are fertilized but are unable to travel to the uterus for implantation, they may implant into the wall of the uterine tube or on the outside of structures in the abdominal cavities. These pregnancies are called ectopic pregnancies and 0% of them are viable. The developing embryo and its placenta must be removed before they cause damage to the non-uterine structure they've implanted in. If the placenta proceeds in development, bleeding, sometimes lethal bleeding, can occur. In addition to ectopic pregnancy, PID and its resulting scar tissue can cause infertility by preventing sperm from reaching released eggs. Another consequence to the loss of motility can be endometriosis, where sloughed off endometrial tissue travels up through the uterine tubes instead of out through the vagina. The endometrial tissue can become lodged in the tubes or on the outside of structures in the abdominal cavity and can remain there for years.

(page intentionally left blank)

Activity 33.2 Histology of Gametogenesis

This activity targets LOs 33.8, 33.9, 33.10, and 33.11.

Most cells in the human body have 46 chromosomes. There are 23 pairs of chromosomes, and one in each pair is inherited from each parent. Figure 33.6 shows all 46 chromosomes isolated from a human patient and arranged in pairs. You are not an exact copy of either of your parents, and your siblings (if you have any) are not genetically identical to you (unless you have an identical twin) because of the way that chromosomes are mixed during gametogenesis and reproduction.

Figure 33.6	Humans Have 46 Individual Chromosomes

There are 23 unique pairs, in each pair, one chromosome was inherited via sperm and the other was inherited via the egg. Each of the two chromosomes in a pair contain the same genes, but may contain different versions of those genes. The last pair, the 23rd pair, can either contain two X chromosomes or one X chromosome and one Y chromosome.

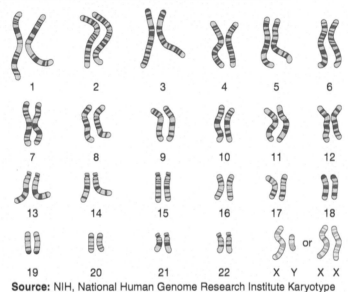

Source: NIH, National Human Genome Research Institute Karyotype

Gametes, eggs and sperm, are produced through the process of meiosis. Meiosis separates the pairs of chromosomes so that each of the gametes that results has only one copy of each chromosome (Figure 33.7). Because the chromosomes sort into different combinations each time meiosis occurs, no two gametes are exactly identical.

Meiosis

Meiosis is summarized in Figure 33.8. Meiosis begins with a stem cell that has a full set of (46) chromosomes. The term **diploid**, or **2n**, refers to a set of chromosomes that contains two of each pair. A cell that has a single representative chromosome of each pair, for example one chromosome 1, one chromosome 2, and so on, is called **haploid** or **1n**. While most of the cells in your body (called somatic cells) are diploid, eggs and sperm are haploid. The two chromosomes within a pair contain the same genes, and yet are slightly different from each other. Therefore, we refer to them as homologous, just as we use that word to describe anatomical structures that have the parts arranged slightly differently like the scrotum and labia.

The process of meiosis can be divided into two halves, called meiosis I and meiosis II. Each half has the same phases as mitosis, prophase, metaphase, anaphase, telophase, and cytokinesis. In the first half, meiosis I, a diploid cell begins by replicating its chromosomes, just as it would in mitosis. The identical, freshly replicated chromosomes are held together in an X-shape and are paired with their homologue. Each of the homologous pairs exchanges a bit of the chromosome, swapping chunks that contain the same genes. This process of swapping between homologues is called **crossing over**. The resulting chromosomes are now a blend of genes from the original chromosomes. Crossing over provides infinite recombination potential, ensuring that no two gametes produced by one individual

Figure 33.7 Meiosis Produces Haploid Cells

During the process of meiosis, a stem cell containing all 46 chromosomes divides in such a way that it produces daughter cells that each contain 23 unique chromosomes (no pairs).

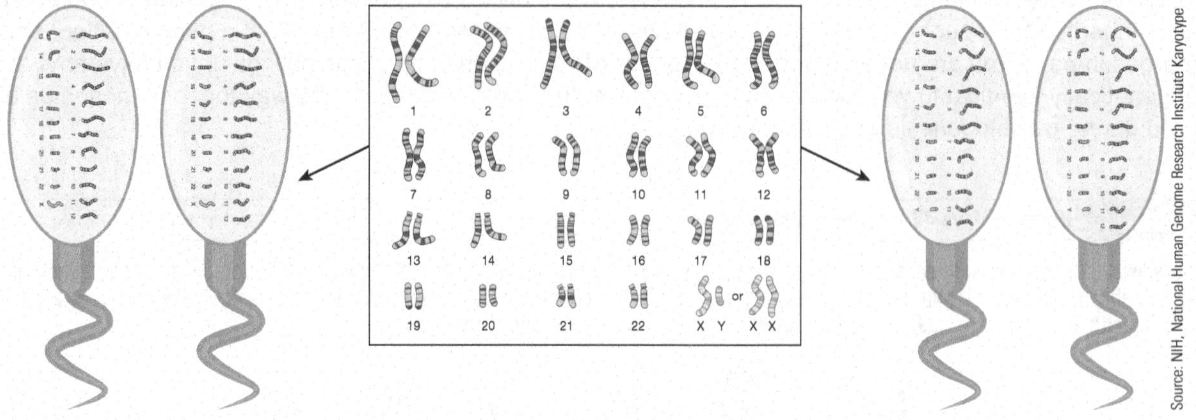

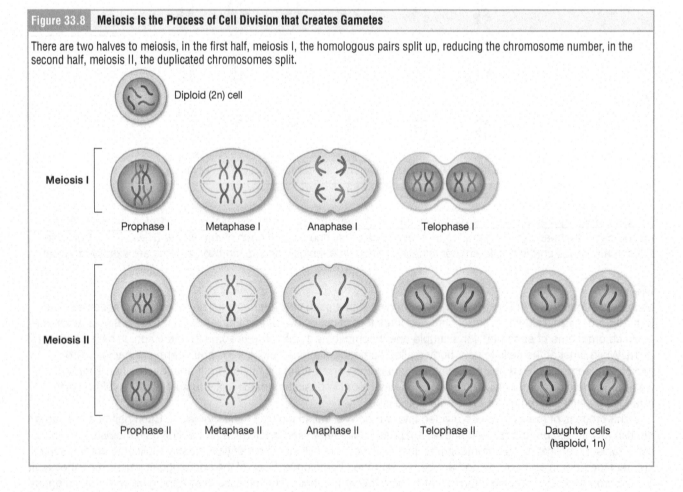

1
2
3
4
5
6
7
8
9
10
11
12
13
14
15
16
17
18
19
20
21
22
X Y
or
X X
Source: NIH, National Human Genome Research Institute Karyotype

Figure 33.8 Meiosis Is the Process of Cell Division that Creates Gametes

There are two halves to meiosis, in the first half, meiosis I, the homologous pairs split up, reducing the chromosome number, in the second half, meiosis II, the duplicated chromosomes split.

Diploid (2n) cell
Meiosis I
Prophase I
Metaphase I
Anaphase I
Telophase I
Meiosis II
Prophase II
Metaphase II
Anaphase II
Telophase II
Daughter cells (haploid, 1n)

Figure 33.9 **Spermatogenesis**

The development of sperm from stem cells takes place in the seminiferous tubules.

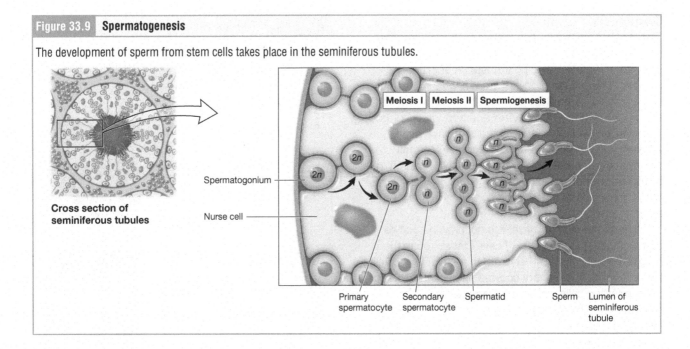

Cross section of seminiferous tubules

Spermatogonium

Nurse cell

Meiosis I | Meiosis II | Spermiogenesis

Primary spermatocyte

Secondary spermatocyte

Spermatid

Sperm

Lumen of seminiferous tubule

will ever be exactly the same. After crossing over, meiosis I proceeds with the separation of the homologues. The homologues separate randomly, insuring a different assortment of chromosomes each time. Each of the two resulting cells now has one chromosome (though it is replicated) from each pair. These cells are haploid. In meiosis II, the haploid cells separate the replicates of their chromosomes.

Spermatogenesis

The development of sperm through meiosis is a process known as **spermatogenesis**. It is summarized in Figure 33.9. The development of eggs through meiosis is a process known as **oogenesis**. It is summarized in Figure 33.10. Spermatogenesis and oogenesis happen in very different ways in humans. Oogenesis begins before birth in fetal ovaries.

Figure 33.10 **Oogenesis**

The mitotic events that generate eggs are completed before birth and all the eggs are arrested in development. At the onset of puberty, meiosis begins again and is only completed in eggs that are fertilized.

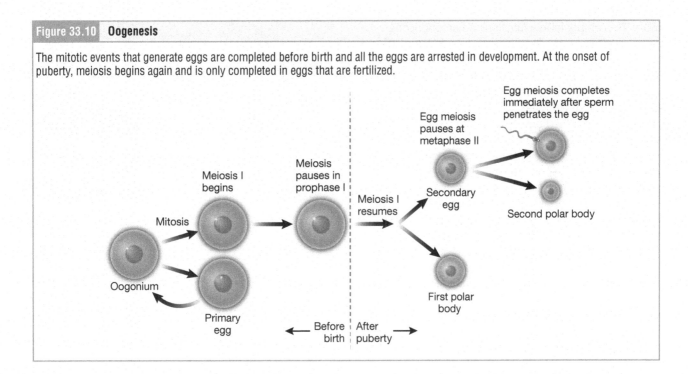

Egg meiosis completes immediately after sperm penetrates the egg

Egg meiosis pauses at metaphase II

Meiosis I begins

Meiosis pauses in prophase I

Meiosis I resumes

Secondary egg

Second polar body

Mitosis

Oogonium

Primary egg

First polar body

Before birth | After puberty

Oogenesis

Oogenesis begins with ovarian stem cells. During the fetal period, prior to birth, the stem cells begin the process of meiosis, forming thousands of **primary oocytes**. The primary oocytes are stopped in this stage of meiosis I. The partially developed primary oocytes are present in the ovary at birth and through childhood. The **ovarian cycle**, which occurs about once every 28 days beginning at puberty and continuing until menopause (the cessation of reproductive function in biological women) will stimulate a small group of these primary oocytes to mature and complete meiosis I and go on to meiosis II.

During the ovarian cycle, it is not only the oocyte that changes, but the group of cells that surround the oocyte, collectively called an **ovarian follicle**. The follicle that surrounds the primary oocytes has a single layer of flat supporting cells, called **granulosa cells**. During the ovarian cycle, the oocyte swells and proceeds further through the stages of meiosis and the granulosa cells proliferate, creating an elaborate structure around the egg. As the granulosa cells divide, the follicles increase in diameter, adding a new outer layer of connective tissue, blood vessels, and **thecal cells**—cells that work with granulosa cells to produce estrogens. This growing follicle is now called a secondary follicle. Deep inside the secondary follicle the oocyte secretes a thin jelly-like coat called the **zona pellucida**. A thick fluid, called **follicular fluid**, gathers within the follicle. Follicles that reach a stage in which the pool of fluid, called the **antrum**, is large and fully formed are called **tertiary follicles**.

Once the egg is released from the ovary, the granulosa cells and thecal cells are left behind in the ovary. This empty follicle functions as an endocrine gland, releasing hormones into the bloodstream. It is called a **corpus luteum**. If the egg is fertilized, the developing embryo will send a hormonal signal to the corpus luteum which encourages it to remain an active endocrine gland, secreting both estrogen and progesterone for the beginning of the pregnancy. If the egg is not fertilized and no signal is sent to the corpus luteum, it will degrade slowly over the course of a few weeks. Figure 33.11 shows examples of the stages of follicle development that you might see in the lab.

In contrast to oogenesis, which happens in a first phase before birth and then small batches throughout the reproductive years, spermatogenesis is an ongoing process that initiates in puberty and continues for decades. Spermatogenesis specifically refers to the meiotic process of creating haploid cells. These haploid cells, though, are not yet ready to fertilize an egg. First, they must mature and differentiate to become cells capable of movement and digestion of the egg's zona pellucida. **Spermiogenesis** is the separate process of

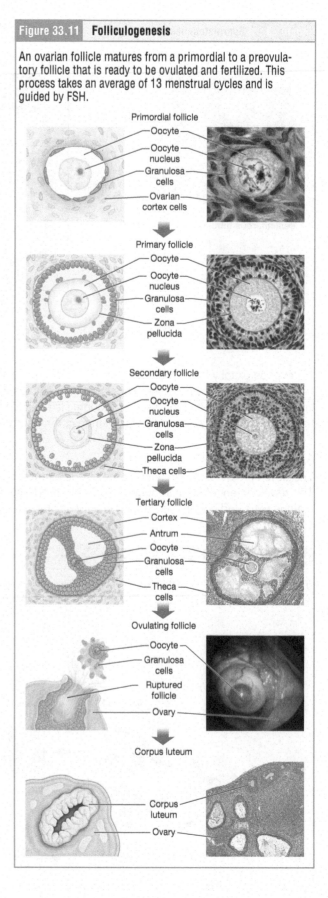

Figure 33.11 **Folliculogenesis**

An ovarian follicle matures from a primordial to a preovulatory follicle that is ready to be ovulated and fertilized. This process takes an average of 13 menstrual cycles and is guided by FSH.

Primordial follicle
- Oocyte
- Oocyte nucleus
- Granulosa cells
- Ovarian cortex cells

Primary follicle
- Oocyte
- Oocyte nucleus
- Granulosa cells
- Zona pellucida

Secondary follicle
- Oocyte
- Oocyte nucleus
- Granulosa cells
- Zona pellucida
- Theca cells

Tertiary follicle
- Cortex
- Antrum
- Oocyte
- Granulosa cells
- Theca cells

Ovulating follicle
- Oocyte
- Granulosa cells
- Ruptured follicle
- Ovary

Corpus luteum
- Corpus luteum
- Ovary

sperm maturation that takes place in the epididymis. In this activity in the lab, we will explore only spermatogenesis, which takes place in the testes.

Two types of cells in the testes collaborate in the making of sperm. **Interstitial cells,** found outside the seminiferous tubules, are responsible for generating androgens—especially testosterone, which drives spermatogenesis and spermiogenesis forward. Inside the seminiferous tubules we can find **nurse cells**, which directly support the process of making sperm. Inside the seminiferous tubules we can also find developing sperm cells. As these cells develop they progress from the basement membrane—at the perimeter of the tubule—toward the lumen (Figure 33.12).

Figure 33.12 | **Interstitial Cells**

There are two types of cells that are central to the functions of the testes. Interstitial cells, which lie between the seminiferous tubules, produce testosterone. Nurse (sustentacular) cells line the walls of the seminiferous tubules and support the maturation of sperm.

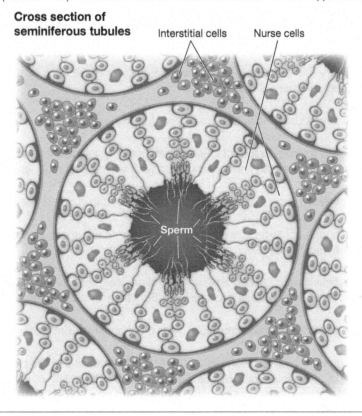

The stem cells that give rise to sperm are **spermatogonia** (singular = spermatogonium). They are found lining the basement membrane inside the tubule (Figure 33.9). Spermatogonia undergo mitosis (not meiosis). Mitosis results in two diploid daughter cells. One of these remains a spermatogonium and will be a stem cell for future use. The other becomes a **primary spermatocyte**, which undergoes meiosis to yield four haploid gametes called **spermatids**. Although they have reduced their chromosome number, early spermatids look very similar to cells in the earlier stages of spermatogenesis, with a round shape, central nucleus, and large amount of cytoplasm. Spermiogenesis transforms these early spermatids into sperm by reducing the amount of cytoplasm and elongating the flagellum in preparation for swimming. At the end of this process the formed sperm can be found within the wall of the seminiferous tubule nearest the lumen. Eventually the sperm will be released into the lumen and are moved along a series of ducts toward the epididymis where they will mature.

Materials

- Model of meiosis
- Slide of ovary
- Slide of seminiferous tubules

Structures Identified in This Activity

Antrum
Corpus luteum
Granulosa cells
Interstitial cells
Nurse cells

Oocyte/egg
Ovarian follicle
Spermatogonia
Zona pellucida

Instructions

1. Examine a model of meiosis or Figure 33.8. Identify the cells in each stage and designate them as haploid (1n) or diploid (2n).

2. Obtain a slide of an ovary and examine it under the lowest power of the microscope. Typically, the majority of follicles will be along the periphery of the ovary.

3. Increase magnification, refocusing each time, to 40× or the highest your microscope can magnify.

4. Using Figure 33.11, identify as many stages of follicle development as you can. You may benefit from examining the microscopes of other students near you and showing them what you have found in yours.

5. If this ovary recently ovulated, a corpus luteum would be visible, can you find one on your ovary slide?

6. In Table 33.3, draw each type of follicle you were able to identify. In the space provided, indicate whether a follicle in this stage is haploid or diploid.

7. Obtain a slide of the seminiferous tubules. Examine the slide using the lowest power objective to see how many cross-sections of tubule you can fit into one view. Find the interstitial cells.

8. Increase magnification, refocusing each time, to 40× or the highest your microscope can magnify.

9. Identify the wall and lumen of the seminiferous tubule. Compare the cells at the periphery of the wall, to those closer to the lumen. Examine their nuclei. Notice the differences in how dark or dense their nuclei appear. Which cells have darker nuclei, those closer to the lumen or the ones closer to the periphery? What might this indicate about their DNA?

10. Compare the size of the cells at the periphery and those closer to the lumen. Where are the larger cells?

Table 33.3 Development of an Oocyte	
Primary Oocyte	**Secondary Oocyte**
Tertiary Oocyte	**Corpus Luteum**

(page intentionally left blank)

Activity 33.3 Hormones of Reproduction

This activity targets LOs 33.12 and 33.13.

The development of follicles, release of the egg in ovulation, the readying of the uterine lining for pregnancy, whether menstruation occurs or not, all of these reproductive functions as well as others are controlled through hormones.

The hypothalamus produces gonadotrophin-releasing hormone (GnRH), a hormone that signals the anterior pituitary gland to produce the hormones follicle-stimulating hormone (FSH) and luteinizing hormone (LH) (Figure 33.13). These hormones leave the pituitary and travel through the bloodstream to the ovaries, where they bind to receptors on the granulosa cells and thecal cells of the follicles. FSH stimulates the follicles to grow, and a handful of tertiary follicles expand in size. The release of LH also stimulates the granulosa cells and thecal cells of the follicles to produce the sex steroid hormone estradiol, a type of estrogen. This phase of the ovarian cycle is known as the follicular phase.

The larger and more mature a follicle is, the more estradiol it will produce in response to LH stimulation. As a result of these large follicles producing large amounts of estradiol, blood levels of estradiol increase. More than one follicle grows and matures at a time, in fact about 15–20 follicles mature during each ovarian cycle, therefore the levels of estradiol during the follicular phase become quite high. These large doses of estradiol in the blood cause the anterior pituitary to release a huge quantity of LH in a surge and this wave of LH triggers the largest of the follicles to release its egg, a process called ovulation. Ovulation occurs in the middle of the menstrual cycle, which would be around day 14 for a 28-day cycle. Note that 28 days is the average, but is by no means typical, menstrual cycles can vary widely among healthy individuals.

Figure 33.13	The Follicular Stage

FSH drives the development of follicles in the ovary. As they mature, they contribute more and more estradiol to the bloodstream.

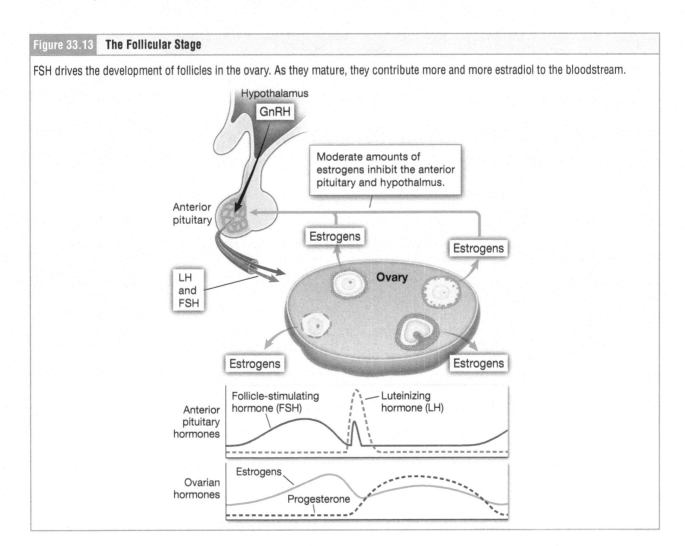

After the egg has been released, most of the granulosa cells and all of the thecal cells remain in the ovary as the corpus luteum, an endocrine gland. They continue to secrete estrogen and now secrete progesterone as well. The corpus luteum will stay in the ovary for about 2 more weeks, secreting hormones. It awaits a signal, the hormone, **human choriogonad-otropin (hCG)**. This hormone is secreted by embryos in the event that the egg is fertilized. In the absence of fertilization, and therefore the absence of hCG, the corpus luteum will begin to degrade and secrete less and less estradiol and progesterone into the bloodstream. As levels of these hormones fall, the endometrium will begin to breakdown and slough off and menstruation occurs.

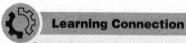
Instructions

Use the graphs in the figure below to answer the following seven questions.

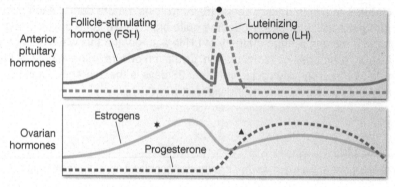

Questions

1. In this figure at the time of the ✳ which structures are secreting estrogens?
   a. Only one corpus luteum
   b. Many corpora lutea
   c. Only one developing follicle
   d. Many developing follicles

2. At the triangle, the concentration of both hormones in the bloodstream are low. What occurs at this time?
   a. Menstruation
   b. Ovulation
   c. Follicle growth and maturation

3. In this figure, at the time marked with the circle, you can see the levels of LH are very high, but then they are low for the remainder of the menstrual cycle. What is the function of LH?
   a. Follicle growth and maturation
   b. Uterine lining growth and maturation
   c. Preparation of the breasts for lactation
   d. Ovulation

4. What causes the increase in LH secretion that occurs at the time marked with the circle?
   a. An increase in progesterone secretion
   b. The presence of hCG in the bloodstream
   c. An increase in estrogen secretion
   d. All of these need to occur simultaneously for the increase in LH to happen

5. If fertilization occurs, the developing embryo secretes a hormone called human choriogonadotropin, hCG. The main target tissue of hCG is the corpus luteum in the ovary. A home pregnancy test looks for the presence of hCG. Examine the graph on page 854. At what time in pregnancy would hCG be highest?

   a. At fertilization (around day 14–21 of the menstrual cycle)

   b. At 2 weeks of pregnancy (around day 28 of the menstrual cycle)

   c. At 6 weeks of pregnancy

   d. At 4 months of pregnancy

6. What would a home pregnancy test—like the one in the figure—show if taken by a person who was 36 weeks pregnant?

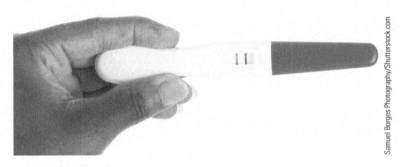

Samuel Borges Photography/Shutterstock.com

   a. It would show a definitively positive test result.

   b. It would show a negative test result or slightly positive test result.

7. In the space provided, draw your own graph based on the following information. Patty is on birth control pills, these pills supply steady levels of estrogen for 21 days of the cycle. On days 22–28 the pills supply no hormones. What would Patty's graphs look like?

Anterior pituitary hormones

Ovarian hormones

(page intentionally left blank)

Activity 33.4 Anatomical Changes during Pregnancy

This activity targets LO 33.14.

At the end of a full-term pregnancy, a fetus may weigh between 5 and 15 pounds (average weights in the United States are between 6 and 8 pounds) and the fetal structures such as the placenta and amniotic fluid lend an additional 3–5 pounds. Other body structures change and swell during pregnancy as well. And yet, all of these new materials must still fit within the pregnant person's body. In order to accommodate the growing fetus and its structures, the anatomy of the pregnant individual changes constantly throughout the pregnancy. In this lab we will investigate some of these changes.

Materials
• Non-pregnant torso model or diagrams provided in this manual
• Pregnant torso model or diagrams provided in this manual

Instructions
1. On each of the models or using Figure 33.14, identify the following structures:

Fetus	Large intestine	Uterus:
Placenta	Rectum	Body
Umbilical cord	Small intestine	Cervix
Vagina	Stomach	Fundus

Figure 33.14 Torso of a Pregnant Person

As the fetus and uterus increase in size and dominate the abdominal cavity, the other organs must move out of the way as seen in anterior (A) and sagittal (B) views.

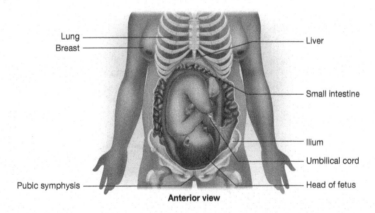

Lung
Breast
Liver
Small intestine
Ilium
Umbilical cord
Head of fetus
Pubic symphysis
Anterior view

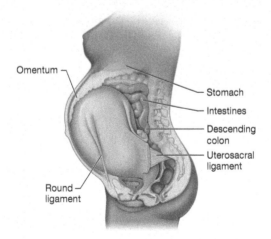

Omentum
Stomach
Intestines
Descending colon
Uterosacral ligament
Round ligament

Questions

1. Mitosis and meiosis are related processes that produce new cells. Which of the following statements is correct about mitosis and meiosis in human reproduction?

   a. Meiosis is involved in both the production of gametes and the increase in uterine size during pregnancy.

   b. Mitosis is involved in both the production of gametes and the increase in uterine size during pregnancy.

   c. Meiosis is responsible for the production of gametes, while mitosis occurs during pregnancy as the uterus expands in size.

   d. Mitosis is responsible for the production of gametes, while meiosis occurs during pregnancy as the uterus expands in size.

2. When the embryo implants into the uterine wall, what layer of the uterus does it first encounter?

   a. Endometrium

   b. Perimetrium

   c. Myometrium

Learning Connection

Quiz Yourself

What are the functions of each of the three layers of the uterine wall?

Clinically, a pregnancy is said to last for an average of 40 weeks and the beginning of the pregnancy is measured from the first day of the last observed menstrual period.

3. Using the figure below, at the beginning of which week of pregnancy is progesterone first secreted?

   a. Week 1

   b. Week 3

   c. Week 8

   d. It has been secreted all along, even before the pregnancy

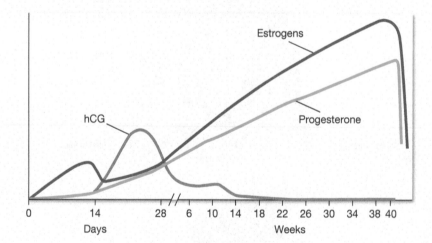

4. On the pregnant model or in the figure below, what region of the uterus is the site of placental attachment?
   a. Fundus
   b. Body
   c. Cervix

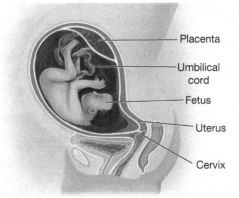

Placenta

Umbilical cord

Fetus

Uterus

Cervix

5. What do you think might happen if the placenta attached lower, as seen in the figure below?
   a. It might grow larger than usual, as the uterus is wider at this location.
   b. It might block the exit of the fetus through the cervix.
   c. There would be no complications from the placenta attaching lower.

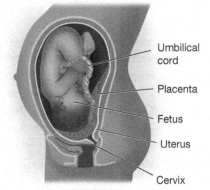

Umbilical cord

Placenta

Fetus

Uterus

Cervix

6. There are some pregnancy complications in which the placenta does not grow to full size (a typical full-term placenta weighs around 1 lb and measures 22 cm in diameter). Which of the following might be complications if the placenta had not reached a typical size?
   a. Lower than typical levels of progesterone
   b. Insufficient surface area for nutrient and waste exchange with the fetus
   c. Lower than typical weight gain during pregnancy
   d. All of these

7. Which of the following correctly describes the blood vessels of the umbilical cord?
   a. One vein and one artery
   b. Two veins and two arteries
   c. One vein and two arteries
   d. Two veins and one artery

8. Umbilical cord prolapse is a condition that can develop during labor, it is illustrated in the figure below. During umbilical cord prolapse, a section of the umbilical cord descends through the cervix, but then gets compressed under the fetal head, and is pinched between the head and the cervix, which stops blood flow through the umbilical vessels. Which of the following is a true statement about the fetus during umbilical cord prolapse.

a. Fetal cells would still have access to oxygen, but would not receive any nutrients.

b. Fetal cells would have oxygen and nutrients but waste products would build up.

c. Fetal cells would lack both oxygen and nutrients, and wastes would build up.

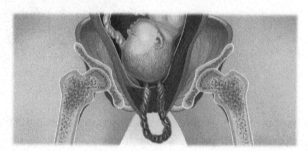

9. A common condition in pregnant people at the end of their pregnancy is constipation. Looking at the model or Figure 33.14, which of the following explains why?

a. Pregnant individuals often eat poorly at the end of pregnancy and their diets may lack fiber.

b. The uterus compresses the rectum during pregnancy, preventing the passage of stool.

c. The uterus compresses the large intestine during pregnancy, slowing down digestion, which allows more water to be extracted and stool to become harder and more difficult to excrete.

Clinical Correlation	Maternal–Fetal Exchange

The function of the placenta is to provide surface area for the fetal circulation to obtain nutrients and oxygen from maternal blood. The fetal blood also releases waste products into maternal blood. However, fetal blood is entirely contained in vessels within the placenta. Maternal blood bathes the surface of the placenta, but does not cross into the fetal circulation. The blood supplies of the two individuals, fetus and mother, never mix. Materials in the maternal circulation may cross the placenta, or may not, depending on their molecular properties. Alcohol, for example, is a very small molecule that diffuses easily, so alcohol in the maternal circulation readily crosses the placenta and enters the fetal circulation. Other drugs, for example insulin, are larger and do not diffuse well, therefore maternal insulin does not cross the placenta and is unavailable to the fetus. Some pathogens cross the placenta and others do not, for example, the virus HIV does not bind to any receptors on the placenta, and therefore it does not cross the placenta. An HIV+ mother can give birth to an HIV− child if care is taken to prevent exchange during birth. Most bacteria are too large to cross the placenta and bacterial diseases rarely infect the fetus.

Lab 33 Post-Lab Quiz

1. Which structure(s) does the surgeon need to cut through in order to get to the ductus deferens during a vasectomy? (Hint: you explored this location in Activity 33.1.)
   a. Scrotal skin
   b. Dartos muscle
   c. Transverse abdominus
   d. Inguinal ligament

2. After the vasectomy, when this person ejaculates what will be released?
   a. Nothing
   b. Secretions from the prostate, bulbourethral glands and seminal vesicles, but no sperm
   c. Secretions from the prostate, bulbourethral glands and seminal vesicles, and sperm
   d. Secretions from the bulbourethral glands and seminal vesicles but not prostate

3. What structure(s) will a surgeon have to cut through in order to reach the uterine tubes during a tubal ligation? (Hint: you explored this location in Activity 33.1.)
   a. Transverse abdominus
   b. Labia
   c. Cervix
   d. fundus of uterus

4. During the course of a menstrual cycle, the endometrium changes. It becomes dramatically thicker through cell proliferation. What process is occurring in the endometrial cells at this time?
   a. Mitosis
   b. Meiosis
   c. Both
   d. Neither

5. During the course of a menstrual cycle, the endometrium changes. It becomes dramatically thicker through cell proliferation. This process is driven by estradiol. What is the main source of this estradiol?
   a. Hypothalamus
   b. Anterior pituitary
   c. Interstitial cells in the ovaries
   d. Developing follicles in the ovaries

6. Which of the following is found within a mature ovarian follicle? (select all that apply)
   a. An oocyte
   b. Estrogen
   c. Follicle-stimulating hormone
   d. An embryo

7. Fluoroacetamide is a drug that inhibits the process of spermiogenesis. Rats that are exposed to fluoroacetamide do not produce functional sperm, but the results are reversible, meaning that after the rats stop ingesting fluoroacetamide they resume making functional sperm. This is because fluoroacetamide does not affect stem cells. Does fluoroacetamide affect haploid or diploid cells?
   a. Diploid (2n)
   b. Haploid (1n)
   c. Both diploid and haploid

8. Robin's last menstrual period was 8 weeks ago. She believes that she might be pregnant. Her doctor performs an ultrasound of her ovaries, which can visualize structures present within the ovaries. If Robin is pregnant, which of the following best describes what her doctor might see during the ultrasound?

    a. Many tertiary follicles

    b. One tertiary follicle and some primordial follicles

    c. Many primordial follicles and a corpus luteum

9. Estrogens, including estradiol, have effects on many tissues throughout the body. One target tissue of estrogens is ligaments, estrogens cause ligaments to loosen. Looser ligaments contribute to the wider pelvises observed in biological females. Understanding this, what would you expect to be true of the pelvis of an individual at 38 weeks of pregnancy compared to their pelvis at 12 weeks of pregnancy.

    a. The pelvis will be the same size and shape at 12 weeks of pregnancy and 38 weeks of pregnancy.

    b. The pelvis will be wider at 12 weeks of pregnancy than at 38 weeks of pregnancy.

    c. The pelvis will be wider at 38 weeks of pregnancy than at 12 weeks of pregnancy.

10. Endometriosis is a condition in which endometrial tissue becomes lodged in the abdominal cavity. Through which of the following structures did the endometrial tissue travel to reach the abdomen?

    a. Cervix

    b. Uterine tubes

    c. Vagina

    d. Ovary

Lab 34 Development

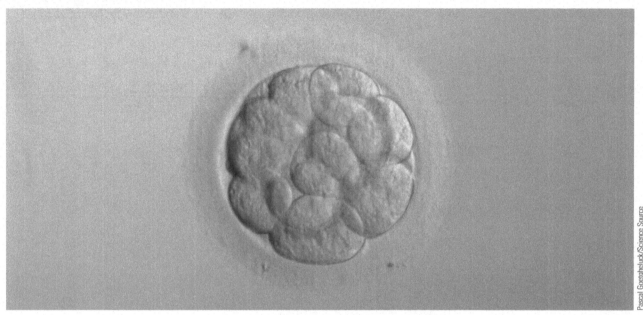

Learning Objectives: After completing this lab, you should be able to:

34.1 Describe the process and events of fertilization.

34.2 Define the preembryonic period, embryonic period, and fetal period, and describe the main events that occur in each.

34.3 Explain the process and timing of implantation.

34.4 Describe the major events of the fetal period.

34.5 List the maternal and fetal components of the placenta.

34.6 Trace the pathway of blood flow from the placenta, through the fetal heart and body, and back to the placenta.

Introduction

Fertilization is the event in which a sperm encounters and penetrates an egg, resulting in the combining of their genetic material. This single cell, now with a full complement of 46 chromosomes, is called a **zygote**. After fertilization, the zygote will begin to rapidly expand its cell number by repeated mitotic divisions that eventually result in a ball of cells. This preembryo then attaches to the uterine wall during **implantation** so that a **placenta** can be established. The placenta creates a location where fetal capillaries and maternal capillaries travel closely next to each other, exchanging gases, nutrients, and wastes. Following implantation, the embryo further develops into a **fetus**. The fetus' remaining time in the womb is dedicated to continued mitotic division for growth of the organism and specialization of the tissues and organs of the body.

The Human Anatomy and Physiology Society includes more than 1,700 educators who work together to promote excellence in the teaching of this subject area. The HAPS A&P Learning Outcomes measure student mastery of the content typically covered in a two-semester Human A&P curriculum at the undergraduate level. The full Learning Outcomes are available at https://www.hapsweb.org.

Lab 34 Pre-Lab Quiz

This quiz will strengthen your background knowledge in preparation for this lab. For help answering the questions, use your resources to deepen your understanding. The best resource for help on the first five questions is your text, and the best resource for help answering the last five questions is to read the introduction section of each lab activity.

1. Anatomy of the placenta and extraembryonic membranes. Identify the following structures on the figure:

Amnion	Chorion	Placenta
Amniotic fluid	Fetus	Umbilical cord

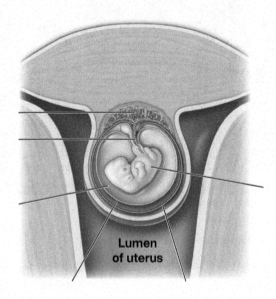

Lumen
of uterus

2. Match the root next to its definition:

Root	Meaning
	lead or direct
	skin or membrane
	long slender hair
	membrane around the fetus
	to sprout

List of Roots

amnion

blasto

chorion

duct

villus

3. Which of the following structures must a sperm pass through to fertilize an ovulated egg? Choose all correct answers.

a. The acrosome

b. The zona pellucida

c. The corona radiata (granulosa cells)

4. Implantation of the preembryo can only occur once it has developed a/an:

a. Trophoblast

b. Morula

c. Acrosome

5. The trophoblast of the preembryo secretes human chorionic gonadotropin (hCG). This hormone targets the:

   a. Endometrium of the uterus
   b. Corpus luteum
   c. Vagina

6. The direct product of the process of fertilization is a/an:

   a. Zygote
   b. Embryo
   c. Fetus

7. The final result of the process of cleavage is a:

   a. Morula
   b. Blastocyst
   c. Zygote

8. The portion of the blastocyst that will develop into the chorion is the:

   a. Central cavity
   b. Trophoblast
   c. Inner cell mass

9. The structure that delivers oxygen- and nutrient-rich blood to the fetus is the:

   a. Umbilical artery
   b. Umbilical vein

10. The temporary fetal structure that allows blood from the umbilical vein to enter the fetal cardiovascular system is the:

    a. Foramen ovale
    b. Ductus arteriosus
    c. Ductus venosus

(page intentionally left blank)

Activity 34.1 Fertilization

This activity targets LO 34.1.

The process of fertilization occurs when the nucleus of a sperm enters an ovulated egg, but the ability of a sperm to reach an egg relies on the environment within the reproductive tract through which it travels on its way to that egg. The fluids of the vagina and uterus contain enzymes that alter the cell membrane of sperm, a process called **sperm capacitation**. In sperm capacitation, three functional changes to the sperm occur. The sperm become chemotactically active, meaning that they can follow chemical signals from the egg and the cells of the corona radiata surrounding the egg. In addition, the beating of the sperm flagella becomes more active to propel the sperm more rapidly. Finally, capacitation thins the **acrosome**, a cap at the tip of the sperm's head, so that upon contact with the layers that surround the egg, it will rupture more readily to release the enzymes within.

As the capacitated sperm make their way through the vagina and then uterus, they enter the uterine tubes. If an egg has recently been ovulated, the sperm is likely to encounter it here; the ampulla of the uterine tubes is the most likely location of fertilization. Following ovulation, the egg is able to be fertilized for a maximum of 24 hours. After this time, if it is not fertilized the egg will degenerate and leave the uterus through the vagina during menstruation.

Examine Figure 34.1 to visualize the process of fertilization described here. The arriving sperm push through the **granulosa cells** of the **corona radiata** due to the increased activity of their flagella. As the sperm encounter the **zona pellucida** deep to the corona radiata their acrosomal caps rupture to release enzymes, chipping away at the zona pellucida. The acrosomal enzymes of a single sperm are not enough to fully penetrate the barrier of the zona pellucida, but the contribution of each sperm paves the way deeper through the zona pellucida for other sperm that follow. Eventually, one sperm merges its membrane with the now-revealed membrane of the egg, pushes its nucleus into the egg cytoplasm, and fertilizes it. Immediately following the fusion of the cell membranes of the sperm and egg, a cortical reaction occurs throughout the zona pellucida. This reaction hardens the zona pellucida into an impenetrable coating around the egg, preventing any other sperm from being able to merge their membranes or inject their nucleus into the now-fertilized egg.

The result of this process is a zygote. This is a single cell with the full complement of 46 chromosomes: 23 each from the sperm and egg. Only the 23 chromosomes from the sperm are transferred to the egg; therefore, all other cell organelles (mitochondria, lysosomes, cytoskeleton, etc.) are from the mother's egg.

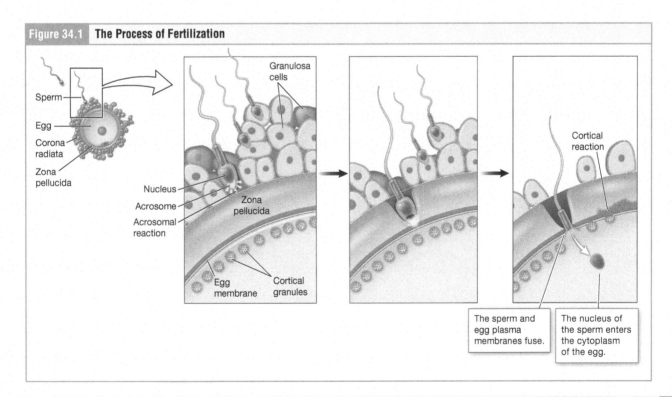

Figure 34.1 **The Process of Fertilization**

The sperm and egg plasma membranes fuse.

The nucleus of the sperm enters the cytoplasm of the egg.

Instructions

1. Using the introduction, Figure 34.1, and any other resources provided by your instructor, create a flow chart to represent the steps that are necessary to lead to fertilization. A possible start to this flow chart has been provided to help you get started. Use a pencil to develop the flow chart so that you can edit as necessary in step 2.

2. Once you have completed your flow chart, share your work with your lab partner and discuss the following prompts.

   a. Are there any missing steps? Add as necessary.

   b. Does the flow chart contain enough detail to explain the process?

   c. A flow chart also shows what will happen if something does NOT occur at critical moments. Does the flow chart account for what will happen if:

      i. Sperm capacitation does not occur?

      ii. An egg has not been ovulated in the last 24 hours?

      iii. Not enough sperm reach the egg?

      iv. The cortical reaction does not occur?

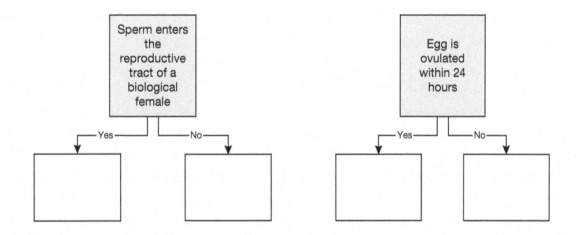

(page intentionally left blank)

Activity 34.2 Cleavage, Blastocyst Formation, and Implantation

This activity targets LOs 34.2, 34.3, and 34.4.

Once fertilization occurs, the zygote will begin a series of rapid mitotic divisions. These divisions, shown in Figure 34.2, increase the number of cells to two, then four, then eight, and so on. But because there is little pause for growth of the resulting cells, each round of mitosis results in smaller and smaller cells that together take up the same amount of space as the single fertilized zygote did originally. Once the divisions have produced a solid ball of eight cells, the structure is called a **morula**. It takes approximately four days for these divisions, called **cleavage**, to eventually produce a morula of 16 to 32 cells. Concurrently, this mass of dividing cells has been traveling through the remaining sections of the uterine tube to reach the uterus. As the cells of the morula complete more mitotic divisions, the cells shift their arrangement to create a hollow ball of cells called a **blastocyst**. This is the final step of cleavage. The blastocyst consists of an outer ring of cells (the **trophoblast**) that will develop into the **chorion,** an inner cell mass that will develop into the embryo, and a fluid-filled central cavity. The early blastocyst departs the uterine tube and approximately 7 days after fertilization it is floating in the lumen of the uterus. At this time the blastocyst can also be called a preembryo, as it has not yet implanted in the uterine wall.

Between days 7 and 12, the late blastocyst begins the process of implantation into the apical surface of the uterine wall—the **endometrium**. Remember that the hormones of the female reproductive cycle have prompted the development of the endometrium into a dense layer of tissue with nutrient-rich mucus and abundant vasculature in the form of **spiral arteries**.

Figure 34.2	The Processes of Cleavage, Blastocyst Formation, and Implantation

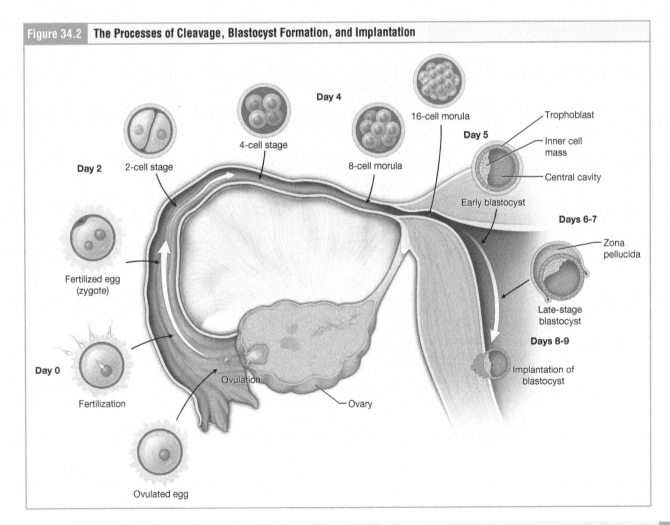

The cells of the trophoblast perform several critical functions at this time. They begin to secrete **human chorionic gonadotrophin (hCG)** to signal the corpus luteum to stay, grow, and continue to secrete progesterone. Recall that the corpus luteum is in the ovary and is the remnant of the ovarian follicle once the egg has been ejected via ovulation. This progesterone from the corpus luteum, among other actions, promotes the maintenance of the endometrial layer. As implantation begins, the trophoblast also secretes enzymes to weaken a small area of endometrium so that the blastocyst can burrow into the tissue and reach the spiral arteries. Finally, the cells of the trophoblast develop into the **chorion**—an elaborately folded layer of tissue that will become the fetal portion of the placenta.

As implantation occurs and the chorion develops, the structure can now be called an **embryo**. Examine Figure 34.3. The shape of the newly implanted embryo has also been changing into a **gastrula**. One end of the hollow ball looks as if it has been indented or pushed inward toward the central cavity to eventually create three layers of cells: the endoderm, mesoderm, and ectoderm. These three germ layers will begin to differentiate and give rise to specific tissues and organs of the body over the next six weeks. At the base of this indentation, the **amniotic cavity** begins to develop and will eventually surround the developing embryo.

By week 9, the embryo has developed all organ systems. At this time, the embryo is called a fetus. The remaining time, from week 9 to birth, will see the continued growth and development of the needed organs and tissues of the body. Rapid mitotic division to produce growth in size is characteristic of this time, with both weight and length of the fetus increasing dramatically. While a 9-week fetus measures approximately 9 cm and weighs an average of 28 grams, a full-term fetus measures approximately 53 cm and usually weighs between 2.5 to 4.5 kilograms.

Figure 34.3 | **Germ Layers**

Formation of the three primary germ layers occurs during the first 2 weeks of development. The embryo at this stage is only a few millimeters in length.

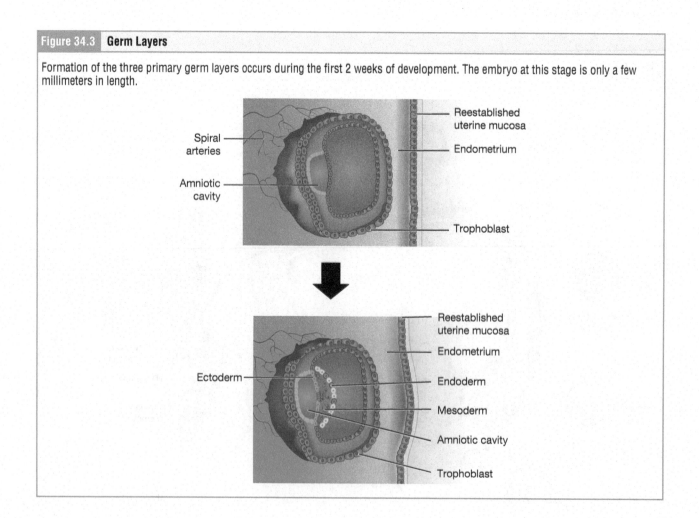

Materials

- Compund microscope (if slides are used)
- Microscopic slides of embryonic development of the sea urchin and/or color copies of the embryonic development of the sea urchin

Instructions

1. Obtain the slides or color copies of the embryonic development of the sea urchin.

2. If you are using slides and the microscope, refresh your understanding of microscope use as necessary before you begin.

3. You will identify and sketch four different stages of development: zygote, morula, blastula (called a blastocyst in the human), and gastrula.

4. For each stage, draw the structure in the space provided and complete the requested labeling and notes alongside your drawing.

Slide 1: Zygote

Label the zona pellucida on your drawing.
How many cells are present?

This structure is present at day _____
after fertilization.
In the human, where would the zygote be found?

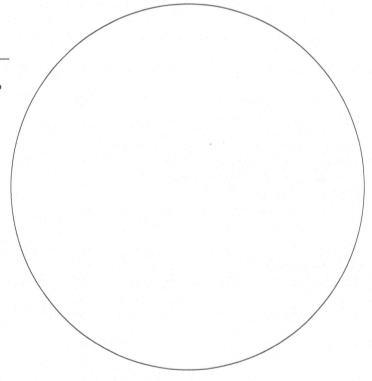

Slide 2: Morula

How many cells are present? (this may be an educated guess, depending on your slide or image)

This structure is present at approximately days

_____ to _____ after
fertilization.

In the human, where would the morula be found?

Slide 3: Blastula (Blastocyst in the Human)

Label the location of the central cavity on your drawing.
If visible, label the trophoblast and the inner cell mass on your drawing.
This structure is present at approximately days

_____ to _____ after
fertilization.
In the human, where would an early-stage blastocyst be found?

Slide 4: Gastrula

This structure is present at approximately days

_____ to _____ after

fertilization.

In the human, where would the gastrula be
found?

Learning Connection

Organizing Your Ideas

Remembering which developmental stages occur in which developmental periods (preembryonic, embryonic, and fetal) is easier
when you organize your ideas.

- On separate sticky notes, write out all of the developmental stages: zygote, morula, blastocyst, gastrula, and fetus.
- Divide a piece of paper into thirds, labeling a section with: preembryonic, embryonic, and fetal periods.
- Place the sticky notes into the correct developmental periods. Arrange them in order within the periods.

(page intentionally left blank)

Name: _____   Section: _____

Activity 34.3 The Placenta and Umbilical Cord

This activity targets LOs 34.5 and 34.6.

The placenta is a large, flat structure that is a nest of fetal capillaries. The placenta develops from the trophoblast of the embryo to interface with the maternal uterine wall. Examine Figure 34.4 and notice how the blood from the mother's uterine spiral arteries bathes the surface of the placenta. While maternal and fetal blood never mix, the nutrients, gases, and wastes can diffuse between the two circulation systems. As the portion of the trophoblast facing deep into the maternal endometrium develops into the chorion, the surface of the tissue begins to develop elaborate folded projections called **chorionic villi (placental villi)**. These villi create a large surface area where the fetal tissue meets the **placental sinus**, a broad space where maternal blood arriving from the spiral arteries pools in close proximity to fetal venules and arterioles.

Fetal venules converge to emerge from the placenta as the **umbilical vein**, which carries oxygen-rich fetal blood from the placenta to the fetal vena cava via the **ductus venosus**. A pair of **umbilical arteries** eventually carry oxygen-depleted fetal blood back to the placenta where they branch into fetal arterioles.

Materials
- Fresh or preserved placenta with umbilical cord, if available

Instructions
1. Observe the fresh or preserved placenta, if available. Alternately, examine the placenta in Figure 34.4.

| Figure 34.4 | The Placenta |

(A) A full-term placenta is slightly larger then a dinner plate and weighs about two pounds. (B) The structure of the placenta keeps fetal blood within capillary beds while maternal blood bathes the outer surface of the capillaries. Thus, fetal and maternal blood never mix, but gas and nutrient exchange is possible.

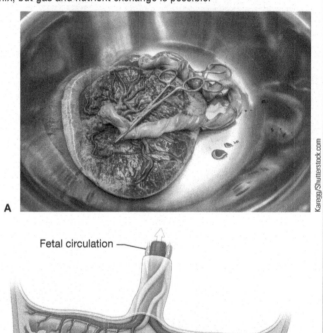

2. The placenta is flattened, somewhat like a thick plate. Distinguish between the maternal surface of the placenta and the fetal surface. The maternal surface is rough, while the fetal surface is shiny, with the umbilical cord attached somewhat centrally.

   a. Label the maternal and fetal surfaces on Figure 34.4 (only a small edge of one of the two surfaces is visible in the figure).

   b. Why is the maternal surface rough? In other words, what small fetal structures project from this side?

   c. What function is served by these small projections?

3. Observe the umbilical cord.

a. What three structures are contained in the umbilical cord?

b. Label these structures on Figure 34.4.

c. What are the characteristics of the blood in these structures? (For example, is the blood oxygen-rich or -poor? What about nutrients and wastes?) Add this information to your labels in Figure 34.4.

Activity 34.4 Fetal Circulation

This activity targets LO 34.5.

As the fetal blood arrives through the umbilical vein, it is rich in oxygen and nutrients that have been passed from maternal blood to fetal blood at the placenta. However, at the fetal heart this arriving blood will mix with fetal blood from the inferior vena cava and then the superior vena cava. This blood is usually called "mixed," because it is a combination of oxygen- and nutrient-rich blood from the placenta and oxygen-poor and waste-containing blood from the fetus' venous circuit. The mixed blood transits through the rest of the fetal circulatory system. A high proportion of fetal hemoglobin in the blood and high cardiac output in the fetus allow for adequate oxygen delivery despite the lower oxygen content of this blood.

   The structure of the fetal cardiovascular system differs from that of a newborn in three important areas. These structural differences provide three shunts to bypass two relatively unneeded organs in the fetus: the liver and the lungs. The shunts re-direct blood to other areas of the fetus that are actively growing and in need of a robust blood supply. Find these three shunts in Figure 34.5.

1. The *foramen ovale* is an opening in the interatrial septum of the heart that allows blood to flow from the right atrium to the left atrium, thus bypassing the lungs.

2. The *ductus arteriosus* is a short muscular vessel that connects the pulmonary trunk to the aorta. Most of the blood pumped from the right ventricle into the pulmonary trunk is thereby diverted into the aorta and bypasses the lungs.

3. The *ductus venosus* acts as a temporary blood vessel that reaches from the umbilical vein to the inferior vena cava, so the liver is bypassed.

Materials
• Colored pencils in blue, pink or red, and purple

Figure 34.5 **Fetal Circulation**

Three major features of the fetal circulatory system that shunt blood around the developing lungs and liver.

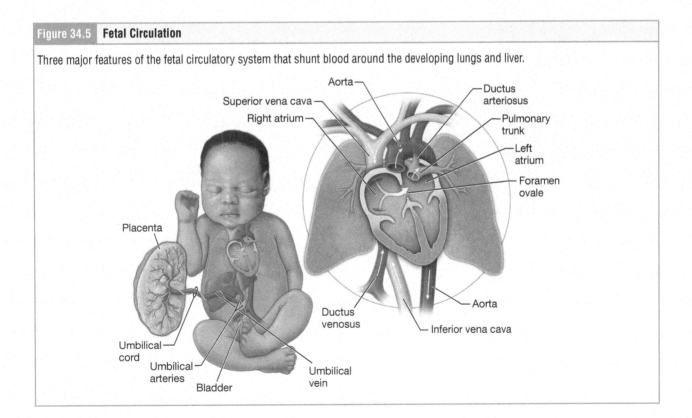

Instructions

1. Use the introduction and Figure 34.5 to label and color the drawing of fetal circulation below.

   a. Label the following structures on the drawing:

Umbilical vein	Foramen ovale
Umbilical arteries	Inferior vena cava
Ductus venosus	Internal iliac arteries
Ductus arteriosus	

   b. Add to your labels to include what structures are bypassed by the three shunts:

      i. At the ductus venosus: what structure will be bypassed?

      ii. At the ductus arteriosus: what structures will be bypassed?

      iii. At the foramen ovale: what structures will be bypassed?

   c. Starting with the umbilical vein and ending with the umbilical arteries, draw arrows to show the flow of blood through the entire fetal circulation. At both the foramen ovale and the ductus arteriosus, blood can take two paths. Use two different-sized arrows to indicate the path the *majority* of the blood will take.

   d. Use blue, pink or red, and purple colored pencils to color in all of the blood vessels shown in the image.

      i. Blue vessels are those carrying oxygen-poor and waste-containing blood.

      ii. Pink or red vessels are those carrying oxygen-rich and nutrient-rich blood.

      iii. Purple vessels are those carrying mixed blood.

   e. Share your completed work with your lab partner and resolve any differences between your drawings.

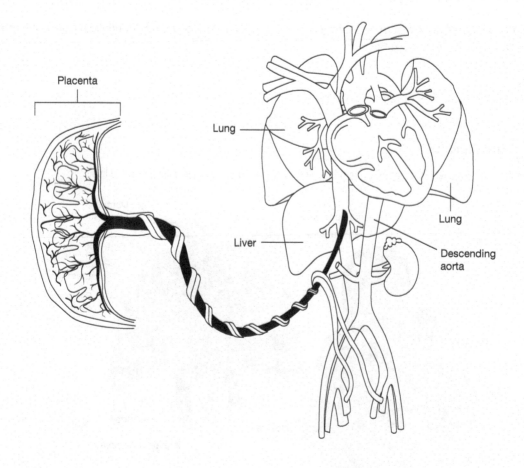

2. Answer the following questions:

   a. The introduction describes how the shunts re-direct blood from the lungs and liver. You identified on the drawing which shunts bypass which structures. Are these structures fully bypassed? In other words, does no blood flow to the fetal liver and lungs? Explain.

   b. Post-birth the umbilical cord is cut near to the *umbilicus* (belly button) to separate the newborn from the placenta. After this cut is made, what happens to the portion of umbilical cord that traveled within the fetus' body? This is not covered in the introduction, but you and your lab partner should be able to puzzle it out.

Digging Deeper:
Home Pregnancy Tests

Human chorionic gonadotropin (hCG) is released by the trophoblast cells of the developing blastocyst as they begin their migration through the endometrium; this can be as early as day 5. The hCG must travel through the bloodstream to reach the target cells of the corpus luteum. The amount of hCG is quite small at this point, but the trophoblast continues to develop and grow into a chorion, secreting more and more hCG as it does. By approximately day 10, levels of hCG in the blood are high enough to be detected by a blood test; by approximately days 12 to 14 there is enough hCG in the blood that it can be detected in the urine as it is removed at the kidneys.

The presence of hCG in the urine is what is detected in an at-home pregnancy test such as the one seen here. Such tests are usually accurate when hCG levels in the blood reach 20–50 mU/mL, which corresponds to approximately 4 weeks since the individual's last menstrual period. Levels of hCG increase rapidly and consistently from here, reaching upwards of 26,000 to 290,000 mU/mL in the 9th to 12th weeks since the individual's last menstrual period. When used correctly, at-home pregnancy tests are considered quite accurate 4 weeks after the individual's last menstrual period. If used earlier than that, an at-home test may give a false negative (read "not pregnant" when the individual is pregnant but not enough hCG is present to be detected by the test yet).

An at-home pregnancy test

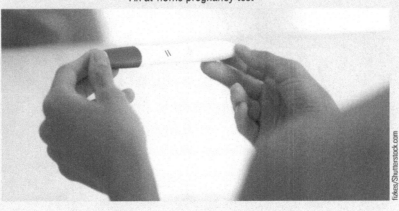

fizkes/Shutterstock.com

(page intentionally left blank)

Activity 34.5 Case Study: Patent Ductus Arteriosus

Case Study

This activity targets LOs 34.2 and 34.4.

Camila is 40 weeks pregnant with her second child. She has had an uneventful pregnancy with regular check-ups with her midwife. When labor begins Camila calls the midwife; several hours later Camila delivers the baby, whom she names Juan. At birth Juan appears healthy, though the midwife does notice that Juan has a higher rate of breathing than she is used to seeing. She asks Camila to pay attention to Juan's respiration rate in the coming days.

When Juan is two weeks old, Camila brings the baby to visit the pediatrician for the first time. She shares that Juan continues to have a high respiration rate and that he is not feeding as well as Camila's first child did. The pediatrician notices that Juan has not gained any weight since birth. In listening to Juan's lungs, the pediatrician notices some congestion.

The pediatrician tells Camila that she suspects that Juan's lung problems may be due to a problem that has developed with Juan's heart. She orders an echocardiogram, an ultrasound scan of the heart. Camila is nervous, but the pediatrician assures her that an echocardiogram is non-invasive, using sound waves to create a visual picture of the heart on the computer screen so that the structures of Juan's heart can be examined. She is especially interested in seeing the echocardiogram images of the pulmonary artery and the aortic arch.

1. You can deduce that the pediatrician suspects a problem with what fetal heart structure?

    a. Ductus venosus

    b. Ductus arteriosus

    c. Foramen ovale

2. The pediatrician suspects a heart problem has led to congestion in the lungs. If blood is being delivered to the lungs at a greater rate than it is being removed, lung congestion can occur. This would suggest that the _____ side of the heart is delivering more blood to the lungs than can be received back on the other side.

    a. left

    b. right

Camila returns to visit the pediatrician when the results of the echocardiogram are back. The pediatrician shares the results, which can be seen here, with Camila. The results show that Juan has a *patent ductus arteriosus (PDA)*. (The echocardiogram shows the PDA as a small dark opening and it shows the blood flow through the PDA colored as red and orange.) The pediatrician explains that this means that the fetal structure of the ductus arteriosus, which should have closed off at birth, remains open and blood is traveling through it in the opposite direction that it moved in the fetus. This PDA is described as a left-to-right shunt.

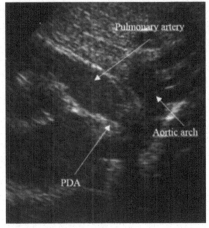
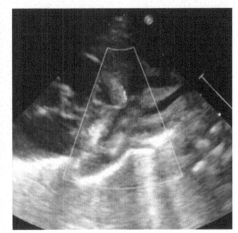

Source: Arlettaz R (2017) Echocardiographic Evaluation of Patent Ductus Arteriosus in Preterm Infants. *Front. Pediatr.* 5:147. doi: 10.3389/fped.2017.00147

3. What was the purpose of the ductus arteriosus in Juan's heart?

   a. In the fetus, the ductus arteriosus allows blood from the right atrium to move to the left atrium, bypassing the body.

   b. In the fetus, the ductus arteriosus allows blood from the aortic arch to move to the right ventricle, bypassing the pulmonary artery.

   c. In the fetus, the ductus arteriosus allows blood from the pulmonary artery to move to the aortic arch, bypassing the lungs.

4. Why does this fetal bypass occur?

   a. The fetal lungs do not provide oxygen to the blood.

   b. The fetal heart cannot work that hard.

   c. The umbilical arteries send fetal blood to the lungs, not the fetal heart.

5. If the PDA left-to-right shunt goes uncorrected, what effect does this have on the blood flowing through the heart?

   a. Already oxygenated blood that should be going to the body will be returned to the lungs.

   b. Low-oxygen blood that should be going to the lungs will be sent to the body.

The pediatrician thinks that Juan's PDA is not serious and asks to see Juan back in two more weeks to assess his symptoms. At that checkup the pediatrician is pleased to tell Camila that the symptoms that they noticed earlier have reduced and that aggressive treatment of the PDA is not necessary. She prescribes an anti-inflammatory to reduce the lung congestion. At a follow-up echocardiogram at 6 months, Juan's PDA has closed on its own.

Lab 34 Post-Lab Quiz

1. The period of development characterized by growth of a placenta is the _____ period.

   a. preembryonic
   b. embryonic
   c. fetal

2. Examine the image. This structure is most accurately called a:

   a. Preembryo
   b. Embryo
   c. Fetus

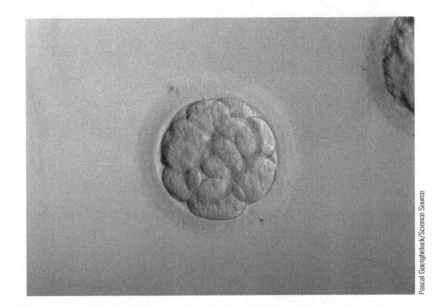

3. If the cortical reaction of a fertilized egg did not occur immediately after penetration by a sperm, the resulting zygote would most likely contain ___ chromosomes.

   a. 23
   b. 46
   c. 69

4. Diseases caused by mutations in mitochondrial DNA are inherited from one's:

   a. Mother
   b. Father
   c. Either mother or father

5. In comparing two fertilized eggs, one that has been fertilized as it enters the uterus and one that has been fertilized in the ampulla of the uterine tube, the egg fertilized in the uterus will be _____ likely to implant into the endometrium.

   a. less
   b. just as
   c. more

6. You are provided with the slide seen here. Considering the developmental stage of this structure, you can assume that this would be found in the:

   a. Ovary
   b. Uterine tube
   c. Lumen of the uterus
   d. Endometrium of the uterus

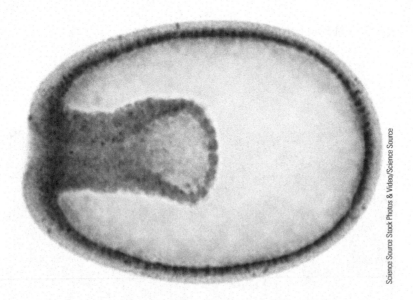

Science Source Stock Photos & Video/Science Source

7. Which of the following components of the placenta will contain fetal blood? Choose all correct answers.

   a. Placental sinus
   b. Umbilical veins
   c. Umbilical arteries

8. At birth, the umbilical cord is usually pinched closed and cut. If you examined the blood that was produced when the cord is cut, you would find that it contains:

   a. Highly oxygenated blood only
   b. Low-oxygen blood only
   c. A mixture of high-oxygen and low-oxygen blood

9. The blood within the umbilical arteries will pass next into the:

   a. Placenta
   b. Ductus venosus
   c. Internal iliac arteries

10. If a newborn infant's foramen ovale remained open following birth, you could expect that the infant's ability to _____ would be reduced.

    a. process dietary nutrients
    b. transport glucose
    c. exchange oxygen and carbon dioxide

Glossary

A

abdominal aorta the largest artery in the abdomen, a continuation of the descending aorta. **617**

abduction (Latin: *abductio* - movement away from center) motion in the coronal plane that pulls a structure away from the midline of the body. **280**

acetabulum (Latin: *acetabulum* - a shallow cup) deep, cup-shaped cavity on the lateral side of the hip bone; formed by the convergence of the ilium, pubis, and ischium. **243**

acetylcholine (ACh) a neurotransmitter used in the central nervous system and in neuromuscular junctions; binds to receptors on the motor end-plate to trigger depolarization. **523**

acinus a berry-shaped group of cells, such as the alveolar sac. **761**

acromial end of the clavicle (aka lateral end of the clavicle) (Greek: *akron* - tip; *ōmos* – shoulder) lateral end of the clavicle; articulates with the acromion of the scapula. **231**

acromion (Greek: *akron* - tip; *ōmos* - shoulder) broad, flat projection that extends laterally from the scapular spine to form the tip of the scapula. **231**

acrosome (Greek: *akro* - at the end; *soma* - the body) an enzyme filled cap that covers the extremely compact nucleus of the nuclear head. **863**

action the particular movement of a muscle. **309, 329**

action potentials (pl: action potentials) (Latin: *actio* - to do; *potentia* - power) electrical signal unique to neurons and muscle fibers; caused by a change in voltage across a cell membrane in response to a stimulus. These electrochemical impulses allow the neurons to communicate throughout the body. **360**

acuity (French: *acutus* - sharpness) a term used to describe how accurately our brain can understand a stimulus; often applied to the sense of vision. **421, 441**

adaptation (Latin: *adapto* - to adjust) a process by which a sensory neuron stops or slows its rate of neurotransmitter release even though it is still receiving graded potentials. **392, 438**

adduction (Latin: *adduco* - to bring toward) motion in the coronal plane that brings a structure toward or past the midline of the body. **280**

adenosine triphosphate (ATP) chemical compound that provides energy for cellular processes. **51**

adipocytes (sing: adipocyte) (aka fat cells) (Latin: *adip* - fat; Greek: *kytos* - cell) specialized lipid storage cells. **129**

adrenal arteries (Latin: *ad* - to; *ren* - kidney) an artery that branches off of the abdominal aorta, and provides blood to the adrenal (suprarenal) glands. **620**

adrenal cortex (aka suprarenal cortex) (Latin: *ad* - to; *ren* – kidney) the outer layer of the adrenal gland, which secretes several hormones called corticoids; these steroid hor-mones regulate glucose balance, electrolyte balance, and some reproductive function. **549**

adrenal glands (*aka* suprarenal glands) (Latin: *ad* - to; *ren* - kidney; *glans* – acorn) endocrine glands composed of the adrenal cortex and adrenal medulla; located on top of the kidneys; cortex secretes steroid hormones and medulla secretes epinephrine and norepinephrine. **549**

adrenal medulla (aka suprarenal medulla) (Latin: *ad* - to; *ren* - kidney; *medulla* – marrow) the internal portion of the adrenal gland; secretes the hormones epinephrine and norepinephrine. **531, 549**

adrenal vein (Latin: *ad* - to; *ren* - kidney) a vein that drains blood from the adrenal (suprarenal) glands; the right adrenal vein drains directly into the inferior vena cava, and the left adrenal vein drains into the left renal vein. **620**

adrenocorticotropic hormone (ACTH) (*aka* corticotropin) (Latin: *ad* - to; *ren* - kidney; Greek: *trophē* - nurturing) a hormone produced by the anterior pituitary gland; stimulates the adrenal cortex to secrete corticosteroid hormones, such as cortisol. **556**

adventitia the outermost layer of tubular organs of the GI tract that lie outside of the abdominopelvic cavity, such as the pharynx and esophagus; consists of dense connective tissue. **755, 799**

aerobic cellular respiration (*aka* oxidative phosphorylation) (Greek: *aēr* - air; Latin: *respiratio* – breathe) breakdown of pyruvate in the presence of oxygen; results in ATP formation. **712**

afferent (Latin: *afferens* - to bring to) nerves that bring input toward the central nervous system. **363**

afferent arterioles (Latin: *arteriola* - an artery) arterioles that transport oxygen-rich blood from the cortical radiate (intralobular) arteries to the glomerulus in each nephron. **795**

afferent lymphatic vessels (Latin: *lympha* - water) lymphatic vessels that drain lymph into a lymph node to be filtered. **671**

afferent signals signals traveling along an afferent neuron. **480**

agglutination (Latin: *agglutinare* - to glue) the clumping of erythrocytes containing certain antigens by antibodies that specifically bind to those antigens. **581**

agonist (aka prime mover) (Greek: *agōn* – contest) prime mover whose action is responsible for a particular movement. **329**

ala (pl: alae) (Latin: *alaris* - wing) small, rounded, cartilaginous structure that forms the lateral wall of a nostril. **687**

aldosterone a mineralocorticoid hormone produced by the zona glomerulosa of the adrenal cortex; involved in regulation of sodium and potassium levels in the blood and of blood pressure. **549**

alpha cells a pancreatic islet cell type that produces the hormone glucagon. **549**

alveolar ducts (Latin: *alveolus* - basin) a small tubular organ that branches off of a respiratory bronchiole and opens into a cluster of alveoli called an alveolar sac. **700**

alveolar macrophages phagocytic cells found in the alveoli of the lungs, which engulf and destroy pathogens and debris. **703**

alveolar sacs (Latin: *alveolus* - basin) a cluster of alveoli that branches from an alveolar duct. **700**

alveoli (Latin: *alveolus* - basin) rounded sacs found in clusters within the mammary lobules of the breast; composed of cells that secrete milk to nourish a baby; (Note: the term *alveoli* also refers to saclike structures in the lungs). **684**

amniotic cavity the space within the amniotic membrane containing the amniotic fluid and developing embryo. **872**

amphiarthrosis (aka amphiarthrotic joint) (Greek: *amphi* - both sides; *arthrōsis* - joint) joint at which slight motion is possible. **265**

amylases a class of enzymes that catalyze the hydrolysis of starch into sugars. **767**

anaphase (Greek: *ana* - up; *phasis* - appearance) third stage of mitosis (and meiosis), during which sister chromatids separate and move toward the poles of the dividing cell. **56**

anatomical position (Greek: *ana* - up; *tomia* - cutting) standard position to reference when describing locations and directions on the human body. **5**

anemia (Greek: *an* - lack of; *haima* - blood) a deficiency in the oxygen-carrying capacity of the blood, due to a lack of sufficient red blood cells or hemoglobin. **577**

angle of the rib (Latin: *angulus* - angle) portion of the rib body with the greatest curvature. **222**

antagonist muscle whose action is the opposite of an agonist. **329, 309**

antebrachium (Latin: *ante* - before; *brachium* - arm) the portion of the upper limb that is between the elbow and wrist joint. **235**

anterior arch (Latin: *anterior* - earlier; *arcus* - bow) anterior portion of the ring-shaped C1 vertebra. **215**

anterior cerebral artery (Latin: *ante* - before; *cerebro* - brain) an artery that branches from the internal carotid artery; supplies blood to portions of the frontal and parietal lobes of the cerebrum. **617**

anterior chamber the space between the iris and the endothelium of the cornea, filled with aqueous humor. **415**

anterior columns (*aka* ventral columns, anterior funiculi) (Latin: *ante* - front) white matter located between the anterior horns of the spinal cord; composed of groups of axons of both ascending and descending tracts. **483**

anterior communicating artery (Latin: *ante* - before) an artery that unites the anterior cerebral arteries in the cerebral arterial circle; supplies blood to portions of the frontal lobe of the cerebrum. **617**

anterior cruciate ligament (Latin: *ante* - in front; *cruciatus* - resembling cross; *ligamentum* - a band) intracapsular ligament that extends from anterior, superior surface of the tibia to the inner aspect of the lateral condyle of the femur. **275**

anterior horn (*aka* ventral horn) (Latin: *ante* - front) a region of gray matter in the spinal cord, which contains multipolar motor neurons. **483**

anterior inferior iliac spine (aka AIIS) (Latin: *ante* - before; *inferior* - lower; *spina* – backbone) small protuberance on the anterior margin of the ilium, inferior to the anterior superior iliac spine. **243**

anterior interventricular artery (*aka* left anterior descending artery, LAD) (Latin: *anterior* - before; Greek: *cardiac* - *kardia* meaning heart) major branch of the left coronary+D4vessels that parallel the small cardiac arteries and drain the anterior surface of the right ventricle; bypass the coronary sinus and drain directly into the right atrium. **597**

anterior interventricular sulcus (Latin: *anterior* - before; Greek: *cardiac* - *kardia* meaning heart) the groove in the heart wall in which the anterior interventricular artery sits. **597**

anterior median fissure (Latin: *ante* - front; *medianus* - middle; *fissura* - cleft) a deep midline groove in the anterior spinal cord, which marks the separation between the right and left sides of the cord. **483**

anterior ramus the anterior division of a spinal nerve supplying the lateral trunk and limbs. **484**

anterior root (aka ventral root) (Latin: *ante* - front) a bundle of axons of multipolar motor neurons that transmits nerve impulses from the anterior horn of the spinal cord to the skeletal muscles; briefly runs through a spinal nerve along with a posterior root. **483**

anterior superior iliac spine (aka ASIS) (Latin: *ante* - before; *superior* - upper; *spina* - backbone) rough, rounded projection on the anterior end of the iliac crest. **243**

anterior tibial artery (Latin: *ante* - before; *tibia* - shinbone) an artery that branches off of the popliteal artery; provides blood to the muscles and skin of the anterior and lateral areas of the leg, and continues as the dorsalis pedis artery in the foot. **623**

anterior tibial vein the anterior of the two tibial veins, which drains blood from the dorsal venous arch in the foot and empties it into the popliteal vein. **623**

antibodies (*aka* immunoglobulins, gamma globulins) (Greek: *anti* - opposite) proteins produced by B lymphocytes that defend the body from infection and disease, by acting against specific antigens on pathogens such as bacteria, viruses, and fungi via direct attack, inflammation, or complement activation. **579**

antidiuretic hormone (ADH) (aka ADH, vasopressin) (Greek: *anti* - against; *diouretikos* - promoting urine) a peptide hormone, produced by the hypothalamus and released by the posterior pituitary gland, that promotes the reabsorption of water in the distal convoluted tubules and collecting ducts of the nephrons; this decreases the urine volume and increases the blood volume and blood pressure. **548**

antigens (Greek: *anti* - opposite; *genes* - produced) a molecule that evokes an immune response in the body, since it is recognized as being "non-self" or foreign by the immune system. **579**

antrum (Greek: *antron* - a cave) a cavity inside a mature (tertiary) ovarian follicle that is filled with fluid. **848**

anus the exterior opening of the rectum. **750**

aorta the largest artery of the systemic circuit and the body, which transports blood from the left ventricle to all body tissues except the respiratory tissues of the lungs. **615**

aortic arch the curved portion of the aorta, which lies between the ascending aorta and the descending aorta. **615**

aortic semilunar valve (aka aortic valve) (Latin: *semi* - half; *lunaris* - moon) valve in the heart located between the left ventricle and the aorta; prevents backflow of blood from the aorta into the left ventricle while the heart is relaxing. **602**

apex (Latin: *apex* - summit or tip) the "point" of the heart at the inferior aspect 000; the tip of the external nose, which protrudes more anteriorly than any other portion of the nose 000; the pointed region at the superior tip of each lung. **598, 687**

apical (Latin: *apex* - summit or tip) generally, the part of a cell or tissue facing an external space or internal environment. **115**

aquaporins channels that transport water through the plasma membrane of a cell; activated in the collecting ducts of the kidney by ADH. **84**

aqueous humor (Latin: *aqua* - water; *umor* - liquid) watery fluid that fills the anterior cavity of the eye. **15**

arachnoid mater (Greek: *arachnē* - cobweb; *eidos* - resemblance; Latin: *mater* - mother) the middle layer of the three meninges; a thin, weblike membrane that forms a loose sac around the brain and spinal cord. **453**

arbor vitae (Latin: *arbor* - tree; *vita* - life) the inner white matter tracts that form a treelike shape in the cerebellum. **459**

arcuate arteries (Latin: *arcuatus* - bent like a bow) arteries that transport blood from the interlobar arteries to the cortical radiate (intralobular) arteries in the kidney; form arches around the bases of the renal pyramids. **795**

arcuate line a curvilinear structure. **243**

arcuate veins (Latin: *arcuatus* - bent like a bow) a vein that transports oxygen-poor blood from the cortical radiate (intralobular) veins to the interlobar veins in the kidney; arches around the base of each renal pyramid. **795**

arrhythmia irregularities in the beating of the heart. **655**

arteries (*sing*: artery) (Greek: *arteria* - pipe) blood vessels that carry blood away from the heart. **611**

arterioles (*aka* resistance vessel) a blood vessel with a small diameter that transports blood from an artery to a capillary. **611**

artery blood vessel that carries blood away from the heart. **615**

articular capsule (Latin: *articulo* - to articulate; *capsula* - box) structure made of fibrous connective tissue that forms the walls of the joint cavity of a synovial joint. **271**

articular cartilage (Latin: *articulo* - to articulate; *cartilago* - gristle) thin layer of hyaline cartilage covering the epiphyses, or articulating surfaces of the bones, at a synovial joint; reduces friction and absorbs shocks. **271**

articular surface of the lateral condyle the location at which the lateral condyle meets the articular surface of another bone. **247**

articular surface of the medial condyle the location at which the medial condyle meets the articular surface of another bone. **247**

arytenoid small, paired cartilages of the larynx that help to control the tension of the vocal cords. **694**

ascending aorta (Latin: *ascendere* - to climb up) the first portion of the aorta, which extends upward out of the left ventricle; transports blood between the left ventricle and the aortic arch. **615**

ascending colon (Latin: *ascendere* - to climb up) the region of the colon between the cecum and the transverse colon, which runs upward in the right side of the abdominopelvic cavity. **750**

ascending limb (Latin: *ascendere* - to climb up) the portion of a nephron loop (loop of Henle) that transports filtrate from the descending limb of the nephron loop to the distal convoluted tubule (DCT) in each nephron. **791**

astrocytes (Greek: *astron* - star; *kytos* - cell) star-shaped glial cells that maintain concentration of chemicals in the extracellular space, remove excess signaling molecules, and react to tissue damage. **369**

atlas (Greek: *Atlas* - the mythic titan who supported the heavens on his shoulders) first cervical vertebra. **215**

auditory tube the tube linking the nasopharynx to the middle ear; also called the Eustachian tube. **425, 691**

auricle (*aka* auricula) (Latin: *auris* - an ear) (1) fleshy portion of the external ear; located on the lateral aspect of the head; (2) extension of an atrium visible on the superior surface of the heart. **425, 597**

auricular surface (Latin: *auris* - an ear) rough area on the posterior, medial side of the ilium of the os coxae. **243**

auscultation the act of listening to sounds of the body during a physical examination. **637**

autonomic nervous system (ANS) functional division of the nervous system responsible for involuntary control of the body, usually to maintain homeostasis. **364**

avascular without blood supply. **115**

axial skeleton (Greek: *skeletos* - skeleton) the bones of the head, neck, torso, and back. **193**

axillary artery (Latin: *axilla* - armpit) a continuation of the subclavian artery as it passes behind the clavicle; runs between the subclavian artery and the brachial artery, and supplies blood to the shoulder, upper arm, and chest. **615**

axillary vein (Latin: *axilla* - armpit) a vein in the axillary region that drains blood from the upper limb; arises from the merging of the brachial and basilic veins and continues to become the subclavian vein. **617**

axis the second cervical vertebra. **215**

axon (pl: axons) (Greek: *axōn* - axis) part of the neuron that extends from the cell body to another cell. **127, 365**

axon hillock (Greek: *axon* - axis) tapered region of the cell body of the neuron where the axon emerges; location where signals from different dendrites converge to become a single signal. **365**

axon terminals (*aka* synaptic boutons, terminal boutons, end feet) (Greek: *axon* - axis;

Latin: *terminus* - a limit) the end of the axon; usually includes several branches that extend toward the target cell. **365**

B

ball-and-socket joint synovial joint formed by the articulation between the rounded head of one bone and the bowl-shaped socket of the adjacent bone. **267**

basal generally, the part of a cell or tissue facing the extracellular matrix. **115**

basal epithelial cells (aka basal cells) stem cells inferior to the taste buds that can reproduce to replace the taste receptor cells if they age or are damaged. **400**

basal nuclei (aka basal ganglia) (Latin: *nucleus* - a little nut) nuclei of the cerebrum (with a few components in the upper brainstem and diencephalon) responsible for cognitive processing, planning and initiation of movements. **457**

base (Latin: *basis* - bottom) (1) compound that accepts hydrogen ions (H+) in solution 00; (2) bottom of a structure 000; (3) the flat portion at the posterior of the heart formed by the merging of the pulmonary veins with the left atrium. **598**

basement a skin or membrane that covers a part of your body) a thin layer of fibrous material that supports and anchors the basal epithelium; made up of the lamina lucida and the lamina densa. **115**

basilar artery an artery formed from the union of the vertebral arteries at the base of the brain; supplies blood to several regions of the brain. **617**

basilar membrane a membrane that forms the floor of the cochlear duct and the roof of the scala tympani. **426**

basilic vein a superficial vein that courses up the medial side of the arm; originates from the palmar and dorsal venous arches and merges with the brachial vein in the upper arm; empties into the axillary vein. **617**

basophils (Latin: *basis* - base; *philus* - loving) granulocytes that secrete heparin and histamine to enhance inflammation; make up the smallest percentage of leukocytes, and are stained by basic (alkaline) dyes. **568**

beta cells a pancreatic islet cell type that produces the hormone insulin. **549**

biaxial type of joint that allows for motion within two planes. **266**

bicipital groove (aka intertubercular groove, sulcus intertubercularis) (Latin: *bi* - two; *caput* - head) narrow groove between the greater and lesser tubercles of the humerus. **235**

bile duct (Latin: *bilis* - fluid secreted by the liver) a green, alkaline mixture of substances produced by the liver, which functions in lipid emulsification and excretion of wastes, such as bile pigments, from the body. **759**

biliary apparatus (Latin: *bilis* - fluid secreted by the liver; *apparatus* - equipment)

a network of interconnected ducts between the pancreas, liver, gallbladder, and duodenum, which functions in the secretion and excretion of bile. **749**

bipolar cells (Latin: *bi* - two) cell type in the retina anterior to the photoreceptor cells; connect the photoreceptors to the ganglion cells. **416**

blastocyst a stage of embryo development that includes the trophectoderm and an inner cell mass. **871**

blood pressure force or hydrostatic pressure applied by the blood against the wall of a blood vessel; expressed in units of mm Hg. **645**

body the main and middle region of the stomach. **202, 213, 749, 833**

body of the pancreas the longest portion of the pancreas; located posterior to the stomach. **750**

body of the rib the angle and anterior surface of the rib, excluding the head, neck, and tubercle. **222**

bone markings surface features on bones, such as depressions and projections, used to identify the locations of other body structures. **193**

Bowman's capsule (aka capsula glomeruli, glomerular capsule) the cup-shaped beginning of a renal tubule; along with the glomerulus, makes up the renal corpuscle; receives the filtrate from the glomerulus during glomerular filtration and transports it to the proximal convoluted tubule. **791**

brachial artery (Latin: *brachium* - arm) a continuation of the axillary artery as it passes into the upper arm; provides blood to the upper arm, and terminates in the elbow region by dividing into the radial and ulnar arteries. **615**

brachial vein (Latin: *brachium* - arm) a deep vein of the arm that arises from the merging of the radial and ulnar veins in the forearm; empties into the axillary vein. **617**

brachiocephalic trunk one of the great vessels of the aortic arch; supplies blood to the head and neck. **615**

brachium (Latin: *brachium* - arm) the portion of the upper limb located between the shoulder and elbow joints. **235**

brainstem region of the brain including the medulla oblongata, pons and midbrain. **460**

bridge the region of the external nose that lies between the root and the lower portion of the nose; formed by the nasal bones. **687**

bronchial artery (Greek: *bronkhos* - windpipe) an artery that branches off of the thoracic aorta, and is part of the systemic circuit; supplies blood to the tissues of the bronchi, lungs, and pleural membranes. **617**

bronchial tree (Greek: *bronkhos* - windpipe) the name for the upside-down branching structure of the respiratory system; the trachea represents the tree trunk, the bronchi represent the tree limbs, and the bronchioles represent the smaller branches of the tree. **684, 699**

bronchial vein (Greek: *bronkhos* - windpipe) a vein of the systemic circuit that drains blood from the lungs, bronchi, and pleural membranes; empties into the azygos vein. **620**

bronchi (Greek: *bronkhos* - wind pipe) tubular organs of the respiratory system; the primary bronchi branch directly off the trachea, the secondary bronchi branch off the primary bronchi, and the tertiary bronchi branch off the secondary bronchi. **664**

bronchoconstriction (Greek: *bronkhos* - windpipe; Latin: *constrictionem* - drawing together) a decrease in the diameter of the bronchi and bronchioles, due to contraction of the smooth muscle in their walls. **705**

bronchodilation (Greek: *bronkhos* - windpipe; Latin: *dilatare* - enlarge) an increase in the diameter of the bronchi and bronchioles, due to relaxation of the smooth muscle in their walls. **705**

bronchus-associated lymphatic tissue (BALT) (Greek: *bronkhos* - wind pipe; Latin: *lympha* - water) lymphoid nodules found in the mucosa of the respiratory tract, especially the bronchi; a type of MALT (mucosa-associated lymphoid tissue), which protects against infection by pathogenic microorganisms entering the body in inhaled air. **663**

brush border enzymes (*aka* striated border; brush border membrane) the luminal surface of the proximal convoluted tubules in the kidney, formed by microvilli on the surface of the simple cuboidal epithelial cells; greatly increases surface area for reabsorption and secretion of various substances. **767**

bulbourethral glands (*aka* Cowper's glands) (Latin: *bulbus* - bulb-shaped; *urethra* - passage for urine) small glands of the reproductive system of a biological male that secrete a mucus-containing fluid that lubricates the penis during sexual intercourse and becomes part of the semen during ejaculation. **841**

bursa (pl: bursae) (Latin: *bursa* - purse) small sac of synovial fluid surrounded by connective tissue; prevents friction between adjacent structures by separating structures and providing cushioning. **272**

C

calcaneal tendon (aka Achilles tendon) (Latin: *calcaneum* - heel) strong tendon that inserts into the calcaneal bone of the ankle; site of insertion of the superficial muscles in the posterior compartment of the lower leg. **248**

calcaneus (aka heel bone) (Latin: *calcaneum* – heel) large posterior, inferior tarsal bone; forms the heel of the foot. **248**

calcitonin a peptide hormone produced by the parafollicular cells of the thyroid gland; functions to decrease blood calcium levels when they are too high. **549**

callus (Latin: *callus* - hard skin) a fibrocartilaginous template for later mineralization formed during bone growth. **186**

calyces (*sing*: calyx) (Greek: *kalyx* - seed pod) cuplike drainage tubes that receive urine from the collecting ducts and transport it to the renal pelvis of the kidney. **791**

canaliculi (sing: canaliculus) (Latin: *canaliculus* - a small channel) channels within the bone matrix. **180**

capillaries (Latin: *capillaris* - hairlike) the smallest of the blood vessels, which exchange oxygen, carbon dioxide, nutrients, and wastes between the blood and the cells. **611**

capitate (Latin: *caput* - head) from the lateral side, the head-shaped third of the four distal carpal bones; articulates with the scaphoid and lunate proximally, the trapezoid laterally, the hamate medially, and primarily with the third metacarpal distally. **236**

capitulum (aka lateral condyle of the humerus) (Latin: *caput* - head) knoblike bony structure of the lateral, distal end of the humerus; located lateral to the trochlea. **235**

capsular space within Bowman's capsule, the space between the visceral and parietal layers. **791**

capsule tough, fibrous tissue that surrounds the kidney. **671**

cardia (aka cardiac region) (Greek: *cardia* – heart) the region of the stomach that encircles the opening through which food enters the stomach from the esophagus; named for its proximity to the heart. **749**

carina a carrtilaginous ridge located at the base of the trachea; separates the openings of the right and left bronchi. **699**

carotid canals (Greek: *karotides* - great arteries of the neck; Latin: *canal* - channel) the zig-zag passageway through which the internal carotid artery enters the skull; enters on the inferior aspect, anteromedial to the styloid process and opening to the middle cranial cavity near the posterior-lateral base of the sella turcica. **199**

carpal bones (Greek: *karpus* - wrist) one of the eight bones that form the wrist and base of the hand. **235**

cartilaginous joint (Latin: *cartilago* - gristle) joint at which the adjacent bones are united by hyaline cartilage or fibrocartilage. **265**

cauda equina (Latin: *cauda* - tail; *equina* - horse) a long bundle of spinal nerve roots descending below the first lumbar vertebra, encased within the inferior portion of the spine. **487**

caudate lobe the posterosuperior surface of the right lobe of the liver; bounded on the right by the fossa of the inferior vena cava, the left by the fossa of the ductus venosus and the ligamentum venosum, and is superior to the porta hepatis. **750**

cecum (aka caecum) (Latin: *caecum* - gut) a pouchlike structure that forms the beginning of the large intestine; lies between the ileum of the small intestine and the ascending colon of the large intestine. **750**

celiac trunk (aka celiac artery) (Latin: *coeliacus* - pertaining to bowels) a single artery that arises from the abdominal aorta, and branches into the left gastric artery, splenic artery, and common hepatic artery. **618**

cell cycle (pl: cell cycles) (Latin: *cella* - chamber; Greek: *kyklos* - circles) life cycle of a single cell, until its division into two new daughter cells. **55**

cell membrane (pl: cell membranes) (aka plasma membrane) (Latin: *cella* - chamber; *membrana* - a skin that covers part of the body) membrane surrounding all cells, composed primarily of phospholipids. **46**

cementum a calcified type of connective tissue, similar to bone, that helps to hold each tooth in place; connects the root of the tooth to its periodontal ligament. **745**

central adaptation sensory adaptation occuring in the brain. **438**

central canal (aka Haversian canal) (Latin: *canal* - a small channel) 1) longitudinal channel in the center of the osteon that contains blood vessels, nerves, and lymphatic vessels; 2) a long tubular channel running down the center of the spinal cord; circulates cerebrospinal fluid. **179, 455**

central nervous system (CNS) anatomical division of the nervous system; includes the brain and spinal cord. **364, 480**

centrioles (*pl*: centrioles) (Greek: *kentron* - center) small organelle is the origin of the microtubule spindle that moves chromosomes during cell replication. **51**

cephalic vein (Latin: *cephalicus* - pertaining to head) a superficial vein that courses up the lateral side of the arm; originates from the dorsal venous arch and empties into the axillary vein. **617**

cerebellum (Latin: *cerebellum* - little brain) a region of the adult brain that develops from the metencephalon; responsible for balance, coordination, and complex motor functions. **459**

cerebral aqueduct (aka aqueduct of Sylvius) (Latin: *cerebrum* - brain; *aquaeductus* - canal) a tubular channel that transports cerebrospinal fluid between the third and fourth ventricles; located in the midbrain. **455**

cerebral arterial circle (*aka* circle of Willis) (Latin: *cerebrum* - brain) a circular arterial anastomosis located at the base of the brain, composed of arteries that supply blood to the brain; consists of many branches entering and exiting the circle; resembles a traffic circle. **617**

cerebral cortex (Latin: *cerebrum* - brain) the outer layer of the cerebrum; consists of gray matter and is marked by ridges and indentations known as gyri and sulci. **452**

cerebrospinal fluid (CSF) (Latin: *cerebrum* - brain; *spinalis* - spine) fluid within the subarachnoid space that provides a liquid cushion and nutrition to the brain and spinal cord. **370, 455**

cerebrum (Latin: *cerebrum* - brain) a large region of the adult brain that develops from the telencephalon; responsible for higher neurological functions including memory, language, and consciousness. **452**

cervical enlargement the section of the spinal cord from C5 to T1; spinal cord expansion correspoding to the arms. **488**

cervical vertebrae the seven vertebrae (C1-C7) of the neck. **214**

cervix (Latin: *cervix* - neck) the inferior portion of the uterus, that connects to the vagina. **841**

cheeks the area of the face below the eyes, bordered laerally by the nose and ear. **743**

chemical digestion the breakdown of food via the action of enzymes on specific chemical bonds within the food molecules. **743**

chondrocytes (Greek: *chondrion* - cartilage; *kytos* - cell) cartilage cells; cells that are differentiated from chondroblasts and occupy lacunae. **129**

chordae tendineae (*aka* tendinous cords) (Latin: *chord* - string or rope; *tendo* - stretch) stringlike extensions of tough connective tissue that extend from the flaps of the atrioventricular valves to the papillary muscles. **601**

chorion the outermost membrane; with the amnion, forms the amniotic sac. **867**

chorionic villi (placental villi) finger-like projections of the placenta that increase placental surface area for nutrient and gas exchange with maternal blood. **877**

choroid (aka choroida, choroid coat) (Greek: *chorioeidēs* - like a membrane) highly vascular loose connective tissue in the wall of the eye that provides blood supply to the outer retina; forms the middle layer of the wall of the eye. **416**

choroid plexus (Greek: *choroeidēs* - membrane; Latin: *plexus* - braid) a structure within the ventricles of the brain containing blood capillaries lined by ependymal cells that filter blood to produce CSF. **455**

chromaffin cells (Greek: *chrōma* - color; Latin: *affinis* - affinity) neuroendocrine cells that make up the adrenal medulla; produce epinephrine and norepinephrine. **531**

chromatophilic substance granules found in the cytoplasm of nerve cells; stained by basic dyes. **365**

chromosomes (*pl*: chromosomes) (Greek: *chrōma* - color; *sōma* - body) condensed version of chromatin. **55**

cilia (sing: cilum) (Latin: *cilium* - eyelid) small appendages on some cells that include microtubules; capable of a sweeping motion across the cell surface. **53**

ciliary body (aka ciliary bodies) (Latin: *ciliaris* - relating to the eye) a mass of smooth muscle attached to the lens of the eye by suspensory ligaments; controls lens shape through the zonule fibers. **416**

circular layer circular muscle fibers that surround organs including the stomach and intestines. **757**

circumduction (Latin: *circum* - around; *ductus* - to draw) circular movement; involves a sequential combination of flexion, adduction, extension, and abduction. **280**

circumferential lamellae layers of bone tissue that extend around the surface of the bone. **180**

circumflex artery (Latin: *circum* - all around; Greek: *arteria* - pipe) branch of the left coronary artery that follows the coronary sulcus. **597**

cisterna chyli (aka receptaculum chyli) (Latin: *cisterna* - underground water reservoir) a saclike vessel that forms the beginning of the thoracic duct; formed from the merging of the two lumbar trunks and the intestinal trunk. **668**

clavicle (aka collarbone) an S-shaped bone that connects the sternum to the scapula. **231**

clavicular notches shallow depressions located on the superior-lateral sides of the sternal manubrium. **222**

cleavage a series of mitotic cell divisions that occur in the egg following fertilization. **878**

clitoris (aka glans clitoris) (Greek: *kleis* - a key) a small mass of erectile tissue that lies anterior to the vestibule in biological females; contains many nerves and is sensitive to sexual stimulation. **841**

clonus rhythmic, involuntary muscle contractions. **507**

coarse focus a control on the microscope that controls relatively large changes in the height of the microscope stage. **69**

coccyx (aka tailbone) (Greek: *kokkyx* - the coccyx) bone located at inferior end of the adult vertebral column; formed by the fusion of four coccygeal vertebrae during adulthood. **225**

cochlea (Latin: *cochlea* - snail) fluid-filled tube located in the inner ear; contains structures to detect and transduce sound stimuli. **426**

cochlear duct (aka scala media) (Latin: *cochlea* - snail) space within inner ear containing the spiral organ (organ of Corti); adjacent to the scala tympani and scala vestibuli. **426**

collateral ganglia (*aka* prevertebral ganglia) (Latin: *prae* - before; *vertebral* - joint) sympathetic ganglia, which lie anterior to

the vertebral column; consist of the celiac, superior mesenteric, and inferior mesenteric ganglia; run between the sympathetic chain ganglia and abdominal organs that they regulate. **525**

collecting ducts (*aka* duct of Bellini) the most distal portions of the renal tubules, which collect filtrate from the distal convoluted tubules (DCTs) of several nephrons. **795**

colloid (Greek: *kolla* - glue; *eidos* - appearance) 1) liquid mixture in which the solute particles consist of clumps of molecules large enough to scatter light; 2) a viscous fluid filling the central cavity of thyroid follicles; precursor to thyroid hormones. **551**

colon (aka large intestine) the main part of the large intestine, which lies between the cecum and rectum; consists of ascending, transverse, descending, and sigmoid portions. **750**

common bile duct (aka bile duct) a duct formed by the convergence of the common hepatic duct and the cystic duct; transports bile from the liver and gallbladder to the duodenum via the hepatopancreatic sphincter. **750**

common hepatic artery (Latin: *hepaticus* - pertaining to the liver) an artery that branches off the celiac trunk, and supplies blood to the liver, the gallbladder, portions of the stomach, and the pancreas. **618**

common iliac arteries (Latin: *iliacus* - pertaining to colic) a terminal branch of the abdominal aorta that divides into the internal and external iliac arteries; supplies blood to the pelvis, the lower abdominal area, and the lower limbs. **615**

Compact bone dense, strong bone type; located under the periosteum, provides support and protection. **173**

compound microscope a microscope composed of both an objective lens and an eyepiece. **67**

concentration gradient (pl: concentration gradients) (Latin: *con* - together; *centrum* - center) difference in the concentration of an ion or molecule between two locations. **81**

concentric lamella (Latin: *lamina* - leaf) concentric rings of calcified matrix that support osteons. **179**

condenser a lens that focuses light on the stage of a microscope. **69**

conducting zone the structures that provide passageways for air to travel to and from the lungs, including the nose, pharynx, larynx, trachea, bronchi and bronchioles. **684**

condyloid synovial joint formed by the articulation of a shallow depression at the end of one bone and a rounded end from a second bone. **267**

cones photoreceptor cells located in the retina; responsive to bright light. **416**

conjunctiva (Latin: *conjugo* - to join together) a layer of stratified columnar epithelium which forms the inner lining of the eyelid, and folds back to form a covering over the sclera. **413**

connective tissue (French: *tissu* - woven) type of tissue that serves to hold in place, connect, and integrate the body's organs and systems. **109**

contractile (Latin: *contractus* - to draw together) the ability to shorten with force by pulling on attachment points. **123**

contralateral reflexes (Latin: *latus* - side; *reflexus* - a bending back) reflexes that begin and end on opposite sides of the body. **498**

coracoid process (Greek: *korakōdēs* - like a raven's beak) lateral, hook-shaped anterior projection from the superior margin of the scapula. **231**

cornea (Latin: *corneus* - horny) clear portion of the fibrous layer of the eye; lies anterior to the iris and pupil and helps to refract light rays entering the eye. **416**

corniculate cartilages small, paired cartilages of the larynx that help control the tension of the vocal cords. **694**

coronal suture (Latin: *corona* - crown) joint that joins the frontal bone parietal bones across the top of the skull. **198, 211**

corona radiata a white matter sheet carrying neurons from the cerebral cortex. **863**

coronary sinus (Latin: *corona* - crown) large, thin-walled vein on the posterior surface of the heart that lies within the atrioventricular sulcus and drains the heart myocardium directly into the right atrium. **598**

coronary vessels (Latin: *corona* - crown) the general term for both arteries and veins that carry blood to and from the heart wall. **597, 637**

coronoid fossa (Greek: *korōnē* - a crow; *eidos* - resembling; Latin: *fossa* - a trench) depression located on the anterior surface of the humerus superior to the trochlea; accommodates the coronoid process of the ulna when the elbow is maximally flexed. **235**

coronoid process of the mandible (Greek: *korōnē* - a crow; *eidos* - resembling; Latin: *processus* - advance; *mando* - to chew) anterior projection from the ramus; site of attachment for one of the biting muscles. **235**

coronoid process of the ulna (Greek: *korōnē* - a crow; Greek: *ōlenē* - ulna) projecting bony lip on the anterior, proximal ulna; forms the anterior edge of the trochlear notch. **202**

corpus luteum (Latin: *corpus* - body; *luteus* - yellow) an endocrine structure formed by the remnants of the ovarian follicle after ovulation; secretes estrogen, progesterone, and inhibin. **848**

cortex (*pl:* cortices) (Latin: *cortex* - bark of a tree) in hair, the second or middle layer of compressed keratinocytes 000; the outer region of a lymph node, which consists of follicles containing dividing B cells. **671, 787**

cortical nephrons (Latin: *cortex* - bark of a tree; Greek: *nephros* - kidney) nephrons that reside mainly in the renal cortex, whose nephron loops (loops of Henle) extend a small distance into the renal medulla. **795**

corticotropin-releasing hormone (CRH) a peptide hormone which stimulates the synthesis of adrenocorticotropic hormone in the pituitary gland. **556**

cortisol a glucocorticoid produced by the zona fasciculata of the adrenal cortex; promotes the breakdown of glucose and other stored nutrients in response to stress. **549**

costal cartilages (Latin: *costa* - ribs; *cartilago* - gristle) strips of hyaline cartilage that attach the ribs to the sternum. **222**

costal facet the point of articulation between a vertebra and a rib, located on the vertebrae. **216**

costal groove (Latin: *costa* - ribs) shallow groove along the inferior margin of the body of a rib that allows blood vessels and a nerve to pass from the spinal cord to the thoracic wall. **222**

cranial bones the 8 bones that form the neurocranium, including the ethmoid, sphenoid, frontal, occipital, right and left parietal and right and left temporal bones. **197**

cranial nerves twelve pairs of nerves that are directly connected to the brain; responsible for the sensory and motor functions of the head, neck, thoracic and abdominal organs. **461**

cranial reflexes (Greek: *kranion* - skull; Latin: *reflexus* - a bending back) reflexes that involve the brain. **497**

craniosacral system (*aka* parasympathetic system) the parasympathetic division of the autonomic nervous system; axons emerge from the central nervous system through the brainstem and sacral portion of the spinal cord; associated with maintaining homeostasis while a person is in a "rest and digest" or relaxed state. **525**

cremaster muscle muscles that descend from the internal oblique muscle of the abdominal wall and cover each testis like a muscular net that is responsible for elevating the testes. **841**

crenation a shape with a scalloped or round-toothed edge. **97**

cribriform plates (Latin: *cribrum* - a sieve; *forma* - form; Greek: *platys* - flat) small, flattened areas with many tiny holes located on the sides of the crista galli on the ethmoid bone. **200**

cricoid cartilage (Latin: *cartilago* - gristle) a ring-shaped mass of hyaline cartilage that lies between the thyroid cartilage and the trachea in the larynx. **693**

crista galli (Latin: *crista* - crest; *galli* - rooster) small bony projection of the ethmoid bone located at the midline of the cranial floor; provides an anterior attachment point for a covering layer of the brain. **200**

crossed-extensor reflex (Latin: *reflexus* - a bending back) contralateral reflexes that activate the opposite side of the body, usually due to painful stimulus that needs to be avoided. **499**

crossing over the exchange of genetic material between non-sister chromatids of two homologous chromosomes; occurs between prophase I and metaphase I of meiosis. **845**

crown the region of a tooth that protrudes externally from the gums (gingiva). **743**

cuboid a three dimensional shape with rectangular faces. **248**

cuneiform cartilages small, paired cartilages of the larynx that help control the tension of the vocal cords. **694**

cystic artery (Greek: *kystis* - bladder) an artery that branches off of the common hepatic artery, and provides blood to the gallbladder. **618**

cystic veins veins that drain the gallbladder. **622**

cytokinesis (Greek: *kytos* - cell; *kinesis* - movement) final stage in cell division, where the cytoplasm divides to form two separate daughter cells. **55, 56**

cytoplasm (Greek: *kytos* - a hollow cell; *plasma* - thing formed) internal material of the cell including the water-based fluid called cytosol, and all the other organelles and solutes. **49**

cytoskeleton (Greek: *kytos*: a hollow cell; *skeletos* - skeleton) "skeleton" of a cell; formed by fibrous proteins that support the cell's shape and enable movement. **51**

D

dartos muscle the muscles that make up the subcutaneous muscle layer of the scrotum. **841**

deep describes a position farther from the surface of the body. **24, 312**

deep brachial arteries (Latin: *brachio* - arm) arteries that provide blood to the posterior surface of the arm. **615**

deep femoral artery (*aka* deep artery of the thigh) (Latin: *femur* - thigh) an artery that branches off of the femoral artery; supplies blood to the thigh muscles and branches to form the lateral and medial circumflex arteries. **623**

deep femoral vein (*aka* profunda femoris vein) (Latin: *femur* - thigh) a deep vein that drains the deep regions of the thigh, and empties to the femoral vein. **623**

deltoid tuberosity (Greek: *deltoeidēs* - resembling the shape of a delta; Latin: *tuberosus* - full of lumps) roughened protrusion on the lateral mid-shaft of the humerus. **235**

denaturing (Latin: *de* - from; Greek: *hydr* - water; *syn* - together; *thesis* - arranging) change in the structure of a molecule most often a protein. **771**

dendrites (sing: dendrite) (aka branches) (Greek: *dendrites* - related to a tree) part of the neuron that receives signals from neighboring cells. **127, 365**

dens bony projection that extends upward from the body on the anterior side of the second cervical vertebra. **215**

dentin (Latin: *dens* - tooth) a hard bone-like material that lies just under the enamel of a tooth crown; makes up a large majority of the tooth. **745**

depression (Latin: *depressio* - to press down) inferior movement of the mandible or scapula. **281**

dermal papilla (pl: dermal papillae) (Greek: *derma* - skin) projections of the papillary layer of the dermis that increase surface contact between the epidermis and dermis. **157**

dermis (Greek: *derma* - skin) layer of tissue between the epidermis and hypodermis; composed mainly of connective tissue; contains blood vessels, lymph vessels, nerves, and other structures. **153**

descending moving or sloping downwar. **615**

descending aorta (Latin: *descendere* - come down) the portion of the aorta that extends inferiorly from the aortic arch to the termination point, where it splits into the common iliac arteries; divided into thoracic and abdominal segments. **615**

descending colon the portion of the colon that runs down the left side of the body; connects the transverse colon to the sigmoid colon. **750**

descending limb (aka nephron loop) (Latin: *descendere* - come down) the portion of a nephron loop (loop of Henle) that transports filtrate from the proximal convoluted tubule (PCT) to the ascending limb of the nephron loop in each nephron. **795**

desmosomes (sing: desmosome) (Greek: *desmos* - a band; *soma* - body) a type of anchoring junction that unites two neighboring cells structurally. **301**

diaphragma sellae (Greek: *diaphragma* - a partition wall) one of the cranial dural septa; a fold of dura mater that contains an opening for the pituitary stalk. **455**

diaphragm (Greek: *diaphragma* - barrier) a flat, dome-shaped muscular organ that separates the thoracic cavity from the abdominopelvic cavity; one of the main muscles of inspiration. **70, 684, 716**

diarthrosis (pl: diarthroses) (Greek: *di* - two; *arthron* - joint) joint at which a lot of motion is possible; freely mobile joint. **265**

diastole (Greek: *diastole* - dilation/relaxation) relaxation of the heart wall, also the term for the phases of the cardiac cycle in which relaxation occurs. **642**

diffuse lymphatic tissue unencapsulated lymphocytes located in the lamina of organs, associated with the epithelium. **660**

diffusion (Latin: *difusus* - to pour in different directions) the movement of molecules in response to concentration gradients. **81**

digital artery (Latin: *digitus* - finger or toe) arteries that arise from the superficial and deep palmar arches; provide blood to the digits (fingers). **615**

digital veins (Latin: *digitus* - finger or toe) veins that drain the digits (fingers or toes) and empty blood into the palmar arches (hand) or dorsal venous arch (foot). **617, 623**

diploid (aka 2n) (Greek: *diploos* - double; *eidos* - resemblance) condition of a cell having two sets of chromosomes, one set inherited from the sperm and one inherited from the egg. **845**

distal convoluted tubule (Latin: *distantem* - remote; *convolutus* - to roll together) the part of the nephron distal to the nephron loop, which transports filtrate between the nephron loop and the collecting ducts; conducts reabsorption and secretion of several substances. **791**

dorsal arch (aka arcuate arch, arcuate artery) (Latin: *dorsalis* - back) an arterial arch formed by the merging of the dorsalis pedis artery with the medial and plantar arteries; provides blood to the metatarsal region of the foot and the digits (toes). **623**

dorsalis pedis artery (Latin: *dorsalis* - back; *pedalis* - foot) an artery that branches off of the anterior tibial artery; provides blood to the ankle and dorsal portion of the foot. **623**

dorsal venous arch (Latin: *dorsalis* - back) a curved superficial vein that arises from the dorsal metatarsal and dorsal digital veins, and drains into the great and small saphenous veins and the tibial veins. **623**

dorsiflexion (Latin: *dorsalis* - back; *flecto* - to bend) motion in the ankle that lifts the top of the foot toward the anterior leg. **281**

dorsum nasi (Latin: *dorsalis* - of the back) the long, middle region of the external nose that lies between the bridge and the apex. **687**

dual innervation (Latin: *duo* - two) the regulation of organ function by both the sympathetic and parasympathetic divisions of the autonomic nervous system. **520**

ductus deferens (aka vas deferens) a tube that carries sperm from the epididymis to the ejaculatory duct. **836**

ductus venosus (Latin: *ductus* - pipe) a short vessel that transports oxygen-rich blood from the umbilical vein to the inferior vena cava in the fetus, bypassing the fetal liver. **877**

duodenum (Latin: *duodecim* – twelve) the first portion of the small intestine, which runs between the pyloric sphincter and the jejunum; receives chyme from the stomach and mixes it with bile and pancreatic juice. **749**

dura mater (Latin: *mater* - mother) the outermost layer of the three meninges; a thick, tough layer that surrounds and protects the brain and spinal cord. **453**

E

ECG surface recording of the electrical activity of the heart that can be used for diagnosis of irregular heart function. **637**

echocardiogram (Greek: *echo* - a returned sound; *cardiac* - kardia meaning heart) an ultrasound of the heart that allows blood flow to be visualized. **637**

effectors (Latin: *effector* - producer) an organ or muscle that responds to a signal. **363, 497**

efferent arterioles (Latin: *effere* - to carry away; *arteriola* - an artery) an arteriole that transports blood from the glomerulus to the peritubular capillaries in each nephron. **795**

efferent (Latin: *effere* - to bring out) a neuron that brings input away from the central nervous system. **363**

efferent signals signals traveling along an efferent neuron. **480**

electrocardiogram (ECG or EKG) (Greek: *echo* - a returned sound; *cardiac* - kardia meaning heart) surface recording of the electrical activity of the heart that can be used for diagnosis of irregular heart function. **637**

electrocardiologists cardiologist specializing in the electrical system of the heart. **637**

elevation (Latin: *elevo* - to lift up) superior movement of the mandible or scapula. **281**

embryo the developing organism from fertilization until the ninth week after conception. **872**

emulsify the proces of mixing two or more normally immiscilbe liquids. **777**

enamel the hard-bonelike covering over the crown of a tooth; overlies the dentin. **745**

endocardium (Greek: *endo* - within, inside) innermost layer of the heart lining the heart chambers and heart valves; composed of endothelium reinforced with a thin layer of connective tissue that binds to the myocardium. **601**

endocrine of or relating to hormones. **542**

endometrium (Greek: *endo* - within; *meter* - mother) the inner layer of the wall of the uterus; thickens during the proliferative and secretory phases of the uterine cycle and is shed during the menstrual phase. **841, 871**

endomysium (Greek: *endo* - within; *mys* - muscle) thin layer of collagen and reticular fibers that surrounds each muscle fiber within the fascicle; contains extracellular fluid and nutrients. **297**

endoplasmic reticulum (ER) (pl: endoplasmic reticula) (Greek: *endo*: within; plastos: formed; Latin: *reticulum* - a small net) cellular organelle that consists of connected membranous sacs. When associated with ribosomes, it is called rough ER, when not associated with ribosomes, called smooth ER. **49**

endosteum (Greek: *endo* - within; *osteon* - bone) the membranous lining of the medullary cavity of a bone. **179**

eosinophils (Greek: *eos* - dawn; Latin: *philus* - loving) granulocytes that secrete anti-inflammatory chemicals and substances that destroy parasitic worms; make up a small percentage of leukocytes, and are stained by the dye, eosin. **568**

ependymal cells (Greek: *ependyma* - an upper garment) glial cell type in the CNS that filters blood to produce cerebrospinal fluid. **370**

epicardium (Greek: *epi* - upon) the outermost layer of the heart wall. **601**

epididymis (*pl*: epididymides) (Greek: *epi* - on; *didymos* - testicle) a tubular organ of the reproductive system of a biological male, in which sperm reside and mature for about two weeks after leaving the testis. **841**

epigastric region the upper, central region of the abdomen, located between the costal margins and subcostal plane. **15**

epiglottis (Greek: *epi* - upon; *glossa* - tongue) an oval mass of elastic cartilage in the larynx; remains open to allow air to flow into the larynx and trachea and closes during swallowing to prevent food from entering these organs of the airway. **691**

epimysium (Greek: *epi* - upon; *mys* - muscle) sheath of dense, irregular connective tissue that surrounds an entire skeletal muscle. **297**

epinephrine (*aka* adrenaline) (Greek: *epi* - upon; *nephros* - kidney) signaling molecule (hormone) secreted by the adrenal medulla as part of the sympathetic response. **549**

epiphyseal line (Greek: *epi* - upon; *physis* - growth) remnant of the epiphyseal plate following ossification. **173**

epiphysis (Greek: *epi* - upon; *physis* – growth) wide sections at each end of a long bone. **173**

epithelial tissues (Greek: *epi* - upon, *thēlē* - nipple; French: *tissu* - woven) highly cellular tissue that serves primarily as a covering or lining of body parts. **109**

equilibrium (pl: equilibria) (Latin: *aequus* - equal) a dynamic state at which molecules or ions move randomly, but there is no net movement of molecules or ions in either direction of the cell membrane. **82**

erythrocytes (*aka* red blood cells, RBCs) (Greek: *erythro* - red; *kytos* - cell) blood cells that transport oxygen and some carbon dioxide through the blood; make up the vast majority of the formed elements of the blood. **129, 563**

esophageal artery (Greek: *oisophagos* - passage for food) an artery that branches off of the thoracic aorta; provides blood to a portion of the esophagus. **617**

esophageal hiatus (Greek: *oisophagos* - passage for food; Latin: *hiare* - gape) an opening in the diaphragm through which the esophagus passes from the thoracic cavity into the abdominopelvic cavity. **747**

esophageal vein (Greek: *oisophagos* - passage for food) a vein that drains blood from the inferior part of the esophagus, and empties into the azygos vein. **620**

esophagus (Greek: *oisophagos* - passage for food) a tubular organ of the digestive system, which transports food from the pharynx to the stomach via peristaltic waves. **747**

estrogen a class of sex hormones responsible for development of the femael reproductive system and secondary sex characteristics. **549**

ethmoid bone (Greek: *ēthmos* - sieve) unpaired bone located between the two eye orbits; forms a barrier between the nose and the brain. **200**

ethmoid sinuses small spaces separated by very thin bony walls located in the lateral aspects of the ethmoid bone. **200, 688**

eversion (Latin: *everto* - to overturn) the turning of the foot to angle the bottom of the foot away from the midline of the body. **281**

exhalation (*aka* expiration) (Latin: *exhalare* - to breathe out) the process of moving air out of the lungs into the atmosphere. **712**

expiratory reserve volume (ERV) the additional amount of air that can be exhaled after a normal exhalation. **719**

extension (Latin: *extendo* - to stretch out) movement in the sagittal plane that increases the angle of a joint. **280**

external acoustic canal a canal in the center of the temporal bone; tunnel through which sound reaches the inner ear. **425**

external acoustic meatus (Latin: *externus* - outside; *meatus* - to pass; Greek: *akoustikos* - sound) a canal in the center of the temporal bone; tunnel through which sound reaches the inner ear. **199**

external carotid artery (Latin: *externus* - outwards; Greek: *karotides* - great arteries of the neck) an artery that branches from the common carotid artery and supplies blood to various structures around the face, scalp, lower jaw, neck, esophagus, and larynx. **617**

external elastic membrane (Latin: *externus* - outside) a layer of elastic fibers that lies between the tunica media and the tunica externa in large arteries. **628**

external iliac artery (Latin: *externus* - outside; *iliacus* - pertaining to colic) an artery that branches off of the common iliac artery and provides blood to the lower limbs and a few structures of the anterior wall of the abdominal cavity. **622**

external iliac vein (Latin: *externus* - outside; *iliacus* - pertaining to colic) a continuation of the femoral vein as it enters the trunk of the body; drains blood from the legs and empties into the common iliac vein. **623**

external intercostal muscles (Latin: *externus* - outside; *inter* - between; *costa* – rib) short muscles between the ribs, which expand the thoracic cavity for inhalation. **716**

external jugular vein (Latin: *externus* - outside; *iugulum* - collarbone) a vein found in the superficial portion of the neck; empties blood from the superficial regions of the head and scalp; drains into the subclavian vein. **617**

external occipital protuberance (Latin: *externus* - outside) small bump at the midline on the posterior skull on the occipital bone; attachment site for a ligament of the posterior neck. **199**

extracellular fluid (ECF) (Latin: *extra* - outside; Greek: *kytos* - a hollow cell; Latin: *fluo* - to flow) fluid exterior to cells; includes the interstitial fluid, blood plasma, and fluid found in other reservoirs in the body. **80**

extrinsic (Latin: *extrinsecus* - from without) muscle that originates outside of the area in which it acts. **311, 329**

extrinsic ligaments ligament located on the outside of the articular capsule. **271**

F

facial bones fourteen bones that form the structures of the face, upper and lower jaws, and the hard palate. **197**

facial nerve seventh cranial nerve; regulates the muscles of facial expressions, part of the sense of taste and the production of saliva. **526**

facilitated diffusion (aka mediated diffusion) (Latin: *facilis* - easy; *diffuses* - to pour in different directions) process by which a substance moves from areas where it is more concentrated to less concentrated, but requires a membrane protein to move. **82**

falciform ligament a ligament connecting the liver to the anterior body wall; divides the liver into the left and right lobes. **750**

false ribs ribs 8–12; these ribs do not attach directly to the sternum. **222**

falx cerebelli (Latin: *falx* - sickle) one of the cranial dural septa that form a partition between the two halves of the cerebellum. **455**

falx cerebri (Latin: *falx* - sickle) the largest of the cranial dural septa; forms a partition between the right and left cerebral hemispheres. **455**

fat pad small masses of adipose tissue found at many joints, which cushion the bones of the joint. **272**

fauces (Latin: *fauces* – gullet) the passageway from the posterior portion of the oral cavity to the pharynx. **743**

femoral artery (Latin: *femur* - thigh) a continuation of the external iliac artery as it enters the thigh; supplies blood to the structures of the thigh, and is renamed to the popliteal artery as it runs down the posterior side of the knee. **623**

femoral vein (Latin: *femur* - thigh) a continuation of the popliteal vein in the thigh, which empties into the external iliac vein in the trunk of the body; drains blood from the thigh; tributaries include the great saphenous, the deep femoral, and the femoral circumflex veins. **623**

femur (aka thigh bone) (Latin: *femur* - thigh) long, single bone of the thigh. **247**

fertilization (Latin: *fertilis* - fruitful) The event in which a sperm encounters and penetrates an egg, resulting in the combining of their genetic material. **863**

fetus developing offspring from the ninth week after fertilization until birth. **863**

fibroblasts the most abundant cell type in connective tissue, secrete protein fibers that form connective tissue. **129**

fibrous capsule tough, fibrous tissue that surrounds the kidney. **787**

fibrous joint (Latin: *fibra* - fiber) joint at which the adjacent bones are directly connected by fibrous connective tissue. **265**

fibrous layer (Latin: *fibra* - fiber) the outer layer of the eye, which consists of dense connective tissue and provides a strong framework for the eye. **416**

fibrous pericardium one of two sublayers of the pericardium. **595**

fibrous skeleton of the heart the collective term for the valves of the heart, which are fibrous and nonconductive in composition. **601**

fibula (Latin: *fibula* - brooch) thin, lateral bone of the lower leg. **247**

fibular collateral ligament (Latin: *fibula* - brooch) extrinsic ligament on the lateral side of the knee joint that runs from the lateral epicondyle of the femur to the head of the fibula. **276**

fibular vein (Latin: *fibula* - brooch) a vein that drains blood from the muscles and skin in the region of the fibula, and empties into the popliteal vein. **623**

filtrate the fluid that occupies the renal tubules of the nephron; produced from the small molecules of the blood plasma that are filtered out of the glomerulus in the renal corpuscle. **791**

filtration slits long, narrow gaps in the pedicels (foot processes) of podocytes, the cells that surround the glomeruli; allow substances filtered by the glomeruli to enter the glomerular (Bowman's) capsule. **791**

filum terminale a thin strand of fibrous tissue extending inferiorly from the conus medullaris to the coccyx; inferior anchor of the spinal cord and spinal meninges. **488**

fine focus knob a precise control for the objective height of a microscope. **69**

fissures narrow cracks, grooves or openings. **452**

flagellum (pl: flagella) (Latin: *flagrum* - whip) appendage on certain cells formed by microtubules and modified for movement. **53**

flexion (Latin: *flecto* - to bend) movement in the sagittal plane that decreases the angle of a joint. **280**

floating ribs ribs 11–12; these ribs do not attach to the sternum or to the costal cartilage of another rib. **222**

follicle-stimulating hormone (FSH) (Latin: *folliculus* - a small bag) an anterior pituitary hormone, which stimulates the production and maturation of sex cells in the ovary or testis. **558**

follicular cells cells of the thyroid that secrete thyroid hormone. **549**

follicular fluid liquid surrounding the ovum in an ovarian follicle within the follicular antrum. **848**

foot terminal portion of a limb that bears weight and allows for locomotion. **247**

foramen lacerum (Latin: *foramen* - opening; *lacero* - to tear to pieces) irregular opening at the base of the skull, immediately inferior to the exit of carotid canal. **206**

foramen magnum (Latin: *foramen* - opening; *magnus* - large) large opening in the occipital bone of the skull through which the spinal cord emerges and the vertebral arteries enter the cranium. **199**

foramen ovale (Latin: *foramen* – opening; *ovum* - egg) 1) large opening in the floor of the middle cranial fossa; allows for the passage of the spinal cord through the skull; 2) an opening between the right and left atria in the fetal heart, which transports oxygen-rich blood from the right atrium directly into the left atrium; this bypasses the fetal lungs, as they are nonfunctional. **201, 601**

foramen rotundum (Latin: *foramen* - opening) round opening in the floor of the middle cranial fossa, inferior to the superior orbital fissure; exit point for sensory nerve supplying the cheek, nose, and upper teeth. **201**

foramen spinosum (Latin: *foramen* - opening; *spina* - thornlike projection) small opening in the floor of the middle cranial fossa, located posterior-lateral to the foramen ovale; entry point for artery supplying covering layers that surround the brain. **201**

formed elements the cells and fragments of the blood; consist of erythrocytes, all five types of leukocytes, and platelets; make up about 45 percent of the blood volume. **563**

fossa ovalis (Latin: *fossa* - ditch) oval-shaped depression in the interatrial septum that marks the former location of the foramen ovale. **601**

fourth ventricle (Latin: *ventriculus* - belly) one of the chambers in the brain that produces and circulates cerebrospinal fluid; lies between the cerebellum and the brainstem; opens into the subarachnoid space through the median and lateral apertures. **455**

fovea capitis (Latin: *fossa* - trench; *caput* - head) indentation on the head of the femur; serves as the site of attachment for a ligament. **247**

fovea centralis (Latin: *fovea* - a pit; *centralis* - central) an indented area in the center of the macula lutea that contains the highest density of cones and almost no rods. **421**

frontal bone unpaired bone that forms forehead, roof of orbit, and floor of anterior cranial fossa. **197**

frontal lobe a large region of the cerebrum that lies directly beneath the frontal bone; responsible for planning and initiating muscle contraction, including contractions for speech and eye movements, and higher-order cognitive behaviors. **457**

frontal plane (aka coronal plane) two-dimensional, vertical plane that divides a structure into anterior and posterior halves. **9**

frontal sinuses air-filled spaces within the frontal bone; just above the eyebrows. **197, 688**

functional residual capacity (FRC) the amount of air left in the lungs after a normal exhalation; can be calculated by taking the sum of the ERV and RV. **719**

functional unit conduct the filtration, reabsorption, and secretion processes of urine formation; consist of the renal corpuscles and renal tubules; empty urine into the collecting ducts. **292**

fundus (Latin: *fundus* - bottom) the rounded portion of the stomach, located above the cardiac region; often traps swallowed air 000; the rounded region of the uterus that extends superiorly to the points of entry of the uterine tubes. **749, 841**

G

G1 phase (*aka* gap one phase) first phase of the cell cycle, after a new cell is born. **55**

G2 phase (*aka* gap two phase) third phase of the cell cycle, after the DNA synthesis phase. **55**

gallbladder (Greek: *gaster* - stomach) an accessory organ of the digestive system, which stores and concentrates bile, and secretes it into the common bile duct via the cystic duct. **749**

gametes (Greek: *gamete* - a wife) sex cells (sperm or egg) that contain the genetic material to produce offspring; haploid cells. **838**

gametogenesis the formation of haploid gametes; the processes of sperm and egg formation. **837**

ganglia a group of neuron cell bodies in the peripheral nervous system. **520**

ganglion cells (Greek: *ganglion* - a swelling) cells that form the anterior layer of the retina, which synapse with the bipolar cells and leave the eye to become the optic nerve. **416**

gas exchange across the respiratory membrane the movement of respiratory gases (oxygen and carbon dioxide) between the lungs and the blood. **712**

gas exchange between the blood and tissues the movement of respiratory gases (oxygen and carbon dioxide) throughout the body within the blood. **712**

gas transport the movement of respiratory gases (oxygen and carbon dioxide) throughout the body within the blood. **712**

gastric veins veins that drain the stomach. **622**

gastrula a multi-layered structure formed in early embryonic development; includes ectoderm, endoderm and mesoderm. **872**

gender social, cultural and behavioral aspects of identity as a man, woman or other gender identity, which may or may not correspond to biological sex. **834**

general senses senses including touch, pain, temperature, proprioception, vibration and pressure; receptors are distributed throughout the body. **391, 408, 437**

genicular artery (Latin: *geniculatus* - having knots) an artery that branches off of the femoral artery; provides blood to the knee area. **623**

gingivae (sing: gingiva) (Latin: *gingivae* - gums) the gums that surround the neck of a tooth. **743**

glabella (Latin: *glabellus* - hairless) slight depression at the anterior midline of the frontal bone. **197**

glenoid cavity (aka glenoid fossa) (Greek: *glēnē* - socket of joint; Latin: *cavus* - hollow) shallow depression on the lateral scapula, forms the corner between the superior and lateral borders; articulates with the head of the humerus. **231**

glial cells (*aka* neuroglial cells) (Greek: *glia* - glue) various cell types that make up neural tissue excluding neurons; cells are responsible for maintenance and support of neurons. **127, 480**

glial (neuroglial) cells various cell types that make up neural tissue excluding neurons; cells are responsible for maintenance and support of neurons. **360**

glomerulus (*pl*: glomeruli) (Latin: *glomerationem* - ball) a cluster of capillaries encased in a glomerular (Bowman's) capsule; a renal corpuscle consists of a glomerulus and a glomerular capsule; filters small molecules out of the blood. **795**

glossopharyngeal nerve ninth cranial nerve; responsible for controlling muscles in the oral cavity and upper throat, as well as part of the sense of taste and the production of saliva. **526**

glottis (Greek: *glossa* – tongue) the opening between the vocal folds (vocal cords) through which air passes for sound production. **693**

glucagon (Greek: *glyco* - sweet; *agon* - push forward) a hormone secreted by the pancreas that acts upon the liver cells to inhibit glycogen synthesis and stimulate glycogen breakdown, releasing glucose into the blood. **549**

gluteal tuberosity (Greek: *gloutos* - buttock; Latin: *tuberosus* - full of lumps) roughened area of the proximal femur; extends inferiorly from the base of the greater trochanter. **247**

glycosidic bond an ether bond between a carbohydrate and another group. **767**

goblet cells (sing: goblet cell) (Greek: *kytos* - cell) mucous-secreting, unicellular exocrine glands found in columnar epithelium of mucous membranes. **115, 754**

golgi apparatus (*pl*: apparati) (Latin: *apparatus*: preparation) cellular organelle formed by a series of flattened, membrane-bound sacs that functions in protein modification, tagging, packaging, and transport. **50**

gomphosis (Greek: *gomphos* - bolt) fibrous joint that anchors the root of a tooth into its bony jaw socket via strong periodontal ligaments. **266**

gonadal arteries arteries that branch off of the abdominal aorta, and supply blood to the gonads; in females, supplies blood to the ovary and uterine tube, and is also called the ovarian artery; in males, supplies blood to the testis, and is also called the testicular artery. **620**

gonadal veins (Greek: *gonos* - child) a vein that drains the gonads; drains the ovary in the female and is also called the ovarian vein; in the male, drains the testis and is also called the testicular vein. **622**

gonadotropin-releasing hormone (GRH) peptide hormone released from the hypothalamus responsible for releasing follicle-stimulating hormone and luteinizing hormone from the anterior pituitary. **558**

gonads (Greek: *gonos* - child) the primary reproductive organs that produce sex cells (sperm or eggs) and sex hormones. **837**

granules vesicles in the cytoplasm of leukocytes that contain chemicals used in the immune response; used to destroy pathogenic microorganisms or to regulate the inflammatory response. **567**

granulosa cells (Latin: *granulum* - granule) the cells that surround and nourish the oocyte in an ovarian follicle. **848, 863**

great cardiac vein (Greek: *cardiac* - *kardia* meaning heart) vessel that follows the interventricular sulcus on the anterior surface of the heart and flows along the coronary sulcus into the coronary sinus on the posterior surface; parallels the anterior interventricular artery and drains the areas supplied by this vessel. **598**

greater curvature (aka curvatura ventriculi major) the broad inferior convex curve of the stomach. **749**

greater omentum (aka great omentum, omentum majus, gastrocolic omentum, epiploon) a large mesenterial sheet that hangs down from the inferior surface of the stomach like a curtain covering the small intestine. **747**

greater sciatic notch (Latin: *sciaticus* - corruption of; Greek: *ischiadikos* - hip joint) inverted U-shaped indentation on the posterior margin of the ilium, superior to the ischial spine. **243**

greater trochanter (Greek: *trochantēr* - a runner) on the lateral side of the neck of the femur; bony expansion of the femur that projects superiorly from the base of the femoral neck. **247**

greater tubercle (Latin: *tuberosus* - full of lumps) large prominence on the lateral side of the proximal humerus. **235**

greater wings of the sphenoid bone (Latin: *sphenoides* - resembling a wedge) lateral projections of the sphenoid bone that extend from the sella turcica, forming the base of the skull. **201**

great saphenous vein a large superficial vein that travels up the medial side of the leg and thigh; arises from the dorsal venous arch of the foot, and empties into the femoral vein; drains blood from the superficial regions of the leg and thigh; longest vein in the body. **623**

ground substance fluid or semifluid portion of the extracellular matrix of connective tissue. **129**

growth hormone (GH) a protein hormone produced by the anterior pituitary, which promotes cell growth and tissue building and influences nutrient metabolism. **555**

growth hormone-inhibiting hormone (GHIH) peptide hormone secreted by delta cells that regulates the endocrine system. **556**

growth hormone-releasing hormone (GHRH) peptide hormone produced by the arcuate nucleus of the hypothalamus that releases growth hormone. **556**

gyrus (Greek: *guros* - a ring) a raised ridge on the surface of the cerebrum. **452**

H

hair cell mechanoreceptor cells of the inner ear that transduce sound stimuli. **425**

hair follicle (Latin: *folliculus* - a small bag) cavity or sac from which hair originates; originates in the epidermis but partially located in the dermis. **157**

hair root portion of hair follicle that is between the hair bulb and hair shaft, anchored to the follicle. **157**

hair shaft externally visible portion of hair; composed of keratin. **157**

hallux (aka big toe, great toe) (Latin: *hallex* - great toe) digit 1 of the foot; the big toe. **248**

hamate (Latin: *hamatus* - a hook) from the lateral side, the hook-shaped fourth of the four distal carpal bones; articulates with the lunate and triquetrum proximally, the fourth and fifth metacarpals distally, and the capitate laterally. **236**

hand region of the upper limb distal to the wrist. **235**

haploid (Greek: *haplous* - single; *eidos* - appearance) homologous pairs of chromosomes, usually. **23, 845**

hard palate bony plate at the roof of the mouth; composed of the palatine process of the maxilla and the horizontal plate of the palatine bone. **688, 743**

head of the femur (Latin: *femur* - thigh) rounded, proximal end of the femur; articulates with the acetabulum of the hip bone. **247**

head of the fibula (Latin: *fibula* - brooch) knob-shaped, proximal end of the fibula. **248**

head of the humerus (Latin: *humerus* - shoulder) smooth, rounded region located on medial side of the proximal humerus; forms the glenohumeral (shoulder) joint with the glenoid fossa of the scapula. **235**

head of the radius (Latin: *radius* - ray) disc-shaped proximal end of the radius; articulates with the capitulum at the elbow joint, and with the radial notch of the ulna at the proximal radioulnar joint. **235**

head of the rib posterior end of a rib that articulates with the costal facet of the thoracic vertebra. **222**

head of the ulna (Greek: *ōlenē* - ulna) distal end of the ulna; articulates with the ulnar notch of the distal radius. **235**

hematoma mostly clotted blood that pools in a space within the body. **186**

hemoglobin (Greek: *haima* - blood; Latin: *globus* - ball) a protein that makes up a portion of erythrocytes; binds to and transports oxygen and some carbon dioxide molecules through the blood. **567**

hepatic artery (Latin: *hepaticus* - pertaining to the liver) the artery that delivers oxygen-rich blood to the liver. **618, 750, 759**

hepatic portal vein (Latin: *hepaticus* - pertaining to the liver) the vein that transports oxygen-poor blood, containing absorbed nutrients, to the liver. **622, 750, 759**

hepatic vein (Latin: *hepaticus* - pertaining to the liver) a vein that drains blood from the liver and drains into the inferior vena cava. **622**

hepatocytes (Latin: *hepaticus* - pertaining to the liver; Greek: *kytos* - cell) liver cells, which produce bile and adjust nutrient concentrations in the blood as it flows through the liver. **759**

hepatopancreatic ampulla (aka ampulla of Vater, hepatopancreatic duct) (Latin: *hepaticus* - pertaining to the liver; *ampulla* - flask) a rounded structure in the wall of the duodenum, located at the site at which the main pancreatic duct and the common bile duct merge; transports bile and pancreatic juice through the major duodenal papilla into the duodenum. **750**

hilum the indented entry/exit point on the medial surface of each lung, through which nerves, blood and lymphatic vessels, and a primary bronchus enters/exits the lung. **787**

hilus the indented entry/exit point on the medial surface of each lung, through which nerves, blood and lymphatic vessels, and a primary bronchus enters/exits the lung. **671**

hinge joint synovial joint formed by the articulation of the convex surface of one bone with the concave surface of a second bone. **267**

Histology (Greek: *histos* - tissue; *logos* - study) microscopic study of tissue appearance, organization, and function. **74**

homologous structures structures that share anatomcial features, indicating a common developmental origin. **837**

horizontal plates extension of the palatine bone that joins with a paired horizontal plate to form the posterior hard palate. **204**

human chorionic gonadotrophin (hCG) (Greek: *khorion* - membrane closing the fetus; *gonos* - child; *tropikos* - pertaining to a change) a hormone that is secreted as a signal from the developing embryo. **872**

humerus (Latin: *humerus* - shoulder) single long bone of the brachium. **235**

hydrophilic (Greek: *hydōr* - water; *philos* - loving) describes a substance or structure attracted to water. **80**

hydrophobic (Greek: *hydōr* - water; *phobos* - fear) describes a substance or structure repelled by water. **80**

hyoid bone (Greek: *hyoeidēs* - shaped like U) small, independent, U-shaped bone located in upper neck; serves as an anchor for some muscles of the tongue; does not contact any other bone. **204**

hyperextension (Greek: *hyper* - over; Latin: *extendo* - to stretch out) an increase in the joint angle beyond 180 degrees. **280**

hyperosmotic a solution with a greater concentration of non-permeating solutes relative to another solution. **93**

hypertonic (Greek: *hyper* - over; *tonos* - tension) describes a solution concentration that is higher than a reference concentration. **97**

hypodermis (aka subcutaneous layer) (Greek: *hypo* - under; *derma* - skin) connective tissue connecting the skin to the underlying fascia. **153**

hypogastric region the lower central region of the abdomen, located between the umbilicus and the pubis bone. **15**

hypoosmotic a solution with a lower concentration of non-permeating solutes relative to another solution. **93**

hypophyseal portal veins vessels that drain the primary capillary plexus and lead to the secondary plexus of the brain. **548**

hypothalamic neurons hormone-secreting neurons that extend from the hypothalamus to the posterior pituitary. **548**

hypothalamus (Greek: *hypo* - under) a structure of the diencephalon inferior and anterior to the thalamus; has both neural and endocrine functions, and is responsible for monitoring many homeostatic functions, such as hunger, body weight, thirst, body temperature, heart rate, and pituitary gland activity. **547**

hypotonic (Greek: *hypo* - under; *tonos* - tension) describes a solution concentration that is lower than a reference concentration. **97**

I

ileocecal sphincter (aka ileocecal valve) (Latin: *ileum* - intestines; *sphincter* - contractile muscle) a ring of muscle that regulates the flow of chyme from the ileum of the small intestine into the cecum of the large intestine. **750**

ileum (Latin: *ileum* - intestines) the final and longest segment of the small intestine, which runs between the jejunum and the cecum of the large intestine. **749**

iliac crest (Latin: *ilium* - groin; *crista* - crest) arched, superior ridge of the ilium. **243**

ilium (Latin: *ilium* - groin) large, superior region of the hip bone. **243**

implantation (French: *implanter* - implant) The process in which the preembryo attaches and embeds in the endometrium; can only occur when the preembryo has developed a trophoblast and endometrium is ready. **863**

incus (aka anvil) (Latin: *incus* - anvil) ossicle that connects the malleus to the stapes. **425**

inferior angle (Latin: *inferior* - below; *scapulae* - shoulder blades) corner of the scapula located between the medial and lateral borders. **231**

inferior articular processes (Latin: *articulo* - to articulate) flat, downward-facing bony process that extends from the vertebral arch of a vertebra to the superior articular process of the next lower vertebra. **213**

inferior colliculi (Latin: *inferior* - below) two of the four structures of the corpora quadrigemina, which lie along the anterior surface of the midbrain; part of the brainstem auditory pathway. **460**

inferior mesenteric artery (Latin: *inferior* - lower; *mesenterie* - middle of intestines) an artery that branches off of the abdominal aorta, and provides blood to the portion of the large intestine between the transverse colon and the sigmoid colon, and the rectum. **620**

inferior mesenteric vein a vein that drains blood from the large intestine. **622**

inferior nasal conchae (Latin: *nasus* - nose; *concha* – shell) one of two paired bones that project from the lateral walls of the nasal cavity. **204**

inferior orbital fissure opening formed by the sphenoid bone and maxilla; located at the inferior and lateral aspect of the orbit. **201**

inferior phrenic arteries arteries that branch off of the abdominal aorta, and provide blood to the inferior portion of the diaphragm. **620**

inferior ramus of the pubis structure forming the border of the urogenital region of the pelvic outlet; between the pubic tubercle and ischial tuberosity. **244**

inferior vena cava (Latin: *inferior* - lower; *vena* - vein; *cava* - hollow) one of the largest veins in the systemic circuit; formed by the merging of the common iliac veins; drains blood from most body regions inferior to the diaphragm, such as the lower limbs and the inferior portion of the trunk; empties into the right atrium. **620**

infraorbital foramina (Latin: *infra* - below; *orbis* - circle; *foramen* - opening) opening located on anterior skull, below the orbit. **203**

infraspinous fossa (Latin: *infra* - below; *spina* - backbone; *fossa* - trench) broad depression on the posterior scapula, inferior to the spine. **231**

infundibulum (aka pituitary stalk) (Latin: *infundibulum* – funnel) 1) a long, narrow stem of tissue connecting the hypothalamus to the posterior pituitary; contains the axons

of neurons that secrete oxytocin and antidiuretic hormone (ADH) 2) the widest part of the uterine tube, which lies closest to the ovary; contains the fimbriae which surround the end of the ovary that releases oocytes. **547**

inguinal canal (Latin: *inguinalis* - groin) an opening in the wall of the abdominal cavity through which either the spermatic cord or round ligament runs. **841**

inhalation (aka inspiration) (Latin: *inhalare* - inhale) the process of moving air into the lungs from the atmosphere. **712**

inner ear structure within the bony labyrinth of the temporal bone; composed of the cochlea, the vestibule, and the semicircular canals. **425**

insertion (Latin: *insertio* - to plant in) moveable end of a muscle; attaches to the structure being pulled during muscle contraction. **309, 329**

inspiratory capacity (IC) (Latin: *inspirare* - breathe upon) the maximal amount of air that can be inhaled after a normal exhalation; can be calculated by taking the sum of the TV and IRV. **719**

inspiratory reserve volume (IRV) (Latin: *inspirare* - breathe upon) the additional amount of air that can be inhaled after a normal inhalation. **719**

insula (Latin: *insula* - island) a pancreatic hormone produced by beta cells that enhances the cellular uptake and utilization of glucose; its systemic effect is to decrease the blood glucose level. **457**

insulin (Latin: *insula* - island) a pancreatic hormone produced by beta cells that enhances the cellular uptake and utilization of glucose; its systemic effect is to decrease the blood glucose level. **549**

insulin-like growth factors (IGFs) proteins that enhance cellular proliferation, inhibit apoptosis, and stimulate the cellular uptake of amino acids; produced by the liver and other tissues when stimulated by GH. **556**

integrating center a region in the brain that signals the body to respond to a stimulus. **497**

interatrial septum cardiac septum located between the two atria; contains the fossa ovalis after birth. **601**

intercalated disks (Latin: *intercalatus* - interposed) physical junction between adjacent cardiac muscle cells; consisting of desmosomes, specialized linking proteoglycans, and gap junctions that allow passage of ions between the two cells. **123**

intercondylar eminences (Latin: *eminentia* - raised surface) irregular, elevated area of the superior end of the tibia, located between the articulating surfaces of the medial and lateral condyles. **247**

intercondylar fossa (Latin: *fossa* - trench) deep depression located on the posterior side of the distal femur; separates the medial and lateral condyles. **247**

intercostal arteries (Latin: *interi* - between; *costa* - rib) branches of the

thoracic aorta, which supply blood to the structures of the vertebral column, segments of the spinal cord, and the skin and muscles of the back. **615, 617**

intercostal vein (Latin: *interi* - between; *costa* - rib) a vein that drains the muscles of the wall of the thorax, and empties into the azygos vein. **620**

interlobar arteries (Latin: *inter* - between) arteries that transport oxygen-rich blood from the segmental arteries to the arcuate arteries in the kidney; run through the renal columns between the renal pyramids. **795**

interlobar veins (Latin: *inter* - between) veins that transport oxygen-poor blood from the arcuate veins to the renal vein in the kidney; run through the renal columns between the renal pyramids. **795**

interlobular arteries (aka cortical radiate arteries) arteries that transport oxygen-rich blood from the arcuate arteries to the afferent arterioles of the renal corpuscles; also supply blood to the tissue of the renal cortex. **795**

interlobular veins (aka cortical radiate vein) (Latin: *inter* - between) a vein that transports oxygen-poor blood from the peritubular capillaries of the nephrons to the arcuate vein in the kidney. **795**

intermediate cuneiform (Latin: *inter* - between; *medius* - middle; *cuneus* - wedge) center of the three cuneiform tarsal bones; articulates posteriorly with the navicular bone, medially with the medial cuneiform bone, laterally with the lateral cuneiform bone, and anteriorly with the second metatarsal bone. **248**

intermediate (Latin: *inter* - between; *medius* - middle) muscles located intermediate to the thenar and hypothenar muscles. **312**

intermediate filaments (*pl*: filaments) (Latin: *inter* - between; *medius* - middle; *filamentum* - thread) type of cytoskeletal filament made of keratin, characterized by an intermediate thickness, and playing a role in resisting cellular tension. **51**

internal acoustic meatus opening in the temporal bone; provides passage for the nerve from the hearing and equilibrium organs of the inner ear. **200**

internal carotid artery (Latin: *internus* - internal; Greek: *karotides* - great arteries of the neck) an artery that branches from the common carotid artery, and joins with the vertebral artery at the base of the brain to form the cerebral arterial circle; provides the main blood supply to the brain. **617**

internal elastic membrane (aka internal elastic lamina) (Latin: *internus* - internal) a layer of elastic fibers that lies between the tunica intima and the tunica media in large arteries. **628**

internal iliac artery (Latin: *internus* - internal; *iliacus* - pertaining to colic) an artery that branches off of the common iliac artery; provides blood to pelvic organs, such as the urinary bladder, rectum, uterus (females) and ductus deferens (males), and to the external genitalia. **622**

internal intercostals (Latin: *internus* - internal; *inter* - between; *costa* - rib) one of 11 pairs of intermediate intercostal muscles; draw the ribs together, resulting in the contraction of the rib cage. **716**

internal jugular vein (Latin: *internus* - internal; *iugula* - collarbone) a vein that arises from the dural venous sinuses of the brain, travels through the jugular foramen of the skull, and empties blood into the subclavian vein; drains blood from a large portion of the brain; considered to be the counterpart of the carotid artery. **617**

Interphase (Latin: *inter* - between) entire life cycle of a cell, excluding mitosis. **55**

Interstitial cells (aka cells of Leydig) (Latin: *interstitium* - interval) cells that produce testosterone and are located between the coiled seminiferous tubules rather than within them. **847**

intertrochanteric crest (Latin: *inter* - between; Greek: *trochanter* - a runner) ridge on the posterior side of the proximal femur; runs between the greater and lesser trochanters. **247**

intertrochanteric line (Latin: *inter* - between; Greek: *trochanter* - a runner) rough ridge running between the greater and lesser trochanters on the anterior side of the femur. **247**

interventricular foramen (aka foramen of Monro) (Latin: *inter* - between; *ventriculus* - diminutive; *foramen* - an opening) a small opening between each of the lateral ventricles and the third ventricle in the brain; allows for the passage of CSF. **455**

interventricular septum (Latin: *inter* - between; *ventriculus* - diminutive; *septum* - partition) the wall between the right and left ventricles of the heart. **601**

intervertebral discs (Latin: *inter* - between; *vertebra* - joint) fibrocartilaginous structure located between adjacent vertebrae that joins the vertebrae. **213**

intervertebral foramina the openings through which spinal nerves exit the vertebral column; located between adjacent. **213**

intracapsular ligaments (Latin: *intra* - within; *capsula* - box; *ligamentum* - a band) ligament located within the articular capsule. **271**

intracellular fluid (ICF) (Latin: *intra* - within; Greek: *kytos* - cell) the fluid within the cells. **80**

intramural ganglia (Latin: *intra* - within; *muralis* - pertaining to a wall; Greek: *ganglion* - knot) terminal ganglia of the parasympathetic system; located within the walls of the target or effector organs. **525**

intrapulmonary pressure (Latin: *intra* - within; *pulmonis* - lung) the pressure within the alveoli. **715**

intrinsic (Latin: *intrinsecus* - on the inside) muscle that originates and is located inside the area in which it acts. **311, 329, 497**

intrinsic ligaments (Latin: *intrinsecus* - on the inside; *ligamentum* - a band) ligament that is fused to or incorporated into the wall of the articular capsule. **271**

Inversion (Latin: *inverto* - to turn upside down) the turning of the foot to angle the bottom of the foot toward the midline of the body. **281**

involuntary without conscious control. **123**

Ipsilateral reflexes (Latin: *ipse* - same; *latus* - side; *reflexus* - a bending back) reflex that begins and ends on the same side of the body. **498**

Iris (Latin: *iris* - rainbow) colored portion of the anterior eye; composed of smooth muscle that surrounds the pupil and regulates the amount of light entering the eye. **416**

ischial ramus (Greek: *ischion* - hip joint) narrow anterior, superior projection from the ischial tuberosity. **244**

ischial spine (Greek: *ischion* - hip joint; Latin: *spina* - backbone) projection of the posterior margin of the ischium; separates the greater sciatic notch and lesser sciatic notch. **244**

ischial tuberosity (aka "sit bone") (Greek: *ischion* - hip joint; Latin: *tuberosus* - full of lumps) large, roughened area of the inferior ischium. **244**

ischium (Greek: *ischion* - hip joint) posteroinferior region of the hip bone. **244**

isosmotic (Greek: *isos* - equal; *ōsmos* - impulsion) two solutions that have the same concentration of solutes. **93**

isotonic (Greek: *isos* - equal; *tonos* - tension) a solution outside a cell with the same concentration of solutes as the intracellular fluid. **97**

J

jejunum (Latin: *ieiunus* - hungry) the middle portion of the small intestine, which lies between the duodenum and the ileum; performs most of the final digestion and absorption of dietary nutrients. **749**

joint a structure in which adjacent bones and/or cartilage come together. **261**

joint cavity (aka synovial cavity) (Latin: *cavus* - hollow) fluid-filled space formed by the articular capsule of a synovial joint; houses the articulating surfaces of adjacent bones. **261**

jugular foramina irregularly shaped opening located in the lateral floor of the posterior cranial cavity. **206**

jugular notch a shallow U-shaped border on the top of the manubrium. **222**

juxtaglomerular apparatus (Latin: *iuxta* - beside; *glomerationem* - ball; *apparatus* - tools) a structure that lies at the point at which the distal convoluted tubule passes between the afferent and efferent arterioles of the nephron; functions in the regulation

of blood pressure, electrolyte balance, glomerular filtration rate, and blood flow to the kidney. **791, 808**

juxtaglomerular cells (Latin: *iuxta* - beside; *glomerationem* - ball) a specialized smooth muscle cell found in the wall of the afferent ateriole; secretes the enzyme renin in response to a decrease in blood pressure. **808**

juxtamedullary nephrons (Latin: *iuxta* - beside; *medulla* - marrow; Greek: *nephros* - kidney) nephrons whose renal corpuscles lie close to the border of the renal cortex and medulla, which contain very long nephron loops that extend deep into the renal medulla. **791**

K

keratin (Greek: *keras* - horn) an intracellular fibrous protein that is found in the dead cells on the apical surface; gives skin, hair, and nails their hard, water-resistant properties. **115**

keratinocytes (Greek: *keras* - horn; *kytos* - cell) epidermal cell that produces keratin. **151**

kinetic energy (Greek: *kinētos* - moving; *energeia* - in work) energy that matter possesses because of its motion. **81**

knob a rounded protuberance **69**

L

labia (*sing*: labium) (aka lips) (Latin: *labia* - lips) the lips, which help in eating, facial expression, and closing the mouth to keep out objects. **743, 841**

lacrimal bones (Latin: *lacrima* - tear) small, rectangular paired bone that contribute to the anterior-medial wall of each orbit; contain a tunnel between the inner corner of the eye and the nasal cavity. **204**

lacrimal glands (Latin: *lacrima* - tear; *glans* - an acorn) glands located just above the lateral side of the upper eyelid; produce tears. **413**

lacunae (*sing*: lacuna) (Latin: *lacus* - a hollow lake) small spaces in bone or cartilage tissue occupied by osteocytes or chondrocytes, respectively. **129**

lacuna (*sing*: lacuna) (Latin: *lacus* - a hollow lake) small spaces in bone or cartilage tissue occupied by osteocytes or chondrocytes, respectively. **179**

lambdoidal suture (Latin: *lambdoid* - resembling Greek letter lambda (λ) in form) V-shaped juncture that joins the occipital bone and the right and left parietal bones on the posterior skull. **198, 211**

lamina (*sing*: = lamina) (Latin: *lamina* - layer) portion of the vertebral arch, which extends between the transverse and spinous process. **213**

large intestine (aka colon) the portion of the of the GI tract, or alimentary canal, that lies between the small intestine and the rectum; absorbs water, vitamins, and electrolytes, and produces the feces. **750**

laryngeal prominence (aka Adam's apple) (Greek: *larynx* - upper windpipe; Latin: *prominentia* - projection) the point of fusion of the two hyaline cartilage plates of the thyroid cartilage; this protruding area, which can be seen from the outside, is called the "Adam's apple." **693**

laryngopharynx (aka hypopharynx) (Greek: *larynx* - upper windpipe; *pharynx* - throat) the most inferior part of the pharynx. Lies posteriorly to the larynx, and connects the oropharynx to the esophagus and trachea; transports respiratory gases to and from the larynx and transports food to the esophagus. **691, 747**

larynx (aka voice box) (Greek: *larynx* - upper windpipe) an oval cartilaginous structure that lies anterior to the esophagus and between the pharynx and trachea; contains the vocal cords, produces vocal sounds, protects the trachea from food entry, and controls the amount of air entering or leaving the lungs. **684**

lateral border lateral margin of the scapula. **231**

lateral circumflex artery (Latin: *lateralis* - side; *circumflex* - bent around) an artery that branches off of the deep femoral artery; surrounds the neck of the femur, and provides blood to the some of the deep thigh muscles and nearby skin. **623**

lateral columns (aka lateral funiculi) (Latin: *lateralis* – side) masses of white matter on both sides of the spinal cord between the posterior horn and the anterior horn; composed of many different groups of axons of both ascending and descending tracts. **483**

lateral condyle of the femur (Latin: *latus* - side; *femur* - thigh; Greek: *kondylōma* - resembling knuckle) smooth, articulating surface of the lateral expansion of the distal femur. **247**

lateral condyle of the tibia (Latin: *latus* - side; *tibia* - shin bone; Greek: *kondylōma* - resembling knuckle) smooth, flat, proximal surface on the lateral side of the tibia. **247**

lateral cuneiform (Latin: *latus* - side; *cuneus* - wedge) most lateral of the cuneiform tarsal bones; articulates posteriorly with the navicular bone, medially with the intermediate cuneiform bone, laterally with the cuboid bone, and anteriorly with the third metatarsal bone. **248**

lateral epicondyle of the femur site of attachment of the fibular collateral ligament of the knee joint. **247**

lateral epicondyle of the humerus (Latin: *latus* - side; *epi* - on; *kondyloma* - resembling knuckle; *humerus* - shoulder) small bony projection on the lateral side of the distal humerus. **235**

lateral horns (Latin: *lateralis* - side) masses of gray matter in the thoracic and upper lumbar regions of the spinal cord; contain neurons of the autonomic nervous system. **483**

lateral inhibition the capacity of an excited neuron to inhibit the activity of nearby neurons. **438**

lateral malleolus (Latin: *latus* - side; *malleus* - mallet) expanded bump formed by the distal end of the fibula. **248**

lateral meniscus (Latin: *latus* - side; Greek: *mēniskos* - crescent) c-shaped articular disc located between the lateral condyle of the femur and the lateral condyle of the tibia. **276**

lateral plantar artery (Latin: *lateralis* - side; *planta* - sole of the foot) an artery formed by the splitting of the posterior tibial artery in the foot; provides blood to the lateral plantar region of the foot. **623**

lateral rotation (aka external rotation) (Latin: *latus* - side; *rotatio* - rotation) movement that rotates the limb so the anterior surface moves away from the midline of the body. **281**

lateral ventricles (Latin: *lateralis* - side; *ventriculus* - belly) paired superior chambers housed within the cerebral hemispheres, which produce and circulate CSF. **455**

learned not innate. **497**

left atrioventricular valve (aka bicuspid valve, mitral valve) (Latin: *atrium* - entry hall) valve located between the left atrium and ventricle; consists of two flaps of tissue. **601**

left atrioventricular valve valve located between the left atrium and ventricle; consists of two flaps of tissue.

left atrium upper-left chamber of the heart; receives blood from the pulmonary circuit that flows into the left ventricle. **597**

left common carotid artery an artery that supplies blood to the head, neck, and brain; a branch of the aortic arch. **615**

left common iliac artery a terminal branch of the abdominal aorta that divides into the internal and external iliac arteries; supplies blood to the pelvis, the lower abdominal area, and the lower limbs. **622**

left coronary artery (Latin: *corona* - crown) one of two arteries that branch off the aorta to bring oxygenated blood to the heart wall; distributes blood to the left atrium, the left ventricle, and the intraventricular septum. **597**

left gastric artery (Greek: *gaster* - stomach) an artery that branches off of the celiac trunk, and provides blood to the stomach and a portion of the esophagus. **618**

left lobe lobe of the liver located in the epigastric an dleft hyypochondriac regions of the abdomen. **750**

left pulmonary artery artery supplying deoxygenated blood from the heart to the left lung. **624**

left subclavian artery supplying blood to the left arm. **615**

left ventricle one of the primary pumping chambers of the heart located in the lower portion of the heart; the left ventricle is the major pumping chamber on the lower left side of the heart that ejects blood into the systemic circuit via the aorta and receives blood from the left atrium. **597**

lens clear disc through which light enters the eye; helps to focus light on the fovea centralis. **416**

lens paper soft paper used for cleaning microscope lenses. **67**

lesser curvature the superior concave curve of the stomach. **749**

lesser sciatic notch (Latin: *sciaticus* - corruption of; Greek: *ischiadikos* - hip joint) slightly curved posterior margin of the ischium, superior to the ischial tuberosity. **244**

lesser trochanter (Greek: *trochantēr* - a runner) small, bony prominence on the medial aspect of the femur, at the base of the femoral neck. **247**

lesser tubercle (Latin: *tuberosus* - full of lumps) small prominence on anterior side of the proximal humerus. **235**

lesser wings paired lateral extensions of the sphenoid bone; form the ridge separating the anterior and middle cranial fossae. **201**

leukocytes (aka white blood cells, WBCs) (Greek: *leukos* - white; *kytos* - cell) blood cells that defend the body from infection and disease; consisting of neutrophils, eosinophils, basophils, monocytes, and lymphocytes, they make up a small percentage of the formed elements of the blood. **129, 563**

ligaments fibrous tissue that connects bone to bone. **271**

linea aspera (aka rough line) (Latin: *linea* - line) roughened ridge that runs longitudinally along the posterior mid-femur. **247**

lingual frenulum (Latin: *lingua* - tongue) a small fold of tissue that connects the tongue to the floor of the oral cavity; helps to hold the tongue in place and aids in speech and eating. **743**

lingual tonsils (Latin: *Iinguae* - of the tongue) a mass of lymphoid tissue that lies on the posterior side of the tongue; helps remove and destroy pathogens in incoming air and food. **663, 691**

liver an accessory organ of the digestive system; secretes bile and adjusts nutrient concentrations in the blood; the largest internal organ of the human body; also accomplishes many types of blood and metabolic regulation. **750**

lobules substructures within testis that are divided by septa. **759**

longitudinal layer middle layer of muscle in the wall of the stomach. **757**

lower esophageal sphincter (aka cardiac sphincter) (Greek: *oisophagos* - passage for food; Latin: *sphincter* - contractile muscle) a ring of smooth muscle that regulates the passage of food from the esophagus into the stomach. **747**

lower leg the region of the lower limb that is made up of the two bones that run parallel to each other and sit between the knee joint and the ankle joint. **247**

lower respiratory tract the portion of the respiratory tract that lies within the thorax; consists of the trachea, bronchi, bronchioles, alveolar ducts, alveolar sacs, and alveoli. **684**

lumbar arteries (Latin: *lumbaris* – loin) four pairs of arteries that branch off of the abdominal aorta; provide blood to the muscles and skin of the lumbar region, spinal cord, lumbar vertebrae, and the posterior portion of the abdominal wall. **620**

lumbar enlargement attachment point on the spinal cord of nerves that supply the lower limbs. **488**

lumbar veins (Latin: *lumbaris* - loin) veins that drain blood from the lumbar region of the abdominal wall and a portion of the spinal cord; empty into either the hemiazygos or the azygos vein, and eventually into the superior vena cava. **620**

lumbar vertebrae (Latin: *lumbus* - a loin) five vertebrae (L1–L5) located in the lumbar region of the vertebral column; characterized by large, thick vertebral bodies. **217**

lumen (Latin: *lumen* - light) 1) an enclosed space, lined with epithelia; 2) the interior of a tubular organ or blood vessel through which a substance is transported; 3) the inside of a tubular or hollow body structure. **115**

lunate (Latin: *luna* - moon) from the lateral side, the moon-shaped second of the four proximal carpal bones; articulates with the radius proximally, the capitate and hamate distally, the scaphoid laterally, and the triquetrum medially. **236**

luteinizing hormone (LH) a hormone secreted by the anterior pituitary gland; triggers ovulation and formation of the corpus luteum in females; stimulates the production of ovarian hormones in females and testosterone in males. **558**

lymphatic capillaries (Latin: *lympha* - water) the smallest vessels of the lymphatic system; pick up interstitial fluid from the tissue spaces and transport it through the lymphatic system, to eventually return it to the bloodstream. **667**

lymphatic trunks (Latin: *lympha* - water) large lymphatic vessels that receive lymph from smaller lymphatic vessels and empty it into lymphatic ducts. **667**

lymph nodes (Latin: *lympha* - water) a small, bean-shaped organ of the lymphatic system, which lies along a lymphatic vessel; filters lymph, removing pathogenic organisms and engaging in antigen recognition and attack. **660**

lymphocytes (Latin: *lympha* - water; Greek: *kytos* - cell) agranular leukocytes (white blood cells), which function in adaptive (specific) immunity; types of lymphocytes include B-lymphocytes, T-lymphocytes, and NK (natural killer) cells; comprise approximately 25 to 33 percent of the leukocyte population. **568**

lymphoid nodule (Latin: *lympha* - water) unencapsulated masses of lymphoid tissue found mainly in the walls of organs of the respiratory and digestive tracts; protect the body from infection by pathogens entering the body in the air or food. **664**

Lysosomes (*pl*: lysosomes) (Greek: *lysis* - loosening: *soma* - body) membrane-bound cellular organelle originating from the Golgi apparatus and containing digestive enzymes. **50**

M

macrophages (*pl*: **macrophages**) (Greek: *makros* - large; *phago* - to eat) large phagocytic cell, found in various tissues of the body, which arises from the differentiation of monocytes after they leave the bloodstream. **568**

macula densa cells (Latin: *macula* - spot) a group of cells found in the portion of the distal convoluted tubule that makes up the juxtaglomerular apparatus; detects changes in Na+ concentration in the filtrate, as a measure of electrolyte concentration in the blood. **791, 808**

malleus (aka hammer) (Latin: *malleus* - hammer) ossicle that is attached to the tympanic membrane that articulates with the incus. **425**

mandible (Latin: *mando* - to chew) single bone that forms the lower jaw; the only moveable bone of the skull. **202**

mandibular angles inferior termination of the ramus of the mandible. **202**

mandibular condyle bony projection from the ramus of the mandible; articulates with the temporal bone. **202**

mandibular fossa (Latin: *mando* - to chew; *fossa* - trench) deep, oval-shaped depression located on the external base of the skull; site of joint between mandible and skull. **199**

mandibular notch (Latin: *mando* - to chew) broad U-shaped curve located between the condylar process and coronoid process of the mandible. **202**

manubrium (Latin: *manubrium* - handle) wider, superior portion of the sternum. **222**

marginal arteries Greek: *arteria* - pipe) branches of the right coronary artery that supply blood to the superficial portions of the right ventricle. mass amount of matter contained within an object. **597**

mastoid process (Greek: *mastos* - the breast; *eidos* - resemblance) large bony prominence of the temporal bone; located on the inferior, lateral skull, just behind the earlobe. **199**

maxilla paired bone that forms the upper jaw, part of the eye orbit, the lateral base of the nose, and anterior portion of the hard palate. **203**

maxillary sinuses (Latin: *maxilla* - jawbone; *sinus* - cavity) large, air-filled space located within the maxillary bone; largest of the paranasal sinuses. **688**

meatuses a group of three narrow air passages (superior, middle, and inferior) in the nasal cavity, formed by the conchae, which increase the surface area to warm, filter, and humidify incoming air. **688**

mechanical digestion the process of breaking down food into smaller particles to prepare it for enzymatic digestion; includes chewing, stomach churning, and intestinal segmentation. **743**

medial border (Latin: *medialis* - middle; *scapulae* - shoulder blades) medial margin of the scapula. **231**

medial condyle of the femur (Latin: *medialis* - middle; *femur* - thigh; Greek: *kondylōma* - resembling knuckle) smooth, articulating surface of the medial expansion of the distal femur. **247**

medial condyle of the tibia (Latin: *medialis* - middle; *tibia* - shin bone; Greek: *kondylōma* - resembling knuckle) smooth, flat, proximal surface on the medial side of the tibia. **247**

medial cuneiform (Latin: *medialis* - middle; *cuneus* - wedge) most medial of the cuneiform tarsal bones; articulates posteriorly with the navicular bone, laterally with the intermediate cuneiform bone, and anteriorly with the first and second metatarsal bones. **248**

medial epicondyle of the femur (Latin: *medialis* - middle; *epi* - on; *kondyloma* - resembling knuckle; *femur* - thigh) rough area on the medial side of the medial condyle of the distal femur. **247**

medial epicondyle of the humerus (Latin: *medialis* - middle; *epi* - on; *kondyloma* - resembling knuckle; *humerus* - shoulder) prominent bony projection on the medial side of the distal humerus. **235**

medial malleolus (Latin: *medialis* - middle; *malleus* - mallet) distal medial expansion of the tibia. **247**

medial meniscus (Latin: *medialis* - middle; Greek: *mēniskos* - crescent) C-shaped articular disc located between the medial condyle of the femur and medial condyle of the tibia. **276**

medial plantar artery (Latin: *medialis* - middle; *planta* - sole of the foot) an artery formed by the splitting of the posterior tibial artery in the foot; provides blood to the medial plantar region of the foot. **623**

medial rotation (aka internal rotation) (Latin: *medialis* - middle; *rotatio* - rotation) movement that brings the anterior surface of a limb towards the midline of the body. **281**

medial sacral crest series of tubercles found on the lateral aspect of the sacral groove. **218**

median antebrachial vein (Latin: *medianus* - middle; *ante* - before; *brachium* - arm) a vein that runs parallel to the ulnar vein, and lies between the radial ulnar veins; arises from the palmar venous arches and empties into the basilic vein. **617**

median cubital vein (Latin: *medianus* - middle; *cubitus* - elbow) a superficial vein of the antecubital region, that connects the cephalic vein to the basilic vein; often used for drawing blood. **617**

mediastinal artery an artery that branches off of the thoracic aorta; provides blood supply to the posterior portion to the mediastinum, such as lymph nodes and a portion of the esophagus. **617**

mediastinum (Latin: *medius* - middle) the subdivision in the center of the thoracic cavity; contains the heart and lungs. **595, 663**

medulla (Latin: *medulla* - marrow) in hair, the innermost layer of keratinocytes and the spaces between them 000; the inner portion of the kidney, which consists of the renal pyramids; *aka* renal medulla. **671, 787**

medulla oblongata the lower portion of the brain stem; responsible for autonomic functions. **526**

medullary sinuses vessles that drain lymph from the trabecular sinuses; contain reticular cells and histocytes. **671**

megakaryocytes (Greek: *mega* - great; *karyon* - nut; *kytos* - cell) very large cells of the bone marrow, that release small cytoplasmic fragments that become platelets. **567**

Melanin (Greek: *melan* - black) one of the pigments found in the skin; determines the color of hair and skin and protects cells from UV radiation damage. **151**

Melanocyte (Greek: *melan* - black; *kytos* - cell) cell found in the stratum basale that produces the pigment melanin. **151**

melatonin hormone secreted by the pineal gland in response to low-light as detected by the eye. **548**

membrane thin sheets of tissue that surround structures within the body. **14, 115**

meninges (Greek: *mēninx* - membrane) the three membranes that surround and protect the brain and spinal cord; consist of the dura mater, the arachnoid mater, and the pia mater. **453, 487**

meniscus (Greek: *mene* - moon; *mēniskos* - crescent) c-shaped articular disc found between articulating surfaces in some joints; located between the femoral and tibial condyles in the knee joint; composed of fibrocartilage. **272**

mental foramina two openings located on the anteior surface of the mandble. **202**

mental protuberance a triangular protrusion on the symphysis of the mandible. **202**

Merkel cell (pl: Merkel cells) (aka Merkel-Ranvier cell, tactile epithelial cell) sensory receptor cell of the stratum basale connected to sensory nerves; responds to touch. **151**

metacarpal bones (Greek: *meta* - between; *karpos* - wrist) one of the five bones that form the palm of the hand. **235**

metaphase (Greek: *meta* - between; *phasis* - an appearance) the stage of mitosis (and meiosis), characterized by the alignment of chromosomes in the center of the cell. **55**

metatarsal one of the five bones that form the anterior region of the foot. **248**

metatarsal bones (Greek: *meta* - between; *tarsos* - tarsus) one of the five bones that form the anterior region of the foot. **247**

methylene blue methylthioninium chloride; salt used as a dye or a drug to reduce ferric iron. **73**

microfilaments (*pl*: microfilaments) (Greek: *mikros* - small; Latin: *filamentum* - thread) the thinnest of the cytoskeletal filaments; composed of actin subunits that function in muscle contraction and cellular structural support. **51**

microglia (Greek: *mikros* - small; *glia* - glue) glial cell type in the CNS that serves as the resident immune cells. **370**

microtubules (*pl*: microtubules) (Greek: *mikros* - small; Latin: *tubus* - tube) the thickest of the cytoskeletal filaments, composed of tubulin subunits that function in cellular movement and structural support. **51**

microvilli (sing: microvillus) (Greek: *mikros* - small; Latin: *villi* - shaggy hair) small projections on the surface of cells that function to increase surface area. **51**

micturition (aka urination) (Latin: *micturitium* - to desire to urinate) urination or emptying of the urinary bladder. **799**

midbrain (aka mesencephalon) the middle region of the adult brain. **460, 525**

middle cardiac vein (Greek: *cardiac* - *kardia* meaning heart) vessel that parallels and drains the areas supplied by the posterior interventricular artery; drains into the great cardiac vein. **598**

middle cerebral arteries (Latin: *cerebro* - brain) an artery that branches from the internal carotid artery; supplies blood to portions of the frontal, temporal, and parietal lobes of the cerebrum. **617**

middle ear space spanned by the ossicles. **425**

middle nasal conchae (Latin: *medialis* - middle; *nasus* - nose; *concha* - shell) thin, curved projection of the ethmoid bone; located between the superior and inferior conchae within the nasal cavity. **200**

midsagittal plane two-dimensional, vertical plane that divides the body or organ into roughly equal right and left portions. **9**

minute ventilation (aka minute respiratory volume) the total volume of inhaled air that enters or exits the lungs in one minute; can be calculated by multiplying the number of breaths per minute by the tidal volume. **720**

mitochondria (*pl*: mitochondria) (Greek: *mitos* - thread; *chondros* - granule) one of the cellular organelles bound by a double lipid bilayer that function primarily in the production of cellular energy (ATP). **50**

mitosis (Greek: *mitos* - thread) division of genetic material, during which the cell nucleus breaks down and two new, fully functional, nuclei are formed. **55**

mitotic spindle (*pl*: spindles) (Greek: *mitos*: thread) network of microtubules, originating from centrioles, that arranges and pulls apart chromosomes during mitosis. **55**

monocytes (Greek: *mono* - one; *kytos* - cell) agranular leukocytes that conduct phagocytosis of pathogens; once they leave the bloodstream, they differentiate into macrophages. **568**

monomers (Greek: *mono* – single; *meros* - part) individual units that make up a larger molecule. **763**

monosynaptic reflex arc (Greek: *monos* - single; *syn* - together; Latin: *reflexus* - a bending back) a reflex that only contains one synapse, which is found between the sensory and motor neurons. **498**

morula stage of embryo development from 16-32 cells. **871**

motor neuron (aka efferent neurons) (Latin: *motor* - to move; Greek: *neuro* – nerve) nerve fibers that carry signals that emanate from the central nervous system and connect to effectors. **497**

mucosa-associated lymphatic tissue (MALT) (Latin: *mucosus* - mucous; *lympha* – water) lymphoid nodules found in the mucosa of the digestive, respiratory, genital, and urinary tracts, as well as the breast, eye, and skin, which protect against infection by pathogenic organisms entering the body from the external environment. **663, 754**

mucosa (Latin: *mucosus* – mucous) the innermost layer of the GI tract; consists of an epithelial layer, a lamina propria (connective tissue) and a muscularis mucosae (smooth muscle). **754, 799**

multiaxial (Latin: *multus* – much) joint that allows for motion within several planes. **266**

muscarinic receptors (pl: muscarinic receptors) one type of acetylcholine receptor protein; also binds to the mushroom poison muscarine. **525**

muscle organ a group of muscle fibers, blood vessels, nerve fibers and connective tissue, colletively referred to as a muscle organ. **297**

muscle spindle bundle of muscle cells that are wrapped with sensory neuron dendrites. **499**

muscle tissue (Latin: *musculus* - muscle) contractile tissue capable of generating tension in response to stimulation; produces movement. **109**

muscularis the muscle layer of the wall of the GI tract; consists of smooth muscle in most areas, but skeletal muscle in a few areas; lies between the submucosa and the serosa. **754, 799**

myelin insulating layer of lipids that surround the axons of some neurons, formed by glial cells. **360, 365**

myelination the formation of a myelin sheath surrounding the axon of a neuron. **360**

myelin sheath fatty layer of insulation that surrounds some axons; facilitates the transmission of electrical signals down the axon. **373**

myocardium (Greek: *cardiac - kardia* meaning heart) thickest layer of the heart composed of cardiac muscle cells built upon a framework of primarily collagenous fibers and blood vessels that supply it and the nervous fibers that help to regulate it. **601**

myofibrils (Greek: *mys* - muscle) the contractile machinery of muscle cells; long cylinders of contractile proteins that shorten during muscle contraction. **292, 297**

myometrium (Greek: *mys* - muscle) the thick muscle layer of the wall of the uterus; contractions of this muscle layer excrete the menses each month, and also expel the fetus during childbirth. **841**

N

nail bed epidermal structure that produces the nail body. **159**

nail body (pl: nail bodies) hard, bladelike keratinous plate that forms the nail. **159**

nail root protected region at the proximal side of the nail bed from which the nail body grows. **159**

nares the opening into the nostrils. **687**

nasal bones (Latin: *nasus* - nose) small, paired bones that articulate to form the bridge of the nose. **204**

nasal cavity the air-filled space behind and above the nose. **684**

nasal conchae (Latin: *nasus* - nose; *concha* - shell) mucous membrane-covered bony plates that project from the lateral walls of the nasal cavity; function to subdivide the nasal cavity. **688**

nasal septum (Latin: *nasus* - nose; *saeptum* - a partition) flat structure that

divides the nasal cavity into halves; formed by the perpendicular plate of the ethmoid bone and the vomer bone. **687**

nasal vestibules the anterior section of the nasal cavity. **687**

nasolacrimal duct duct between the lacrimal sac of the eye to the nasal cavity. **413**

nasopharynx (Latin: *nasus* - nose; Greek: *pharynx* - throat) the most superior part of the pharynx, which lies posteriorly to the nasal cavity and transports air from the nasal cavity to the oropharynx. **747**

nasopharynx (Latin: *nasus* - nose; Greek: *pharynx* - throat) the most superior region of the pharynx; lies posterior to the nasal cavity, transports only air and is involved in the breathing process. **691**

navicular (Latin: *navis* - ship) last bone in the proximal row of tarsal bones; articulates posteriorly with the talus bone, laterally with the cuboid bone, and anteriorly with the medial, intermediate, and lateral cuneiform bones. **248**

neck of the femur (Latin: *femur* - thigh) narrowed region of the femur; immediately inferior to the head of the femur. **247**

neck of the rib narrowed region of a rib; located next to the rib head. **222**

nephron loop (*aka* Loop of Henle) (Greek: *nephros* - kidney) the portion of the nephron that extends into the renal medulla of the kidney; transports filtrate between the proximal convoluted tubule and the distal convoluted tubule; performs reabsorption of NaCl and water. **791**

nerve a bundle of axons in the peripheral nervous system. **491**

nervous tissue (French: *tissu* - woven) tissue that is capable of sending and receiving impulses throughout the body using electrochemical communication. **109**

neurilemma cells (*aka* Schwann cell) (Greek: *neuro* - nerve; *lemma* - husk) a type of glial cell that insulates axons with myelin in the peripheral nervous system. **370**

neurofibril nodes (*aka* nodes of Ranvier) each gap along the axon that contributes to saltatory conduction. **365**

neurons (*sing*: neuron) (Greek: *neuro* – nerve) neural cells responsible for communication; generate and propagate electrical signals into, within, and out of the nervous system. **127, 360, 480**

neurosecretory cells neurons that produce hormones. **531**

neutrophils (Latin: *neuter* - neither) the most abundant of the five types of leukocytes (white blood cells) in the body; strong phagocytic cells that engulf mainly bacteria; circulate in the blood, but also leave the bloodstream to migrate to infection sites via chemotaxis. **568**

nicotinic receptors (pl: nicotinic receptors) one type of chemically gated ion channel

acetylcholine receptor protein; characterized by also binding to nicotine. **523**

norepinephrine (NE) signaling molecule released as a neurotransmitter by most postganglionic sympathetic fibers; also secreted as an amine hormone by the chromaffin cells of the adrenal medulla in response to short-term stress; prepares the body for fight-or-flight responses, such as an increase in heart rate, breathing rate, and blood pressure, and dilation of the pupils and airways. **523, 549**

nurse cells (*aka* sustentacular cells, Sertoli cells) cells that make up the wall of the seminiferous tubule and surround the developing sperm through all stages of their development. **847**

O

objective lens lens nearest to the stage on a microscope, gathesr and focuses light. **68**

oblique layer layer of muscle that makes up the wall of the stomach. **757**

obturator foramina (Latin: *obturo* - to occlude; *foramen* - an opening) large opening in the anterior hip bone formed at the junction of the rami of the pubis and ischium. **243**

occipital bone single bone that forms the posterior skull and the posterior base of the cranial cavity. **199**

occipital condyles paired, oval-shaped bony knobs located on the inferior skull, on either side of the foramen magnum; form joints with the first cervical vertebra. **199**

occipital lobe a region of the cerebrum beneath the occipital bone of the cranium; responsible for processing visual stimuli. **457**

ocular lens the eyepiece of a microscope, the lens closest to the eye. **68**

oculomotor nerve third cranial nerve; responsible for eye movements by controlling four of the extraocular muscle. **525**

olecranon fossa (Greek: *ōlenē + kranion* - head of the ulna; Latin: *fossa* - trench) large depression on the posterior side of the distal humerus; accommodates the olecranon process of the ulna when the elbow is fully extended. **235**

olecranon process (Greek: *ōlenē + kranion* - head of the ulna) curved extension of the ulna, formed by the posterior and superior portions of the proximal ulna; fits into the olecranon fossa of the humerus when the elbow is extended. **235**

olfactory bulbs (Latin: *olfacio* - to smell) structures that lie at the anterior inferior portion of the brain, that transmit nerve impulses from the olfactory receptor cells to the brain; participate in the sense of smell and in limbic system function. **395**

oligodendrocytes (Greek: *oligos* - few; *dendron* - tree; *kytos* - cell) glial cell type in the CNS that insulates axons in myelin. **370**

oogenesis (Greek: *oon* – egg; Latin: *genesis* - origin) the process of oocyte production and development in the biological female; includes the production and mitosis of oogonia and development of primary oocytes by meiosis during fetal development; further development into secondary oocytes occurs only upon fertilization. **833**

opposition (Latin: *oppono* - oppose) movement of the thumb that brings the tip into contact with the tip of a finger. **281**

optic canals (Greek: *optikos* - eyes; *canal* – channel) opening located at the anterior lateral corner of the sella turcica; provides passage of the optic nerve to the orbit. **201**

organelle (Greek: *organon* - a tool) tiny functioning units within a cell. **46**

origin (Latin: *origio* - source) fixed (unmovable) end of a muscle; generally attached to a bone. **309, 329**

oropharynx (Latin: *oris* - mouth; Greek: *pharynx* - throat) the middle part of the pharynx; lies posteriorly to the oral cavity and runs between the nasopharynx and the laryngopharynx; a passageway for air and food. **691, 747**

osmosis (Greek: *osmos* - pushing fluid) the diffusion of water from regions of higher water concentration to regions of lower water concentration across a selectively permeable membrane. **84**

ossicles (Latin: *ossiculum* - bone) three tiny bones of the middle ear, called the malleus, incus, and stapes, which transmit sound vibrations from the tympanic membrane to the fluid of the inner ear. **425**

osteocytes (Greek: *osteon* - bone; *kytos* - cells) most abundant cell in mature bone; maintain the mineral concentration in the matrix via secretion of enzymes. **129**

osteon (aka Haversian system) (Greek: *osteon* - bone) microscopic structural unit of compact bone; composed of concentric layers of lamellae. **179**

outer ear the external structures of the ear, including the auricle and ear canal. **425**

oval window membrane-covered opening between the middle ear and the inner ear. **425, 426**

ovarian cycle the selection, growth and maturation of an oocyte within the ovary. **848**

ovarian follicle (Latin: *folliculus* - a little bag) a structure in the ovary consisting of an oocyte and the granulosa cells that surround and nourish it. **848**

ovaries (*sing*: **ovary**) (Latin: *ovum* - egg) the gonads of biological females; produce sex hormones and gametes (eggs). **549, 836**

oxytocin (OT) (Greek: *oxus* - swift; *tokos* - childbirth) a peptide hormone produced by the hypothalamus and stored in the posterior pituitary gland; stimulates uterine contractions during labor, milk ejection during breastfeeding, and feelings of attachment. **548**

P

palatine bones paired, irregularly shaped bones form small parts of the nasal cavity and the medial wall of the orbit, and the posterior hard palate. **204**

palatine tonsils (Latin: *palatum* - roof of mouth) paired lymphoid structures in the posterior portion of the oropharynx, that remove and destroy pathogens in incoming air and food. **663, 691**

palmar arches (Latin: *palma* - palm of hands) arterial arches formed by the merging of the radial and ulnar arteries; provide blood to the hand and give off branches that supply the fingers; consist of superficial and deep arches. **615**

palmar venous arches (Latin: *palma* - palm of hands) curved veins that drain blood from the hands and fingers, and empty into the radial and ulnar veins. **617**

palpation the technique of using the hands to determine the size, shape, firmness and location of part of the body. **24**

pancreas (Greek: *pankreas* - sweetbread) an organ with both endocrine and exocrine functions; secretes hormones that regulate blood glucose and appetite and produces digestive enzymes and buffers; located posterior to the stomach. **749**

pancreas (Greek: *pankreas* - sweetbread) one of the accessory digestive organs, which lies posterior to the stomach; the exocrine portion secretes pancreatic juice, and the endocrine portion secretes insulin and glucagon. **549**

pancreatic duct the duct that transports pancreatic juice from the pancreas into the duodenum. **750**

pancreatic islets (*aka* islets of Langerhans) (Greek: *pankreas* - sweetbread) small groups of endocrine cells in the pancreas that produce the hormones insulin. **761**

papillae (*sing*: papilla) (Latin: *papilla* - nipple) small, fingerlike projections of body structures; lingual papillae house taste buds on the tongue; the duodenal papilla connects the common bile duct and main pancreatic duct to the duodenum. **400**

papillary layer (Latin: *papilla* - nipple) superficial layer of the dermis, made of areolar connective tissue; contains fibroblasts and small blood vessels, along with other structures. **153**

papillary muscle (Latin: *papilla* - nipple) extension of the myocardium in the ventricles to which the chordae tendineae attach. **601**

parafollicular cells calcitonin-secreting cells located in the thyroid. **549**

paranasal sinuses (*pl*: **paranasal sinuses**) (Greek: *para* - near; Latin: *nasus* - nose; *sinus* - fold) one of the chambers within a skull bone (frontal, maxillary, sphenoid, and ethmoid bones) that opens into the nasal cavity; serves to resonate the voice and decrease the weight of the skull. **206, 688**

parasagittal plane off-center, two-dimensional, vertical plane that divides the body or organ into unequal right and left portions. **9**

parasympathetic division (Greek: *para* - near; *sympathētikos* - to feel with) division of the autonomic nervous system responsible for restful and digestive functions; dominant between emergency situations. **520**

parathyroid glands (Greek: *para* - near; *thyreoeidēs* - shield-shaped; *glans* - acorn) small, oval glands embedded in the posterior surface of the thyroid gland; secrete parathyroid hormone. **549**

parathyroid hormone (PTH) (Greek: *para* - near; *thyreoeidēs* - shield-shaped) a peptide hormone produced by the parathyroid glands, which increases the blood calcium level. **549**

parietal relating to or attached to the wall of the body or a cavity. **13**

parietal bones (Latin: *parietalis* – wall) paired bones that form the superior lateral sides of the skull. **198**

parietal lobes (Latin: *parietalis* - wall) a region of the cerebrum that lies superior to the lateral sulcus, directly beneath the parietal bone of the cranium; associated with somatosensation. **457**

parietal pericardium the portion of the pericardium that lines the superficial interior pericardial sac. **595**

parotid salivary glands (Greek: *para* - beside; *ot* - ear) the largest pair of salivary glands in the body, found inferior and anterior to the ears; secrete serous saliva containing amylase into the roof of the mouth. **743**

passive transport (Latin: *passivus* - to endure; *transporto* - to carry over) form of transport across the cell membrane that does not require input of cellular energy. **84**

patella (aka kneecap) (Latin: *patella* - small, shallow dish) sesamoid bone of the knee; articulates with the distal femur. **247**

patellar ligament (Latin: *patella* - small, shallow dish; *ligamentum* - a band) inferior end of the quadriceps tendon, continuation of the patellar tendon; located between the patella and the tibia, just below the knee. **276**

patellar surface (Latin: *patella* - small, shallow dish) wide groove on the anterior side of the distal femur, at the meeting of the medial and lateral condyles; site of articulation for the patella. **247**

pectinate muscles (Latin: *pectinatus* - resembling a comb) muscular ridges seen on the anterior surface of the right atrium. **601**

pectineal line (Latin: *pectineus* - comblike) narrow ridge that runs along the superior margin of the superior pubic ramus. **244**

pectoral appendages limbs attached to the pectoral girdle. **35**

pectoral girdle the region formed by the scapula and clavicle, which attaches each upper limb to the axial skeleton. **228**

pedicles (Latin: *pediculus* - footstalk) one of the lateral sides of the vertebral arch, attached to the vertebral body. **213, 791**

pelvic appendages (Latin: *pediculus* - footstalk) one of the lateral sides of the vertebral arch, attached to the vertebral body. **37**

pelvic brim (aka pelvic inlet) (Latin: *pelvis* - basin) the dividing line between the greater and lesser pelvic regions; formed anteriorly by the superior margin of the pubic symphysis and posteriorly by the pectineal lines of each pubis, the arcuate lines of each ilium, and the sacral promontory. **243**

pelvic girdle the portion of the pelvic including the ilium, ischium and pubis; connects the pelvic spine to the lower limbs. **228**

pelvic splanchnic nerves (Greek: *splankhna* - inward parts) parasympathetic nerves that arise from the S2-S4 spinal nerves; innervate organs such as the distal portion of the large intestine, the urinary bladder, and some of the reproductive organs. **526**

pelvis (Latin: *pelvis* - basin) the right and left hip bones, sacrum, and coccyx. **243**

penis (Latin: *penis* - penis) an organ of the external genitalia in biological males; participates in sexual intercourse and houses the urethra through which both semen and urine are transported to the outside of the body through the urethra. **833**

pepsinogen (Greek: *pepsis* - digestion; *genesis* - origin) the inactive form of the enzyme pepsin, secreted by the chief cells of the gastric mucosa; converted into pepsin by hydrochloric acid. **771**

pericardial artery (Greek: *peri* - around; Latin: *cardiacus* - heart) an artery that branches off of the thoracic aorta; provides blood to the pericardium, the serous membrane that surrounds the heart. **617**

pericardial cavity (Greek: *peri* - around; *cardiac* - *kardia* meaning heart) cavity surrounding the heart filled with a lubricating serous fluid that reduces friction as the heart contracts. **595**

pericardium (*aka* pericardial sac) (Greek: *peri* - around; *cardiac* - *kardia* meaning heart) membrane that separates the heart from other mediastinal structures; consists of two distinct, fused sublayers: the fibrous pericardium and the parietal pericardium. **13, 595**

perimetrium Greek: *peri* - around;) the outermost layer of the wall of the uterus. **841**

perimysium (Greek: *peri* - around; *mys* - muscle) layer of connective tissue that closely surrounds a fascicle. **297**

perinephritic fat (Greek: *peri* - around; *nephros* - kidney) a shock-absorbing layer of adipose tissue that surrounds the kidney; lies between the fibrous capsule of the kidney and the renal fascia. **787**

periodontal ligaments (Greek: *peri* - around; *odous* - tooth; Latin: *ligamentum* - a band) one of the short bands of dense connective tissue that anchors the root of a tooth into its socket in the maxilla or mandible. **743**

periosteum (Greek: *peri* - around; *osteon* - bone) membrane covering the outer surface of bone. **179**

peripheral adaptation adaptation to a stimulus due to synaptic fatigue. **438**

peripheral nervous system (PNS) anatomical division of the nervous system including all parts except the brain and spinal cord; largely located outside of the brain and vertebral column. **364, 480**

peritoneum (Greek: *peri* - around; *tonos* - stretched) a double-layered serous membrane that surrounds the abdominopelvic organs and lines the abdominopelvic cavity. **13, 747**

peritubular capillaries (Greek: *peri* - around) the second capillary system of the nephrons; surround the renal tubules, and participate in the reabsorption and secretion process of urine formation. **795**

peroxisomes (*pl*: peroxisomes) (Greek: *sōma* - body) membrane-bound organelle that contains enzymes primarily responsible for detoxifying harmful substances. **50**

perpendicular plate the largest extension of the ethmoid bone; a downward, midline extension that forms the superior portion of the nasal septum. **200**

Peyer's patches lymphoid follicles located in the lower portion of the small intestine. **759**

phalanges the bones that form the fingers or toes. **235, 247**

pharyngeal tonsils (*aka* adenoid) (Greek: *pharynx* - throat) a lymphoid structure in the posterior wall of the nasopharynx, which removes and destroys pathogens in inhaled air. **663, 691**

pharynx (aka throat) (Greek: *pharynx* - throat) a tubular passageway in the conducting zone that lies posteriorly to the nasal and oral cavities and the larynx; lined mainly with pseudostratified ciliated columnar epithelium; transports air toward the larynx and food toward the esophagus; also called the throat. **684, 747**

phasic receptors (Latin: *recipere* - to receive) the sensory receptors that adapt quickly to repeated or prolonged stimuli. **443**

phrenic vein (Greek: *phrenicus* – diaphragm) the vein that drains venous blood from the diaphragm. **622**

pia mater (Latin: *pia* - tender; *mater* - mother) the thin, transparent, innermost membrane of the meninges, which directly covers the brain and spinal cord. **453**

pineal gland (Latin: *pineus* - relating to pine; *glans* - acorn) a small gland within the epithalamus that helps to regulate wake-sleep cycles, via the secretion of the hormone melatonin. **548**

pisiform (Latin: *pisum* - pea; *forma* - appearance) from the lateral side, the pea-shaped fourth of the four proximal carpal bones; articulates with the anterior surface of the triquetrum. **236**

pituitary gland (Latin: *pituita* - phlegm; *glans* - acorn) bean-sized organ suspended from the hypothalamus that produces hormones in response to hypothalamic stimulation; composed of the anterior and posterior pituitary glands. **459, 547**

pivot joint (aka trochoid joint) synovial joint at which the rounded portion of a bone is enclosed within a ring formed by a ligament and an articulating bone; the bone rotates within the ring. **266**

placenta (Latin: *placenta* - a cake) a fetal structure that provides an exchange site between the maternal and fetal circulation. **863**

placental sinus location on the placenta wehre maternal blood pools in close proximity to fetal venules and arterioles. **877**

plane imaginary two-dimensional surface that passes through the body. **9**

plane joint (aka gliding joint) synovial joint formed between articulation of the flat surfaces of adjacent bones. **267**

plantar arch (Latin: *planta* – sole of the foot) an arterial arch formed by the merging of the dorsalis pedis artery with the medial and plantar arteries; provides blood to the distal portion of the plantar surface of the foot and the digits (toes). **623**

plantar flexion (Latin: *planta* - sole of the foot; *flecto* - to bend) motion in the ankle that lifts the heel of the foot from the ground; pointing the toes downward. **281**

plantar veins (Latin: *planta* - sole of the foot) veins that drain blood from the plantar (inferior) region of the foot, and empty into the plantar venous arch and eventually into the posterior tibial vein. **623**

plantar venous arch (Latin: *planta -* sole of the foot) a curved deep vein that arises from the plantar veins, and drains into the anterior and posterior tibial veins. **623**

plasma (Greek: *plasma* - something formed) the fluid component of the blood; considered to be the matrix of the blood. **563**

platelets (*aka* thrombocytes) (Greek: *platys* - flat) small cytoplasmic fragments of megakaryocytes that function in hemostasis (the stoppage of bleeding); make up a small percentage of the formed elements of the blood. **563**

pleura (Greek: *pleura* - a rib) serous membrane that lines the pleural cavity and covers the lungs. **13**

podocytes (Greek: *pod* – foot; *kytos* - cells) specialized cells of the inner layer of the glomerular (Bowman's) capsule, which closely surround the glomerular capillaries; the filtration slits in the pedicels of these cells permit small molecules to enter the glomerular capsule. **791**

polar having a separation of electric charge. **80**

pollex (aka thumb) (Latin: *pollex* - thumb) digit 1 of the hand; the thumb. **237**

polymer (Greek: *polys* – many) larger molecule made up of smaller monomers. **763**

polysynaptic reflex arc Greek: *polys* - many; *syn* - together; Latin: *reflexus* - a bending back) a reflex that contains more than one synapse and therefore involves interneurons in addition to sensory and motor neurons. **498**

pons (Latin: *pons* - bridge) the bulging middle portion of the brainstem, which connects the cerebellum to the brainstem and the spinal cord to some of the higher brain centers; functions in information relay and in the control of respiration. **526**

popliteal artery (Latin: *popliteus* - back of knee joint) a continuation of the femoral artery as it passes posteriorly to the knee joint; divides into the anterior and posterior tibial arteries. **623**

popliteal vein (Latin: *popliteus* - back of knee joint) a vein that arises from the merging of the anterior and posterior tibial veins in the posterior knee region; drains blood from the lower leg and posterior knee areas, and continues as the femoral vein. **623**

portal triad a group of three vessels that runs through each corner of the hepatic lobules; consists of a branch of the bile duct, the hepatic artery, and the hepatic portal vein. **759**

posterior arch posterior portion of the ring-shaped C1 vertebra. **215**

posterior cardiac vein (Greek: *cardiac* - *kardia* meaning heart) vessel that parallels and drains the areas supplied by the marginal artery branch of the circumflex artery; drains into the great cardiac vein. **598**

posterior cerebral arteries (Latin: *cerebro* - brain) an artery that branches off of the basilar artery, and forms part of the cerebral arterial circle; supplies blood to portions of the occipital and temporal lobes of the cerebrum. **617**

posterior chamber the space between the peripheral iris and anterior to the suspensory ligament of the lens. **415**

posterior columns (*aka* posterior funiculi) (Latin: *posterus* - following) masses of white matter of the spinal cord located between the posterior horns; composed of ascending tracts that carry sensory information to the brain. **483**

posterior communicating artery (Latin: *posterus* - following) an artery that connects the posterior cerebral artery to the middle cerebral artery, and forms part of the cerebral arterial circle; supplies blood to the posterior portions of the cerebrum. **617**

posterior cruciate ligament (Latin: *cruciatus* - resembling cross; *ligamentum* - a band) intracapsular ligament that extends from the posterior, superior surface of the tibia to the inner aspect of the medial condyle of the femur. **275**

posterior horn (Latin: *posterus* - following) a gray matter region of the spinal cord containing the axons of sensory neurons. **483**

posterior inferior iliac spine (aka PIIS) projection on the inferior margin of the auricular surface on the posterior ilium. **243**

posterior interventricular artery (*aka* posterior descending artery) branch of the right coronary artery that runs along the posterior portion of the interventricular sulcus toward the apex of the heart and gives rise to branches that supply the interventricular septum and portions of both ventricles. **597**

posterior interventricular sulcus sulcus located between the left and right ventricles on the anterior surface of the heart. **597**

posterior median sulcus (Latin: *posterus* - following) a midline groove on the posterior side of the spinal cord; separates the right and left sides of the spinal cord. **483**

posterior ramus the posterior division of a spinal nerve supplying the lateral trunk and limbs. **483**

posterior root a group of cell bodies of sensory neurons that extend into the spinal cord through the posterior root. **483**

posterior root ganglion (*aka* dorsal root ganglion) (Latin: *posterus* - following; Greek: *ganglion* - knot) a group of cell bodies of sensory neurons that extend into the spinal cord through the posterior root. **483**

posterior superior iliac spine (aka PSIS) rounded area on the posterior end of the iliac crest. **243**

posterior tibial artery (Latin: *posterior* - behind; *tibia* - shinbone) an artery that branches off of the popliteal artery; provides blood to the lateral and posterior areas of the leg and the bottom of the foot, and branches to form the fibular (peroneal) artery and the medial and lateral plantar arteries. **623**

posterior tibial vein (Latin: *posterior* – behind; *tibia* - shinbone) a vein that arises from the dorsal venous arch and empties into the popliteal vein; drains blood from the posterior tibial region. **623**

preganglionic neuron (Latin: *prae* - before; Greek: *ganglion* - knot) a neuron that transmits neural information from the spinal cord to an autonomic ganglion. **523**

primary bronchi bronchi that branch directly from the trachea. **699**

primary oocytes (Greek: *oon* - egg; *kytos* - cell) immature eggs formed by meiosis. **848**

primary spermatocyte Diploid cells formed following mitosis of the spermatogonium. **847**

prime mover (aka agonist) principal muscle responsible for a movement. **309, 329**

progesterone steroid hormone inovlved in embryogenesis, pregnancy and the menstrual cycle. **549**

prolactin-inhibiting hormone (PIH) hormone released by the hypothalamus to regulate the anterior pituitary. **558**

prolactin (PRL) a hormone produced by the anterior pituitary gland, which promotes development of the mammary glands and the production of breast milk in lactating individuals. **558**

prolactin-releasing hormone (PRH) peptide hormone synthesized in the hypothalamus that stimulates prolactin. **558**

pronation (Latin: *pronus* - bent forward) positioning the forearm so that the palm of the hand faces posteriorly; if the arm is outstretched, positioning of the forearm so that the palm faces inferiorly. **281**

prone (Latin: *pronus* - bent forward) anterior side of the body down so that the posterior is viewable/up. **5**

prophase (Greek: *pro* - before; *phasis*: appearance) first stage of mitosis (and meiosis), characterized by breakdown of the nuclear envelope and condensing of the chromatin to form chromosomes. **55**

prostate gland a ring-shaped gland at the base of the urinary bladder in biological males; surrounds the prostatic urethra and secretes an alkaline fluid that becomes part of the semen during ejaculation. **841**

proton pump inhibitor (PPI) class of drug that irreversibly inhibits the H^+/K^+ ATPase proton pumps in the stomach. **779**

protraction (Latin: *protractus* - to draw forth) anterior movement of a portion of the body, such as forward thrusting of the mandible or scapula. **281**

proximal convoluted tubule (Latin: *convolutus* - to roll together) a coiled tubule of the nephron, which transports filtrate from the glomerular capsule to the nephron loop; performs a significant portion of reabsorption and secretion for the nephron. **791**

pterygoid processes paired projections of the sphenoid bone located on the inferior skull; forms the posterior portion of the nasal cavity. **201**

pubic symphysis (Latin: *pubes* - groin; Greek: *symphysis* - growing together) articulation between the pubic bodies of the two ossa coxae. **243**

pubic tubercle (Latin: *pubes* - groin; *tuberosus* - full of lumps) small bump on the superior aspect of the pubic body. **244**

pubis (Latin: *pubes* - groin) anterior region of the hip bone. **244**

pulmonary semilunar valve (*aka* pulmonary valve, pulmonic valve, right semilunar valve) (Latin: *pulmonarius* - lungs; Latin: *semi* - half; *luna* - moon) valve at the base of the pulmonary trunk that prevents backflow of blood into the right ventricle; consists of three flaps. **601**

pulmonary surfactant (Latin: *pulmonarius* - lungs) a mixture of phospholipids and proteins, secreted by the type II alveolar cells, which decreases the surface tension between water molecules in the alveoli. **703**

pulmonary trunk (Latin: *pulmonarius* - lungs) a short, large-diameter arterial blood vessel that transports blood ejected from the right ventricle, and soon divides into the right and left pulmonary arteries. **624**

pulmonary veins (Latin: *pulmonarius* - lungs) veins that carry highly oxygenated blood into the left atrium, which pumps the blood into the left ventricle, which in turn pumps oxygenated blood into the aorta and to the many branches of the systemic circuit. **624**

pulp cavity the inner part of the crown of a tooth, which contains nerves and blood vessels; continues into the root of the tooth as the root canal. **743**

pyloric sphincter (Greek: *pyloros* - gate; Latin: *sphincter* - contractile muscle) a ring of smooth muscle that lies between the stomach and the duodenum, which regulates the rate of gastric emptying into the duodenum. **749**

pylorus (Greek: *pyloros* - gate) the J-shaped curve at the distal end of the stomach, which lies between the body of the stomach and the duodenum. **749**

Q

quadrate lobe lobe of the liver located on the undersurface of the left lobe **750**

quadriceps femoris tendon (Latin: *quattuor* - four; *caput* - head; *femur* - thigh) four muscles in the anterior compartment of the thigh; extend and stabilize the knee. **276**

R

radial artery one of the terminal branches of the brachial artery; runs along the radius on the lateral side of the forearm; in the hand, merges with the ulnar artery to form the palmar arches; provides blood to the forearm and hand. **615**

radial fossa (Latin: *radius* - ray; *fossa* - trench) small depression on the anterior humerus superior to the capitulum; provides space for the head of the radius when the elbow is maximally flexed. **235**

radial notch (Latin: *radius* - ray; Greek: *ōlenē* - ulna) lateral side of the proximal ulna; articulates with the head of the radius at the proximal radioulnar joint. **235**

radial tuberosity (Latin: *radius* - ray; *tuberosus* - full of lumps) rough protuberance on the medial side of the proximal radius; serves as a point of attachment for muscles. **235**

radial vein a vein that originates from the palmar venous arches and runs parallel to the radius on the lateral side of the forearm; empties into the brachial vein; drains blood from the hand and forearm. **617**

radius (Latin: *radius* - ray) bone located on the lateral side of the antebrachium of each upper limb. **235**

rami portions of a bone. **202**

receptive field (*aka* sensory space) (Latin: *recipere* - to receive) a region of tissue space within which a receptor receives information. **441**

reciprocal activation reflex in which antagonistic muscles are stimulated to contract. **501**

reciprocal inhibition reflex in which antagonistic muscles are inhibited from contracting. **501**

rectum The final portion of the large intestine. **750**

red bone marrow connective tissue in the interior cavity of a bone; site of hematopoiesis. **174, 660**

red pulp an area in the spleen that consists mainly of red blood cells; removes damaged or worn-out red blood cells from the circulation. **664**

reflex arc a neural pathway that controls a reflex **493**

reflexes (Latin: *reflexus* - a bending back) connections between the sensory and motor neurons that do not include the higher brain centers or include conscious or voluntary aspects of movement. **493, 480**

renal arteries (Latin: *renes* - kidneys) the major artery that transports oxygen-rich blood to the kidney; branches directly off the abdominal aorta. **620, 795**

renal columns (Latin: *renes* - kidneys) connective tissue bands running between the renal pyramids, that transport blood vessels between the renal cortex and the renal medulla. **787**

renal corpuscle (Latin: *renes* - kidneys; *corpuscle* - small particle) the beginning of a nephron, which performs the process of filtration during urine formation; consists of a glomerulus and a glomerular capsule. **791**

renal papillae (Latin: *renes* - kidneys; *papilla* - nipple) the pointed tips of the renal pyramids that face the renal pelvis; site at which urine is drained into the collecting ducts into the minor calyces. **787**

renal pelvis (Latin: *renes* - kidneys) the funnel-shaped region of the kidney that collects all of the urine for each kidney; formed by the merging of the major calyces; after leaving the kidney at the hilum, narrows to become the ureter. **787**

renal pyramids (Latin: *renes* - kidneys) cone-shaped units of the renal medulla, which consist of the collecting ducts and the long nephron loops of juxtamedullary nephrons. **787**

renal veins (Latin: *renes* - kidneys) the major vein that transports the filtered, oxygen-poor blood out of the kidney; drains into the inferior vena cava. **620, 795**

reposition movement that returns the thumb from opposition to anatomical position. **281**

residual volume (RV) the amount of air that always remains in the lungs, even after forced exhalation. **719**

respiratory bronchioles (Latin: *respirare* - breathe; Greek: *bronkhos* - windpipe) a bronchiole that branches off of a terminal bronchiole and leads into an alveolar duct; the first structure of the bronchial tree to be able to perform gas exchange, due to the presence of alveoli budding from its walls. **700**

respiratory zone the portion of the respiratory tract including the bronchioles, alveolar ducts and alveoli. **684**

reticular layer (Latin: *reticulum* - a small net) layer of connective tissue underlying the papillary layer; reticulated appearance due to the abundant collagen and elastin fibers. **153**

retina (Latin: *rete* - net) the innermost layer of the wall of the eyeball; contains the rods and cones (the photoreceptor cells for the sense of vision). **416**

retraction (Latin: *retraho* - to draw back) posterior movement of the mandible or scapula. **281**

ribosomes cellular machinery composed of ribosomal RNA and proteins; responsible for protein synthesis. **50**

right and left hypochondriac regions the upper right region of the abdomen, inferior to the thoracic cage, that includes the liver; the upper left region of the abdomen, inferior to the thoracic cage, that includes part of the liver, the spleen and part of the stomach. **15**

right and left iliac regions the lower right region of the abdomen, located below the right hypochondriac region; the lower left region of the abdomen, located below the left hypochondriac region. **15**

right atrioventricular valve (*aka* tricuspid valve) (Latin: *atrium* - entry hall; *ventriculum* - belly) valve located between the right atrium and ventricle; consists of three flaps of tissue. **601**

right atrium upper-right chamber of the heart; receives blood from the system circuit that flows into the right ventricle. **597**

right common carotid artery vessel that branches from the brachiocephalic artery. **615**

right common iliac artery a terminal branch of the abdominal aorta that divides into the internal and external iliac arteries; supplies blood to the pelvis, the lower abdominal area, and the lower limbs. **622**

right coronary artery (Latin: *corona* - crown) one of two arteries that branch off the aorta to bring oxygenated blood to the heart wall; distributes blood to the right atrium, portions of the left and right ventricles, and the heart conduction system. **597**

right gastric artery an artery that branches off of the common hepatic artery, and supplies blood to a portion of the stomach. **618**

right lobe lobe of the liver located on the right hypochondrium. **750**

right lymphatic duct (Latin: *lympha* - water) the smaller of the two lymphatic collecting ducts; drains lymph from the right side of the head, upper limb, and chest into the right subclavian vein. **667**

right pulmonary artery artery supplying deoxygenated blood to the right lung. **624**

right subclavian artery artery supplying blood to the right arm. **615**

right ventricle the major pumping chamber on the lower right side of the heart that ejects blood into the pulmonary circuit via the pulmonary trunk and receives blood from the right atrium. **597**

rods photoreceptor cells located in the retina; responsive to low light. **416**

root 1) the upper portion of the external nose, which runs between the eyebrows; 2) the region of a tooth which lies below the gum line, and anchors the tooth to the alveolar process of the maxilla or mandible; 3) the internal part of the penis. **687, 743**

root canal the passageway through which blood vessels and nerves within the pulp cavity of a tooth run through the root into the maxilla or mandible. **745**

rootlets nerve bundles that make up the anerior and posterior roots. **483**

rotation (Latin: *rotare* - to rotate) twisting movement of the body or a limb around the long axis. **280**

Rough ER (RER) an organelle in a cell that functions to synthesize and modify proteins. **50**

round window opening between the middle ear and inner ear; covered by the secondary tympanic membrane. **427**

rugae (*pl*: **rugae**) (Latin: *ruga* - wrinkle) a fold in the inner lining (mucosa and submucosa) of certain organs to allow for expansion; rugae are visible in the stomach when it is empty but disappear when it is full following a meal; similarly, rugae in the vagina permit it to stretch during childbirth and sexual intercourse. **749, 800**

S

sacral canal (Latin: *sacral* - sacred; *canalis* - channel) bony tunnel that passes inferiorly through the sacrum. **218**

sacral foramina (Latin: *sacral* - sacred; *foramen* - an opening) series of paired openings through which nerves exit the anterior (ventral) and posterior (dorsal) aspects of the sacrum. **218**

sacral hiatus (Latin: *sacral* - sacred; *hiatus* - an opening) the inferior opening of the sacral canal; located near the inferior tip of the sacrum. **218**

sacral promontory (Latin: *sacral* - sacred; *promontorium* - projection) anterior lip of the body of the first sacral vertebra. **218**

sacroiliac joint (aka SI joint) (Latin: *sacrum* - sacred; *ilium* - groin) largely immobile articulation between the auricular surfaces of the sacrum and ilium. **243**

sacrum (Latin: *sacrum* - sacred) single bone forming part of the posterior portion of the pelvis; located near the inferior end of the adult vertebral column; formed by the fusion of the sacral vertebraes. **218**

saddle joint (aka sellar joint) synovial joint formed by the articulation of the convex and concave surfaces of two saddle-shaped bones. **267**

sagittal plane (Latin: *sagitta* - an arrow) two-dimensional, vertical plane that divides the body or organ into right and left portions. **9**

sagittal suture (Latin: *sagitta* - an arrow; *sutura* - seam) joint that unites the parietal bones at the superior midline along the top of the skull. **198, 211**

sarcomere (Greek: *sarx* - flesh; *meros* - part) repeating contractile unit of skeletal muscle; consists of overlapping thick and thin myofilaments. **292**

sarcoplasmic reticulum (SR) (Greek: *sarx* - flesh; *plasma* - a thing formed; Latin: *rete* - net) specialized smooth endoplasmic reticulum of a skeletal muscle cell, which stores and releases calcium ions for muscle contraction. **297**

satellite cells (Latin: *satelles* - attendant) glial cell type in the PNS that surrounds and supports the cell bodies of sensory neurons. **370**

scaphoid (Greek: *skaphē* - boat; *eidos* - resemblance) from the lateral side, the boat-shaped first of the four proximal carpal bones; articulates with the radius proximally, the trapezoid, trapezium, and capitate distally, and the lunate medially. **236**

scapula (pl: scapulae) (aka shoulder blade) (Latin: *scapulae* - shoulder blades) flat, triangular bone located on the posterior side of the shoulder. **231**

sclera (Greek: *sklēros* - hard) white portion of the fibrous layer of the eye. **416**

scrotum (Latin: *scrotum* - a skin) a sac of skin and muscle tissue that contains the testes and helps with temperature regulation of the developing sperm. **837**

sebaceous gland (Latin: *sebum* - oily; *glans* - acorn) oil gland found in the dermis, usually associated with a hair follicle; secretes sebum to lubricate and waterproof skin and hair. **158**

sebum (Latin: *sebum* - oily) oily mixture of lipids that lubricates the skin, hair, and stratum corneum. **158**

secondary bronchi bronchi that branch from the primary bronchi. **699**

section in anatomy, a single flat surface that results when a three-dimensional structure has been cut. **9**

segmental arteries (Latin: *segmentum* - a piece cut off) arteries that branch from the renal artery as it enters the kidney at the hilum; branch into the interlobar arteries. **795**

selectively permeable (Latin: *seligo* - to select; *permeao* - to pass through) feature of any barrier that allows certain substances to cross but excludes others. **80**

sella turcica (aka Turkish saddle) (Latin: *sella* - saddle; *turcica* - Turkish) bony region of the sphenoid bone located at midline of the middle cranial fossa. **201**

semicircular canals three fluid-filled tubes within the inner ear; involved in balance. **426**

seminal vesicle (*aka* seminal glands) (Latin: *seminalis* - seed) glands that contribute approximately 60 percent of the semen volume. **834**

sensory neuron (*aka* afferent neurons) (Latin: *sensatio* - feeling; Greek: *neuro* - nerve) neurons that collect information about the internal and external environment and send it toward the central nervous system. **497**

sensory receptor (Latin: *sensatio* - feeling; *recipere* - to receive) cells or structures that detect sensation. **392, 497**

septum (*pl*: septa) (Latin: *septum* - fence, partition) a wall or partition that divides the heart into chambers. **597**

serosa the outermost layer of the wall of the GI tract; continuous with the visceral peritoneum of organs of the abdominopelvic cavity. **755**

serous (*pl*: serous glands) (Latin: *glans* - acorn) gland that produces enzyme-rich secretions derived from blood plasma. **14**

serous membranes (*sing*: serous membrane) (*aka* serosa) (Latin: *membrana* - membrane) epithelial membranes lining body cavities that do not open to the outside; lining both the cavity wall and organ surface; 2) membrane that covers organs and releases a fluid to reduce friction. **13**

serous pericardium the inner layer of the pericardium, composed of the parietal and visceral serious pericardium. **595**

sex considerations of a person's anatomy, chromosomes, or hormones. **834**

shaft of the humerus (Latin: *humerus* - shoulder) cylindrical central region of the humerus. **235**

sigmoid colon (Greek: *sigmoeidēs* - S-shaped) the final region of the colon, which connects the descending colon to the rectum. **750**

simple diffusion movement of a substance from an area of higher concentration to one of lower concentration. **82**

small cardiac vein (Greek: *cardiac - kardia* meaning heart) parallels the right coronary artery and drains blood from the posterior surfaces of the right atrium and ventricle; drains into the great cardiac vein. **598**

small intestine a major organ of the GI tract, which accomplishes most of the digestion and absorption of nutrients. **749**

small saphenous vein a vein that arises from the dorsal veinous arch and dorsal vein of the little toe, and empties into the popliteal vein; travels up the lateral side of the leg, and drains the superficial areas of the foot and lower leg. **623**

Smooth ER (SER) an organelle that functions to synthesize and modify carbohydrates as well as serve some detoxification functions. **50**

soft palate (*aka* velum, palatal velum, muscular palate) (Latin: *palatum* - roof of mouth) the muscular portion of the palate, which separates the oral cavity from the nasal cavity; lies posterior to the hard (bony) palate, and contains the uvula, which closes the entrance to the nasal cavity during swallowing. **688, 743**

solute (Latin: *solutus* - dissolved) substance that is put into a solution to dissolve. **89**

solvent (Latin: *solvere* - to loosen) a substance in a solution that dissolves a solute. **89**

somatic nervous system (SNS) (Greek: *sōma* – body) functional division of the nervous system concerned with voluntary motor responses and conscious perception. **364**

somatic reflexes (Greek: *sōma* - body; Latin: *reflexus* - a bending back) reflexes where the motor response is carried out by skeletal muscles. **497**

special senses senses with specialized organs; includes vision, hearing, balance, smell and taste. **391, 408**

specific gravity the ratio of the density of a substance to the density of a reference material (often water). **819**

spermatids (Greek: *sperma* - seed) a cell that arises from a secondary spermatocyte during meiosis I, and will eventually mature into a sperm. **847**

spermatogenesis (Greek: *sperma* - seed; *genesis* - origin) the process of sperm production, which occurs in the seminiferous tubules of the testes. **833**

spermatogonia (*sing*: spermatogonium) (Greek: *sperma* - seed) the most primitive form of spermatogenic cells; eventually develop into primary and secondary spermatocytes, spermatids, and finally sperm. **849**

sperm capacitation (Latin: *capacitas* - make capable) the process in which the sperm penetrates the outer layers of the egg. **867**

spermiogenesis (Greek: *sperma* - seed) the last stage of spermatogenesis, in which spermatids mature into sperm (or spermatozoa). **848**

S phase stage of the cell cycle during which DNA replication occurs. **55**

sphenoid bone (Latin: *sphenoides* - resembling a wedge) single bone that forms the central base of skull and articulates with almost every other bone in the skull. **201**

sphenoid sinus (Latin: sphenoides - resembling a wedge; sinus - cavity) air-filled space located within the body of the sphenoid bone; most posterior of the paranasal sinuses. **201**

sphenoid sinus (Latin: *sphenoides* - resembling a wedge; *sinus* - cavity) air-filled space located within the body of the sphenoid bone; most posterior of the paranasal sinuses. **688**

sphygmomanometer an instrument used to measure the blood pressure, consisting of an inflatable cuff, bulb (for inflating the cuff), and a pressure gauge. **645**

spinal nerve the 31 pairs of mixed nerves attached to the spinal cord. **461, 483**

spinal reflexes (Latin: *reflexus* - a bending back) reflexes that involve the spinal cord. **497**

spine of the scapula (*aka* scapular spine) (Latin: *spina* - backbone; *scapulae* - shoulder blades) posterior ridge passing mediolaterally along the scapular surface. **231**

spinous process single bony process that projects posteriorly from the midline of the back; functions as a site of muscle attachment. **213**

spiral arteries small arteries that supply blood to the endometrium during the luteal phase of the menstrual cycle. **865**

spirometer (Latin: *spira* - a coil; Greek: *metron* - a measure) a device that measures the volume of air that a person inhales or exhales, as a result of different degrees of effort in the breathing process; used to test various respiratory volumes in patients with certain respiratory conditions.sss. **719**

spleen one of the secondary lymphoid organs, and the largest lymphatic organ in the body; removes pathogenic organisms from the blood, produces B lymphocytes (white pulp) for the adaptive immune response, and removes worn-out or damaged red blood cells and platelets from the bloodstream (red pulp). **660**

splenic artery an artery that branches off of the celiac trunk, provides blood to the spleen and pancreas. **618**

splenic veins vein of the heptic portal system; drains from the spleen, fundus of the stomach and the pancreas. **622**

spongy bone (aka cancellous bone) bone containing lacunae and osteocytes arranged in trabeculae. **173**

squamous suture (Latin: *squamosus* - scaly; *sutura* – seam) joint that unites the parietal bone to the squamous portion of the temporal bone. **198, 211**

stapes (aka stirrup) ossicle that is attached to the oval window of the inner ear. **425**

sternal angle (Greek: *sterno* - chest) junction between the manubrium and body of the sternum; site of attachment of the second rib to the sternum. **222**

sternal body the longest portion fo the sternum; inferior to the manubrium and superior to the xiphoid process. **222**

sternal end (aka medial end of the clavicle) (Greek: *sterno* - chest) triangle-shaped medial end of the clavicle; articulates with the manubrium of the sternum. **231**

sternum (Greek: *sterno* - chest) elongated, flat bony structure located at the center of the anterior chest. **222**

stimulus an environmental change that initiates a response. **480**

stratum basale (Latin: *stratum* - layer– basis - foundation) the deepest layer of the epidermis, composed of epidermal stem cells; connects the epidermis to the basement membrane. **151**

stratum corneum (Latin: *stratum* - layer; *corneus* - horn) most superficial layer of the epidermis; composed of layers of dead cells. **153**

stratum granulosum (Latin: *stratum* - layer; *granulum* - grain) layer of the epidermis superficial to the stratum spinosum; characterized by the presence of intracellular granules. **152**

stratum lucidum (Latin: *stratum* - layer; *lucidus* – clear) layer of the epidermis between the stratum granulosum and stratum corneum, found in the skin of the palms and soles of the feet. **153**

stratum spinosum (Latin: *stratum* - layer; *spinosus* - thorny) layer of the epidermis superficial to the stratum basale; composed of football-shaped keratinocytes, joined together by desmosomes. **152**

stretch reflexes (Latin: *reflexus* - a bending back) type of reflex which responds to the activation of muscle spindle stretch receptors; results in contraction of the muscle to maintain a constant length. **499**

styloid process (Greek: *stylos* - pillar; *eidos* - resemblance) elongated, downward bony projection, located posterior to the mandibular fossa on the external base of the skull. **199**

styloid process of the radius (Greek: *stylos* - pillar; *eidos* - resemblance; Latin: *radius* - ray) delicate inferior projection on the lateral end of the distal radius. **235**

styloid process of the ulna (Greek: *stylos* - pillar; *eidos* - resemblance; *ōlenē* – ulna) delicate inferior projection located on the medial end of the distal ulna. **235**

subcapsular sinus (Latin: *sub* - under; *sinus* - a fold) the region of a lymph node through which lymph enters; lies just inside of the capsule; filters the lymph, bringing pathogens and debris into contact with phagocytic cells that can engulf and destroy them. **671**

subclavian veins a continuation of the axillary vein as it passes behind the clavicle; drains blood from the upper limb and scapular region; drains into the brachiocephalic vein. **617**

sublingual salivary glands (Latin: *sub* - under; *lingualis* - tongue) one of the major pairs of salivary glands; lies under the tongue; secretes mucous saliva into the floor of the mouth. **743**

submandibular glands (Latin: *sub* - under; *mandibula* – jaw) one of the major pairs of salivary glands, which lies deep to the mandible; secretes mixed saliva (part serous and part mucous) into the floor of the mouth. **743**

submucosa (Latin: *sub* - under) the layer of areolar connective tissue in the wall of the GI tract that is located between the mucosa and the muscularis. **754**

subscapular fossa (Latin: *sub* - under; *scapulae* - shoulder blades; *fossa* - trenches) broad depression on the anterior surface of the scapula. **231**

subscapular vein a vein that drains the subscapular area, and drains into the axillary vein. **617**

sulci (sing: sulcus) (Latin: *sulcus* - furrow, wrinkle) grooves or indentations on the surface of the cerebrum; lie between gyri (raised ridges). **452**

sulcus (*pl*: **sulci**) (Latin: *sulcus* - groove) fat-filled groove visible on the surface of the heart; coronary vessels are also located in these areas. **597**

superficial (Latin: *superficialis* - surface) describes a position nearer to the surface of the body. **24, 312**

superior angle (Latin: *scapulae* - shoulder blades) corner of the scapula located between the superior and medial borders. **231**

superior articular processes flat projection extending upward from the vertebral arch of a vertebra; articulates with the inferior articular process of the adjacent vertebra. **213, 218**

superior border (Latin: *scapulae* - shoulder blades) superior margin of the scapula. **231**

superior colliculi (Latin: *superiorem* - higher) two of the four structures of the corpora quadrigemina of the midbrain; responsible for combining sensory information with motor output. **460**

superior mesenteric artery an artery that branches off of the abdominal aorta, and provides blood to the small intestine, the pancreas, and the proximal portion of the large intestine. **620**

superior mesenteric vein vessel that drains from the small intestine; meets the splenic vein to form the portal vein. **622**

superior nasal conchae (*pl*: conchae) (Latin: *nasus* - nose; *concha* - shell) thin, curved projection of the ethmoid bone; smallest and most superiorly located of the nasal conchae. **200**

superior nuchal line (Latin: *nucha* - nape) paired bony lines on the posterior skull that extend laterally from the external occipital protuberance; sites of attachment for muscles of the neck. **199**

superior orbital fissure (Latin: *superiorem* - higher; *fissura* - cleft) irregularly shaped opening located laterally to the optic canal; provides passage for the artery supplying the eyeball. **201**

superior phrenic artery an artery that branches off of the thoracic aorta and provides the blood supply to the superior portion of the diaphragm. **617**

superior ramus of the pubis segment of bone that passes laterally from the pubic body to the ilium. **244**

superior vena cava one of the largest veins of the systemic circuit; formed by the merging of the left and right brachiocephalic veins. **617**

supination (Latin: *supinare* - to bend backwards) positioning the forearm so that the palm of the hand faces anteriorly; if the arm is outstretched, positioning the forearm so that the palm faces superiorly. **281**

supine (Latin: *supinare* - to bend backwards) position of the body resting on the posterior so that anterior structures are visible and accessible. **5**

supraorbital foramina (Latin: *supra* - above; *foramen* - opening) small opening located at the superior margin of the orbit; provides passage for the sensory nerve of the forehead. **197**

supraorbital margin (Latin: *supra* - above) feature of the frontal bone, superior rim of the orbit. **197**

supraspinous fossa (Latin: *supra* - above; spina - spine; fossa – trench) narrow depression on the posterior scapula, superior to the spine. **231**

sutures (Latin: *sutura* - seam) immobile joint at which adjacent bones of the skull are connected by fibrous connective tissue. **266**

sympathetic division (Greek: *sympathētikos* - to feel with) portion of the autonomic nervous system associated with preservation of life in emergency situations. **520**

symphysis (Greek: *symphysis* - growing together) cartilaginous joint in which bones are joined by a fibrocartilage pad; amphiarthrotic. **266**

synapses (Greek: *syn* - with; *hapto* - to clasp) narrow junction between two neurons across which a chemical signal passes. **366**

synarthrosis (Greek: *syn* - together; *arthrosis* - joints) joint at which no motion is possible. **265**

synchondrosis (Greek: *syn* - together; *chondros* - cartilage) cartilaginous joint in which bones are joined together by hyaline cartilage, or where bone is joined to hyaline cartilage. **266**

syndesmoses (Greek: *syndesmos* - a fastening) type of fibrous joint in which two parallel bones are united by fibrous connective tissue. **266**

synergists (Greek: *synergia* - together) muscle that contributes to the movement of a prime mover; often by flexing a joint or stabilizing the insertion. **309, 329**

synovial fluid (Greek: *syn* - together; Latin: *ovum* - egg; *fluidus* - to flow) thick, slimy fluid that is released into a synovial joint; lubricates the joint and nourishes the cells of cartilage within the joint. **271**

synovial joint joint at which the articulating surfaces of the bones are not directly connected, but are located within a cavity formed by an articular capsule. **265**

synovial membrane (Greek: *syn* - together; Latin: *ovum* - egg; *membrana* - membrane) connective tissue membrane that lines the cavities of a freely movable joint; composed of cells that produce synovial fluid. **271**

systole (Greek: *systole* - contraction) period of time when the heart muscle is contracting. **642**

T

tachycardia a heart rate that is faster than normal resting rate. **653**

tail of the pancreas the portion of the pancreas that is closest to the spleen. **750**

talus (Latin: *talus* - ankle) most superior tarsal bone; articulates superiorly with the tibia and fibula at the ankle joint, inferiorly with the calcaneus bone and anteriorly with the navicular bone. **248**

tarsal bones (Greek: *tarsos* - ship) one of the seven bones that make up the posterior foot and ankle. **247**

tastants a chemical that stimulates the sensory cell of the taste bud. **399**

taste buds structures within the papilla of the tongue that contain gustatory receptor cells. **400**

taste receptors cells (*aka* gustatory cells) (Latin: *recipere* - to receive) specialized epithelial cells that are the receptors and transducers of taste stimuli. **400**

tectorial membrane a membrane in the cochlea of the inner ear. **426**

telophase (Greek: *telos* - end; *phasis* - appearance) final stage of mitosis (and meiosis), preceding cytokinesis, characterized by the formation of two new daughter nuclei. **56**

temporal bones (Latin: *tempus* - temple) one of a pair of bones that forms the lateral, inferior portion of the skull; subdivided into squamous, mastoid, and petrous regions. **199**

temporal lobes (Latin: *temporalis* - temples) a region of the cerebral cortex directly beneath the temporal bone of the cranium; functions in the understanding of language, memory, processing of auditory information, and some visual perception. **457**

temporal processes (Latin: *temporalis* - temple; Greek: *zygon* - yoke) short projection of the zygomatic bone; forms the anterior portion of the zygomatic arch. **203**

tendon reflex (Latin: *reflexus* - a bending back) a reflex that serves to protect the muscles during contraction by stretching a tendon during muscle contraction. **501**

tendon sheath (Latin: *tendo* - to stretch) elongated structure made of connective tissue that surrounds a tendon where the tendon crosses a joint; contains synovial fluid to reduce friction and allow for smooth movement of the tendon. **272**

tendons (Latin: *tendo* - to stretch) structure made of dense connective tissue that attaches muscle to bone. **271**

tentorium cerebelli (Latin: *tentorium* - tent; *cerebrum* - brain) one of the four cranial dural septa; provides a roof over the cerebellum, separating the cerebellum from the rest of the brain. **455**

tertiary follicles (*aka* antral follicles, mature follicles, mature antral follicles, Graafian follicles) (Latin: *tertiarius* - third; *folliculus* - little bag) ovarian follicles consisting of a primary or secondary oocyte, a zona pellucida, a single fluid-filled antrum, and several layers of granulosa cells; ready for ovulation. **848**

testes (*sing*: testis) (Latin: *testis* - testicle) the gonads that produce sperm and sex hormones including testosterone in biological males. **549, 837**

testosterone steroid sex hormone; involved in the development of male reproductive structures and secondary sex characteristics. **549**

thecal cells cells that surround developing ovarian follicles in the ovary; cells of the theca interna produce androgens, and cells of the theca externa from a connective tissue sheath around the follicles. **848**

the pulmonary circuit the pathway that blood follows as it travels from the heart to the lungs, and then back to the heart; functions in gas exchange between the blood and the lungs. **612**

the systemic circuit the pathway that blood follows as it travels from the heart to the body cells, and then back to the heart; functions in gas and nutrient exchange between the blood and the cells of the body. **612**

thigh area of the lower limb between the hip and knee joints. **247**

third ventricle (Latin: *ventriculus* - belly) one of the four chambers in the brain that produces and circulates cerebrospinal fluid; located in the diencephalon. **455**

thoracic cage (Greek: *thōrax* - chest) forms the chest; consists of the sternum, 12 pairs of ribs and their costal cartilages; protects the heart and lungs. **221**

thoracic duct (Latin: *thorax* - chest) the larger of the two lymphatic collecting ducts; drains lymph from the lower limbs, abdomen, left upper limb, and the left side of the head and chest, and empties it into the left subclavian vein. **667**

thoracic vertebrae (Greek: *thōrax* - chest; Latin: *vertebra* - joint of the spine) twelve vertebrae, located in the thoracic region of the vertebral column. **216**

thymosins (Greek: *thumos* - warty excrescence) hormones produced and secreted by the thymus; contribute to the development and differentiation of T cells. **549**

thymus (Greek: *thumos* - a warty excrescence) one of the primary lymphoid organs; responsible for the maturation and proliferation of T lymphocytes received from the red bone marrow. **549, 660**

thyroid cartilage (Greek: *thyreoiedes* - shield-shaped) the largest cartilage of the larynx; consists of two plates of hyaline cartilage that fuse in the middle to form the laryngeal prominence or "Adam's apple." **693**

thyroid gland (Greek: *thyreoiedes* - shield-shaped; *glans* – acorn) a large, butterfly-shaped endocrine gland located over the anterior surface of the trachea, inferior to the larynx; responsible for the synthesis of thyroid hormones and calcitonin. **549**

thyroid hormone (TH) any hormone produced by the thyroid gland; includes triiodothyronine and thyroxine. **549**

thyroid-stimulating hormone (TSH) (*aka* thyrotropin) (Greek: *thyreoiedes* - shield-shaped) an anterior pituitary hormone that stimulates the secretion of thyroid hormones. **556**

thyrotropin-releasing hormone (TRH) (aka thyrotropin) (Greek: *thyreoiedes* - shield-shaped) an anterior pituitary hormone that stimulates the secretion of thyroid hormones. **556**

tibia (aka shin bone) the larger and more medial of the two bones of each lower leg; articulates with the femur to form the knee joint; important weight-bearing bone. **247**

tibial collateral ligament (Latin: *tibia* - shin bone; *ligamentum* - a band) extrinsic ligament on the medial knee that runs from the medial epicondyle of the femur to the medial tibia. **276**

tibial tuberosity (Latin: *tibia* - shin bone; *tuberosus* - full of lumps) prominent elevated area located on the anterior surface of the proximal tibia. **247**

tidal volume (TV) the amount of air that enters or leaves the lungs during one normal breath. **719**

tongue one of the accessory organs of the digestive system, which occupies a large portion of the oral cavity; composed of skeletal muscle; active in the chewing and swallowing processes. **743**

tonicity the concentration of solute in the extracellular fluid compared to the concentration of solute in the intracellular fluid. **97**

tonic receptors (Greek: *tonikos* - tone; Latin: *recipere* - to receive) receptors that do not adapt, or adapt very slowly, even when the stimulus remains constant. **443**

tonsillar crypts deep indentations of the epithelium of a tonsil, which accumulate materials entering the body through eating and breathing; contain macrophages and leukocytes, which interact with incoming materials to destroy pathogens. **664, 691**

tonsils masses of lymphoid nodules associated with the pharynx; consist of the pharyngeal tonsil (adenoid), the palatine tonsils, and the lingual tonsil. **660**

total lung capacity (TLC) the maximal amount of air that the lungs can hold; can be calculated by taking the sum of the TV, ERV, IRV, and RV. **719**

trabeculae carneae (Latin: *trabecula* - a small beam) ridges of muscle covered by endocardium located in the ventricles. **601**

trabeculae (*sing*: trabecula) (Latin: *trabs* - a beam) lattice-like beams making up the matrix of spongy bone. **671, 173**

trabecular sinus sinus within a lymph node that drains into the medullary sinus. **671**

trachea (aka windpipe) (Greek: *trakheia* – windpipe) a tubular organ that runs vertically from the larynx to the primary bronchi; contains cartilaginous C-shaped incomplete rings that keep it open at all times; provides the air passageway between the larynx and the lungs. **684, 747**

tracheal cartilage c-shaped rings of cartilage that surround the trachea. **699**

transverse colon the portion of the colon that lies between the ascending colon and the descending colon; runs across the upper portion of the abdominopelvic cavity from right to left. **750**

transverse costal facet the point of artiulation between a vertebra and a rib, located on the vertebrae. **216**

transverse foramina (Latin: *transversus* - crosswise; *foramen* - opening) opening located in the transverse processes of the cervical vertebrae. **214**

transverse plane (Latin: *transversus* - crosswise) two-dimensional, horizontal slice that divides the body or organ into superior and inferior portions. **9**

transverse processes (Latin: *transversus* - crosswise) paired bony processes that extend laterally from the junction between the pedicle and lamina of the vertebra. **213**

trapezium (Greek: *trapezion* - table) from the lateral side, the table-shaped first of the four distal carpal bones; articulates with the scaphoid proximally, the first and second metacarpals distally, and the trapezoid medially. **236**

trapezoid (Greek: *trapeza* - a table; *eidos* - resembling) from the lateral side, the table-shaped second of the four distal carpal bones; articulates with the scaphoid proximally, the second metacarpal distally, the trapezium laterally, and the capitate medially. **236**

triquetrum (Latin: *triquetrus* - three-cornered) from the lateral side, the three-cornered third of the four proximal carpal bones; articulates with the lunate laterally, the hamate distally, and has a facet for the pisiform. **236**

trochlea (Greek: *trochileia* - a pulley) pulley-shaped region of the distal end of the humerus; articulates at the elbow with the trochlear notch of the ulna. **235**

trochlear notch (Latin: *trochlea* - a pulley) C-shaped depression on the anterior side of the proximal ulna; articulates with the trochlea of the humerus. **235**

trophoblast (Greek: *trophe* – nourishment) a structure that will yield the placenta and other structures required for pregnancy that are not part of the fetus. **865**

tropic hormone (Greek: *trope* - turning) hormones that act on a specific endocrine gland to trigger the release of another hormone from that gland. **555**

true ribs ribs 1–7; attach directly to the sternum via the costal cartilage. **222**

T-tubule deep invaginations in the sarcolemma that transmit electrical signals into the sarcoplasmic reticulum of muscle cells. **297**

tubercle of the rib (Latin: *tuberculum* - a knob) small bump located at the posterior end of a rib; articulates with the transverse process of a thoracic vertebra. **222**

tunica intima (*aka* tunica interna) (Latin: *tunica* - coat; *intima* - innermost) the innermost layer of the wall of a blood vessel. **628**

tunica media (Latin: *tunica* - coat; *media* - middle) the middle layer of the wall of a blood vessel; not present in capillaries. **628**

type I alveolar cells (Latin: *alveolus* - basin) a simple squamous epithelial cell that makes up the alveolar walls; performs rapid diffusion of respiratory gases with the pulmonary capillaries, due to its flattened shape and single-layered structure. **703**

type II alveolar cells (Latin: *alveolus* - basin) a cuboidal epithelial cell found in the walls of the alveoli; secretes pulmonary surfactant to decrease the surface tension between water molecules in the alveoli, allowing them to expand easily during inhalation. **703**

U

ulna (Greek: *ōlenē* - ulna) bone located on the medial side of the antebrachium of each upper limb. **235**

ulnar artery one of the terminal branches of the brachial artery; runs along the ulna on the medial side of the forearm; in the hand, merges with the radial artery to form the palmar arches; supplies blood to the forearm and hand. **615**

ulnar collateral arteries arteries that supply blood to structures in the the elbow area; arise from the brachial artery. **615**

ulnar vein (Greek: *ōlenē* - ulna) a vein that originates from the palmar venous arches and runs parallel to the ulna on the medial side of the forearm; empties into the brachial vein; drains blood from the hand and forearm. **617**

umbilical arteries two arteries that transport oxygen-poor blood and wastes from the fetus to the placenta; arise from the internal iliac arteries of the fetus, travel through the umbilical cord, and empty into the placenta. **877**

umbilical region the central region of the abdomen surrounding the umbilicus, that includes part of the stomach, intestine, and kidneys. **15**

umbilical vein a single vein that transports blood containing oxygen and nutrients from the placenta to the fetus; blood travels from the placenta into the umbilical cord, then into the fetal liver or ductus venosus, then into the inferior vena cava, and finally into the right atrium of the fetal heart. **877**

uniaxial (Latin: *unus* - one) joint that allows for motion within a single plane. **266**

upper respiratory tract the portion of the respiratory tract that lies in the head and neck; consists of the nasal cavity, pharynx, and larynx. **684**

urethra (Greek: *ourethra* - passage for urine) a tubular organ that transports urine from the urinary bladder to the outside of the body. **833**

uterine tubes (*aka* fallopian tube, oviduct) (Latin: *uterinus* - pertaining to the womb) tubular organ that transports ovulated oocytes from the ovary to the uterus; the site of fertilization of an oocyte by a sperm. **838**

uterus (Latin: *uterus* - womb) the highly muscular organ which houses a developing embryo and fetus, and propels the fetus to the outside world during childbirth. **838**

uvula (aka palatine uvula) (Latin: *uvula* - bunch of grapes) a fingerlike projection of soft tissue that hangs down from the center of the soft palate into the throat; prevents food from entering the nasal cavity during swallowing. **743**

V

vagina (Latin: *vagina* - sheath) a tubular organ in biological females that connects the uterus to the outside of the body; receives semen for fertilization, forms part of the birth canal, and excretes the menstrual flow. **833**

vagus nerve (Latin: *vagus* - wandering) tenth cranial nerve; responsible for contributing to homeostatic control of the organs of the thoracic and upper abdominal cavities. **526**

valve in the cardiovascular system, a specialized structure located within the heart or vessels that ensures one-way flow of blood. **601**

vasa recta (*aka* vasa rectae renis) (Latin: *vas* - vessel; *recta* - right) long straight capillaries that run parallel to the nephron loops of the juxtamedullary nephrons; function in urine concentration. **795**

vascular layer (Latin: *vasculum* - a small vessel) the middle layer of the eye that is primarily connective tissue but is rich with blood vessels and intrinsic muscles. **416**

vasoconstriction contraction of the smooth muscle tissue in the wall of a blood vessel, resulting in a decrease in the diameter of the vessel. **628**

vasodilation (Latin: *vas* - a vessel; *dilato* - to spread out) relaxation of the smooth muscle tissue in the wall of a blood vessel, resulting in the widening of blood vessels. **628**

vein (*pl*: **veins**) (Latin: *vena* - vein) a blood vessel that transports blood toward the heart. **611**

ventilation (aka pulmonary ventilation) (Latin: *ventilatus* - a breeze) the process of breathing, or moving of air into and out of the lungs; consists of inhalation and exhalation. **712**

ventricles (*pl*: **ventricles**) (Latin: *ventriculus* - belly) (1) one of the primary pumping chambers of the heart located in the lower portion of the heart; the left ventricle is the major pumping chamber on the lower left side of the heart that ejects blood into the systemic circuit via the aorta and receives blood from the left atrium; the right ventricle is the major pumping chamber on the lower right side of the heart that ejects blood into the pulmonary circuit via the pulmonary trunk and receives blood from the right atrium. (2) space within the brain where cerebrospinal fluid is produced. **370, 455**

venules a small vessel that transports blood from a capillary into a vein. **611**

vertebrae (Latin: *vertebrum* - joint) irregular bone located in the neck and back regions of the vertebral column; typically consist of a body, vertebral arch, and processes. **213**

vertebral arch (Latin: *vertebrum* - joint) posterior portion of the vertebra surrounding the spinal cord; consists of two pedicles and two laminae. **213**

vertebral arteries (Latin: *vertebrum* - joint) an artery that branches off of the subclavian artery and passes through the transverse foramina of the cervical vertebrae, continuing to the brain; joins the internal carotid artery to form the cerebral arterial circle; supplies blood to the central nervous system. **617**

vertebral foramen (Latin: *vertebrum* - joint; *foramen* - opening) opening within a vertebra, enclosed by the vertebral arch; provides passage for the spinal cord. **213**

vestibular folds (Latin: *vestibulum -* entrance) folded regions of mucous membrane that lie superior to the vocal folds in the larynx; do not produce sound, but help support the epiglottis during the swallowing process and to clear debris from the larynx by triggering the cough reflex. **693**

vestibule central part of the inner ear, medial to the tympanic membrane; responsible for the sense of balance. **426**

vestibulocochlear nerve (*aka* auditory nerve, acoustic nerve) eighth cranial nerve; responsible for the senses of hearing and equilibrium (balance). **427**

vibrissae stiff hair used by mammals to sense the enivronment; also known as whiskers. **687**

visceral relating to the internal organs of the body. **14**

visceral pericardium the outermost layer of the heart wall. **595**

visceral reflexes (Latin: *viscera -* internal organs; *reflexus -* a bending back) a reflex where the motor response involves smooth or cardiac muscle or a gland. **497**

vital capacity (VC) the amount of air that can be forcefully exhaled after a forced inhalation; can be calculated by taking the sum of the TV, ERV, and IRV. **719**

vitreous humor (Latin: *vitreus -* glassy) jellylike fluid that fills the posterior cavity of the eye. **415**

vocal folds (*aka* vocal cords, true vocal cords) white, membranous folds in the larynx, which produce sound when air is forced through them in their closed position, causing them to vibrate. **693**

vocal ligaments ligaments that extend from the thyroid cartilage to the vocal processes of the arytenoid vartilages. **693**

voluntary with conscious control. **123**

vomer single bone that forms the inferior and posterior parts of the nasal septum. **204**

vulva (*aka* pudendum) (Latin: *vulva -* womb) a group of external reproductive organs in biological females that includes the labia majora, labia minora, clitoris, mons pubis, vestibule, and vestibular glands. **840**

W

white matter a mass of myelinated axons in the central nervous system. **373**

white pulp the portion of the spleen that contains B cells and T cells, and macrophages; filter pathogens from the blood and engage in antigen recognition and attack. **664**

withdrawal reflex (Latin: *reflexus -* a bending back) a reflex in which the nociceptors that are activated by a painful stimulus activate the motor neurons responsible for contraction of a muscle to withdraw the body part from the painful stimulus. **498**

working distance the distance between the front edge of the objective lens and the surface of the specimen when the specimen is in focus. **71, 75**

X

xiphoid proces (Greek: *xiphos -* sword) the inferior tip of the sternum; cartilaginous early in life, but ossifies beginning in middle age. **222**

Y

yellow bone marrow (Greek: *zygon -* yoke; Latin: *temporalis -* temple) anterior projection of the temporal bone; forms the posterior portion of the zygomatic arch. **174**

Z

zona pellucida an extracellular matrix surrounding the oocyte. **848, 867**

zygomatic bones (Greek: *zygon -* yoke) paired bones that contribute to the lateral wall of the orbit and the anterior zygomatic arch. **203**

zygomatic process (Greek: *zygon -* yoke; Latin: *temporalis -* temple) anterior projection of the temporal bone; forms the posterior portion of the zygomatic arch. **199**

zygote (Greek: *zygōtós -* yoked) a fertilized egg. **863**

Index

Note: Page numbers followed by *f* and *t* represent figures and tables respectively.

cardiopulmonary resuscitation (CPR)
chest compressions during, 224, 224*f*
cardiovascular system, 592. *See also*
blood; heart
aorta, 615, 615*f*, 616*f*
arteries
of abdomen, 617–618, 619*f*, 620
of head, 617, 618*f*
of lower limbs, 622–623, 622*f*, 623*t*
of neck, 617, 619*f*
of thorax, 617–618, 619*f*, 620
of upper limbs, 615, 616*f*, 617*t*
circulation pathways, 612, 612*f*
fetal, 879, 879*f*
pulmonary circuit, 624, 625*f*
veins
of abdomen, 620–622, 620*f*
of head, 617, 619*f*
of lower limbs, 623, 623*t*, 624*f*
of neck, 617, 619*f*
of thorax, 620–622, 620*f*
of upper limbs, 617, 617*f*, 617*t*
carotid canals, 199, 207*f*
carotid pulse point, 32
carpal bones, 235, 235*f*, 236, 238*f*
carpal region, 30*f*
cartilaginous joints, 265, 266, 267*f*
casts, 813, 814*f*
cauda equina, 487
caudate lobe, 745, 746*f*
causative agent of disease, 76
cavities, body, 13
cecum, 733*f*, 745
celiac trunk, 618, 619*f*
cell cycle, 55–56, 55*f*, 56*f*
cell junctions, 301, 301*f*
cell membranes, 46, 49, 64, 80
diffusion across, 82*f*, 93–95
hydrophobic, 80
osmosis across, 93–95
phospholipid bilayer of, 49, 49*f*
proteins, 80
selective permeability, 80
cells
alpha, 549
appendages, 51, 53, 53*f*, 54*t*
bacterial, 54
beta, 549
bipolar, 416, 416*f*
of bones, 179, 179*t*
cardiac muscle, 293, 301
chromaffin, 531
cytoplasm, 49
cytoskeleton, 51, 52*f*

diploid (2n), 845, 846*f*
endothelial, 163
enteroendocrine, 750
epithelial, 813, 814*f*
follicular, 549, 551
ganglion, 416, 416*f*
glial (neuroglial), 127, 360, 369–371,
371*t*, 480
of CNS, 369–370, 369*f*
of PNS, 370, 370*f*
goblet, 115, 750
granulosa, 848, 848*f*, 853, 867,
867*f*
haploid, 845, 846*f*, 848
human, 49, 49*f*
interstitial, 849, 849*f*
juxtaglomerular, 808
keratinocytes, 151
macula densa, 808
mitosis, 55–58, 56*f*
neurilemma, 370, 370*f*, 373
neurosecretory, 531
nurse, 849, 849*f*
organelles, 49
parafollicular, 549
parts of, 49–54. *See also specific*
parts
photoreceptor, 416, 416*f*, 421
red blood. *See* erythrocytes
replication, 55
satellite, 370, 370*f*
skeletal muscle, 292
smooth muscle, 301
somatic, 845
staining, 74
stem, 845, 849
of stomach, 753, 753*f*
structures, 54*t*
taste, 401*f*
in testes, 849, 849*f*
thecal, 848, 848*f*
type I alveolar, 703, 703*f*
type II alveolar, 703, 703*f*
white blood. *See* leukocytes
cementum, 740*f*, 741
central adaptation, 438
central canal, 180, 455, 456*f*
central nervous system (CNS), 364,
480
brain. *See* brain
glial cells of, 369–370, 369*f*
spinal cord. *See* spinal cord
centrioles, 51
centrosome, 51

cephalic region, 26*f*, 30*f*
cephalic vein, 617, 617*f*
cerebellum, 459, 460*f*
cerebral aqueduct, 455, 456*f*, 460
cerebral arterial circle, 617, 618*f*
cerebral cortex, 452
cerebrospinal fluid (CSF), 104, 369*f*,
370, 455, 456*f*, 488
cerebrum, 452, 453*f*
lobes of, 457, 457*f*, 459
cervical enlargement, 488
cervical region, 26*f*, 30*f*
cervical vertebrae, 214–216, 214*f*, 215*f*
cervix, 838, 838*f*
chambers, of heart, 595*f*, 597
cheek bones. *See* zygomatic bones
cheeks, 739
chemical digestion, 739
chemical senses, 391–392
alterations in COVID-19, 403–404,
403*f*
gustation, 399–402, 399*f*–401*f*, 402*t*
olfaction, 395–397, 395*f*, 396*f*
chemotherapy, 59–60
chest compressions, during
cardiopulmonary resuscitation,
224, 224*f*
chin. *See* mental protuberance
cholinergic synapse, 523
chondrocytes, 129
chordae tendineae, 601, 601*f*, 603*f*
chorion, 871, 872
chorionic villi/placental villi, 877, 877*f*
choroid, 415*f*, 416
choroid plexus, 455, 456*f*
chromaffin cells, 531
chromatid, 57
chromatophilic substance, 365
chromosomes, 55–56, 845, 845*f*, 863
chronic kidney disease, 827–828
chymotrypsin, 771
cigarette smoking. *See* smoking
cilia, 53, 53*f*
ciliary body, 415*f*, 416
circular folds, 754
circular layer, 753
circulatory system. *See* cardiovascular
system
circumduction, 279*f*, 280, 344*f*
circumferential lamellae, 180
circumflex artery, 597, 599*t*
clavicle, 231, 232*f*, 233*t*
clavicular notches, 221*f*, 222
clean catch urine sample, 811

ethmoid sinuses, 200, 200*f*, 206, 206*f*, 688, 688*f*
eversion, 280*f*, 281
exhalation/expiration, 712
  forced, 716
  muscles of, 716, 716*f*
  quiet, 716
expiratory reserve volume (ERV), 719, 719*f*
extension, 279*f*, 280, 339*f*, 344*f*, 348*f*
extensor carpi radialis brevis muscle, 341*t*, 343*f*
extensor carpi radialis longus muscle, 340*t*, 343*f*
extensor carpi ulnaris muscle, 341*t*, 343*f*
extensor digiti minimi muscle, 341*t*, 343*f*
extensor digitorum brevis, 351*f*, 352*t*
extensor digitorum muscle
  foot, 351*f*, 352*t*
  forearm, 341*t*, 343*f*
  lower leg, 350*f*, 352*t*
extensor hallucis longus muscle, 350*f*, 351*f*, 352*t*
extensor indicis muscle, 341*t*, 343*f*
extensor pollicis brevis muscle, 341*t*, 343*f*
extensor pollicis longus muscle, 341*t*, 343*f*
extensors, 344*f*
external abdominal oblique muscle, 323*t*
external acoustic canal, 425
external acoustic meatus, 199, 200*f*
external carotid artery, 617, 618*f*
external elastic membrane, 628
external genitalia, 840
external iliac artery, 622, 622*f*, 623
external iliac vein, 623, 624*f*
external intercostal muscles, 716, 716*f*
external intercostals muscle, 321*f*, 324*t*
external jugular vein, 617, 619*f*
external occipital protuberance, 199, 199*f*
external rotation, shoulder, 339*f*
extracellular fluid (ECF), 80–81, 89
extrinsic ligaments, 271
extrinsic muscles
  appendicular, 329, 330*f*
  axial, 311, 311*f*
eyeball, 417*f*
eyes, 391
  anatomy of, 415, 415*f*
  conjunctiva, 413, 414*f*

external features of, 414*f*
light path through, 421, 422*f*
models, 418*f*
muscles of, 413, 413*f*
structures of, 417*t*–418*t*
tissue layers of, 415, 415*f*, 416

F

facial bones, 197, 197*f*, 202, 208*t*
  inferior nasal conchae, 204, 204*f*
  lacrimal bones, 204, 204*f*, 208*t*
  mandible, 202, 202*f*, 208*t*
  maxilla, 203, 203*f*, 208*t*
  nasal bones, 204, 204*f*, 208*t*
  palatine bones, 204, 205*f*, 208*t*
  vomer, 204, 204*f*
  zygomatic bones, 203, 203*f*, 208*t*
facial expression, muscles of, 315, 315*f*, 315*t*–316*t*
facial paralysis, 325–326, 325*f*
facial region, 30*f*
facial VII nerve, 461*f*, 461*t*, 526
facilitated diffusion, 82
falciform ligament, 745, 746*f*
false ribs, 221*f*, 222
falx cerebelli, 455, 456*f*
falx cerebri, 455, 455*f*
fascia, 148*f*
fascicles, 297
FAST acronym, 473*f*
fat pad, 272
fats
  emulsification, 773, 774*f*
  perinephritic, 787
fatty acids, 734
fatty acid tails, 80
fauces, 739, 739*f*
females
  clitoris, 840
  embryos in, 840
femoral artery, 622*f*, 623
femoral region, 26*f*, 30*f*
femoral vein, 623, 624*f*
femur, 247, 247*f*, 248*f*, 251*t*
fertilization, 863, 867–868, 867*f*
fetal circulation, 879–881, 879*f*
fetal hemolytic newborn disease, 587–588
fetal venules, 877
fetus, 863
  maternal-fetal exchange, 860
fiber, 769
fibroblasts, 129
fibrous capsule, 787

fibrous joints, 265
  function of, 266
  subcategories of, 266, 266*f*
fibrous layer/fibrous tunic layer, 415*f*, 416
fibrous pericardium, 595, 596*f*
fibrous skeleton of heart, 601
fibula, 247, 247*f*, 248, 249*f*, 251*t*
fibular collateral ligament, 275*f*, 276
fibularis brevis muscle, 350*f*, 351*f*, 352*t*
fibularis longus muscle, 350*f*, 351*f*, 352*t*
fibularis tertius muscle, 351*f*, 352*t*
fibular nerve damage, 355–356
fibular vein, 623, 624*f*
Fick's law, 84
field diameter, estimation of, 75–76, 75*f*, 76*t*
filtrate, 791
filtration rate, kidneys, 808
filtration slits, 791, 791*f*
filum terminale, 487*f*, 488
fine focus knob, 69
fingers, movements of, 344*f*
fissures, 452
flagellum, 53, 53*f*
flat bones, 171, 171*t*
flexion, 279*f*, 280, 339*f*, 344*f*, 348*f*
flexor carpi radialis muscle, 340*t*, 343*f*
flexor carpi ulnaris muscle, 330, 340*t*, 343*f*
flexor digiti minimi muscle, 342*t*, 345*f*
flexor digitorum longus, 350*f*, 352*t*
flexor digitorum profundus muscle, 340*t*, 343*f*
flexor digitorum superficialis muscle, 340*t*, 343*f*
flexor hallucis longus, 350*f*, 352*t*
flexor pollicis brevis muscle, 341*t*, 345*f*
flexor pollicis longus muscle, 340*t*, 343*f*
flexor reflex, 498
flexors, 344*f*
floating ribs, 221*f*, 222
focus knobs, 68–69
follicles, 551
follicle-stimulating hormone (FSH), 558, 853, 853*f*
follicular cells, 549, 551
follicular fluid, 848, 848*f*
follicular phase, 853, 853*f*
folliculogenesis, 848, 848*f*

foot
  bones of, 247, 247f, 248, 250f, 251t
  movements, 280f, 281
  muscles of, 350, 351f, 352t
foramen lacerum, 206, 207f
foramen magnum, 199, 199f
foramen ovale, 201, 207f, 601, 879, 879f
foramen rotundum, 201, 207f
foramen spinosum, 201, 207f
forced expiration, 716
forced inspiration, 716
forearm. *See* antebrachium/forearm
  movements of, 344f
  muscles of, 339, 339t–342t, 343f
formed elements, 563, 567
fossa ovalis, 601
fourth ventricle, 455, 456f
fovea capitis, 247, 248f
fovea centralis, 421, 422, 422f
fractures, 185–186, 185f
FRC. *See* functional residual capacity (FRC)
frontal bone, 197, 198f, 208t
frontal lobe, 457, 457f
frontal plane, 9, 9f
frontal region, 30f
frontal sinuses, 197, 206, 206f, 688, 688f
FSH. *See* follicle-stimulating hormone (FSH)
functional magnetic resonance imaging (fMRI), 378, 452, 452f, 460
functional magnetoencephalography (fMEG), 378, 378f–385f
functional residual capacity (FRC), 719
functional unit, 292
fundus, 744, 745f, 838, 838f

G

gallbladder, 733f, 744, 746f
gametes, 837, 837f, 838. *See also* sperms
gametogenesis, 837
  histology of, 845–850
ganglia, 520
ganglion cells, 416, 416f
  axons of, 421
gas exchange
  across the respiratory membrane, 712
  at alveoli, 723, 723f

between blood and tissues, 712, 723–725, 723f, 724f, 726f
gas transport, 712
gastric bypass, 748
gastric epithelium, 777–779, 777f
gastric pits, 753, 753f
gastric veins, 621f, 622
gastrocnemius muscle, 350f, 352t, 507
gastrointestinal (GI) tract, 733, 743
  epithelium of, 750, 750f
  human torso model, 748f
  organs
    anus, 733f, 745
    esophagus, 733f, 744, 744f
    histology of, 753–759, 753f–756f
    large intestine, 733f, 745
    mouth, 733f, 739–741, 739f–742f
    pharynx, 733f, 743, 743f
    rectum, 733f, 745
    small intestine, 733f, 744
    stomach, 733f, 744, 745f
  tissue layers, 749–751, 749f
    mucosa, 749f, 750
    muscularis, 749f, 750
    serosa, 749f, 751
    submucosa, 749f, 750
gastroparesis, 537–538
gastrula, 872
gatorade, 103
gender, 834
general senses, 391, 408
  adaptation. *See* adaptation
  overview of, 437–438
  tactile localization, 445–446, 445f, 446t
  two-point discrimination, 441–442
genicular artery, 622f, 623
germ layers, 872f
GH. *See* growth hormone (GH)
GHIH. *See* growth hormone-inhibiting hormone (GHIH)
GHRH. *See* growth hormone-releasing hormone (GHRH)
Giemsa stain, 74
gingivae/gums, 739
GI tract. *See* gastrointestinal (GI) tract
glabella, 197, 198f
glands, 158, 158f, 159f
  adrenal, 542, 547f, 549, 550f, 551
  bulbourethral, 840
  parathyroid, 547f, 549
  pineal, 459, 459f, 547f, 548
  pituitary, 459, 459f, 547, 548f
  prostate, 839f, 840

salivary, 733f, 739, 740f
  thyroid, 547f, 549, 551
glasses, need for, 422
glenoid cavity/fossa, 231, 232f
glial (neuroglial) cells, 127, 360, 369–371, 371t, 480
  of CNS, 369–370, 369f
  of PNS, 370, 370f
glomerulus, 791, 791f, 795
glossopharyngeal IX nerve, 461f, 461t, 526
glottis, 693, 694f
glucagon, 549, 551
gluteal region, 30f
gluteal tuberosity, 247, 248f
gluteus maximus muscle, 347f, 348t
gluteus medius muscle, 347f, 348t
gluteus minimus muscle, 347f, 348t
glycosidic bonds, 767, 771
GnRH. *See* gonadotrophin-releasing hormone (GnRH)
goblet cells, 115, 750
Golgi apparatus, 50, 51f
Golgi reflex, 501, 501f
gomphosis, 266, 266f
gonadal arteries, 619f, 620
gonadal veins, 620f, 622
gonadotrophin-releasing hormone (GnRH), 853
gonadotropin-releasing hormone (GRH), 558
gonads, 622, 837, 837f
goniometer, 283, 283f
G_1 phase, 55
G_2 phase, 55
gracilis muscle, 347f, 349t
granules, 567
granulosa cells, 848, 848f, 854, 867, 867f
gray matter, of gray matter, 483, 483f
great cardiac vein, 597f, 598, 598f, 599t
greater curvature, 744
greater omentum, 743
greater sciatic notch, 243, 244f
greater trochanter, 247, 248f
greater tubercle, 235, 236f
greater wings of sphenoid bone, 201, 201f
great saphenous vein, 623, 624f
greenstick fracture, 185, 185f
GRH. *See* gonadotropin-releasing hormone (GRH)
ground substance, 129
growth hormone (GH), 173, 555–556, 555f

growth hormone-inhibiting hormone (GHIH), 556
growth hormone-releasing hormone (GHRH), 555–556
gums, 739
gustation, 399–402, 399f–401f, 402t
gyrus, 452, 453f

H

hair, 157, 157f
hair bulb, 157f
hair cells, 425, 426–427
hair follicle, 157, 160f
hair root, 157, 157f
hair shaft, 157, 157f
hallux, 248, 250f
hallux region, 30f
hamate, 236, 238f
hands
 bones of, 235, 235f, 236–237, 238f, 239t
 digits and phalanges of, 36
 movements, 280f, 281, 344f
 muscles of, 339, 339t–342t, 345f
 palm, bones of, 237, 238f
haploid cells, 845, 846f, 848
hard palate, 688, 739, 739f
hCG. See human choriogonadotropin (hCG)
HCl. See hydrochloric acid (HCl)
head
 arteries of, 617, 618f
 blood vessels of, 620t
 sagittal model of, 741f
 surface anatomy of, 31–32, 32t
 veins of, 617, 619f
head of the femur, 247, 248f
head of the fibula, 248, 249f
head of the humerus, 235, 236f
head of the radius, 235, 237f
head of the rib, 222, 222f
head of the ulna, 235, 237f
hearing, 425–433. See also ear
 tests, 431–433, 431f, 432t
heart, 15f, 592, 637
 anterior view of, 597f
 blood flow through, 605
 chambers of, 595f, 597
 dissection, 597
 external features of, 599t
 frontal section, 601f, 602f
  within pericardium, 596f

internal structures of, 601–603, 601f, 603f
murmurs, 642
position of, 595–596, 595f, 596f
posterior view of, 598f
sounds, 641–643
superficial structures of, 597–600
transverse section of, 641
valves, 601, 601f, 641, 641f
veins, 598, 599t
vessels, 597, 598f, 599t, 600f, 600t
wall, 301
 layers of, 601
heat stroke, 101–102
hematocrit, 577–578, 577t, 578f
hematoxylin, 74, 144
hemodialysis, 794
hemoglobin, 567, 577–578, 577t, 724, 724f
hemoglobin and hematocrit (H&H), 578
hemolysis, 103
hepatic artery, 618, 619f, 745, 746f, 755, 756
hepatic portal vein, 621f, 622, 745, 746f, 755–756
hepatic veins, 620f, 622
hepatocytes, 755
hepatopancreatic ampulla, 745, 746f
hilum/hilus, 671, 671f, 787, 787f
hinge joint, 267, 268f
hips
 joints, and osteoarthritis, 288
 movements of, 348f
 muscles of, 347f, 348t–349t
histology, 74
holes, in skull, 206, 207f
homeostasis, 520
home pregnancy tests, 881
homologous structures, 837
horizontal plates, 204, 205f
hormones, 542
 aldosterone, 549
 antidiuretic, 548
 calcitonin, 549, 551
 corticoids, 549
 cortisol, 549
 epinephrine, 549, 551
 estrogen, 549
 follicle-stimulating, 853, 853f
 glucagon, 549, 551
 gonadotrophin-releasing, 853
 growth, 173
 insulin, 542, 551

 luteinizing, 853, 853f
 melatonin, 548
 norepinephrine, 549, 551
 oxytocin, 548
 parathyroid, 549
 progesterone, 549, 872
 of reproduction, 853–855
 testosterone, 549, 849
 thymosins, 549
 thyroid, 549
 tropic. See tropic hormone
horns, 483, 483f
human cells, 49, 49f
human choriogonadotropin (hCG), 854, 872, 881
humerus, bones of, 235, 235f, 236f, 239t
hydrochloric acid (HCl), 771
hydrogen ions (H$^+$), 400
hydrophilic, 80
hydrophobic cell membrane, 80
hyoid bone, 204, 205f
hyperextension, 279f, 280, 344f
hyperosmotic solution, 93
hypertonic solution, 97, 97f
hypochondriac regions, 15
hypodermis, 148, 148f, 153
hypogastric region, 15
hypoglossal XII nerve, 461f, 461t
hypoosmotic solution, 93
hypophyseal portal veins, 548
hypothalamic neurons, 548
hypothalamus, 459, 459f, 547, 548, 548f, 558, 853
hypotonic solution, 97, 97f

I

IC. See inspiratory capacity (IC)
IGF. See insulin-like growth factor (IGF)
ileum, 733f, 744
iliac crest, 243, 244f
iliac regions, 15
iliacus muscle, 323f, 323t
iliocostalis muscle, 317f, 318t
ilium, 243, 243f, 244f
immune system, 660
impacted fracture, 185f
impetigo, 148
implantation, 863, 871, 871f, 872
incus, 425, 425f
inferior, directional term, 7t
inferior angle, 231, 232f

M

macrophages, 567f, 568
macula densa cells, 791, 808
magnetic resonance imaging (MRI)
  functional, 378
magnetoencephalography (MEG)
  functional, 378, 378f–385f
magnification power, of microscope
  lenses, 69f, 76t
males
  embryos in, 840
  glands, 840
  penis, 840
  reproductive anatomy of, 838,
    839f
malleus, 425, 425f
mammary region, 30f
mandible, 202, 202f, 208t
mandibular angles, 202, 202f
mandibular condyle, 202, 202f
mandibular fossa, 199, 200f
mandibular notch, 202, 202f
manual region, 30f
manubrium, 221f, 222
Marfan syndrome, 143–144
marginal arteries, 597, 597f, 599t
Masson's trichrome, 74
mastoid process, 199, 200f
maternal-fetal exchange, 860
maxilla, 203, 203f, 208t
maxillary sinuses, 206, 206f, 688,
  688f
mechanical digestion, 739
medial, directional term, 7t
medial border, of scapula, 231, 232f
medial condyle of the femur, 247,
  248f
medial condyle of the tibia, 247,
  249f
medial cuneiform, 248, 250f
medial epicondyle of the femur, 247,
  248f
medial epicondyle of the humerus,
  235, 236f
medial malleolus, 247, 249f
medial meniscus, 275f, 276
medial plantar artery, 622f, 623
medial rotation, 281
medial sacral crest, 218, 218f
medial styloid process, 235
median antebrachial vein, 617,
  617f
median cubital vein, 617, 617f

mediastinal artery, 617, 619f
mediastinum, 595, 595f, 663
medulla, 460, 460f, 671, 671f, 787,
  787f, 795
medulla oblongata, 526
medullary sinuses, 671, 671f
MEG. See magnetoencephalography
  (MEG)
megakaryocytes, 567, 567f
meiosis, 55, 845, 846f, 847
melanin, 151–152
melanocyte, 151
melanoma, 163
melatonin, 548
meninges, 453, 454f, 455, 455f–456f,
  487
meniscus, 272
mental foramina, 202, 202f
mental protuberance, 202, 202f
mental region, 30f
Merkel cell, 151
Merkel cell carcinoma, 163
metacarpal bones, 235, 235f, 237,
  238f
metaphase, 56, 56f
metatarsal bones, 247f, 248, 250f
methylene blue, 64, 73
  rate of diffusion, 89, 89t
microanatomy, of bones, 179–181
microfilaments, 51, 52f
microglia, 369f, 370
microorganisms, 811, 812f
microscope
  anatomy of, 65, 65f
  compound, 67, 69f
  for disease diagnosis, 76
  field diameter, estimation of,
    75–76, 75f, 76t
  focus knobs, 68–69
  invention of, 63
  lenses of, 68, 68t, 69f
  lens paper for cleaning of, 67,
    67f
  light, 69–70
  parts of, 67–70
  slide preparation, 73–74, 73f
  usage of, 71–72
  way to carry, 67, 67f
microtubules, 51, 52f
microvilli, 53, 53f, 755
micturition, 797
midbrain, 460, 460f, 525
middle cardiac vein, 598, 598f,
  599t

middle cerebral arteries, 617,
  618f
middle ear, 425, 425f
middle meatuses, 688
middle nasal conchae, 200, 200f
middle phalanx bone
  of foot, 248, 250f
  of hand, 237, 238f
midsagittal plane, 9
midsagittal section of skull, 204,
  205f
minute ventilation, 720
mitochondria, 50–51, 52f
mitosis, 55–58, 849
  stages of, 55–56, 56f
mitotic spindle, 55
molecules
  kinetic energy, 81, 81f, 84
  phospholipid, 80, 80f
  properties of, 89t
  rate to cross membrane, 84
  water, 82–84, 83f
monocytes, 567f, 568
monomers, 734, 763–764
monosaccharides, 400, 734, 767
monosynaptic reflex arc, 498,
  498f
motor information, 391
motor neurons, 497
mouth, 733f, 739–741, 739f–742f
movements
  fingers, 344f
  foot, 280f, 281
  forearm, 344f
  hands, 280f, 281
  hips, 348f
  shoulder, 339f
  at synovial joints, 279–281,
    279f–280f
  wrist, 344f
MRI. See magnetic resonance
  imaging (MRI)
MS. See multiple sclerosis (MS)
mucociliary escalator, smoking and,
  707–708
mucosa, 749f, 750, 753, 799
mucosa-associated lymphoid tissue
  (MALT), 663, 664, 676, 676f,
  750, 751f, 755
multiaxial synovial joint, 266
multiple sclerosis (MS), 515–516
muscarinic receptors, 525
muscle fibers, 292
muscle organ, 297

olfactory receptor cells, 391
oligodendrocytes, 369f, 370, 373
oocytes, 837
  development of, 851t
  primary, 848, 848f
oogenesis, 847, 847f, 848–849
open fracture, 185f
openings
  in skull, 206, 207f
  in sphenoid bone, 210, 210f
opponens digiti minimi muscle, 342t,
    345f
opponens pollicis muscle, 341t, 343f
opposition, 280f, 281
optic canals, 201, 201f
optic II nerve, 461f, 461t
oral cavity, 739. See also mouth
oral region, 30f
orbicularis oculi muscle, 315f, 315t
orbicularis oris muscle, 315t
organelles, 46, 49
origin, muscles, 309, 329, 330
oropharynx, 691, 692f, 743, 743f
os coxae bones, 243, 243f
osmosis, 82, 83f, 84
  across cell membranes, 93–95
ossicles, 425, 425f, 426f
osteoarthritis, 288
osteoblasts, 179, 179t, 188
osteoclasts, 179–180, 179t, 188
osteocytes, 129, 179, 179t, 183, 188
osteons, 180
osteopetrosis, 187–188
OT. See oxytocin (OT)
otic region, 30f
outer ear, 425, 425f
oval window, 425, 425f, 426
ovarian cycle, 848
ovarian follicle, 848, 848f
ovaries, 547f, 549, 837
ovulation, 853
oxygen, 684
oxytocin (OT), 548

P

palatine bones, 204, 205f, 208t
palatine tonsils, 663, 663f, 691
palm, bones of, 237, 238f
palmar arches, 615, 616f
palmar interossei muscle, 342t,
    345f
palmaris longus muscle, 340t,
    343f

palmar region, 30f
palmar venous arches, 617, 617f
palpation, 24
pancreas, 547f, 549, 551, 733f, 745
  body of, 745
  histology of, 756, 756f
  tail of, 745
pancreatic duct, 745, 746f
pancreatic islets, 547f, 549, 551, 756,
    756f
papillary layer, of dermis, 153, 153f,
    156f
papillary muscle, 601, 601f
papillary muscles, 603f
parafollicular cells, 549
paralysis, facial, 325–326, 325f
paranasal sinuses, 206, 206f, 688, 688f
parasagittal plane, 9
parasympathetic division, of ANS, 520,
    523, 524t
  anatomy of, 525–527, 525f, 527f
  physiology of, 533, 534t–535t
  sympathetic division vs., 523, 524f
parathyroid glands, 547f, 549
parathyroid hormone (PTH), 549
parietal bones, 198, 198f, 208t
parietal layer, of serous membrane,
    14, 14f
parietal lobes, 457, 457f
parietal pericardium, 595, 596f
parotid salivary glands, 733f, 739, 740f
passive transport, 84
Patapoutain, Ardem, 438
patella/kneecap, 247, 247f, 275
patellar ligament, 275f, 276
patellar reflex, 507
patellar region, 30f
patellar surface, 247, 248f
patent ductus arteriosus (PDA),
    883–884
pathology, 76
PDA. See patent ductus arteriosus
    (PDA)
pectinate muscles, 601, 601f
pectineus muscle, 347f, 349t
pectoral appendages, 35
pectoralis minor muscle, 337t, 338f
pectoral/shoulder girdle, 228
  bones of, 231–233, 231f–232f, 233t
    clavicle, 231, 232f, 233t
    scapula, 231, 232f, 233t
pedal region, 30f
pedicles, 213, 214f, 791
pelvic appendages, 37

pelvic brim, 243, 243f
pelvic cavity, 13f
pelvic girdle, 228
  bones of, 243–253, 243f–250f
pelvic inflammatory disease (PID), 843
pelvic region, 26f, 30f
pelvic splanchnic nerves, 526
pelvis, 243, 243f
penis, 840
pepsin, 771
pepsinogen, 771
peptidases, 771
peptide bonds, 771
pericardial artery, 617, 619f
pericardial cavity, 13f, 595, 596f
pericardium, 13, 595
  frontal section of heart within, 596f
  layers of, 595, 596f
perimetrium, 838, 838f
perimysium, 297
perinephritic fat, 787
periodontal ligaments, 739, 740f
periosteum, 179
peripheral adaptation, 438
peripheral nervous system (PNS), 364,
    480
  afferent division, 391
  cranial nerves, 461, 461f, 461t
  efferent division, 391
  glial cells of, 370, 370f
  motor information, 391
  sensory information, 391, 392
peritoneum, 13, 743, 743f
peritubular capillaries, 795
peroneal nerve
  deep, 355
  superficial, 355
peroxisomes, 50
perpendicular plate, 200, 200f
Peyer's patches, 676, 755
pH, phenol red as indicator of, 774,
    774f
phalanges, 36, 235, 235f
phalanx bone
  of foot, 248, 250f
  of hand, 237, 238f
pharyngeal tonsils, 663, 663f, 691
pharynx, 684, 733f, 743, 743f
  anatomy of, 691–692, 692f
  laryngopharynx, 691, 692f, 743,
    743f
  nasopharynx, 691, 692f, 743, 743f
  oropharynx, 691, 692f, 743, 743f
phasic receptors, 443

radius, 235, 235f, 237f, 239t, 344f

rami/ramus, 202, 202f

range of motion (ROM), 283–284, 283f, 285t

receptive field, 441, 441f

receptors

itch, 447–448

muscarinic, 525

nicotinic, 523

phasic, 443

sensory, 392, 497, 497f

sweet, 400

taste, 400

tonic, 443

reciprocal activation, 501

reciprocal inhibition, 501

rectum, 733f, 745

rectus abdominis muscle, 323t

rectus femoris muscle, 347f, 349t

red blood cell (RBC). *See* erythrocytes

red bone marrow, 174, 174f, 660, 663

red pulp, 664

reflex arc, 501

components of, 497, 497f

defined, 493

monosynaptic, 498, 498f

polysynaptic, 498, 498f

reflexes, 493

autonomic, 497, 503–505, 503f

classifications of, 502t

contralateral, 498, 499f

cranial, 497

crossed-extensor, 499

intrinsic, 497

ipsilateral, 498, 499f

learned, 497

plantar, 511, 513f

somatic, 497

spinal, 497

stretch, 499, 500f, 501, 507–508, 509t

superficial, 511–514, 512t

tendon, 501, 501f

visceral/autonomic, 497

withdrawal, 498, 500f, 501

reflex hammer, 516, 516f

regional terms, 29

regions

abdominopelvic, 15

of brachium/upper arm, 235, 235f

of head and neck, 31f

human body, 26, 26f, 30f

hypochondriac, 15

iliac, 15

of lower limbs, 247, 247f

of stomach, 744, 745f

of upper limb, 35f

of uterus, 834, 834f

renal arteries, 619f, 620, 795

renal columns, 787, 787f

renal corpuscle, 791, 791f

renal papillae, 787, 787f, 791

renal pelvis, 787, 787f

renal pyramids, 787, 787f

renal system. *See* urinary system/renal system

renal veins, 620, 620f, 795

reposition, 281

reproduction

gametogenesis, histology of, 845–850

hormones of, 853–855

reproductive structures on anatomical models, identifying, 837–842, 837f–840f, 841t–843t

residual volume (RV), 719, 719f

respiration, 712

respiratory bronchioles, 700, 700f

respiratory system

bronchi, anatomy of, 699, 700f

larynx, anatomy of, 693–695, 693f–694f

lungs, anatomy of, 699, 699f, 700f

nose, anatomy of, 687–690, 687f, 688f

overview of, 684

pharynx, anatomy of, 691–692, 692f

trachea, anatomy of, 699–700, 699f, 700f

respiratory tract, 684

conducting zone, 684

lower, 684

bronchi, anatomy of, 699, 700f

lungs, anatomy of, 699, 699f, 700f

trachea, anatomy of, 699–700, 699f, 700f

respiratory zone, 684

structural differences along, 705, 705t–706t

upper, 684

larynx, anatomy of, 693–695, 693f–694f

nose, anatomy of, 687–690, 687f, 688f

pharynx, anatomy of, 691–692, 692f

respiratory zone, 684, 700, 700f

restrictive disease, 727. *See also* tuberculosis

reticular fibers, 129, 129f

reticular layer, of dermis, 153, 153f, 156f

retina, 415f, 416

histology of, 419, 419f

retraction

of mandible, 280f, 281

shoulder movements, 339f

rheumatic heart disease, 653–654

Rh group/factor, 579

matching of, 583–585, 583f, 584t, 586t

Rho(D) immune globulin (RhIG/RhoGAM shot), 588

rhomboid major muscle, 337t, 338f

rhomboid minor muscle, 337t, 338f

rib cage. *See* thoracic cage

ribosomes, 50

ribs, 194f, 221f, 222

false, 221f, 222

floating, 221f, 222

parts of, 222, 222f

true, 221f, 222

right and left hypochondriac regions, 15

right and left iliac regions, 15

right atrioventricular valve, 601, 601f

right atrium, 595f, 597

interior walls of, 601, 601f

right bronchomediastinal trunk, 667

right common carotid artery, 615, 615f

right common iliac artery, 622, 622f

right coronary artery, 597, 597f, 599t

right gastric artery, 618, 619f

right jugular trunk, 667

right lobe, 745, 746f

right lumbar trunks, 667

right lymphatic duct, 667

right pulmonary artery, 624, 625f

right subclavian artery, 615, 615f

right subclavian trunk, 667

right ventricle, 595f, 597

Rinne test, 431, 431f, 432t

rods, 416, 416f, 421

ROM. *See* range of motion (ROM)

root

of nose, 687, 687f

tooth, 739, 740f

root canal, 740f, 741

rootlets, 483, 484f

rotation, 279f, 280–281, 339f

rough ER (RER), 50, 50f

round window, 427

rugae, 744

RV. *See* residual volume (RV)

S

sacral canal, 218, 218f
sacral foramina, 218, 218f
sacral hiatus, 218, 218f
sacral promontory, 218, 218f
sacral region, 30f
sacroiliac joint, 243, 243f
sacrum, 218, 218f, 243, 243f
saddle joint, 267, 268f
sagittal plane, 9, 9f
sagittal suture, 198, 211
salivary glands, 733f, 739, 740f
sarcolemma, 297, 297f
sarcomeres, 292
sarcoplasmic reticulum (SR), 297, 297f, 298
sartorius muscle, 347f, 349t
satellite cells, 370, 370f
scalenes muscle, 317f, 318t
scaphoid, 236, 238f
scapula, 231, 232f, 233t
Schwann cells, 370
sclera, 415f, 416
scrotum, 837, 837f, 840
sebaceous gland, 158, 158f, 159f
sebum, 158, 158f
secondary bronchi, 699, 700f
section
  defined, 9
  planes of, 9–11, 9f
sediment analysis of urine sample, 813–816, 814f, 815f
segmental arteries, 795
selective permeability, cell membranes, 80
sella turcica, 201, 201f
semen, 840
semicircular canals, 425f, 426, 426f, 427
semimembranosus muscle, 347f, 349t
seminal vesicle, 839f, 840
semitendinosus muscle, 347f, 349t
sensation-integration-response cycle, of nervous system, 363, 363f
senses
  general, 391, 408
    adaptation. See adaptation
    overview of, 437–438
    tactile localization, 445–446, 445f, 446t
    two-point discrimination, 441–442
  special, 391, 408
    equilibrium, 425–433
    gustation, 399–402, 399f–401f, 402t

hearing, 425–433
    olfaction, 395–397, 395f, 396f
    vision, 413–423, 413f–419f, 417t–418t
sensory information, 391, 392
sensory neurons, 497, 497f
  receptive field of, 441, 441f
sensory receptor, 392, 497, 497f
sensory stimuli, localization of, 445–446, 445f
sensory system
  general senses, 391, 408, 437–438
  special senses, 391, 408
    equilibrium, 425–433
    gustation, 399–402, 399f–401f, 402t
    hearing, 425–433
    olfaction, 395–397, 395f, 396f
    vision, 413–423, 413f–419f, 417t–418t
sentinel lymph node biopsy (SLNB), 674
septum, 597
serosa, 749f, 751
serous membranes, 14, 14f
serous pericardium, 595, 596f
serratus anterior muscle, 337t, 338f
sex, 834
  hormones, 173
shaft of the humerus, 235, 236f
sheep brain
  dissection, 469–470, 470t
  inferior view of, 466f, 469f, 470t
  lateral view of, 470t
  midsagittal view of, 470t
  superior view of, 466f, 470t
short bones, 171, 171t
shoulder
  movements, 339f
  muscles of, 337, 337t–338t, 338f
shoulder flexion
  range of motion for, 284
shoulder girdle. See pectoral/shoulder girdle
shoulder hyperextension
  range of motion for, 284
sigmoid colon, 733f, 745
simple diffusion, 82
sinuses, of skull, 206, 206f
skeletal muscle, 123, 292, 293f
  anatomy of, 297–298, 297f
  cells, 292
  contraction, 305–306
  histology, 299f

skeletal system, 193. See also appendicular skeleton; axial skeleton
skin, 148
  accessory structures of. See accessory structures of skin
  infections and illnesses, 148
  layers of, 148, 148f, 151–156, 155t. See also dermis; epidermis; hypodermis
  thick vs. thin, 151, 152f
skin cancer, 163–164, 164f
skull, 194f
  bones of, 197, 197f
    cranial, 197–201, 197f–201f, 208t
    facial, 197, 197f, 202–204, 202f–205f, 208t
  holes/openings in, 206, 207f
  midsagittal section of, 204, 205f
  sinuses of, 206, 206f
  sutures of, 211, 211f
slide preparation, microscope, 73–74, 73f
small cardiac vein, 598, 599t
small intestine, 733f, 744
  histology of, 754–755, 754f
  length of, 755
  lumen and walls of, 754f
small saphenous vein, 623, 624f
smell. See olfaction
smoking, and mucociliary escalator, 707–708
smooth ER (SER), 50, 50f
smooth muscle, 123, 292, 293, 293f
  cells, 301
  histology, 301, 303f
SNS. See somatic nervous system (SNS)
sodium ions (Na$^+$), 400
soft palate, 688, 739, 739f
soleus muscle, 350f, 352t
solute, 89
solution
  hyperosmotic, 93
  hypertonic, 97, 97f
  hypoosmotic, 93
  hypotonic, 97, 97f
  isosmotic, 93
  isotonic, 97, 97f
solvent, 89
somatic cells, 845
somatic nervous system (SNS), 364, 520

somatic reflexes, 497
  autonomic reflexes *vs.*, 503, 503*f*
sounds, heart, 641–643
special senses, 391, 408
  equilibrium, 425–433
  gustation, 399–402, 399*f*–401*f*,
    402*t*
  hearing, 425–433
  olfaction, 395–397, 395*f*, 396*f*
  vision, 413–423, 413*f*–419*f*,
    417*t*–418*t*
specific gravity analysis of urine
    sample, 819–820, 820*f*
spermatids, 849
spermatogenesis, 847, 847*f*, 848–849
spermatogonia, 849
sperm capacitation, 867
sperms, 840
  acrosomal enzymes of, 867
  capacitated, 867
  nucleus of, 867
S phase, 55
sphenoid bone, 201, 201*f*, 208*t*
  openings in, 210, 210*f*
sphenoid sinus, 201, 206, 206*f*, 688,
    688*f*
sphygmomanometer, 645, 648*f*
spinal cord
  cross section of, 483, 483*f*
  functions, 480*f*
  gross anatomy of, 487–488, 487*f*,
    488*t*
  injury, 480, 488, 489
  length of, 487
  longitudinal view of, 487*f*
  microanatomy of, 483–485, 483*f*,
    484*f*, 485*t*
  nerves associated with, 483, 484*f*
  overview, 480
  structures seen in cross section,
    485*t*
spinalis muscle, 317*f*, 318*t*
spinal nerves, 461, 483, 484*f*
spinal reflexes, 497
spine of the scapula, 231, 232*f*
spinous process, 213, 214*f*
spiral arteries, 871
spiral fracture, 185*f*
spirometer, 719, 719*f*, 720
spleen, 660, 663
  anatomy of, 675, 675*f*
splenic artery, 618, 619*f*
splenic veins, 621*f*, 622
splenius capitus muscle, 317*f*, 318*t*

splenius cervicis muscle, 317*f*, 318*t*
spongy bone, 173
sprained knee, 287–288, 287*f*
squamous cell carcinoma, 163–164
squamous sutures, 198, 211
SR. *See* sarcoplasmic reticulum (SR)
staining, cells, 74
stapes, 425, 425*f*
starches, 767
  Lugol's iodine tests for presence of,
    767, 768*f*
stem cells, 849
sternal angle, 221*f*, 222
sternal body, 221*f*, 222
sternal end, 231, 232*f*
sternocleidomastoid muscle, 317*f*,
    318*t*
sternum, 194*f*, 221*f*, 222
stethoscope, 645
  placement for heart auscultation,
    642, 643*f*
  position for wearing, 642*f*
stiffening, blood vessels, 627
stimulus, 480, 480*f*
stomach, 733*f*, 744
  cells of, 753, 753*f*
  histology of, 753, 753*f*
  human model, 751*f*
  lumen, 777
  regions of, 744, 745*f*
stones, kidneys, 829–830
stratum basale, 151–152, 151*f*, 156*f*
stratum corneum, 151*f*, 153, 156*f*
stratum lucidum, 151*f*, 153
stratum spinosum, 151*f*, 152, 156*f*
stretch reflexes, 499, 500*f*, 501,
    507–508, 509*t*
stroke, 387, 473–475, 473*f*–475*f*
styloid process
  of radius, 235, 237*f*
  of ulna, 235, 237*f*
subcapsular sinus, 671, 671*f*
subclavian artery, 615, 616*f*
subclavian veins, 617, 617*f*, 619*f*
subclavius muscle, 337*t*, 338*f*
sublingual salivary glands, 733*f*, 739,
    740*f*
submandibular glands, 733*f*, 739, 740*f*
submucosa, 749*f*, 750
subscapular fossa, 231, 232*f*
subscapularis muscle, 337*t*
subscapular vein, 617, 617*f*
sugar, 767
sulci, 452, 453*f*

sulcus, 597
superficial, directional term, 7*t*, 24
superficial layer, 312, 312*f*
superficial peroneal nerve, 355
superficial reflexes, 511–514, 512*t*
superficial structures, of heart,
    597–600
superior, directional term, 7*t*
superior angle, 231, 232*f*
superior articular processes, 213, 214*f*,
    218, 218*f*
superior border, of scapula, 231, 232*f*
superior colliculi, 460
superior gemellus muscle, 347*f*, 348*t*
superior meatuses, 688
superior mesenteric artery, 619*f*, 620
superior mesenteric vein, 621*f*, 622
superior nasal conchae, 200, 200*f*
superior nuchal line, 199, 199*f*
superior orbital fissure, 201
superior phrenic artery, 617, 619*f*
superior ramus of the pubis, 244, 245*f*
superior vena cava, 595*f*, 597*f*, 598*f*,
    600*t*, 617
supination, 279*f*, 281, 344*f*
supinator muscle, 340*t*, 343*f*
supine, 5
supraorbital foramina, 197, 198*f*
supraorbital margin, 197, 198*f*
supraspinatus muscle, 337*t*
supraspinous fossa, 231, 232*f*
sural region, 30*f*
surface anatomy
  of head and neck, 31–32, 32*t*
  of lower limb, 37–38, 37*f*, 38*t*
  of trunk, 33–34, 33*f*, 34*t*
  of upper limb, 35–36, 35*f*, 36*t*
sutures, 211, 211*f*, 266, 266*f*
sweat, 95
  production of, 101, 101*f*
sweat glands, 158, 158*f*, 159*f*
sweet receptors, 400
sympathetic division, of ANS, 520,
    523, 524*t*
  anatomy of, 529–532, 529*f*–530*f*, 532*f*
  parasympathetic division *vs.*, 523,
    524*f*
  physiology of, 533, 534*t*–535*t*
symphysis, 266, 267*f*
synapses, 365*f*, 366
synarthrosis, 265, 265*f*
synchondrosis, 266, 267*f*
syndesmoses, 266, 266*f*
synergists, 309, 310*f*, 329, 330*f*

synovial fluid, 271

synovial joints, 265

  anatomy of, 271–272, 271f–272f

  ligaments at, 271

  movements at, 279–281, 279f–280f

  range of motion, 283–284, 283f, 285t

  types of, 266–267, 268f

synovial membrane, 271

systemic circuit, 612, 612f

systole, 627, 642

systolic murmurs, 642

T

tachycardia, 653

tactile (Meisner's) corpuscles, 153

tactile localization, 445–446, 445f, 446t

tail of pancreas, 745

talus, 248, 250f

tarsal bones, 247f, 248, 250f

tarsal region, 30f

tastants, 399

taste. *See* gustation

taste buds, 400, 400f

taste cells, 401f

taste information, pathway of, 399, 399f

taste receptors, 400

taste receptors cells, 400, 400f

Taxol, 59

tectorial membrane, 426, 427f

telophase, 56, 56f

temporal bones, 199, 200f, 208t

temporal lobes, 457, 457f

temporal processes, 203, 203f

temporomandibular joint, 32, 199, 268, 268f

tendon/Golgi reflex, 501, 501f

tendons, 271, 309

  knee, 275f, 276

tendon sheath, 272

tensor fascia latae muscle, 347f, 348t

tentorium cerebelli, 455, 455f

teres major muscle, 338t

teres minor muscle, 338t

tertiary bronchi, 700, 700f

tertiary follicles, 848, 848f

testes, 547f, 549, 837, 837f, 838, 839f

  cells in, 849, 849f

testosterone, 173, 549, 849

TH. *See* thyroid hormone (TH)

thalamus, 459, 459f

thecal cells, 848, 848f

thermoreceptors, 438, 438t

thick skin, 151, 152f

thigh/upper leg

  bone of, 247, 247f, 248f

  muscles of, 347f, 348t–349t

thin skin, 151, 152f

third ventricle, 455, 456f

thoracic aorta, 615, 616f

thoracic cage, 194f

  bones of, 221–222, 221f, 222f

thoracic cavity, 13f, 321

thoracic duct, 667

thoracic region, 26f, 30f

thoracic vertebrae, 216, 216f–217f

thoracolumbar system, 529. *See also* sympathetic division, of ANS

thorax

  arteries of, 617–618, 619f, 620

  blood vessels of, 623t

  muscles of, 321–322, 321f–322f, 323f, 323t–324t

  veins of, 620–622, 620f

throat. *See* pharynx

thymosins, 549

thymus, 549, 660, 663

thyrocervical artery, 615

thyroid cartilage, 693, 693f, 696, 696f

thyroid gland, 547f, 549, 551

thyroid hormone (TH), 549

thyroid-stimulating hormone (TSH), 556

thyrotropin-releasing hormone (TRH), 556

tibia, 247, 247f, 249f, 251t, 275

tibial collateral ligament, 275f, 276

tibialis anterior muscle, 350f, 351f, 352t

tibialis posterior, 350f, 352t

tibial tuberosity, 247, 249f

tidal volume (TV), 719, 719f

tissue layers

  of eyes, 415, 415f, 416

  of GI tract, 749–751, 749f

    mucosa, 749f, 750

    muscularis, 749f, 750

    serosa, 749f, 751

    submucosa, 749f, 750

tissues

  connective, 105

    structure of, 129–135, 129f

  epithelial, 105

    structure of, 115–121, 115f

    terminology associated with, 116t

  gas exchange between blood and, 712, 723–725, 723f, 724f, 726f

  making histology slide, 144

muscle, 105

  quality of, 292

  structure of, 123–125, 123t

  types of, 292, 293f

nervous, 105

  physiology of, 375–377, 375f

  structure of, 127–128

overview of, 105

TLC. *See* total lung capacity (TLC)

T lymphocytes, 568, 663

tongue, 739, 739f

tonicity, 97

tonic receptors, 443

tonsillar crypts, 664, 691

tonsils, 660, 663, 663f, 664, 676

  lingual, 691

  palatine, 663, 663f, 691

  pharyngeal, 663, 663f, 691

tooth

  anatomy of, 739, 740f, 741

  crown, 739, 740f

  model of, 741f

  root, 739, 740f

total lung capacity (TLC), 719, 719f

trabecula, 173

trabeculae, 671, 671f

trabeculae carneae, 601, 601f, 603f

trabecular sinus, 671, 671f

trachea, 684, 693, 743

  anatomy of, 699–700, 699f, 700f

tracheal cartilage, 699, 699f

transverse abdominis muscle, 323t

transverse colon, 733f, 745

transverse costal facet, 216

transverse foramina, 214, 214f

transverse fracture, 185f

transverse plane, 9, 9f

transverse processes, 213, 214f

transversospinalis muscle, 318t

trapezium, 236, 238f

trapezius muscle, 337t, 338f

trapezoid, 236, 238f

TRH. *See* thyrotropin-releasing hormone (TRH)

triceps brachii muscle, 329, 340t, 343f

  stretch reflex of, 507

trigeminal V nerve, 461f, 461t

triglycerides, 773

triquetrum, 236, 238f

trochlea, 235, 236f

trochlear IV nerve, 461f, 461t

trochlear notch, 235, 237f

trophoblast, 867, 867f, 868

tropic hormone
  defined, 555
  pathway physiology
    cortisol hormone, 556, 557f
    growth hormone, 555–556, 555f
    thyroid hormone, 556, 556f
true ribs, 221f, 222
trunks
  brachiocephalic, 615, 615f
  celiac, 618, 619f
  intestinal, 667
  left bronchomediastinal, 667
  left jugular, 667
  left lumbar, 667
  left subclavian, 667
  lymphatic, 667, 668f
  pulmonary, 597f, 600t, 624, 625f
  right bronchomediastinal, 667
  right jugular, 667
  right lumbar, 667
  right subclavian, 667
  surface anatomy of, 33–34, 33f, 34t
trypsin, 771
TSH. *See* thyroid-stimulating hormone
    (TSH)
T-tubule, 297, 297f
tubercle of the rib, 222, 222f
tuberculosis, 727–728
tunica externa, 628, 628f
tunica intima, 628, 628f
tunica media, 628, 628f
tunics, 628, 628f
TV. *See* tidal volume (TV)
tympanic membrane, 425, 425f
type I alveolar cells, 703, 703f
type II alveolar cells, 703, 703f

U

ulna, 235, 235f, 237f, 344f
ulnar artery, 615, 616f
ulnar collateral arteries, 615, 616f
ulnar deviation, 344f
ulnar vein, 617, 617f
umami, 400
umbilical arteries, 877
umbilical cord, 877–878, 877f
umbilical/navel region, 15, 30f
umbilical vein, 877, 879
uniaxial synovial joint, 266
upper leg. *See* thigh/upper leg
upper limbs. *See also* antebrachium/
    forearm
  arteries of, 615, 616f, 617t
  blood vessels of, 617t

bones of, 235–238, 235f–238f, 239t
  muscles of
    brachial, 337, 337t–338t
    forearm, 339, 339t–342t, 343f
    hand, 339, 339t–342t, 345f
    shoulder, 337, 337t–338t, 338f
    wrist, 339, 339t–342t
  regions of, 235, 235f
  surface anatomy of, 35–36, 35f, 36t
  veins of, 617, 617f, 617t
upper respiratory tract, 684
  larynx, anatomy of, 693–695,
    693f–694f
  nose, anatomy of, 687–690, 687f,
    688f
  pharynx, anatomy of, 691–692,
    692f
urea, 103
ureters, histology of, 799
urethra, 837f, 838
  histology of, 799
urinalysis, 807
  tests, interpretation of results from,
    823–824, 825t
  urine sample
    color and transparency of, 817
    dipstick analysis of, 821–822,
      821t
    obtaining, 811
    sediment analysis of, 813–816,
      814f, 815f
    specific gravity analysis of,
      819–820, 820f
urinary bladder, histology of, 799–802,
    800f
urinary incontinence, 803
urinary system/renal system
  accessory structures of, 799t
  anatomy of, 784
  kidneys
    blood supply to, 795–796, 797f
    and dialysis, 794, 794f
    gross anatomy of, 787–789, 787f
    nephrons, anatomy of, 791–792,
      791f, 793f
  overview of, 784
  ureters, histology of, 799
  urethra, histology of, 799
  urinary bladder, histology of,
    799–800, 800f
urine
  average amount of, 807
  composition, 807
  production of, 807

sample
  color and transparency of, 817
  dipstick analysis of, 821–822, 821t
  obtaining, 811
  sediment analysis of, 813–816, 814f,
    815f
  specific gravity analysis of,
    819–820, 820f
uterine tubes, 838, 838f
uterus, 838, 838f
  layers of, 838, 838f
  regions of, 838, 838f
uvula, 739, 739f

V

vagina, 840
vagus X nerve, 461f, 461t, 526
valves, heart, 601, 601f, 641, 641f
vasa recta, 795
vascular layer/vascular tunic layer,
    415f, 416
vascular stiffening, 627
vasoconstriction, 628
vasodilation, 628
vastus intermedius muscle, 347f, 349t
vastus lateralis muscle, 347f, 349t
vastus medialis muscle, 347f, 349t
VC. *See* vital capacity (VC)
veins, 611, 612, 628
  of abdomen, 620–622, 620f
  arcuate, 795
  arteries *vs.*, 627, 627f
  coronary, 597f, 598, 598f, 599t
  of head, 617, 619f
  heart, 598, 599t
  interlobar, 795
  interlobular, 795
  of lower limbs, 623, 623t, 624f
  of neck, 617, 619f
  renal, 795
  of thorax, 620–622, 620f
  umbilical, 877, 879
  of upper limbs, 617, 617f, 617t
ventilation, 712
  minute, 720
ventilator, 718
ventral, directional term, 7t
ventral gluteal injection site, 38
ventricles, 369f, 370
  of brain, 455, 456f
  of heart, 642
    internal structures, 603f
    left, 595f, 597
    right, 595f, 597

venules, 611, 628
vertebrae, 213, 214*f*
  cervical, 214–216, 214*f*, 215*f*
  lumbar, 217, 217*f*
  thoracic, 216, 216*f*–217*f*
vertebral arch, 213, 214*f*
vertebral arteries, 617, 618*f*
vertebral artery, 615
vertebral cavity, 13*f*
vertebral column, 194*f*
  bones of, 213–219, 213*f*–218*f*
    cervical vertebrae, 214–216, 214*f*,
      215*f*
    coccyx, 218, 218*f*
    lumbar vertebrae, 217, 217*f*
    sacrum, 218, 218*f*
    thoracic vertebrae, 216, 216*f*–217*f*
    vertebrae, 213, 214*f*
vertebral foramen, 213, 214*f*
vessels, blood. *See* blood vessels
vestibular folds, 693, 693*f*
vestibules, 425*f*, 426, 426*f*, 427
  nasal, 687, 687*f*
vestibulocochlear nerve, 427
vestibulocochlear VIII nerve, 461*f*,
  461*t*
vibrissae, 687, 687*f*
villi/villus, 754–755, 754*f*

visceral layer, of serous membrane,
  14, 14*f*
visceral pericardium, 595, 596*f*
visceral reflexes, 497
vision, 413–423, 413*f*–419*f*, 417*t*–418*t*.
  See also eyes
  tests, 421–423, 421*f*, 422*f*
visual acuity. *See* acuity, visual
vital capacity (VC), 719, 719*f*
vitreous humor, 415
vocal folds/cords, 693, 693*f*
vocal ligaments, 693
voicebox. *See* larynx
voluntary muscle, 123
vomer, 204, 204*f*
vulva, 840, 840*f*

W

water
  lipid molecules in, 773*f*
  molecules, 82–84, 83*f*
weight loss surgery, 748
wet mount, preparation of, 73, 73*f*
white blood cell (WBC). *See* leukocytes
white matter, 373
white pulp, 664
windpipe. *See* trachea

withdrawal/flexor reflex, 498, 500*f*, 501
working distance, 71, 71*f*, 75
wrist
  bones of, 236, 238*f*
  movements of, 344*f*
  muscles of, 339, 339*t*–342*t*
wrist flexion, range of motion for, 284
wrist hyperextension, range of motion
  for, 284

X

X chromosomes, 845
xiphoid process, 221*f*, 222
X-rays, 256, 256*f*

Y

Y chromosomes, 845
yellow bone marrow, 174, 174*f*

Z

zona pellucida, 848, 848*f*
zygomatic bones, 203, 203*f*, 208*t*
zygomatic process, 199, 200*f*
zygomaticus major muscle, 315*f*, 316*t*
zygomaticus minor muscle, 315*f*, 316*t*
zygote, 863, 867, 871